HEALTH RISKS
FROM EXPOSURE TO
LOW LEVELS OF
IONIZING
RADIATION

BEIR VII PHASE 2

Committee to Assess Health Risks from Exposure to Low Levels of Ionizing Radiation

Board on Radiation Effects Research
Division on Earth and Life Studies

NATIONAL RESEARCH COUNCIL
OF THE NATIONAL ACADEMIES

THE NATIONAL ACADEMIES PRESS
Washington, D.C.
www.nap.edu

THE NATIONAL ACADEMIES PRESS   500 Fifth Street, N.W.   Washington, DC 20001

NOTICE: The project that is the subject of this report was approved by the Governing Board of the National Research Council, whose members are drawn from the councils of the National Academy of Sciences, the National Academy of Engineering, and the Institute of Medicine. The members of the committee responsible for the report were chosen for their special competences and with regard for appropriate balance.

This study was supported by Environmental Protection Agency Grant #X-826842-01, Nuclear Regulatory Commission Grant #NRC-04-98-061, and U.S. Department of Commerce, National Institute of Standards and Technology Grant #60NANB5D1003. Any opinions, findings, conclusions, or recommendations expressed in this publication are those of the author(s) and do not necessarily reflect the views of the organizations or agencies that provided support for the project.

Library of Congress Cataloging-in-Publication Data

Health risks from exposure to low levels of ionizing radiation : BEIR VII, Phase 2 / Committee to Assess Health Risks from Exposure to Low Levels of Ionizing Radiation, Board on Radiation Effects, Research Division on Earth and Life Studies, National Research Council of the National Academies.
     p. cm.
   This is the seventh in a series of reports from the National Research Council prepared to advise the U.S. government on the relationship between exposure to ionizing radiation and human health.
   Includes bibliographical references and index.
   ISBN 0-309-09156-X (pbk.) — ISBN 0-309-53040-7 (pdf) 1. Ionizing radiation—Toxicology.
2. Ionizing radiation—Physiological effect. 3. Ionizing radiation—Dose-response relationship.
I. National Research Council (U.S.). Committee to Assess Health Risks from Exposure to Low Levels of Ionizing Radiation.
   RA1231.R2H395 2006
   363.17′99—dc22
                           2006000279

Additional copies of this report are available from the National Academies Press, 500 Fifth Street, N.W., Lockbox 285, Washington, DC 20055; (800) 624-6242 or (202) 334-3313 (in the Washington metropolitan area); Internet, http://www.nap.edu.

THE NATIONAL ACADEMIES
Advisers to the Nation on Science, Engineering, and Medicine

The **National Academy of Sciences** is a private, nonprofit, self-perpetuating society of distinguished scholars engaged in scientific and engineering research, dedicated to the furtherance of science and technology and to their use for the general welfare. Upon the authority of the charter granted to it by the Congress in 1863, the Academy has a mandate that requires it to advise the federal government on scientific and technical matters. Dr. Ralph J. Cicerone is president of the National Academy of Sciences.

The **National Academy of Engineering** was established in 1964, under the charter of the National Academy of Sciences, as a parallel organization of outstanding engineers. It is autonomous in its administration and in the selection of its members, sharing with the National Academy of Sciences the responsibility for advising the federal government. The National Academy of Engineering also sponsors engineering programs aimed at meeting national needs, encourages education and research, and recognizes the superior achievements of engineers. Dr. Wm. A. Wulf is president of the National Academy of Engineering.

The **Institute of Medicine** was established in 1970 by the National Academy of Sciences to secure the services of eminent members of appropriate professions in the examination of policy matters pertaining to the health of the public. The Institute acts under the responsibility given to the National Academy of Sciences by its congressional charter to be an adviser to the federal government and, upon its own initiative, to identify issues of medical care, research, and education. Dr. Harvey V. Fineberg is president of the Institute of Medicine.

The **National Research Council** was organized by the National Academy of Sciences in 1916 to associate the broad community of science and technology with the Academy's purposes of furthering knowledge and advising the federal government. Functioning in accordance with general policies determined by the Academy, the Council has become the principal operating agency of both the National Academy of Sciences and the National Academy of Engineering in providing services to the government, the public, and the scientific and engineering communities. The Council is administered jointly by both Academies and the Institute of Medicine. Dr. Ralph J. Cicerone and Dr. Wm. A. Wulf are chair and vice chair, respectively, of the National Research Council.

www.national-academies.org

Preface

BACKGROUND

This is the seventh in a series of reports from the National Research Council (NRC) prepared to advise the U.S. government on the relationship between exposure to ionizing radiation and human health. In 1996 the National Academy of Sciences (NAS) was requested by the U.S. Environmental Protection Agency to initiate a scoping study preparatory to a new review of the health risks from exposure to low levels of ionizing radiations. The main purpose of the new review would be to update the Biological Effects of Ionizing Radiation V (BEIR V) report (NRC 1990), using new information from epidemiologic and experimental research that has accumulated during the 14 years since the 1990 review. Analysis of those data would help to determine how regulatory bodies should best characterize risks at the doses and dose rates experienced by radiation workers and members of the general public. BEIR VII—Phase 1 was the preliminary survey to evaluate whether it was appropriate and feasible to conduct a BEIR VII—Phase 2 study. The Phase 1 study determined that it was appropriate and feasible to proceed to Phase 2. The Phase 1 study, *Health Effects of Exposure to Low Levels of Ionizing Radiations: Time for Reassessment?*, published in 1998, also provided the basis for the Phase 2 Statement of Task that follows.

BEIR VII—PHASE 2 STATEMENT OF TASK

The primary objective of the study is to develop the best possible risk estimate for exposure to low-dose, low linear energy transfer (LET) radiation in human subjects. In order to do this, the committee will (1) conduct a comprehensive review of all relevant epidemiologic data related to the risk from exposure to low-dose, low-LET radiation; (2) define and establish principles on which quantitative analyses of low-dose and low-dose-rate effects can be based, including requirements for epidemiologic data and cohort characteristics; (3) consider relevant biologic factors (such as the dose and dose-rate effectiveness factor, relative biologic effectiveness, genomic instability, and adaptive responses) and appropriate methods to develop etiologic models (favoring simple as opposed to complex models) and estimate population detriment; (4) assess the current status and relevance to risk models of biologic data and models of carcinogenesis, including critical assessment of all data that might affect the shape of the response curve at low doses, in particular, evidence for or against thresholds in dose-response relationships and evidence for or against adaptive responses and radiation hormesis; (5) consider, when appropriate, potential target cells and problems that might exist in determining dose to the target cell; and (6) consider any recent evidence regarding genetic effects not related to cancer. In performing the above tasks, the committee should consider all relevant data, even if obtained from high radiation exposures or at high dose rates.

With respect to modeling, the committee will (1) develop appropriate risk models for all cancer sites and other outcomes for which there are adequate data to support a quantitative estimate of risk, including benign disease and genetic effects; (2) provide examples of specific risk calculations based on the models and explain the appropriate use of the risk models; (3) describe and define the limitations and uncertainties of the risk models and their results; (4) discuss the role and effect of modifying factors, including host (such as individual susceptibility and variability, age, and sex), environment (such as altitude and ultraviolet radiation), and life-style (such as smoking history and alcohol consumption) factors; and (5) identify critical gaps in knowledge that should be filled by future research.

WHAT HAS CHANGED SINCE THE LAST BEIR REPORT ON THE HEALTH EFFECTS OF LOW LEVELS OF LOW-LET IONIZING RADIATION

In the 15 years since the publication of the previous BEIR report on low-LET radiation (BEIR V), much new information has become available on the health effects of ionizing radiation. Since the 1990 BEIR V report, substantial new information on radiation-induced cancer has become available from the Hiroshima and Nagasaki survivors, slightly less than half of whom were alive in 2000. Of special importance are cancer incidence data from the Hiroshima and Nagasaki tumor registries. The committee evaluated nearly 13,000 incidences of cancer and approximately 10,000 cancer deaths in contrast to fewer than 6000 cancer deaths available to the BEIR V committee. Also, since completion of the 1990 report, additional evidence has emerged from studies of the Hiroshima and Nagasaki atomic bomb survivors suggesting that other health effects, such as cardiovascular disease and stroke, can result from radiation exposure.

A major reevaluation of the dosimetry at Hiroshima and Nagasaki has recently been completed that lends more certainty to dose estimates and provides increased confidence in the relationship between radiation exposure and the health effects observed in Japanese A-bomb survivors. Additional new information is also available from radiation worker studies, medical radiation exposures, and populations with environmental exposures.

Although the cancer risk estimates have not changed greatly since the 1990 report, confidence in the estimates has risen because of the increase in epidemiologic and biological data available to the committee.

Progress has also been made since the 1990 report in areas of science that relate to the estimation of genetic (hereditary) effects of radiation. In particular, (1) advances in human molecular biology have been incorporated into the conceptual framework of genetic risk estimation, and (2) it has become possible to project risks for all classes of genetic diseases (*i.e.*, those with more complex as well as simple patterns of inheritance).

Advances in cell and molecular biology have also contributed new information on the mechanisms through which cells respond to radiation-induced damage and to the close associations between DNA damage response and cancer development.

ORGANIZATION OF THE STUDY

The NRC appointed a committee comprised of scientists and educators. Some had particular expertise in conducting research on ionizing radiation, while others were experienced in fields relevant to the committee's charge. The NRC vetted all potential members to ensure that each was free from any apparent or potential conflict of interest. The work of the committee was conducted with the assistance of the Board of Radiation Effects Research of the Division on Earth and Life Sciences.

The committee held 11 meetings over a period of 4.5 years. The long duration of the committee was due largely to a period of reduced activity while awaiting completion of the update of the dosimetry and exposure estimates to atomic bomb survivors of Hiroshima and Nagasaki, Japan (the so-called DS02: Dosimetry System 2002).

Six of the meetings included participation of the public for a portion of the meeting, and five of the meetings were conducted exclusively in executive session. Each meeting included extensive deliberations involving the committee as a whole; in addition, two major subcommittees were formed that were termed "biology" and "epidemiology." Dr. Monson convened the epidemiology sessions and Dr. Cleaver convened the biology sessions. Also, a number of loosely organized and nonpermanent working groups were formed to discuss the many issues before the committee. This enabled biologists and nonbiologists to work together and evaluate each other's work.

ORGANIZATION OF THE REPORT

As noted under its STATEMENT OF TASK, the committee's focus was to develop the best possible risk estimate for exposure to low-dose, low-LET radiation in human subjects. Accordingly, Chapters 1–4 discuss basic aspects of radiation physics and radiation biology, including the known interaction between radiation exposure and genetic material, cellular structures, and whole organisms. Chapters 5–9 discuss basic principles of epidemiology as well as substantive data relating to exposure from the atomic bombs, medical radiation, occupational radiation, and environmental radiation. Chapters 10–12, to the extent possible, integrate the information from biology and epidemiology and develop risk estimates based on this information. Three summary sections provide different levels of description of the report. Chapter 13 is an overall scientific summary and lays out the research needs identified by the committee. The Executive Summary is an abbreviated and reorganized version of Chapter 13 that provides an overview of the report. The Public Summary addresses the findings of the committee and the relevance of the report to public concerns about exposure to ionizing radiation.

Reviewers

This report has been reviewed in draft form by persons chosen for their diverse perspectives and technical expertise in accordance with procedures approved by the National Research Council's Report Review Committee. The purposes of this review are to provide candid and critical comments that will assist the institution in making the published report as sound as possible and to ensure that the report meets institutional standards of objectivity, evidence, and responsiveness to the study charge. The review comments and draft manuscript remain confidential to protect the integrity of the deliberative process. We wish to thank the following for their participation in the review of this report:

Seymour Abrahamson, University of Wisconsin, Madison, WI

John F. Ahearne, Sigma Xi, The Scientific Research Society, Research Triangle Park, NC

Allan Balmain, University of California, San Francisco, CA

Michael Cornforth, University of Texas, Galveston, TX

James F. Crow, University of Wisconsin, Madison, WI

John Easton, University of Chicago Hospitals, Chicago, IL

Eric J. Hall, Columbia University College of Physicians and Surgeons, New York, NY

Richard D. Hichwa, University of Iowa, Iowa City, IA

Hedvig Hricak, Memorial Sloan-Kettering Cancer Center, New York, NY

Glenn F. Knoll, University of Michigan, Ann Arbor, MI

Jack S. Mandel, Emory University Rollins School of Public Health, Atlanta, GA

John P. Murnane, University of California, San Francisco, CA

Hooshang Nikjoo, National Aeronautics and Space Administration, Houston, TX

Jonathan M. Samet, Johns Hopkins University, Baltimore, MD

Susan S. Wallace, University of Vermont, Burlington, VT

Chris G. Whipple, ENVIRON International Corporation, Emeryville, CA

Although the reviewers listed above have provided many constructive comments and suggestions, they were not asked to endorse the conclusions or recommendations, nor did they see the final draft of the report before its release. The review of this report was overseen by George M. Hornberger, Ernest H. Ern Professor of Environmental Sciences and Associate Dean for the Sciences, University of Virginia, and John C. Bailar III, Professor Emeritus, University of Chicago. Appointed by the National Research Council, they were responsible for making certain that an independent examination of this report was carried out in accordance with institutional procedures and that all review comments were carefully considered. Responsibility for the final content of this report rests entirely with the authoring committee and the National Research Council.

GENERAL ACKNOWLEDGMENTS

The committee thanks the directors and staff of the Radiation Effects Research Foundation (RERF), Hiroshima, Japan, for providing the most current Life Span Study data on the Japanese atomic bomb survivors. These data continue to be the primary source of epidemiologic information on the relationship between exposure to ionizing radiation and its effects on human health. In particular, Dr. Donald Pierce was especially helpful in communication between RERF and the committee; he also added his insightful experience to the work of the committee.

The committee was aided in the consideration of its charge not only by comments from the public but also by formal presentations by experts from a number of fields. The following presentations were made as part of the public portion of the meetings (in order of appearance):

Presentations by Sponsors

Jerome Puskin, Ph.D.
Environmental Protection Agency

Vincent Holahan, Ph.D.
U.S. Nuclear Regulatory Commission

Bonnie Richter, Ph.D.
U.S. Department of Energy

Scientific Speakers

John Boice, Ph.D.
International Epidemiology Institute
Epidemiology that should be considered by BEIR VII

Charles Waldren, Ph.D.
Colorado State University
Adaptive effects, genomic instability, and bystander effects

John Ward, Ph.D.
University of California, San Diego
Differences between ionizing radiation-induced DNA
    damage and endogenous oxidative damage

Antone Brooks, Ph.D.
Washington State University Tri-cities
Overview of projects funded by the Department of Energy
    low-dose program

Charles Land, Ph.D.
National Institutes of Health (NIH)
National Cancer Institute's update of the 1985 NIH
    Radioepidemiologic Tables

L.B. Russell, Ph.D.
Oak Ridge National Laboratory
Early information derived from radiation-induced
    mutations in mice

R. Chakraborty, Ph.D.
University of Texas School of Public Health
Mini- and microsatellite mutations and their possible
    relevance for genetic risk estimation

Allan Balmain, Ph.D.
University of California, San Francisco
High- and low-penetrance genes involved in cancer
    incidence

Al Fornace, Ph.D.
Harvard School of Public Health
Functional genomics and informatics approaches to
    categorize radiation response

Steve Wing, Ph.D.
University of North Carolina
Relevance of occupational epidemiology to radiation risk
    assessment

Edward Calabrese, Ph.D.
University of Massachusetts
Radiation hormesis

David Utterback, Ph.D.
National Institute of Occupational Safety and Health
Exposure assessment and radiation worker studies

Sharon Dunwoody, Ph.D.
University of Wisconsin
Challenges in the communication of scientific uncertainties

Suresh Moolgavkar, Ph.D., M.B.B.S.
School of Public Health and Community Medicine,
    University of Washington and Fred Hutchinson
    Cancer Research Center
Biology-based models

We thank these presenters and all other members of the public who spoke on issues related to ionizing radiation.

The committee thanks Dr. Isaf Al-Nabulsi for her assistance at the beginning of this study and Doris Taylor and Cathie Berkley for their administrative assistance in assuring that its members showed up at the right place at the right time. The committee was also aided in its work by a talented group of program assistants. We thank Courtney Gibbs for her assistance in the preparation of this manuscript. We thank Courtney Slack, a Christine Mirzayan Science and Technology Policy Graduate Fellow, who provided additional valuable assistance to NRC staff.

We thank Dr. Evan Douple for pulling us in and holding us together. His wise and patient counsel along with his gentle encouragement, when needed, kept the committee focused on its charge.

Finally, special thanks are due to Dr. Rick Jostes, the study director. His scientific expertise, persistence, equanimity, and organizational skills were essential to our staying the course.

RICHARD MONSON, *Chairman*

Units Used to Express Radiation Dose

Radiation exposures are measured in terms of the quantity *absorbed dose*, which equals the ratio of energy imparted to the mass of the exposed body or organ. The unit of absorbed dose is joules per kilogram (J/kg). For convenience this unit has been given the special name gray (Gy).

Ionizing radiation can consist of electromagnetic radiation, such as X-rays or gamma rays (γ-rays), or of subatomic particles, such as protons, neutrons, and α-particles. X- and γ-rays are said to be sparsely ionizing, because they produce fast electrons, which cause only a few dozen ionizations when they traverse a cell. Because the rate of energy transfer is called *linear energy transfer* (*LET*), they are also termed *low-LET* radiation; low-LET radiations are the subject of this report. In contrast, the heavier particles are termed *high-LET* radiations because they transfer more energy per unit length as they traverse the cell.

Since the high-LET radiations are capable of causing more damage per unit absorbed dose, a weighted quantity, *equivalent dose*, or its average over all organs, *effective dose*, is used for radiation protection purposes. For low-LET radiation, *equivalent dose* equals *absorbed dose*. For high-LET radiation—such as neutrons, α-particles, or heavier ion particles—*equivalent dose* or *effective dose* equals the *absorbed dose* multiplied by a factor, the *quality factor* or the *radiation weighting factor* (see Glossary), to account for their increased effectiveness. Since the weighting factor for radiation quality is dimensionless, the unit of *equivalent dose* is also joules per kilogram. However, to avoid confusion between the two dose quantities, the special name sievert (Sv) has been introduced for use with *equivalent dose* and *effective dose*.

Although the BEIR VII report is about low-LET radiation, the committee has had to consider information derived from complex exposures—especially from atomic bomb radiation—that include a high-LET contribution in addition to low-LET radiation. A *weighted dose*, with a weight factor

that differs from the quality factor and the radiation weighting factor, is employed in these computations. The unit sievert is likewise used with this quantity.

Whenever the nature of the quantity is apparent from the context, the term *dose* is used equally in this report for *absorbed dose*, *equivalent dose*, *effective dose*, and *weighted dose*. With regard to risk assessment, reference is usually to the *equivalent dose* to specified organs or to the *effective dose*. The unit sievert is then used, although absorbed dose and equivalent dose are equal for low-LET radiation. In experimental radiation biology and radiotherapy, exact specification of absorbed dose is required and the dose values are frequently larger than in radiation protection considerations. With reference to those fields, therefore, use is made of absorbed dose and the unit is gray.

The Public Summary refers to radiation protection, and the dose therefore is given as sieverts throughout that chapter (for a more complete description of the various dose quantities and units used in this report, see the Glossary and the table below).

TABLE 1 Units of Dose

Unit[a]	Symbol	Conversion Factors
Becquerel (SI)	Bq	1 disintegration/s = 2.7×10^{-11} Ci
Curie	Ci	3.7×10^{10} disintegrations/s = 3.7×10^{10} Bq
Gray (SI)	Gy	1 J/kg = 100 rads
Rad	rad	0.01 Gy = 100 erg/g
Sievert (SI)	Sv	1 J/kg = 100 rem
Rem	rem	0.01 Sv

NOTE: Equivalent dose equals absorbed dose times Q (quality factor). Gray is the special name of the unit (J/kg) to be used with absorbed dose; sievert is the special name of the unit (J/kg) to be used with equivalent dose.

[a]International Units are designated SI.

Contents

Public Summary

INTRODUCTION

The health effects of low levels of ionizing radiation are important to understand. Ionizing radiation—the sort found in X-rays or gamma rays[1]—is defined as radiation that has sufficient energy to displace electrons from molecules. Free electrons, in turn, can damage human cells. One challenge to understanding the health effects of radiation is that there is no general property that makes the effects of man-made radiation different from those of naturally occurring radiation. Still another difficulty is that of distinguishing cancers that occur because of radiation exposure from cancers that occur due to other causes. These facts are just some of the many that make it difficult to characterize the effects of ionizing radiation at low levels.

Despite these challenges, a great deal about this topic is well understood. Specifically, substantial evidence exists that exposure to high levels of ionizing radiation can cause illness or death. Further, scientists have long known that in addition to cancer, ionizing radiation at high doses causes mental retardation in the children of mothers exposed to radiation during pregnancy. Recently, data from atomic bomb survivors suggest that high doses are also connected to other health effects such as heart disease and stroke.

Because ionizing radiation is a threat to health, it has been studied extensively. This report is the seventh in a series of publications from the National Academies concerning radiation health effects, referred to as the Biological Effects of Ionizing Radiation (BEIR) reports. This report, BEIR VII, focuses on the health effects of low levels of low linear energy transfer (LET) ionizing radiation. Low-LET radiation deposits less energy in the cell along the radiation path and is considered less destructive per radiation track than high-LET radiation. Examples of low-LET radiation, the subject of this report, include X-rays and γ-rays (gamma rays). Health effects of concern include cancer, hereditary diseases, and other effects, such as heart disease.

This summary describes:

- how ionizing radiation was discovered,
- how ionizing radiation is detected,
- units used to describe radiation dose,
- what is meant by low doses of ionizing radiation,
- exposure from natural "background" radiation,
- the contribution of man-made radiation to public exposure,
- scenarios illustrating how people might be exposed to ionizing radiation above background levels,
- evidence for adverse health effects such as cancer and hereditary disease,
- the BEIR VII risk models,
- what bodies of research the committee reviewed,
- why the committee has not accepted the view that low levels of radiation might be substantially more or less harmful than expected from the model used in this BEIR report, and
- the committee's conclusions.

HOW IONIZING RADIATION WAS DISCOVERED

Low levels of ionizing radiation cannot be seen or felt, so the fact that people are constantly exposed to radiation is not usually apparent. Scientists began to detect the presence of ionizing radiation in the 1890s.[2] In 1895, Wilhelm Conrad Roentgen was investigating an electrical discharge generated in a paper-wrapped glass tube from which most of the air had been evacuated. The free electrons generated in the "vacuum tube," which were then called cathode rays, were

---

[1]X-rays are man-made and generated by machines, whereas gamma rays occur from unstable atomic nuclei. People are continuously exposed to gamma rays from naturally occurring elements in the earth and outer space.

[2]Health Physics Society. *Figures in Radiation History, http://www.hps.org.* September 2004.

1

in themselves a form of radiation. Roentgen noted that when the electrons were being generated, a fluorescent screen on a nearby table began to glow. Roentgen theorized that invisible emissions from the cathode-ray tube were causing the fluorescent screen to glow, and he termed these invisible emissions X-rays. The electrons produced by the electrical discharge had themselves produced another form of radiation, X-rays. The next major discovery occurred when Henri Becquerel noted that unexposed photographic plates stored in a drawer with uranium ore were fogged. He concluded that the fogging was due to an invisible emission emanating from the uranium atoms and their decay products. This turned out to be naturally occurring radiation emanating from the uranium. Marie and Pierre Curie went on to purify radium from uranium ore in Becquerel's laboratory, and in subsequent years, many other forms of radiation including neutrons, protons, and other particles were discovered. Thus, within a period of several years in the 1890s, man-made and naturally occurring radiation were discovered.

Roentgen's discovery of X-rays resulted in the eventual invention of X-ray machines used to image structures in the human body and to treat health conditions. Adverse health effects of high levels of ionizing radiation exposure became apparent shortly after these initial discoveries. High doses to radiation workers would redden the skin (erythema), and this rough measure of radiation exposure was called the "skin erythema dose." The use of very large doses, primitive dosimetry (dose measurement) such as the skin erythema dose, and the fact that many of these early machines were not well shielded led to high radiation exposures both to the patients and to the persons administering the treatments. The development of chronic, slow-healing skin lesions on the hands of early radiologists and their assistants resulted in the loss of extremities in some cases. These incidents were some of the first indications that radiation delivered at high doses could have serious health consequences. Subsequent studies in recent years have shown that early radiologists had a higher mortality rate than other health workers. This increased mortality rate is not seen in radiologists working in later years, presumably due to vastly improved safety conditions resulting in much lower doses to radiologists.

The early indications of health effects after high radiation exposures are too many to chronicle in this Public Summary, but the committee notes one frequently cited example. In 1896, Thomas Edison developed a fluoroscope that consisted of a tapered box with a calcium tungstate screen and a viewing port by which physicians could view X-ray images. During the course of these investigations with X-rays, Clarence Dally, one of Edison's assistants, developed a degenerative skin disease, that progressed into a carcinoma. In 1904, Dally succumbed to his injuries in what may have been the first death associated with man-made ionizing radiation in the United States. Edison halted all of his X-ray research noting that "the x rays had affected poisonously my assistant, Mr.

Dally . . ."[3] Today, radiation is one of the most thoroughly studied potential hazards to humans, and regulatory standards have become increasingly strict over the years in an effort to protect human health.

HOW IONIZING RADIATION IS DETECTED

The detection of ionizing radiation has greatly improved since the days of Roentgen, Becquerel, and the Curies. Ionizations can be detected accurately by Geiger counters and other devices. Because the efficiency of the detector is known, one can determine not only the location of the radiation, but also the amount of radiation present. Other, more sophisticated detectors can evaluate the "signature" energy spectrum of some radiations and thus identify the type of radiation.

UNITS USED TO DESCRIBE RADIATION DOSE

Ionizing radiation can be in the form of electromagnetic radiation, such as X-rays or γ-rays, or in the form of subatomic particles, such as protons, neutrons, alpha particles, and beta particles. Radiation units can be confusing. Radiation is usually measured in dose units called grays (Gy) or sieverts (Sv), which are measures of energy deposited in living tissue. X- and γ-rays are said to have low LET. Low-LET radiation produces ionizations sparsely throughout a cell; in contrast, high-LET radiation transfers more energy per unit length as it traverses the cell and is more destructive per unit length.

Although this BEIR VII report is about low-LET radiation, the committee has considered some information derived from complex exposures that include radiation from high-LET and low-LET sources. High-LET or mixed radiations (radiation from high-LET and low-LET sources) are often described in units known as sievert. The units for low-LET radiation can be sievert or gray. For simplicity, all dose units in the Public Summary are reported in sieverts (Sv). For a more complete description of the various units of dose used in this report, see "Units Used to Express Radiation Dose" which precedes the Public Summary, as well as the terms Gray, Sievert, and Units in the glossary.

WHAT IS MEANT BY LOW DOSES OF IONIZING RADIATION

For this report, the committee has defined low dose as doses in the range of near zero up to about 100 mSv (0.1 Sv) of low-LET radiation. The committee has placed emphasis on the lowest doses where relevant data are available. The annual worldwide background exposure from natural sources of low-LET radiation is about 1 mSv.

---

[3]Health Physics Society. Figures in Radiation History, *http://www.hps.org.* September 2004.

EXPOSURE FROM NATURAL BACKGROUND RADIATION

Human beings are exposed to natural background radiation every day from the ground, building materials, air, food, the universe, and even elements in their own bodies. In the United States, the majority of exposure to background ionizing radiation comes from exposure to radon gas and its decay products. Radon is a colorless, odorless gas that emanates from the earth and, along with its decay products, emits a mixture of high- and low-LET radiation. Radon can be hazardous when accumulated in underground areas such as poorly ventilated basements. The National Research Council 1999 report, *Health Effects of Exposure to Radon* (BEIR VI), reported on the health effects of radon, and therefore those health effects are not discussed in this report. Average annual exposures worldwide to natural radiation sources (both high and low LET) would generally be expected to be in the range of 1–10 mSv, with 2.4 mSv being the present estimate of the central value.[4] Of this amount, about one-half (1.2 mSv per year) comes from radon and its decay products. Average annual background exposures in the United States are slightly higher (3.0 mSv) due in part to higher average radon levels. After radon, the next highest percentage of natural ionizing radiation exposure comes from cosmic rays, followed by terrestrial sources, and "internal" emissions. Cosmic rays are particles that travel through the universe. The Sun is a source of some of these particles. Other particles come from exploding stars called supernovas.

The amount of terrestrial radiation from rocks and soils varies geographically. Much of this variation is due to differences in radon levels. "Internal" emissions come from radioactive isotopes in food and water and from the human body itself. Exposures from eating and drinking are due in part to the uranium and thorium series of radioisotopes present in food and drinking water.[5] An example of a radioisotope moving through the food chain would be carbon-14 (^{14}C), a substance found in all living things. ^{14}C is created when cosmic rays collide with nitrogen atoms. ^{14}C combines with oxygen to create carbon dioxide gas. Plants absorb carbon dioxide during photosynthesis, and animals feed on those plants. In these ways, ^{14}C accumulates in the food chain and contributes to the internal background dose from ionizing radiation.

As mentioned previously, possible health effects of low-dose, low-LET radiation are the focus of this BEIR VII report. Because of the "mixed" nature of many radiation sources, it is difficult to estimate precisely the percentage of natural background radiation that is low LET. Figure PS-1 illustrates the approximate sources and relative amounts of high-LET and low-LET radiations that comprise the natural background exposure worldwide. This figure illustrates the relative contributions of three natural sources of high-LET radiation and three natural sources of low-LET radiation to the global population exposure. The smaller, detached segment of the chart represents the relative contribution of low-LET radiation sources to the annual background exposure. The total average annual population exposure worldwide due to low-LET radiation would generally be expected to be in the range of 0.2–1.0 mSv, with 0.9 mSv being the present estimate of the central value.

CONTRIBUTION OF MAN-MADE RADIATION TO PUBLIC EXPOSURE

In addition to natural background radiation, people are also exposed to low- and high-LET radiation from man-made sources such as X-ray equipment and radioactive materials used in medicine, research, and industry. A 1987 study[6] of ionizing radiation exposure of the population of the United States estimated that natural background radiation comprised 82% of the annual U.S. population exposure, while man-made sources contributed 18% (see Figure PS-2, pie chart in the lower left portion of the figure).

In Figure PS-2, the man-made radiation component (upper right portion of the figure) shows the relative contributions of the various types of man-made radiation to the U.S. population.[7] Medical X-rays and nuclear medicine account for about 79% of the man-made radiation exposure in the United States. Elements in consumer products, such as tobacco, the domestic water supply, building materials, and to a lesser extent, smoke detectors, televisions, and computer screens, account for another 16%. Occupational exposures, fallout, and the nuclear fuel cycle comprise less than 5% of the man-made component and less than 1% of the combined background and man-made component. Additional small amounts of exposure from background and man-made radiation come from activities such as traveling by jet aircraft (cosmic radiation—add 0.01 mSv for each 1000 miles traveled), living near a coal-fired power plant (plant emissions—add 0.0003 mSv), being near X-ray luggage inspection scanners (add 0.00002 mSv), or living within 50 miles of a nuclear power plant (add 0.00009 mSv).[8]

---

[4]United Nations Scientific Committee on the Effects of Atomic Radiation (UNSCEAR). 2000. Sources and Effects of Ionizing Radiation, Volume 1: Sources. New York: United Nations. Table 31, p. 40.

[5]UNSCEAR. 2000. Sources and Effects of Ionizing Radiation. Report to the General Assembly, with scientific annexes. New York: United Nations.

[6]National Council on Radiation Protection and Measurements (NCRP). 1987. Ionizing Radiation Exposure of the Population of the United States. Washington, DC: NCRP, No. 93.

[7]National Council on Radiation Protection and Measurements. 1987. Ionizing Radiation Exposure of the Population of the United States. Washington, DC: NCRP, No. 93.

[8]National Council on Radiation Protection and Measurements Reports #92-95 and #100. Washington, DC: NCRP.

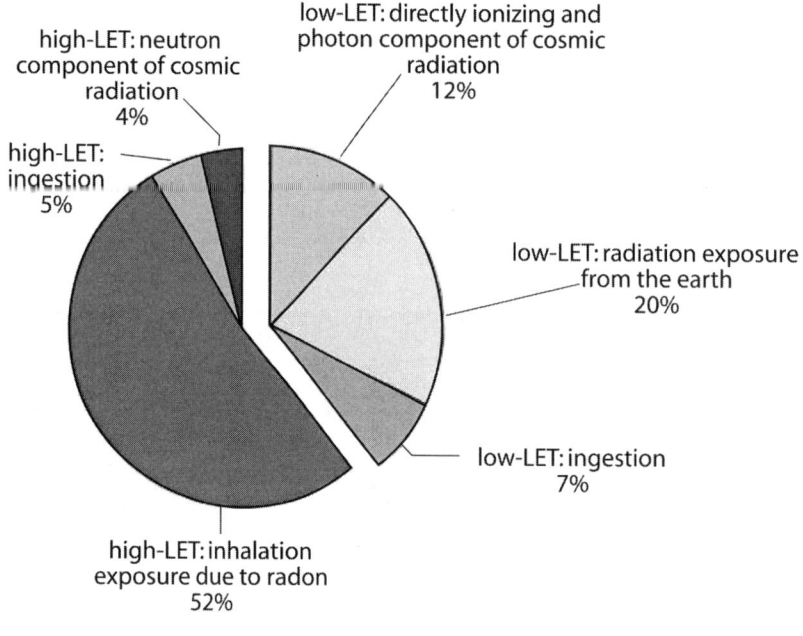

high-LET: neutron component of cosmic radiation 4%

high-LET: ingestion 5%

low-LET: directly ionizing and photon component of cosmic radiation 12%

low-LET: radiation exposure from the earth 20%

low-LET: ingestion 7%

high-LET: inhalation exposure due to radon 52%

**Worldwide background radiation
2.4 mSv/year**

FIGURE PS-1 Sources of global background radiation. The pie chart above shows the relative worldwide percentage of all sources of natural background radiation (low and high LET). Because this report evaluates the health effects of low-LET radiation, the low-LET portion of the pie chart is separated to illustrate the relative contributions of the three major sources of low-LET radiation exposure. SOURCE: Data from UNSCEAR 2000a.

There are many ways in which an individual's exposure to ionizing radiation could vary from the averages. Factors that might increase exposure to ionizing radiation include (1) increased uses of radiation for medical purposes, (2) occupational exposure to radiation, and (3) smoking tobacco products.[9] Factors that might decrease radiation exposure include living at lower altitudes (less cosmic radiation) and living and working in the higher floors of a building (less radon).

SCENARIOS ILLUSTRATING HOW PEOPLE MIGHT BE EXPOSED TO IONIZING RADIATION ABOVE BACKGROUND LEVELS

This section provides three scenarios illustrating how some people might be exposed to ionizing radiation above background levels. These examples are for illustration purposes only and are not meant to be inclusive.

Whole-Body Scans

There is growing use of whole-body scanning by computed tomography (CT) as a way of screening for early signs

of disease among asymptomatic adults.[10] CT examinations result in higher organ doses of radiation than conventional single-film X-rays. This is because CT scanners rotate around the body, taking a series of cross-sectional X-rays. A computer compiles these X-ray slices to produce a three-dimensional portrait. According to Brenner and Elliston, who estimated both radiation dose and risks from such procedures, a single full-body scan results in a mean effective radiation dose of 12 mSv.[11] These authors write, "To put this (dose) in perspective, a typical mammogram . . . has an effective dose of 0.13 mSv—a factor of almost 100 times less." According to Brenner and Elliston's calculations, "a 45-year-old adult who plans to undergo 30 annual full-body CT examinations would potentially accrue an estimated lifetime cancer mortality risk of 1.9% (almost 1 in 50). . . . Correspondingly, a 60-year-old who plans to undergo 15 annual full-body CT examinations would potentially accrue an estimated lifetime cancer mortality risk of one in 220." Citing a National Vital Statistics Report, Brenner and Elliston note, for comparison that, "the lifetime odds that an individual born in the United States in 1999 will die in a traffic accident

[9]National Council on Radiation Protection and Measurements. 1987. Radiation exposure of the U.S. population from Consumer Products and Miscellaneous Sources. Bethesda, MD: NCRP, Report No. 95.

[10]Full-Body CT Scans: What You Need to Know (brochure). U.S. Department of Health and Human Services. 2003. Accessed at *www.fda.gov/cdrh/ct.*

[11]Brenner, D.J., and C.D. Elliston. 2004. Estimated radiation risks potentially associated with full-body CT screening. Radiology 232:735–738.

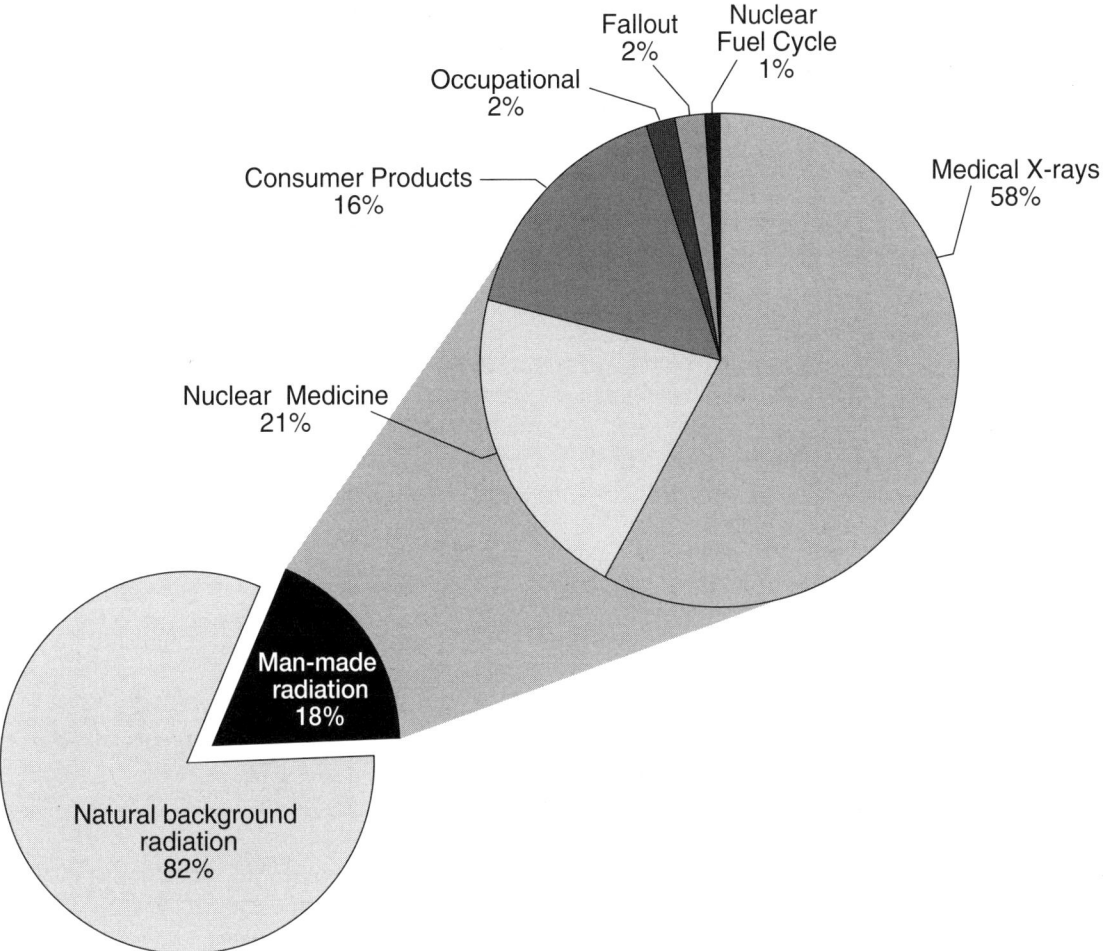

FIGURE PS-2 The pie chart in the lower left portion of the figure shows the contribution of man-made radiation sources (18%) relative to natural background radiation (82%) exposure of the population of the United States. Sources of man-made radiation are detailed in the upper right portion of the pie chart. SOURCE: Data from NCRP 1987.

are estimated to be one in 77."[12] Further information on whole-body scans is available from the U.S. Food and Drug Administration web site.[13]

CT Scans Used in Diagnostic Procedures

The use of CT scans in adults experiencing symptoms of illness or injury is widely accepted, and CT scan use has increased substantially in the last several decades. The BEIR VII committee recommends that in the interest of radiological protection, there be follow-up studies of cohorts of persons receiving CT scans, especially children. In addi-

tion, the committee recommends studies of infants who experience diagnostic radiation exposure related to cardiac catheterization and of premature infants who are monitored with repeated X-rays for pulmonary development.

Working near Ionizing Radiation

People who work at medical facilities, in mining or milling, or with nuclear weapons are required to take steps to protect themselves from occupational exposures to radiation. The maximum amount of radiation that workers are allowed to receive in connection with their occupations is regulated. In general these limits are 50 mSv per year to the whole body, with larger amounts allowed to the extremities. The exposure limits for a pregnant worker, once pregnancy is declared, are more stringent. In practice the guidelines call for limiting exposures to as low as is reasonably achievable.

Combined analyses of data from nuclear workers offer an opportunity to increase the sensitivity of such studies and to

---

[12]Hoyert, D. L., E. Arias, B.L. Smith, S.L. Murphy, and K.D. Kochanek. 2001. Deaths: Final data for 1999. National Vital Statistics Report USA 49:1–113.

[13]Full-Body CT Scans: What You Need to Know (brochure), U.S. Department of Health and Human Services. 2003. Accessed at *www.fda.gov/cdrh/ct.*

provide direct estimates of the effects of long-term, low-dose, low-LET radiation. It should be noted however that even with the increased sensitivity, the combined analyses are compatible with a range of possibilities, from a reduction of risk at low doses to risks twice those on which current radiation protection recommendations are based.

Veterans Exposed to Radiation Through Weapons Testing

An example of man-made radiation exposures experienced by large numbers of people in the past is the experience of the U.S. atomic veterans during and after World War II. From 1945 to 1962, about 210,000 military and civilian personnel were exposed directly at a distance to aboveground atomic bomb tests (about 200 atmospheric weapons tests were conducted in this period).[14] In general, these exercises, conducted in Nevada, New Mexico, and the Pacific, were intended to familiarize combat teams with conditions that would be present during a potential war in which atomic weapons might be used. As an example, in the series of five atmospheric tests conducted during Operation UPSHOT-KNOTHOLE, individual battalion combat teams experienced low-LET γ-ray doses as low as 0.4 mSv and as high as 31 mSv. This range of exposures would correspond to the equivalent of about five chest X-rays for the lowest-exposed combat team to approximately 390 chest X-rays for the highest-exposed combat team (by assuming a dose from one chest X-ray to be about 0.08 mSv).

EVIDENCE FOR ADVERSE HEALTH EFFECTS SUCH AS CANCER AND HEREDITARY DISEASE

The mechanisms that lead to adverse health effects after exposure to ionizing radiation are not fully understood. Ionizing radiation has sufficient energy to change the structure of molecules, including DNA, within the cells of the human body. Some of these molecular changes are so complex that it may be difficult for the body's repair mechanisms to mend them correctly. However, the evidence is that only a very small fraction of such changes would be expected to result in cancer or other health effects. Radiation-induced mutations would be expected to occur in the reproductive cells of the human body (sperm and eggs), resulting in heritable disease. The latter risk is sufficiently small that it has not been detected in humans, even in thoroughly studied irradiated populations such as those of Hiroshima and Nagasaki.

As noted above, the most thoroughly studied individuals for determination of the health effects of ionizing radiation are the survivors of the Hiroshima and Nagasaki atomic bombs. Sixty-five percent of these survivors received a low dose of radiation (less than 100 mSv; the definition of low dose used by this BEIR VII report). A dosage of 100 mSv is equivalent to approximately 40 times the average yearly background radiation exposure worldwide from all sources (2.4 mSv) or roughly 100 times the worldwide background exposure from low LET radiation, the subject of this report. At dose levels of about 100 to 4000 mSv (about 40 to 1600 times the average yearly background exposure), excess cancers have been observed in Japanese atomic bomb survivors. Excess cancers represent the number of cancers above the levels expected in the population. In the case of *in utero* exposure (exposure of the fetus during pregnancy), excess cancers can be detected at doses as low as 10 mSv.[15] For the radiation doses at which excess cancers occur in the Hiroshima and Nagasaki studies, solid cancers[16] show an increasing rate with increasing dose that is consistent with a linear association. In other words, as the level of exposure to radiation increased, so did the occurrence of solid cancers.

Major advances have occurred during the last decade in several key areas that are relevant to the assessment of risks at low radiation doses. These advances have contributed to greater insights into the molecular and cellular responses to ionizing radiation and into the nature of the relationship between radiation exposure and the types of damage that underlie adverse health outcomes. Also, more data on radiation-induced cancers in humans have become available since the previous BEIR report on the health effects of low-dose, low-LET radiation, and those data are evaluated in this report.

THE BEIR VII RISK MODELS

Estimating Cancer Risk

An important task of the BEIR VII committee was to develop "risk models" for estimating the relationship between exposure to low levels of low-LET ionizing radiation and harmful health effects. The committee judged that the linear no-threshold model (LNT) provided the most reasonable description of the relation between low-dose exposure to ionizing radiation and the incidence of solid cancers that are induced by ionizing radiation. This section describes the LNT; the linear-quadratic model, which the committee adopted for leukemia; and a hypothetical linear model with a threshold. It then gives an example derived from the BEIR VII risk models using a figure with closed circles representing the frequency of cancers in the general population and a star representing estimated cancer incidence from ra-

---

[14]National Research Council. 2003. A Review of the Dose Reconstruction Program of the Defense Threat Reduction Agency. Washington, DC: National Academies Press, *http://www.nap.edu/catalog/10697.html*.

[15]Doll, R., and R. Wakeford. 1997. Risk of childhood cancer from foetal irradiation. Brit J Radiol 70:130–139.

[16]Solid cancers are cellular growths in organs such as the breast or prostate as contrasted with leukemia, a cancer of the blood and blood-forming organs.

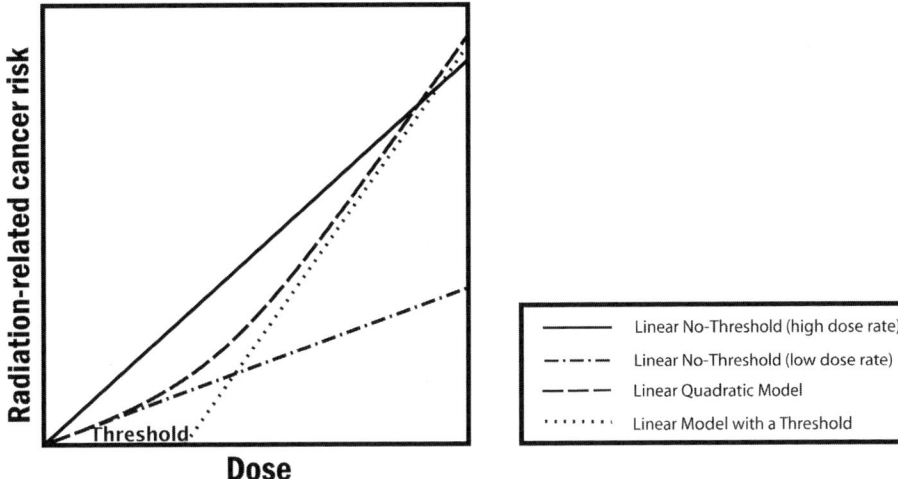

FIGURE PS-3 The committee finds the linear no-threshold (LNT) model to be a computationally convenient starting point. Actual risk estimates improve upon this simplified model by using a dose and dose-rate effectiveness factor (DDREF), which is a multiplicative adjustment that results in downward estimation of risk and is roughly equivalent to using the line labeled "Linear No-Threshold" (low dose rate). The latter is the zero-dose tangent of the linear-quadratic model. While it would be possible to use the linear-quadratic model directly, the DDREF adjustment to the linear model is used to conform with historical precedent dictated in part by simplicity of calculations. In the low-dose range of interest, there is essentially no difference between the two. Source: Modified from Brenner and colleagues.[17]

diation exposure using the BEIR VII risk models. Next, the section explains how the absence of evidence for induced adverse heritable effects in the children of survivors of atomic bombs is consistent with the genetic risk estimated through the use of the doubling dose method in this report.

At doses less than 40 times the average yearly background exposure (100 mSv), statistical limitations make it difficult to evaluate cancer risk in humans. A comprehensive review of the biology data led the committee to conclude that the risk would continue in a linear fashion at lower doses without a threshold and that the smallest dose has the potential to cause a small increase in risk to humans. This assumption is termed the "linear no-threshold model" (see Figure PS-3).

The BEIR VII committee has developed and presented in Chapter 12 the committee's best risk estimates for exposure to low-dose, low-LET radiation in human subjects. An example of how the data-based risk models developed in this report can be used to evaluate the risk of radiation exposure is illustrated in Figure PS-4. This example calculates the expected cancer risk from a single exposure of 0.1 Sv. The risk depends on both sex and age at exposure, with higher risks for females and for those exposed at younger ages. On

FIGURE PS-4 In a lifetime, approximately 42 (solid circles) of 100 people will be diagnosed with cancer (calculated from Table 12-4 of this report). Calculations in this report suggest that approximately one cancer (star) per 100 people could result from a single exposure to 0.1 Sv of low-LET radiation above background.

[17]Brenner, D.J., R. Doll, D.T. Goodhead, E.J. Hall, C.E. Land, J.B. Little, J.H. Lubin, D.L. Preston, R.J. Preston, J.S. Puskin, E. Ron, R.K. Sachs, J.M. Samet, R.B. Setlow, and M. Zaider. 2003. Cancer risks attributable to low doses of ionizing radiation: Assessing what we really know. P Natl Acad Sci USA 100:13761–13766.

average, assuming a sex and age distribution similar to that of the entire U.S. population, the BEIR VII lifetime risk model predicts that approximately 1 person in 100 would be expected to develop cancer (solid cancer or leukemia) from a dose of 0.1 Sv above background, while approximately 42 of the 100 individuals would be expected to develop solid cancer or leukemia from other causes. Lower doses would produce proportionally lower risks. For example, the committee predicts that approximately one individual per thousand would develop cancer from an exposure to 0.01 Sv. As another example, approximately one individual per hundred would be expected to develop cancer from a lifetime (70-year) exposure to low-LET, natural background radiation (excluding radon and other high-LET radiation). Because of limitations in the data used to develop risk models, risk estimates are uncertain, and estimates that are a factor of two or three larger or smaller cannot be excluded.

Health Effects Other Than Cancer

In addition to cancer, radiation exposure has been demonstrated to increase the risk of other diseases, particularly cardiovascular disease, in persons exposed to high therapeutic doses and also in A-bomb survivors exposed to more modest doses. However, there is no direct evidence of increased risk of noncancer diseases at low doses, and data are inadequate to quantify this risk if it exists. Radiation exposure has also been shown to increase risks of some benign tumors, but data are inadequate to quantify this risk.

Estimating Risks to Children of Parents Exposed to Ionizing Radiation

Naturally occurring genetic (*i.e.*, hereditary) diseases contribute substantially to illness and death in human populations. These diseases arise as a result of alterations (mutations) occurring in the genetic material (DNA) contained in the germ cells (sperm and ova) and are heritable (*i.e.*, can be transmitted to offspring and subsequent generations). Among the diseases are those that show simple predictable patterns of inheritance (which are rare), such as cystic fibrosis, and those with complex patterns (which are common), such as diabetes mellitus. Diseases in the latter group originate from interactions among multiple genetic and environmental factors.

Early in the twentieth century, it was demonstrated that ionizing radiation could induce mutations in the germ cells of fruit flies. These findings were subsequently extended to a number of other organisms including mice, establishing the fact that radiation is a mutagen (an agent that can cause mutations in body cells); human beings are unlikely to be exceptions. Thus began the concern that exposure of human populations to ionizing radiation would cause an increase in the frequency of genetic diseases. This concern moved to center stage in the aftermath of the detonation of atomic weapons over Hiroshima and Nagasaki in World War II. Extensive research programs to examine the adverse genetic effects of radiation in the children of A-bomb survivors were soon launched. Other studies focusing on mammals that could be bred in the laboratory—primarily the mouse—were also initiated in different research centers around the world.

The aim of the early human genetic studies carried out in Japan was to obtain a *direct* measure of adverse effects in the children of A-bomb survivors. The indicators that were used included adverse pregnancy outcomes (*i.e.*, stillbirths, early neonatal deaths, congenital abnormalities); deaths among live-born infants over a follow-up period of about 26 years; growth and development of the children; chromosomal abnormalities; and specific types of mutations. Specific genetic diseases were *not* used as indicators of risk, because not enough was known about them when the studies began.

The *initial* goal of the mouse experiments was to examine the effects of different doses, types, and modes of delivery of radiation on mutation frequencies and the extent to which the germ cell stages in the two sexes might differ in their responses to radiation-induced mutations. As it turned out, however, the continuing scarcity of data on radiation-induced mutations in humans and the compelling need for quantitative estimates of genetic risk to formulate adequate measures for radiological protection necessitated the use of mouse data for indirect prediction of genetic risks in humans.

As in previous BEIR reports, a method termed the "doubling dose method," is used to predict the risk of inducible genetic diseases in the children of people exposed to radiation using naturally occurring genetic diseases as a framework. The doubling dose (DD) is defined as the amount of radiation that is required to produce as many mutations as those occurring spontaneously in *one generation*. The doubling dose is expressed as a ratio of *mutation rates*. It is calculated as a ratio of the average spontaneous and induced mutation rates in a set of genes. A large DD indicates small relative mutation risk, and a small doubling dose indicates a large relative mutation risk. The DD used in the present report is 1 Sv (1 Gy)[18] and derives from human data on spontaneous mutation rates of disease-causing genes and mouse data on induced mutation rates.[19] Therefore, if three mutations occur spontaneously in 1 million people in one generation, six mutations will occur per generation if 1 million people are each exposed to 1 Sv of ionizing radiation, and three of these six mutations would be attributed to the radiation exposure.

More than four decades have elapsed since the genetic studies in Japan were initiated. In 1990, the final results of

----

[18]For the purposes of this report, when low-LET radiation is considered, 1 Gy is equivalent to 1 Sv.

[19]UNSCEAR. 2001. Hereditary Effects of Radiation. Report to the General Assembly. New York: United Nations.

those studies were published. They show (as earlier reports published from time to time over the intervening years showed) that there are no statistically significant adverse effects detectable in the children of exposed survivors, indicating that at the relatively low doses sustained by survivors (of the order of about 400 mSv or less), the genetic risks, as measured by the indicators mentioned earlier, are very low. Other, mostly small-scale studies of the children of those exposed to high doses of radiation for radiotherapy of cancers have also shown no detectable increases in the frequencies of genetic diseases.

During the past 10 years, major advances have occurred in our understanding of the molecular nature and mechanisms underlying naturally occurring genetic diseases and radiation-induced mutations in experimental organisms including the mouse. These advances have shed light on the relationships between spontaneous mutations and naturally occurring genetic diseases and have provided a firmer scientific basis for inferences on the relationships between induced mutations and diseases. The risk estimates presented in this report have incorporated all of these advances. They show that at low or chronic doses of low-LET irradiation, the genetic risks are very small compared to the baseline frequencies of genetic diseases in the population. Additionally, they are consistent with the lack of significant adverse effects in the Japanese studies based on about 30,000 children of exposed survivors. In other words, given the BEIR VII estimates, one would not expect to see an excess of adverse hereditary effects in a sample of about 30,000 children (the number of children evaluated in Hiroshima and Nagasaki). One reason that genetic risks are low is that only those genetic changes compatible with embryonic development and viability will be recovered in live births.

RESEARCH REVIEWED BY THE COMMITTEE

The committee and staff ensured that the conclusions of this report were informed by a thorough review of published, peer-reviewed materials relevant to the committee's formal Statement of Task. Specifically, the sponsors of this study asked for a comprehensive review of all relevant epidemiologic data (*i.e.*, data from studies of disease in populations) related to health effects of low doses of ionizing radiation. In addition, the committee was asked to review all relevant biological information important to the understanding or modeling of those health effects. Along with the review of these bodies of literature and drawing on the accumulated knowledge of its members, the committee and staff also considered mailings, publications, and e-mails sent to them. Data on cancer mortality and incidence from the Life Span Study cohort of atomic bomb survivors in Hiroshima and Nagasaki, based on improved dose estimates, were used by the committee. The committee also considered radiation risk information from studies of persons exposed for medical, occupational, and environmental reasons. Models for breast and thyroid cancer drew directly on medical studies. Further information was gathered in open sessions of the committee held at meetings in Washington, D.C., and Irvine, California. Questions and concerns raised in open sessions were considered by committee members in writing this report.

Why Has the Committee Not Accepted the View That Low Doses Are Substantially More Harmful Than Estimated by the Linear No-Threshold Model?

Some of the materials the committee reviewed included arguments that low doses of radiation are more harmful than a LNT model of effects would suggest. The BEIR VII committee has concluded that radiation health effects research, taken as a whole, does not support this view. In essence, the committee concludes that the higher the dose, the greater is the risk; the lower the dose, the lower is the likelihood of harm to human health. There are several intuitive ways to think about the reasons for this conclusion. First, any single track of ionizing radiation has the potential to cause cellular damage. However, if only one ionizing particle passes through a cell's DNA, the chances of damage to the cell's DNA are proportionately lower than if there are 10, 100, or 1000 such ionizing particles passing through it. There is no reason to expect a greater effect at lower doses from the physical interaction of the radiation with the cell's DNA.

New evidence from biology suggests that cells do not necessarily have to be hit directly by a radiation track for the cell to be affected. Some speculate that hit cells communicate with nonhit cells by chemical signals or other means. To some, this suggests that at very low radiation doses, where all of the cells in the body are not hit, "bystander" cells may be adversely affected, resulting in a greater health effect at low doses than would be predicted by extrapolating the observed response at high doses. Others believe that increased cell death caused by so-called bystander effects might lower the risk of cancer by eliminating cells at risk for cancer from the irradiated cell population. Although additional research on this subject is needed, it is unclear at this time whether the bystander effect would have a net positive or net negative effect on the health of an irradiated person.

In sum, the total body of relevant research for the assessment of radiation health effects provides compelling reasons to believe that the risks associated with low doses of low-LET radiation are no greater than expected on the basis of the LNT model.

Why Has the Committee Not Accepted the View That Low Doses Are Substantially Less Harmful Than Estimated by the Linear No-Threshold Model?

In contrast to the previous section's subject, some materials provided to the committee suggest that the LNT model exaggerates the health effects of low levels of ionizing radiation. They say that the risks are lower than predicted by the

LNT, that they are nonexistent, or that low doses of radiation may even be beneficial. The committee also does not accept this hypothesis. Instead, the committee concludes that the preponderance of information indicates that there will be some risk, even at low doses. As the simple risk calculations in this Public Summary show, the risk at low doses will be small. Nevertheless, the committee's principal risk model for solid tumors predicts a linear decrease in cancer incidence with decreasing dose.

Before coming to this conclusion, the committee reviewed articles arguing that a threshold or decrease in effect does exist at low doses. Those reports claimed that at very low doses, ionizing radiation does not harm human health or may even be beneficial. The reports were found either to be based on ecologic studies or to cite findings not representative of the overall body of data.

Ecologic studies assess broad regional associations, and in some cases, such studies have suggested that the incidence of cancer is much higher or lower than the numbers observed with more precise epidemiologic studies. When the complete body of research on this question is considered, a consensus view emerges. This view says that the health risks of ionizing radiation, although small at low doses, are a function of dose.

Both the epidemiologic data and the biological data are consistent with a linear model at doses where associations can be measured. The main studies establishing the health effects of ionizing radiation are those analyzing survivors of the Hiroshima and Nagasaki atomic bombings in 1945. Sixty-five percent of these survivors received a low dose of radiation, that is, low according to the definition used in this report (equal to or less than 100 mSv). The arguments for thresholds or beneficial health effects are not supported by these data. Other work in epidemiology also supports the view that the harmfulness of ionizing radiation is a function of dose. Further, studies of cancer in children following exposure *in utero* or in early life indicate that radiation-induced cancers can occur at low doses. For example, the Oxford Survey of Childhood Cancer found a "40 percent increase in

the cancer rate among children up to [age] 15."[20] This increase was detected at radiation doses in the range of 10 to 20 mSv.

There is also compelling support for the linearity view of how cancers form. Studies in radiation biology show that "a single radiation track (resulting in the lowest exposure possible) traversing the nucleus of an appropriate target cell has a low but finite probability of damaging the cell's DNA."[21] Subsets of this damage, such as ionization "spurs" that can cause multiple damage in a short length of DNA, may be difficult for the cell to repair or may be repaired incorrectly. The committee has concluded that there is no compelling evidence to indicate a dose threshold below which the risk of tumor induction is zero.

CONCLUSIONS

Despite the challenges associated with understanding the health effects of low doses of low-LET radiation, current knowledge allows several conclusions. The BEIR VII committee concludes that current scientific evidence is consistent with the hypothesis that there is a linear dose-response relationship between exposure to ionizing radiation and the development of radiation-induced solid cancers in humans. The committee further judges it unlikely that a threshold exists for the induction of cancers but notes that the occurrence of radiation-induced cancers at low doses will be small. The committee maintains that other health effects (such as heart disease and stroke) occur at high radiation doses, but additional data must be gathered before an assessment can be made of any possible connection between low doses of radiation and noncancer health effects. Additionally, the committee concludes that although adverse health effects in children of exposed parents (attributable to radiation-induced mutations) have not been found, there are extensive data on radiation-induced transmissible mutations in mice and other organisms. Thus, there is no reason to believe that humans would be immune to this sort of harm.

---

[20]As noted in Cox, R., C.R. Muirhead, J.W. Stather, A.A. Edwards, and M.P. Little. 1995. Risk of radiation-induced cancer at low doses and low dose rates for radiation protection purposes. Documents of the [British] National Radiological Protection Board, Vol. 6, No. 1, p. 71.

[21]As noted in Cox, R., C.R. Muirhead, J.W. Stather, A.A. Edwards, and M.P. Little. 1995. Risk of radiation-induced cancer at low doses and low dose rates for radiation protection purposes. Documents of the National Radiological Protection Board, Vol. 6, No. 1, p. 74.

Executive Summary

INTRODUCTION

This report, prepared by the National Research Council's Committee on the Biological Effects of Ionizing Radiation (BEIR), is the seventh in a series that addresses the health effects of exposure of human populations to low-dose, low-LET (linear energy transfer) ionizing radiation. The current report focuses on new information available since the 1990 BEIR V report on low-dose, low-LET health effects.

Ionizing radiation arises from both natural and man-made sources and at very high doses can produce damaging effects in tissues that can be evident within days after exposure. At the low-dose exposures that are the focus of this report, so-called late effects, such as cancer, are produced many years after the initial exposure. In this report, the committee has defined low doses as those in the range of near 0 up to about 100 milligray (mGy) of low-LET radiation, with emphasis on the lowest doses for which meaningful effects have been found. Additionally, effects that may occur as a result of chronic exposures over months to a lifetime at dose rates below 0.1 mGy/min, irrespective of total dose, are thought to be most relevant. Medium doses are defined as doses in excess of 100 mGy up to 1 Gy, and high doses encompass doses of 1 Gy or more, including the very high total doses used in radiotherapy (of the order of 20 to 60 Gy).

Well-demonstrated late effects of radiation exposure include the induction of cancer and some degenerative diseases (*e.g.*, cataracts). Also, the induction of mutations in the DNA of germ cells that, when transmitted, have the potential to cause adverse health effects in offspring has been demonstrated in animal studies.

EVIDENCE FROM BIOLOGY

There is an intimate relationship between responses to DNA damage, the appearance of gene or chromosomal mutations, and multistage cancer development. Molecular and cytogenetic studies of radiation-associated animal cancers and more limited human data are consistent with the induction of a multistage process of cancer development. This process does not appear to differ from that which applies to spontaneous cancer or to cancers associated with exposure to other carcinogens.

Animal data support the view that low-dose radiation acts principally on the early stages of tumorigenesis (initiation). High-dose effects on later stages (promotion or progression) are also likely. Although data are limited, the loss of specific genes whose absence might result in animal tumor initiation has been demonstrated in irradiated animals and cells.

Adaptation, low-dose hypersensitivity, bystander effect, hormesis, and genomic instability are based mainly on phenomenological data with little mechanistic information. The data suggest enhancement or reduction in radiation effects and in some cases appear to be restricted to special experimental circumstances.

Radiation-Induced Cancer: Mechanisms, Quantitative Experimental Studies, and the Role of Molecular Genetics

A critical conclusion about mechanisms of radiation tumorigenesis is that the data reviewed greatly strengthen the view that there are intimate links between the dose-dependent induction of DNA damage in cells, the appearance of gene or chromosomal mutations through DNA damage misrepair, and the development of cancer. Although less well established, the available data point toward a single-cell (monoclonal) origin of induced tumors. These data also provide some evidence on candidate radiation-associated mutations in tumors. These mutations include loss-of-function DNA deletions, some of which have been shown to be multigene deletions. Certain point mutations and gene amplifications have also been characterized in radiation-associated tumors, but their origins and status are uncertain.

One mechanistic caveat explored was that novel forms of cellular damage response, collectively termed induced genomic instability, might contribute significantly to radiation cancer risk. The cellular data reviewed in this report identified uncertainties and some inconsistencies in the expression of this multifaceted phenomenon. However, telomere-associated mechanisms[1] did provide a coherent explanation for some *in vitro* manifestations of induced genomic instability. The data did not reveal consistent evidence for the involvement of induced genomic instability in radiation tumorigenesis, although telomere-associated processes may account for some tumorigenic phenotypes.

Quantitative animal data on dose-response relationships provide a complex picture of low-LET radiation, with some tumor types showing linear or linear-quadratic relationships, while studies of other tumor types are suggestive of a low-dose threshold, particularly for thymic lymphoma and ovarian cancer. However, the induction or development of these two cancer types is believed to proceed via atypical mechanisms involving cell killing; therefore it was judged that the threshold-like responses observed should not be generalized. Adaptive responses for radiation tumorigenesis have been investigated in quantitative animal studies, and recent information is suggestive of adaptive processes that increase tumor latency but do not affect lifetime risk.

The review of cellular, animal, and epidemiologic or clinical studies of the role of genetic factors in radiation tumorigenesis suggest that many of the known, strongly expressing, cancer-prone human genetic disorders are likely to show an elevated risk of radiation-induced cancer, probably with a high degree of organ specificity. Cellular and animal studies suggest that the molecular mechanisms that underlie these genetically determined radiation effects largely mirror those that apply to spontaneous tumorigenesis and are consistent with the knowledge of somatic mechanisms of tumorigenesis. In particular, evidence has been obtained that major deficiencies in DNA damage response and tumor-suppressor-type genes can serve to elevate radiation cancer risk.

A major theme developing in the study of cancer genetics is the interaction and potential impact of more weakly expressing variant cancer genes that may be relatively common in human populations. Knowledge of such gene-gene and gene-environment interactions, although at an early stage, is developing rapidly. The animal genetic data provide proof-of-principle evidence of how such variant genes with functional polymorphisms can influence cancer risk, including limited data on radiation tumorigenesis.

Given that the functional gene polymorphisms associated with cancer risk may be relatively common, the potential for significant distortion of population-based risk was explored with emphasis on the organ specificity of genes of interest. A preliminary conclusion is that common polymorphisms of DNA damage response genes associated with organ-wide

---

[1]Mechanisms associated with the structure and function of telomeres, which are the terminal regions of a chromosome that include characteristic DNA repeats and associated proteins.

radiation cancer risk would be the most likely source of major interindividual differences in radiation response.

ESTIMATION OF HERITABLE GENETIC EFFECTS OF RADIATION IN HUMAN POPULATIONS

In addition to the induction of cancers in humans by radiation, there is evidence for the heritable genetic effects of radiation from animal experiments. It is now possible to estimate risks for all classes of genetic diseases. The advances that deserve particular attention are the following: (1) introduction of a conceptual change for calculating the doubling dose (from the use of mouse data for *both* spontaneous and induced mutation rates in 1990 to the use of human data on spontaneous mutation rates and mouse data on induced mutation rates now; the latter was the procedure used in the 1972 BEIR report); (2) elaboration of methods to estimate mutation component (*i.e.*, the relative increase in disease frequency per unit relative increase in mutation rate) and use of estimates obtained through these methods to assess the impact of induced mutations on the incidence of Mendelian and chronic multifactorial diseases; (3) introduction of an additional factor, the "potential recoverability correction factor," in the risk equation to bridge the gap between the rates of radiation-induced mutations estimated from mouse data and the predicted risk of radiation-inducible heritable diseases in humans, and (4) introduction of the concept that multisystem developmental abnormalities are likely to be among the principal phenotypes of radiation-induced genetic damage in humans.

The risk estimates presented in this report incorporate all of the above advances. They show that at low or chronic doses of low-LET irradiation, the genetic risks are very small compared to the baseline frequencies of genetic diseases in the population.

The total risk for *all* classes of genetic diseases estimated in this report is about 3000 to 4700 cases per million first-generation progeny per gray. These figures are about 0.4 to 0.6% of the baseline risk of 738,000 cases per million (of which chronic diseases constitute the predominant component—namely, 650,000 cases per million). The BEIR V risk estimates (which did not include chronic diseases) were <2400 to 5300 cases per million first-generation progeny per gray. Those figures were about 5 to 14% of the baseline risk of 37,300 to 47,300 cases per million.

EVIDENCE FROM EPIDEMIOLOGY

Studies of Atomic Bomb Survivors

The Life Span Study (LSS) cohort of survivors of the atomic bombings in Hiroshima and Nagasaki continues to serve as a major source of information for evaluating health risks from exposure to ionizing radiation and particularly for developing quantitative estimates of risk. The advantages of this population include its large size (slightly less than half of the survivors were alive in 2000); the inclusion of both sexes and all ages; a wide range of doses that have been estimated for individual subjects; and high-quality mortality and cancer incidence data. In addition, the whole-body exposure received by this cohort offers the opportunity to assess risks for cancers of a large number of specific sites and to evaluate the comparability of site-specific risks. Special studies of subgroups of the LSS have provided clinical data, biological measurements, and information on potential confounders or modifiers.

Mortality data for the period 1950–1997 have been evaluated in detail. Importantly, cancer incidence data from both the Hiroshima and the Nagasaki tumor registries became available for the first time in the 1990s. These data not only include nonfatal cancers, but also offer diagnostic information that is of higher quality than that based on death certificates, which is especially important when evaluating site-specific cancers. The more extensive data on solid cancer that are now available have allowed more detailed evaluation of several issues pertinent to radiation risk assessment. Analyses evaluating the shape of the dose-response and focusing on the large number of survivors with relatively low doses (less than 0.5 Sv) generally confirm the appropriateness of linear functions to describe solid cancer risks. Both excess relative risk and excess absolute risk models have been used to evaluate the modifying effects of sex, age at exposure, and attained age.

Health end points other than cancer have been linked with radiation exposure in the LSS cohort. Of particular note, a dose-response relationship to mortality from nonneoplastic disease has been demonstrated with statistically significant associations for the categories of heart disease; stroke; and diseases of the digestive, respiratory, and hematopoietic systems. However, noncancer risks at the low doses of interest for this report are especially uncertain, and the committee has not modeled the dose-response for nonneoplastic diseases, or developed risk estimates for these diseases.

Medical Radiation Studies

Published studies on the health effects of medical exposures were reviewed to identify those that provide information for quantitative risk estimation. Particular attention was focused on estimating risks of leukemia and of lung, breast, thyroid, and stomach cancer in relation to radiation dose for comparison with the estimates derived from other exposed populations, in particular atomic bomb survivors.

For lung cancer, the excess relative risk (ERR)[2] per gray from the studies of acute or fractionated high dose-rate ex-

posures are statistically compatible and in the range 0.1–0.4 per Gy. For breast cancer, both the ERR and the excess absolute risk (EAR) appear to be quite variable across studies. A pooled analysis of A-bomb survivors and selected medically exposed cohorts indicated that the EAR for breast cancer was similar (about 10 per 10^4 person-years ([PY]) per gray at age 50) following acute and fractionated moderate to high-dose-rate exposure despite differences in baseline risks and dose rate. Women treated for benign breast conditions appeared to be at higher risk, whereas the risk was lower following protracted low-dose-rate exposures in hemangioma cohorts.

For thyroid cancer, all of the studies providing quantitative information about risks are studies of children who received radiotherapy for benign conditions. For subjects exposed below the age of 15, a linear dose-response was seen, with a leveling or decrease in risk at the higher doses used for cancer therapy (10+ Gy). An ERR of 7.7 per gray and an EAR of 4.4 per 10^4 PY per gray were derived from pooled analyses of data from medical exposures and atomic bomb survivors. Both estimates were significantly affected by age at exposure, with a strong decrease in risk with increasing age at exposure and little apparent risk for exposures after age 20. The ERR appeared to decline over time about 30 years after exposure but was still elevated at 40 years. Little information on thyroid cancer risk in relation to medical iodine-131 (^{131}I) exposure in childhood was available. Studies of the effects of ^{131}I exposure later in life provide little evidence of an increased risk of thyroid cancer.

For leukemia, ERR estimates from studies with average doses ranging from 0.1 to 2 Gy are relatively close, in the range 1.9 to 5 per gray, and are statistically compatible. Estimates of EAR are also similar across studies, ranging from 1 to 2.6 per 10^4 PY per gray. Little information is available on the effects of age at exposure or of exposure protraction.

For stomach cancer, the estimates of ERR per gray range from negative to 1.3. The confidence intervals are wide however, and they all overlap, indicating that these estimates are statistically compatible. Finally, studies of patients having undergone radiotherapy for Hodgkin's disease or breast cancer suggest that there may be some risk of cardiovascular morbidity and mortality for very high doses and high-dose-rate exposures. The magnitude of the radiation risk and the shape of the dose-response curve for these outcomes are uncertain.

Occupational Radiation Studies

Numerous studies have considered the mortality and incidence of cancer among various occupationally exposed groups in the medical, manufacturing, nuclear, research, and aviation industries.

The most informative studies are those of nuclear industry workers (including the workers of Mayak in the former Soviet Union), for whom individual real-time estimates of

---

[2]The ERR is (the rate of disease in an exposed population divided by the rate of disease in an unexposed population) minus 1.0. The EAR is the rate of disease in an exposed population minus the rate of disease in an unexposed population.

doses have been collected over time with the use of personal dosimeters. More than 1 million workers have been employed in this industry since its beginning in the early 1940s. Studies of individual worker cohorts are limited, however, in their ability to estimate precisely the potentially small risks associated with low levels of exposure.

Combined analyses of data from multiple cohorts offer an opportunity to increase the sensitivity of such studies and provide direct estimates of the effects of long-term, low-dose, low-LET radiation. The most comprehensive and precise estimates to date are those derived from the UK National Registry of Radiation Workers and the Three-Country Study (Canada-United Kingdom-United States), which have provided estimates of leukemia and all cancer risks. In these studies, the leukemia risk estimates are intermediate between those derived using linear and linear-quadratic extrapolations from the A-bomb survivors' study. The estimate for all cancers is smaller, but the confidence intervals are wide and consistent both with no risk and with risks up to twice the linear extrapolation from atomic bomb survivors.

Because of the remaining uncertainty in occupational risk estimates and the fact that errors in doses have not formally been taken into account in these studies, the committee concluded that the risk estimates from occupational studies, although directly relevant to the estimation of effects of low-dose protracted exposures, are not sufficiently precise to form the sole basis for radiation risk estimates.

Environmental Studies

Ecological studies of populations living around nuclear facilities and of other environmentally exposed populations do not contain individual estimates of radiation dose or provide a direct quantitative estimate of risk in relation to dose. This limits the interpretation of such data. Several cohort studies have reported health outcomes among persons exposed to environmental radiation. No consistent or generalizable information is contained in these studies.

Results from environmental exposures to ^{131}I have been inconsistent. The most informative findings are from studies of individuals exposed to radiation after the Chernobyl accident. Recent evidence indicates that exposure to radiation from Chernobyl is associated with an increased risk of thyroid cancer and that the relationship is dose dependent. The quantitative estimate of excess thyroid cancer risk is generally consistent with estimates from other radiation-exposed populations and is observed in both males and females. Iodine deficiency appears to be an important modifier of risk, enhancing the risk of thyroid cancer following radiation exposure.

INTEGRATION OF BIOLOGY AND EPIDEMIOLOGY

The principal conclusions from this work are the following:

• Current knowledge of cellular or molecular mechanisms of radiation tumorigenesis tends to support the application of models that incorporate the excess relative risk projection over time.

• The choice of models for the transport of cancer risk from Japanese A-bomb survivors to the U.S. population is influenced by mechanistic knowledge and information on the etiology of different cancer types.

• A combined Bayesian analysis of A-bomb epidemiologic information and experimental data has been developed to provide an estimation of the dose and dose-rate effectiveness factor (DDREF) for cancer risk estimates reported in this study.

• Knowledge of adaptive responses, genomic instability, and bystander signaling among cells that may act to alter radiation cancer risk was judged to be insufficient to be incorporated in a meaningful way into the modeling of epidemiologic data.

• Genetic variation in the population is a potentially important factor in the estimation of radiation cancer risk. Modeling studies suggest that strongly expressing mutations that predispose humans to cancer are too rare to distort appreciably population-based estimates of risk, but are a significant issue in some medical radiation settings.

• Estimation of the heritable effects of radiation takes advantage of new information on human genetic disease and on mechanisms of radiation-induced germline mutation. The application of a new approach to genetic risk estimation leads the committee to conclude that low-dose induced genetic risks are very small when compared to baseline risks in the population.

• The committee judges that the balance of evidence from epidemiologic, animal, and mechanistic studies tends to favor a simple proportionate relationship at low doses between radiation dose and cancer risk. Uncertainties in this judgment are recognized and noted.

Each of the above points contributes to refining earlier risk estimates, but none leads to a major change in the overall evaluation of the relation between exposure to ionizing radiation and human health effects.

ESTIMATING CANCER RISKS

As in past risk assessments, the LSS cohort of survivors of the atomic bombings in Hiroshima and Nagasaki plays a principal role in the committee's development of cancer risk estimates. Risk models were developed primarily from cancer incidence data for the period 1958–1998 and based on DS02 (Dosimetry System 2002) dosimetry, the result of a major international effort to reassess and improve survivor dose estimates. Data from studies involving medical and occupational exposure were also evaluated. Models for estimating risks of breast and thyroid cancer were based on pooled analyses that included data on both the LSS cohort and medically exposed persons.

To use models developed primarily from the LSS cohort for the estimation of lifetime risks for the U.S. population, it was necessary to make several assumptions that involve uncertainty. Two important sources of uncertainty are (1) the possible reduction in risk for exposure at low doses and dose rates (*i.e.*, the DDREF) and (2) the use of risk estimates based on Japanese atomic bomb survivors for estimating risks for the U.S. population.

The committee has developed and presented its best possible risk estimates for exposure to low-dose, low-LET radiation in human subjects. As an example, Table ES-1 shows the estimated number of incident cancer cases and deaths that would be expected to result if each individual in a population of 100,000 persons with an age distribution similar to that of the entire U.S. population was exposed to a single dose of 0.1 Gy, and also shows the numbers that would be expected in the absence of exposure. Results for solid cancers are based on linear models and reduced by a DDREF of 1.5. Results for leukemia are based on a linear-quadratic model.

The estimates are accompanied by 95% subjective confidence intervals (*i.e.*, random as well as judgmental) that reflect the most important sources of uncertainty—namely, statistical variation, uncertainty in the factor used to adjust risk estimates for exposure at low doses and dose rates, and uncertainty in the method of transport. In this report the committee also presents example estimates for each of several specific cancer sites and other exposure scenarios, although they are not shown here.

In general the magnitude of estimated risks for total cancer mortality or leukemia has not changed greatly from estimates in past reports such as BEIR V and recent reports of the United Nations Scientific Committee on the Effects of Atomic Radiation and the International Commission on Radiological Protection. New data and analyses have reduced sampling uncertainty, but uncertainties related to estimating risk for exposure at low doses and dose rates and to transporting risks from Japanese A-bomb survivors to the U.S. population remain large. Uncertainties in estimating risks of site-specific cancers are especially large.

As an illustration, Figure ES-1 shows estimated excess relative risks of solid cancer versus dose (averaged over sex and standardized to represent individuals exposed at age 30 who have attained age 60) for atomic bomb survivors, with doses in each of 10 dose intervals less than 2.0 Sv. The figure in the insert represents the ERR versus dose for leukemia. This plot conveys the overall dose-response relationship for the LSS cohort and its role in low-dose risk estimation. It is important to note that the difference between the linear and linear-quadratic models in the low-dose ranges is small relative to the error bars; therefore, the difference between these models is small relative to the uncertainty in the risk estimates produced from them. For solid cancer incidence the linear-quadratic model did not offer a statistically significant improvement in fit, so the linear model was used. For leukemia, a linear-quadratic model (insert in Figure ES-1) was used since it fitted the data significantly better than the linear model.

CONCLUSION

The committee concludes that current scientific evidence is consistent with the hypothesis that there is a linear, no-threshold dose-response relationship between exposure to ionizing radiation and the development of cancer in humans.

RECOMMENDED RESEARCH NEEDS

A more detailed listing of the BEIR VII recommended research needs can be found at the end of Chapter 13.

Research Need 1: Determination of the level of various molecular markers of DNA damage as a function of low-dose ionizing radiation

Currently identified molecular markers of DNA damage and other biomarkers that can be identified in the future should be used to quantify low levels of DNA damage and to identify the chemical nature and repair characteristics of the damage to the DNA molecule.

TABLE ES-1 The Committee's Preferred Estimates of the Lifetime Attributable Risk of Incidence and Mortality for All Solid Cancers and for Leukemia

	All Solid Cancers		Leukemia	
	Males	Females	Males	Females
Excess cases (including nonfatal cases) from exposure to 0.1 Gy	800 (400, 1600)	1300 (690, 2500)	100 (30, 300)	70 (20, 250)
Number of cases in the absence of exposure	45,500	36,900	830	590
Excess deaths from exposure to 0.1 Gy	410 (200, 830)	610 (300, 1200)	70 (20, 220)	50 (10, 190)
Number of deaths in the absence of exposure	22,100	17,500	710	530

NOTE: Number of cases or deaths per 100,000 exposed persons.

[a]95% subjective confidence intervals.

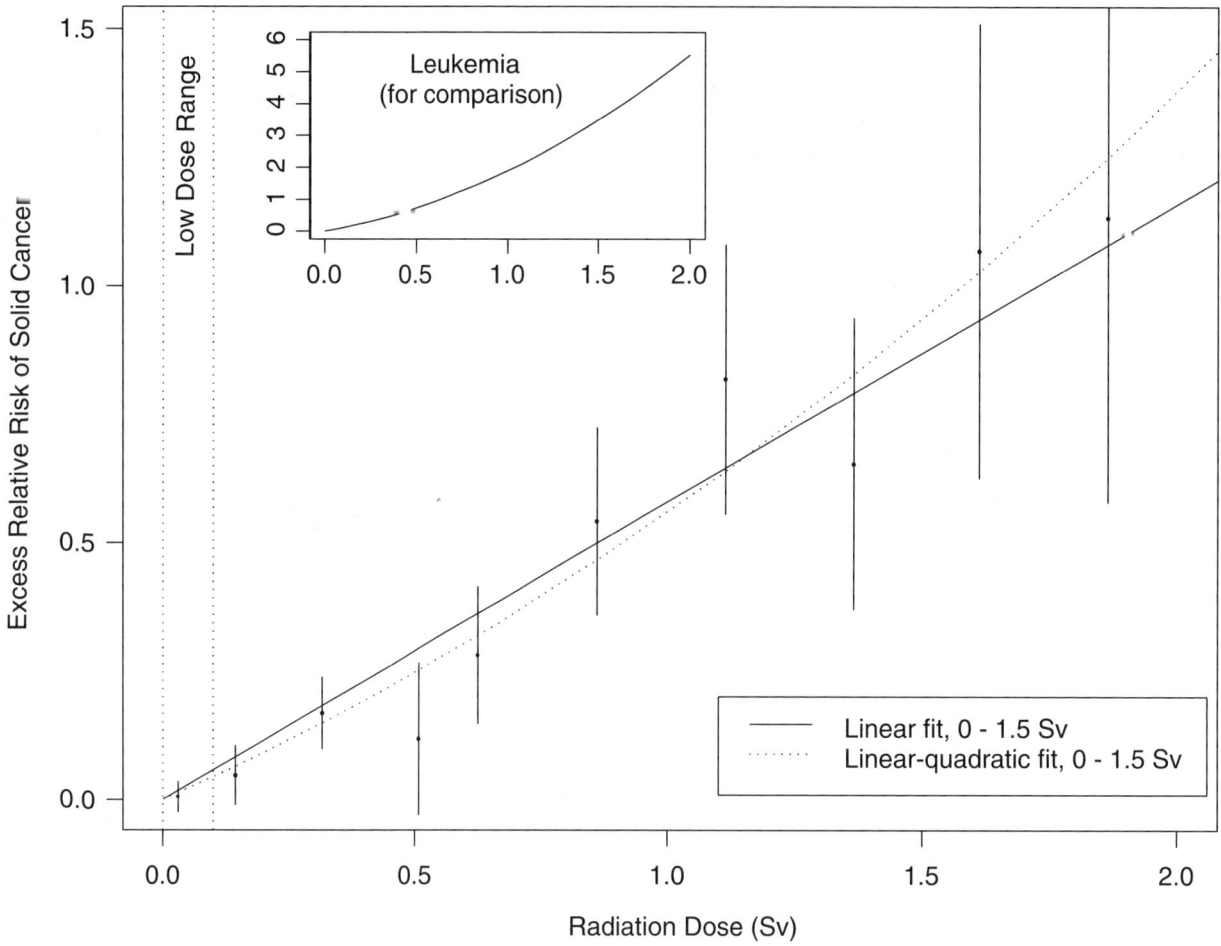

FIGURE ES-1 Excess relative risks of solid cancer for Japanese atomic bomb survivors. Plotted points are estimated excess relative risks of solid cancer incidence (averaged over sex and standardized to represent individuals exposed at age 30 who have attained age 60) for atomic bomb survivors, with doses in each of 10 dose intervals, plotted above the midpoints of the dose intervals. If R(d) is the age-specific instantaneous risk at some dose d, then the *excess relative risk* at dose d is [R(d) – R(0)]/R(0) (which is necessarily zero when the dose is zero). Vertical lines represent approximate 95% confidence intervals. Solid and dotted lines are estimated linear and linear-quadratic models for excess relative risk, estimated from all subjects with doses in the range 0 to 1.5 Sv (not estimated from the points, but from the lifetimes and doses of individual survivors, using statistical methods discussed in Chapter 6). A linear-quadratic model will always fit the data better than a linear model, since the linear model is a restricted special case with the quadratic coefficient equal to zero. For solid cancer incidence however, there is no *statistically significant* improvement in fit due to the quadratic term. It should also be noted that in the low-dose range of interest, the difference between the estimated linear and linear-quadratic models is small relative to the 95% confidence intervals. The insert shows the fit of a linear-quadratic model for leukemia to illustrate the greater degree of curvature observed for that cancer.

Research Need 2: Determination of DNA repair fidelity, especially with regard to double and multiple strand breaks at low doses, and whether repair capacity is independent of dose

Repair capacity at low levels of damage should be investigated, especially in light of conflicting evidence for stimulation of repair at low doses. In these studies the accuracy of DNA sequences rejoined by these pathways must be determined, and the mechanisms of error-prone repair of radiation lesions have to be elucidated.

Research Need 3: Evaluation of the relevance of adaptation, low-dose hypersensitivity, bystander effect, hormesis, and genomic instability for radiation carcinogenesis

Mechanistic data are needed to establish the relevance of these processes to low-dose radiation exposure (*i.e.*, <100 mGy). Relevant end points should include not only chromosomal aberrations and mutations but also genomic instability and induction of cancer. *In vitro* and *in vivo* data are needed for delivery of low doses over several weeks or months at very low dose rates or with fractionated exposures. The cumulative effect of multiple low doses of less than 10 mGy delivered over extended periods has to be explored further. The development of *in vitro* transformation assays utilizing nontransformed human diploid cells is judged to be of special importance.

Research Need 4: Identification of molecular mechanisms for postulated hormetic effects at low doses

Definitive experiments that identify molecular mechanisms are necessary to establish whether hormetic effects exist for radiation-induced carcinogenesis.

Research Need 5: Tumorigenic mechanisms

Further cytogenetic and molecular genetic studies are necessary to reduce current uncertainties about the specific role of radiation in multistage radiation tumorigenesis.

Research Need 6: Genetic factors in radiation cancer risk

Further work is needed in humans and mice on gene mutations and functional polymorphisms that influence radiation response and cancer risk.

Research Need 7: Heritable genetic effects of radiation

Further work should be done to establish (1) the potential roles of DNA double-strand break repair processes in the origin of deletions in irradiated stem cell spermatogonia and oocytes (the germ cell stages of importance in risk estimation) in mice and humans and (2) the extent to which large radiation-induced deletions in mice are associated with multisystem development defects. In humans, the problem can be explored using genomic databases and knowledge of mechanisms of origin of radiation-induced deletions to predict regions that may be particularly prone to radiation-inducible deletions.

With respect to epidemiology, studies on the genetic effects of radiotherapy for childhood cancer should be encouraged, especially when they can be coupled with modern molecular techniques (such as array-based comparative genomic hybridization).

Research Need 8: Future medical radiation studies

Most studies of medical radiation should rely on exposure information collected prospectively, including cohort studies as well as nested case-control studies. Future studies should continue to include individual dose estimation for the site of interest, as well as an evaluation of the uncertainty in dose estimation.

Studies of populations with high- and moderate-dose medical exposures are particularly important for the study of modifiers of radiation risks. Because of the high level of radiation exposure in these populations, they are also ideally suited to study the effects of gene-radiation interactions, which may render particular subsets of the population more sensitive to radiation-induced cancer. Genes of particular interest include BRCA1, BRCA2, ATM, CHEK2, NBS1, XRCC1, and XRCC3.

Of concern for radiological protection is the increasing use of computed tomography (CT) scans and diagnostic X-rays. Epidemiologic studies of the following exposed populations, if feasible, would be particularly useful: (1) follow-up studies of persons receiving CT scans, especially children; and (2) studies of infants who experience diagnostic exposures related to cardiac catheterization, those who have recurrent exposures to follow their clinical status, and premature babies monitored for pulmonary development with repeated X-rays.

There is a need to organize worldwide consortia that would use similar methods in data collection and follow-up. These consortia should record delivered doses and technical data from all X-ray or isotope-based imaging approaches including CT, positron emission tomography, and single photon emission computed tomography.

Research Need 9: Future occupational radiation studies

Studies of occupational radiation exposures, in particular among nuclear industry workers, including nuclear power plant workers, are well suited for direct assessment of the carcinogenic effects of long-term, low-level radiation exposure in humans. Ideally, studies of occupational radiation should be prospective in nature and rely on individual real-time estimates of radiation doses. Where possible, national registries of radiation exposure of workers should be established and updated as additional radiation exposure is accumulated and as workers change employers. These registries should include at least annual estimates of whole-body radiation dose from external photon exposure. These exposure registries should be linked with mortality registries and, where they exist, national tumor (and other disease) registries. It is also important to continue follow-up of workers exposed to relatively high doses, that is, workers at the Mayak nuclear facility and workers involved in the Chernobyl cleanup.

Research Need 10: Future environmental radiation studies

In general, additional ecological studies of persons exposed to low levels of radiation from environmental sources are not recommended. However, if there are disasters in which a local population is exposed to unusually high levels of radiation, it is important that there be a rapid response not only for the prevention of further exposure but also for scientific evaluation of possible effects of the exposure. The data collected should include basic demographic information on individuals, estimates of acute and possible continuing exposure, the nature of the ionizing radiation, and the means of following these individuals for many years. The possibility of enrolling a comparable nonexposed population should be considered. Studies of persons exposed environmentally as a result of the Chernobyl disaster or as a result of releases from the Mayak nuclear facility should continue.

Research Need 11: Japanese atomic bomb survivor studies

The LSS cohort of Japanese A-bomb survivors has played a central role in BEIR VII and in past risk assessments. It is important that follow-up for mortality and cancer incidence continue for the 45% of the cohort who remained alive at the end of 2000.

In the near future, an uncertainty evaluation of the DS02 dosimetry system is expected to become available. Dose-response analyses that make use of this evaluation should thus be conducted to account for dosimetry uncertainties.

Development and application of analytic methods that allow more reliable estimation of site-specific estimates is also needed. Specifically, methods that draw on both data for the specific site and data for broader cancer categories could be useful.

Research Need 12: Epidemiologic studies in general

Data from the LSS cohort of A-bomb survivors should be supplemented with data on populations exposed to low doses and/or dose rates, especially those with large enough doses to allow risks to be estimated with reasonable precision. Studies of nuclear industry workers and careful studies of persons exposed in countries of the former Soviet Union are particularly important in this regard.

1

Background Information

This report focuses on the health effects of low-dose, low-LET (low linear energy transfer) radiation. In this chapter the committee provides background information relating to the physical and chemical aspects of radiation and the interaction of radiation with the target molecule DNA. The committee discusses contributions of normal oxidative DNA damage relative to radiation-induced DNA damage and describes the DNA repair mechanisms that mammalian cells have developed to cope with such damage. Finally, this chapter introduces a special subject, the physical characteristics that determine the relative biological effectiveness (RBE) of neutrons, estimates of which are required in the derivation of low-LET radiation risk estimates from atomic bomb survivors.

PHYSICAL ASPECTS OF RADIATION

The central question that must be resolved when considering the physical and biological effects of low-dose ionizing radiation is whether the effects of ionizing radiation and the effects of the free radicals and oxidative reaction products generated in normal cellular metabolism are the same or different. Is ionizing radiation a unique insult to cells, or are its effects lost in the ocean of naturally occurring metabolic reaction products? Can cells detect and respond to low doses of ionizing radiation because of detectable qualitative and quantitative differences from endogenous reaction products?

Different Types of Ionizing Radiation

Ionizing radiation, by definition, contains enough energy to displace electrons and break chemical bonds. Charged particles, such as high-energy electrons, protons, α-particles, or fast heavy ions, are termed *directly ionizing* because, while they traverse the cell, they ionize numerous molecules by direct collisions with their electrons. Electromagnetic radiations, such as X- and γ-rays, consist of photons that can travel relatively large distances in tissue without interaction.

Once an interaction with one of the electrons in the material occurs, part or all of the photon energy is transferred to the electron. The energetic electrons released in this way produce the bulk of ionizations. X- and γ-rays are accordingly termed "indirectly ionizing" radiation. This term is also applied to fast neutrons, because they too traverse large distances in tissue without interaction but can, in occasional collisions, transfer much of their energy to atomic nuclei that in turn produce the main part of the ionizations.

In addition to the distinction between indirectly ionizing and directly ionizing (*i.e.*, uncharged and charged radiation) a distinction is made between sparsely ionizing, or low-LET, and densely ionizing, or high-LET, radiation. The (unrestricted) LET of an ionizing charged particle is defined as the average energy lost by the particle due to electronic interactions per unit length of its trajectory; it is expressed in kiloelectronvolts per micrometer (keV/μm).[1] High-energy electromagnetic radiations, such as X-rays or γ-rays, are sparsely ionizing since, in tissue, they release fast electrons that have low LET. Neutrons are densely ionizing because in tissue they release fast protons and heavier atomic nuclei that have high LET.

Figure 1-1 gives the LET of electrons as a function of their kinetic energy and compares it to the considerably higher LET of protons. It is seen that electrons are generally sparsely ionizing while protons are, at moderate energies, densely ionizing. However it is also noted that very energetic protons, as they occur in altitudes relevant to aviation and in space, are sufficiently fast to be sparsely ionizing.

---

[1] *Restricted* linear energy transfer, L_Δ, results when, within the charged particle tracks, secondary electrons (δ-rays) with energies in excess of Δ are followed separately. It is important to distinguish between *track average LET* and *dose average LET*. Dose average LET represents more realistically the high local energy densities that can occur in a track even for low-LET radiation, and it therefore can assume larger values. For example, the track average of L_{100eV} for cobalt-60 γ-rays is 0.23 keV/μm, and the dose average is 5.5 keV/μm (ICRU 1970).

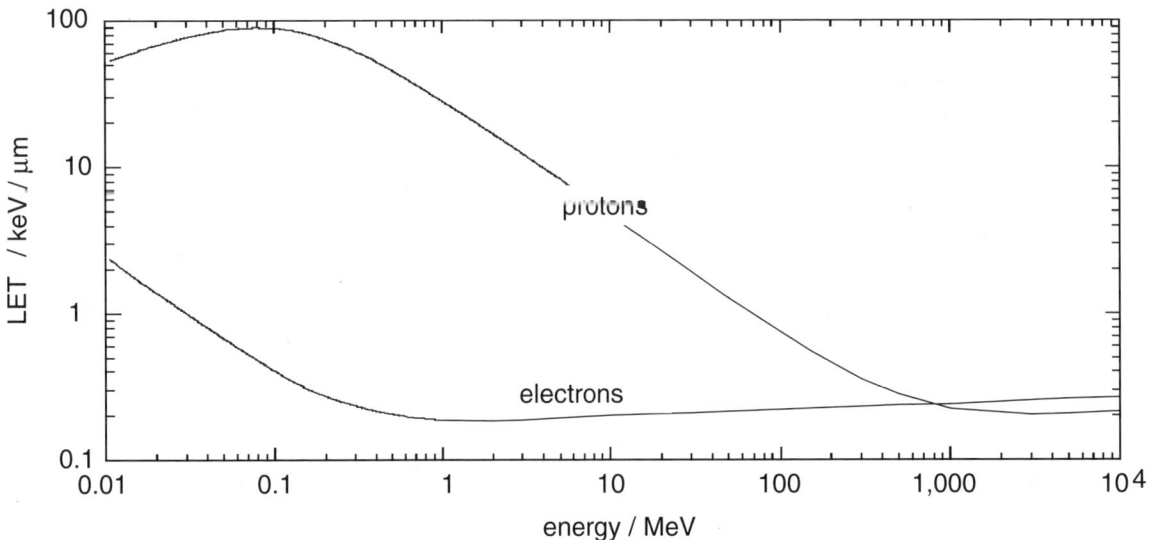

FIGURE 1-1 Linear energy transfer of protons and electrons in water. SOURCE: Data from ICRU (1970).

The effects of high-LET particles (*i.e.*, protons and heavier ions) are outside the scope of this report. However, neutrons and their high relative biological effectiveness must be considered in the context of low-LET risk estimates derived from the observations on delayed health effects among A-bomb survivors. The reason is that a small fraction of the absorbed dose to A-bomb survivors was due not to the predominant high-energy γ-rays, but to fast neutrons. Because of the greater effectiveness of these fast neutrons, this small dose component must be taken into consideration.

Photon Spectral Distributions

The absorption and scattering of photons depends on their energy. The γ-rays from radioactive decay consist of monoenergetic photons that do not exceed several million electronvolts (MeV) in energy; γ-rays that result from the fission of uranium or plutonium have a spectrum of energies with a maximum of 2 MeV. Higher-energy γ-rays, up to 7 MeV, can be generated by inelastic scattering, as occurred in the neutron-nitrogen interaction from the atomic bomb explosions in Hiroshima and Nagasaki.

Artificially produced X-rays have a wide spectrum of energies resulting from the deceleration of electrons as they traverse high-atomic-number materials. A continuous distribution of photon energies is generated, with a mean energy of about one-third the maximal energy of the accelerated electrons. Added filtration selectively removes the "soft" (*i.e.*, less energetic) photon component and, thus, hardens the X-rays. Discrete energy "spikes" also occur in the X-ray spectrum; these spikes originate in the ejection of electrons from atoms of the affected element, which is followed by the transition of electrons from outer shells to inner shells of the

atom releasing photons of discrete energy. Conventional X-rays, used for diagnostic radiology, are commonly produced with accelerating voltages of about 200 kV. For mammography, where high contrast is sought and only a moderate thickness of tissue must be traversed by the X-rays, the low acceleration voltage of 29 kV is usually employed.

There are three different types of energy-transfer processes whereby photons of sufficient energy eject electrons from an atom, which can then interact with other atoms and molecules to produce a cascade of alterations that ultimately lead to observable biological effects. These are the photoelectric process, Compton scattering, and pair production.

At low energies (<0.1 MeV), the photoelectric process dominates in tissue. A photon interacts with and ejects an electron from one of the inner shells of an atom. The photon is extinguished, and most of its energy is imparted to the ejected electron as kinetic energy.

At medium photon energies (about 0.5–3.5 MeV), Compton scattering is the most probable event. Compton scattering occurs when an incoming photon's energy greatly exceeds the electron-binding energy of the affected atom. In this case the energy of the incoming photon is converted into the kinetic energy of an ejected electron and a secondary "scattered" photon. The scattered photon has less energy than the primary photon and can undergo further Compton scattering until its energy is sufficiently degraded for the photoelectric process to occur.

At energies greater than 1.02 MeV, pair production can occur. A photon interacts with an atomic nucleus, and the photon energy is converted into a positron and an electron. The photon energy above 1.02 MeV is converted into the kinetic energy of the newly created particles. The electron and the positron interact with and can ionize other molecules.

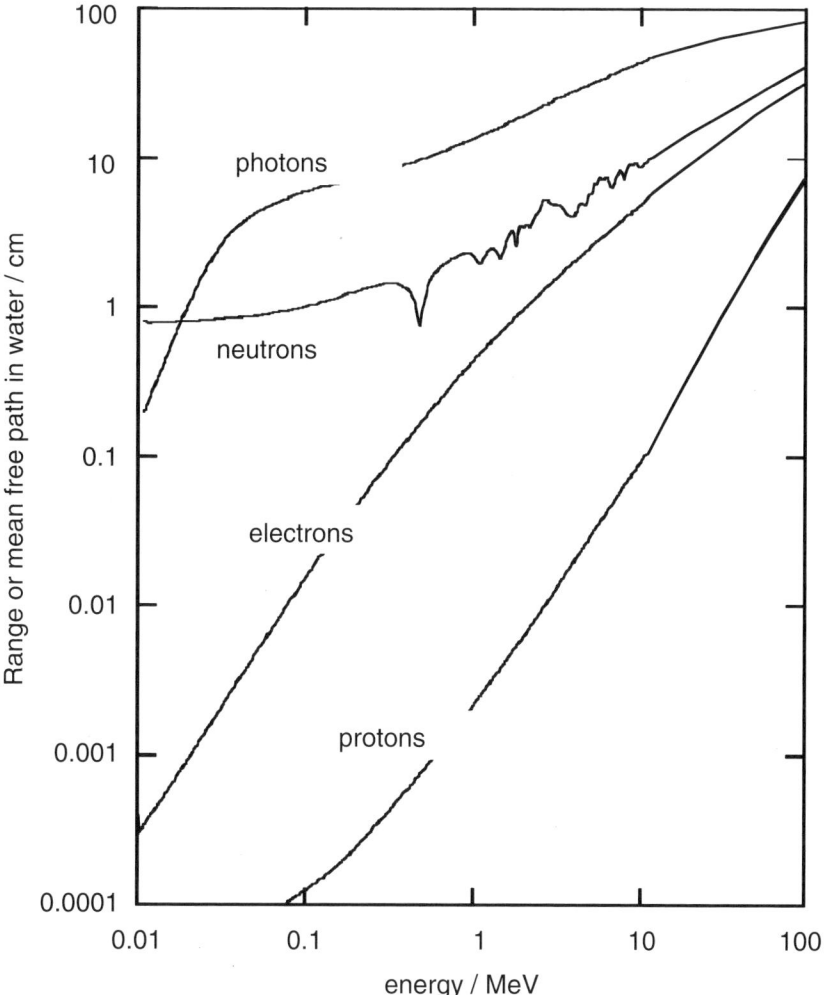

FIGURE 1-2 Mean free path of photons and neutrons in water and range of electrons and protons. SOURCE: Data from ICRU (1970).

The positron ultimately interacts with another electron, and this results in an "annihilation" event in which the mass is extinguished and two 0.51 MeV photons are emitted in opposite directions. The annihilation photons can themselves produce further ionizations.

Figure 1-2 shows the mean free path for monoenergetic photons (*i.e.*, the average distance in water until the photon undergoes an interaction). To compare the penetration depth of photon radiation with that of electron radiation, the mean range of electrons of specified energy is given in the same diagram. It is seen that the electrons released by photons are always considerably less penetrating than the photons themselves.

Figure 1-3 compares in terms of the distributions of photon energy fluence the γ-rays from the A-bomb explosions with the distributions of photon energy for orthovoltage X-rays and low-energy mammography X-rays. These different electromagnetic radiations are all classified as low-LET (*i.e.*,

sparsely ionizing) radiation. There are, nevertheless, differences in effectiveness and possibly also differences in the risk for late effects due to these radiations.

Track Structure

The passage of fast electrons through tissue creates a track of excited and ionized molecules that are relatively far apart. X- and γ-rays produce electrons with relatively low linear energy *transfer*, (*i.e.*, energy loss per unit track length) and are considered low-LET radiation. For example, the track average of unrestricted LET of the electrons liberated by cobalt-60 (^{60}Co) gamma rays is about 0.25 keV/μm, which can be contrasted with an average LET of about 180 keV/μm for a 2 MeV α-particle, a high-LET radiation. LET is an important measure in the evaluation of relative biological effectiveness (ICRU 1970; Engels and Wambersie 1998) of a given kind of radiation.

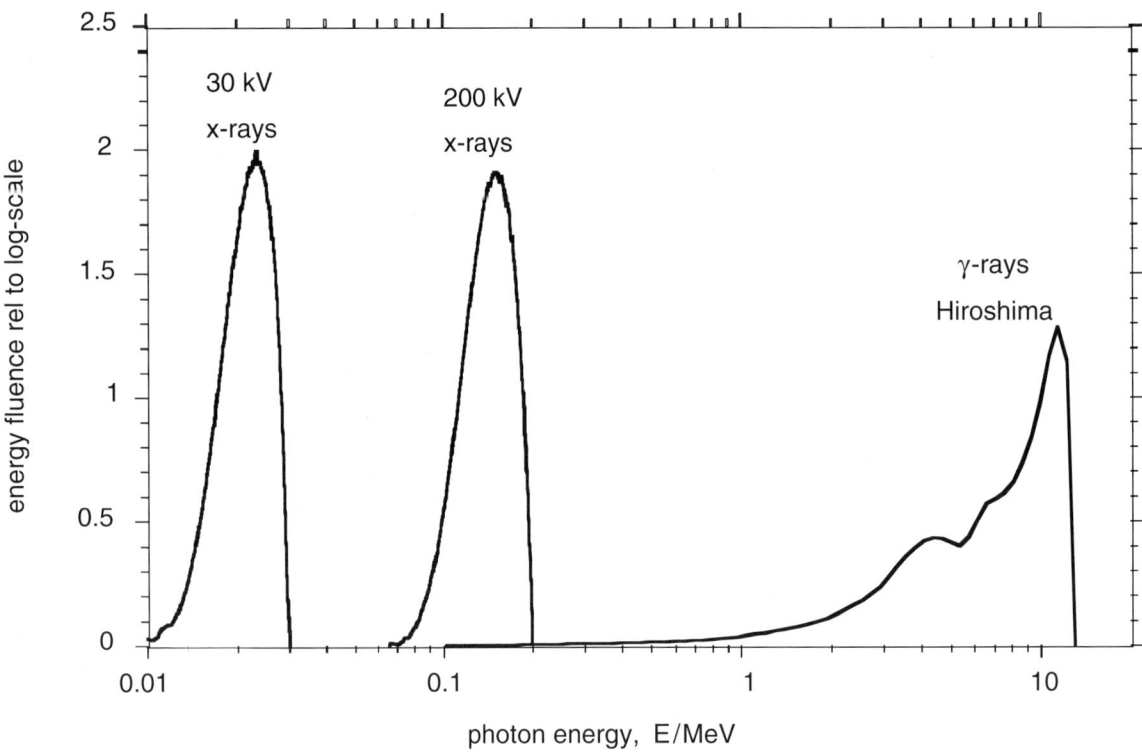

FIGURE 1-3 Distributions of photon energy fluence for mammography X-rays, orthovoltage X-rays, and γ-rays from the atomic bomb explosion in Hiroshima. The distributions of the energy fluence relative to the logarithmic scale of energy are plotted, because they represent roughly the fractional contribution of incident photons of specified energy to the dose absorbed by a person. SOURCE: Data from Seelentag and others (1979) and Roesch (1987).

Different Effectiveness of γ-Rays and X-Rays

LET and Related Parameters of Radiation Quality

While γ-rays and X-rays of various energies are all sparsely ionizing, in the body they generate electrons with somewhat different spectra of LET values (ICRU 1970). To quantify the differences, reference is usually made to the dose average LET or to the mean values of the related microdosimetric parameter dose-averaged linear energy, y.

Figure 1-4 gives the dose average LET values for the electrons released by monoenergetic photons (solid curves) and compares these values to the averages for 29 kV mammography X-rays and 200 kV X-rays (solid circles and squares, respectively; ICRP 2003). In addition to the dose average, L_D, of the unrestricted LET, the diagram contains the dose averages, $L_{D,\Delta}$, of the restricted LET, L_Δ. The restricted LET treats the Δ-rays beyond the specified cutoff energy Δ as separate tracks. This accounts in an approximate way for the increased local energies due to Δ-rays and therefore provides larger values that are more meaningful than those of unrestricted LET.

High-energy photons (*e.g.*, ^{60}Co γ-rays) release Compton electrons of comparatively high energy and correspondingly low LET. Photons of less energy (*e.g.*, conventional 200 kV

X-rays) produce less energetic Compton electrons with higher LET. This explains the substantial difference between the mean LET of high-energy γ-rays and conventional X-rays. For lower-energy X-rays the photon energy is further reduced, and the photo effect (*i.e.*, the total transfer of photon energy to electrons) begins to dominate. Accordingly, the average energy of the electrons begins to increase again, which explains the relatively small difference in average LET between 200 kV X-rays and soft X-rays. At very low photon energies (*i.e.*, less than about 20 keV) the LET values increase strongly, but these ultrasoft X-rays are of little concern in radiation protection because of their very limited penetration depth.

The dose average, $L_{D,\Delta}$, of the restricted LET is a parameter that correlates with the low dose effectiveness of photon or electron radiation. With a cutoff value Δ = 1 keV, the numerical values of $L_{D,\Delta}$ are consistent with a low-dose RBE of about 2 for conventional X-rays versus γ-rays. A similar dependence on photon energy is seen in the related microdosimetric parameter dose lineal energy, y, which has been used as reference parameter by the liaison committee of the International Commission on Radiological Protection (ICRP) and the International Commission on Radiation Units and Measurements (ICRU) in *The Quality Factor in Radiation Protection* (ICRU 1986). Figure 1-5 gives values of its

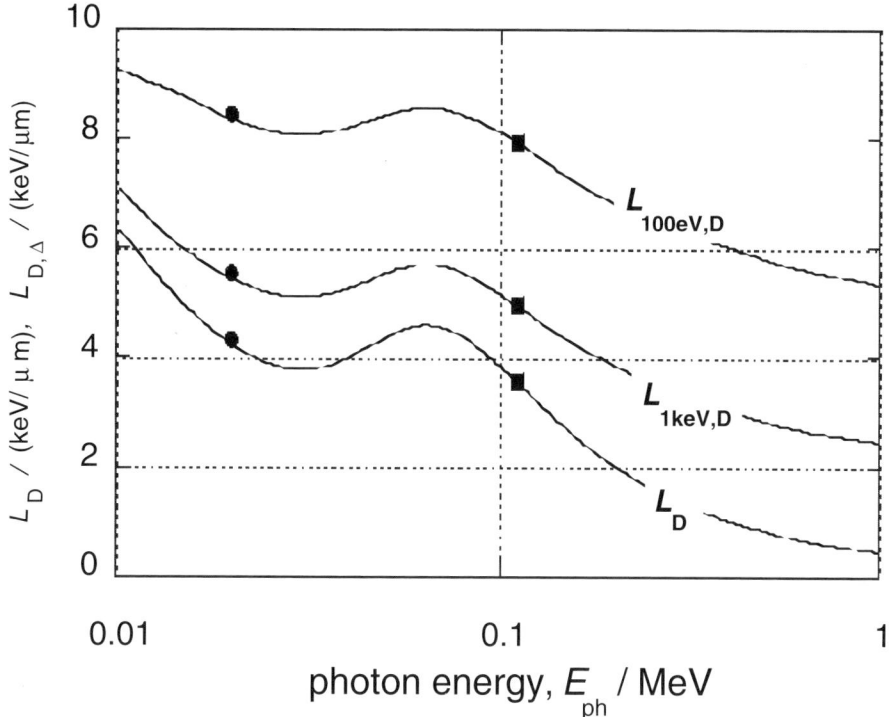

FIGURE 1-4 The dose mean restricted and unrestricted linear energy transfer for electrons liberated by monoenergetic photons of energy E_{ph}. The dots and squares give the values for the 29 kVp and the 200 kVp X-rays. They are plotted at the weighted photon energies of the X-ray spectra. SOURCE: Data from Kellerer (2002).

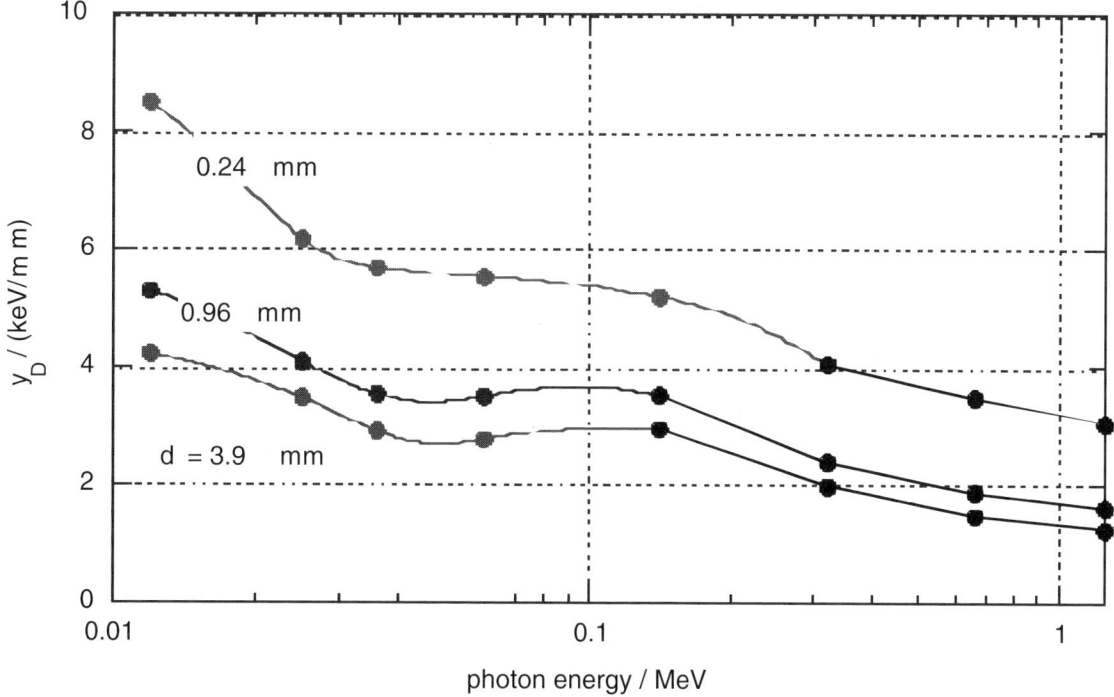

FIGURE 1-5 Measured dose average lineal energy, y_D, for monoenergetic photons and for different simulated site diameters, d. SOURCE: Data from Kliauga and Dvorak (1978).

dose average, y_D, as measured by Kliauga and Dvorak (1978) for various photon radiations and for different simulated site diameters, *d*.

The γ-rays from the atomic bomb explosions had average energies between 2 and 5 MeV at the relevant distances (Straume 1996). Figures 1-4 and 1-5 do not extend to these energies; however, it is apparent from Figures 1-4 and 1-5 that the mean values of the restricted LET or the lineal energy do not decrease substantially beyond a photon energy of 1 MeV. There is, thus, little indication that the hard γ-rays from the atomic bombs should have an RBE substantially less than unity compared to conventional ^{60}Co γ-rays.

Information from In Vitro Studies

It has long been recognized in experimental radiobiology that low-*LET* radiations do not all have the same effectiveness at low doses. With regard to mutations in *Tradescantia*, aberrations in human lymphocytes, and killing of mouse oocytes (Bond and others 1978), conventional 200 kV X-rays have been found to be about twice as effective at low doses as high-energy γ-rays. Fast electrons may be even less effective than γ-rays. These differences are most clearly documented in cell studies and, especially, in studies on chromosome aberrations (Sinclair 1985; ICRU 1986). The most reliable and detailed data on photon RBE exist for chromosome aberrations in human lymphocytes. Edwards and others (1982) have obtained the data for dicentrics in human lymphocytes listed in Table 1-1 for 15 MeV electrons, ^{60}Co γ-rays, and 250 kV X-rays. New data have since confirmed these substantial differences of effectiveness for different types of penetrating low-LET radiations.

Sasaki and colleagues (1989; Sasaki 1991) have determined the yields of dicentrics in human lymphocytes over a broad range of photon energies. The upper panel of Figure 1-6 gives the linear coefficients (and standard errors) from linear-quadratic fits to the dose dependencies. The closed circles relate to γ-rays and to broad X-ray spectra; the squares, to characteristic X-rays and monoenergetic photons

from synchrotron radiation. The lower panel gives analogous data obtained by Schmid and others (2002).

The diagram demonstrates that there is a substantial decrease of the yield of dicentrics from conventional X-rays to γ-rays. The photon energies below 20 keV are of special interest with regard to biophysical consideration, but are less relevant to exposure situations in radiation protection. They are included here to show the full trend of the energy dependence.

It is seen that the low-dose RBE for dicentrics for moderately filtered 200 kV X-rays is about 2–3 relative to γ-rays, while the RBE of mammographic X-rays (29 kV) relative to the moderately filtered 200 kV X-rays is somewhat in excess of 1.5.

The data for dicentrics in Figures 1-6 are reasonably consistent with the LET values in Figure 1-4 for a cutoff value in excess of 1 keV. The difference by a factor of 2–3 in the low-dose effectiveness of conventional X-rays and γ-rays has been known and, even if it should apply equally to radiation-induced late effects, would not necessarily require a departure from the current convention for radiation protection, which assigns the radiation weighting factor unity to all photon radiations. However, the difference has to be noted whenever risk estimates are derived from exposures to γ-rays and then applied to X-rays.

Apart from these considerations it is uncertain whether the marked dependence of the low-dose RBE on photon energy for chromosome aberrations also is representative for late radiation effects in man. The dependence of RBE on photon energy for dicentric chromosomes reflects the fact that the dose dependencies have large curvature for ^{60}Co γ-rays (α/β = 0.2 Gy in the data reported by Schmid and others 2002), but little curvature for 29 kV X-rays (α/β = 1.9 Gy). If there were no curvature below 1 Gy in the dose relations for chromosome aberrations, the low-dose RBE of 29 kV X-rays would be only 1.65 compared to ^{60}Co γ-rays. Since the dose dependence for solid tumors among A-bomb survivors indicates little curvature, the dependence of risk on photon energy may be similarly weak for tumor induction in man. It is of interest to compare the biophysical information and the experimental results to the radioepidemiologic evidence for health effects.

Information from Radioepidemiology

Numerous epidemiologic studies on medical cohorts have provided risk estimates that exhibit considerable variation. Many of these studies on patients relate to X-ray exposures, but there is no consistent epidemiologic evidence for higher risk factors from X-rays than from γ-rays. In fact, while the risk estimates from medical studies are not inconsistent with those for atomic bomb survivors, they tend to be, as a whole, somewhat lower (UNSCEAR 2000b). The radiation-related increase in breast cancer incidence can serve as an example because it has been most thoroughly studied.

TABLE 1-1 Low-Dose Coefficients (and standard errors) for Induction of Chromosome Aberrations in Human Lymphocytes by Low-LET Penetrating Radiation

Radiation Type	Dicentrics per Cell per Gray
15 MeV electrons	0.0055 (± 0.011)
^{60}Co γ-rays	0.0157 (± 0.003)
250 kV X-rays	0.0476 (± 0.005)

NOTE: The low-dose coefficients represent the linear component of a linear-quadratic fit to the data. SOURCE: Data from Edwards and others (1982).

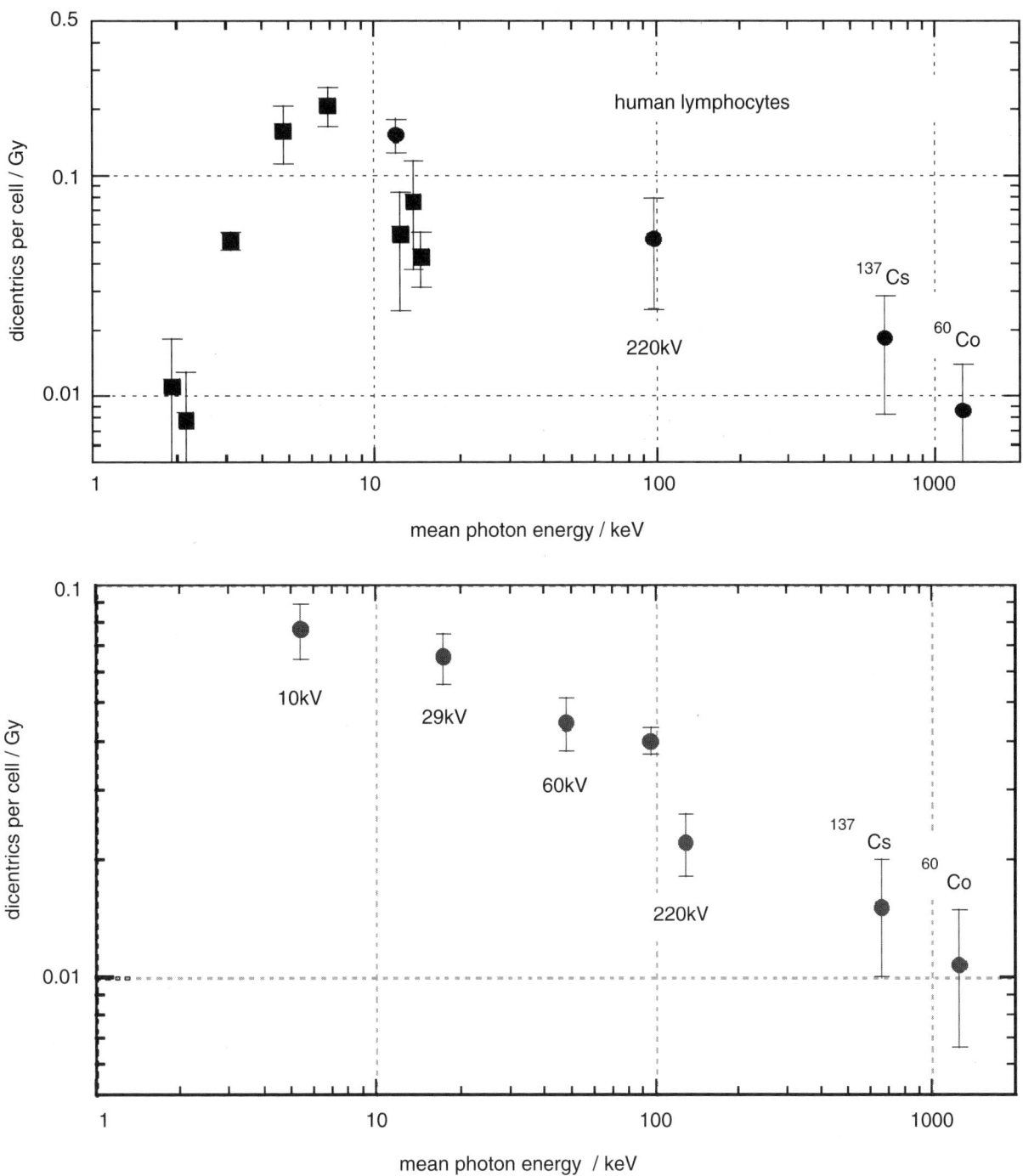

FIGURE 1-6 Data points are linear coefficients (and standard errors) of the dose dependence for dicentric chromosomes in human peripheral blood lymphocytes. Squares are for monoenergetic photons; circles are X-ray spectra or γ-rays. The two data points in the lower panel labeled 220 kV both had 220 kV generating voltage, but the filtration was different. SOURCE: *Upper panel:* Data from Sasaki and others (1989; Sasaki 1991). *Lower panel:* Data from Schmid and others (2002).

Figure 1-7 gives risk estimates from major studies on radiation-induced breast cancer. The estimated risk coefficients (and 90% confidence intervals) are expressed in terms of the excess relative risk (ERR) per gray and the excess absolute risk (EAR) per gray per 10,000 person-years (PY).

The uncertainties are large, and the risk estimates vary widely because the patient treatment regimes differed not only in the type of radiation but also in the various exposure modalities, such as acute, fractionated, or protracted exposure; whole- or partial-body exposure; exposure rate; and

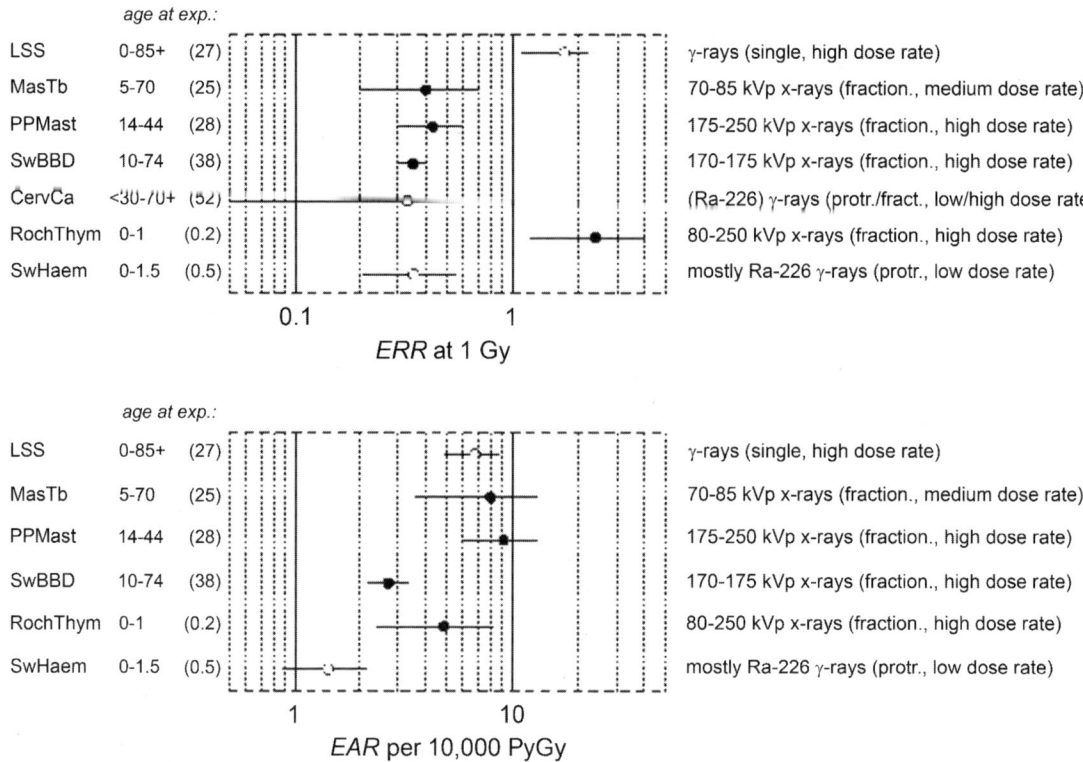

FIGURE 1-7 Excess relative risk (and 90% confidence interval) from various epidemiologic studies of breast cancer. The upper panel shows the excess relative risk per gray, the lower panel, the absolute risk per 10,000 person-years per gray. (For the description of individual studies, see UNSCEAR 2000b and Preston and others 2002a.) The confidence limit for the study of cervical carcinoma patients is recalculated. *Cohorts:* LSS: Life Span Study of atomic bomb survivors; MasTb: Massachusetts tuberculosis patients; PPMast: New York postpartum mastits patients; SwBBD: Swedish benign breast disease patients; CervCa: cervical cancer patients (case-control study); RochThym: Rochester infants with thymic enlargement; SwHem: Swedish infants with skin hemangioma.

magnitude of the exposure. Furthermore, there are ethnic differences, including those related to life-style, that are associated with greatly different background rates of breast cancer. Populations with low spontaneous rates tend to exhibit comparatively high ERR, while their EAR tends to be low. This complicates the comparison of risk estimates, since it remains uncertain whether relative or absolute excess incidence is the more relevant measure of risk.

The various exposed cohorts also differ considerably in the duration of follow-up and, especially, the age at exposure. The last two studies (RochThym, SwHem) relate to exposures in childhood, while the remainder refer to exposures at intermediate or higher ages. The last factor is especially critical, because both ERR and lifetime integrated EAR decrease substantially with increasing age at exposure.

The dominant influence of the various modifying factors makes it impossible on the basis of epidemiologic data to confirm the difference in effectiveness between γ-rays and X-rays or the difference between X-rays of different energies. Studies related to other types of cancer are even further removed from providing an answer. Thus, although cell stud-

ies and biophysical considerations suggest a low-dose RBE for conventional X-rays versus hard γ-rays of about 2–3, this difference cannot be confirmed at present through epidemiologic investigations.

Effects of Radiation on DNA, Genes, and Chromosomes

The probability that a low-LET primary electron will interact with a DNA molecule along its track is low, but a direct interaction of this sort is possible (Nikjoo and others 2002). Along the primary electron track, secondary electrons with lower energies are also formed, producing clusters of ionizations (see Figure 1-8, panel A). If such an ionization cluster occurs near a DNA molecule, multiple damages can occur in a very localized segment of the DNA (Figure 1-8, panel B). These clusters have been referred to as as clustered-damage or locally multiply damaged sites (LMDS) (Ward and others 1985; Goodhead 1994).

Figure 1-8 illustrates two typical structures of electron tracks produced by low-LET photons (*e.g.*, γ-rays). The wavy lines outside the sphere represent primary and second-

Panel A

Low-LET tracks
in cell nucleus
for example,
from gamma rays

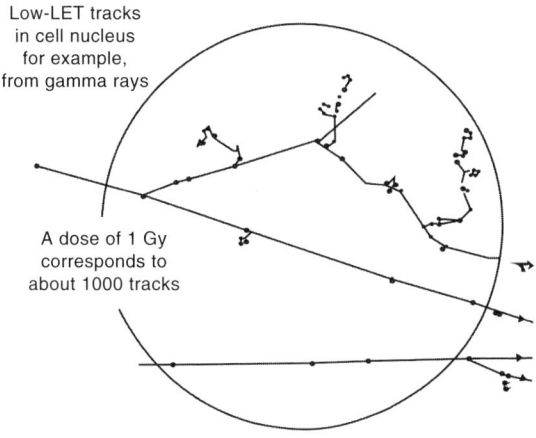

A dose of 1 Gy
corresponds to
about 1000 tracks

Panel B

Tracks in chromatin fiber

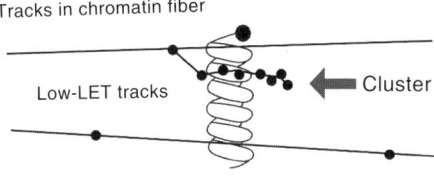

Low-LET tracks

← Cluster

FIGURE 1-8 *Panel A:* Illustration of primary and secondary electron tracks producing clusters of ionization events. The calculated number of tracks is based on a nucleus with a diameter of 8 μm. The track size is enlarged relative to the nucleus to illustrate the theoretical track structure. *Panel B:* Illustration of clustered damage. The arrow identifies an ionization cluster near a DNA molecule to represent the possibility of locally multiply damaged sites. Only a segment of the electron track is illustrated in Panel B.

ary photons; the straight lines represent the paths of ejected electrons. For clarity of presentation, the size of the tracks is increased relative to the cell and is not drawn to scale. As the energetic electron interacts with atoms of the material, secondary electrons are produced and kinetic energy is lost. Such collisions can result in deflection of the primary electron from its original path (Figure 1-8, panel A). Important components of the track structure are the clusters of secondary ionizations that occur in a very small volume (see Figure 1-8, panel B). These clusters, acting directly or indirectly on the DNA molecule, may produce clustered damage, LMDS, that may in turn be refractory to repair. The likely site of health effects of low-dose radiation is the genetic

material, which directs the structure and function of the organism. This genetic material is made up of DNA organized into genes and chromosomes (for a brief description, see Appendix A). Radiation can damage DNA as described in this chapter, and the damage can result in cell lethality, impaired cell function, or may produce damage involved in the carcinogenic process. Radiation has also been shown to produce heritable gene mutations in animals. For a basic description of gene mutations, see Appendix A.

Relative Biological Effectiveness of Neutrons

This report assesses the biological effects of low-LET radiation, that is, photons and electrons. It does not deal with densely ionizing radiation, such as heavy ions (including α-particles) and fast neutrons. Although neutrons need not be considered here on their own account, they must be accounted for in the analysis of the most important source of information on radiation risks, observations on the atomic bomb survivors of Hiroshima and Nagasaki. Such analysis requires consideration of the relative biological effectiveness of neutrons. The following remarks deal with the RBE of neutrons in general terms.

According to the 1986 dosimetry system, DS86, only a small fraction of the absorbed dose to atomic bomb survivors was due to neutrons—about 2% in Hiroshima in the most relevant dose range and 0.7% in Nagasaki (Roesch 1987). The current reevaluation of the Hiroshima and Nagasaki dosimetry, DS02, is in general agreement with these observations. However, although the absorbed dose fraction of neutrons was small in both cities, it is known from a multitude of radiobiological investigations that the RBE of small neutron doses can be large enough for even the small absorbed dose fraction to add appreciably to the late health effects among atomic bomb survivors.

Fast neutrons interact with exposed tissue predominantly by releasing recoil nuclei. At neutron energies up to a few million electronvolts, the energy transfer is predominantly to protons. On the average, a neutron transfers half its energy to a recoil proton in a collision. Neutrons of 1 MeV therefore produce recoil protons with an average initial energy of 500 keV. At a neutron energy of 0.4 MeV, the typical recoil proton energy is 200 keV, enough to allow the proton to go through its maximal LET of about 100 keV/μm, which is reached at its Bragg peak energy of 0.1 MeV. The ionization density in such proton tracks is far greater than that in an electron track, as depicted in Figure 1-1. It is evident that the resulting high local energy concentration will produce far more clusters of closely spaced ionizations than do low-LET photons and thus more LMDS (clustered damage) that may remain unrepaired or misrepaired. In addition, recoil protons have track lengths of a few micrometers, so critical damage can, with fairly high probability, be caused in neighboring chromosomal structures. The interaction of closely spaced chromosomal damage has long been noted to be a critical factor in the production of chromosomal aberrations (Lea 1946).

Recoil protons with energy of a few hundred kiloelectronvolts appear, in line with the above biophysical considerations, to be the particles that produce maximal cellular damage per unit energy imparted. This is confirmed by various experimental studies that consistently demonstrate the maximal effectiveness of neutrons at a neutron energy of about 0.4 MeV (Kellerer and Rossi 1972b).

The dose-effect relationship, $E(D)$, for photons can in many radiobiological investigations be described as a linear quadratic function of absorbed dose:

$$E(D) = aD_\gamma + bD_\gamma{}^2. \qquad (1\text{-}1)$$

In experiments with fast neutrons, the effect is typically proportional to the absorbed dose, D_n, of neutrons over a variable dose range depending on the tissue and effect:

$$E(D) = a_n D_n. \qquad (1\text{-}2)$$

The linear dose coefficient, a_n, for neutrons is always substantially larger than the linear dose coefficient, a, for photon radiation. The RBE of neutrons is defined as the ratio of a γ-ray dose to the neutron dose that produces the same effect:

$$\text{RBE} = D_\gamma / D_n, \text{ with: } E(D_\gamma) = E(D_n). \qquad (1\text{-}3)$$

In terms of Equations (1-1) and (1-2), RBE can be expressed as a function of the neutron dose or the photon dose. The latter expression is somewhat simpler:

$$RBE(D_\gamma) = a_n /(a + bD_\gamma). \qquad (1\text{-}4)$$

This implies that RBE assumes its maximal value, $\text{RBE}_{max} = a_n/a$, at low doses, whereas it decreases with increasing dose and then tends to be inversely proportional to the photon dose.

Experimental Observations

Indeed, numerous experimental investigations of chromosomal aberrations, cellular transformations, and cell killing have confirmed that maximal RBE values of neutrons occur at low doses and that, at somewhat higher doses, RBE varies inversely with increasing reference dose (*i.e.*, the photon dose). The same has been observed for more complex effects such as opacification of the lens and, more important in the context of risk assessment, induction of tumors in animals. A synopsis of such findings was provided in the context of the microdosimetric interpretation of the neutron RBE (Kellerer and Rossi 1972b).

Although the general features of the dependence of neutron RBE on dose are brought out consistently in experimental studies, the numerical values of RBE vary, and the variation appears to be largely a matter of the different magnitude of the linear dose component for photon radiation.

Cell survival curves usually exhibit pronounced initial slopes, and the observed maximal neutron RBE rarely exceeds a factor of about 10. For dicentric chromosomal aberrations in human lymphocytes, values of about 70 are obtained for the maximal RBE of 0.5 MeV neutrons against γ-rays (Dobson and others 1991; Schmid and others 2000). This large maximal value might be seen as an indication of an exceptionally high effectiveness of neutrons at low doses. In fact the dose-effect relationship for neutrons is simply linear, and the high maximal RBE of neutrons is merely a reflection of the very shallow and imprecisely known (standard error, 30–40%) initial slope in the dose-effect relationship for γ-rays. The RBE of neutrons versus a γ-ray dose of 1 Gy is only about 12 (Bauchinger and others 1983; Schmid and others 2000).

In the context of risk estimation, the major interest is in neutron RBEs that have been evaluated in animal experiments with regard to tumor induction. A multitude of results have been reported in the literature for many tumor systems (NCRP 1990). Experiments with rodents show considerable variation, especially in female mice and rats, and this variation reflects the decisive influence of hormonal status. In experiments with female Sprague-Dawley rats, Shellabarger and others (1980) found that 4 mGy of fast neutrons produced as many mammary neoplasms as 0.4 Gy of X-rays, which implied an RBE of 100. Broerse and Gerber (1982) used female Sprague-Dawley rats, which have a much lower spontaneous incidence, and found substantially lower values of neutron RBE. However, considerable differences in neutron RBE at higher doses were observed for different tumor types. As an extreme example, one may refer to lung adenomas in female RFM mice, in which there is a clear reduction in age-adjusted incidence after γ-ray exposures up to about 2 Gy, but neutron doses of 0.2 Gy cause a substantial increase (Ullrich and others 1976). The simple assumptions made in the calculation of RBE do not seem to be applicable in such a case.

In view of this complexity, it appears best to refer to experiments with male mice or rats that determine the overall incidence of solid tumors. In an extensive series of studies of the French Commissariat a l'Energie Atomique using male Sprague-Dawley rats, a fission neutron dose of 20 mGy was consistently found to be equivalent to an acute γ-ray dose of 1 Gy with regard to both nonlethal tumors (Lafuma and others 1989) and lethal tumors (Wolf and others 2000). This comparison corresponds to a neutron RBE of 50 against a reference γ-ray dose of 1 Gy. When the experiments were evaluated in terms of life shortening as a proxy for tumor mortality, the inferred RBE was closer to 30 (Wolf and others 2000). Smaller values of the RBE—around 20 compared to a γ-ray dose of 1 Gy and about 15 compared to X-rays—are suggested by major studies with mice that were evaluated in terms of life shortening, again as a reflection of increased mortality from tumors (Storer and others 1988; Carnes and others 1989; Covelli and others 1989).

In all experimental studies with rodents, it was difficult or impossible to determine excess tumor rates at γ-ray doses substantially less than 1 Gy. For the purpose of risk estimation, it is therefore assumed in this report that the relevant animal experiments with rodents indicate a neutron RBE for solid tumors of 20–50 compared to a reference γ-ray dose of 1 Gy. Experimental evidence suggests lower neutron RBEs for leukemia; in experiments with RFM mice (Ullrich and Preston 1987), an RBE of about 3 was seen versus a γ-ray dose of 0.5 Gy; at lower γ-ray doses, statistical uncertainty did not permit the specification of a neutron RBE.

CHEMICAL ASPECTS OF RADIATION

Electron Ionization of Water Molecules and Indirect Effects on DNA

As previously described, free electrons can be produced by X- and γ-ray interactions with atoms in tissue. These electrons can then interact with the DNA molecule and create damage in the form of strand breaks or damaged bases; these are known as *direct effects*. *Indirect effects* can occur after a photon interacts with a water molecule. Water molecules make up 70% of human tissue. Ejection of an electron from a water molecule by an incoming photon produces an ionized water molecule, H_2O^+. Trapping of the electron by polarizing water molecules produces a so-called hydrated electron, e^-_{aq}. When the ionized water molecule collides with another water molecule, it reacts to produce a highly reactive hydroxyl radical, $OH^\bullet$, according to the reaction

$$H_2O^+ + H_2O \rightarrow OH^\bullet + H_3O^+.$$

Other reactions produce a hydrogen radical ($H^\bullet$), hydrogen peroxide, and water. Thus, these reactions produce three important reactive species—e^-_{aq}, $H^\bullet$, and $OH^\bullet$, which have initial relative yields of about 45%, 10%, and 45%, respectively, in the case of γ-radiation. The reactive species can damage DNA, and such damage is termed an *indirect effect*.

The relatively long-lived (about 10^{-5} s) $OH^\bullet$ radical is believed to be the most effective of the reactive species; as an oxidizing agent, it can extract a hydrogen atom from the deoxyribose component of DNA, creating a DNA radical. Early experiments demonstrated that about 70% of the DNA damage can be prevented by the addition of $OH^\bullet$ scavengers (Roots and Okada 1972). Because $OH^\bullet$ is so highly reactive, it has been estimated that only the radicals formed within about 3 nm of DNA can react with it (Ward 1994). Although DNA is deemed the most important target for biological damage that leads to health effects, other sites—such as the nuclear membrane, the DNA-membrane complex, and the outer cell membrane—may also be important for some biological effects. Signal transduction from cell membrane phospholipids damaged by free radicals and oxidizing reactions is an important natural process. This is one set of biochemical pathways by which the effects of ionizing radiation may overlap with the effects of endogenous processes, such as macrophage oxidative bursts. These processes may underlie those seen in irradiated cells that have been characterized as "bystander effects" and "adaptation" (see Chapter 2).

Nikjoo and colleagues (1997, 2002) have modeled the probability of electron and $OH^\bullet$ radical interaction with DNA. In a 1997 publication, they modeled the spectrum of DNA damage (direct energy deposition and reactions with diffusing $OH^\bullet$ radicals) induced by low-energy secondary electrons (0.1–4.5 keV). They note that to extrapolate available epidemiologic and experimental data from high-dose and high-dose-rate studies to the relevant low levels of single isolated tracks, it is essential to develop a more molecular and mechanistic approach based on the amounts, types, and repairability of the early molecular damage that results from the initial physical and chemical processes. Their calculations for secondary electrons show that most (about 66–74%) low-energy electron interactions in DNA "do not lead to damage in the form of strand breaks and when they do occur, they are most frequently single strand breaks" (SSBs). Although the data are complex, SSB percentages in their study range from about 22 to 27% in the electron energy range of 0.1–4.5 keV and double-strand break (DSB) percentages range from about 1.4–2.4% in the same energy range. However, more than 30% of DSBs are of a more complex form; these complex breaks are somewhat analogous to LMDS, but Nikjoo and colleagues do not include base damage in their model. Their calculations also indicate that the DNA damage tends to be along short lengths of DNA: 1–34 base pairs (bp) for 0.3 and 1.5 keV electrons. The authors conclude that the large deletions seen in radiation-induced mutations may have other mechanisms, such as nonhomologous recombination (Nikjoo and others 1997).

In the case of energetic electron interactions with DNA (0.1 eV to 100 keV electrons), Nikjoo and others (2002) estimate that more than 80% of the interactions do not cause damage in the form of DNA SSBs. Of the interactions that do cause strand breaks, the authors calculate that a small percentage (about 0.5–1.4%) produce DSBs. They note, however, that there is still a considerable contribution (>20%) to the DSB yield from complex DSBs in which a simple DSB is accompanied by at least one additional strand break within 10 bp. As in the low-energy study just described, this model does not include any contribution to the yield of strand breaks from damaged bases.

Another recent study suggests that single low-energy electrons can produce DNA SSBs and DSBs at energies below ionization thresholds (Boudaiffa and others 2000). The authors speculate that these breaks are initiated as direct damage by resonant electron attachment to DNA components followed by bond dissociation. The breaks were produced in DNA in a vacuum, so the relevance of the resonance phenomenon to DNA breaks in the intracellular aqueous environment is open to question. Hanel and colleagues (2003) have shown that electrons at energies below the threshold for electronic excitation (<3 eV) can decompose gas-phase uracil to generate a mobile hydrogen radical. The relevance of this observation to DNA damage *in vivo* awaits further experimentation.

Spontaneous DNA Damage Relative to Radiation-Induced DNA Damage

DNA is an unstable chemical entity under *in vitro* conditions because it is the target of a variety of reactive small molecules. DNA undergoes degradative reactions caused by active hydrolysis that result in depurination and deamination, and it undergoes base adduct formation by reactions with metabolites and coenzymes, damage by reactive oxygen species generated by "leakage" from mitochondria, lipid peroxidation, and many other sources of spontaneous damage (Lindahl 1993; Marnett and Burcham 1993; Beckman and Ames 1997; Lindahl and Wood 1999; see Table 1-2).

More than 90% of naturally occurring oxidation in a cell originates in the mitochondria, and oxidative nuclear damage occurs only for reactive products that can migrate far enough to enter the nucleus and react with DNA. The cell nucleus consequently is almost anoxic (Joenje 1989), and oxidative damage is quenched about fiftyfold by histones and by suppression of Fenton oxidants. However, the nucleus is not radiobiologically hypoxic (<8 μmol/L). The superoxide radical (O_2^-) formed by one-electron reduction of molecular oxygen is generated in all aerobic cells. Chemical or enzymatic dismutation of O_2^- produces hydrogen peroxide, H_2O_2. Although proteins and small molecules, such as glutathione, serve as scavengers for reactive oxygen and thus protect the nucleic acids, there is a considerable amount of oxidative DNA base damage per cell per day (Saul and Ames

1986). However, the steady-state level of DNA damage is low, so most of the spontaneous and metabolically generated damage is apparently repaired efficiently and correctly. Although DNA in cells is basically unstable, the instability is counteracted by DNA repair processes.

Strong evidence pointing to differences between X-ray damage and oxidative damage has come from studies in the yeast *Saccharomyces cerevisiae*. A genome-wide collection of nearly 5000 deletion mutants in all nonessential genes is now available for this species. Using this collection, all genes that were required for resistance to the lethal effects of X-rays and hydrogen peroxide were determined (Birrell and others 2001, 2002). Of those that were resistant to either agent, few genes were in common and their rankings were different. Of the top 100 genes conferring resistance to X-rays, only 35 were in the top 100 that were sensitive to hydrogen peroxide (see Annex A-1). These rankings indicate that the types of damage caused by X-rays and hydrogen peroxide were significantly different and required different mechanisms for repair. In another study using these deletion mutants, the oxidative damage caused by five different oxidants was found to differ significantly, indicating an unexpected complexity for oxidative damage (Thorpe and others 2004). Despite these differences, all of the oxidants caused predominantly protein damage, and few of the genes involved in DNA repair were involved in resistance to damage caused by any of these oxidants. These studies indicate that DNA damage is a more significant factor in resistance to X-ray damage than to oxidative damage. These studies also showed that the genes whose expression was induced by X-rays or hydrogen peroxide were not the genes required for resistance to these agents; few of the X-ray DNA repair genes in particular were inducible by damage (Birrell and others 2002).

Background Radiation

Added to the sources of spontaneous damage and metabolically produced oxidative DNA damage is background radiation, which includes radon, cosmic rays, terrestrial γ-radiation, and natural radioisotopes in the human body. Collectively, background radiation is responsible for delivering an average effective dose per person worldwide of about 2.4 mSv per year (typical range, 1–10 mSv; UNSCEAR 2000b). This background value includes radon exposure, the health effects of which are not evaluated in this report. Medical sources of radiation (diagnostic X-rays, nuclear medicine, and so on) can substantially increase a person's yearly radiation exposure.

Ionizing radiation produces several kinds of damage in DNA, including SSBs and DSBs in DNA chains, DNA-DNA covalent cross-links, and DNA-protein covalent cross-links and a large variety of oxidative changes in the nucleotide bases (Hutchinson 1985; Ward 1988). The identified oxidative base products of ionizing radiation are chemically iden-

TABLE 1-2 Rates of Production and Steady-State Levels of Spontaneous DNA Damage in Mammalian Cells[a]

Result of Damage	Production Rates	Steady-State Levels[b]
Depurination	9000–10,000 per day	<100
Deamination	100–500 per day	<100
3-Methyladenine	600 per day	<50
8-Hydroxyguanine[c]	500–1000 per day	100 (15,000)

[a]For comparison, background radiation of 5 mGy produces an average of about 1 electron track per cell resulting in 5–10 damaged bases, 2.5–5.0 SSBs, and 0.25 DSBs.

[b]Values are for repair-proficient normal cells. Value in parentheses is for repair-deficient liver cells.

[c]Best estimate of 8-hydroxyguanine values, disregarding reports of high values where chemical oxidation occurred during sample preparation.

tical with those produced by other oxidizing agents, such as H_2O_2 in the presence of iron or copper ions or those resulting from the normal metabolic production of free radicals that are by-products of the transport of electrons to oxygen in mitochondria (Dizdaroglu and others 1987, 1991; Gajewski and others 1990; Nackerdien and others 1991; Dizdaroglu 1992; Beckman and Ames 1997). It has been argued in the scientific press and the lay press that the quantity of spontaneous and metabolically generated DNA damage is many orders of magnitude greater than that resulting from low, protracted doses of radiation from environmental sources. This argument implies that the contribution from low doses of ionizing radiation is trivial and can be ignored (Billen 1990; Beckman and Ames 1997)—in other words, that the DNA damage produced by background radiation and the low doses of radiation to which some workers are exposed does not add appreciably to the extensive spontaneous and metabolic damage. However, measurement of naturally produced DNA damage generated by reactive oxygen species is difficult, and some early estimates of DNA products of spontaneous damage, such as 8-hydroxyguanine, are not likely to be accurate estimates, but rather to be overestimates due to chemical oxidation after extraction. An additional consideration is that the distribution of oxidative events produced by radiation may, in some cases, have a unique impact on DNA.

Locally Multiply Damaged Sites

Accumulated evidence shows that the products of ionizing radiation may differ from chemically generated oxidation products in their microdistribution rather than in the chemistry of individual lesions (Ward 1981, 1988, 1994). A portion of the energy of ionizing radiation, primarily that from secondary electrons, is deposited in large enough packets to produce clusters of OH• radicals. Because OH• has a very short range, owing to its high reactivity, it can produce a cluster of damage within a few base pairs of DNA if the cluster is generated within 3 nm of the DNA. Ward and colleagues (1985) have referred to such lesions as LMDS. The probability of clustered damage or LMDS increases with dose and LET but is independent of dose rate because it results from the passage of a single particle track (Prise and others 1994; Holley and Chatterjee 1996; Rydberg 1996; Nikjoo and others 2001). A DSB resulting from a single energy deposition is the most obvious example of an LMDS, but other combinations of strand breaks, cross-links, and base or sugar products can also occur (Ward 1994). Furthermore, both direct interactions of radiation with DNA and reactions of OH• contribute to the complexity of LMDS (clustered damage; Nikjoo and others 1997).

A second property of ionizing radiation that might distinguish it from chemically generated oxidation products is the extensive production of peroxyl radicals due to initial radical damage to molecules other than DNA (Floyd 1995; Milligan and others 1996). Peroxyl radicals produce oxidized

bases, but not DNA strand breaks, and might account for the greater-than-expected yield of base damage, as opposed to strand breaks, observed in irradiated cells (Nackerdien and others 1992). Peroxyl radicals might also account for the production of double base lesions by single radicals that have been observed in irradiated oligonucleotides (Box and others 1995).

Ward and colleagues (1985) have calculated that 5 μM H_2O_2 can produce 15 Gy-equivalents of SSBs in mammalian cell DNA in 30 min through OH• generation catalyzed by iron ions bound to DNA; on the basis of these SSB yields, it takes 1000 Gy-equivalents to kill cells. At the D_{37} dose of cell killing, it has been calculated that each cell will have sustained 2.5 million SSBs from H_2O_2. In contrast, the D_{37} dose for low-LET ionizing radiation produces only 1000 SSBs and 40 DSBs—damage that is not characteristic of lethal doses of H_2O_2. Such data suggest that DSBs and other LMDS (clustered damage) produced by ionizing radiation and a few radiomimetic chemicals are the primary lethal lesions. A recent study that used the phosphorylation of the histone protein H2AX as a marker for DSBs suggests that the yield of DSBs as a function of dose is linear down to as low a dose as 3 mGy (Rothkamm and Lobrich 2003). The fraction of the energy deposited that can yield LMDS increases with LET, and LMDS are generally thought to explain the increased biological effectiveness of high-LET radiation in inducing DNA damage. Whether such LMDS are poorly repaired is still a matter of conjecture, especially in view of the multiple homologous and nonhomologous mechanisms of repair of DNA breaks. At the least, clustering will create complex DSBs within up to 10 bp or so (Ward and others 1985; Holley and Chatterjee 1996; Nikjoo and others 2001). Because of the wrapping of DNA around nucleosomes and the organization of the chromatin fiber, some clusters might include DSBs at two or more sites that are several thousand base pairs apart or even removed from each other by the distance of a chromosomal loop (about 100 kbp; Lobrich and others 1996; Rydberg 1996). For cells to survive without mutations, DNA damage must be faithfully repaired. Yet because large regions of the genome in somatic cells do not contain active genes or contain genes that are not expressed, inaccurate repair that simply restores the integrity of the DNA may be sufficient to produce viable cells that have minimal alterations in function. Conversely, it has been argued that whereas spontaneous damage is readily repaired in repair-competent cells, the DSBs and clustered lesions produced by even low-LET radiation are likely to be repaired with difficulty or incorrectly, if at all (Ward 1994). However, detailed experimental comparisons between the biological effects associated with the repair of spontaneous damage versus damage due to ionizing radiation have yet to be made.

In summary, LMDS (clustered damage) may be viewed as complex lesions associated with ionizing radiation and not with normal endogenous oxidative processes. If they are refractory to repair, the risk to humans posed by ionizing radiation may be viewed as greater than that posed by endogenous oxidative stress.

MOLECULAR MECHANISMS OF DNA REPAIR

Ionizing radiation can cause a wide array of damage to individual DNA bases and SSBs and DSBs resulting from deoxyribose destruction (for basic biological and genetic concepts, see Appendix A). Damaged bases are repaired by mechanisms that involve excision and replacement of individual damaged bases (base-excision repair) or of larger oligonucleotide fragments (nucleotide-excision repair). SSBs are repaired in a process similar to base-excision repair with some of the same enzymatic components. DSBs potentially involve a number of repair processes, especially because organisms require the ability to distinguish between breaks caused by damage and those associated with normal processes, such as recombination, telomere maintenance, DNA replication, and processing of genes encoding antibodies. Some DSBs are simply rejoined end to end in a process called nonhomologous end joining (NHEJ). Others are repaired by a process of homologous recombination (HR) in which the broken strand is repaired by crossing over with an adjacent identical DNA sequence; this generally occurs only during or after chromosome duplication and before chromosome segregation. Damage, especially DSBs, also elicits a signal transduction process that uses a cascade of kinase and other protein modifications and changes in gene transcription, all of which contribute to a cellwide response to DNA damage.

Base-Excision Repair

Release of altered bases by base-excision repair (BER) is initiated by DNA glycosylases that hydrolytically cleave the base-deoxyribose glycosyl bond of a damaged nucleotide residue (Figure 1-9). A present estimate would be that human cell nuclei have ten to twelve different DNA glycosylases, which have varied but overlapping specificities for different base damage. BER has two main pathways that result in replacement of the damaged base with either a short or a long patch.

A common strategy for DNA glycosylases, deduced largely from structural studies, appears to be facilitated diffusion along the minor groove of DNA until a specific type of damaged nucleotide is recognized. The enzyme then kinks the DNA by compression of the flanking backbone in the same strand as the lesion, flips out the abnormal nucleoside residue to accommodate the altered base in a specific recognition pocket, and mediates cleavage (Parikh and others 1998). The DNA glycosylase then may remain clamped to the damaged site until displaced by the next enzyme in the BER pathway, APE1 (also called HAP1), which has greater affinity for the abasic site. This strategy (Parikh and others 1998; Waters and others 1999) protects the cytotoxic abasic residue and may delay the rearrangement of the base-free deoxyribose into a reactive free-aldehyde conformation that could cause cross-linking and other unwanted side effects.

The main human apurinic-apyrimidinic (AP) endonuclease, APE1, occupies a pivotal position in BER of anomalous residues, recognizing and cleaving at the 5′ side of abasic sites generated by spontaneous hydrolysis, reactive oxygen species, and DNA glycosylases. Abasic sites generated by nonenzymatic depurination probably outnumber those generated by all of the DNA glycosylases; consequently, APE1 and subsequent key proteins in the BER pathway (XRCC1 and polymerase β) are essential, whereas mice with knockouts of various DNA glycosylases so far investigated have been viable (Wilson and Thompson 1997). In a substrate recognition process similar to DNA glycosylases, APE1 flips out the base-free deoxyribose residue from the double helix before chain cleavage (Gorman and others 1997; Parikh and others 1998). When bound to DNA, the APE1 protein interacts with the next enzyme in the BER pathway, POL β, and recruits the polymerase to the site of repair (Bennett and others 1997). POL β has two distinct domains that are well suited for DNA gap filling during BER. The larger domain is the polymerase domain itself; a small basic NH_2-terminal domain contains an AP lyase activity that excises the abasic sugar-phosphate residue at the strand break (Matsumoto and Kim 1995; Sobol and others 1996). POL β also interacts with the noncatalytic XRCC1 subunit of the XRCC1-DNA ligase III heterodimer. Consequently, XRCC1 acts as a scaffold protein by bringing the polymerase and ligase together at the site of repair and interacts with poly(ADP-ribose) polymerase and polynucleotide kinase (Whitehouse and others 2001); further stabilization of the complex may be achieved by direct binding of the NH_2-terminal region of XRCC1 to the DNA SSB (Kubota and others 1996; Marintchev and others 1999). XRCC1 contributes to the normal X-ray resistance of mammalian cells, and mutant cells with a defective XRCC1 protein are hypersensitive to ionizing radiation.

When the terminal sugar-phosphate residue has a more complex structure that is relatively resistant to cleavage by the AP lyase function of POL β, DNA strand displacement may occur instead—involving either POL β or a larger polymerase such as POL δ—for filling in gaps a few nucleotides long (Fortini and others 1998; Dianov and others 1999). The FEN1 structure-specific nuclease removes the displaced flap, and the PCNA protein stimulates these reactions (Wu and others 1996; Klungland and Lindahl 1997), acting as a scaffold protein in this alternative pathway in a way similar to that of XRCC1 in the main pathway. Another replication factor, DNA ligase I (LIG1), then completes this longer-patch form of repair. An important property of FEN1 here, in addition to processing the 5′ ends of Okazaki fragments during lagging-strand DNA replication, is to minimize the possibility of hairpin-loop formation and slippage during

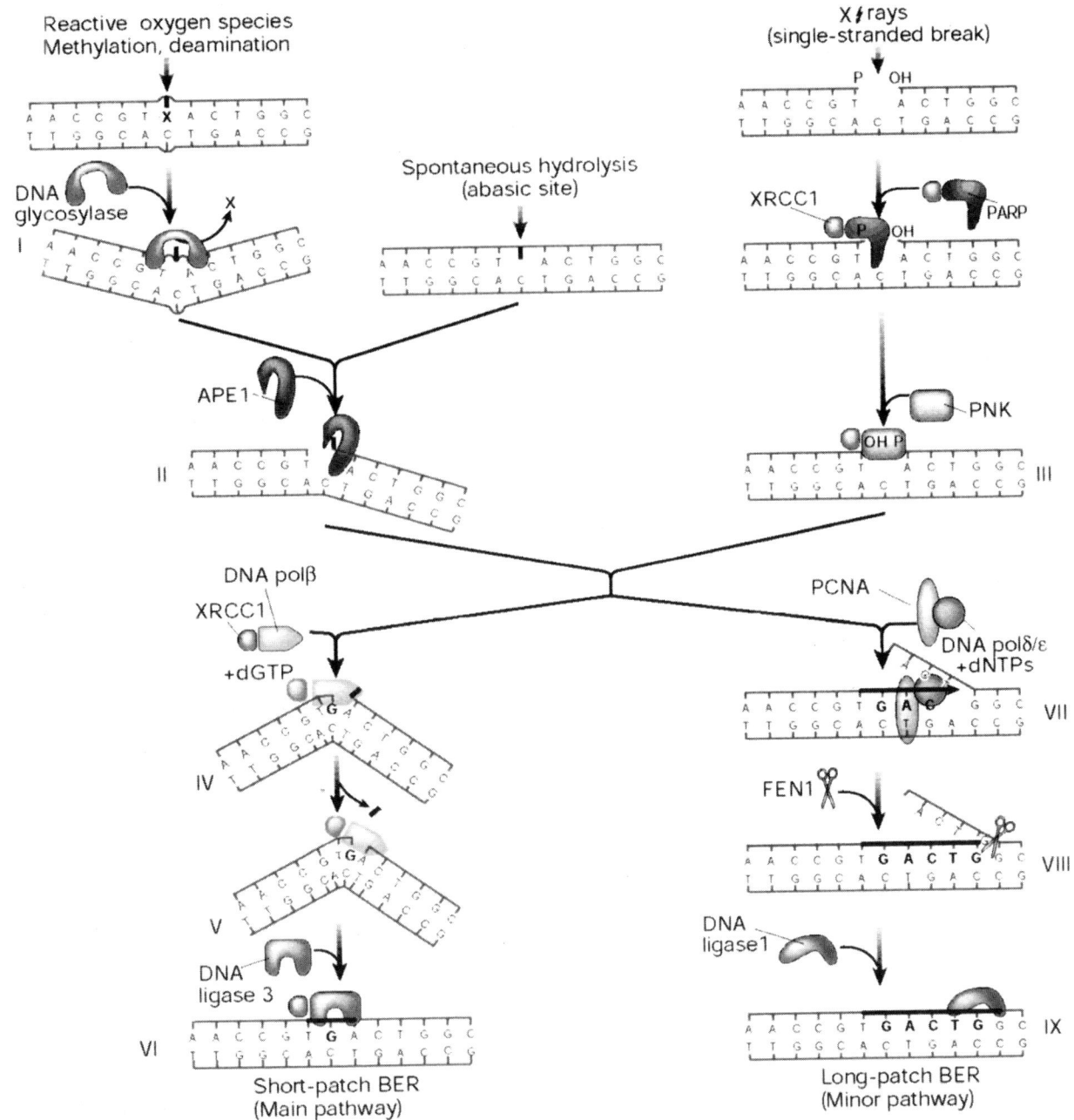

FIGURE 1-9 Base-excision repair. This pathway repairs single-base damage (from X-rays, reactive oxygen species, methylation, or deamination), apurinic sites, and SSBs (from X-rays). A damaged base is removed by glycosylases, leaving an apurinic site that is a substrate for apurinic endonuclease (APE1), which converts it into a SSB. X-ray breaks are modified by XRCC1, polynucleotide kinase (PNK), and poly(ADP-ribose) polymerase (PARP) to produce a cleaved substrate with 3′ and 5′ termini similar to those produced by APE1. The break is then patched by short- or long-patch BER. The short-patch pathway predominates in mammalian cells, and involves polymerase β, which can remove a 5′-deoxyribose moiety by its lyase activity and then insert a single base patch that is sealed by DNA ligase III. The long-patch pathway involves polymerase δ or ε, which is anchored to DNA by a PCNA collar and carries out strand displacement synthesis. The displaced flap is cleaved by the structure-specific endonuclease FEN1, and the patch is sealed by ligase I. XRCC1 is a nonenzymatic scaffold protein that interacts with many of the participants of BER and anchors them to the substrate and hands on repair intermediates through successive stages of BER. SOURCE: Reproduced with permission from J.H. Hoeijmakers (2001).

strand displacement and subsequent DNA synthesis, which might otherwise result in local expansion of sequence repeats (Tishkoff and others 1997; Freudenreich and others 1998). The temporary inefficiency of this process during early mammalian development could explain the origin of several human syndromes that are associated with expansion of triplet repeats in relevant genes.

A series of pairwise interactions between the relevant proteins in BER seem to occur in most cases without any direct strong protein-protein interactions in the absence of DNA. The XRCC1-LIG3 heterodimer is the only preformed complex, and no large preassembled multiprotein BER complex is likely to exist. Nevertheless, the consecutive ordered interactions may protect reaction intermediates and ensure efficient completion of the correction process after initial DNA damage recognition.

Nucleotide-Excision Repair of Cyclodeoxynucleosides

The great majority of endogenous DNA lesions produced by reactive oxygen species are corrected by the BER pathway, and the contributions of the different pathways of nucleotide-excision repair (NER) and mismatch repair are very minor. However, exposure of DNA or cells to ionizing radiation under hypoxic conditions causes the formation of 5′, 8-purine cyclodeoxynucleosides. This chemically stable and distorting form of DNA damage, in which the purine is attached by two covalent bonds to the sugar-phosphate backbone, can be removed only by NER (Heyer and others 2000; Kuraoka and others 2000). Similarly, a major lipid peroxidation product, malondialdehyde, reacts with G to produce an exocyclic pyrimidopurinone (M_1G) that requires NER for repair. These are not the major mutagenic or cytotoxic lesions that occur as a consequence of exposure to ionizing radiation, but they could be critical in individuals with impaired ability to perform NER.

Repair of Single-Strand Breaks

Reactive oxygen species cause DNA strand breaks by destroying deoxyribose residues. Such SSBs are processed and repaired by the same enzymes responsible for the later stages of BER, sometimes with the additional steps of exonucleolytic removal of base pairs and phosphorylation of 5′ termini by DNA kinase. In contrast to the continuous protection of DNA reaction intermediates when an altered base residue is replaced however, the initial strand break is fragile and attracts unwelcome recombination events. An abundant nuclear protein, poly(ADP-ribose) polymerase-1 (PARP1), appears to have as its main role the temporary protection of DNA single-strand interruptions (Le Rhun and others 1998; Lindahl and Wood 1999). PARP1 rapidly shuttles strand breaks in DNA on and off, with NAD-dependent synthesis of poly(ADP-ribose) as its release mechanism. PARP1 knockout mice are viable but show increased numbers of spontaneous sister-chromatid exchanges and sensitivity to ionizing radiation. Extracts of cells from such mice contain low concentrations of other PARP enzymes, which may have distinct unknown roles but could also have backup functions. Crossing PARP1 knockout mice with severe combined immunodeficient disease knockout mice that lack DNA-dependent protein kinase, which is required for VDJ recombination during lymphocyte development, alleviates the DNA-processing defect in the latter and allows some low-fidelity recombination (Morrison and others 1997). PARP1 plays no clear role in the BER process itself, as POL β and LIG3 do, but it interacts with the scaffold protein XRCC1 and may in this way accelerate the recruitment of these repair enzymes for strand interruptions (Mackey and others 1999).

Repair of Double-Strand Breaks

Exposure of DNA to ionizing radiation produces about 5–7% as many DSBs as SSBs (*e.g.*, see earlier discussion of Nikjoo and others 1997, 2000). DSBs are sites at which a surprisingly large number of proteins can bind, carry out strand-break repair, and initiate a complex series of cellular signals that regulate cell cycle progression and the induction and activation of many downstream genes. Cells often encounter DNA DSBs under natural circumstances. These include termini (*e.g.*, telomeres at chromosome ends); recombination intermediates; and immunoglobulin rearrangement during the processing of antibody genes (which leads to increased versatility in the repertoire of immature immunocytes), during the processing of stalled or collapsed replication forks arrested by damage on the template strand and during topoisomerase action on DNA. DSB repair enzymes have been suggested as playing an essential role in telomere maintenance in normal undamaged cells (Blackburn 2000).

One critical difference between metabolically generated DSBs and those generated by ionizing radiation is that some fraction of the latter contain complex radiochemical damage that results in LMDS. LMDS (clustered damage) involve frank breaks, radiolytic fragments as termini, and base damage that is processed into breaks by cellular glycosylases (Blaisdell and Wallace 2001). DSBs thus are not inherently novel, although substantial differences between natural and radiation-induced breaks are likely. Cells contain many genes that code for DNA-binding proteins and signal transduction pathways that respond specifically to DNA double-strand breakage. Consequently, cells can distinguish between a naturally occurring end of DNA at a telomere or recombination structure, for example, and a DSB at an unusual location with atypical chemistry. This suggests that metabolic responses to DSBs and LMDS are highly evolved in most cell types and that cells are not completely unprepared and unequipped for these kinds of lesions, but are in fact able to exercise considerable discrimination in their detection and repair. Cells can also repair damage by novel chemicals, such as cisplatinum, which was newly synthesized in the twentieth century, an indication that novelty or uniqueness is no barrier to the repair of DNA damage.

Repair of DSBs involves a number of biochemically distinct processes. Direct rejoining of the broken ends occurs by several mechanisms, generally described as NHEJ. A fast NHEJ process involves end-binding proteins (Ku70, Ku80, and DNA-PK; Baumann and West 1998; Critchlow and Jackson 1998; Zhao and others 2000), and a slower process involves the hMre11/hRad50/Nbs1 DNA-binding and exonuclease complex that appears to act on refractory, complex breaks (Haber 1998; Petrini 1999). A more complicated rejoining process—homologous recombination—depends on matching damaged DNA with its identical sequence in a sister chromatid after DNA replication or in the homologous chromosome in diploid cells. This process depends on the hRad51 protein, which facilitates homologous pairing, and accessory proteins, such as hRad52, hRad54, XRCC2, and XRCC3 (Thompson 1996). How cells coordinate these processes and determine which should be used under various circumstances is unknown. Coordination may be under the control of the Brca1 and Brca2 proteins. Brca1 binds to unusual DNA structures (Parvin 2001) and is found in a large complex that contains many repair and replication proteins (Wang and others 2000).

The proteins directly involved in DNA strand-break repair do not appear to be inducible (Tusher and others 2001) or to be strongly influenced by p53 functions, except where recombination is involved. Radiation-induced genes represent predominantly cellular signaling molecules, particularly those induced by transactivation by p53. Radiation does, however, activate a series of protein kinases, of which ATM (ataxia-telangiectasia-mutated) is the most prominent, that modify the activity of many other proteins in the repair pathways (Bakkenist and Kastan 2003).

Nonhomologous End Joining—Fast Reaction

DSBs begin to rejoin rapidly after irradiation, with half-times of about 10 min or less (Ward and others 1991). This rapid rejoining involves accumulation of the end-binding proteins Ku70 and Ku80, DNA-PK kinase, the DNA ligase IV-XRCC4 heterodimer, PARP, and others (Figure 1-10). The same factors are also an integral part of the normal process of immunologic rearrangement (Labhart 1999). Conceivably, if the LMDS contains damaged bases, the ends will also require repair steps involving glycosylases, apurinic endonuclease, and DNA polymerase β. Attempted repair by these BER enzymes can enhance DSB formation and loss of base pairs, which then must be repaired by NHEJ (Blaisdell and Wallace 2001). Attempted BER of LMDS in human lymphoblastoid cells produces lethal and mutagenic DSBs (Yang and others, 2004). Small deletions associated with NHEJ have been mapped by sequencing techniques and range up to about 10 nucleotides (Daza and others 1996).

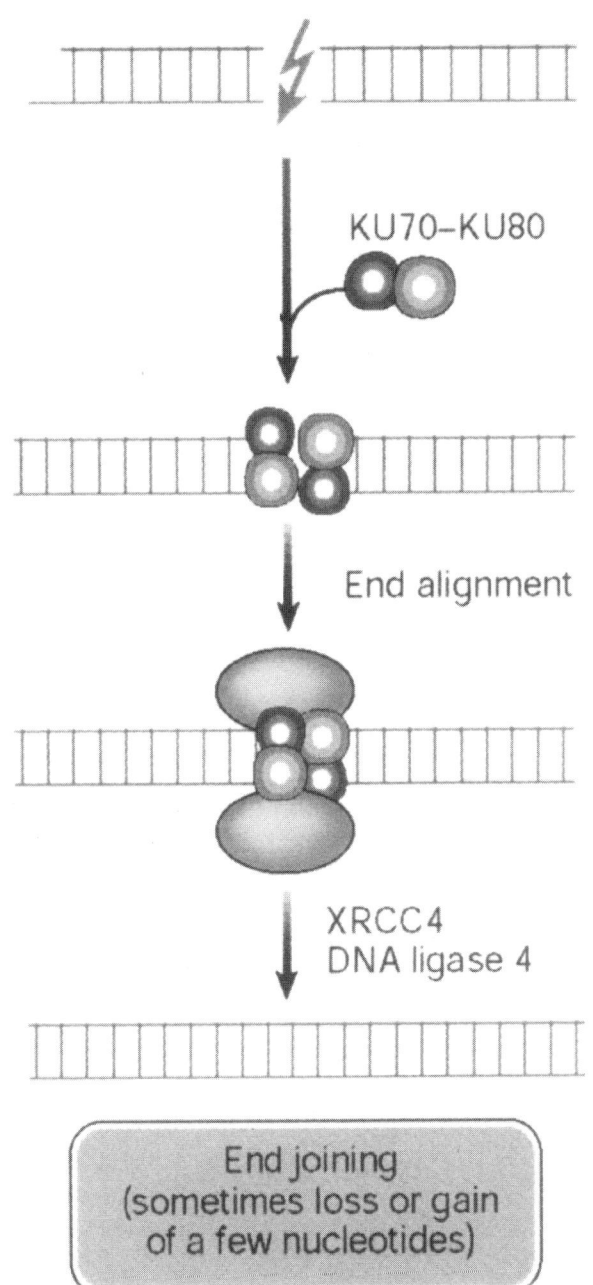

FIGURE 1-10 Nonhomologous end joining: this repair pathway re-ligates DNA DSBs by using the end-binding proteins Ku70 and Ku80 to maintain alignment, and p450 kinase acts as a binding factor. The region across the break is then sealed by ligase IV and its cofactor XRCC4. The sealed break often gains or loses a few nucleotides, especially if the break is an LMDS. In some cases, nonhomologous end joining appears to be responsible for large DNA deletions and chromosome aberrations. In these cases, considerably more than a few nucleotides can be lost. SOURCE: Reproduced with modifications and with permission from Hoeijmakers (2001).

The histone protein H2AX is phosphorylated rapidly over large regions of DNA around sites of DSBs by ATM kinase (Burma and others 2001). Loss of H2AX phosphorylation occurs rapidly with the repair of DSBs, but the biochemical details of dephosphorylation remain to be ascertained. A recent study showed that in human cells, a background level of H2AX phosphorylation occurred in about 5% of the cells. After low doses of X-rays that initially increased the level to 10%, most cells eliminated this phosphorylation, except for a small fraction in which it persisted unless the cell entered DNA synthesis (Rothkamm and Lobrich 2003). Whether this means that a small fraction of cells cannot repair some classes of LMDS or that dephosphorylation of H2AX can be slower than repair itself in a subset of cells remains to be determined.

The DNA-PK kinase is a member of a class of phosphatidyl-3-inosityl enzymes that includes ataxia-telangiectasia-mutated (ATM) and ataxia-telangiectasia-related (ATR) kinases, all of which are involved in signaling the presence of DNA damage (Shiloh 2001, 2004; Figure 1-11). Although DNA-PK kinase can phosphorylate many proteins *in vitro*, it is unclear which proteins it usually phosphorylates *in vivo*. Early cytologic evidence of X-ray damage is phosphorylation of a histone protein to create γ-H2AX foci that are visible microscopically within minutes of irradiation.

Nonhomologous End Joining—Slow Reaction

After the rapid phase of rejoining is complete, the repair of DSBs slows to a second phase with a half-time of several hours. Foci containing the hMre11/hRad50/Nbs complex form or persist and reach a maximum at about 4–6 h. Because this complex has endonuclease and DNA-binding activity, it may be involved in the slower repair of refractory DSBs that cannot be repaired by the earlier, fast mechanism. The complex is not active unless the Nbs1 protein is phosphorylated on several sites by ATM kinase (Figure 1-11), which is itself activated by DNA breaks (Shiloh 2001; Bakkenist and Kastan 2003). The precise DNA structures involved in these refractory breaks are unknown. However, one model suggests that nuclease action by the Mre11 complex resects single DNA strands and that short regions of sequence identity (microhomologies) can be used for alignment and rejoining of DNA strands (Figure 1-12).

Homologous Recombination

Repair of a DSB by HR involves matching the two broken ends of a DNA strand with identical sequences of intact DNA (Figure 1-12). The broken and intact molecules are aligned according to their sequences and encompassed by a toroid of hRad51 molecules that facilitate repair by having DNA single strands invade their homologues, producing an X-shaped four-armed structure called a Holliday junction. Resolution of this structure by specific junction nucleases produces two intact double-strand DNA molecules with or without exchanges according to the orientation of the resolution nuclease actions. The activity of hRad51 is enhanced by other factors, such as hRad52, XRCC2, and XRCC3, and suppressed by p53, which binds to both Holliday junctions and hRad51 (Buchhop and others 1997). HR is much more efficient and important for repair in yeast and somatic chick cells than in normal (nonmalignant) mammalian (human) somatic cells, where NHEJ is the dominant mechanism for DSB repair (Sonoda and others 1998). However, there are exceptions, and there may be times in the cell cycle, such as late S, when HR assumes greater importance because of the proximity of sister chromatids (Thompson 1996). The low level of sister-chromatid exchange, a form of HR, induced by X-rays and high-LET radiation indicates that, in absolute terms, HR remains a minor pathway for the repair of damage caused by ionizing radiation in somatic cells.

There is some question about the source of an identical matching sequence for repair by HR in somatic human cells. A homologous sequence may be the other allele on a chromosome of a recently replicated sister-chromatid sequence on a daughter chromatid or a similar sequence in a repetitive region along the same chromosome. In the latter case the sequences may not be identical over long regions, and the mechanism is known as "homeologous" recombination. Recombination between alleles on separate chromosomes occurs at much lower frequency than between identical sequences on sister chromatids or arranged in tandem on the same chromosome. In general, HR between sister chromatids may occur at higher frequencies late in the cell cycle (*e.g.*, late S; Thompson and Schild 1999), and homeologous recombination is likely to result in the loss of intervening sequences with the production of deletion mutations.

The HR involving hRad51 can be visualized immunohistochemically: foci containing hRad51, Brca1, and other proteins can be seen microscopically soon after irradiation (Scully and others 1997). Cells generally exhibit either hRad51 foci or hMre11/hRad50/Nbs foci, but not both, and the choice of which of the mutually exclusive pathways an irradiated cell follows may be determined by Brca1 (Parvin 2001).

DSB Signal Transduction and Inducible Repair

Bacteria live in a highly variable environment and have evolved efficient inducible DNA repair processes to deal with sudden challenges of DNA damage from oxygen free radicals, ionizing radiation, chemicals, and ultraviolet radiation. These inducible repair pathways are now mechanistically well understood. In *Escherichia coli*, the regulatory genes *sox*R, *ada*, and *lex* control transcription of DNA repair functions, and increased amounts of relevant DNA repair enzymes can be produced in response to environmental challenges. In mammalian cells, the same types of DNA damage are recognized by similar DNA repair enzymes. However, a

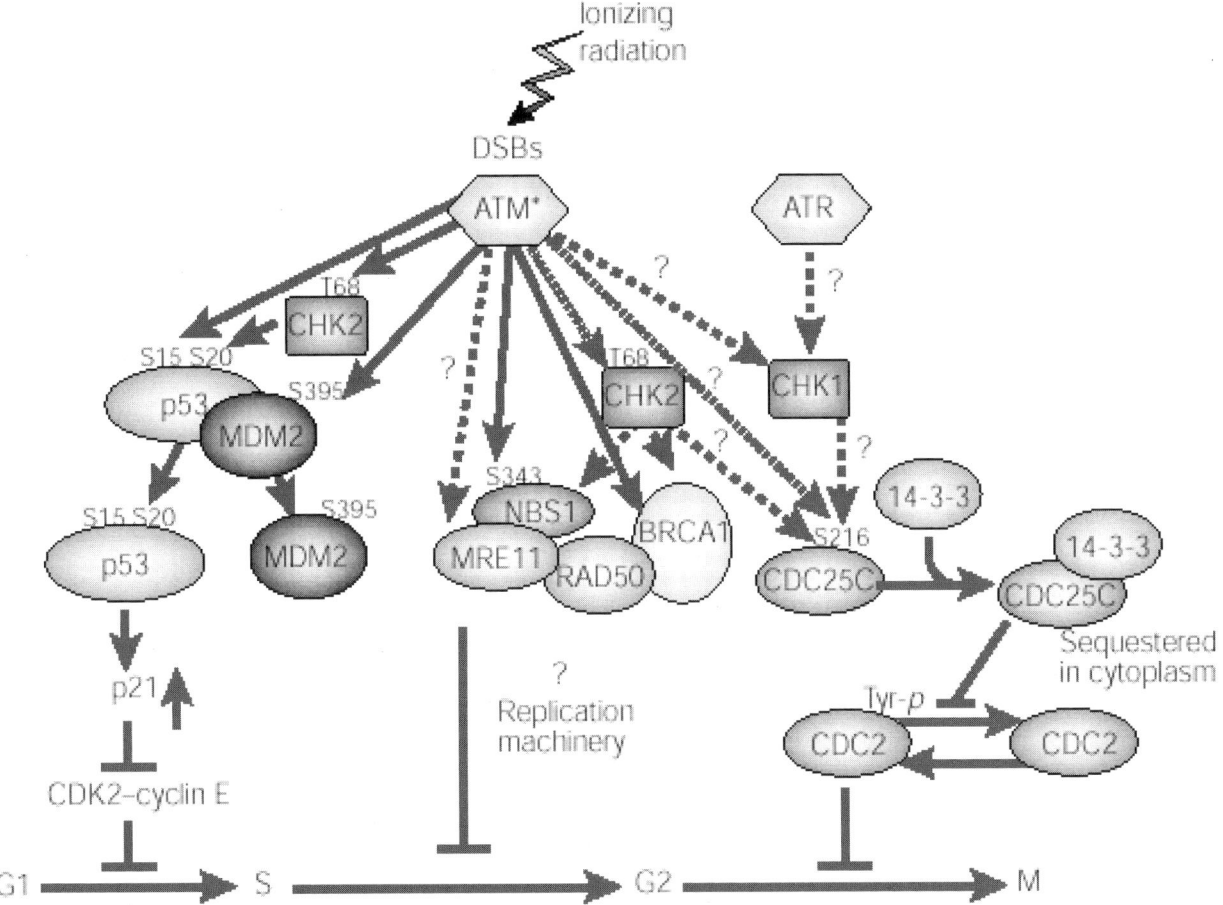

FIGURE 1-11 Network of protein kinases activated by DNA DSBs. ATM is the primary kinase that phosphorylates downstream kinases. The specific activity of ATM is increased after introduction of DSBs in DNA through ionizing radiation or other means; this then activates other proteins by phosphorylation (denoted by amino acid symbol and number) and in a cell cycle-specific manner. G_1 *phase:* Activated ATM (ATM*) directly phosphorylates three proteins involved in controlling p53 functions or levels—p53 (serine 15), CHK2 (threonine 68), and MDM2 (serine 395). CHK2 kinase may also be activated by ATM and in turn phosphorylate p53 on serine 20. This phosphorylation event and the phosphorylation of MDM2 seem to inhibit binding of MDM2 to p53 and should result in an increase in p53 protein. The increased p53 protein transcriptionally induces p21, which inhibits CDK2-cyclin E and causes arrest in the G_1 phase of the cycle. *S phase:* Activated ATM also phosphorylates NBS1 (serine 343), and this phosphorylation event is required for the ionizing radiation-induced S-phase arrest. NBS1 exists in a complex with MRE11, RAD50, and BRCA1. The potential role of these proteins in S-phase arrest remains to be clarified; CHK2 may also be involved in this pathway, after activation by ATM, through phosphorylation of BRCA1 or NBS1. G_2 phase: Details of the downstream targets of ATM at the G_2 checkpoint have not been determined. CHK2 and CHK1 may be targets for ATM and ATR in the G_2-M checkpoint pathway, respectively. CDC25C and 14-3-3 have been implicated in regulation of CDC2 kinase and progression through G_2. Dashed arrows and question marks represent possible signaling steps; solid arrows represent reported phosphorylation events. SOURCE: Reproduced with permission from Kastan and Lim (2000).

major difference from microorganisms is that mammalian enzymes are constitutively expressed. Thus, there are no transcription control or mammalian counterparts of *sox*R, *ada*, and *lex*. This situation presumably reflects the much greater constancy of cellular environment in complex multi-cellular organisms. Therefore, the work on inducible DNA repair in bacteria offers no direct guidelines for the relative resistance of human cells repeatedly exposed to DNA-damaging agents.

Many reports have appeared about adaptive responses involving increased resistance or hypersensitivity in mammalian cells in response to single or multiple doses of ionizing radiation (adaptive effects). There are also reports that the effects of radiation on single cells can influence the response of adjacent nonirradiated cells (bystander effect). These reports are discussed specifically in Chapter 2, but this chapter describes the general stress response and signal transduction pathways that are known to occur after exposure to radiation.

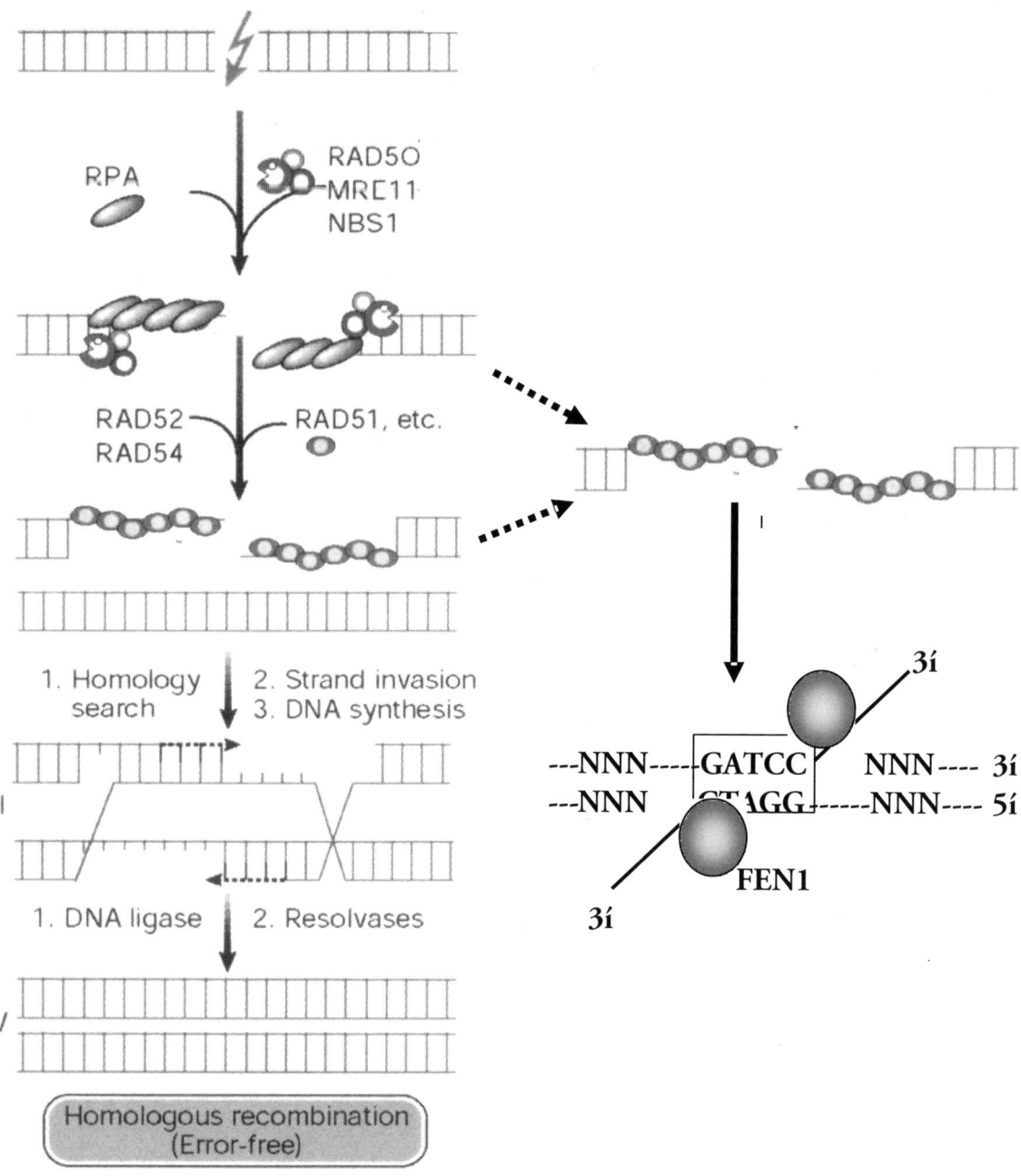

FIGURE 1-12 HR- and microhomology-mediated DSB repair. These two pathways for repair of DSBs are driven by stretches of homologous DNA. HR requires an identical sequence spanning the part of the DNA molecule containing the break and extensive remodeling of the broken DNA termini. Mre11/Rad50/Nbs1 resects individual strands by its 5′- to 3′-exonuclease activity and binds homologous double-stranded DNA by the Rad50 moiety. Exposure of single-stranded regions with only small regions of homology flanking the original break can allow microhomology-mediated strand-break rejoining coupled by cleavage of overhanging strands by FEN1 and resynthesis of any resulting gaps. The repair will, at the least, result in loss of one of the regions of microhomology. Exposure of single-stranded regions homologous to adjacent double-stranded DNA can lead to strand invasion and HR. Single-stranded regions are coated with single-strand binding protein (RPA); homology search and strand invasion are mediated by Rad52, 54, Brca 1 and 2, and Rad51. The complex structure produced forms a Holliday junction that is cleaved by junction-specific nucleases (resolvases), and associated polymerase and ligases complete an error-free exchange of DNA strands. SOURCE: Modified reproduction and reproduced with permission of J. Hoeijmakers (2001).

Damage to cells elicits increases and decreases in the expression of many genes. Recent microarray analysis has shown that these changes can involve hundreds of genes and that different stresses can invoke both a common set of genes and genes that are peculiar to particular kinds of stress (Amundson and others 1999a, 1999b). Despite the large number of affected genes, none appears to be directly involved in repair of DSBs (Tusher and others 2001). Central to most damage responses is stabilization of the tumor-suppressor gene p53, which occurs as a result of posttranslational phosphorylation or acetylation of the protein (Blattner and others 1999; Figure 1-11). Multiple potential serine and threonine residues in p53 are capable of being phosphorylated by different kinases in response to cellular stress, and several thousand combinations of modifications are possible in an irradiated cell. Resolving the functional role of any particular site can be difficult (Blattner and others 1999). The kinases include ATM, ATR, Chk1, Chk2, DNA-dependent protein kinase, and casein kinase I and II (Blattner and others 1999; Chehab and others 2000). (For the role that p53, pRb, cdc25C, chk1, chk2, 14-3-3 proteins, bub1, and the various cyclins and cyclin-dependent kinases play in radiation-induced checkpoints in G_1, G_2, and mitosis, see Little 1994; Jacks and Weinberg 1998; Lengauer and others 1998; Schmidt-Kastner and others 1998; Chan and others 1999; Ford and Pardee 1999; White and Prives 1999).

ATM is a centrally important kinase for X-ray damage that is activated by DNA DSBs (Bakkenist and Kastan 2003; Figure 1-11). In X-irradiated cells, phosphorylation of serine 15 and 37 interferes with the association of p53 with another protein mdm2 that also becomes phosphorylated and normally causes degradation of p53, extending its lifetime. The increased stability of p53 in irradiated cells permits it to form a tetramer and then act as a transactivating factor, increasing the expression of many other genes. Clearly, this will result in large-scale alterations of the gene expression pattern of irradiated cells that can influence their behavior. One downstream target for p53 is the cell cycle regulator protein p21; increased transcription of p21 due to p53 results in delays in the onset of DNA synthesis (the G_1 checkpoint) and reduced DNA synthesis due to p21 binding the replication factor PCNA. The major response of cells to ionizing radiation is a reduction in initiation of the S phase and of replication origins during S. Another important radiation-responsive gene is *GADD45*; both this and p21 showed a linear dose-response relation for induction from 20 to 500 mGy with no indication of a threshold (Amundson and others 1999b).

Most of the members of the signal transduction pathways including ATM, p53, Chk1, Chk2, Brca1, and hMre11/hRad50/Nbs1 are protein products of tumor-suppressor genes. Loss of function of these members can result in genomic instability and in some instances may contribute to a series of events resulting in malignancy. They influence cell cycle checkpoints, DNA replication, DNA repair, and recombination. Thus, it is possible for a single DNA DSB to

activate ATM and p53 and create a cell-wide response through this cascade of protein modifications and alterations in gene expression.

These signal transduction pathways are also activated by extracellular signals working through specific receptors on the cell membrane that then activates kinases, such as MAPKs, which phosphorylate p53. Irradiated cells also generate extracellular signals that resemble cytokines released during normal *in vivo* cell-cell communication processes (Herrlich and others 1992). These can, through receptors on adjacent cells or gap junctions, result in activation of the signal transduction pathways in nearby cells. These multiple intracellular and extracellular pathways of protein modification and signal transduction may constitute the mechanisms by which many of the transient alterations in cellular metabolism occur after exposure to ionizing radiation (Blattner and others 1994).

Some responses observed in particular regimes of exposure to ionizing radiation and given unique names (*e.g.*, adaptive response, bystander effect, genomic instability) may constitute particular manifestations of these general stress responses and signal transduction pathways. These apparently distinct radiation responses have been described mainly in cell biology experiments, and in no case do they have solid biochemical support or mechanistic understanding. In addition to controversy among laboratories, some of the responses described appear to be valid only within a limited dose range and under particular experimental conditions. It is also unclear whether different types of cells, such as epithelial cells, fibroblasts, and lymphoid cells, respond similarly or differently in this regard. Some of the inducible responses appear to be complex in that they depend on participation of intercellular gap junctions in communicating radiation responses to neighboring cells. Work on this subject is in the preliminary, descriptive stage, and there is no understanding of what compounds or factors would be transferred between cells in the gap junction. Therefore, it is difficult to evaluate whether the phenomena are of any general physiologic significance.

SUMMARY

In this chapter the committee has provided background information relating to the physical and chemical aspects of radiation and the interaction of radiation with the target molecule DNA. The chapter describes the physics of electrons and beta particles, which are important contributors to direct DNA damage after ionizing radiation exposure, and introduces a special subject—the effect that neutron RBEs have on low-LET radiation risk estimates. Radical formation by ionizing radiation and its contribution to DNA damage are also described. The committee has discussed the contributions of normal oxidative DNA damage relative to radiation-induced DNA damage and described the DNA repair mechanisms that mammalian cells have developed to cope with such damage. Modeling of electron interactions with DNA suggests that when more than one strand break occurs due to an electron interaction, approximately 30% of the breaks will be multiple events (three or more) that occur over a very small distance. These multiple events, sometimes referred to as LMDS, would be expected to occur at the same average rate per electron traversal of the DNA, whether the overall dose is high or low. It is reasonable to expect that multiple lesions of this sort would be more difficult to repair or might be prone to misrepair. This may explain the apparent inconsistency between the lethality and mutagenicity of agents that principally cause DNA single-strand breaks and ionizing radiation, which also produces double-strand breaks and LMDS. Furthermore, modeling of multiple damages in a small length of DNA suggests that the normal cellular oxidative damage of DNA may differ qualitatively from that due to ionizing radiation. Recent information is presented as an annex to this chapter, about a significant disparity in the genes that repair oxidative damage in yeast DNA and genes that repair radiation damage.

ANNEX 1A: IONIZING RADIATION AND OXIDATIVE DAMAGE—A VIEWPOINT FROM *SACCHAROMYCES CEREVISIAE*

Approximately 4800 deletion mutations have been made in all the nonessential genes in the yeast *Saccharomyces cerevisiae*. These have been used by two groups of investigators to identify the genes responsible for resistance against ionizing radiation, ultraviolet light, cisplatin, and a number of different oxidizing agents (hydrogen peroxide, diamide, linoleic acid 13-hydroperoxide, menadione, and cumene hydroperoxide; Birrell and others 2001, 2002; Game and others 2003; Thorpe and others 2004; Wu and others 2004). The set of genes required for resistance against a particular agent is an indication of the nature of the cellular biochemical pathways required to restore viability and, indirectly, of the kind of damage generated by the agent. If a common set of genes is required for several different agents, these will point to a common or overlapping chemical nature of the damage. The striking observation about the results in *S. cerevisiae* is that the sets of genes required for resistance against each agent differed significantly from each other. When pairwise comparisons were made between ionizing radiation and each oxidant, the overlap was low: less than half of the genes required for resistance against ionizing radiation were also required for resistance to oxidative damage (Figures 1A-1, and 1A-2).

Large numbers of genes not obviously involved in DNA repair fall within the list of sensitive mutants to ionizing radiation and oxidants. Several genes whose deletion produced sensitivity to radiation and oxidants were involved in DNA replication and recombination, suggesting that this process was vulnerable to all kinds of cellular damage in yeast. In contrast, the most important genes in human cells for repair by NHEJ were not represented among the sensitive mutants because this is a minor pathway in yeast. An additional observation is that the set of genes whose expression was induced by damage differed from the genes required for resistance against each agent, implying that repair genes were not among those induced by damage (Birrell and others 2002).

The committee carried out a detailed comparison of the genes reported by each group, using publicly available data sets. One group (Birrell and others 2001, 2002; Game and others 2003) reported the response of the complete set of 4800 genes and ranked them in sequence, from most sensitive to least sensitive. About 10% of all genes (470) showed some degree of sensitivity to ionizing radiation. The other group (Thorpe and others 2004) reported only those genes that showed sensitivity to at least one oxidant (approximately 675 genes) and ranked them in categories 1–7, with the most sensitive in category 1.

Comparison between these data sets is complicated by different methods of reporting and different technical approaches to determining sensitivity. Comparisons were therefore made in general terms rather than gene by gene, and only those genes were considered that were reported by both groups. The committee first compared the genes required for resistance against hydrogen peroxide as reported by two independent research groups, to establish the consistency of the data (Figure 1A-1). A set containing about 200 genes was common to both groups as necessary for resistance to hydrogen peroxide. Of these, 150 were also sensitive to ionizing radiation. Since different methods were used to detect sensitivity and rank the strains, some differences are not surprising. The common set of 150 genes required for resistance to both ionizing radiation and hydrogen peroxide included those involved in postreplication repair and recombination, but the genes that ranked among the most sensitive toward ionizing radiation were ranked lower on the list for hydrogen peroxide (Birrell and others 2002).

The committee then compared the genes required for resistance to different oxidizing agents with those required for resistance to X-rays (Figure 1A-2). The overlap was small in comparison to the number of genes required for resistance to ionizing radiation; conversely, more than half of the genes required for resistance to each oxidant were also required for resistance to ionizing radiation. However, the same genes were not involved for each oxidant.

The implication of these results is that each agent that is toxic to *S. cerevisiae* produces a unique spectrum of cellular damage, with some overlap. The relevance of these comparisons to this report lies in the attempts that have been made to explain low-dose ionizing radiation as no more than a special case of oxidative damage (Pollycove and Feinendegen 2003). If this were true, low doses of ionizing radiation would be insignificant compared to the levels of naturally occurring reactive oxygen species and could therefore be ignored as having no detrimental health effects. How-

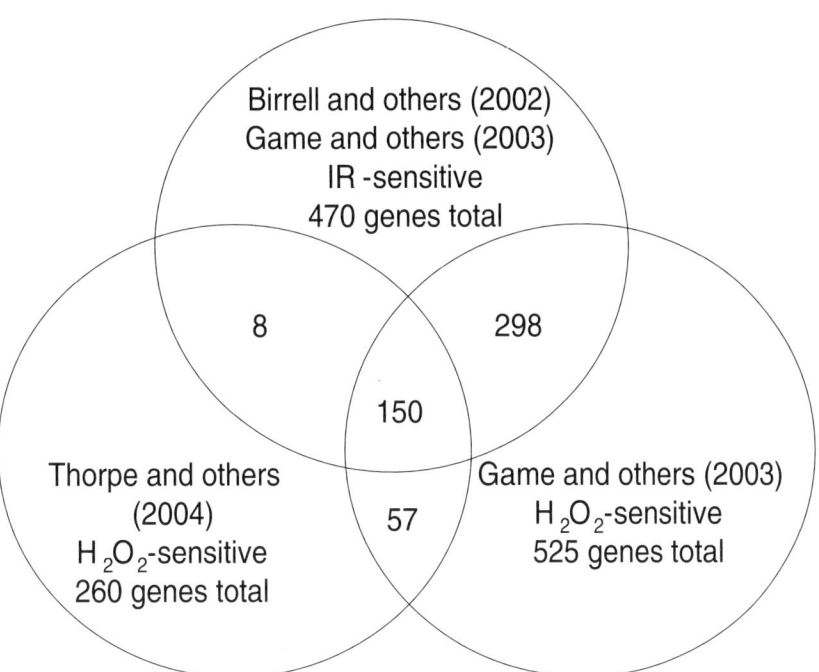

FIGURE 1A-1 Venn diagram representing the overlap among genes involved in resistance against ionizing radiation and hydrogen peroxide as indicated in the reports cited. Numbers in regions of overlap represent the number of genes responsible for resistance against two agents as reported by one or another group.

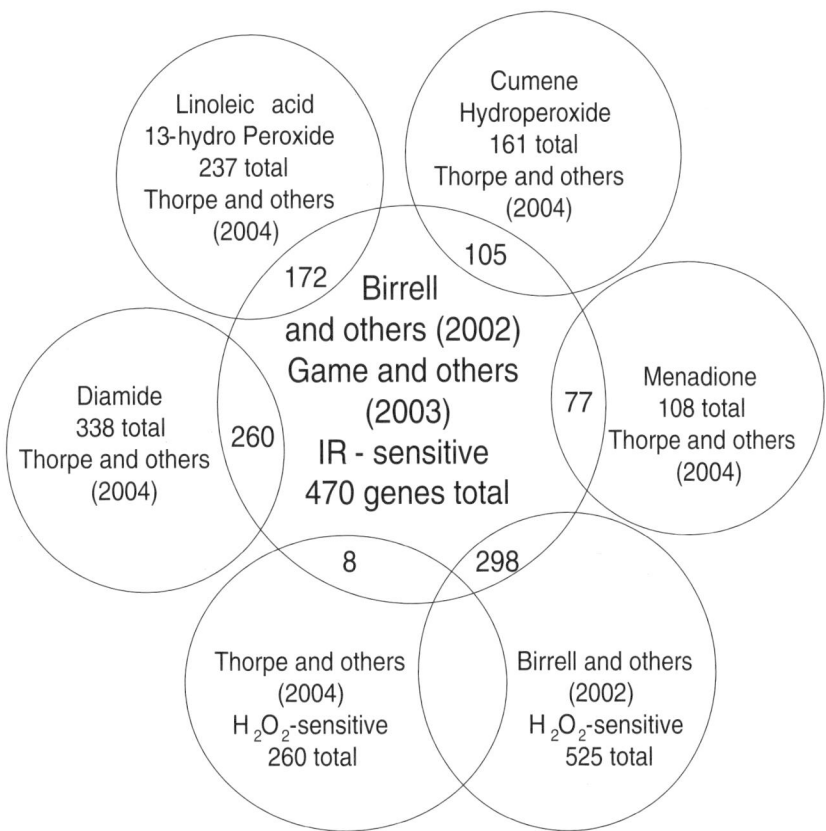

FIGURE 1A-2 Venn diagram representing the overlap among genes involved in resistance against ionizing radiation and various oxidizing agents as indicated in the reports cited. Numbers in regions of overlap represent the number of genes responsible for resistance against two agents as reported by one or another group.

ever, each oxidizing agent involved a significantly different set of genes, which also differed from those required for protection against X-rays, indicating that oxidative damage cannot be considered a single entity, but is dependent on the chemical source of the oxidation. Mutants sensitive to hydrogen peroxide included an overrepresentation of mitochondrial respiratory functions, but those sensitive to diamide encompassed genes involved in vacuolar protein sorting. This makes it especially difficult to predict what kinds of damage would result from endogenous reactive oxidative species. Endogenous damage could present its own unique spectrum of genes required for resistance, different from each of the exogenous sources as well as from ionizing radiation.

These results must be confirmed and extended to human cells, because the genes known to be involved in repair of DNA DSBs by NHEJ (Ku70, Ku80, and DNA-PK) were rarely found among those involved in resistance to ionizing radiation or oxidative damage in yeast, where they play a very minor role. The majority of genes required for resistance to oxidative damage were, however, considered by one set of authors (Thorpe and others 2004) as more representative of damage to the protein components of the cell than to DNA. These included genes required for transcription, protein trafficking, and vacuolar function.

These damage responses in *S. cerevisiae* are, however, dominated by the efficient homologous recombination that plays a major role in response to DNA damage (Kelley and others 2003). Homologous recombination may therefore mask some of the effects caused by loss of genes on pathways that may be minor in yeast but more important in mammalian cells (Swanson and others 1999; Gellon and others 2001; Morey and others 2003). For example, mice that are defective in apurinic endonuclease are embryonic lethals, and blastocysts derived from these nulls are radiosensitive (Xanthoudakis and others 1996; Ludwig and others 1998). RNAi ablation of a pyrimidine-specific DNA glycosylase in mice confers radiosensitivity (Rosenquist and others 2003). Although the results described in yeast do indicate differences between ionizing radiations and oxidizing agents, the extent of differences or of overlap may not be the same in mammalian cells.

These results in *S. cerevisiae*, however, provide *no* support for the attempts to equate low-dose ionizing radiation with endogenous oxidative reactions. The committee would expect even greater divergence between ionizing radiation and oxidative damage in human cells because of the higher ratio of cytoplasmic and nuclear proteins to DNA than in *S. cerevisiae* and the greater role of NHEJ.

2

Molecular and Cellular Responses to Ionizing Radiation

Since the early years of radiobiology the cellular effects of ionizing radiation have been studied in the context of induced chromosomal aberrations, and early models of radiation action were largely based upon such studies (Savage 1996). In the 1970s, somatic cell genetic techniques were developed to allow the quantification and characterization of specific gene mutations arising in irradiated cultures of somatic cells. In more recent years, findings of persistent postirradiation genomic instability, bystander effects, and other types of cellular response have posed additional questions regarding the mechanisms underlying the cytogenetic and mutagenic effects of radiation and their potential to contribute to radiation tumorigenesis.

This chapter considers the general aspects of dose-response relationships for radiobiological effects and subsequently reviews the largely cellular data on a range of radiobiological end points. The main focus of the review is the issue of cellular effects at low doses of low-LET (linear energy transfer) radiation. Many of the conclusions reached from this review, when aggregated with those of Chapters 1 and 3, contribute to the judgments made in this report about human cancer risk at low doses and low dose rates.

GENERAL ASPECTS OF DOSE-RESPONSE RELATIONSHIPS

Any effect of radiation exposure must be quantified in relation to the effect observed in a control population. In this way the dose to an irradiated population is considered in the context of, among other factors, the natural background radiation received. For low-LET radiation an absorbed dose of 1 Gy[1] (1000 mGy) corresponds to an equivalent dose of 1 Sv (1000 mSv). Because this report focuses on low-LET radiation, reference is mostly to grays and not to sieverts. Low-

LET background radiation worldwide is responsible for an average annual effective dose per person of about 0.9 mGy per year (UNSCEAR 2000b). This includes an estimated 0.48 mGy from external terrestrial radiation to the body, 0.28 mGy from cosmic radiation (excluding the neutron component), and 0.17 mGy from radioisotopes in the body. For the purposes of this report, it does not include background radiation of about 1.2 mSv delivered to the lungs from radon and radon progeny or other high-LET radiation. Radon is the subject of the BEIR VI report (NRC 1999).

The maximal permissible levels that are recommended in the United States by the National Council on Radiation Protection and Measurements (NCRP) for people exposed to radiation other than background radiation and from medical applications are 1 mSv per year for the general population and 50 mSv per year for radiation workers employed by nuclear-related industries (Federal Register 1987). Considering the levels of background radiation, the maximal permissible levels of exposure of radiation workers now in effect, and the fact that much of the epidemiology of low-dose exposures includes people who in the past have received up to 500 mSv, the BEIR VII committee has focused on evaluating radiation effects in the low-dose range <100 mGy, with emphasis on the lowest doses where relevant data are available. Effects that may occur as the radiation is delivered chronically over several months to a lifetime are thought to be most relevant.

An effect (E) (for example chromosomal aberrations, mutations, or animal carcinogenesis) induced by an acute dose of low-LET radiation delivered over a few minutes has been described by the relationship $E = \alpha D + \beta D^2$, where D = dose; this is a linear-quadratic dose-response relationship curving upward (Lea 1946; Cox and others 1977). Theoretically, the α term represents the single-hit intratrack component, and β represents the two-hit intertrack component. An alternative interpretation is that the D^2 term may arise from multiple tracks that would increase the overall burden of damage in a cell and thereby partially saturate a repair

---

[1]Because the older dose term "rad" is used in some figures, the committee notes here that 1 Gy = 100 rads.

system and reduce the probability of repair of particular damage from a track (UNSCEAR 1993). However, there is no experimental evidence to support this model. As the dose is reduced, the β term becomes less important, and the dose-response relationship approaches linearity with a slope of α. For doses delivered in multiple fractions or at low dose rates, in which case the effects during the exposure period are independent and without additive or synergistic interactions, the dose-response relationship should also be linear with a slope of α. Theoretically, the value of α should be the same for high and low dose rates and for single or multiple doses, and there should be a limiting value, α_1, so that reducing the dose rate further would not reduce the α term (see Figure 2-1 for an illustration of these concepts).

For extrapolating data from acute high-dose-rate experiments to results expected for low doses and low-dose-rate experiments, the dose and dose-rate effectiveness factor, DDREF, is estimated (see Figure 2-1). The DDREF is estimated by comparing the linear extrapolation (curve B) of the induced incidence for a set of acute dose points (curve A) with the linear curve (D) for low dose rate. The DDREF is equal to the slope α_L for curve B divided by the slope α_1 for curve D. If only acute high-dose data are available, the slope (α_1) for the linear extrapolation of the data for acute doses that approach zero (tangent to curve A) is used. This is the dose effectiveness factor (DEF), which is assumed and shown (Cornforth and others 2002) to be equal to the dose-rate effectiveness factor (DREF). Therefore, the term DDREF is used to estimate effects for either low doses or low dose rates. This value for DDREF can be estimated from a fit of the acute data using the relationship described above (*i.e.*, $E = \alpha D + \beta D^2$). Thus, the DDREF = $[(\alpha D + \beta D^2)/D] / (\alpha D/D) = (\alpha D + \beta D^2)/\alpha D$, which equals $1 + D\beta/\alpha$ or $1 + D/(\alpha/\beta)$. D is the dose at which the response for acute irradiation is divided by the response for low-dose-rate irradiation to obtain the DDREF, and the relationship shows that DDREF will increase with the dose at which the curves A and D are compared. Note, the contribution from the β term (βD^2) equals the contribution from the α term (αD) (*i.e.*, $\beta D^2 = \alpha D$, when $D = \alpha/\beta$). For this dose equal to α/β, the incidence for curve D is equal to the difference between the incidence for curve A and the incidence for curve D; thus, curve A intersects the linear curve B at the dose equal to α/β. For example, if α/β equals 1 Gy, the DDREF for a dose of 1 Gy would theoretically equal (1 + 1/1) or 2; for a dose of 0.5 Gy, the DDREF would equal 1.5, and for a dose of 2 Gy, it would equal 3. If α/β equals 2 Gy, curves A and B would intersect at 2 Gy where the DDREF equals 2; at doses less than or greater than 2 Gy, the DDREF would be less than or greater than 2, respectively. This concept is illustrated with experimental data in Figure 2-8; for the induction of HPRT (hypoxanthine-guanine phosphoribosyl transferase) muta-

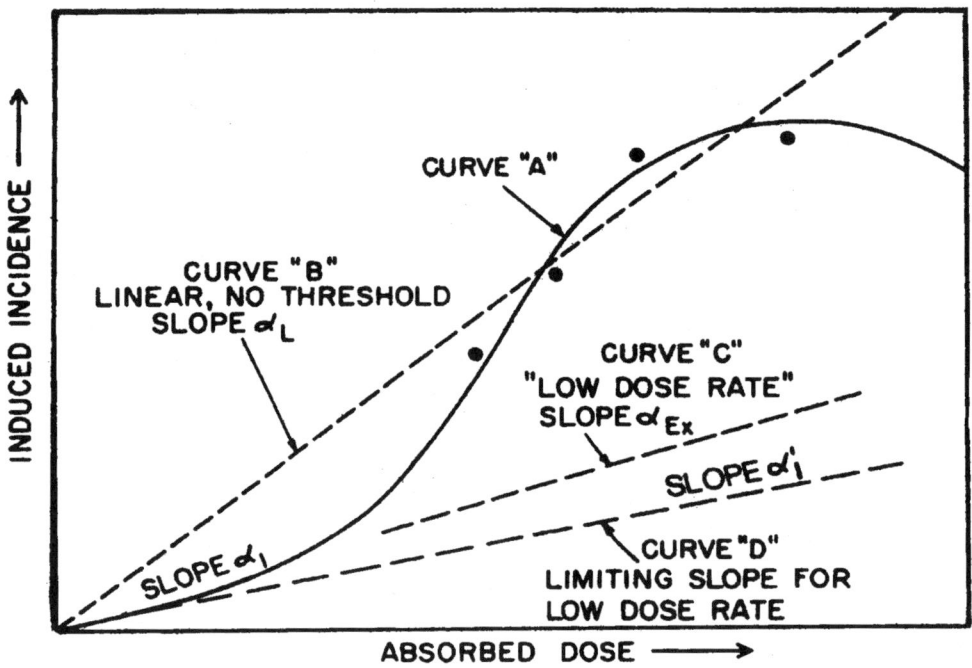

FIGURE 2-1 Schematic curves of incidence versus absorbed dose. The curved solid line for high absorbed doses and high dose rates (curve A) is the "true" curve. The linear, no-threshold dashed line (curve B) was fitted to the four indicated "experimental" points and the origin. Slope α_1 indicates the essentially linear portion of curve A at low doses. The dashed curve C, marked "low dose rate," slope α_{Ex}, represents experimental high-dose data obtained at low dose rates. SOURCE: Reproduced with permission of the National Council on Radiation Protection and Measurements, NCRP Report No. 64 (NCRP 1980).

tions in mouse splenic T lymphocytes, the DDREF was ~1.5 at 1 Gy and ~4 at 5 Gy. Also, in Figure 10-2, dose-response curves for the incidence of solid cancers in Japanese A-bomb studies were constructed over the dose range of 0–1.5 Sv, assuming α/β = 1.45 Sv and α/β = 3.33 Sv, and DDREF values were calculated by dividing the slope of curve B by the slope of curve D. These slope ratios give DDREF values of 1.8 for α/β = 1.45 Sv and 1.3 for α/β = 3.33 Sv.

Several factors may affect the theoretical dose-response relationships described above, namely: variations in radiosensitivity during the cell cycle; induction of an adaptive response to an initial exposure, which can reduce the effect of later exposures; a bystander effect that causes an irradiated cell to have an effect on a nearby unirradiated cell; the induction of persistent genomic instability; and hyper-radiation sensitivity in the low-dose region. Except for the cell cycle, these factors have been identified and studied since the BEIR V report (NRC 1990). These factors, together with data on the induction of gene/chromosomal mutations in somatic cells are discussed in subsequent sections of this chapter.

INDUCTION OF CHROMOSOME ABERRATIONS

Early studies on the mechanisms of chromosome aberration induction summarized by Savage (1996) lead to the following conclusions: Primary radiation-induced break-type lesions can (1) reconstitute without morphological change to chromosomes; (2) rejoin illegitimately with another break close in time and space to produce an intra- or interchromosomal aberration visible at the subsequent mitosis; or (3) remain "open," leading to a simple break at mitosis. These early conclusions, based primarily on work with plant cells, are supported by subsequent studies with mammalian cells. The quantitative cytogenetic systems developed over the years, particularly in G_0 human lymphocytes, have been utilized in studies on the effects of dose, dose rate, and radiation quality. From a mechanistic viewpoint there is compelling evidence that the induction and interaction of DNA double-strand breaks (DNA DSBs or, more correctly, double-stranded lesions) is the principal mechanism for the production of chromosome aberrations. The fundamental arguments supporting this widely accepted conclusion have been discussed in depth (Bender and others 1974; Scott 1980; Cornforth and Bedford 1993; Natarajan and Obe 1996). Of particular note are the data showing excess aberrations following the introduction of DNA DSB-inducing restriction endonucleases into cells (Bryant 1984; Obe and others 1985; Morgan and Winegar 1990). The increased chromosomal radiosensitivity in cells genetically deficient in processes associated with DNA DSB repair, reviewed by ICRP (1998), also supports this conclusion.

The biophysical modeling of the dose-response and LET dependence for chromosome aberration induction has been a major focus in radiobiological research for many years. In the following paragraphs, a brief outline is provided of the current state of knowledge of the mechanisms that are believed to play a role in the induction of chromosomal aberrations (see Bedford and Dewey 2002 for a detailed discussion). Aberrations formed following irradiation of cells in the G_0/G_1 phase of the cell cycle are dicentric exchanges, centric rings, and monocentric exchanges (translocations). The vast majority of studies show that the dose-response for low-LET radiation is curvilinear and fits well to the equation $\alpha D + \beta D^2$. At high doses, saturation effects occur, and the dose-response tends to turn down; for human lymphocytes, saturation occurs at doses greater than 4–5 Gy. The linear coefficient α, representing the initial slope of the dose-response, increases with the LET of the radiation, reaches a maximum at ~70 keV μm^{-1}, and then falls. The quadratic coefficient β is approximately constant up to around 20 keV μm^{-1} but reduces at higher LET (>100 keV μm^{-1}). A reduction in low-LET dose rate reduces aberration yields in a dose-dependent manner; the value of α is unaffected, but the value of β decreases (Edwards and others 1989).

A current explanation of the above dose-response characteristics is that DNA DSBs are the principal causal events for aberration induction and that these are induced with linear kinetics at around 30 DNA DSBs Gy^{-1}. Correct repair and misrepair processes operate in competition for these DNA DSBs, with the majority of breaks restituting correctly and a small fraction taking part in misrepair-mediated chromosomal exchanges (Hlatky and others 1991). The fraction of misrepair events is suggested to be dose dependent, with the close proximity of DNA DSBs promoting exchanges and thereby imposing curvature on the low-LET dose response. The two-track component of DNA lesion production and interaction increases as a quadratic function of dose and produces biophysical curvature on the dose-response. However, the concept of proximity-promoted interaction of lesions gives more weight to lesions arising along the path of single tracks. Such proximity effects have been reviewed (Sachs and others 1997). Modeling procedures of this type, while providing a coherent explanation of low-LET dose-response, are insufficient to account fully for high-LET effects (Moiseenko and others 1997). An additional factor considered in some modeling of dose- and LET-dependent responses is the possibility that some exchanges might involve interaction of a DNA DSB with an undamaged DNA site (*i.e.*, recombinational-like DNA misrepair). It seems likely that a variety of repair and misrepair options are available to the cell and that their relative importance is LET dependent; this feature may relate to the complexity of a significant fraction of initial DNA DSBs (see Chapter 1).

Dose and LET dependence also apply to the morphological complexity of the induced chromosomal aberrations themselves. The development of fluorescence *in situ* hybridization (FISH) methods of chromosome painting has allowed aberration complexity to be studied in detail. In brief, aberration complexity reflects the number of DNA DSBs involved in a given chromosomal exchange event; not surprisingly, aberration complexity becomes most apparent at high doses of low-LET radiation and at all doses of high-LET radiation (Finnon and others 1995, 1999; Griffin and others 1995; Anderson and others 2000). The precise mechanism of formation of these complexes remains uncertain, but multiple pairwise exchanges involving the same chromosomes play some part (Edwards and Savage 1999). However, cyclic exchanges involving three and four breaks are not uncommon, implying that the interaction of multiple DNA DSBs can occur. Recent studies using multicolor mFISH analyses further emphasize the complexity of many radiation-induced chromosomal exchanges produced after high acute doses of radiation (Loucas and Cornforth 2001). These mFISH analyses also show that even after exposure at very low dose rates, the formation of complex chromosomal exchanges is not completely eliminated (Loucas and others 2004).

Combining FISH painting and premature chromosome condensation techniques (Darroudi and others 1998) has also facilitated studies on the rate of formation of aberrations. In these studies (Darroudi and others 1998; Greinert and others 2000) a substantial portion of exchanges have been shown to form rapidly, although some require several hours. There is some evidence that those aberrations forming rapidly tend to be incomplete exchanges, which suggests a time dependence for pairwise exchange (Alper and others 1988) of DNA DSBs. The general picture that emerges from these biophysical studies is that the misrepair events of radiation-induced DNA DSBs that lead to chromosome aberrations are probably associated with the dominant postirradiation function of the nonhomologous end joining (NHEJ) repair processes described in Chapter 1.

Overall, biophysical approaches to the modeling of dose-response for chromosomal aberrations, although not without some uncertainties on mechanisms, imply that the single-track component of radiation action will dominate responses at low doses and low dose rates (*i.e.*, the dose-response for all forms of aberrations will be linear at low doses and low dose rates). Considerable effort has been expended to test this proposition, and in a very large multicenter study using assays of dicentric aberrations in human lymphocytes, the linearity of the response was evident down to at least 20 mGy of low-LET radiation (Lloyd and others 1992), which is illustrated in Figure 2-5. Below that dose, the statistical power of the data was not sufficient to exclude the theoretical possibility of a dose threshold for radiation effects.

Another important feature of the chromosomal response to radiation is the postirradiation period during which initial DNA damage is fixed and then expressed in the form of aberrations such as dicentric chromosomes. On the basis of direct observation and theory, the conventional cytogenetic view is that all such chromosomal damage sustained within a given cell cycle will be fixed and then expressed at the first postirradiation mitosis. Accordingly, Carrano and Heddle

(1973) predicted that the dicentric aberration frequency will fall by a factor of around 2 per cell division on the basis that at each mitotic anaphase, a given dicentric has an equal chance of falling free or producing a lethal anaphase bridge. This prediction has been tested as part of a recent study (Pala and others 2001) that showed dicentric yields falling by up to a factor of 4 between the first and second postirradiation cell division. It seems therefore that the vast majority of initial unrepaired and misrepaired lesions are expressed as chromosomal damage at the first division. Cells carrying unbalanced chromosomal exchanges (dicentrics) or substantial chromosomal losses are not expected to contribute to the viable postirradiation population. By contrast, cells carrying small deletions or balanced exchanges such as reciprocal translocations are likely to remain viable, and some may have the potential to contribute to tumor development.

Later in the chapter this conventional view is contrasted with data implying that in some circumstances, a certain fraction of irradiated cells can express chromosomal damage over many cell cycles (*i.e.*, persistent genomic instability). The proposition that this induced instability phenotype can contribute to tumorigenesis is explored in Chapter 3.

INDUCTION OF GENE MUTATIONS IN SOMATIC CELLS

Ionizing radiation is known to induce a broad range of potentially mutagenic lesions in DNA ranging from damaged DNA bases to frank DNA breaks and chemically complex lesion clusters (see Chapter 1). Not unexpectedly, molecular analyses of radiation-induced somatic mutations at a number of loci provide evidence of induction of point mutations in single genes and of small and large deletions that may encompass a number of physically linked genes (Sankaranarayanan 1991; Thacker 1992). An important factor in the induction and recovery of deletion-type, multilocus mutations is the degree to which multiple gene loss may be tolerated by the cell. There is good evidence that such tolerance is highly dependent on the genetic context of the mutation (*i.e.*, its position in respect to essential genes and, for autosomal loci, the genetic status of the second gene copy on the homologous chromosome). These issues are discussed in depth elsewhere (Thacker 1992); here it is sufficient to note that genetic context can result in up to a twentyfold change in induced mutation frequencies in autosomal genes (Bradley and others 1988; Amundson and Liber 1991). There is strong molecular evidence that in most circumstances, a DNA deletion mechanism dominates mutagenic response after ionizing radiation (Sankaranarayanan 1991; Thacker 1992), and it is for this reason that the genetic context of the mutation is of great importance. In illustration of this, radiation mutagenesis in cells hemizygous (one gene copy deleted) for autosomal APRT (adenine phosphoribosyltransferase) is constrained by the proximity of an essential sequence; induced mutation frequencies are relatively low, and only ~20% of induced mutations are of the deletion or rearrangement type (Miles and others 1990)—many deletions will have led to cell death. By contrast, radiation mutagenesis at the X-linked HPRT gene is much less constrained by neighboring sequence; induced mutation frequencies are substantially higher, and ~70% of induced mutations show HPRT deletion or rearrangement (Thacker 1986)—many more will have been tolerated (Bedford and Dewey 2002). Stated simply, gene loss mutations are characteristic of radiation, but their recovery in viable cells can be a major limiting factor. Also, gene amplification can result from the process of DSB repair (Difilippantonio and others 2002). As shown later, these features are important for consideration of carcinogenic mechanisms and are also discussed in respect of germline mutagenesis.

Deletion and rearrangement of APRT, HPRT, and other target genes do occur spontaneously but are generally less frequent than point mutation; in the case of most chemical mutagens, there is a strong bias toward the induction of point mutations (Thacker 1986; Miles and others 1990; Sankaranarayanan 1991).

Studies of the effect of radiation quality on the induction of gene mutations show a relationship similar between relative biological effectiveness (RBE) and LET to that noted for chromosome aberration induction. Mutagenic effectiveness peaks at a LET of 100–200 keV μm^{-1}, with maximum RBE values usually in the range of 7–10 based largely on initial slopes of the dose-response (Cox and Masson 1979; Thacker and others 1979; Thacker 1992). Molecular analyses broadly suggest that a DNA deletion mechanism predominates for all radiation qualities (Thacker 1986; Gibbs and others 1987; Aghamohammadi and others 1992; Jostes and others 1994), but there are some conflicting data on this issue.

DNA sequence data for radiation-induced intragenic deletions in APRT and larger deletions encompassing HPRT indicate the frequent involvement of short direct or inverted DNA repeats at deletion breakpoints (Miles and others 1990; Morris and Thacker 1993). The presence of these short repeats is highly suggestive of an important role for illegitimate recombination processes in mutagenesis and, as for chromosome aberration induction, the involvement of DNA DSBs and error-prone NHEJ repair. Evidence for a close relationship between gene mutations and chromosome aberrations is that several induced gene mutations are associated with macroscopic region-specific chromosomal deletions or rearrangements (Cox and Masson 1978; Thacker and Cox 1983; Morris and Thacker 1993).

If, as molecular data suggest, error-prone NHEJ repair of DNA DSBs is the principal source of radiation-induced gene mutations, then a linear dose-response would be anticipated at low doses. For technical reasons, dose-response relationships for gene mutations are far less precise than those for chromosome aberrations. In general, however, a linear or linear-quadratic relationship provides a satisfactory description of the dose-response down to ~200 mGy (Thacker 1992) and, from limited data, at lower doses. The exceptions to this are the data from a particularly sensitive in vivo system that scores reversion mutations (as hair color changes) at the pink-eyed unstable (Bonassi and others 1995) locus in the mouse. Using this system, a linear nonthreshold low-LET dose response has been obtained at doses down to 10 mGy (Schiestl and others 1994), but as discussed later in this chapter, that system is probably reflecting a mutagenic component from the induction of genomic instability.

Studies of radiation-induced gene mutation in radiosensitive mutant cell lines indicate that increased mutability can be associated not only with defective repair of DNA DSB but also with processes that affect the regulation of DNA repair (Thacker and others 1994). Finally, in studies on the effects of low-dose-rate, low-LET radiation and other cellular repair-related factors (Thacker 1992), there is consistent evidence for potentially increased efficiency of repair of premutagenic lesions at low dose rates, but none of these studies specifically suggest the presence of a low-dose threshold. The following sections consider specific aspects of cellular response relating to cell cycle effects, adaptive responses to radiation, the transfer of damage signals between cells (bystander effects), induced and persistent genomic instability, low-dose hyper-radiation sensitivity, and other aspects of dose-response.

RADIATION-INDUCED GENOMIC INSTABILITY

Radiation-induced genomic instability has been defined as the manifestation of genetic damage in a certain fraction of irradiated cells over many cell cycles after they were irradiated (Little 2003). This persistent instability is expressed as chromosomal rearrangements, chromosomal bridge formation, chromatid breaks and gaps, and micronuclei (Grosovsky and others 1996; Murnane 1996; Poupon and others 1996; Limoli and others 1997a; Suzuki and others 1998) in the progeny of cells that survive irradiation. Reduction in cell cloning efficiency several generations after irradiation is called delayed lethality; it is supposedly a manifestation of genomic instability associated with an increase in lethal mutations (Seymour and Mothersill 1997). Also, gene mutations, such as HPRT mutations, that arise de novo several generations after irradiation are thought to be another manifestation of genomic instability. The spectrum of these de novo mutations resembles that of spontaneous mutations (i.e., primarily point mutations instead of deletions that are induced directly by irradiation; Little and others 1997). There is controversy, however, as to whether all of these different end points represent the same fundamental chromosomal alterations that result in genomic instability (Chang and Little 1992; Morgan and others 1996; Limoli and others 1997a; Little 1998; Mothersill and others 2000a). However, the similarity in the frequencies of genomic instability induced in X-irradiated cells, (3 to 19) × 10^{-5} per cell/mGy, and the frequencies of chromosomal aberrations induced directly by irradiation may suggest that the induction of chromosomal aberrations is a primary event that plays a major role in radiation-induced genomic instability (data presented in section "Observed Dose-Response Relationships at Low Doses").

There is controversy concerning the fundamental radiation target and lesions that result in genomic instability. Evidence that the nucleus is the target (Limoli and others 1997b; Kaplan and Morgan 1998) is that [125]IdU (iododeoxyuridine) disintegrations in the DNA resulted in chromosomal instability, whereas ^{125}I disintegrations in the cytoplasm and cellular membrane did not. Furthermore, incorporation of BrdU (bromodeoxyuridine) into DNA increased the amount of radiation-induced chromosomal instability (Limoli and others 1999), which argues for DNA as the target. However, since restriction enzymes that produced DSBs in DNA (Kinashi and others 1995), mutations (Phillips and Morgan 1994), and chromosomal aberrations (Bryant 1984) did not induce chromosomal instability (Limoli and others 1997b), the hypothesis was presented that DSBs themselves are insufficient and that complex clustered damage in the DNA, such as that from ^{125}I disintegrations, is required. There is also some evidence that genomic instability results from complex chromosomal abnormalities created *de novo* by rearrangements that generate unstable combinations of DNA sequences (Murnane 1990), such as inverted repeats or associations of euchromatin with heterochromatin (Grosovsky and others 1996). Nevertheless, since the amount of instability induced by ^{125}I disintegrations in the DNA was relatively low (maximum of 4–9% unstable clones; Kaplan and Morgan 1998; Griffin and others 2000), the possibility was suggested that targets in addition to DNA might be involved (Limoli and others 2001). At the least, damage and/or error-prone repair in DNA is probably involved in radiation-induced genomic instability because mutant cells deficient in the repair enzymes needed for NHEJ are most sensitive to the induction of radiation-induced instability (Little 2003) and especially genomic instability induced by DNA DSBs (Difilippantonio and others 2002).

There are also data indicating that reactive oxygen species (Limoli and others 2001; Little 2003), potentially persistent over several generations, may play an important role in ongoing genomic instability. In addition, alterations in signal transduction pathways may be involved (Morgan and others 1996), and alterations in nucleotide pools have been shown to lead to genomic instability (Poupon and others 1996). Another possibility is that damage to centrosomes might be an important target because centrosome defects are thought to result in genomic instability through missegregation of chromosomes (Pihan and others 1998; Duensing and others 2001) that would result in aneuploidy (Duensing and Munger 2001). However, as reported recently (Hut and others 2003), centrosomal damage can result from incompletely replicated or damaged DNA.

Because chromosomal instability has been associated with breakage-fusion-bridge (B/F/B) cycles (Fouladi and others, 2000; Gisselsson and others 2000; Lo and others, 2002a, 2002b; Little 2003), the roles of telomeres may be particularly relevant. See Mathieu and colleagues (2004) and Murnane and Sabatier (2004) for reviews. Chromosome instability can also be initiated by DSBs that result in the loss of a telomere that protects the chromosome end and prevents chromosome fusion. A single DSB introduced at a telomere with the I-SceI endonuclease in mouse embryonic stem (ES) cells (Lo and others, 2002a) and spontaneous telomere loss in a human tumor cell line (Fouladi and others, 2000; Lo and others, 2002b) were found to result in sister-chromatid fusion and chromosome instability. Chromosome instability can be associated with prolonged B/F/B cycles; these cycles arise as a consequence of breakage of fused sister chromatids when their centromeres are pulled in opposite directions during anaphase, with subsequent re-fusion in the next cell cycle. These B/F/B cycles result in extensive DNA amplification and cease only when the chromosome acquires a new telomere, often by nonreciprocal translocations from another chromosome. However, because the nonreciprocal translocations provide telomeres that stabilize the marker chromosome, those chromosomes that donate the nonreciprocal translocations can become unstable due to the loss of their telomeres. Then, a subsequent nonreciprocal translocation can serve to transfer instability to another chromosome (Murnane and Sabatier 2004; Sabatier and others 2005). Thus, the loss of a single telomere can result in transfer of instability from one chromosome to another, leading to extensive genomic instability.

The importance of telomere loss as a mechanism for chromosome instability through B/F/B cycles in cancer has been emphasized by the demonstration that telomerase-deficient mice that are also deficient in p53 have a high cancer incidence (Artandi and others 2000; Chang and others 2001; Rudolph and others 2001). The analysis of the tumor cells from these mice demonstrated the presence of chromosome rearrangements typical of B/F/B cycles, including gene amplification and nonreciprocal translocations commonly seen in human cancer. It is possible that the genomic instability observed for chromosomal aberrations, HPRT mutations, and longer telomere terminal restriction fragment lengths in X-irradiated CHO cells (Romney and others 2001) is also a manifestation of nonreciprocal translocations that lead to telomere loss.

A question that has to be addressed is the relevance of radiation-induced genomic instability for radiation-induced cancer, and a corollary of this question is the relationship among expression of p53, radiation-induced apoptosis, and radiation-induced genomic instability. The "guardian-of-the-genome" hypothesis postulates that either cell cycle arrest allows additional time for repair of DNA damage or, alternatively, apoptosis eliminates damaged cells, thereby preventing progeny from manifesting genomic instability and ultimately carcinogenesis (Lane 1992; Kemp and others 1994; White and others 1994; Levine 1997; Lengauer and others 1998). Evidence has been presented that radiation-induced apoptosis can occur via p53-dependent and p53-independent mechanisms (Strasser and others 1994) initiated by damage in the nucleus (Guo and others 1997) or cytoplasm-mem-brane (Haimovitz-Friedman 1998). This damage results in cells undergoing apoptosis either during interphase without attempting division (Endlich and others 2000), several hours after they have divided a few times (Forrester and others 1999), or during an aberrant mitosis (Endlich and others 2000). The signal transduction pathways (White and Prives 1999) resulting in radiation-induced apoptosis involve the nucleus and cytoplasm with alterations in mitochondrial electron transport (Voehringer and others 2000) and release of cytochrome c from the mitochondria, which initiates caspase cleavage (Finucane and others 1999) and terminates in activation of a nuclease responsible for internucleosomal digestion of DNA (Wyllie 1998).

In accord with the guardian-of-the-genome hypothesis, mouse tumors undergoing apoptosis in a p53-independent manner contained abnormally amplified centrosomes, aneu-ploidy, and gene amplification (Fukasawa and others 1997). Also, a decrease in radiation-induced apoptosis associated with nonfunctional p53 or expression of Bcl2 correlated with an increase in mutagenesis (Xia and others 1995; Cherbonnel-Lasserre and others 1996; Yu and others 1997). However, the latter correlation might be due not to p53-mediated's enhancement of radiation-induced apoptosis (Xia and others 1995) but instead to p53-mediated's suppression of homologous recombination (Sturzbecher and others 1996), which in turn might suppress genomic instability and a hypermutable phenotype. However, there is evidence that radiation-induced genomic instability is independent of p53 expression (Kadhim and others 1996). Furthermore, when the guardian-of-the-genome hypothesis was tested in lymphocyte cultures that were irradiated under different dose-rate and mitogen-treatment conditions, postradiation incubation allowing apoptotic processes to remove damaged cells did not prevent the development of chromosomal instability during long-term cell proliferation over 51–57 days (Holmberg and others 1998). Thus, the relationship between radiation-induced genomic instability, radiation-induced apoptosis, and radiation-induced cancer is uncertain (discussed at length in Chapter 3). Furthermore, radiation-induced genomic instability could not be induced in normal diploid human fibroblasts (Dugan and Bedford 2003) and may be related to confounding *in vitro* stress factors (Bouffler and others 2001) or to the cells being partially transformed. Finally, as discussed in Chapter 3, it may be that genomic instability plays a more important role in tumor progression than in tumor initiation.

Data are critically needed for the definition of molecular targets and processes responsible for genomic instability in order to define and understand the dose-response relation-ship for genomic instability and especially why, in some cellular systems, the induction frequency saturates with only about 10–30% of the surviving cells manifesting genomic instability (Little 1998; Limoli and others 1999) (data presented in Table 2-1). It may be that only a certain fraction of the cells, or those in a certain part of the cell cycle, are susceptible to radiation-induced genomic instability. Until the molecular mechanisms responsible for genomic instability and its relationship to carcinogenesis are understood, the extrapolation of dose-response data for genomic instability to radiation-induced cancers in the low-dose range <100 mGy is not warranted.

CELL CYCLE EFFECTS

In a number of mammalian cell lines, cells irradiated in mitosis or late G_2 are most susceptible, cells in G_1 are intermediate in susceptibility, and cells in middle to late S phase and early G_2 are most resistant to the induction of cell lethality, chromosomal aberrations, and mutations (Sinclair and Morton 1963; Terasima and Tolmach 1963; Dewey and others 1970; Burki 1980; Jostes and others 1980; Watanabe and Horikawa 1980; Chuang and Liber 1996; Leonhardt and others 1997). Also, cells irradiated at the G_1/S transition are often observed to be more radiosensitive than cells in G_1 or S. However, exceptions have been observed, such as little variation in radiosensitivity during the cell cycle (Henderson and others 1982) and greater sensitivity of cells in late S than of cells in G_1 (Thompson and Humphrey 1968; Guo and others 1997; Furre and others 1999). Since radioresistance during late S phase has been attributed to error-free repair of DNA DSBs by homologous recombination when sister chromatids have been replicated (Rothkamm and Lobrich 2003; Rothkamm and others 2003), the lack of radioresistance during late S phase in some cell lines may be attributed to their inability to carry out repair by homologous recombination. Those effects have been observed in connection with relatively high acute doses of 1.5–10 Gy (1500-10,000 mGy), but how such variations in radiosensitivity during the cell cycle may affect responses to low doses up to 100 mGy is not known. Also, there are no reports of studies to determine whether there may be variations in radiosensitivity during the cell cycle for induction of genomic instability. However, studies with cell lines have indicated that cells are most susceptible to malignant transformation *in vitro* when they are irradiated with high-LET radiation or low-energy X-rays in late G_2/M (Cao and others 1992, 1993; Miller and others 1992).

The inverse dose-rate effect (Crompton and others 1990; Amundson and Chen 1996), in which cells at first become more radioresistant and then more radiosensitive again as the dose rate of low-LET radiation is decreased below about 1–10 mGy/min, has been attributed to the arrest of cells in a radiosensitive G_2 phase of the cycle (Mitchell and others 1979; Furre and others 1999). However, evidence has been presented that the inverse dose-rate effect can be observed when cells do not arrest in G_2 and, instead, correlates with low-dose hyper-radiation sensitivity (HRS; Mitchell and others 2002). This conclusion may be consistent with recent results from the same research group (Marples and others 2003), which reported that HRS for acute radiation doses was attributed to cells in radiosensitive G_2 failing to arrest before mitosis. For high-LET radiation, the inverse dose-rate effect has been attributed to the traversal of cells through a radiosensitive G_2 phase (Brenner and others 1996; Elkind 1996; Tauchi and others 1999). Such an inverse dose-rate effect has been reported for cell lethality and mutations induced by low-LET radiation and for transformation induced by high-LET radiation.

Vilenchik and Knudson (2000) hypothesized that the increase in mutability observed below a dose rate of 1 mGy/min for mouse spermatogonia and 10 mGy/min for cells *in vitro* is not caused by variations in radiosensitivity during the cell cycle but rather by a diminished activation of error-free repair at very low dose rates inasmuch as the rate of induced DNA damage (signal) is lower than the background rate of spontaneous DNA damage (noise). This interpretation of the data remains controversial, particularly since there is evidence that argues against the inducibility of DNA repair genes. However, Collis and colleagues (2004) reported recently that DNA damage introduced at a very low dose rate of 0.33 or 1.5 mGy/min produced less activation of the radiation damage sensor ATM (ataxia-telangiectasia-mutated), as detected by H2AX foci, than activation at a high dose rate of 750 mGy/min. Furthermore, this reduction of ATM activation was observed after irradiation in G_0/G_1, S, and G_2/M, and correlated with enhanced cell killing. For a discussion of the expression of particular genes involved in DNA repair and controlling checkpoints in the cell cycle, see "DSB Signal Transduction and Inducible Repair" in Chapter 1, along with Figure 1-10.

Although some small transient effects on cell cycle progression have been reported for doses of 20–100 mGy (Puck and others 1997; Amundson and others 1999b), no inverse dose-rate effect would be expected at these dose levels (Brenner and others 1996), and if it did exist, it would be difficult to demonstrate. However, at approximately 100 mGy, an inverse dose-rate effect of fission-spectrum neutrons has been observed between 4 and 100 mGy/min for neoplastic transformation of C3H 10T1/2 cells (Hill and others 1982, 1984) and between 10 versus 250 mGy/min and 0.0083 versus 0.083 mGy/min for induction of lung adenocarcinomas and mammary adenocarcinomas in mice (Ullrich 1984). Apparently, these inverse dose-rate effects could not be explained by perturbations in the cell cycle, and for mammary tumors, the effect was associated with an increased probability of progression of carcinogen-altered cells rather than an increased number of initiated cells (Ullrich 1986). Furthermore, an inverse dose-rate effect was not observed for the induction of ovarian tumors, for which the response

to dose at low dose rates was much lower than that at high dose rates (Ullrich 1984). How these data on high-LET fission neutrons can be extrapolated to low-LET radiation is unknown, especially because the RBE for these carcinogenic effects has been estimated to be as high as 10 or more. This means that the equivalent doses and equivalent dose rates mentioned above, when expressed in millisieverts, would be at least 10 times greater than the values expressed in milligrays.

Furthermore, when the same tumors were induced in mice by low-LET radiation at doses of 0.1–6.0 Gy, no inverse dose-rate effect was observed between 0.04 and 0.6 mGy/min; these low dose rates always had a dose-response relationship significantly below that observed for acute high-dose-rate irradiation (Ullrich and others 1976, 1987; Ullrich and Storer 1979a, 1979b, 1979c; Ullrich 1983). Similar observations were reported for neoplastic transformation of C3H 10T1/2 cells by low-LET radiation, for which the dose-response relationship for a low dose rate of 1 mGy/min was much below that observed for an acute high dose rate of 1.0 Gy/min (Han and others 1980). The lack of a low-LET inverse dose-rate effect for tumor induction and neoplastic transformation *in vitro* contrasts with the inverse dose-rate effect seen for cell killing and induction of mutations that is sometimes attributed to perturbations in cell cycle progression. However, results obtained with mammalian cell lines, in particular those for neoplastic transformation, should be interpreted with great caution if they are to be used in estimating radiation risk to humans.

ADAPTIVE RESPONSE

Organisms, such as bacteria, that live in a highly changeable environment have multiple mechanisms for adapting to environmental stress. The bacterium *Escherichia coli* has two distinct, inducible, redox-regulated transcriptional switches involving the soxRS and oxyR transcription factors, which respond to exposure to superoxide and hydrogen peroxide, respectively (Demple 1991; Choi and others 2001). After exposure to ionizing radiation, these factors reprogram the cellular transcription pattern with increased expression of proteins that inactivate reactive oxygen species and some DNA repair enzymes that process oxidative DNA damage. As a consequence, *E. coli* cells exhibit a distinct adaptive response to oxidative stress: exposure to a low dose of active oxygen makes the cells more resistant to later exposures for some finite period. In that situation, there is a clear threshold value for deleterious effects of ionizing radiation. However, the soxRS and oxyR gene regulons have not been conserved during evolution, and human cells, which exist in a much more stable cellular environment than bacteria, do not appear to have counterparts. Thus, humans do not have an adaptive response to oxidative damage similar to the well-characterized systems in bacteria.

A broad perturbation of DNA transcription is observed in human cells after exposure to ionizing radiation; it involves the activation of transcription factors, such as NF-kappaB and c-jun/c-fos. After exposure of human lymphoblastoid cells to 5 Gy of radiation, 2–3% of the genes exhibit more than a 50% change in induction or repression (Tusher and others 2001). These genes include several involved in cell cycle control. No genes involved in repair of DNA DSBs generated by ionizing radiation were induced (Tusher and others 2001; Wood and others 2001). It should also be noted that the base-excision repair enzymes involved in the removal of oxidative damage are not induced by low doses of ionizing radiation in human cells (Inoue and others 2004). These studies have provided no support for a general adaptive repair response in human cells to counteract DNA DSB formation that can result in cell death or mutagenesis.

A different type of apparent adaptive response has been well documented for the induction of chromatid-type breaks and mutations in human lymphocytes stimulated to divide. In most studies, a priming or adaptive dose of about 10 mGy significantly reduces the frequency of chromosomal aberrations (Shadley and others 1987; Wolff 1992a, 1996) and mutations (Kelsey and others 1991) induced a few hours later by 1–3 Gy. However, when the priming dose was 10 mGy, the adaptive response for chromosomal aberrations was reduced significantly as the priming dose rate was reduced from 50 mGy/min to 6.4 mGy/min (Shadley and Wiencke 1989). Adaptive responses of this type were reviewed by UNSCEAR (1994).

Although alterations in cell cycle progression have been implicated in the mammalian cell adaptive phenomenon (Aghamohammadi and Savage 1991), carefully controlled studies indicate that the priming dose induces radioresistance for induction of chromosomal aberrations in human lymphocytes (Wolff 1996); priming doses less than 5 mGy, or greater than about 200 mGy, yield very little if any adaptation (Wolff 1992b). The induction and magnitude of the adaptive response in human lymphocytes are highly variable among people (Bose and Olivieri 1989; Sankaranarayanan and others 1989; Shadley and Wiencke 1989; Hain and others 1992; Vijayalaxmi and others 1995; Upton 2000), and the adaptive response could not be induced when lymphocytes were given the priming dose during G_0 (Shadley and others 1987). Although inhibitor and electrophoretic studies (Youngblom and others 1989; Wolff 1992b) suggest that alterations in transcribing messenger RNA and synthesis of proteins are involved in the adaptive response in lymphocytes, no specific signal transduction or repair pathways have been identified. Finally, humans exposed occupationally (Barquinero and others 1995) or to iodine-131 (^{131}I) for treatment of thyroid disease (Monsieurs and others 2000) or as children after Chernobyl (Tedeschi and others 1995) varied in their ability to demonstrate an apparent adaptive response for chromosomal aberrations (Padovani and others 1995; Tedeschi and others 1996). This variability may relate to the genetic variation reported for radiation-induced transcriptional changes (Correa and Cheung 2004).

Adaptive responses to radiation observed in other cellular systems for induction of cell lethality, chromosomal aberrations, mutations (Zhou and others 1993; Rigaud and others 1995), and defects in embryonic development provide little information that can be used to suggest that the dose-response curve in the dose range 0–100 mGy will be less steep than that described by the limiting value of α mentioned above. When mouse embryos were exposed to a priming dose of about 10 mGy and evaluated for chromosomal aberrations or defects in development induced by a challenge dose several hours later, the results were highly variable for the induction of an adaptive response (Muller and others 1992; Wojcik and others 1992; Wolff 1996; Wang and others 1998). Studies of radiation-induced mutagenesis also had variable results. Adaptation not only decreases the frequency of mutants induced by a challenge dose but also appears to alter the types of mutants. Adaptation of human lymphoblastoid cells to a challenge dose of 4 Gy 6 h after 20 mGy decreased the proportion of HPRT mutants of the deletion type relative to small point mutations (Rigaud and others 1995). In contrast, adaptation of human-hamster hybrid A_L cells to a challenge dose of 3 Gy after a priming dose of 40 mGy increased the proportion of complex unstable mutations (Ueno and others 1996). An extensive study (Sasaki 1995) of chromosomal aberrations, HPRT mutations, and cell killing demonstrated adaptation in quiescent cultured m5S mouse embryonic skin cells preexposed in G_1 to 10–50 mGy; cells exposed 4 h later to doses greater than 2 Gy were significantly more resistant than nonadapted cells for all three end points (see Figure 2-2 for cell-killing results). The adaptation phenomenon appeared to involve a protein kinase C signaling pathway. In addition, the lack of an adaptive response in a tumorigenic variant, clone 6110, and restoration of the adaptive response obtained by introducing human chromosome 11 (five other chromosomes had no effect) further suggested that interference of signaling pathways may alter adaptive responses in malignant cells. The observation (Broome and others 2002) that a priming dose as low as 1 mGy induced an adaptive response in a nontransformed human fibroblast cell line for micronuclei induced by a challenge dose of 2 Gy has to be confirmed for other systems and end points, such as mutation induction. Also, the large variation in adaptive response for radiation-induced micronuclei in human lymphoblastoid cell lines must be considered (Sorensen and others 2002). Most important, the adaptive response has to be demonstrated for both priming and challenging doses in the low-dose range <100 mGy, and an understanding of the molecular and cellular mechanisms of the adaptive response is essential if it is to have relevance for risk assessment.

Studies of adaptation for malignant transformation *in vitro* provide conflicting information and might not be relevant to malignant transformation *in vivo*. Although the

X-RAY DOSE (Gy)

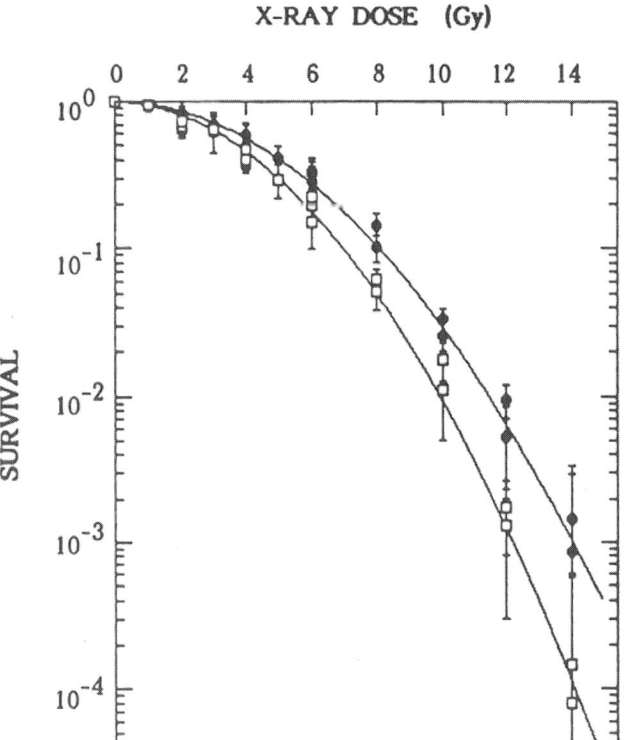

FIGURE 2-2 Effects of preirradiation on clonogenic survival of mouse m5S cells. Closed symbols represent results in cells in G_1 preirradiated with 20 mGy of X-rays 5 h before graded doses of acute radiation. Open symbols represent results in cells in G_1 given graded doses of acute radiation only. Statistical errors are standard errors of the mean based on variation in the number of recovered colonies in irradiated dishes (this does not include propagation of error in plating efficiency of nonirradiated controls). SOURCE: Sasaki (1995).

morphologic transformation frequency of m5S adapted mouse embryonic skin cells that had received 20 mGy was about half the spontaneous frequency of 3×10^{-5} observed in nonprimed cells, the adapted cells exposed 5 h later to a challenge dose of 1 Gy were more susceptible to morphologic transformation than the nonadapted cells (Sasaki 1995). These transformation results, however, contrast with results in mouse C3H 10T1/2 cells that were exposed in plateau phase to a challenge dose of 4 Gy 5 h after a priming dose of 100 or 670 mGy (*i.e.*, adapted cells were more resistant to malignant transformation than nonadapted cells; Azzam and others 1994). Furthermore, the priming dose of 100 or 670 mGy caused an increase by a factor of 2–5 in the transformation frequency relative to the frequency of about 3×10^{-4} observed for nonirradiated cells. When the same group of investigators exposed the same C3H 10T1/2 cells in plateau phase to priming doses of 1, 10, or 100 mGy, the neoplastic transformation frequency was lower by a factor of 3–4 than the spontaneous frequency (Azzam and others 1996).

The reduction was observed only when the cells were trypsinized and replated 24 h after irradiation for the transformation assay; trypsinization and replating immediately after irradiation did not alter the frequency. Similar results have been reported by Redpath and coworkers (Redpath and Antoniono 1998; Redpath and others 2001): the malignant transformation frequency was reduced by about half when human hybrid cells approaching confluence were trypsinized and replated 24 h after a priming dose of 10 mGy; again, no statistically significant reduction in transformation frequency was observed when the cells were trypsinized and replated immediately after irradiation.

The validity of extrapolating any of the results from *in vitro* neoplastic transformation systems to malignant transformation *in vivo* may be questioned for the following reasons. First, the effects associated with variations in time of trypsinization and replating after irradiation must be understood (Schollnberger and others 2002). Second, the measured neoplastic transformation frequency depends on both the density of viable cells plated (Bettega and others 1989) and the number of generations before the cells become confluent (Kennedy and others 1980). Third, when priming doses of 1–100 mGy resulted in a decrease in the neoplastic transformation frequency, the spontaneous transformation frequency was unusually high in one case (Azzam and others 1994), and a Hela X skin fibroblast human hybrid cell system was used in the other (Redpath and Antoniono 1998). Fourth, studies of malignant transformation in immortalized (already-transformed) cell lines may have little relevance to malignant transformation of normal nonimmortalized cells, especially *in vivo*, where complex interactive processes can occur (Harvey and Levine 1991; Kamijo and others 1997).

For several mammalian cell lines in culture, adaptive responses for cell lethality after doses of 200–600 mGy (Marples and Joiner 1995; Joiner and others 1996; Marples and Skov 1996; Wouters and others 1996; Skov 1999) and for enhanced removal of thymine glycols after a dose of 2 Gy (Le and others 1998) have been observed 4–6 h after a priming dose of 200 mGy. In Chinese hamster V79 cells, the rate of repair of DNA DSBs induced by 1.5 or 5.0 Gy was increased 4 h after a priming dose of 50 mGy (Ikushima and others 1996).

The adaptive responses of mammalian cells described above, at least for cell survival and repair of DNA strand breaks (Robson and others 2000), may be associated in part with the downregulation of a gene DIR1 90 min after doses of 50–1000 mGy. This gene codes for proteins (Robson and others 1997, 1999, 2000) similar to a family of heat shock-related proteins (HSPs) known as immunophilins with tetrapeptide repeats (TPRs). TPR-containing proteins, such as cell cyclin proteins cdc23, cdc27, and cdc16, have been reported to form complexes *in vivo*, and the TPR domain is thought to be involved in binding HSP90 and HSP70. Less binding of HSP70 and the induction of other members of the HSP70 family by low doses of radiation (Sadekova and others 1997) might result in adaptation through the same mechanisms.

The recent microarray expression studies (Yin and others 2003) that demonstrated downregulation of the large HSPs 30 min after irradiating the mouse brain with 100 mGy may support these conjectures. Also, the radiation-induced downregulation of CDC16, which belongs to the anaphase-promoting complex, was enhanced by an adaptive dose of 20 mGy (Zhou and Rigaud 2001). In fact, regulation of repair and cell cycle progression may be achieved by differential complex formation (Eckardt-Schupp and Klaus 1999). For instance, PCNA (proliferating cell nuclear antigen) expression, which is modulated by p53 in response to radiation, may play an important role in regulating and coordinating cell cycle progression, DNA replication, translesion synthesis, and DNA excision repair, depending on its partner proteins. Within minutes after ionizing radiation, the immediate-response genes transcription factors such as c-jun, c-fos, and NF-kB are turned on, possibly thwarting the general downregulation of transcription after irradiation and allowing privileged transcription of special genes. The sensors for these fast responses are in membranes, and they initiate signal transduction by several cascades of protein kinases (Eckardt-Schupp and Klaus 1999) that may involve reactive oxygen intermediates (Mohan and Meltz 1994; Hoshi and others 1997). Therefore, adaptation in mammalian cells probably involves induction of signal transduction pathways (Stecca and Gerber 1998) rather than induction of DNA repair enzymes.

There is much variability and heterogeneity in the ability to induce adaptive responses that usually require a priming dose of 10–200 mGy and a large challenge dose of 1–2 Gy. Challenge doses of this magnitude probably have little relevance to risk assessment for low radiation doses of 1–100 mGy. Furthermore, the molecular pathways associated with the phenomenon have not been delineated. Available data indicate that the adaptive response results from DNA damage that can be induced by ^{3}HTdR (triliated thymidine) incorporated into DNA, by H_2O_2, and by restriction enzymes (Wolff 1992b; Sasaki 1995; Belyaev and Harms-Ringdahl 1996). The ability to induce an adaptive response appears to depend on the genotype (Wojcik and others 1992), which may relate to genetic variation reported for radiation-induced transcriptional changes (Correa and Cheung 2004). In fact, the effect of the genotype on the adaptive response has been demonstrated most conclusively in *Drosophila melanogaster* (Schappi-Bushi 1994).

A priming dose has been reported to reduce chromosomal damage in some chromosomes and increase it in others (Broome and others 1999). Data are needed, particularly at the molecular level, on adaptation induced when both priming and challenging doses are in the low-dose range <100 mGy; relevant end points should include not only chromosomal aberrations and mutations but also genomic instability and, if possible, tumor induction. *In vitro* and *in vivo*

data are needed on delivery of the priming and challenge doses over several weeks or months at very low dose rates or with fractionated exposures.

Finally, we should be concerned about the cumulative effect of multiple low doses of less than 10 mGy. Such data have not yet been obtained, in particular those explaining the molecular and cellular mechanisms of the adaptive response. Therefore, it is concluded that any useful extrapolations for dose-response relationships in humans cannot be made from the adaptive responses observed in human lymphocytes or the other cellular systems mentioned above. In fact, a study (Barquinero and others 1995) reporting that an average occupational exposure of about 2.5 mGy per year over 7–21 years resulted in a variable adaptive response for chromosomal aberrations induced in human lymphocytes by a large challenge dose of 2 Gy also reported that the incidence of spontaneous aberrations was increased significantly by the occupational exposure. Barquinero and colleagues (1995) also cite six reports indicating that basal rates of chromosomal abnormalities are in general higher in exposed human populations; recent papers (Tanaka and others 2000; Tawn and others 2000a, 2004; Burak and others 2001; Liu and others 2002; Maffei and others 2004) present similar information. Therefore, based on current information, the assumption is unwarranted that any stimulatory effects of low doses of ionizing radiation substantially reduce long-term deleterious radiation effects in humans.

BYSTANDER EFFECTS

A factor that could have a significant effect on the dose-response relationship is the bystander effect that irradiated cells have on nonirradiated cells. Recent comprehensive reviews of bystander effects observed *in vitro* (Morgan 2003a) and *in vivo* (Morgan 2003b) emphasized their possible mechanisms, implications, and variability. In addition, reviews have been published recently on the relationship between the bystander effect, genomic instability, and carcinogenesis (Little 2003; Lorimore and others 2003). Observations that irradiated cells or tissues could have deleterious effects on nonirradiated cells or tissues were reported many years ago (Bacq and Alexander 1961) and were termed abscopal effects. As an example of such an effect, plasma from patients who underwent localized radiation therapy induced chromosomal aberrations in lymphocytes from nonirradiated patients (Hollowell and Littlefield 1968; Littlefield and others 1969). A bystander effect has been demonstrated conclusively for cells in culture exposed to high-LET radiation, usually α-particles. Little and colleagues estimated that a single α-particle traversing a cell can induce HPRT mutations (Nagasawa and Little 1999), sister-chromatid exchanges (Nagasawa and Little 1992), upregulation of p21 and p53, and downregulation of cyclin B1, cdc2, and rad51 (Azzam and others 1998) in unirradiated cells. At least for the bystander effect on signal transduction pathways and induction of mutations, the irradiated and nonirradiated cells had to be in contact with each other through gap junctions. Hall and colleagues demonstrated the same bystander phenomenon for cell killing, induction of mutations (Zhou and others 2000), micronuclei formation (Hall 2000), and malignant transformation (Sawant and others 2001a); the magnitude of the bystander effect increased with the number of α-particles traversing the nuclei (Sawant and others 2001a).

For malignant transformation, the frequency when only 10% of the cells were traversed by an α-particle was as great as when every cell was exposed to an α-particle; furthermore, nonirradiated cells did not have to be in contact with irradiated cells. However, the same group subsequently reported that gap junctions appeared to be required for another bystander effect resulting in cell lethality in nonhit cells (Sawant and others 2002). The group also showed that irradiating the cytoplasm with α-particles (Wu and others 1999) induced mutations (small deletion and base-pair alteration mutations) that resembled those occurring spontaneously, not the larger deletions observed when the nucleus was irradiated to induce mutations in both irradiated and non-irradiated cells (Zhou and others 2000). Lorimore and colleagues (1998) have observed a similar bystander effect: delayed chromosomal aberrations associated with genomic instability when cells were exposed to α-particles. Prise and colleagues (1998) have observed a bystander effect for genomic instability associated with the formation of micronuclei 20–30 generations after individual cells were irradiated with a charged-particle microbeam. Their subsequent studies with primary human fibroblasts (Belyakov and others 2001) showed that even though a single cell had been targeted, an additional 80–110 cells had micronuclei; the yield of cells that had excess micronuclei was independent of the number of charged particles delivered to the targeted cell.

The molecular mechanisms proposed for the bystander effects described above are speculative (see Chapter 1 "DSB Signal Transduction and Inducible Repair" for a discussion of possible repair and signal transduction pathways that may be involved). Activation of the p53-mediated DNA damage response pathway in bystander cells has led to speculation (Grosovsky 1999) that reduced replication fidelity or increased recombinational activity might lead to the genetic effects that occur in these cells. However, α-particle-induced chromosomal instability was reported to be independent of the p53 status of the cells (Kadhim and others 1996). The bystander phenomenon may involve the diffusion of cytokines or long-lived reactive oxygen species (ROS; Narayanan and others 1997, 1999; Lorimore and others 1998; Wu and others 1999; Azzam and others 2002; Morgan 2003a, 2003b) including any products formed by reaction with hydrated electrons or OH• radicals (Ward 2002). Also, the diffusion of paracrine proapoptotic or antiapoptotic factors induced by upregulation of p21 (Chang and others 2000) may be involved. Because CDC2 is downregulated by α-

particles, there may be reduced phosphorylation of connexin 43 by CDC2 and thus increased membrane permeability (Azzam and others 1998). This hypothesis is supported by the observation that membrane signaling is involved in the bystander effect for sister-chromatid exchanges and HPRT mutations induced indirectly by α-particles (Nagasawa and others 2002).

Regardless of the molecular mechanisms involved, the bystander effects observed with high-LET particles may have important implications for low doses of high-LET radiation. According to Sawant and others (2001a), "These results, if applicable *in vivo*, would have significant consequences in terms of radiation risk extrapolation to low doses, implying that the relevant target for radiation oncogenesis is larger than an individual cell, and that the risk of carcinogenesis would increase more slowly, if at all, at higher doses— an effect seen *in vivo*, as well as epidemiologically. Thus, a simple linear extrapolation of radiation risk from high doses (where they can be measured) to lower doses (where they must be inferred) would be of questionable validity." In other words, it is speculated that there could be a convex, downward-curving dose-response relationship at low doses, and that extrapolation of data from high doses could lead to an underestimate of the effect at low doses of high-LET radiation.

A most critical question, however, is whether these types of bystander effects exist for low-LET radiation doses <100 mGy, which are the focus of this report. For α-particles and other high-LET radiation used in bystander studies, the dose to the nucleus was calculated to be 130– 500 mGy per α-particle traversal, depending on the size and shape of the cell and its nucleus (Azzam and others 1998); that is, a flattened cell nucleus would have a much lower dose from high-LET radiation than a spherical rounded cell nucleus because of the geometry of the nucleus in relation to the radiation source (Clutton and others 1996a, 1996b). For low-LET radiation (assuming an RBE of 3), the dose corresponding to that from the high-LET radiation would be 0.39– 1.5 Gy. Because the bystander effect resulting from an α-particle traversal through an irradiated cell was lower by a factor of 3–5 than the direct effect on the irradiated cell and because the magnitude of the bystander effect appeared to increase as the number of traversals through the cell increased (Sawant and others 2001a), one might expect that the same type of bystander effect would not be observed in the low-dose range <100 mGy for low-LET radiation. In fact, data indicate that the bystander effect for induced expression of p53 was much greater and persisted much longer after α-irradiation than after X-irradiation (Hickman and others 1994).

In human keratinocytes, a bystander effect for cell lethality that required cell-to-cell contact with gap junctions has been reported for γ-ray doses of 500 mGy and above (Mothersill and Seymour 1997). In the same dose range, a bystander effect that did not require cell-to-cell contact was observed when cell culture medium from irradiated cells was added to nonirradiated cells (Mothersill and Seymour 1998a). The observed bystander effect is specific for keratinocytes because it was not observed for fibroblasts. The effect is eliminated by heating the medium at 70°C for 30 min, and there is some evidence that an alteration in energy metabolism and induction of apoptosis are involved (Mothersill and others 2000b). Furthermore, the bystander effect from transfer of medium varies among cell lines (Mothersill and others 2000b; Seymour and Mothersill 2000), and its contribution to cell lethality has been reported either to plateau with about 40% of human keratinocytes killed at 30–60 mGy (Seymour and Mothersill 2000) or to increase at doses over 1 Gy delivered to CHO (Chinese hamster ovary) cells (Mothersill and others 2000b). Finally, bystander cell killing reported for a dose as low as 10 mGy appears to be greater for delayed cell lethality quantified by cloning efficiency at about 14 d after irradiation than for initial cell lethality quantified by cloning efficiency determined immediately after irradiation (Seymour and Mothersill 2000). Delayed lethality is supposedly a manifestation of genomic instability associated with an increase in lethal mutations in cells that survive irradiation (Seymour and Mothersill 1997).

In another study, a low-LET radiation bystander effect that required gap junctions was observed in a three-dimensional Chinese hamster culture model (Bishayee and others 1999). The bystander effect that caused cell lethality in the nonirradiated cells became apparent only after the irradiated cells had undergone 1000–2000 disintegrations of ^{3}HTdR in the DNA, that is, at a very high dose of about 2.5–5.0 Gy (Dewey and others 1965).

Several issues should be considered in relation to the bystander effect. First, in contrast with the results summarized above that involved enhancement of damage, a bystander effect was reported to increase survival (Dent and others 1999) when medium from γ-irradiated mammary carcinoma cells was transferred to nonirradiated cells 120 min after a dose of 2 Gy. Apparently, the soluble TGF-α (transforming growth factor-α) that was released induced secondary activation of EGFR (epidermal growth factor receptor), MAPK (mitogen-activated protein kinase), and JNK (c-jun N-terminal kinase), which resulted in an increase in survival. Thus, as reviewed by Waldren (2004) both beneficial and detrimental effects may result from the bystander effect. A similar observation was reported for normal human diploid lung fibroblasts exposed to low doses of α-particles; the observed enhancement of cell growth was hypothesized to result from an ROS-caused increase in TGF-β (Iyer and Lehnert 2000). Second, there is a suggestion that an adaptive response induced by a priming dose of 1 mGy for reducing radiation-induced micronuclei was due in part to a bystander effect (Broome and others 2002). However, the bystander effect of a priming dose has not been found to induce a radioprotective or adaptive response for chromosomal aberra-

tions or cell killing (Wolff 1992b; Mothersill and Seymour 1998a). Third, an adaptive response induced by irradiating a cell directly may cancel out at least part of the bystander effect; this was observed for cell lethality when mouse C3H 10T1/2 cells were irradiated with 20 mGy of X-rays 6 h before α-particle irradiation (Sawant and others 2001b). Fourth, molecular mechanisms responsible for the bystander effect of low-LET radiation, as well as high-LET radiation, that may include genetic variation in transcriptional response to radiation exposure (Correa and Cheung 2004), have not been delineated. Fifth, recent results (Prise and others 2003) suggest that a bystander effect for cell lethality from soft X-ray irradiation (LET of 25–30 keV/μ) might be observed down to 50 mGy but not below. Sixth, until molecular mechanisms of the bystander effect are elucidated, especially as related to an intact organism, and until reproducible bystander effects are observed for low-LET radiation in the dose range of 1–5 mGy, where an average of about one electron track traverses the nucleus, a bystander effect of low-dose, low-LET radiation that might result in a dose-response curving either upwards or downwards should not be assumed.

HYPER-RADIATION SENSITIVITY AT LOW DOSES

Another factor that can cause the dose-response to deviate from the alpha-beta model is HRS that has been reported for cell lethality induced by low-LET radiation at doses up to 200 mGy (Joiner and others 1996; Skov 1999; Figure 2-3). In this dose range, survival can decrease to 85–90%, depending on the cell line, which is significantly lower than survival predicted by the value of α determined from survival values above 1–2 Gy. HRS might be associated with a bystander effect, but a recent study (Mothersill and others 2002) suggests that it is not. Although the magnitude of HRS varies, there is some evidence that it also occurs for fractionated doses of about 400 or 500 mGy both *in vitro* (Smith and others 1999; Short and others 2001) and *in vivo* for kidney and skin (Joiner and others 1996) and for glioma cell lines irradiated with multiple fractions of 700–800 mGy (Beauchesne and others 2003). Furthermore, an observed inverse dose-rate effect was attributed to HRS seen for low acute doses (Mitchell and others 2002), and recent cell cycle studies (Mitchell and others 2002; Marples and others 2003; Short and others 2003) suggest that HRS may be related to cells not arresting in radiosensitive G_2. Since a high proportion of the target stem-like cells in humans would be noncycling G_0 cells (see Chapter 3, "General Aspects of Dose-Response"), the last two observations, if generally true, would suggest that neither HRS nor the inverse dose-rate phenomenon should have any significant effect on the dose-response for cancer induction in humans.

Molecular mechanisms involved in HRS have been described in only a preliminary way. However, HRS for cell lethality up to 200 mGy was not observed in radiosensitive

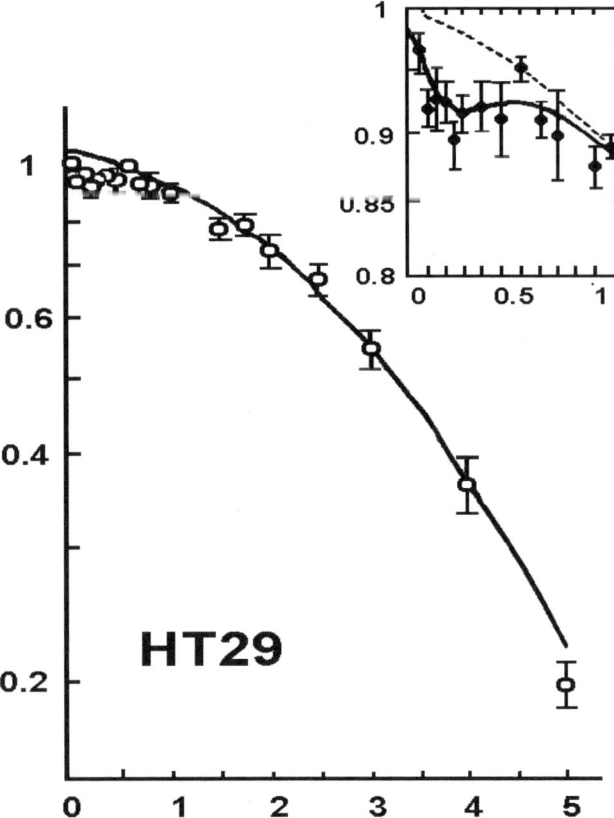

FIGURE 2-3 Illustrative example of hyper-radiation sensitivity for low doses. Example is from HT29 cells given graded doses of X-rays. SOURCE: From Joiner and colleagues (1996).

AT and XR-V15B cell lines (Skov 1999) or with high-LET radiation (Lambin and others 1993). For doses above 50–200 mGy, for which HRS is no longer observed, the flattening of the survival curve between 500 and 1000 mGy may be related to DNA PKcs activity (Marples and others 2002) or to the downregulation of the DIR1 gene (Robson and others 1997, 1999); this downregulation has been shown to correlate with an increase in rate of repair of DNA single-strand breaks (Robson and others 2000; Collis and others 2004; Marples and others 2004).

DNA damage introduced at a very low rate may not activate the radiation damage sensor ATM (Collis and others 2004). Consequently, exposure to low levels of chronic radiation may cause more cell damage than estimated from extrapolation of higher doses. This hypersensitivity to killing could serve to eliminate cells that have received DNA damage and potentially carcinogenic changes to their genome. Alternatively, it cannot presently be excluded that some of these cells may survive and proliferate as clones of mutated cells. It is important to note that the effect of cellular hypersensitivity to killing by very low chronic doses of ionizing radiation is a modest effect that has been detected only in some, but not all, human cell lines investigated.

Studies of other end points have provided some additional evidence of HRS. In a signal transduction study that used γ-ray doses of 20, 50, 100, 250, and 500 mGy, there was a suggestion of HRS up to 200 mGy for radiation-induced transcription of MDM2, ATF3, and BAX in a human myeloid tumor line (Amundson and others 1999b; Figure 2-4).

Similar observations over the same dose range were reported (Yang and others 2000) for X-ray induction of protein-8 (XIP8) in human MCF-7:W58 breast cancer cells; this protein as it complexes with Ku70/Ku80 appears to be an important cell-death signal. HRS was also observed in mice as gene deletions that reverted unstable mutations in melanocytes exposed to 10 mGy of X-radiation (Schiestl and others 1994); that is, there was a threefold effect at 10 mGy and a twelvefold effect at 1 Gy. The frequency of gene deletions was about 100 times higher than the frequency of other

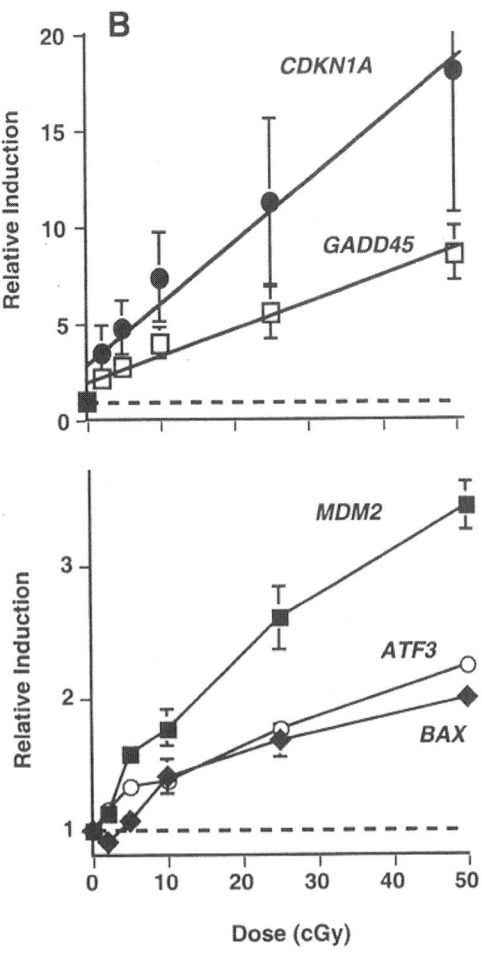

FIGURE 2-4 Maximal induction of CDKN1A (●), GADD45 (□), MDM2 (■), ATF3 (○), and BAX (◆) by low doses of γ-rays. Points are averages of four independent experiments; error bars are standard errors. Dashed line indicates basal level in untreated controls; solid lines were fitted by linear regression through the data. SOURCE: From Amundson and colleagues (1999b).

recessive mutations at other coat color loci; therefore, the authors speculated that the deletions resulted from non-targeted effects, such as increased recombination frequencies (*i.e.*, genomic instability) in the proliferating melanocytes.

In summary, there are data suggesting HRS for cell lethality and signal transduction up to 200 and some data suggesting HRS for mutagenesis or genomic instability at up to 50 mGy. However, it is not clear (Malaise and others 1994; Skov 1999) whether HRS for cell lethality would cause an increase in deleterious effects in surviving cells or would actually decrease deleterious effects by increased killing of damaged cells. Also, it is not known what effect HRS for signal transduction pathways (such as that illustrated in Figure 2-4) will have in mitigating or increasing deleterious effects. Most important, it is not known if HRS plays a role when radiation doses <100 mGy are delivered over weeks to months, which could be relevant for low doses of low-LET radiation delivered to radiation workers.

OBSERVED DOSE-RESPONSE RELATIONSHIPS AT LOW DOSES

At the time of publication of the BEIR V report (NRC 1990) and during the next several years, dose-response relationships for induction of chromosomal aberrations and gene mutations by acute doses of low-LET X-irradiation were described quite satisfactorily down to ~200 mGy by the linear quadratic (alpha-beta) relationship discussed earlier. In general, low dose rates and fractionated doses reduced the induction frequencies by factors of about 2 or more, but the results were variable and included a few reports of inverse dose-rate effects (Thacker 1992). In this section, more recent experiments conducted with mammalian cellular systems that have measured frequencies of various events resulting from relatively low doses and low dose rates of X-rays or γ-rays are reviewed (Table 2-1). The objective is to summarize data acquired primarily since the 1990 BEIR V report that provide information on the shape of the dose-response curve down to 100 mGy. Whenever possible, these data will be related to human exposures, although caution should be exercised whenever attempting to extrapolate from *in vitro* systems to the human.

Normal human fibroblasts irradiated in plateau phase with doses of 109–6000 mGy gave a linear dose-response relationship for the induction of chromosomal aberrations detected by premature chromosomal condensation immediately after irradiation (Darroudi and others 1998); the slope was 6 × 10^{-3} fragments per cell per milligray (Cornforth and Bedford 1983). When the cells entered metaphase and were scored for chromosomal dicentrics and rings after repair or misrepair of DNA damage had occurred (released from confluence after potentially lethal damage repair had occurred), a β-component was apparent, and the α-component decreased to 5.8 × 10^{-5} aberrations per cell per milligray (Cornforth and Bedford 1987). Six laboratories collaborated

in quantifying the frequency of chromosomal aberrations in human lymphocytes exposed to eight different acute doses from 3 to 300 mGy; a linear dose-response relationship was observed above 20 mGy, with a slope of 2.9 × 10^{-5} chromosomal aberrations per cell per milligray (Lloyd and others 1992; Figure 2-5).

Below 20 mGy, however, the data could not distinguish between a linear and a threshold model. When immortalized human lymphocytes were irradiated in G$_2$ with four different doses from 50 to 500 mGy, a linear dose-response relationship was observed, with a slope of 2.5 × 10^{-5} chromosomal aberrations per cell per milligray (Puck and others 1997). These results are similar to those obtained with primary human skin fibroblasts (Cornforth and others 2002), irradiated while the cells were arrested in G$_0$. For total aberrations per cell, an α-component of (5.8 ± 2.4) × 10^{-5}/mGy for acute radiation corresponded to a linear dose-response relationship of (4.9 ± 2.0) × 10^{-5}/mGy for low-dose-rate irradiation (0.5 or 1 mGy/min) between 300 and 6000 mGy. For dicentrics, the frequency was (1.9 ± 1.2) × 10^{-5}/mGy for the low dose rate (LDRs). These LDR coefficients correspond to the limited slope (curve D) in Figure 2-1.

An extensive aberration study was conducted in which mice were irradiated daily for 21, 42, or 63 d at doses of 6.4, 18.5, or 55 mGy; lymphocyte cultures set up two weeks after irradiation was completed yielded a linear dose-response (1.2 × 10^{-5} chromosomal translocations per cell per milligray), with no evidence of either an adaptive or a supralinear response (Tucker and others 1998) (see Figure 2-6 in which the frequencies determined for painted chromosomes were corrected for the whole genome). The DDREF for acute exposures of 1–3 Gy was about 4–6 (see "General Aspects of Dose-Response Relationships" for definition of DDREF). Most important, the induced frequency of chromosomal translocations was not significantly different from that reported in workers at the Sellafield Nuclear Facility who were occupationally exposed to lifetime cumulative doses of more than 500 mSv, that is, (1.0 ± 0.25) × 10^{-5}/mGy for smokers (Tawn and others 2000a). A subsequent analysis by Tawn and colleagues (2004) reported a linear dose-response between 50 and 1000 mSv of (1.11 ± 0.19) × 10^{-5} translocations per cell per millisievert.

In addition, the α-component is 1.9 × 10^{-5}/mGy for the frequency of chromosomal translocations in lymphocytes of cleanup workers of the Chernobyl nuclear accident who received an estimated average dose of 95 mGy over 6–13 years (Jones and others 2002). These values are similar to the frequency of dicentrics (1.4 × 10^{-5}/mGy) observed in people who were exposed to 100 ± 124 mGy of cobalt-60 over about 10 years (Liu and others 2002) and for Mayak nuclear workers exposed over 1–5 years (0.5–0.9 × 10^{-5}/mGy for translocations; Burak and others 2001). Note that in the seven studies above, the dose-response relationships are consistent with a linear no-threshold model in which the aberration frequencies per milligray are similar.

TABLE 2-1 Dose-Response Relationships at Relatively Low Doses

System (including exposure conditions and acute α^a or LDR^b)	End Point	Dose Range, mGy	Curve Shape	Frequency of Events per Viable Cell per Milligray	Comments and References
Human fibroblasts in $\bar{G}_0$	Immediate PCC fragments	109–6000 (acute)	Linear	8×10^{-3}	LNT^e extrapolates to 5 mGy (Cornforth and Bedford 1983)
Human fibroblasts in G_0 α-component-metaphase	Chromosome dicentrics and rings	1000–12,000 (acute)	Upward curvature	5.8×10^{-5}	(Cornforth and Bedford 1987)
Immortal human lymphocytes in G_2	Chromatid gaps	50–500 (acute)	Linear	2.5×10^{-5}	LNT > ~50 mGy (Puck and others 1997)
Human lymphocytes in G_0 (six laboratories)	Chromosome dicentrics	3–300 (acute)	Linear	2.9×10^{-5}	LNT > ~20 mGy (Lloyd and others 1992)
Human primary fibroblasts in G_0 (acute α-component)	Chromosome aberrations	1000–6000 (acute)	Upward curvature	5.8×10^{-5}	α-Component for acute corresponds to linear dose-response for LDR (Cornforth and others 2002)
Human primary fibroblasts in G_0-0.5 or 1 mGy/min	Chromosome aberrations	300–6000 (LDR)	Linear	4.9×10^{-5}	LNT > 300 mGy (Cornforth and others 2002)
Mice—daily doses of 6.4, 18.5, or 55 mGy for 21, 42, or 63 d, respectively	Chromosome translocations	100–3500 (LDR)	Linear	1.2×10^{-5}	LNT > ~100 mGy DDREF^d of 4–6 for 1–2 Gy acute exposure (Tucker and others 1998)
Nuclear workers at Sellafield— lymphocyte cultures	Chromosome translocations	50–1000 (LDR)	Linear	1.1×10^{-5}	LNT > 50 mGy (Tawn and others 2000a, 2004)
Cleanup workers at Chernobyl— lymphocyte cultures	Chromosome translocations	~95 (LDR)	?	1.9×10^{-5}	Increase of 30% (10–53% p < .002) relative to controls (Jones and others 2002)
Chinese hamster cells with human chromosome 11	Loss of antigen on chromosome 11	250–1500 (acute)	Linear	7×10^{-6}	LNT > ~ 250 mGy (Puck and Waldren 1987)
TK6 human lymphoblasts—daily doses of 10, 25, 50, or 100 mGy for 1 month	HPRT mutations	50–2000 (LDR)	Linear	6×10^{-9}	LNT > ~50 mGy (Grosovsky and Little 1985)
Mice—T lymphocytes in spleen— chronic at 0.69 mGy/min or 0.1 mGy/min	HPRT mutations	300–6000 (LDR)	Linear	3×10^{-9}	LNT > ~300 mGy DDREF of ~1.5 for acute <2 Gy (Lorenz and others 1994)
Cleanup workers at Chernobyl— lymphocyte cultures	HPRT mutations	~95 (LDR)	?	5×10^{-8}	Increase of 41% (19–66% p < .001) relative to controls (Jones and others 2002)
Chinese hamster cells with human chromosome 11	Genomic instability Translocations on chromosome 11	1000–10,000 (acute)	Linear	3×10^{-5}	Based on percent unstable clones with BrdU saturates at 30% (Limoli and others 1999)
Chinese hamster cells (CHO)	Genomic instability *de novo* HPRT mutations	2000	?	5×10^{-5}	Based on percent unstable clones; from 4 to 12 Gy saturates at 20% (Little 1998)
Melanocytes in irradiated mice	Genomic instability gene deletions	10–1000	Linear	8×10^{-5}	LNT > 10 mGy, but supralinear from 0 to 10 mGy (Schiestl and others 1994)

continues

TABLE 2-1 Continued

System (including exposure conditions and acute α^a or LDR^b)	End Point	Dose Range, mGy	Curve Shape	Frequency of Events per Viable Cell per Milligray	Comments and References
Human blood lymphocytes stimulated with PHA	Genomic instability chromosomal aberrations	1000–3000 (acute)	?	$(3–10) \times 10^{-5}$	Analyzed at 51–57 d after irradiation (Holmberg and others 1998)
Hamster embryo cells	Malignant transformation	30–1500 (acute)	Linear	4×10^{-6}	LNT > ~30 mGy (Borek and others 1983)
C3H 10 T1/2 mouse cells (six labs)	Malignant transformation	250–5000 (acute)	Linear	8×10^{-8}	LNT > ~250 mGy (Mill and others 1998)
Hela X skin fibroblast human hybrid cell system	Malignant transformation	0–1000 (acute)	Sigmoid	4×10^{-8}	Threshold at ~300 mGy dependent on time of trypsinization after irradiation (Redpath and others 2001)

NOTE: LDR = low dose rate; PCC = premature chromosome condensation; PHA = phytohemagglutinin.

^{a}Acute indicates that doses were delivered at high dose rate (*e.g.*, 0.1 Gy/min.), and α-component signifies the value of α in the linear-quadratic relationship.

^{b}LDR indicates that the doses were delivered at low dose rates less than 0.01 Gy/min.

^{c}LNT signifies a linear, no-threshold dose-response relationship.

^{d}DDREF is defined and illustrated in Figures 2-1 and 2-8.

Three mutation experiments have yielded a linear dose-response relationship. First, the loss of an antigen marker on human chromosome 11 integrated in Chinese hamster cells and exposed to four different doses from 250 to 1500 mGy yielded a linear dose-response relationship with a slope of 7×10^{-6} mutants per viable cell per milligray (Puck and Waldren 1987). The relatively high frequency is due to the large target size because of the large distance between the antigen marker and essential genes on chromosome 11 (see "Induction of Gene Mutations in Somatic Cells").

Second, human lymphoblast cells (TK6) exposed to one acute dose or to daily doses of 10, 25, 50, or 100 mGy for up to one month, with samples taken every 5 d, yielded a linear dose-response relationship for induction of HPRT or TK mutations (Figure 2-7). Over a total dose range of 50–2000 mGy, the slope for HPRT mutations was 6×10^{-9} mu-

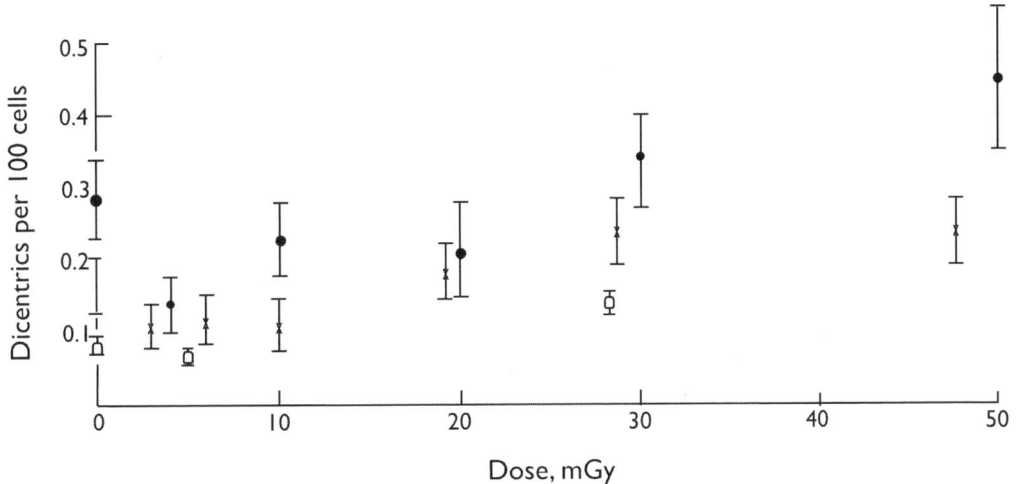

FIGURE 2-5 Dicentric yields as a function of dose; ●, Pohl-Ruling and others (1983); ✖, Lloyd and others (1992), experiment 1; □ experiment 2. SOURCE: From Lloyd and colleagues (1992).

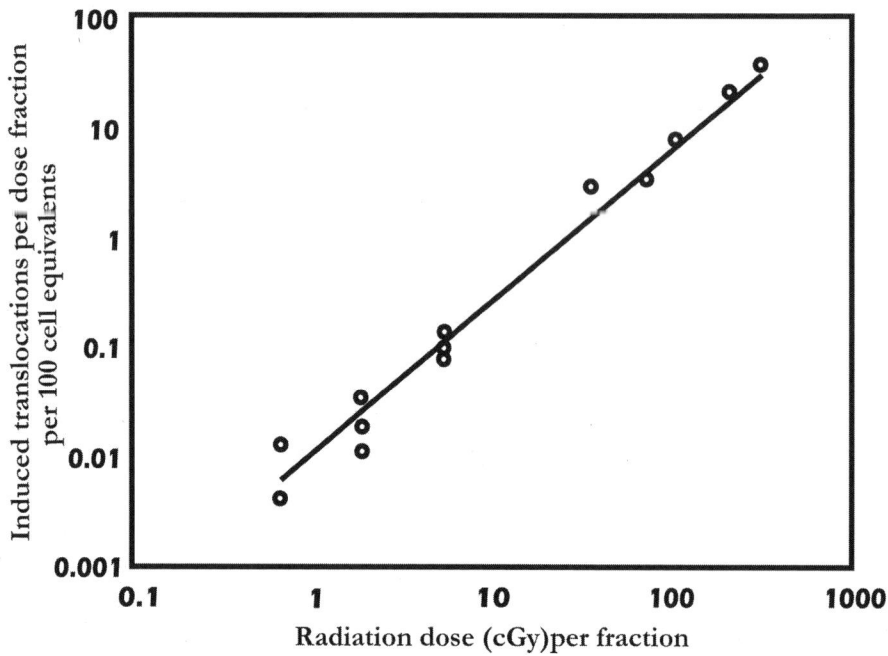

FIGURE 2-6 Induced translocations (observed frequency less control value) per dose fraction as a function of radiation dose per fraction. The line is the least-squares regression fit, with $Y = 0.0121X^{1.367}$; $R^2 = 0.98$. Five points on the upper part of the line represent the acute exposures (DDREF of 4–6), and the three sets of values on the lower portion of the line are from mice that received fractionated exposures. SOURCE: From Tucker and others (1998).

tants per viable cell per milligray (Grosovsky and Little 1985).

Third, mice were irradiated with total doses of 300–6000 mGy applied at an acute dose rate of 500 mGy/min or at a low dose rate of 1000 mGy/d (0.69 mGy/min) or 1000 mGy per week (0.1 mGy/min; Lorenz and others 1994). At 8–10 weeks after irradiation was completed, the frequency of HPRT mutants in splenic T lymphocytes for the LDRs was described by a linear dose-response relationship that had a slope of 3×10^{-9} mutants per viable cell per milligray. This is about one-tenth the frequency of HPRT mutants observed in lymphocytes of cleanup workers of the Chernobyl nuclear accident who received an estimated average dose of 95 mGy over 6–13 years (Jones and others 2002). An interesting observation in the mouse experiments was that an inverse dose-rate effect was not observed; the mutation frequency for 0.1 mGy/min was the same as that for 0.69 mGy/min. From a summary of data for radiation-induced mutations as function of dose rate (Vilenchik and Knudson 2000), an inverse dose-rate effect would not be expected if the induction of HPRT mutations in T lymphocytes in the spleen corresponded to the induction of specific locus mutations in spermatogonia. However, if they corresponded to the induction of HPRT mutations in cells *in vitro*, the mutation frequency for 0.1 mGy/min should have been about half that for 0.69 mGy/min.

By dividing the HPRT mutation frequencies for acute irradiation by the frequencies for LDR irradiation (obtained in the mouse T-lymphocyte experiment of Lorenz and others 1994 described above), the DDREF was 3–5 for acute doses greater than 3 Gy and about 1.5 for acute doses less than 2 Gy (Figure 2-8). The DDREF points (averages of 1.0) plotted for each of the LDRs were obtained by dividing the mutation frequencies for each total dose by the product of 3×10^{-6} mutants per viable cell per gray (value for LDRs) times the total dose. The range of DDREF values for acute doses are similar to those obtained for the same dose ranges in transformation *in vitro* (Han and others 1980) and animal carcinogenesis and life-shortening experiments (Ullrich and Storer 1979a; Ullrich and others 1987). (For definition and illustration of DDREF, see "General Aspects of Dose-Response Relationships.")

Overall, the dose-response for radiation-induced genomic instability is quantitatively similar to that for radiation-induced chromosomal aberrations, with the exception that the frequency for genomic instability saturates between 4 and 12 Gy (Little 1998; Limoli and others 1999), while the frequency for chromosomal aberrations continues to increase with dose. After 10 Gy, 30% of the CHO clones were unstable for chromosomal aberrations, which was the saturation level reached after 4 Gy when the cells had incorporated BrdU (Limoli and others 1999). Furthermore, when induc-

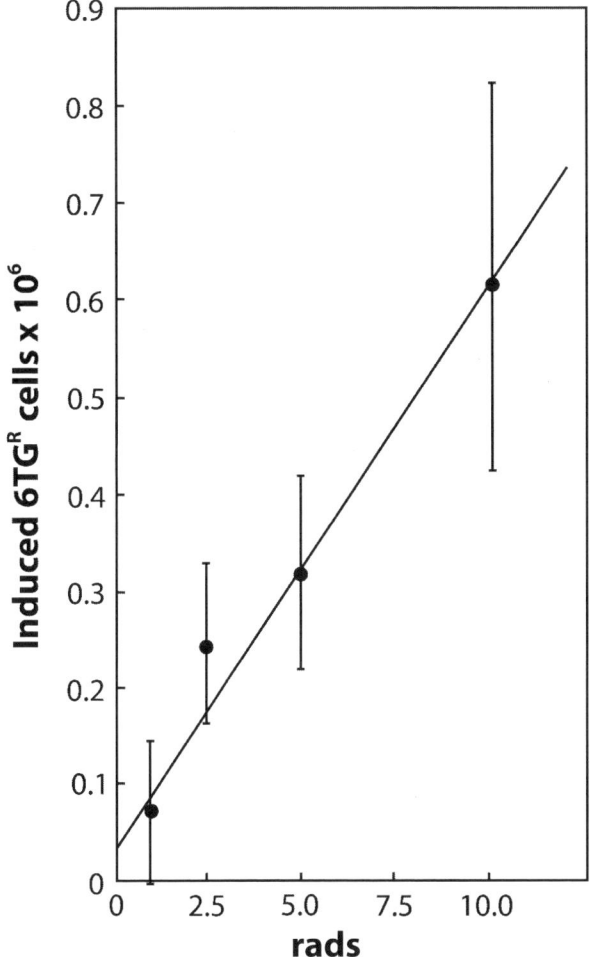

FIGURE 2-7 Frequency of 6TG^R cells induced by 1–10 rads (0.01–0.1 Gy) of X-rays in TK6 human lymphoblastoid cells. Data points (with standard deviations) are from regression analyses of mutations induced per day at various dose rates (1–10 rads/d; 0-30 d) as described in Grosovsky and Little (1985).

human blood lymphocytes stimulated with PHA (phytohemagglutinin) and analyzed 51–57 d after irradiation, the frequency of *de novo* aberrations was (3 to 10) × 10^{-5} chromosomal aberrations per cell per milligray (Holmberg and others 1998).

Genomic instability was also observed in mice as gene deletions in melanocytes exposed to X-irradiation (Schiestl and others 1994a), with a threefold increase at 0.01 Gy and a twelvefold increase at 1.0 Gy. The frequency of gene deletions was about 100 times higher than mutation frequencies; therefore, the authors speculated that the deletions resulted from nontargeted effects, such as an increased recombination frequency or genomic instability in the proliferating melanocytes. The dose-response was linear between 0.01 and 1.0 Gy and had a slope of 8 × 10^{-5} events per cell per milligray. Note that the three values listed above for the frequencies of radiation-induced instability (3–10 × 10^{-5} events per cell per milligray) are of the same order of magnitude as the frequency of chromosomal aberration induced directly by irradiation (1–4 × 10^{-5} events per cell per milligray; Table 2-1).

A malignant transformation experiment with primary hamster embryo cells exposed to five different doses from 0.03 to 1.5 Gy yielded a linear dose-response curve that had a slope of 4 × 10^{-6} transformants per viable cell per milligray (Borek and others 1983). An extensive collaborative study involving six laboratories that quantified malignant transformation of immortalized mouse C3H 10T1/2 cells exposed to

tion of genomic instability was assayed as chromosomal aberrations in mammary epithelial cells at 25 population doublings after the cells had been irradiated *in vitro* or *in vivo* (Ullrich and Davis 1999), a downward-curving dose-response curve was observed between 0 and 0.25 Gy, with the response saturating between 1 and 3 Gy at about 0.35 aberration per cell. The percentage of CHO clones (containing a human chromosome 4) that were stable for chromosomal translocations in chromosome 4, had a linear dose-response of 3 × 10^{-5} events per irradiated cell per milligray between 1 and 10 Gy (Limoli and others 1999). For HPRT mutations in CHO cells, the percentage of clones that were unstable for *de novo* HPRT mutations was 5 × 10^{-5} events per irradiated cell per milligray, based on 10% being unstable after 2 Gy (Little 1998). Between 4 and 12 Gy, the percentage of unstable clones remained the same at 10–20%. For irradiated

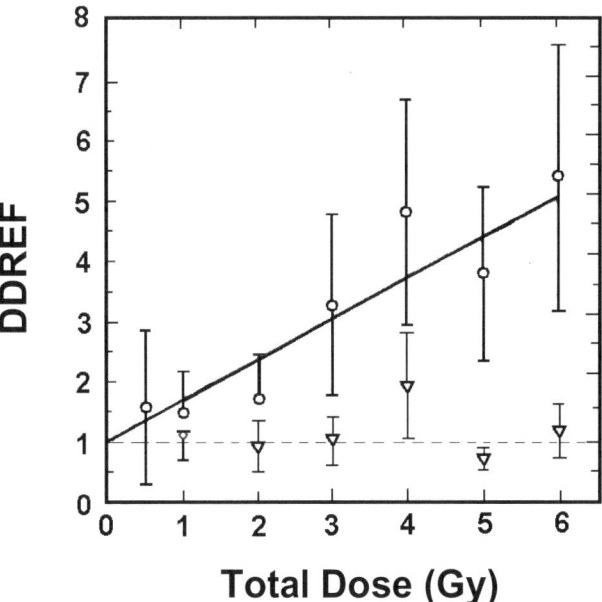

FIGURE 2-8 DDREF for low-LET ^{137}Cs γ-rays: (○) dose rates 0.5 Gy/min; (▽) dose rates 1 Gy/d (0.69 mGy/min) to 1 Gy per week (0.10 mGy/min). SOURCE: From Lorenz and colleagues (1994).

seven different doses from 0.25 to 5.0 Gy reported a linear dose-response with a slope of 8×10^{-8} transformants per viable cell per milligray (Mill and others 1998). A study conducted with a human Hela hybrid cell system (Redpath and others 2001) reported a frequency of 4×10^{-8} transformants per viable cell per milligray beyond a threshold of ~0.3 Gy; however, the results were greatly dependent on the time the cells were trypsinized and plated after irradiation for the transformation analysis. Note that these results for transformation are quite variable and that the frequencies are ten- to a thousandfold lower than the frequencies for radiation-induced genomic instability. However, as discussed earlier under adaptive response, studies of malignant transformation in immortalized (already-transformed) cell lines may have little relevance to malignant transformation of normal nonimmortalized cells, especially *in vivo* where complex interactive processes can occur.

In summary, results of experiments that quantified chromosomal aberrations, malignant transformations, or mutations induced by relatively low total doses or low doses per fraction suggest that the dose-response relationship over a range of 20–200 mGy is generally linear and not affected significantly by either an adaptive or a bystander effect (Table 2-1). No data are available in this dose range for radiation-induced genomic instability. The question of the shape of the dose-response relationship up to about 20 mGy remains, although several of the dose-response relationships described above appear to be consistent with extrapolation linearly down to about 5 mGy. As has been pointed out (Cornforth and Bedford 1983), a macroscopic X-ray dose of about 5 mGy would, on the average, result in one to two electron tracks crossing the nucleus of each cell. Since the tracks are produced randomly, the proportion of nuclei traversed by zero, one, or two electron tracks would be about 0.37, 0.37, and 0.18, respectively. For lower doses, a larger and larger proportion of cell nuclei would receive no dose (track) at all. The nuclei that would receive a track would all receive (on the average) the same dose because the proportion receiving two or more tracks would diminish rapidly. Therefore, unless interactions among neighboring or surrounding cells influence the response, if 5 mGy produces an effect and if the effect is linear above 5 mGy, the dose-response curve must also be linear from 0 to 5 mGy. In addition to the existence of biological information at these very low dose levels, the committee concluded that the biophysical characteristics of the interaction of low-LET radiation with DNA, coupled with the characteristics of DNA repair, argue for a continuation of the linear response at lower doses. However, if a single electron track traversing a cell's nucleus could induce an adaptive or bystander effect, the dose-response relationship below 5 mGy might deviate from linearity depending on whether cellular effects are decreased or increased. In the committee's judgment, there is no evidence for either an adaptive response or a bystander effect for doses below 5 mGy.

Furthermore, the calculated value of 5 mGy for an average of one electron track per nucleus depends on the size and shape of the nucleus, as well as on the energy of the radiation (Rossi and Zaider 1996; Edwards and Cox 2000). For example, the calculated doses for an average of one electron track per nucleus are as follows: about 5 mGy for 60 keV and a 6-μm diameter sphere, about 4 mGy for 60 keV and a 7-μm sphere, about 3 mGy for 300 keV and a 6-μm sphere, and about 2 mGy for 300 keV and a 7-μm sphere. For the very low doses for which important signal transduction events may result from ionizations in either the nucleus or the cytoplasm, the volume of the whole cell might be most appropriate for these types of calculations. Possibly, the shape of the dose-response relationship up to 5 mGy might be determined with *in vitro* and *in vivo* experiments in which multiple doses of about 1–5 mGy are delivered over a long period. However, the question must be addressed rigorously by defining the molecular processes responsible for the end points in question at these very low doses.

SUMMARY

This chapter discusses the biological effects of the ranges of radiation dose that are most relevant for the committee's deliberations on the shapes of dose-response relationships. Considering the levels of background radiation, the maximal permissible levels of exposure of radiation workers now in effect, and the fact that much of the epidemiology of low-dose exposures includes people who in the past have received up to 500 mGy, the committee has focused on evaluating radiation effects in the low dose range of <100 mGy, with emphasis on the lowest doses when relevant data are available. Effects that may occur as the radiation is delivered chronically over several months to a lifetime are thought to be most relevant.

Considerable emphasis has been placed on the dose-response and mechanisms for inducing chromosomal aberrations and gene mutations because, as discussed in Chapter 3, there is evidence that the induction of cancer is associated with these cellular responses. The general pictures that emerge from biophysical studies is that the misrepair of radiation-induced DNA DSBs that lead to chromosome aberrations are probably associated with the dominant postirradiation function of nonhomologous end joining repair processes described elsewhere is this report. Overall, biophysical approaches to the modeling of dose-response for chromosome aberrations, although not without some uncertainties on mechanisms, imply that the single-track α-component of radiation action will dominate at low doses and LDRs (*i.e.*, the dose-response for all forms of aberrations will be linear at low doses and LDRs). Also, as observed, the response at LDRs and low doses, or after fractionated doses, should be lower by a DDREF; then the response to a single acute high-dose-rate exposure for which the two-hit β-component becomes important. In certain cases, an inverse dose-rate effect for cell lethality and mutations has been reported for which the effect at very low dose rates is as high or higher than for single, acute, high-dose-rate exposures. The ability to demonstrate this phenomenon, however, is variable, and no mechanisms have been clearly identified to explain such effects.

Several factors may affect the theoretical dose-response relationships described above: variations in radiosensitivity during the cell cycle; induction of an adaptive response to an initial exposure, which can reduce the effect of later exposures; a bystander effect that causes an irradiated cell to have an effect on a nearby unirradiated cell; the induction of persistent genomic instability; and HRS in the low-dose region. Except for the cell cycle, these factors have been identified and studied since the BEIR V report (NRC 1990). These factors together with quantitative data on the induction of gene or chromosomal mutations in somatic cells are discussed.

Radiation genomic instability has been demonstrated by the manifestation of chromosomal damage in a certain fraction of irradiated cells over many cell cycles after they were irradiated. Data are critically needed for the definition of molecular targets and processes responsible for genomic instability in order to define and understand the dose-response relationship, and especially why the induction frequency saturates with only about 10–30% of the surviving cells manifesting genomic instability. A possibility that has not been investigated is that only a certain fraction of the cells, such as those in a certain part of the cell cycle, are susceptible to radiation-induced genomic instability. Because chromosomal instability has been associated with breakage-fusion-bridge cycles, the role of telomeres may be particularly relevant. Chromosome instability can also be initiated by DSBs that result in the loss of a telomere that protects the chromosome end and prevents chromosome fusion. Furthermore, from limited data, the similarity in the frequencies of genomic instability induced in X-irradiated cells and the frequencies of chromosomal aberrations induced directly by irradiation may suggest that the induction of chromosomal aberrations is a primary event that plays a major role in radiation-induced genomic instability. There is also some evidence that reactive oxygen species may play a role. However, until the molecular mechanisms responsible for genomic instability and its relationship to carcinogenesis are understood, extrapolation of the limited dose-response data for genomic instability to radiation-induced cancers in the low-dose range <100 mGy is not warranted.

An apparent adaptive response has been well documented for cell lethality, chromosomal aberrations, mutations, and *in vitro* transformation. The phenomena are illustrated by a reduction in response to a challenge dose of about 1 Gy delivered a few hours after a low priming dose of about 10–20 mGy. There is much variability in the ability to demonstrate the adaptive response, however. Data are needed, particularly at the molecular level, on adaptation induced when both priming and challenging doses are in the low-dose range <100 mGy; relevant end points should include not only chromosomal aberrations and mutations but also genomic instability and, if possible, tumor induction. Studies of the adaptive response for malignant transformation in immortalized (already-transformed) cell lines may have little relevance to malignant transformation of normal non-immortalized cells, especially *in vivo*, where complex interactive processes can occur. *In vitro* and *in vivo* data are needed on the delivery of priming and challenge doses over several weeks or months at very low dose rates or with fractionated exposures. Specifically, an adaptive response resulting from the cumulative effect of multiple low doses of less than 10 mGy should be determined. Such data have not yet been obtained, particularly those explaining the molecular and cellular mechanisms of the adaptive response. Thus, it is concluded that any useful extrapolations for dose-response relationships in humans cannot be made from the adaptive responses observed in human lymphocytes or other mammalian cellular systems. Therefore, at present, the assumption that any stimulatory effects of low doses of ionizing radiation substantially reduce long-term deleterious radiation effects in humans is unwarranted.

A bystander effect in which an irradiated cell induces a biological response in a neighboring unirradiated cell has been observed with high-LET radiation for inducing cell lethality, chromosome aberrations, sister-chromatid exchanges, mutations, genomic instability, signal transduction pathways, and *in vitro* transformation. There is some evidence that long-lived reactive oxygen species or the diffusion of cytokines plays a role in the bystander effect. For low-LET radiation, the bystander effect has been limited to cell lethality and lethal mutations associated with reduced cloning efficiency. Recent results suggest that a bystander effect for cell lethality from soft X-ray irradiation might be observed down to 50 mGy but not below. Until molecular mechanisms of the bystander effect are elucidated, especially as related to an intact organism, and until reproducible bystander effects are observed for low-LET radiation in the dose range of 1–5 mGy, where an average of about one electron track traverses the nucleus, a bystander effect of low-dose, low-LET radiation that might result in modification of the dose-response should not be assumed.

HRS is a phenomenon for which doses less than about 200 mGy produce a dose-response for cell lethality that is steeper than that predicted from the classic $D + D^2$ model. There are data suggesting HRS for cell lethality and signal transduction at up to 200 mGy and some data suggesting HRS for mutagenesis or genomic instability at up to 50 mGy. Furthermore, from limited data from only one laboratory, an observed inverse dose-rate effect for cell lethality was attributed to HRS seen for low acute doses, and cell cycle analysis suggested that HRS may be related to cells not arresting in radiosensitive G_2. Since a high proportion of the target stem-like cells in humans would be noncycling, the last two observations, if generally true, would suggest that neither HRS nor the inverse dose-rate phenomenon should have any significant effect on the dose-response for cancer induction in humans. Furthermore, molecular mechanisms associated with the two phenomena have not been delineated, and it is not known whether HRS for cell lethality would cause an increase in deleterious effects in surviving cells or would actually decrease deleterious effects by increased killing of damaged cells. Also, it is not known what effect HRS for signal transduction pathways will have in mitigating or increasing deleterious effects. Most important, it is not known if HRS plays a role when radiation doses <100 mGy are delivered over weeks to months, which could be relevant for low doses of low-LET radiation. Finally, until the molecular mechanisms responsible for HRS are understood, its role in low-dose radiation carcinogenesis is uncertain.

Results of experiments that quantified chromosomal aberrations, malignant transformation *in vitro*, or mutations induced by relatively low total doses or low doses per fraction indicate that the dose-response relationship over a range of 20–100 mGy is most likely to be linear and not affected significantly by either an adaptive or a bystander effect. No data are available in this dose range for radiation-induced genomic instability. Furthermore, as stated previously, studies of malignant transformation in immortalized (already-transformed) cell lines may have little relevance to malignant transformation of normal nonimmortalized cells, especially *in vivo* where complex interactive processes can occur. However, the results from these *in vitro* transformation studies may have relevance for effects involved in promoting the immortalization process, possibly through the induction of genomic instability. Thus, the question of the shape of the dose-response relationship up to about 20 mGy remains, although several of the dose-response relationships described above appear to be consistent with extrapolation linearly down to about 5 mGy. The shape of the dose-response relationship up to 5 mGy might be determined with *in vitro* and *in vivo* experiments in which multiple doses of about 1–5 mGy are delivered over a long period. However, this question should be addressed rigorously by defining the molecular processes responsible for the end points in question at these very low doses.

3

Radiation-Induced Cancer: Mechanisms, Quantitative Experimental Studies, and the Role of Genetic Factors

INTRODUCTION

The process of cancer development (tumorigenesis) is recognized to involve multiple changes in genes involved in cell signaling and growth regulation, cell cycle checkpoint control, apoptosis, differentiation, angiogenesis, and DNA damage response or repair. Changes in these genes can involve (1) gene mutations or DNA rearrangements, which result in a gain of function as in the case of the conversion of proto-oncogenes to oncogenes; (2) mutations or DNA deletions or rearrangements, which result in loss of gene function as in the case of tumor-suppressor genes (Kinzler and Vogelstein 1998).

The long latent period between radiation exposure and cancer development together with the multistage nature of tumorigenesis make it difficult to distinguish radiation-induced changes from those alterations that occur once the process has been initiated. Radiation-induced cancers do not appear to be unique or specifically identifiable (UNSCEAR 2000b). The mutations in tumors and their growth characteristics are not readily distinguishable from those in spontaneously occurring tumors of the same site or from tumors at the same site induced by other carcinogenic agents. Attempts to identify radiation-specific changes in human tumors have not been particularly successful despite fairly extensive investigation (UNSCEAR 1993, 2000b). There are, however, clues to possible underlying mechanisms of radiation-induced cancer that emerge from epidemiologic and experimental investigations.

Based mainly on experimental studies, it is generally believed that complex forms of DNA double-strand breaks are the most biologically important type of lesions induced by ionizing radiation, and these complex forms are likely responsible for subsequent molecular and cellular effects (see Chapters 1 and 2). Attempts to repair complex DNA double-strand lesions are judged to be error prone, and there is evidence that this error-prone repair process can lead to gross chromosomal effects and mutagenesis. Molecular analyses

of radiation-induced mutations have found a full range of mutations including base-pair substitutions, frameshift mutations, and deletions. Importantly, the most common radiation-induced mutations are deletions rather than base-pair changes in genes (point mutations; Chapters 1 and 2). Therefore, theories of radiation-induced cancer have generally centered on postirradiation tumor-suppressor gene inactivation that would be expected to occur through DNA deletion rather through the induction of point mutations. Oncogene activation through specific forms of induced chromosome translocation is also a candidate radiation-associated event, particularly for leukemia and lymphoma (UNSCEAR 2000b). Thus, mechanisms involving gene and/or chromosome rearrangements and loss of heterozygosity (signaling specific regions of DNA loss) are considered the most likely radiation-induced events that contribute to cancer development (UNSCEAR 2000b).

More recently, experimental studies have questioned whether the initiating events produced by radiation are indeed direct effects on specific genes (e.g., Little 2000). Rather, it has been proposed that the gene or chromosomal mutations involved in radiation tumorigenesis arise indirectly as a consequence of persistent genomic instability (Chapter 2) induced by the radiation exposure.

This chapter focuses first on studies relevant to mechanisms of radiation-induced tumorigenesis, with particular emphasis on the potential implications for low-dose risks. Subsequently, experimental studies addressing the quantitative relationship between radiation dose and cancer development are reviewed with particular regard to their consistency with proposed underlying mechanisms and the overall implications for cancer risk at low doses.

Advances in human and animal genetics have also highlighted the contribution made to cancer risk by heritable factors (Ponder 2001). Much of the available information concerns germline genes that influence the risk of spontaneous cancer and the mechanisms through which they act. However, evidence is also emerging on the impact of such genes on radiation cancer risk (ICRP 1998). Relevant data on genetic susceptibility to cancer are reviewed in the final section of this chapter, and some interim judgments are developed about their implications for radiation cancer risk in the population.

MECHANISMS OF TUMORIGENESIS

Gene and Chromosomal Mutations in Spontaneously Arising Human Tumors

Studies on the cellular and molecular mechanisms of tumorigenesis have in recent years cast much light on the complex multistep processes of tumorigenesis and its variation among tumor types. There is a vast literature on tumor biology and genetics (Bishop 1991; Loeb 1991, 1994; Hartwell 1992; Levine 1993; Vogelstein and Kinzler 1993; Hinds and Weinberg 1994; Weinberg 1994; Boland and others 1995; Karp and Broder 1995; Levine and Broach 1995; Skuse and Ludlow 1995; Kinzler and Vogelstein 1998; Rabes and others 2000; Khanna and Jackson 2001; Balmain and others 2003), and it is sufficient to highlight the principal points of current fundamental knowledge that may serve to guide judgments on the impact of ionizing radiation on cancer risk.

Tumor development is generally viewed as a multistep clonal process of cellular evolution that may be conveniently but imprecisely divided into a number of overlapping phases: (1) *tumor initiation*, which represents the entry via mutation of a given normal somatic cell into a potentially neoplastic pathway of aberrant development; cellular targets for this process are generally held to have stem cell-like properties; (2) *tumor promotion*, which may now be viewed as the early clonal development of an initiated cell; cell-cell communication, mitogenic stimulation, cellular differentiating factors, and mutational and nonmutational (epigenetic) processes may all play a role in this early pre-neoplastic growth phase; (3) *malignant conversion*, which represents the tumorigenic phase where the evolving clonal population of cells becomes increasingly committed to malignant development; mutation of genes that control genomic stability is believed to be particularly important; and (4) *malignant progression*, which is itself multifaceted, is a relatively late tumorigenic phase during which neoplastic cells become increasingly autonomous and gain a capacity for invasion of surrounding normal tissue and spread to distant sites (metastasis); the development of tumor vasculature is important for the development of solid cancers (Folkman 1995). In addition, there is evidence that inflammatory processes and the microenvironment in which tumors develop are important cofactors for malignant progression (Coussens and Werb 2002). Overall, it is clear that only a small fraction of cells that enter tumorigenic pathways complete the above sequence that results in overt malignancy (Rabes and others 2000), and that the whole process can take many years.

The balance of evidence suggests that sequential gene and chromosomal mutations act as the principal driving force for tumorigenic development, with phase transitions being dependent on the selection and overgrowth of clonal neoplastic variants best fitted for the prevailing *in vivo* conditions. Although there are exceptions, the consensus view is that tumor initiation or promotion is a monoclonal process having its origin in the appearance of a single aberrant cell (Levy and others 1994; Rabes and others 2000).

The tumor initiation phase is most difficult to study directly, but in recent years it has become evident that a relatively tissue-specific set of so-called *gatekeeper* genes (Kinzler and Vogelstein 1997; Lengauer and others 1998) may be critical mutational targets for cellular entry into neoplastic pathways. Table 3-1 provides examples of such genes and their principal associated neoplasms. These gatekeepers are frequently involved in intracellular biochemical signaling pathways, often via transcriptional control, and are subject primarily to productive loss-of-function mutations. They fall into the tumor-suppressor gene category consistent with the germline role of many of these genes in autosomal dominant familial cancer (see "Genetic Susceptibility to Radiation-Induced Cancer," later in this chapter). The somatic loss of function associated with gatekeeper gene inactivation can arise by point mutation (often of the chain-terminating type), intragenic deletion, or gross chromosomal loss events (Sidransky 1996; Kinzler and Vogelstein 1997, 1998). For some genes, epigenetic silencing events may also be important (Jones and others 1992; Feinberg 1993, 2004; Ranier and others 1993; Merlo and others 1995; Issa and Baylin 1996; Roth 1996).

It is evident from Table 3-1 that the gatekeeper gene hypothesis applies principally to the genesis of solid tumors. For lymphomas and leukemia a somewhat different mechanism appears to apply. In these neoplasms, the early productive events often involve chromosomally mediated gain-of-function mutations in tissue-specific proto-oncogenes (*i.e.*, gene activation or intragenic fusion involving juxtaposition of DNA sequences by specific chromosomal exchange; Rabbitts 1994; Greaves and Wiemels 2003). In many instances, these leukemia- or lymphoma-associated chromosomal events involve the DNA sequences (TCR [T cell receptor] and IG [immunoglobin]) involved in immunological

TABLE 3-1 Examples of Human Tumor-Suppressor Genes of the Gatekeeper Type

Gene	Principal Cancer Type	Mode of Action
APC	Colon carcinoma	Transcriptonal regulator
NF1	Neurofibromas	GTPase-activator
VHL	Kidney carcinoma	Transcriptional regulator
WT-1	Nephroblastoma	Transcription factor
PTCH	Skin (basal cell)	Signaling protein

NOTE: GTPase = guanosine 5′-triphosphatase.

response (Rabbitts 1994). Tumorigenic chromosomal exchange events are less well characterized in solid tumors but do occur in certain sarcomas and in thyroid tumors (Rabbitts 1994; Mitelman and others 1997). However, in accord with data from solid tumors, gene deletion and other loss-of-function mutations are not uncommon in lymphohemopoietic tumors (Rabbitts 1994; Mitelman and others 1997).

In relation to tumorigenesis in general, a second broad category of so-called *caretaker* genes has also been identified, although it is important to stress that the distinction between gatekeeper and caretaker genes is somewhat artificial—there are examples of genes that fulfill both criteria. Caretaker genes are those that play roles in the maintenance of genomic integrity (Kinzler and Vogelstein 1997, 1998). Table 3-2 provides examples of such tumor genes and their associated neoplasms. In such cases, mutational loss of function can lead to deficiency in DNA damage response and repair, repair or recombination, chromosomal segregation, cell cycle control, and/or apoptotic response (Loeb 1991; Hartwell and others 1994; Fishel and Kolodner 1995; Kinzler and Vogelstein 1996, 1998). Almost irrespective of the specific nature of the tumor gene in question, the net result of caretaker gene mutation is to elevate the frequency of gene or chromosomal mutations in the evolving neoplastic clone, and there is evidence that in some tumors this phenotype can arise at a relatively early point in neoplastic growth (Schmutte and Fishel 1999). This increased mutation frequency can be seen to provide the high level of dynamic clonal heterogeneity characteristic of tumorigenesis, thereby facilitating the selection of cellular variants that have gained the capacity to evade or tolerate antitumorigenic defenses (Tomlinson and Bodmer 1999). These defenses would include cell-cell communication, apoptosis, terminal differentiation, cell senescence, and immune recognition (Rabes and others 2000). Gene and chromosomal mutations conferring enhanced tumor cell survival or growth characteristics have been identified in a range of malignancies (Greenblatt and others 1994; Branch and others 1995; Kinzler and Vogelstein 1998; Greider 1996; Orkin 1996).

In summary, gene and chromosomal mutations of the general types induced by ionizing radiation are known to play a role throughout the multistep development of tumors. Loss of function of gatekeeper genes may be of particular importance in the initiation of common solid tumors, while gain-of-function chromosomal exchanges and gene loss events can arise early in lymphoma and leukemia. The relatively early spontaneous development of genomic instability via specific mutation of caretaker genes is believed to be important for tumorigenesis in many tissues, but epigenetic gene silencing or activation events have also been characterized. The emphasis placed here on early events in tumorigenesis derives from the prevailing view from epidemiologic and animal studies that ionizing radiation acts pri.cipally as a tumor-initiating agent.

Mechanisms of Radiation Tumorigenesis

Data from quantitative animal tumorigenesis (UNSCEAR 1988; Rabes and others 2000) and human epidemiologic studies (UNSCEAR 1994) imply that low-LET (linear energy transfer) ionizing radiation acts principally as a tumor-initiating agent. Specifically, in humans and animals, single acute doses of low-LET radiation produce a dose-dependent increase in cancer risk with evidence that chronic and fractionated exposures usually decrease that risk. Also, experimental animal data show that radiation only weakly promotes the development of chemically initiated tumors, and the generally greater tumorigenic sensitivity of humans to acute irradiation at young ages is more consistent with effects on tumor initiation than with promotional effects that accelerate the development of preexisting neoplasms.

In this section, molecular and cytogenetic data on radiation-associated human and animal tumors are summarized in the context of the mutagenic and tumorigenic mechanisms discussed previously. Particular attention is given to the proposition, based on somatic mutagenesis data, that early arising, radiation-associated events in tumors will tend to take the form of specific gene or chromosomal deletions or rearrangements.

Gene and Chromosomal Mutations in Radiation-Associated Human Tumors

The acquisition of data on *TP53* tumor-suppressor gene mutational spectra in human tumors associated with ultra-

TABLE 3-2 Examples of Human Tumor Genes of the Caretaker Type

Gene	Principal Cancer Type	Mode of Action
TP53	Multiple types	Transcription factor (DNA damage response)
ATM	Lymphocytic leukemia	PI-3 kinase (DNA damage response)
MSH2, MLH1, PMS	Colon or endometrial carcinoma	DNA mismatch repair
BRCA1/2	Breast or ovarian carcinoma	Transcription factor (DNA damage response)
XPA-G	Squamous, basal cell carcinoma, melanoma	Nucleotide excision repair
MYH	Familial adenomatous polyposis in families that lack the inherited mutation in the APC gene	Removes adenines misincorporated opposite the mutagenic lesion 8-oxoguanine

violet radiation (UVR) and chemical exposures was followed by searches for potential *TP53* mutational signatures in excess lung tumors arising in Japanese A-bomb survivors and radon-exposed uranium miners (Vahakangas and others 1992; UNSCEAR 1993; Taylor and others 1994b; Venitt and Biggs 1994; Bartsch and others 1995; Lo and others 1995; Rabes and others 2000). Subsequently, attention was also given to *TP53* mutations in liver tumors arising in excess in patients receiving the alpha-emitting radiographic contrast agent Thorotrast (Iwamoto and others 1999). Interpretation of these data are problematical, and although one study of lung tumors from uranium miners was suggestive of a possible codon-specific mutational signature of radiation (Taylor and others 1994b), this finding was not confirmed by others (Venitt and Biggs 1994; Bartsch and others 1995; Lo and others 1995). The studies on liver tumors from Thorotrast patients provide some comment on secondary *TP53* mutation and possible instability effects but, overall, the studies cited above do not give consistent evidence that *TP53* is a primary target for ionizing radiation.

A cytogenetic-molecular data set is available on papillary thyroid cancer (PTC) (Bongarzone and others 1997) arising in excess in ^{131}I-exposed children in areas contaminated by the Chernobyl accident (UNSCEAR 2000a). These mechanistic studies were guided by the knowledge that chromosomally mediated rearrangement and activation of the *ret* proto-oncogene is a frequently early arising feature of PTC (Richter and others 1999). Three different forms of *ret* gene rearrangement have been characterized at the cytogenetic and molecular levels (*i.e.*, *ret/PTC1*, *ret/PTC2*, and *ret/PTC3*), and the prevalence of these events has been investigated in post-Chernobyl childhood PTC (Klugbauer and others 1995; Bongarzone and others 1997; Williams 1997; Smida and others 1999a, 1999b). As expected, *ret* activation events were found to be recurrent in Chernobyl-associated childhood PTC, and a similarly high frequency has been reported in adult thyroid cancer of patients with a history of radiation (Bounacer and others 1997). These studies suggest that the spectra of *ret* mutations differ between tumors of adults and children. Some investigations suggest that *ret/PTC3* events in post-Chernobyl childhood cases are more frequent than expected. However this view is questioned by the study of 191 cases by Rabes and colleagues (2000), which provides evidence that the spectrum of *ret* rearrangements may be dependent on postirradiation latency, degree of tumor aggression, and possibly, dose to the thyroid.

At present, causal relationships between *ret* gene rearrangement, childhood PTC, and radiation remain somewhat uncertain. However, a possible clue to radiation causation is the finding that breakpoints in the majority of *ret* rearrangements carry microhomologies and short direct or inverted repeats characteristic of the involvement of nonhomologous endjoining (NHEJ) mediated misrepair (Klugbauer and others 2001). Other investigations have reported that *TP53* gene mutation does not play a significant role in the development

of post-Chernobyl PTC (Nikiforov and others 1996; Smida and others 1997).

Some informative molecular data are also available for basal cell skin carcinomas (BCCs) arising in X-irradiated tinea capitis patients (Burns and others 2002). In five out of five tumors analyzed there was evidence of DNA loss events which encompassed the *Ptch* gene (the gatekeeper for BCC development) plus the closely linked *XPA* gene.

Overall, the studies summarized above, together with reports on the cytogenetic characterization of acute myeloid leukemias in A-bomb survivors (Nakanishi and others 1999) and radiotherapy-associated solid tumors (Chauveinc and others 1997) do not provide clear evidence on the causal gene-specific mechanisms of radiation tumorigenesis. In general however, they do support a monoclonal basis for postirradiation tumor development and suggest that the characteristics of induced tumors are similar to those of spontaneously arising neoplasms of the same type. A possible exception to this is that an excess of complex chromosomal events and microsatellite sequence instability was observed in late-expressing myeloid leukemias arising in A-bomb survivors exposed to high radiation doses (Nakanishi and others 1999); these data are discussed later in this chapter.

Gene and Chromosomal Mutations in Animal Tumors

Although radiation-induced tumors from experimental animals have been available for study for many years, it is only through advances in cytogenetics, molecular biology, and mouse genetics that it has become possible to investigate early events in the tumorigenic process. The most informative data on such early events derives from studies of tumors induced in F_1 hybrid mice in which specific DNA loss events may be analyzed by loss of heterozygosity for genomically mapped polymorphic microsatellites.

Mouse Lymphoma and Leukemia

Early studies with radiation-induced thymic lymphoma provided evidence of recurrent *RAS* gene activation and some indication that the *RAS* gene mutational spectra differs between X-ray and neutron-induced lymphoma (Sloan and others 1990). Other molecular studies include the finding of recurrent chromosome (chr) 4 deletions in thymic and nonthymic lymphomas (Melendez and others 1999; Kominami and others 2002) and T-cell receptor (*Tcr*) gene rearrangements and chromosomal events in thymic lymphoma. However, the above and other somatic mutations in mouse lymphoma have yet to be specifically associated with initial radiation damage.

The situation in mouse acute myeloid leukemia (AML; Silver and others 1999) is clearer. AML-associated, region-specific deletion of chr2 has been shown by cytogenetic analysis of *in vivo* irradiated bone marrow cell populations to be a direct consequence of radiation damage; clonal pre-neoplastic growth of carrier cells has also been reported (Hayata and others 1983; Breckon and others 1991; Bouffler and others 1997). These deletions, which are characteristic of ~90% of AML induced by various radiation qualities, have been analyzed in detail, and a putative myeloid suppressor gene target was identified within a chr2 interval of ~1 centimorgan (cM; 1 centimorgan equals about 1 million base pairs; Clark and others 1996; Silver and others 1999). Site-specific breakage of chr2 is characteristic of early radiation-induced events in AML, and there are cytogenetic and molecular data that support the involvement of telomere-like repeat (TLR) sequence arrays in chr2 breakage and rearrangement at fragile sites (Finnon and others 2002). Initial hypotheses on this form of postirradiation chromosomal fragility centered on increased recombinational activity of such TLR sequence arrays (Bouffler and others 1997). However, the data of Finnon and colleagues (2002) are more consistent with a mechanism of domain-specific chromosomal rearrangement involving chromatin remodeling that is mediated by TLR-associated matrix attachment sequences.

With regard to radiation-induced osteosarcoma, Nathrath and colleagues (2002) have provided evidence for the involvement of two tumor-suppressor gene loci, but whether these loci are direct targets for radiation remains to be determined.

Mouse genetic models of tumorigenesis have also proved to be instructive about the nature of radiation-associated early events in tumor induction. In these models, the germline of the host mouse carries an autosomal deficiency in a given tumor-suppressor or gatekeeper gene, thus exposing the remaining functional (wild-type) copy to spontaneous or induced mutation and thereby tumor initiation (see "Genetic Susceptibility to Radiation-Induced Cancer"). The nature of these tumor gene-inactivating events has been studied in models of different tumor types.

In mice deficient in the *Trp53* tumor suppressor gene (*Trp53*$^{+/-}$ and *Trp53*$^{-/-}$), quantitative tumorigenesis studies implied that loss of the wild-type (wt) gene of *Trp53*$^{+/-}$ heterozygotes was a critical early event for the radiation induction of lymphoma and sarcoma (Kemp and others 1994). Molecular analysis confirmed the loss of wt *Trp53* from tumors but also showed a high frequency of concomitant duplication of mutant (m) *Trp53*—such duplication was much less frequent in spontaneous tumors (Kemp and others 1994). Subsequent cytogenetic studies showed that *Trp53*$^{+/-}$ mice were highly prone to radiation-induced whole chromosome loss and gain (aneuploidy), and that the molecular data on tumorigenesis could be explained by radiation-induced loss of the whole chromosome (chr11) bearing wt *Trp53*, with duplication of the copy bearing m*Trp53* being necessary to regain cellular genetic balance (Bouffler and others 1995). Thus, in this genetic context, *Trp53* loss and tumorigenesis were relatively high-frequency events dependent upon the cellular tolerance of aneuploidy. However a recent study poses questions about whether *Trp53* is indeed a direct target for radiation tumorigenesis in these knockout

mice (Mao and others 2004). This study has raised the hypothesis that after radiation, the wt *Trp53* gene in +/− mice activates the *Fbxw7* gene, leading to genome instability, aneuploidy, and thereby increased *Trp53* loss.

Radiation-induced intestinal tumorigenesis has been studied in F$_1$ hybrid mice of the *Apc*$^{+/-}$ genotype (Luongo and Dove 1996; van der Houven van Oordt and others 1999; Haines and others 2000). In this mouse model, DNA may be sampled from very small, early arising adenomas, thus focusing attention on early clonal events in tumor development (Levy and others 1994). Loss of wt *Apc* with the whole of the encoding chr18 is a relatively common early event in spontaneous intestinal tumorigenesis in *Apc*$^{+/-}$ mice. However, in tumors arising in low-LET-irradiated mice, the spectrum of wt *Apc* loss events was dominated by interstitial chromosome deletions. One study (Haines and others 2000) implicated a second chr18 locus in these early radiation-associated losses and also identified loss of the *Dpc4* gene as a common secondary event in spontaneous and induced tumors. In some genetic backgrounds, mammary, ovarian, and skin tumors also arise in excess in *Apc*$^{+/-}$ mice (van der Houven van Oordt and others 1999).

The same molecular genetic approach to experimental radiation tumorigenesis has been used in tumor-prone rodents that are heterozygous for the *Ptch* and *Tsc-2* tumor-suppressor genes.

Mice deficient in the patched gene (*Ptch*$^{+/-}$) are susceptible to both spontaneous and radiation-induced BCC and medulloblastoma (Hahn and others 1998; Aszterbaum and others 1999; Pazzaglia and others 2002). Of particular note are the recent data of Pazzaglia and colleagues (2002) showing that neonatal mice are highly susceptible to X-ray-induced medulloblastoma and that the predominant mutational event in these tumors is loss of *Ptch*$^+$.

Loss of *Tsc-2*$^+$ was similarly observed in many X-ray-induced renal carcinomas of *Tsc-2*$^{+/-}$ rats (Hino and others 2002), although intragenic deletions and point mutations were also observed. Importantly, the data available in this rodent genetic model (Hino and others 2002) reveal different spectra of tumor-associated *Tsc-2*$^+$ mutations in spontaneous, X-ray, and ethylnitrosourea (ENU) induced renal carcinomas, which strongly suggests that the wt gene in target kidney cells is a direct target for carcinogens. As predicted from *in vitro* studies on somatic mutagenesis (Thacker 1986), tumors induced by the powerful point mutagen ENU were not characterized by *Tsc-2*$^+$ gene loss events.

Studies with gene knockout mice are providing further evidence on the role of DNA damage response genes in determining the *in vivo* radiosensitivity of cells and tissue, together with the impact on growth or development and spontaneous tumorigenesis (Deng and Brodie 2001; Kang and others 2002; Spring and others 2002; Worgul and others 2002). It is expected that such animal genetic models will, in due course, yield more detailed information on the *in vivo* mechanisms of radiation tumorigenesis.

In summary, although studies with radiation-associated human tumors have yet to yield unambiguous data on the nature of causal gene and chromosomal mutations, animal studies are providing valuable guidance on the issue. Three principal points may be made. First, mechanistic studies on murine[1] AML, lymphoma or sarcoma in $Tp53^{+/-}$ mice, intestinal adenoma in $Apc^{+/-}$ mice, medulloblastoma in $Ptch^{+/-}$ mice, and renal carcinoma in $Tsc-2^{+/-}$ rats all argue that the induction of critical cellular events by radiation occurs early in the tumorigenic process—a conclusion that is consistent with previous judgments on the issue. Second, the cytogenetic and molecular data cited for AML and intestinal tumors provide evidence for early monoclonal development of characteristic radiation-induced pre-neoplastic changes implying an initial, single-cell target. Third, for induction of AML and intestinal, medulloblastoma, and renal tumors, the radiation-associated events are predominantly DNA losses targeting specific genomic regions harboring critical genes. This *in vivo* DNA deletion mechanism is consistent with that understood in greater detail from *in vitro* somatic mutation systems. Also, many of the radiation-associated DNA loss events recorded in tumors are of cytogenetic dimensions. It is therefore possible to draw parallels with *in vitro* data on chromosome aberration induction where the predominant importance of DNA DSB induction and postirradiation error-prone NHEJ repair has been used in this report to argue against the proposition of a low-dose threshold in the dose-response.

Evidence on the single-cell origin of radiogenic animal tumors, the *in vivo* gene or chromosomal loss mechanism for tumor initiation that appears to apply, and the close parallels that may be drawn with mechanisms and dose-response for *in vitro* induction of gene or chromosomal mutations argue in favor of a no-threshold relationship between radiation dose and *in vivo* tumor risk. In the examples cited, there is generally concordance between gene loss or mutational events recorded in spontaneous and radiation-associated tumors of a given type; although the data are more limited, such concordance tends to apply to other tumorigenic agents. A degree of gene specificity for different tumor types is also evident.

An obvious caveat to this conclusion is the degree to which these limited mechanistic data provide support for broad judgments about radiation risk at low doses. For example, the data cited on the tolerance of aneuploidy in the bone marrow of irradiated *Trp53*-deficient mice can explain the high-frequency development of lymphoma but may not be wholly relevant to other tissues and/or other genetic settings. Data discussed in the following section on the potentially powerful effects of genetic background on tumorigenic risk in irradiated mice also caution against a dogmatic approach to judgments about low-dose risk that are based

on current mechanistic knowledge. In this respect, the following section summarizes data concerning novel aspects of radiation response that may have relevance to unconventional mechanisms of radiation tumorigenesis.

RADIATION-INDUCED GENOMIC INSTABILITY IN RADIATION TUMORIGENESIS

As noted earlier in this chapter, the spontaneous development of tumors is frequently accompanied by the acquisition of genomic instability phenotypes that serve to promote the mutational evolution of more aggressive neoplastic clones. This form of genomic instability is increasingly well understood, and many of the responsible tumor gene mutations have been identified. Also noted in Chapter 2 is the large body of data showing that initial radiation-induced lesions are processed rapidly and expressed as chromosome aberrations at first postirradiation mitoses. However, during the last decade, evidence has accumulated that under certain experimental conditions, the progeny of cells surviving radiation appear to express an excess of new chromosomal and gene mutations over many postirradiation cell generations. This feature of cellular response (reviewed in Chapter 2) is generically termed radiation-induced persistent genomic instability. There are a variety of different manifestations of this phenomenon, and the developing field has been the subject of a number of recent reviews (Morgan and others 1996; Mothersill and Seymour 1998b; Wright 2000). The available data do not allow for generalizations on the onset and duration of such phenomena. On the basis of these data and previous reports of high-frequency neoplastic cell transformation (Clifton 1996), it has been suggested that epigenetic changes affecting a substantial fraction of irradiated cells can serve to destabilize their genomes and that the elevated postirradiation mutation rates in cell progeny, rather than gene-specific initial mutations, act to drive radiation tumorigenesis (Little 2000; Wright 2000). This section of the chapter focuses attention on *in vivo* studies of induced genomic instability that address the relevance of the phenomenon to radiation tumorigenesis.

Chromatid Instability in Hematopoietic Cells

Radiation-induced genomic instability in hematopoietic cells was first revealed by studies showing a persistent excess of chromatid-type aberrations in the progeny of mouse bone marrow cells irradiated *in vitro* with α-particles and subsequently grown in culture (Kadhim and others 1992). Alpha particles were considered to be substantially more effective than low-LET radiation in inducing this form of genomic instability (Wright 2000), which has also been reported in the progeny of cells that had not been traversed by an α-particle track (*i.e.*, a bystander effect for instability; Lorimore and others 1998). Posttransplantation growth *in vivo* of *in vitro* irradiated bone marrow cells was also re-

---

[1]Genus mus. A rat or mouse.

ported to result in excess chromatid instability (Watson and others 1996). However, on the basis of the data summarized below, the consequences of postirradiation chromatid instability of bone marrow cells for hematopoietic neoplasia remains somewhat doubtful.

Cytogenetic characterization of myeloid leukemia induced in the same mouse strain by α-particles, neutrons, and X-rays did not reveal evidence of the LET-dependent cytogenetic footprint of induced chromatid instability that might be expected from *in vitro* cellular studies with bone marrow cells (Bouffler and others 1996). In addition, the very high α-particle relative biological effectiveness (RBE) for induced genomic instability in bone marrow cells in culture (Kadhim and others 1992) is somewhat inconsistent with the low α-particle RBE suggested to apply to leukemogenic risk *in vivo* (Breckon and Cox 1990; UNSCEAR 2000b).

Early studies of this form of induced instability depended on *in vitro* irradiation. Studies with humans exposed *in vivo* to low- and high-LET radiation (Tawn and others 2000b; Whitehouse and Tawn 2001) have found no evidence of induced chromatid instability in hemopoietic cells. The same negative result was obtained experimentally in the CBA/H mouse strain (Bouffler and others 2001). However Watson and colleagues (2001) provided data that suggested variable expression of *in vivo* induced chromatid instability in the CBH/H mouse strain. Since CBH/H is a highly inbred strain, such variable expression of chromatid instability cannot be ascribed to genetic variation. Experimental factors may therefore be of considerable importance, and relevant to this are the data of Bouffler and colleagues (2001), which indicate the existence of confounding stress factors that may account for *in vitro* and *in vivo* differences in the apparent expression of such instability.

These *in vivo* observations cast considerable doubt on the relevance of radiation-induced chromatid instability for risk of lymphohematopoietic tumors. This view is strengthened by studies showing that the genetic determinants of induced chromatid instability in mouse bone marrow cells differ from those of susceptibility to induced lymphohematopoietic neoplasia (Boulton and others 2001). A similar degree of doubt has been expressed following reanalysis of genomic instability data (Nakanishi and others 1999, 2001) relating to myeloid leukemia arising in A-bomb survivors (Cox and Edwards 2002; Little 2002).

Chromatid Instability in Mouse Mammary Epithelial Cells

Differences in radiosensitivity and susceptibility to radiation induction of specific tumors among inbred mouse strains are well recognized, and there is good evidence that the BALB/c mouse is unusually sensitive to the induction of tissue injury and mammary tumors (Roderick 1963; Storer and others 1988); on these criteria the C57BL/6 mouse falls into the radioresistant category. Initial cytogenetic studies showed that mammary epithelial cells cultured from irradi-

ated BALB/c mice persistently expressed substantially more chromatid aberrations during passage than those derived from irradiated C57BL/6 animals (Ponnaiya and others 1997; Ullrich and Ponnaiya 1998). In follow-up investigations, the chromatid instability phenotype of BALB/c was shown to be associated with a partial deficiency in the NHEJ repair protein DNA-dependent protein kinase catalytic subunit (DNA PKcs) together with compromised postirradiation DNA DSB repair (Okayasu and others 2000). This study, which included an intercomparison of inbred mouse strains, showed deficiency of DNA-PKcs and DNA DSB repair to be restricted to BALB/c suggesting genetic associations with persistent genomic instability and mammary tumor susceptibility. In accord with this, molecular genetic analyses showed BALB/c to carry a rare variant form of the gene (*Prkdc*) encoding DNA-PKcs, and subsequent analysis of recombinant mice provided strong evidence that variant *Prkdc* directly determined DNA-PKcs deficiency and postirradiation chromatid instability in mammary epithelial cells (Yu and others 2001). On the basis of these data it was proposed that induced genomic instability and mammary tumor susceptibility were genetically codetermined. Importantly, these investigations provide genetic evidence that deficiencies in the repair of DNA DSB, rather than as-yet-undefined epigenetic phenomena, are likely to determine persistent chromatid instability in this mouse. The question as to whether such instability is a primary causal element in mammary tumorigenesis or a secondary *in vitro* consequence of DNA repair deficiency and clonal growth selection remains to be resolved.

Recent studies have also suggested a linkage between DNA-PKcs and maintenance of functional telomeres (Bailey and others 2004a, 2004b). As noted elsewhere in this report, the products of telomere dysfunction are dicentric chromosomes created by end-to-end fusion and sister-chromatid fusions, both of which can be associated with breakage-fusion-bridge cycles. More recently, a second product of telomere dysfunction, fusions between telomeres and the ends of broken DNA strands (*i.e.*, DNA DSBs), have been described. Since telomere-DSB fusions have properties that differ from both chromosomal end fusions and ordinary chromosome aberrations, such fusions offer a potentially important new mechanism for induction of instability. These fusions appear to occur only under conditions of telomere dysfunction resulting from defects in the NHEJ pathway (Bailey and others 1999; Mills and others 2004). This suggests that genomic instability as a mechanism in radiation-induced cancer may be limited to specific circumstances in which individuals harbor specific DNA-repair deficiencies.

Telomere-Associated Persistent Chromosomal Instability

Telomeric repeat sequences (Bertoni and others 1994) cap the ends of mammalian chromosomes and serve to protect against replicative erosion and chromosomal fusion; in normal human cells in culture, telomere shortening and instability is a natural feature of replicative cell senescence (Harley and Villeponteau 1995; Bacchetti 1996). In often degenerate forms, telomeric repeats are also found in subtelomeric and interstitial chromosomal locations, and there is some evidence that these loci may act as sites at which radiation-induced and other forms of genomic damage are preferentially resolved (Bouffler 1998).

Early studies of the postirradiation development of chromosomal instability in *in vitro* passaged human diploid fibroblasts were suggestive of instability effects in a high proportion of irradiated cells (Sabatier and others 1992). However, subsequent detailed cytogenetic analyses suggested that passage-dependent instability in cultured human fibroblasts primarily takes the form of telomeric events expressed in cell clones naturally selected by growth rate during passage (Ducray and others 1999). Overall, the data obtained may be interpreted as initial radiation exposure bringing forward in time the natural process of clonal telomeric sequence instability associated with cell senescence and telomere shortening.

A different form of postirradiation telomere-associated instability is expressed in a hamster-human hybrid cell system (Marder and Morgan 1993) where, in some clones, chromosomal instability is persistently expressed at translocations that have telomeric sequences at their junction (Day and others 1998). Similarly, unstable structures have been observed in unirradiated hamster cells undergoing gene amplification (Bertoni and others 1994), and again it may be that radiation is inducing genomic structures that enhance the natural expression of instability.

There is good evidence that telomeric sequence instability is a recurrent feature of tumorigenic development (Bacchetti 1996; Chang and others 2001; Murnane and Sabatier 2004). Of particular relevance to the question of unstable translocation junctions are the so-called segmental jumping translocations that have been well characterized in spontaneously arising human leukemias (Shippey and others 1990). In respect of radiation tumorigenesis, detailed cytogenetic analyses suggest an excess of complex aberrations and segmental jumping translocations in myeloid leukemias arising at old ages in high-dose-exposed atomic bomb survivors (Nakanishi and others 1999). These and other data on excess microsatellite instability in A-bomb myeloid leukemias (Nakanishi and others 2001) have been reanalyzed in respect of dose and probability of tumor causation (Cox and Edwards 2002; Little 2002). These reanalyses largely uncouple the expression of leukemia-associated jumping translocations and microsatellite instability from radiation causation and argue that the potential contribution of induced instability to leukemogenic risk is likely to be small.

Telomeric sequence instability at radiation-associated deletion or translocation breakpoints in mouse myeloid leukemia has also been recorded; this is not a general characteristic of these tumor-associated events, and recent studies argue against the direct involvement of telomeric sequence instability in these events (Bouffler and others 1996; Finnon and others 2002).

In conclusion, although the position regarding radiation-induced persistent genomic instability and its causal association with tumorigenesis is not well understood, a few specific points can be made:

1. In the case of radiation-associated persistent telomeric rearrangement and unstable chromosome translocation junctions, a coherent case can be made that a certain fraction of misrepaired genomic damage after radiation may be prone to ongoing secondary change in clonal progeny. There is evidence that such secondary genomic rearrangement can be a normal component of tumor development, in which case it is reasonable to assume that excess instability of this type could be a feature of some radiation-associated tumors, particularly those arising after high-dose irradiation where multiple or complex rearrangements may be expected.

2. The genetic evidence from mouse studies that postirradiation chromatid instability can be associated with mammary tumor development is also persuasive, although it leaves unanswered questions on the causal role of the excess chromatid damage observed *in vitro*. Thus, in certain genetic settings of DNA repair deficiency, a role for postirradiation chromatid instability in tumorigenesis appears reasonable, and the potential linkage with telomere dysfunction could also be important.

3. Based on the negative or inconsistent data on *in vivo* induced genomic instability in bone marrow cells, the non-sharing of genetic determinants, and the contention on data regarding A-bomb leukemias, induced genomic instability is judged unlikely to impact appreciably on the risk of lymphohematopoietic tumors after low-dose radiation.

There are very few data on radiation-associated human solid tumors from which to assess the potential contribution of induced genomic instability. The central problem is the inherent difficulty in distinguishing this specific radiation-induced phenotype from spontaneously developing genomic instability as a natural consequence of clonal selection during tumor development. Stated simply, does tumor instability correlate with initial radiation damage or with neoplastic phenotype?

This problem is well evidenced by molecular studies on post-Chernobyl (Belarus) childhood thyroid cancer. Initial studies showed evidence of excess microsatellite alterations in these radiation-associated tumors when compared with a reference group of adult thyroid cancers (Richter and others 1999). However, more detailed follow-up studies showed that the principal correlation was between microsatellite alterations and the aggression of early arising tumors. When this factor was taken into account, microsatellite loss or mutation in the early Belarus tumors was shown to be similar to that of the adult reference cases (Lohrer and others 2001).

Based on consideration of the available *in vivo* data it is concluded that, at present, only a weak scientific case can be made for a discernible impact of induced genomic instability on radiation cancer risk. This conclusion is strengthened when account is also taken of the uncertainties noted in Chapter 2 regarding the biological basis and generality of the expression of induced genomic instability in cultured mammalian cells.

QUANTITATIVE STUDIES IN EXPERIMENTAL TUMORIGENESIS

General Aspects of Dose-Response

The preceding discussion of potential mechanisms for radiation-induced cancer has indicated an important role for radiation-induced DNA DSBs, damage response pathways, and gene or chromosomal mutations in the initial events leading to cancer development. On this basis it would be predicted that the form of the dose-response for radiation-induced cancer and the effects of fractionation or reduced dose rate on this dose-response would be compatible with such underlying mechanisms unless factors involved in the expression of initiated cells are limiting in neoplastic development. Such a mechanistic model provides specific predictions with respect to dose-response and time-dose relationships for initial events and provides a framework for prediction of dose-response and time-dose effects for radiation-induced cancer (Ullrich and others 1987). Animal studies can be used to test these predictions. This framework is based on the $\alpha D + \beta D^2$ dose-response model for chromosome aberration induction described in Chapter 2. For single acute exposures the dose-response would be predicted to follow this model such that at low doses the relationship between cancer incidence and dose would be linear, while at higher doses this relationship would follow a function more closely related to the square of the dose. It is unlikely from a statistical standpoint alone that such a function could be proven to hold to the exclusion of all other dose-response models for any set of experimental data.

Because of this, time-dose studies using both fractionated and low-dose-rate exposure regimens are important components in testing mechanistic predictions. On the basis of this model, it would be predicted that the dose-response following low-dose-rate exposures would be linear, with the same slope as the linear portion of the acute dose response model. In other words, at low doses the risk of radiation-induced cancer is independent of the time over which exposure occurs and is a cumulative function of dose. Fractionated exposures can further test these time-dose relationships and also provide information on the kinetics of processes involved. Such kinetic information, while limited, can provide insight into the nature of cellular versus tissue effects as major components in cancer risks in the specific experimental model under study.

Any critical analysis of quantitative data on radiation-induced cancer requires informed selection of data sets. First, the adequacy of a study with respect to statistical power and use of appropriate analytical methodology must be considered. Second, biological factors involved in the pathogenesis of specific neoplasms must be considered with respect to the applicability of the experimental model to carcinogenesis in general and to cancer risk in humans in particular. Given these caveats, there are relatively few studies on animal carcinogenesis where the data are sufficient to address the issue of dose-response relationships or the issue of dose-rate effects and/or fractionation effects. Those studies in which such analyses are possible are limited mainly to rodent studies, principally mice. Biological factors in neoplastic development must also be noted.

As discussed later in this chapter genetic background has a major role in determining neoplastic development at the level of sensitivity to both initiating events and events involved in expression. Therefore even in mouse studies in which there is sufficient statistical power to address questions of low-dose effects and time-dose relationships, the data are limited to mouse strains that are highly susceptible to specific forms of neoplasias. While variations in susceptibility must be considered potential confounding factors in applying animal data to human risks, careful analyses of human and animal data suggest that animal data do in fact have predictive value—for example, they can guide judgments on the choice of cancer risk models (Carnes and others 1998; Storer and others 1988). On the other hand, there are specific murine neoplasms whose pathogenesis appears to be unique to the mouse. In these specific instances it is unlikely that data derived using these systems would be applicable to human risks. These neoplasms are identified in sections below.

Specific Murine Neoplasms

Leukemia and Lymphoma

The induction of leukemia and lymphoma has been examined in a number of murine systems, but the most extensive quantitative data on both dose-effects and time-dose relationships are for myeloid leukemia and thymic lymphoma. The most comprehensive data for myeloid leukemia with respect to dose-response relationships, and fractionation and dose-rate effects are in CBA male mice and RFM male mice (Upton and others 1970; Mole and Major 1983; Mole and others 1983). Interestingly, susceptibility in female mice of the same strains is markedly lower. The CBA mouse has also been used as an important model to dissect underlying radiation-induced molecular events described earlier (Bouffler and others 1991; Clark and others 1996; Silver and others 1999). For both strains, studies have been conducted over the dose range 250–3000 mGy (Upton and others 1970; Mole and Major 1983; Mole and others 1983). Analyses of data sets from both strains have yielded similar conclusions. Briefly, a number of dose-response models were seen to describe the data sets adequately. Data on incidence as a function of dose for both strains could be described by quadratic, linear-quadratic, and simple linear dose-responses with insufficient statistical power to exclude any of these three models on the basis of acute exposure data alone. Fractionation of the dose or low-dose-rate exposures resulted in a linear dose-response consistent with expectations of radiobiological theory in which the dose-response is linear quadratic for acute exposures and linear for low-dose-rate exposures, with the linear slope of the linear quadratic predicting the low-dose-rate and fractionation responses. These results are compatible with the apparent role of alterations in chromosome 2 in initial events for murine myeloid leukemogenesis and consistent with mechanistic predictions of dose and time-dose relationships described previously.

This is not the case for studies on thymic lymphoma. In contrast to myelogenous leukemia, for which male mice are the most sensitive, female RFM mice are significantly more sensitive to the induction of thymic lymphoma following radiation exposures (Ullrich and Storer 1979a). For radiation-induced thymic lymphoma in female RFM mice, the data suggest a more complex relationship between radiation exposure and neoplastic development. Following single acute exposures over the 100–3000 mGy dose range, no simple dose-response model was found to describe the data (Ullrich and Storer 1979a). Low-dose-rate exposures, although significantly less effective with respect to induction of thymic lymphoma than single acute exposures, still resulted in a complex dose-response with a clear suggestion of a large threshold (Ullrich and Storer 1979c). These results should not be unexpected since the development of thymic lymphoma in mice following irradiation is an extremely complex process. The target cells for induction of thymic lymphoma are thought to be in the bone marrow rather than the thymus, and the pathogenesis of the disease appears to be largely mediated through indirect mechanisms with cell killing playing a major role (Kaplan 1964, 1967; Haran-ghera 1976). For example, the expression of thymic lymphoma can be substantially reduced or eliminated by protection of bone marrow stem cells from radiation-induced cell killing. The complex nature of the pathogenesis of this disease and the lack of a comparable counterpart in humans argues against thymic lymphoma as an appropriate model for understanding dose-response and time-dose relationships in humans.

Solid Tumors

Data from experimental studies examining dose-response and time-dose relationships are also available for a limited number of solid cancers in female RFM and BALB/c mice, including pituitary, Harderian gland, lung, and breast cancers (Ullrich and Storer 1979b, 1979c; Ullrich 1983). In a large study examining dose and dose-rate effects in female

RFM mice, increased incidences of pituitary and Harderian gland tumors were reported. In spite of the large numbers of animals used, analyses of the data with respect to dose-response models could not distinguish between linear and linear-quadratic models (Ullrich and Storer 1979b).

However, when the data for low-dose-rate exposures were considered as well, they were most compatible with a linear-quadratic model (Ullrich and Storer 1979c). Importantly, with respect to low-dose effects, these data support a linear response at low doses that is independent of exposure time. Such a response is consistent with predictions of the mechanistic model outlined earlier in this chapter. Although the number of animals used was smaller, a study examining radiation-induced lung and mammary adenocarcinomas in female Balb/c mice reached similar conclusions with respect to dose-response functions and low-dose risks (Ullrich and Storer 1979c; Ullrich 1983). This model was tested further in a series of experiments comparing the effectiveness of single acute exposures, acute fractionated exposures, and low-dose-rate exposures on the induction of lung and mammary tumors in the Balb/c mouse (Ullrich and others 1987). Importantly, in this study the hypothesis of time independence of effects at low doses was critically tested and found to hold. Specifically, similar effects were observed whether the same total dose was delivered as acute low-dose fractions or as low-dose-rate exposures.

While the data for solid tumors described above are compatible with mechanistic models detailed earlier in this chapter, there are data sets that do not support a linear-quadratic dose-response model. Extensive data for mammary cancer induction in the Sprague-Dawley rat appear more consistent with a linear model over a wide range of doses and with linear, time-independent effects at low doses, low-dose fractions, and low dose rates (Shellabarger and others 1980). Although questions have been raised about the applicability of this model system to radiation-induced breast cancer in humans, much of the data from this rat model, from the mouse model in Balb/c mice, and from epidemiologic studies in exposed human populations appear to be consistent with respect to low-dose risk functions (Preston and others 2002b).

In contrast to the data for leukemia and for pituitary, Harderian gland, lung, and mammary cancer described above, data from studies examining radiation-induced ovarian cancer in mice and bone and skin cancer in various animal species are more compatible with threshold dose-response models. In each instance it appears that an important role for cell killing in the process of neoplastic development and progression may explain these observations.

Analysis of the dose-response for radiation-induced ovarian tumors following single acute or low-dose-rate exposures in RFM female mice indicated a marked sensitivity to induction at relatively low radiation doses, but equally importantly the analysis of the data strongly supported a threshold dose-response model (Ullrich and Storer 1979b, 1979c).

In fact, this is one of the few instances for which a linear relationship could be rejected statistically. Studies in other mouse strains, while having less statistical power, also suggest a high sensitivity to induction of ovarian tumors at relatively low doses but with an apparent threshold (Lorenz and others 1947; Ullrich and Storer 1979c). This relatively unusual dose-response combining a threshold with high sensitivity to induction is unique to the mouse. Ovarian cancer in the mouse appears to involve an indirect mechanism for induction involving oocyte cell killing and subsequent alterations in the pituitary ovarian hormonal interactions (Kaplan 1950; Foulds 1975; Bonser and Jull 1977). The hormonal alterations are the proximate cause of tumor formation, with the role of radiation being relatively indirect as a result of its cell-killing effects. Because mouse oocytes are uniquely sensitive to the killing effects of radiation (the LD50 [lethal dose—50%] is ~50 mGy), ovarian tumors occur at very high frequencies following relatively low doses of ionizing radiation (Ullrich and Storer 1979c). A threshold appears to exist because a certain level of oocyte killing is required to cause the hormonal alterations that result in ovarian tumor formation. The principal effect of lowering the dose rate is to increase the threshold. In the RFM mouse, estimates of thresholds were reported as 110 mGy for acute exposures and 700 mGy for low-dose-rate exposures (Ullrich and Storer 1979b, 1979c). In contrast to the mouse, oocytes in humans are relatively resistant, with an LD50 of several grays. This difference in sensitivity is apparently because mouse and human oocytes are at different stages of differentiation in the ovary (Brewen and others 1976). The unique sensitivity of the mouse ovary to radiation makes it unlikely that results using this model system would have general applicability to risks in humans.

Radiation-induced skin cancer has been studied in both mice and rats, although the majority of such studies have focused on the rat model because the rat is significantly more sensitive to skin tumor induction than the mouse (Burns and others 1973, 1975, 1989a, 1989b). In both rats and mice, relatively high total doses are required to induce skin cancer, and there is a clear threshold below which no tumors are seen. Multiple repeated radiation exposures are generally required for tumors to develop in mouse skin, while a single high dose (>10 Gy) is capable of inducing tumors in rat skin. It was for skin tumorigenesis that many of the concepts of multistage carcinogenesis were developed, including concepts related to initiation, promotion, and progression, and it is within this framework that the data for radiation-induced skin tumors are best considered (Jaffe and Bowden 1986; Burns and others 1989b). It appears from a variety of studies that single doses of ionizing radiation are capable of initiating cells with neoplastic potential, but that these cells require subsequent promotion in order to develop into tumors (Hoshino and Tanooka 1975; Yokoro and others 1977; Jaffe and Bowden 1986). Without this promotion these latent initiated cells will not express their neoplastic potential.

Several lines of evidence support this view. Hoshino and Tanooka have demonstrated that small doses of beta irradiation are capable of inducing initiating alterations in mouse skin that required subsequent promotion with 4-nitroquinoline N-oxide (4NQO) for tumors to develop. Jaffe and Bowden (1986) have demonstrated the initiating potential of single doses of electrons when followed by multiple exposures to the tumor-promoting agent TPA (12-O-tetradecanoylphorbol-13-acetate). Fry and his coworkers (1986) have shown that X-ray-initiated cells can be promoted to develop skin tumors by exposure to ultraviolet light. This group has demonstrated further that the apparent threshold dose-response for skin tumorigenesis can be converted to a linear UVR dose-response when promotion is used to maximize the expression of latent initiated cells.

Based on such observations it is logical to speculate that the multiple high-dose fractions of radiation that are generally required to induce skin tumors in mouse skin are acting not only to initiate cells but also to induce tissue damage via cell killing, which in turn acts as a promoting stimulus to facilitate the progression of these initiated cells into skin tumors. Likewise in the rat, the high doses required to produce tumors are likely to produce both transformation of cells and sufficient cell killing to promote the transformed cells. This phenomenon does not appear to be unique to these animal systems. Most evidence suggests that relatively high doses of radiation are necessary to induce skin tumors in humans and that these effects can be enhanced by exposure to UV light from the Sun (Shore 2001). It is also important to note studies by Jaffe and Bowden demonstrating that multiple low doses of radiation to the skin that did not produce tissue damage were not effective in promoting skin tumors initiated by chemical agents (Jaffe and Bowden 1986). These data support the view that the predominant role for low-dose radiation is tumorigenic initiation.

Studies of bone cancer also suggest a threshold response and a requirement for prolonged exposure for tumor development from exposure to low-LET radiation (NCRP 1990). Unfortunately most of the available data have focused on observations of effects rather than dissecting potential underlying mechanisms. Attempts have been made to model bone tumorigenesis however, and these models have again focused on an important role for a mechanism involved in the expression of initiated cells in controlling tumor development (Marshall and Groer 1977). Although speculative, it is likely that mechanisms similar to those proposed for skin tumorigenesis involving the cell-killing effects of radiation are likely involved in producing a threshold response for bone tumors.

Fractionation Kinetics

Studies using fractionation regimens have been useful in addressing issues of time-dose relationships in radiation carcinogenesis. In a few instances, investigators have also used this approach to examine the kinetics of repair of carcinogenic injury. Studies have been conducted examining repair kinetics associated with skin tumorigenesis following localized irradiation of rat skin by Burns and coworkers (1975). In the mouse, repair kinetics were determined by examining tumor development in the mouse ovary and mouse lung following whole-body irradiation (Yuhas 1974; Ullrich 1984; Ullrich and others 1987). The experimental design for these studies has been to compare tumorigenic effects following a single acute exposure with the effects after a similar total dose split into two equal fractions separated in time by hours or days. When there is interaction between the two doses the tumorigenic effectiveness would be predicted to approximate that for the single acute exposure, while if there is recovery from carcinogenic injury, the effectiveness of the split doses would be lower. A simple approach to determining whether cellular-based or tissue-based factors play a limiting role in radiation tumorigenesis is to compare a 24 h fractionation scheme with that in which the time between fractions is much longer and more compatible with tissue kinetics. A convenient time to use has been 30 d between fractions. Not surprisingly, considering the role of cell killing in its pathogenesis, studies examining radiation-induced ovarian tumorigenesis have indicated a recovery time between fractions of 24 h or less (Yuhas 1974). Likewise data for skin tumorigenesis in the rat, for which cell-killing effects appear to play a role in neoplastic development, a recovery time of approximately 4 h has been reported (Burns and others 1973, 1975).

More interesting are data for the induction of lung adenocarcinomas in Balb/c mice (Ullrich and others 1987). Cell killing has not been seen to play a major role in the pathogenesis of this tumor, and the doses used in the fractionation studies are not in the range where cell killing would be likely to produce significant tissue damage (Meyer and others 1980; Meyer and Ullrich 1981). A comparison of the tumorigenic effects of two 1 Gy fractions separated by either 24 h or 30 d with that for a single dose of 2 Gy indicated full recovery by 24 h with no further reduction in tumorigenic effectiveness when the time between fractions was increased to 30 d. Additional studies compared the lung tumorigenic effects produced at a total dose of 2 Gy delivered as a single acute exposure to those of multiple 100 mGy fractions separated by 24 h as well as to continuous low-intensity exposures delivered at a dose rate of 4 mGy/h. The observation of a similar reduction in lung adenocarcinomas following both the low-dose-rate and the fractionated exposure regimens also provides support for recovery kinetics in the range of 24 h or less.

Postirradiation Persistence of Initiated Cells

While fractionation studies suggest that tissues can recover from radiation-induced carcinogenic injury and that this recovery is likely based on kinetics associated with re-

pair of DNA and chromosomal-type damage, another important question is the persistence of radiation-initiated cells once the initial damage has been produced. Two studies using different experimental systems have addressed this issue. Hoshino and Tanooka (1975) examined the persistence of latent carcinogenic damage in irradiated mouse skin. In this study they gave a dose of irradiation that by itself would not result in the development of skin tumors and followed this with promotion using 4NQO over intervals from 11 to 400 days after irradiation. Importantly, they found that radiation-initiated cells could persist as latent carcinogenic damage for up to 400 d. Yokoro and his coworkers (1977), in studies examining the interaction of radiation and hormones in breast cancer development, found that latent radiation carcinogenic damage could be produced in rat mammary glands by a single low dose of radiation and that the expression of this damage could be enhanced by subsequent stimulation with prolactin. As in the Hoshino and Tanooka study, the latent radiation-initiated cells were found to persist for a substantial portion of the rat's lifetime.

Radiation-Induced Life Shortening

It has been known for decades that radiation reduces the life span of animals, and studies in mice and dogs have been conducted using life-span shortening as a means to quantify radiation effects (NCRP 1980; Storer and others 1982; Carnes and Fritz 1991; Carnes and others 2002, 2003). The rationale for such studies has been that life shortening, although a complex end point, can serve as an integrated measure of the deleterious effects of radiation. The degree of life shortening from a specific radiation dose can vary as a function of strain, species, gender, and physiological status of the animals (Storer and others 1979, 1982; Korshurnikova and Shilnikova 1996). This variation is largely a function of the spectrum of spontaneous and induced disease and the age distribution of disease occurrence. For example, a great degree of life shortening is observed in animals susceptible to the induction of thymic lymphoma or myelogenous leukemia, both of which occur relatively early following exposure to ionizing radiation (Storer and others 1979, 1982; Storer and Ullrich 1983).

In contrast, in animals that are not susceptible to such early developing neoplasms, but rather develop late-occurring solid tumors following radiation exposure, substantially less life shortening is observed at the same radiation dose. Regardless of the degree of life shortening observed however, analyses of experimental studies indicate that at low doses of radiation and for radiation delivered at low dose rates, such life shortening is due almost entirely to radiation-induced cancer (Storer and others 1979, 1982; Carnes and others 2002, 2003).

Single acute doses in the range of 500 mGy and higher increased life shortening attributable to nonneoplastic effects, but at lower doses and for a wide range of doses delivered at low dose rates, this nonneoplastic component of life shortening has not been observed (Ullrich and Storer 1979a). A few instances have been reported of apparent radiation-induced life lengthening following exposure to low levels of single or protracted doses of radiation (NCRP 1980). Statistical analyses of the distribution of deaths in these studies indicate that control animals usually show a greater variance around the mean survival time than the groups exposed to low doses of radiation (NCRP 1980). In addition, the longer-living irradiated animals generally have a reduced rate of intercurrent mortality from nonspecific and infectious diseases during their early adult life, followed by a higher mortality rate later in life (NCRP 1980). Since these studies were conducted under conditions in which infectious diseases made a significant contribution to overall mortality, the interpretation of these studies with respect to radiation-induced cancer or other chronic diseases must be viewed with caution.

Experiments designed to address questions of low-dose risk using life shortening have used two different experimental approaches (NCRP 1980). One approach has been to deliver radiation doses at different dose rates over the entire life span of the animals. A second approach has been to develop dose-response relationships following acute, fractionated, and low-dose-rate exposures delivered as defined radiation doses. In such studies, a range of radiation doses have been delivered, generally to young adult animals. In the case of fractionated or low-dose-rate exposure regimens, the exposures were terminated at specific total doses delivered over a well-defined fraction of their life span.

For purposes of understanding risks from low-dose-rate exposures, it is important to make a clear distinction between dose-rate effects (which involve terminated exposures) and protraction effects (which involve radiation exposures over the entire life span). With few exceptions, dose-response relationships derived from life-shortening data following single acute radiation doses, fractionated exposures, and terminated low-dose-rate exposures all suggest linear dose-responses over wide range of doses (NCRP 1980). This apparent linearity in the dose-response for life shortening may reflect the integration of a variety of tumor types whose individual dose-responses may vary widely.

The exceptions are generally related to instances in which a single tumor type is the principal cause of death following radiation exposure. The primary effect of fractionating the radiation dose or reducing the rate at which the dose is delivered is to reduce the slope of the linear response.

Importantly, experiments using multiple, low-dose-rate, terminated exposures suggest a limiting linear slope in all cases (Storer and others 1979; NCRP 1980; Carnes and others 1989). Once this limiting linear response is reached, no further reduction in effect is seen if dose rate is reduced further. However, for protracted exposures that involve irradiation over the entire life span, a further reduction in life shortening per unit dose has been observed (NCRP 1980). This further reduction in slope has been attributed to so-called wasted radiation. According to this concept, radiation injury induced late in life does not have sufficient time to express itself, thereby reducing the slope of any dose-effect relationship.

In fact, both dose-rate effects and protraction effects are more complicated than they appear at first glance. Analysis of cause of death and tumor incidence data indicates that reducing the rate at which a radiation dose is delivered reduces the frequency of radiation-induced tumors and alters the spectrum of neoplastic disease (Storer and others 1979; NCRP 1980). First, the frequencies of early appearing radiation-induced neoplasms such as leukemia and lymphoma are reduced. This effect alone has a major impact on life shortening by switching the spectrum of disease to more late-occurring solid cancers. Second, a reduction in the frequency of late-appearing tumors when compared to animals receiving a single acute exposure is also observed. Depending on the exposure regimen, this effect on solid tumor frequencies may be a result of dose-rate effects in the case of terminated exposures, as well as a protraction effect in the case of lifetime exposures. This duality of effect tends to amplify dose-rate or protraction effects seen for individual tumors. Regardless of the fine structure of dose-rate and protraction effects, it is important to note that all of the data support a linear dose-response for radiation-induced life shortening at low doses and low dose rates over a wide range of doses.

Determining Dose and Dose-Rate Effectiveness Factors from Animal Studies

Application of the linear-quadratic dose dependence, $\alpha D + \beta D^2$, and a wide range of molecular, cellular, and animal data have been used to argue that data on radiation-induced cancer in human populations derived from studies following acute radiation exposures tend to overestimate radiation risks at low doses and low dose rates. In this regard, analyses of the animal studies examining dose-response and dose-rate effects described above have been particularly important. In an attempt to quantify the degree to which extrapolation of acute high-dose data might overestimate risks at low doses and low dose rates, a number of groups have used a similar approach. The approach taken has been relatively simple. Essentially, the effectiveness per unit dose for acute exposures has been determined using a linear interpolation of data in the 2–3 Gy dose range and control data at 0 Gy. The rationale for using only the high-dose data and not data at lower doses was based on the assumption that this would simulate analyses of risks from epidemiologic studies where most of the available data were for single acute exposures at relatively high doses. Except in instances where threshold dose-responses were observed, effects per unit dose following low-dose-rate exposures were derived by calculating the slope of the entire dose-response (not just in the 2–3 Gy dose range).

By dividing the tumorigenic effectiveness per unit dose of acute exposures using the high-dose data and the low-dose-rate exposures, effectiveness ratios were obtained. These ratios have been termed dose and dose-rate effectiveness factors. Since the data from which these ratios are obtained result from comparing high- and low-dose-rate effects, these ratios are literally dose-rate effectiveness factors (DREF). However, since the actual dose-response for most radiation-induced tumors following single acute exposures was found to be linear quadratic, it can be seen from Figure 10-1 that this procedure would tend to overestimate effects for low single acute radiation doses (in the dose range where the response is predominantly linear) as well as for low-dose-rate exposures over a wide range of total doses. Since the ratio should be equally valid for estimating effects at low dose rates (the DREF) and for low single doses, the term dose and dose rate effectiveness factor (DDREF) has commonly been used. This would not be the case if the dose-response following acute exposures is not linear quadratic.

The derivation and application of DDREF must be performed with caution. Tumors for which there is mechanistic knowledge that they are unlikely to be applicable to radiation carcinogenesis in human populations should not be considered. On this basis, quantitative data on dose-rate effects for thymic lymphomas and for ovarian tumors, which have been shown to be highly sensitive to dose-rate effects, should not be used. Likewise, caution should also be exercised when considering data for the induction of pituitary tumors in RFM female mice because of potential effects associated with the sensitivity of the mouse ovary and the subsequent disruption of pituitary and ovarian hormone functions. This leaves a limited data set upon which to base DDREF calculations, which includes data for myeloid leukemia and a few solid tumors including Harderian gland (for which there is no comparable tissue in humans), lung adenocarcinomas, and mammary tumors. Data for myeloid leukemia are available for two mouse strains and from at least three independent studies. All of the data support a reduced effect when comparing high- and low-dose-rate exposures over the 0–3 Gy dose range. Calculation of DDREF values using the procedures described above yields estimates on the order of 2 to 6, with most values in the range of 4–5. For lung adenocarcinomas and Harderian gland tumors, DDREF values of approximately 3 have been calculated over the 0–2 Gy dose range. For mammary tumors, all of the data suggest a DDREF value of less than 2 and closer to a value of 1 when effects of high-dose-rate and low-dose-rate exposures are compared in this 0–2 Gy dose range. Thus, it appears that myeloid leukemia is probably more sensitive to dose-rate effects than are solid tumors.

It should also be pointed out that these values are based on extrapolation of data from acute doses of 2–3 Gy and that extrapolating data from lower doses would result in lower estimates. The impact of dose range must be considered when applying DDREF factors to human risk estimates for which there are good data at and below 1 Gy. Chapters 10 and 12 describe the use of animal data in developing a specific judgment on the value of DDREF to be used in BEIR VII cancer risk estimates.

Adaptive Responses

Human and animal data relating to adaptive responses to radiation and cancer risk have been reviewed by UNSCEAR (1994). That review concluded that the presence of an adaptive response for cancer risk was not readily evident from the results of animal studies and that, for reasons of statistical power, no clear statements were possible from epidemiologic investigations. Since 1994 a number of further animal studies have reported evidence suggestive of some form of adaptive response in the development of certain tumors.

Ishii and colleagues (1996) reported a decreased incidence of thymic lymphoma in AKR mice following chronic fractionated low doses of X-rays. As described in this chapter, the atypical involvement of cell killing in the etiology of murine thymic lymphoma makes interpretation of all data for this tumor type most difficult. On this basis, no great weight can be placed on the data of Ishii and others (1996). Of potentially greater relevance are the adaptive response data on the induction of AML in CBA mice and the development of osteosarcoma or lymphoma in *Trp* 53-deficient mice.

In studies with CBA mice (Mitchel and others 1999), prior exposure to low-dose-rate radiation was shown to change the tumorigenic response of animals receiving a second dose at a higher dose rate delivered one day later. Somewhat surprisingly, the principal effect of the priming dose was not to reduce the lifetime risk of AML but rather to increase tumor latency. Similar delaying effects on tumor latency but not lifetime risk of a low (10 mGy) acute priming dose were subsequently reported for spontaneous development of osteosarcoma and lymphoma in *Trp 53* heterozygotes. The effects of a 100 mGy priming dose differed for osteosarcoma (decreased latency) and lymphoma (increased latency), a result that is suggestive of a mechanism that is dependent on dose and tumor type. These studies are difficult to interpret, particularly since the priming dose appears to influence tumor development rather than initiation.

This result runs counter to expectations from cellular data on adaptive responses (see Chapter 2), which emphasize the potential importance of adaptive DNA damage response processes. To explain the apparent effects of a priming dose on tumor latency it would be necessary to postulate the existence of low-dose-induced physiological signals that have a lifetime of many months. Mitchel and others (2003) suggest that these signals might act via the inhibition of genomic instability, which would then tend to slow tumor development. However given the great uncertainties on the *in vivo* activity of radiation-associated genomic instability already noted in this chapter, the adaptive mechanism suggested by Mitchel and others (2003) is regarded as being highly speculative.

In summary, while these more recent data on adaptive responses for radiation-induced tumorigenesis may act as a focus for further research, they do not provide coherent evidence of the generality of this mechanism and its importance for judgments on low-dose cancer risk.

GENETIC SUSCEPTIBILITY TO RADIATION-INDUCED CANCER

It has been known for many years that there are individuals and families within human populations who carry heritable mutations that can increase their lifetime probability of spontaneously developing cancer. Indeed, family pedigrees providing evidence of strongly expressing predisposition, particularly to colon carcinoma, were published in the early part of the 1900s, but it was not until the development of molecular genetic techniques in the 1970s that the whole field of human cancer genetics began its rapid development.

The primary mechanistic association between heritable cancer in humans and exposure to an environmental carcinogen was made in the late 1960s when Cleaver (1968) demonstrated an excess of skin cancer in sun-exposed DNA, repair-deficient xeroderma pigmentosum (XP) patients (*i.e.*, there was likely to be a direct association between heritable DNA repair or damage response capacity and cancer development). Since the 1970s the generality of this crucial association has been much more firmly established by a combination of clinical, epidemiologic, and molecular genetic approaches. These developments have included the elucidation of two rare human genetic disorders of cancer, ataxia-telangiectasia (AT; Easton 1994) and Nijmegen breakage syndrome (NBS), in which the DNA damage response defects concern the form of DNA damage (Brenner and Ward 1995) critical for cellular response to ionizing radiation (Taylor and others 1994a; Savitsky and others 1995). The DNA damage response defects in these human disorders are considered in depth elsewhere in this report. However, as evident from the data outlined in the following sections, genetic susceptibility to radiogenic tumors extends beyond a simple relationship between DNA damage response deficiency, cellular radiosensitivity, and neoplastic development (ICRP 1998; NRPB 1999).

The first objectives of this section are to outline the data that relate to (1) cancer-prone human genetic disorders determined by strongly expressing genes, (2) less strongly expressing cancer-associated genes, and (3) the evidence available on radiosensitivity and predisposition to radiation tumorigenesis. The principal conclusions from these reviews will then be applied in the development of judgments on the identification of human subgroups having potentially increased cancer risk after radiation and the likely magnitude of that increased risk. In developing these judgments, particular attention will be given to the uncertainties involved.

Cancer-Prone Human Genetic Disorders

The whole field of cancer genetics has expanded dramatically in the last 15 years, and it is appropriate to provide only a brief overview here. Detailed reviews are given elsewhere (Eeles and others 1996; ICRP 1998).

Published genetic catalogs (McKusick 1998; Mulvihill 1999) show that around 6% of recorded human disorders and mutant genes have some degree of association with neoplastic disease. The number of such disorders for which the association is unambiguously strong remains small (less than 50) and tends to be restricted to rare autosomal recessive and autosomal dominant diseases. Highly expressing autosomal dominant diseases usually manifest as familial cancer, often without other major clinical features. As a genetic grouping, these have received much attention in recent years. Autosomal recessive diseases tend to be more rare, and excess cancer is usually accompanied by other characteristic clinical features. Since their manifestation demands a genetic input from both parents, these disorders do not typically express as familial cancer.

Autosomal Recessive Disorders

The majority of human genetic diseases associated with DNA damage response and repair fall into this category. Table 3-3 outlines examples within this category including AT and NBS. There are also examples of autosomal recessive and X-linked disorders of the immune system, which manifest as susceptibility to virally associated neoplasia (ICRP 1998); these are not considered here.

Autosomal Dominant Disorders

In this category are examples of mutations in DNA damage response or repair genes, in proto-oncogenes, and in tumor-suppressor genes. Table 3-4 outlines examples of human disorders that make up this grouping.

In considering the examples given in Tables 3-3 and 3-4, a number of general points can be added to the descriptions. First, there are genetic disorders that might qualify for inclusion in both DNA damage response or repair and tumor-suppressor categories. The prime example is Li-Fraumeni syndrome, which may be ascribed to DNA damage response and tumor suppression activity of the responsible *TP53* gene (ICRP 1998). However, on the basis of their autosomal dominant inheritance and gene loss in tumors, DNA mismatch repair defects in hereditary nonpolyposis colon cancer and, possibly, BRCA-type heritable breast cancer might also be included in the tumor-suppressor category.

Second, there are general clinical and medical genetic features of the cancer-prone disorders of Tables 3-3 and 3-4 that are important for the judgments to be developed. For autosomal dominant human mutations of cancer to be detected readily in the population via family studies, the

TABLE 3-3 Examples of Autosomal Recessive Disorders of DNA Damage Response

Disorder	Genes or Locus	Defect Proposed	Major Clinical Features	Cancer	Approximate Prevalence (per live births)
Xeroderma pigmentosum	*XP-A* to *XP-G* and *XPV*	Excision or postreplication repair	Photosensitivity and cancer of UVR-exposed skin	Squamous cell skin carcinoma, basal cell carcinoma, and melanoma	1 in 250,000
Cockaynes syndrome	*CS-A, CS-B*	Transcribed strand repair	Photosensitivity, dwarfism	No excess	[a]
Trichothiodystrophy	*XP-D*	Excision repair	Photosensitivity, abnormal sulfur-deficient hair	Variable excess (skin)	[a]
Ataxia-telangiectasia	*ATM*	Kinase activity	Radiosensitivity, neuro- and immunodeficiency	Lymphoma	1 in 100,000
Nijmegen breakage syndrome	*NBS*	NHEJ factor (Mrell/RAD50/nbs)	Radiosensitivity, microencephaly, immunodeficiency	Lymphoma	[a]
Fanconi's anemia	*FA-A* to *FA-C*	DNA cross-link repair	Bone marrow deficiency, skeletal abnormalities	Leukemia	1 in 300,000

[a]Less than 1 in 100,000.

TABLE 3-4 Examples of Autosomal Dominant Disorders of Tumor Suppressor Genes, Proto-oncogenes, and DNA Damage Response or Repair Genes

Disorder	Genes or Locus	Defect Proposed	Cancer	Approximate Prevalence (per live births)
Tumor-Suppressor Disorders				
Familial adenomatous polyposis	*APC*	Transcriptional regulation	Colorectal cancer (multiple polyps)	1 in 8000
Von Hippel-Lindau disease	*VHL*	Transcriptional regulation	Renal cancer	1 in 30,000
Denys Drash syndrome	*WT1*	Transcriptional regulation	Nephroblastoma (+ others)	?
Neurofibromatosis type 1	*NF-1*	GTPase regulation	Neurofibroma Schwannoma	1 in 3000
Neurofibromatosis type 2	*NF-2*	Cytoskeletal linkage	Meningioma Neurofibroma	1 in 30,000
Nevoid basal cell carcinoma syndrome	*PTC*	Cellular signaling	Basal cell skin cancer Medulloblastoma	1 in 50,000
Tuberous sclerosis	*TSC1* *TSC2*	Cellular signaling Cellular signaling	Benign lesions of skin, nervous tissue, heart, and kidneys	1 in 20,000
Retinoblastoma	*RB1*	Transcriptional regulation	Retinal tumors, bone or soft-tissue sarcoma, brain cancer, and melanoma	1 in 25,000
Proto-oncogene Disorders				
Multiple endocrine neoplasia (2A and 2B) and familial medullary thyroid cancer	*RET*	Cellular signaling	Thyroid or parathyroid neoplasms	?
DNA Damage Response or Repair Disorders				
Hereditary nonpolyposis colon cancer	*MLH1, MSH2, PMS1, PMS2*	DNA mismatch repair, apoptosis	Colon cancer, endometrial cancer	1 in 2000
Li-Fraumeni syndrome	*TP53* (others?)	DNA damage recognition	Various	1 in 50,000
Heritable breast or ovarian cancer	*BRCA-1* *BRCA-2*	Transcriptional regulation, DNA repair	Breast or ovarian cancer Breast cancer (also male)	1 in 1000

degree of spontaneous tumor risk that is imposed must be sufficient to distinguish that family from others that are non-carriers. Given that, on average, spontaneous cancer incidence in the general population is around 30%, the information currently available is restricted largely to mutations where the cancer in question is expressed at a high relative frequency in gene carriers (*i.e.*, so-called high-penetrance mutations).

Other features of importance are (1) the organ specificity of many cancer-predisposing mutations, (2) the age of onset of given neoplasms in gene carriers that usually occurs at younger ages than in noncarriers, (3) the frequent occurrence of multiple tumors in gene carriers, and (4) the substantial variation for cancer risk between carriers of a given gene mutation, suggestive of major influences from the genetic background and/or life-style of the host. These issues of heritable cancer risk have been summarized by the International Commission on Radiological Proterction (ICRP 1998) and more recently by Ponder (2001). The crucial point, to be developed later, is that current knowledge of heritable cancer susceptibility in humans is restricted largely to relatively rare mutations of high penetrance. Cancer may be regarded as a multifactorial disorder (see Chapter 4), and genetic views developed from the study of other multifactorial conditions, such as coronary heart disease, suggest strongly that there will be many more variant cancer genes having lower penetrance than those listed in Tables 3-1 and 3-2. The current lack of knowledge about the nature, frequency, and impact of such genes imposes fundamental limitations in respect of the objectives stated earlier.

Mechanistic Aspects of Genetically Determined Radiation Response

In making judgments on the radiation response of cancer-prone individuals it is valuable to consider first the theoretical expectations that follow from current knowledge of the cellular mechanisms that are likely to be involved in cancer susceptibility. Germline mutations in DNA damage response or repair genes, tumor-suppressor genes, and proto-oncogenes are considered in turn.

DNA Damage Response-Repair Genes

As outlined in Chapters 1 and 2, different forms of DNA damage are recognized and processed in mammalian cells by different biochemical pathways, which share few genetic determinants. Accordingly, there is no expectation of a global association between DNA damage response or repair deficiency and sensitivity to the tumorigenic effects of radiation. Rather, the expectation is that a deficiency of genes associated with recognition or repair of the form of damage that is critical for cellular response to radiation (*i.e.*, DNA DSB) will be of greatest significance for radiation cancer risk. On this basis the autosomal recessive disorders AT and

NBS in Table 3-3 might be judged to exhibit increased cancer risk after ionizing radiation, whereas XP would not. Stated simply, germline deficiency in the recognition and/or repair of induced DNA damage of specific forms is expected to increase the abundance of genome-wide damage in the somatic cells of body tissues. This increased mutational load will tend to increase cancer risk, albeit with differing degrees of expression among tissues. It is important to recognize, however, that a number of autosomal dominant conditions, particularly Li-Fraumeni syndrome ($TP53^{+/-}$), are determined by genes that play more general roles in the control of stress responses, apoptosis, and/or coordination of the cell reproductive cycle (Chapter 2). Abnormal cellular response or cancer risk in such disorders might be expected for a range of DNA-damaging agents including ionizing radiation.

Tumor-Suppressor Genes

For tumor-suppressor genes such as *VHL* and *NF1* in Table 3-4 there is no specific association with DNA damage response or repair. Accordingly there is no expectation of increased genome-wide sensitivity to the mutagenic effects of radiation. In these instances increased radiation cancer risk may be anticipated on the basis of the now well-supported hypothesis of Knudson (1986). In brief, there is good evidence that many tumor-suppressor type genes act as tissue-specific gatekeepers to neoplastic pathways (Kinzler and Vogelstein 1997). Since loss or mutation of both autosomal copies of such genes from single cells is believed to be rate limiting for the initiation of neoplastic development, tumor initiation in normal individuals is expected to be a rare cellular event.

A carrier of a germline mutation in a given tumor-suppressor gene will however show loss of function of one such gene copy, thus "unshielding" the second copy in all target somatic cells. The lifetime risk of spontaneous loss or mutation of that second copy from any given population of target cells will be relatively high—hence the often dramatic increase in organ-specific cancer risk.

There is also a clear expectation that exposure of the carrier individual to ionizing radiation or indeed other genotoxic carcinogens would, via the same genetic-somatic mechanism, result in a greater-than-normal risk of organ-specific cancer. Stated simply, the enhanced radiation cancer risk in the carrier individual would be driven by a reduction in the target gene number from two to one; in a given disorder the organs at increased risk would tend to be the same as those involved in spontaneous neoplasia.

Proto-oncogenes

There are few well-characterized germline, gain-of-function mutations in proto-oncogenes that have unambiguous associations with cancer risk; a series of characterized *ret* gene mutations are however known to increase the risk of thyroid neoplasia (Table 3-4). As in the case of tumor-suppressor gene loss, germline *ret* mutation may be viewed as removing one early rate-limiting step in multistage thyroid tumorigenesis such that the carrier individual is at increased risk of neoplastic development via the accumulation of further mutations in other genes. Again, greater-than-normal radiation risk to the target organ should be anticipated.

In the following sections, the above propositions are examined on the basis of available cellular, animal, and epidemiologic data.

Cellular Data on Heritable Human Radiosensitivity

Cellular data on heritable radiosensitivity in respect of cell inactivation have been reviewed recently (ICRP 1998). In brief, although there are isolated instances of cancer and/or radiotherapy patients showing clear evidence of radiosensitivity, it is only for AT and NBS that there is unambiguous evidence of profoundly increased radiosensitivity to cell killing associated with known human disorders of DNA damage response or repair and cancer. Claims for increased radiosensitivity in other cancer-prone disorders remain controversial and do not provide clear guidance on radiation cancer risk.

Although sensitivity to cell killing after radiation may at present not be a particularly useful surrogate for cancer risk, there are closer parallels between the induction of chromosome damage and cancer. Although not without some uncertainty, the data accumulating on the patterns of chromosomal radiosensitivity in human cancer-prone disorders are worthy of some attention. These data, considered by Scott and colleagues (1998) and reviewed by the National Radiological Protection Board (NRPB 1999) show that, compared with healthy controls, cells cultured from AT and NBS patients typically exhibit two- to threefold greater chromosomal radiosensitivity, but in some cytogenetic assays, the increased sensitivity can be up to tenfold (Taalman and others 1983; Taylor 1983). The NRPB has summarized a large body of cytogenetic data on which claims of associations between chromosomal radiosensitivity and human cancer susceptibility have been based. As in the case of cell killing, some of these claims remain controversial. More recent studies on the possible radiosensitivity of cells from breast cancer-susceptible BRCA1 and BRCA2 patients have also provided conflicting evidence (Buchholz and others 2002; Trenz and others 2002; Powell and Kachnic 2003). Of additional interest are the data on G_2 cell cycle radiosensitivity, which among other findings suggest that AT heterozygotes are indeed radiosensitive and that up to 40% of unselected breast cancer cases also exhibit modestly elevated radiation-induced chromosome damage (Scott and others 1994; Parshad and others 1996). There is also some evidence of elevated chromosomal radiosensitivity in cells from patients with malignant gliomas (Bondy and others 1996) and colorectal cancer (Baria and others 2001).

In summary the evidence available on human chromosomal radiosensitivity suggests that AT and NBS may be up to tenfold more sensitive than normal; some uncertainty surrounds the chromosomal radiosensitivity of other cancer-prone disorders, but any such increase in sensitivity appears to be modest—not more than two- to threefold. Although critical data are lacking, it is a reasonable assumption that, in general, a heritable increase in chromosomal radiosensitivity would be associated with increased radiation cancer risk, albeit with possible differences in the response of different tissues. Data from G_2 chromosomal radiosensitivity assays are generally supportive of this association, but some data remain controversial.

Animal Data on Radiosensitivity and Tumorigenesis

The experimental data available about the impact of heritable factors on radiosensitivity and tumorigenesis derive principally from studies on the genetic homologues of some of the human disorders listed in Tables 3-3 and 3-4. These studies are summarized in Table 3-5 with references.

Although there are some differences in the patterns of phenotypic expression, in the main the rodent genetic homologues of AT, Li-Fraumeni syndrome (LFS), familial adenomatous polyposes, neroid basal cell carcinoma syndrome (NBCCS), and tuberous sclerosis recapitulate many of the features of their human counterparts. In respect of early responses, $Atm^{-/-}$ mice show extreme radiosensitivity; there is also evidence of moderate *in vivo* radiosensitivity in $Atm^{+/-}$ mice. Studies with $Atm^{+/-}$ knockout mice (Barlow and others 1999) provided evidence of increased *in vivo* radiosensitivity but failed to demonstrate differences in radiation induced tumorigenesis between +/− and +/+ genotypes. However, more recent data on spontaneous tumorigenesis (Spring and others 2002) imply that such studies are best conducted with *Atm* knock-in mice, which recapitulate known human mutations.

Data on BRCA1- and BRCA2-deficient mice have yet to provide clear evidence on the role of these genes in radiation tumorigenesis. The principal benefit of the referenced studies noted in Table 3-5 is the provision of a growing association between the *Brca* genes, *Rad51*, cell cycle perturbation, and DNA damage response.

The most valuable animal genetic data on radiation tumorigenesis have been developed from studies on mice heterozygously deficient in the tumor-suppressor genes *Tp53*, *Apc*, and *Ptch* and in a rat strain (Eker) heterozygously deficient in *Tsc2* (see Table 3-5 for references). In all instances, the germline mutational loss of one copy of the respective tumor-suppressor gene leads not only to an increase in the rate of spontaneous tumorigenesis but also to increased sensitivity to the induction of the same tumor types by whole-body low-LET radiation with doses up to around 5 Gy.

These data provide strong support for the contention, discussed earlier, that the unshielding of tumor-suppressor

TABLE 3-5 Radiation Response and Tumorigenesis in Rodent Homologues of Cancer-Prone Human Genetic Disorders

Genotype		Human Homologue	Radiation Response		Comment	Key References
			Early response	Tumorigenesis		
Mouse	$Atm^{-/-}$	Ataxia-telangiectasia (AT)	Radiosensitive *in vivo* or *in vitro*	May be dependent upon *Atm* genotype (see text)	Defects in meiosis, immunity, and behavior	Barlow and others (1996) Elson and others (1996) Xu and others (1996)
Mouse	$Brca1^{-/-}$ $BRCA^{+/-}$	Heritable breast cancer in heterozygotes	Cellular and embryonic radiosensitivity	No published study identified	Embryolethal; association with $Rad51^{-/-}$ phenotype	Gowen and others (1998) Sharan and others (1997) Mizuta and others(1997) Connor and others (1997)
Mouse	$Tp53^{+/-}$	Li-Fraumeni syndrome (LFS)	Excess aneuploidy and G$_2$/M checkpoint defect in bone marrow cells	Highly sensitive to induction of lymphoma or sarcoma	Tumorigenesis associated with loss of $Tp53^+$	Kemp and others (1994) Bouffler and others (1995)
Mouse	$Apc^{+/-}$	Familial adenomatous polyposis	None reported	Highly sensitive to induction of intestinal adenoma (breast and other cancers in some genetic backgrounds)	Tumorigenesis associated with loss of Apc^+ and other loci	Luongo and Dove (1996) Ellender and others (1997) van der Houven van Oordt and others (1999) Haines and others (2000)
Mouse	$Ptch^{+/-}$	Nevoid basal cell carcinoma syndrome	Some evidence of cellular radiosensitivity	Sensitive to induction of medulloblastoma	Tumorigenesis associated with loss of $Ptch^+$	Hahn and others (1998) Pazzaglia and others (2002)
Rat	$Tsc2^{+/-}$	Tuberous sclerosis	None reported	Sensitive to induction of renal neoplasia	Tumorigenesis associated with loss of $Tsc2^+$	Hino and others (1993, 2002)

genes by germline mutation will lead to a significant increase in individual susceptibility to radiation tumorigenesis. Critical mechanistic support for this hypothesis has been provided by molecular analysis of tumors arising in irradiated $Tp53^{+/-}$, $Apc^{+/-}$, and $Ptch^{+/-}$ mice and $Tsc-2^{+/-}$ rats; as predicted, such analyses strongly suggest that radiation acts by inactivating the wild-type tumor-suppressor gene copy in target somatic cells. These wild-type genes appear to be mutated by radiation through mechanisms principally involving substantial DNA loss events, although there are examples of whole chromosome losses as well as intragenic deletions and point mutations.

Although the above studies provide proof-of-principle experimental evidence of strong genetic effects on radiation tumorigenesis in mammalian species, quantification of the genetically imposed radiation risk is most problematical. An ICRP (1998) Task Group, in reviewing much of the data of Table 3-5, suggested that radiation tumor risk in such suppressor-suppressor gene-deficient mice might be elevated by up to a hundredfold or more but cautioned against firm judgments because of (1) problems associated with experimental design and (2) preliminary evidence that natural variation in the genetic background of host animals can have major modifying effects on tumor yield.

During the last few years the impact of such modifier genes on the expression of tumorigenesis in mice has been demonstrated more clearly (Balmain and Nagase 1998). The principal message from this experimental work is that because of the strongly modifying effects of genetic background, rodent homologues are unlikely to provide a quantitatively reliable representation of radiation tumorigenesis in cancer-prone human genetic disorders. Such genetic modification is to be expected in humans, but the specific nature and impact of the modifier genes are likely to differ among species. The issue of genetic modification of radiation response is considered further in the section of this chapter that deals with cancer-predisposing mutations of low penetrance.

Human Data on Radiosensitivity and Tumorigenesis

As noted earlier in this chapter unambiguous evidence of human genetic disorders showing hypersensitivity to tissue injury after radiation is confined to AT and NBS, where conventional radiotherapy procedures have proved disastrous to patients. Adverse, but less profound, reactions to radiotherapy are however reported to occur in around 5% of cancer patients (Burnet and others 1998). Studies on *in vitro* cellular radiosensitivity in such radiotherapy patients have, so far, failed to reveal evidence of strong correlations between *in vivo* and *in vitro* responses although subsets of these patients do show statistically significant increases in cellular radiosensitivity under some assay conditions (Burnet and others 1998). Similarly limited molecular studies show no correlation between adverse reactions to radiotherapy and heterozygous *ATM* gene mutation (Appleby and others 1997; Burnet and others 1998). The question as to whether adverse tissue reaction to radiotherapy signals potentially increased risk of therapy-related second tumors has yet to be addressed in epidemiologic studies.

Postradiotherapy observations on specific sets of cancer patients have, however, revealed valuable information on genetic associations with risk of second tumors (Meadows 2001). These data are summarized and referenced in Table 3-6. In brief, there is evidence of an excess of radiotherapy (RT)-related tumors in the human cancer-prone conditions heritable retinoblastoma, NBCCS, and LFS plus related conditions, as well as in children from families with a history of early onset cancer. In addition there are reports suggesting that neurofibromatosis is a positive factor for RT-related tumorigenesis (Robison and Mertens 1993). By contrast, a variety of studies discussed by Mark and colleagues (1993) provide no clear evidence that genetic factors are important for RT-related breast cancer. Recent studies provide no evidence that the status of BRCA genes influences postradiotherapy outcomes at 5 years (Pierce and others 2000).

In Table 3-6 the data suggesting that NBCCS and LFS patients have substantial increases in tumorigenic radiosensitivity are in accord with data obtained experimentally with their rodent genetic homologues. For retinoblastoma (RB),

the large size of the U.S.-based epidemiologic studies of Eng and colleagues (1993) and Wong and coworkers (1997a) allows some judgments to be developed on the degree to which this suppressor gene disorder predisposes to (second) radiogenic soft-tissue sarcoma and bone cancer. Although there is a clear dose-response for radiation tumorigenesis, these data imply that excess relative risk (ERR) in heritable RB patients may be lower than in the nonheritable controls.

The background rate of tumorigenesis in RB is, as expected, rather high, and for the purposes of this report, excess absolute risk (EAR) may be a more useful measure of tumorigenic radiosensitivity than ERR. In considering this issue, the ICRP (1998) and NRPB (2000) suggest that the EAR in heritable RB is around fivefold higher than in the nonheritable group. It is notable that low values of ERR for radiogenic cancer in such cancer-prone conditions are consistent with other epidemiologic data on radiation tumorigenesis where high background cancer rates also tend to be accompanied by lower ERRs. Abramson and colleagues (2001) have also reported on third tumors in RB patients after radiotherapy. As might be expected, the sites of these additional tumors generally accorded with the irradiated volume of normal tissue.

In summary, although clinical and epidemiologic data on RT patients are limited, they are sufficient to confirm the view developed from mechanistic knowledge and experimental studies that human genetic susceptibility to spontaneous tumorigenesis is often accompanied by an increase in absolute cancer risk after ionizing radiation. Quantifying that risk is problematical, but the single study on RB patients that has this capacity is suggestive of relatively modest (about fivefold) increases over that of normal individuals. In the

TABLE 3-6 Postradiotherapy Observations on Risk of Second Tumors in Humans

Genetic Disorder or Study Group	First Tumor	Observations	Key References
Retinoblastoma	Retinoblastoma	Excess bone tumors and soft-tissue sarcomas, large cohorts; some dose, dose-response, and risk estimates possible	Tucker and others (1987a) Eng and others (1993) Wong and others (1997a) Abramson and others (2001)
NBCCS	Medulloblastoma	Excess basal cell skin neoplasms and ovarian fibromas, short latency; case reports only	Strong (1977) Southwick and Schwartz (1979)
LFS and related conditions	Various	Follow-up of children developing posttherapy soft-tissue sarcoma, bone tumors, and acute leukemia—linkage with family histories of cancer	Strong and Williams (1987) Heyn and others (1993) Robison and Mertens (1993) Malkin (1993)
Case-control study of therapy-related second tumors	Various	Excess posttherapy tumors in children from non-LFS families with a history of early onset cancer	Kony and others (1997)

future, the growing capacity of molecular screening techniques to detect cancer-susceptible genotypes in the general population will, in principle, allow the radiation risk of such genotypes to be assessed in a number of suitable human cohorts. A summary of such molecular epidemiologic approaches to spontaneous cancer risk is given later in this chapter.

Population Modeling of Radiation Cancer Risk: Impact of Strongly Expressing Genetic Disorders

In conjunction with the work of an ICRP (1998) Task Group, Chakraborty and colleagues (1997, 1998a) have constructed and illustrated the use of a population-based computational model that serves to describe the impact of cancer-susceptible genotypes on radiation cancer risk in the population. For reasons of data sufficiency, breast cancer risk in typical Western populations was considered and illustrated. This approach, which is based on established Mendelian principles, employed best estimates of the prevalence of known, high-penetrance breast cancer-predisposing genes (BRCA1 and BRCA2), the relative risk of spontaneous breast cancer in such genotypes, and a range of factors that describe in a hypothetical fashion the increase in radiation risk imposed by the given gene mutations; the risk of radiogenic breast cancer in normal individuals was based on data from Japanese atomic bomb survivors.

Other issues that were considered included increased gene frequency in certain genetically isolated populations (Ashkenazi Jews) and the influence of reduced penetrance on population risk. The following points summarize the outcome of these modeling exercises.

• Using best estimates of breast cancer gene frequencies, the genetic impact on excess breast cancer in an irradiated Western population would be small even if these mutations were to impose a radiation risk that was as much as a hundredfold greater than that of normal genotypes.
• Using estimates of the higher gene frequencies in Ashkenazi Jewish populations, the genetic impact on radiation-associated breast cancer can become significant but only if the genetically imposed radiation risk is very high.
• The genetic impact of such mutations will be diluted in proportion to decreasing penetrance.

This model and its predictions have been used by the ICRP (1998) and NRPB (1999) to provide interim judgments on the implications of genetic susceptibility to cancer for radiological protection.

Since the overall prevalence of highly penetrant cancer-predisposing mutations in typical human populations is judged to be 1% or less (ICRP 1998) and since available data tend to argue against extreme increases in genetically imposed radiation cancer risk, there is reason to believe that the presence of these rare, highly penetrant mutations will not appreciably distort current estimates of radiation cancer risk in the population. Stated simply, only a very small fraction of excess cancers in an irradiated human population are expected to arise in individuals carrying familial cancer genes.

The ICRP (1998) and NRPB (1999) stressed, however, that this conclusion took no account of the presence of potentially more common cancer genes of low penetrance that do not express familial cancer. The ICRP and NRPB reports also commented on the problems inherent in identifying and making judgments about radiation cancer risk in genetic subgroups carrying such weakly expressing genes and considered the issue of genetically imposed risk to individuals. These matters are discussed in subsequent sections.

Genes of Low Penetrance

As noted earlier in this chapter, knowledge of heritable factors in tumorigenesis stems largely from studies on strongly predisposing autosomal dominant familial traits and autosomal recessive disorders having unambiguous phenotypes. The problem of estimating the heritable impact on cancer risk from weakly expressing genes of low penetrance and other genetic modifiers of the cancer process has been with us for some time. However, not unexpectedly, an understanding of this issue is proving difficult to obtain. To a large measure this is due to the likelihood that, individually, polymorphic variant genes probably contribute small additional cancer risks to each carrier in a largely tissue-specific manner. These will tend to escape detection by conventional medical genetic and epidemiologic studies. A combination of such genes and their interaction with environmental risk factors may, however, provide a substantial genetic component to both spontaneous and radiation-associated risk. The magnitude of this risk in a given human population would then be determined by gene frequencies together with the pattern or strength of gene-gene and gene-environment interactions.

These issues of population cancer risk have been discussed widely in the context of epidemiologic and molecular genetic findings (Hoover 2000; Houlston and Tomlinson 2000; Lichtenstein and others 2000; Peto and Mack 2000; Shields and Harris 2000; Dong and Hemminki 2001; Nathanson and Weber 2001; Ponder 2001). Here it is sufficient to illustrate some of the progress being made in respect of the weakly expressing genetic component of human and animal tumorigenesis. Where possible, emphasis is placed on data having some connection with cancer risk after ionizing radiation.

Human Breast Cancer

BRCA1 and BRCA2 genes have been identified as the principal genetic determinants of the 2–5% of breast cancer that expresses in multiple-case families; other, more weakly expressing genes involved in familial breast cancer remain to be uncovered (Nathanson and Weber 2001; Ponder 2001). However, epidemiologic evidence is highly suggestive of a more extensive genetic component to breast cancer risk (Peto and Mack 2000), and much effort is being expended to identify the functional gene polymorphisms that might be involved. Although some of the evidence remains controversial, Dunning and colleagues (1999) and Nathanson and Weber (2001) note the potential involvement of polymorphic genes that encode steroid hormone receptors and paracrine growth factors (*e.g.*, *AR*, *CYP19*) together with genes involved in the metabolism of chemical species (*e.g.*, *GSTP1*) and in DNA damage response (*e.g.*, *ATM*, *RAD51*, *TP53*). The most persuasive evidence on breast cancer genes other than BRCA1 and BRCA2 concerns the cell cycle checkpoint kinase gene *CHEK2*. A truncating germline deletion of this gene is present in around 1% of healthy individuals and is estimated to result in about a twofold increase of breast cancer risk in women and about a tenfold increase in men (Meijers-Heijboer and others 2002). Two data sets have some association with cancer risk after radiation.

First is the question of breast cancer risk in individuals who are heterozygous carriers of the *ATM* mutation of the highly radiosensitive disorder AT. *ATM* carriers (*ATM$^{+/-}$*) might represent 0.25–1% of the general population, and there is evidence of modestly increased cellular radiosensitivity in *ATM$^{+/-}$* genotypes. It is therefore reasonable to consider an increased risk of radiogenic breast cancer in these carriers. Considerable effort has been expended on molecular epidemiologic analysis of spontaneous breast cancer risk in *ATM$^{+/-}$* women (Bishop and Hopper 1997; ICRP 1998; Broeks and others 2000; Laake and others 2000; Geoffroy-Perez and others 2001; Olsen and others 2001; Teraoka and others 2001). Although the position remains somewhat uncertain, it seems reasonable to conclude that while increased breast cancer risk may be associated with *ATM$^{+/-}$* in some cohorts, the relative risk is likely to be modest (<3), and the overall impact on spontaneous breast cancer risk in the population is rather small. Some data suggest, however, that it is only certain dominant negative missense mutations of *ATM* that predispose to cancer (Khanna 2000; Chenevix-Trench and others 2002), and for these, the relative risk may be substantially higher. The critical question is whether the *ATM$^{+/-}$* genotype may more specifically and significantly increase breast cancer risk after radiation. For good scientific reasons, some early claims on substantial risks at low doses are not regarded as being well founded (see ICRP 1998). While a modestly increased contribution of the *ATM$^{+/-}$* genotype to radiogenic cancer risk should not be discounted, three recent studies on patients developing second cancers after RT argue against a major impact from the *ATM* gene (Nichols and others 1999; Broeks and others 2000; Shafman and others 2000). In total, these studies considered 141 patients with second cancers; the studies of Shafman and colleagues (2000) and Broeks

and colleagues (2000) specifically considered a total of 89 second breast cancer cases. None of the cases studied carried *ATM* mutations.

The second line of evidence concerns the inheritance of chromosomal radiosensitivity and its association with breast cancer risk (Roberts and others 1999). In brief, in studies on cultured blood lymphocytes, up to around 40% of unselected breast cancer cases were shown to exhibit an abnormal excess of chromatid aberrations following X-irradiation in the G_2 phase of the cell cycle. By contrast, this chromosomal trait was seen in only around 5% of age-matched controls. Follow-up family studies provided evidence on the heritability of the trait, which, although not of a simple Mendelian form, could be genetically modeled. As yet there is no evidence on the specific genes involved.

In summary, advances in breast cancer genetics do allow the construction of a general scheme to describe the interactive genetic component of familial risk, including some allowance for common genes of low penetrance (Ponder 2001). Polygenic computational models describing the overall genetic component of spontaneous breast cancer risk in the population are also under development (Antoniou and others 2002). Although gene candidates and cellular phenotypes may prove to be instructive, there is at present little to guide specific conclusions on the question of the common genetic component of radiation-associated cancer risk. The evidence available would tend to argue against a major overall impact on radiation breast cancer risk from the *ATM* gene in its heterozygous form, although specific *ATM* genotypes may, in principle, carry substantially increased risk.

Human Colonic and Other Neoplasms

There is evidence that the genetic component of colonic cancer also includes a significant contribution from genes of low penetrance. In a recent review of 50 studies on the potential impact of common polymorphisms, Houlston and Tomlinson (2001) identified significant associations with risk for *APC-I1307K*, *HRAS1-VNTR*, and *MTHFR-Val/Val*. For *TP53*, *NAT1*, *NAT2*, *GSTM1*, *GSTT1*, and *GSTP1* polymorphisms, the evidence was weaker. Specific data relating to gene polymorphisms and radiation risk are lacking although, as for breast cancer, there is some evidence of an association between colon cancer risk and lymphocyte chromosomal radiosensitivity (Baria and others 2001).

Finally, in illustration of ongoing work, it is relevant to mention polymorphic associations between *GSTP1* and chemotherapy-related leukemia (Allan and others 2001), *MCUL1* and uterine fibroma (Alam and others 2001), *GFRalpha1* and medullary thyroid carcinoma (Gimm and others 2001), *PPARG* and endometrial carcinoma (Smith and others 2001), and *TP53* and adrenal cortical carcinoma (Ribeiro and others 2001). In their review of gene-environment interactions, Shields and Harris (2000) focus on lung cancer risk, and in this area, Bennett and colleagues (1999) have provided evidence on the potential impact of *GSTM1* allelic status on tobacco-related lung cancer risk.

The broad but incomplete picture that emerges from these studies is of some associations between gene polymorphisms and risk for a range of human tumor types, as well as the clear need for larger and more definitive studies.

Human DNA Repair Gene Polymorphisms

It has already been noted that DNA repair genes play a crucial role in cellular responses to radiation and that major germline deficiencies in these genes can lead to heritable predisposition to cancer. Accordingly, considerable effort is being expended in the search for common functional polymorphisms that might act as low-penetrance cancer susceptibility genes.

A series of studies have identified common and less common polymorphisms in around ten DNA repair genes, some of which appear to have cellular consequences (Price and others 1997; Shen and others 1998; Mohrenweiser and Jones 1998; Duell and others 2000). The associations between these polymorphisms and radiosensitivity and/or tumor risk remain unclear, although there are some positive indications (Duell and others 2001; Hu and others 2001). Much of this work has centered on genes involved in base- or nucleotide-excision repair (Miller and others 2001). Studies on genes controlling DNA DSB repair are less well developed. However, there are indications that a relatively common (in ~6% of the population) functional polymorphism in the *XRCC2* gene of the homologous recombinational repair pathway for DNA DSBs associates with a modestly increased risk of breast cancer (Kuschel and others 2002; Rafii and others 2002). A significant association between breast cancer risk and certain polymorphisms of NHEJ DNA repair has also been reported (Fu and others 2003). A recent review of DNA repair gene polymorphisms and cancer risk recommends large, well-designed studies that include consideration of relevant exposures (Goode and others 2002).

Genetic Studies with Animals

The recognized difficulties of resolving the modifying effects of low-penetrance genes on human cancer risk have prompted experimental genetic studies with rodent models in which genetic-environmental interactions can be more closely controlled.

This approach has been applied principally in mice for the study of naturally arising polymorphic variation that influences spontaneous cancer risk and the risk after exposure to chemical carcinogens and, in a few instances, ionizing radiation (Balmain and Nagase 1998). These studies have the capacity to provide proof-of-principle evidence of the impact of such common loci, together with their possible interactions and tissue specificity, as well as the classes of genes and mechanisms involved. Thus, although specific

functional gene polymorphisms identified in mice may not predict those of humans precisely, the overall pattern of cancer risk modification should provide broad guidance on the potential for such effects in humans.

Much of the research on the role of germline polymorphic loci in mouse tumorigenesis has centered on spontaneous and chemically induced neoplasms. These studies include tumors of the skin (*e.g.*, Nagase and others 2001; Peissel and others 2001), lung (*e.g.*, Lee and others 2001; Tripodis and others 2001), and intestinal tract (*e.g.*, van Wezel and others 1996; Angel and others 2000). The most important messages to emerge from these studies are that multiple common loci can exert complex patterns of control over tumor susceptibility and resistance (synergistic and antagonistic interaction), that the loci tend to be relatively tissue specific in their activity, and that genetic determinants of spontaneous and induced tumorigenesis are often shared. A particularly revealing conclusion from the study of Tripodis and colleagues (2001) is that as many as 60 loci may interact to determine the risk of a single tumor type; specific pairwise interaction of a proportion of these loci was also demonstrated.

A second approach used in mouse genetic studies is to seek evidence of natural polymorphic loci that modify the tumorigenic expression of a major cancer-predisposing germline mutation. In this way, evidence has been obtained for substantial genetic modification of tumorigenesis in *Trp53*- (Backlund and others 2001) and *Apc*-deficient mice (van der Houven van Oordt and others 1999; Moser and others 2001). In the case of *Apc*, one of these modifier genes (*Pla2g2a*) has been identified provisionally (Cormier and others 2000). In general, these effects of genetic modifiers are again consistent with the potential interaction of multiple tissue-specific loci, and some of the data relate to tumors induced by ionizing radiation.

Some studies in this area have the specific objective of mapping and characterizing the polymorphic loci that influence tumorigenic radiosensitivity and tumor characteristics. Multiple loci have been shown to influence susceptibility to radiation-induced lymphoma and leukemia (Balmain and Nagase 1998; Szymanska and others 1999; Saito and others 2001; Santos and others 2001). One study of Boulton and colleagues (2001) provided evidence that the AML loci determining leukemia or lymphoma susceptibility were distinct from those that influenced genomic instability in bone marrow cells. However, no candidate genes were identified. Genetic loci influencing the susceptibility of mice to α-particle ([227]Th)-induced osteosarcoma have also been mapped (Rosemann and others 2002), but again, no candidate genes were specifically identified.

By contrast, another set of investigations has associated a strain-specific functional polymorphism of the gene *Prkdc* encoding DNA PKcs with induced genomic instability, DNA DSB repair deficiency, and susceptibility to radiation-induced breast cancer (Okayasu and others 2000; Yu and others 2001). This same *Prkdc* polymorphism has also been implicated in radiation-induced lymphomagenesis, as a modifier of induced intestinal neoplasia in *Apc^min* mice (Degg and others 2003), and as a candidate gene for the *Rapop1* apoptosis-controlling locus (Mori and others 2001). Other tissue-specific loci that control apoptosis have also been genomically mapped (*e.g.*, Weil and others 2001).

With respect to breast cancer susceptibility in mice, it is already clear that loci other than *Prkdc* can be involved (Moser and others 2001). From recent studies, it seems likely that one such gene is *ATM*, which in the heterozygous form can enhance the frequency of both genomic instability and ductal dysplasia of the breast of irradiated mice (Weil and others 2001).

Conclusions

Although much remains to be learned about genetic susceptibility to the tumorigenic effects of radiation, it is possible to frame some interim conclusions of the role it may play in determining radiation cancer risk at the individual and population levels.

The principal point to emphasize is that cancer is a multifactorial set of diseases, and as such, there is expected to be a complex interplay between multiple germline genes and a plethora of other host- and environment-related factors. The data available, although far from complete, tend to support this basic expectation. The key issues and arguments are given here in brief summary.

For rare major gene deficiencies in humans and mice, there can be strong effects on radiation cancer risk, and for individual carriers, it seems likely that the greatest implications may be for the risk of second cancers after RT (see ICRP 1998). Although the data are sparse, such high-dose radiation exposure in childhood may carry the greatest risk. However, due to differences in genetic background, a uniformity of tumorigenic response in RT patients with major gene deficiencies should not be expected.

The fact that strongly expressing cancer-prone disorders are so rare argues against a significant impact and distorting effect on estimates of cancer risk in irradiated populations; population genetic modeling fully supports this view (see ICRP 1998). By contrast, at the level of whole populations it is feasible that certain inherited combinations of common low-penetrance genes can result in the presence of subpopulations having significantly different susceptibilities to spontaneous and radiation-associated cancer. In due course, the accumulation of sufficient molecular epidemiologic data may allow for some meaningful theoretical modeling of the distribution of radiation cancer risk and the possible implications for radiological protection. Irrespective of such modeling, risk estimates based on epidemiologic evaluation of whole populations will encompass this projected genetic heterogeneity of response. Therefore, the key issue is not whether the estimate of overall cancer risk is genetically con-

founded, but rather the extent to which genetic distortion of the distribution of this risk might lead to underprotection of an appreciable fraction of the population. In this respect, some initial guidance for thought is already available from the data discussed in this chapter.

These data suggest large numbers of loci of low penetrance with relatively small individual effects and a significant degree of locus-specific interaction and tissue specificity that may apply to their activity. Projecting this scenario to a range of radiogenic tumors in a genetically heterogeneous human population would tend to lead to a situation in which the balance between a certain set of tumor susceptibility (S) and resistance (R) loci in a given subgroup might serve to emphasize risk in a given set of organs. Equally, however, the balance of additional S and R locus combinations might provide a degree of resistance to the induction and development of cancer in other organs. Thus, with this first genetic scenario, major distortions of the distribution of *overall* cancer risk after radiation might not apply simply because different genetic susceptibilities would tend to "average out" across organs. By contrast, a second hypothetical scenario involves a small subset of common polymorphic loci that exert organ-wide effects on tumor susceptibility or resistance, which might be particularly strong in the specific instance of radiation exposure (*e.g.*, functional polymorphisms for genes involved in initial tissue-wide cellular response to radiation damage). In this instance, genetically determined distortion of the distribution of overall cancer risk might be expected. At present, the data available are insufficient to distinguish the likely contributions from these two genetic scenarios.

Finally, the large study of cancer concordance in 90,000 Nordic twin pairs should be noted. Lichtenstein and colleagues (2000) and Hoover (2000) make some important points about the difficulties that exist in separating the genetic and environmental components of cancer. In essence, Hoover notes that this Nordic study, like others, is consistent with the presence of low-penetrance cancer-predisposing genes in the general population. However, the confidence intervals for the heritable component of cancers at common sites were wide—all ranged from around 5 to 50%. It was also pointed out that for cancer at common sites, the rate of concordance in monozygotic twins was generally less than 15%. Thus, the absolute risk of concordance of site-specific cancer in identical genotypes sharing some common environmental factors is rather low. In addition to this, a study based on the Swedish Family Cancer Database (Czene and others 2002) has provided further information on the genetic component of organ-specific cancer. With the exception of the thyroid, the environment appears to have the principal causal role for cancer at all sites.

One important message that emerges from current data on cancer genes of low penetrance and the overall genetic component of cancer is that predictive genotyping of individuals for the purposes of radiological protection may not be feasible in the medium term. The likely involvement of multiple and relatively organ-specific sets of polymorphisms and gene-gene or gene-environment interactions makes the provision of meaningful judgments on risk most uncertain. For these reasons it may be more realistic at this stage of knowledge to focus attention on general patterns of gene-radiation interactions and their implications for population risk, rather than risk for specific individuals.

SUMMARY

In this chapter, the committee has reviewed cellular-molecular and animal studies relevant to the complex multistage process of radiation tumorigenesis. Attention has also been given to evidence from various studies on the inherited factors that influence radiation cancer risks. The principal objective of this work was to provide judgments on radiation cancer risk of prime importance to radiological protection, particularly where these judgments serve to couple information about the action of radiation on cells (Chapters 1 and 2) with the epidemiologic measures of risk considered in subsequent chapters.

Mechanisms of Radiation Tumorigenesis

A critical conclusion on mechanisms of radiation tumorigenesis is that the data reviewed greatly strengthen the view that there are intimate links between the dose-dependent induction of DNA damage in cells, the appearance of gene or chromosomal mutations through DNA damage misrepair, and the development of cancer. Although less well established, the data available point toward a single-cell (monoclonal) origin for induced tumors and indicate that low-dose radiation acts predominantly as a tumor-initiating agent. These data also provide some evidence on candidate, radiation-associated mutations in tumors. These mutations are predominantly loss-of-function DNA deletions, some of which are represented as segmental loss of chromosomal material (*i.e.*, multigene deletions). This form of tumorigenic mechanism is broadly consistent with the more firmly established *in vitro* processes of DNA damage response and mutagenesis considered in Chapters 1 and 2. Thus, if as judged in Chapters 1 and 2, error-prone repair of chemically complex DNA double-strand damage is the predominant mechanism for radiation-induced gene or chromosomal injury involved in the carcinogenic process, there can be no expectation of a low-dose threshold for the mutagenic component of radiation cancer risk.

One mechanistic caveat explored was that novel forms of cellular damage response, collectively termed induced genomic instability, might contribute significantly to radiation cancer risk. The cellular data reviewed in Chapter 2 identified uncertainties and some inconsistencies in the expression of this multifaceted phenomenon. However, telomere-associated mechanisms did provide a coherent explanation

for some *in vitro* manifestations of induced genomic instability. The data considered in this chapter did not reveal consistent evidence for the involvement of induced genomic instability in radiation tumorigenesis, although telomere-associated processes may account for some tumorigenic phenotypes. A further conclusion was that there is little evidence of specific tumorigenic signatures of radiation causation, but rather that radiation-induced tumors develop in a tumor-specific multistage manner that parallels that of tumors arising spontaneously. However, further cytogenetic and molecular genetic studies are needed to reduce current uncertainties about the specific role of radiation in multistage radiation tumorigenesis; such investigations would include studies with radiation-associated tumors of humans and experimental animals.

Quantitative Studies of Experimental Tumorigenesis

Quantitative animal data on dose-response relationships provide a complex picture for low-LET radiation, with some tumor types showing linear or linear-quadratic relationships while other studies are suggestive of a low-dose threshold, particularly for thymic lymphoma and ovarian cancer. However, since the induction or development of these two cancer types is believed to proceed via atypical mechanisms involving cell killing, it was judged that the threshold-like responses observed should not be generalized.

Radiation-induced life shortening in mice is largely a reflection of cancer mortality, and the data reviewed generally support the concept of a linear dose-response at low doses and low dose rates. Other dose-response data for animal tumorigenesis, together with cellular data, contributed to the judgments developed in Chapters 10 and 12 on the choice of a DDREF for use in the interpretation of epidemiologic information on cancer risk.

Adaptive responses for radiation tumorigenesis have been investigated in quantitative animal studies, and recent information is suggestive of adaptive processes that increase tumor latency but not lifetime risk. However, these data are difficult to interpret, and the implications for radiological protection remain most uncertain.

Genetic Susceptibility to Radiation-Induced Cancer

The review of cellular, animal, and epidemiologic or clinical studies on the role of genetic factors in radiation tumorigenesis shows that there have been major advances in understanding, albeit with some important knowledge gaps. An important conclusion is that many of the known, strongly expressing, cancer-prone human genetic disorders are likely to show an elevated risk of radiation-induced cancer, probably with a high degree of organ specificity. Cellular and animal studies suggest that the molecular mechanisms underlying these genetically determined radiation effects largely mirror those that apply to spontaneous tumorigenesis and are consistent with knowledge of the somatic mechanisms of tumorigenesis reviewed earlier in this chapter. In particular, evidence has been obtained that major deficiencies in DNA damage response and tumor-suppressor-type genes can serve to elevate radiation cancer risk. Limited epidemiologic data from follow-up of second cancers in gene carriers receiving radiotherapy were supportive of the above conclusions, but quantitative judgments about the degree of increased cancer risk remain uncertain. However, since major germline deficiencies in the genes of interest are known to be rare, it is possible to conclude from published analyses that they are most unlikely to create a significant distortion of population-based estimates of cancer risk. The major practical issue associated with these strongly expressing cancer genes is judged to be the risk of radiotherapy-related cancer.

A major theme developing in the whole field of cancer genetics is the interaction and potential impact of more weakly expressing variant cancer genes that may be relatively common in human populations. Knowledge of such gene-gene and gene-environment interactions, although at an early stage, is developing rapidly. The animal genetic data reviewed in this chapter provide proof-of-principle evidence of how such variant genes with functional polymorphisms can influence cancer risk, including limited data on radiation

tumorigenesis. Attention has also been given to recent molecular epidemiology data on associations between functional polymorphisms and cancer risk, particularly with respect to DNA damage response genes. Some issues of study design have been discussed, and although much work has been reported on cancer risk in heterozygous carriers of the *ATM* gene, clear judgments about radiation risks remain elusive.

Given that functional gene polymorphisms associated with cancer risk may be relatively common, the potential for significant distortion of population-based risk was explored, with emphasis on the organ specificity of the genes of interest. A preliminary conclusion is that common polymorphisms of DNA damage response genes associated with organ-wide radiation cancer risk would be the most likely source of major interindividual differences in radiation response.

Although good progress is being made, there are important gaps in understanding the extent of genetic influences on radiation cancer risk. Accordingly, further work is needed in humans and mice on gene mutations and functional polymorphisms that influence radiation response and cancer risk. Human molecular genetic studies should, where possible, be coupled with epidemiologic investigations.

4

Heritable Genetic Effects of Radiation in Human Populations

INTRODUCTION AND BRIEF HISTORY

Naturally occurring mutations in somatic and germ cells contribute respectively to cancers and heritable genetic diseases (*i.e.*, hereditary diseases). The discoveries by Muller (1927) of the mutagenic effects of X-rays in fruit flies (*Drosophila*) and by Stadler (1928a, 1928b) of similar effects in barley and maize, and the subsequent extension of these findings to other types of ionizing radiation (and also to ultraviolet) and other organisms, conclusively established the genetic damage-inducing effects of radiation. However, widespread and serious concern over the possible adverse genetic effects of exposure of large numbers of people to low levels of radiation first arose in the aftermath of the detonation of atomic bombs over Hiroshima and Nagasaki in World War II, some 20 years after the discoveries of the mutagenic effects of X-rays. In June 1947, at the meeting of the Conference on Genetics convened by the Committee on Atomic Casualties of the U.S. National Research Council to assess the program of research on the heritable effects of radiation to be undertaken in Japan, the leading geneticists voted unanimously to record the following expression of their attitude toward the program: "Although there is every reason to infer that genetic effects can be produced and have been produced in man by atomic radiation, nevertheless the conference wishes to make it clear that it cannot guarantee significant results from this or any other study on the Japanese material. In contrast to laboratory data, this material is too much influenced by extraneous variables and too little adapted to disclosing genetic effects. In spite of these facts, the conference feels that this unique possibility for demonstrating genetic effects caused by atomic radiation should not be lost . . ." (NRC 1947). Thus came into existence the genetics program in Hiroshima and Nagasaki under the auspices of the Atomic Bomb Casualty Commission (ABCC), the newly formed joint agency of the Japanese Ministry of Health and Welfare and the U.S. National Academy of Sciences. The ABCC was renamed the Radiation Effects Re-

search Foundation in 1976. In the late 1940s, the mouse was chosen as the primary surrogate for assessing the genetic radiosensitivity of humans, and extensive studies were initiated in different research centers in the United States, England, and Japan.

In the mid-1950s, one major international and several national scientific bodies came into existence, including the United Nations Scientific Committee on the Effects of Atomic Radiation (UNSCEAR), the Committee on the Biological Effects of Atomic Radiation (the BEAR committee; renamed the Committee on the Biological Effects of Ionizing Radiation [BEIR] in 1972) set up by the U.S. National Academy of Sciences, and the Committee of the British Medical Research Council. The UNSCEAR and the BEIR committees have continued their work up to the present, periodically reviewing the levels of radiation to which human populations are exposed and improving assessment of the somatic and genetic risks of radiation exposure (NRC 1972, 1980, 1988, 1990, 1999; UNSCEAR 1993, 2000b, 2001).

From the beginning of these efforts, it was obvious that in the absence of direct human data on radiation-induced germ cell mutations, quantitative estimates of genetic risk could be derived only through a knowledge of the prevalence of naturally occurring hereditary ill health in the population, the role of spontaneous mutations in supporting this burden, and plausible assumptions on the rates of induced germ cell mutations in humans. The methods developed and used by the above committees for risk estimation, therefore, were necessarily indirect. All were geared toward using human data on genetic diseases as a frame of reference, together with mouse data on radiation-induced mutations, to predict the radiation risk of genetic disease in humans. Both the UNSCEAR and the BEIR committees are cognizant of the need to make assumptions given the consequent uncertainties in extrapolating from mouse data on induced mutation rates to the risk of genetic disease in humans.

Details of the genetics program that evolved in Japan and the vast body of data that emerged from these studies have been published in a series of articles. The most relevant ones have now been compiled in a single volume (Neel and Schull 1991). The most important finding of these studies is that there are no statistically demonstrable adverse genetic effects attributable to radiation exposures sustained by the survivors. Although cited and discussed in the UNSCEAR and BEIR reports over the years, these results did not constitute part of the "mainstream thinking" of genetic risk estimators and therefore were not used in risk estimation.

During the past few years, estimates of the baseline frequencies of Mendelian diseases have been revised and mathematical methods have been developed to estimate the impact of an increase in mutation rate (as a result of radiation exposures) on the frequencies of different classes of genetic diseases in the population. Additionally, there have been several advances in our understanding of the molecular basis and mechanisms of origin of human genetic diseases and of radiation-induced mutations in experimental systems. As a result of these developments, it now is possible to reexamine the conceptual basis of risk estimation, reformulate some of the critical questions in the field, and address some of the problems that could not be addressed earlier.

This chapter summarizes the general framework and the methods and assumptions used in risk estimation until the publication of BEIR V (NRC 1990). This is followed by a discussion of the advances in knowledge since that time, their impact on the concepts used in risk estimation, and how they can be employed to revise the risk estimates. Throughout this chapter, the terms "genetic diseases," "genetic effects," and "genetic risks" are used exclusively to mean "heritable genetic diseases," "heritable genetic effects," and "heritable genetic risks," respectively.

GENERAL FRAMEWORK

Goal of Genetic Risk Estimation

The goal of genetic risk estimation, at least as envisioned and pursued by UNSCEAR and the BEIR committees, remains prediction of the additional risk of genetic diseases in human populations exposed to ionizing radiation, over and above that which occurs naturally as a result of spontaneous mutations. The concept of "radiation-inducible genetic diseases," which emerged early on in the field, is based on two established facts and an inference. The facts are that (1) hereditary diseases result from mutations that occur in germ cells and (2) ionizing radiation is capable of inducing similar changes in all experimental systems adequately investigated. The inference, therefore, has been that radiation exposure of human germ cells can result in an increase in the frequency of genetic diseases in the population. Worth noting is the fact that although there is a vast amount of evidence for radiation-induced mutations in diverse biological systems, there is no evidence for radiation-induced germ cell mutations that cause genetic disease in humans.

Germ Cell Stages and Radiation Conditions of Relevance

From the standpoint of genetic risks, the effects of radiation on two germ cell stages are particularly important. In the male, these are the stem cell spermatogonia, which constitute a permanent germ cell population in the testes and continue to multiply throughout the reproductive life span of the individual. In the female, the corresponding cell stages are the oocytes, primarily the immature ones. The latter constitute the predominant germ cell population in the female. Female mammals are born with a finite number of oocytes formed during fetal development. These primordial oocytes, as they are called, grow, and a sequence of nuclear changes comprising meiosis takes place in them. The latter however are arrested at a particular stage until just before ovulation. Because oocytes are not replenished by mitosis during adult life and immature oocytes are the predominant germ cell population in the female, these are clearly the cell stages whose irradiation has great significance for genetic risks.

The radiation exposures sustained by germ cells in human populations are generally in the form of low-LET (linear energy transfer) irradiation (*e.g.*, X-rays and γ-rays) delivered as small doses at high dose rates (*e.g.*, in diagnostic radiology) or are greatly protracted (*e.g.*, continuous exposures from natural and man-made sources). In estimating genetic risks to the population therefore, the relevant radiation conditions are low or chronic doses of low-LET irradiation. As discussed later, most mouse data used for estimating the rates of induced mutations have been collected at high doses and high dose rates. Consequently, assumptions have to be made to convert the rates of induced mutations at high doses and dose rates into mutation rates for radiation conditions applicable for risk estimation in humans.

GENETIC DISEASES

Since the aim of genetic risk estimation is to predict the additional risk of genetic diseases relative to the baseline frequency of such diseases in the population, the concept of genetic diseases and their classification and attributes are considered in this section. The term genetic diseases refers to those that arise as a result of spontaneous mutations in germ cells and are transmitted to the progeny.

Mendelian Diseases

Diseases caused by mutations in single genes are known as Mendelian diseases and are further divided into autosomal dominant, autosomal recessive, and X-linked, depending on the chromosomal location (autosomes or the X chromosome) and transmission patterns of the mutant genes. In an autosomal dominant disease, a single mutant gene (*i.e.*, in the heterozygous state) is sufficient to cause disease. Examples include achondroplasia, neurofibromatosis, Marfan syndrome, and myotonic dystrophy. Autosomal recessive diseases require homozygosity (*i.e.*, two mutant genes at the same locus, one from each parent) for disease manifestation. Examples include cystic fibrosis, phenylketonuria, hemochromatosis, Bloom's syndrome, and ataxia-telangietasia.

The X-linked recessive diseases are due to mutations in genes located on the X chromosome and include Duchenne's muscular dystrophy, Fabry's disease, steroid sulfatase deficiency, and ocular albinism. Some X-linked dominant diseases are known, but for most of them, no data on incidence estimates are currently available. Therefore, these diseases are not considered further in this report. The general point with respect to Mendelian diseases is that the relationship between mutation and disease is simple and predictable.

Multifactorial Diseases

The major burden of naturally occurring genetic diseases in human populations, however, is not constituted by Mendelian diseases, which are rare, but by those that have a complex etiology. The term "multifactorial" is used to designate these diseases to emphasize the fact that there are multiple genetic and environmental determinants in their etiology. Their transmission patterns do not fit Mendelian expectations. Examples of multifactorial diseases include the common congenital abnormalities such as neural tube defects, cleft lip with or without cleft palate, and congenital heart defects that are present at birth, and chronic diseases of adults (*i.e.*, with onset in middle and later years of life) such as coronary heart disease, essential hypertension, and diabetes mellitus.

Evidence for a genetic component in their etiology comes from family and twin studies. For example, first-degree relatives of patients affected with coronary heart disease have a two- to sixfold higher risk of the disease than those of matched controls, and the concordance rates of disease for monozygotic twins are higher (but never 100%) than those for dizygotic twins (Motulsky and Brunzell 1992; Sankaranarayanan and others 1999).

As mentioned earlier, multifactorial diseases are presumed to originate from the joint action of multiple genetic and environmental factors; consequently, the presence of a mutant allele is not equivalent to having the disease. For these diseases, the interrelated concepts of *genetic susceptibility* and *risk factors* are more appropriate. The genetic basis of a common multifactorial disease is the presence of a genetically susceptible individual, who may or may not develop the disease depending on the interaction with other genetic and environmental factors. These concepts are discussed further in Annex 4A. The important general point is that unlike the situation with Mendelian diseases, the relationships between mutations and disease are complex in the case of multifactorial diseases. For most of them, knowledge of the genes involved, the types of mutational alterations, and the nature of environmental factors remains limited. Among the models used to explain the inheritance patterns

of multifactorial diseases and to estimate the recurrence risks in relatives is the multifactorial threshold model (MTM) of disease liability. The MTM, its properties, and its predictions are discussed in Annex 4A.

Chromosomal Diseases

Historically, both UNSCEAR and the BEIR committees have always had an additional class of genetic diseases— "chromosomal diseases"—in their lists that included those that had long been known to arise as a result of gross (*i.e.*, microscopically detectable), numerical (*e.g.*, Down's syndrome, which is due to trisomy of chromosome 21), or structural abnormalities of chromosomes (*e.g.*, cri du chat syndrome, due to deletion of part or the whole short arm of chromosome 5 [5p-]). As discussed later, this is really not an etiological category, and deletions (microscopically detectable or not) are now known to contribute to a number of constitutional genetic diseases grouped under autosomal dominant, autosomal recessive, and X-linked diseases.

RISK ESTIMATION METHODS

In the absence of data on radiation-induced germ cell mutations that can cause genetic disease in humans, all of the methods developed and used for predicting the risk of genetic disease from the mid-1950s to the present are indirect. Their strengths and weaknesses are reviewed in BEIR V (NRC 1990). One such indirect method is the doubling dose method, on which attention is focused in this section. It has been in use since the early 1970s (NRC 1972, 1990; UNSCEAR 1977, 1982, 1986, 1988) and is used in the recent UNSCEAR (2001) report.

The Doubling Dose Method

The doubling dose method enables expressing of the expected increase in disease frequency per unit dose of radiation in terms of the baseline frequency of the disease class. The doubling dose (DD) is the amount of radiation required to produce in a generation as many mutations as those that arise spontaneously. Ideally, it is estimated as a ratio of the average rates of spontaneous and induced mutations in a given set of genes:

$$\text{DD} = \text{average spontaneous mutation rate/average induced mutation rate.} \quad (4\text{-}1)$$

The reciprocal of the DD (*i.e.*, 1/DD) is the relative mutation risk (RMR) per unit dose. Since RMR is the reciprocal of DD, the smaller the DD, the higher is the RMR and *vice versa*. With the doubling dose method, until recently, risk was estimated as a product of two quantities—namely, the baseline disease frequency, P, and 1/DD:

$$\text{Risk per unit dose} = P \times (1/\text{DD}). \quad (4\text{-}2)$$

The population genetic theory that underlies the use of Equation (4-2) is the equilibrium theory that population geneticists use to explain the dynamics of mutant genes in populations. The theory assumes that the stability of mutant gene frequencies (and thus disease frequencies) in a population is the result of the existence of a balance between the rates at which spontaneous mutations enter the gene pool in every generation and the rate at which they are eliminated by natural selection (*i.e.*, through failure of survival or reproduction).

When the mutation rate is increased as a result of radiation in every generation, this balance between mutation and selection is disturbed by the influx of induced mutations, but the prediction is that the population will attain a new equilibrium (over a number of generations) between mutation and selection. The amount of increase in mutation frequency, the time it takes for the population to reach the new equilibrium, and the rate of approach to equilibrium are all dependent on induced mutation rates, the intensity of selection, the type of genetic disease, and whether the radiation exposure occurs in one generation only or generation after generation. It should be noted that since the starting population (before radiation exposure) is assumed to be in equilibrium between mutation and selection, the quantity P in Equation (4-2) represents the equilibrium incidence of the disease, and the product of P and $1/DD$ is the expected increase in disease frequency at the new equilibrium.

Risk Estimation for Different Classes of Genetic Disease

The application of Equation (4-2) to risk estimation is straightforward for autosomal dominant diseases since the relationship between mutation and disease is simple for this class of diseases. Population genetic theory predicts that for these diseases, if there is an *x%* increase in mutation rate in every generation, at the new equilibrium this increase will be reflected as an *x%* increase in the frequency of these diseases. Until recently, estimates of risk for the first, second, or any postradiation generation of interest were obtained through "back calculation" from the predicted new equilibrium incidence using certain assumptions. If the population sustains radiation exposure in one generation only, there will be a transient increase in the mutant frequency in the first postradiation generation, followed by a progressive decline to the "old" equilibrium value.

The method used to predict the risk of X-linked diseases is approximately similar to that for autosomal dominant diseases discussed above. For autosomal recessive diseases, the risk calculation is more involved because when recessive mutations first arise (or are induced), they are present in the heterozygous state and do not precipitate disease in children of the first few postradiation generations. For multifactorial diseases, the situation is complex in that there is no simple relationship between mutation and disease, and as discussed

later, the estimate of risk will depend on the model used for their maintenance in the population.

The Concept of Mutation Component

The concept of mutation component and the statistic MC, which is derived using this concept, help to unify attempts at predicting how the frequencies of different classes of genetic diseases in the population will change as a result of increases in mutation rate. The mutation component is defined as the relative increase in disease frequency (*i.e.*, relative to the baseline frequency) per unit relative increase in mutation rate (*i.e.*, relative to the spontaneous mutation rate). First introduced in BEIR I (NRC 1972) to address the problem of the impact of the radiation risk of multifactorial diseases in the population, and subsequently elaborated by Crow and Denniston (1981, 1985) and Denniston (1983), the concept can be used for all classes of genetic disease as done in BEIR V (NRC 1990). During the past few years, the concept has been developed further with the necessary algebraic formulations, that permit a direct evaluation of the impact of an increase in mutation rate for all classes of genetic disease in any postradiation generation of interest following exposure to radiation in either one generation only or generation after generation (Chakraborty and others 1998a; Denniston and others 1998). These advances are considered in a later section. Suffice to note here that the inclusion of MC in Equation (4-2) yields the revised equation:

$$\text{Risk per unit dose} = P \times (1/DD) \times \text{MC}, \qquad (4\text{-}3)$$

where MC is the disease class and postradiation generation-specific mutation component and the other two quantities are as defined earlier.

RECENT ADVANCES WITH RESPECT TO THE THREE QUANTITIES USED WITH THE DD METHOD OF RISK ESTIMATION

The BEIR V report (NRC 1990) reviewed the advances that occurred from the mid-1950s to 1990 with respect to P, the baseline frequency of genetic disease, DD, and MC, the three quantities considered relevant for risk estimation with the DD method thus far. In the material that follows, attention is focused on progress made since 1990.

Baseline Frequencies of Genetic Diseases

Mendelian Diseases

Estimates of the baseline frequencies of Mendelian diseases used by UNSCEAR since its 1977 report and by the BEIR III and BEIR V committees (NRC 1980, 1990) have been based on the compilations and analysis of Carter (1976a, 1976b) primarily for Western European and Western

European-derived populations. These are the following (all in live births): autosomal dominants, 0.95%; X-linked, 0.05%; and autosomal recessive, 0.25%. Advances in human genetics during the past two decades now permit an upward revision of the above estimates to 1.5% for autosomal dominant diseases, 0.15% for X-linked diseases, and 0.75% for autosomal recessive diseases (Sankaranarayanan 1998). Note that the revised total frequency of Mendelian diseases is thus 2.4%, which is about twice the earlier figure of 1.25%.

Multifactorial Diseases

For multifactorial diseases (which include congenital abnormalities present at birth and chronic diseases), the estimates used by UNSCEAR (1986, 1988, 1993, 2001) derive from data obtained for the population of Hungary (Czeizel and Sankaranarayanan 1984; Czeizel and others 1988). These estimates are 6% of live births for congenital abnormalities and 65% of the population affected by chronic diseases (excluding cancers). Since most chronic diseases have their onset in middle and late ages (published figures pertain to these age groups), data on the distribution of the population in various age intervals (*i.e.*, ages 0, 1, 2, 3–4, 5–9, 10–14, . . . 80–84, 85+, etc.; a total of 21 age intervals) for 1977 to 1981 were used to obtain estimates applicable to the population as a whole. For example, if the published estimate for a given disease pertains to the adult population (*i.e.*, above age 14), the figure was reduced by 21% since the 0–14 year age group constituted 21% of the total population of 10.7 million (Czeizel and others 1988).

For the BEIR V committee (NRC 1990), the starting point for congenital abnormalities was the published data of Czeizel and Sankaranarayanan (1984) and Czeizel and others (1988), which gave an incidence estimate of 6%. This figure was reduced to 2–3% by noting that the 6% figure is ". . . so high, in part, because of the unusually high frequency of congenital dislocation of the hip in Hungary" (Czeizel and Sankaranaryanan 1984). For chronic diseases, the starting point was the estimate of about 60% based on preliminary data of Czeizel and colleagues made available to and used by UNSCEAR in its 1988 report. The BEIR V committee reduced the figure of 60% to 30% by (1) subtracting the estimates for essential hypertension, acute myocardial infarction, other acute and subacute forms of ischemic heart disease, and varicose veins of the lower extremities (together about 25%) and (2) reducing the figure for juvenile osteochondrosis of the spine from 11% (based on radiographic screening) to about 0.5% (on the assumption that only about 5% of the cases identified by radiographic screening may be deemed to be of clinical significance). The resulting adjusted figure of about 30% was given as the estimate for the "selected others" subgroup of "other diseases of complex etiology." Together with the earlier committee's figures for heart disease (60%) and cancer (30%; which were termed "round number approximations"

for all varieties of the above diseases), the total became 120%. Footnote *f* to Table 2-5 of the BEIR V report (NRC 1990) offers the following explanation for the 120% figure: "Includes heart disease, cancer, and other selected disorders . . . . Note that the total exceeds 100%. The genetic component in many of these traits is unknown. To the extent that genetic influences are important, the effects are through genes that have small individual effects but that act cumulatively among themselves and in combination with environment factors to increase susceptibility."

Estimates of Baseline Frequency of Multifactorial Diseases Used in This Report

In examining what would be considered a reasonable estimate of baseline frequency of congenital abnormalities for use in risk estimation, the BEIR VII committee took note of the vast body of data on their prevalence in different parts of the world, including some large-scale studies carried out in North America (Myrianthopoulos and Chung 1974; Trimble and Doughty 1974; Baird and others 1988). The estimates vary over a wide range, from about 1% in live births to a high of about 8.5% in total births (*i.e.*, still- and live births), depending on, among other things, the definition, classification, and diagnostic criteria; entities included; method of ascertainment; duration of follow-up of live-born children; and sample sizes. In one of the largest U.S. studies (Myrianthopoulos and Chung 1974), the overall frequency of major abnormalities was 8.3% (53,257 deliveries of known outcome), which compares favorably with the estimate of about 6% from British Columbia (Baird and others 1988) and of about 6% from Hungary mentioned earlier. This documents the premise that under conditions of good ascertainment, the overall prevalences are similar and are of the order of about 6%. This committee therefore accepts the 6% figure as reasonable for use in risk estimation in this report.

For chronic multifactorial diseases, the committee prefers to use the estimate of 65% obtained by Czeizel and colleagues (1988) in view of the fact that the estimate is based on 26 clear-cut disease entities defined by ICD (International Classification of Diseases) code numbers that were studied epidemiologically in a large population. This estimate was also used by UNSCEAR (1988, 1993, 2001) as the best available overall estimate for chronic diseases as a whole (excluding cancers). Included in the above estimate are heart or blood vessel-related diseases, together, about 25%. For the estimate of 60% mentioned in BEIR V (NRC 1990) under the heading "heart disease" no verifiable source or study is cited. Likewise, for cancers, BEIR V cites an estimate of 30%, again with no citation of the source or the types of cancers included. As mentioned earlier, both of these numbers represent round number approximations.

In the view of the BEIR VII committee, the inclusion of cancers in estimating the *heritable* risks of radiation is not meaningful at the present state of knowledge.

Estimates of Baseline Frequency of Chromosomal Disease

The BEIR V report (NRC 1990) and the UNSCEAR (1993) report assessed the baseline prevalence of chromosomal diseases to be of the order of about 0.4% in live births. The present committee sees no reason to alter this estimate.

Summary of Current Estimates of Baseline Frequencies of Genetic Diseases and Comparison with Those in BEIR V

Table 4-1 presents these comparisons showing that the current estimates for Mendelian diseases are higher than those used in 1990, while those for the other classes remain essentially unchanged.

The Doubling Dose

As discussed earlier DD is one of the important quantities used in the equation for the doubling dose method of risk estimation. Although the DD concept was formulated by Muller (1951, 1954, 1959) in the 1950s and several possible estimates and/or ranges of DDs were discussed in the BEAR report (NRC 1956), in UNSCEAR (1962), and in Lüning and Searle (1971), actual use of the method to obtain quantitative estimates of risk began only in 1972 (NRC 1972). Changes in the conceptual basis and database used for DD estimates from the mid-1950s to the early 1990s have recently been reviewed (Sankaranarayanan and Chakraborty 2000a).

TABLE 4-1 Estimates of the Baseline Prevalences of Genetic Diseases Used in BEIR VII and BEIR V

Disease Class	Baseline Prevalence Estimates per 10^6 Live Births	
	BEIR VII	BEIR V
Mendelian		
Autosomal dominant	15,000	10,000
X-linked	1500	400
Autosomal recessive	7500	2,500
Chromosomal	~4000	~4000
Multifactorial		
Congenital abnormalities	60,000	20,000–30,000
Chronic multifactorial	650,000	[a]
Other Disorders of Complex Etiology		
Heart disease	[b]	600,000
Cancer	[c]	300,000
Selected others	[b]	300,000

[a]BEIR V included these diseases under "other disorders of complex etiology."

[b]Included under chronic multifactorial diseases in BEIR VII.

[c]Not specifically considered in this chapter.

SOURCE: Table reproduced with permission from Chakraborty and others (1998b).

Table 4B-1 (see Annex 4B) summarizes the important developments. As evident from that Table, with one exception, most of the DD estimates used in risk estimation by UNSCEAR and the BEIR committees were based on data on both spontaneous *and* induced mutation rates in mice. The one exception was BEIR I (NRC 1972), which used data on spontaneous rate of mutations of human genes and induced rate of mutations in mouse genes. As discussed below, re-evaluation of the assumptions underlying the use of mouse data on spontaneous mutation rate for DD calculations has shown that these are incorrect and that the use of human data on spontaneous mutation rates along with mouse data on induced rates is correct.

Incorrectness of the Assumption of Similarity of Spontaneous Mutation Rates in Mice and Humans—The Need to Use Human Spontaneous Mutation Rates for DD Calculations

Extrapolation of the mouse-based DD to humans for risk estimation implies the assumption that both the spontaneous and the induced rates of mutations are similar in the two species. The assumption of similarity of induced rates of mutations in both species is defensible on the grounds of generally similar gene organization, 70–90% homology in DNA sequence of genes, and substantial conservation of synteny for many chromosomal regions between humans and mice. However, the situation is different with respect to spontaneous mutations.

The reasons spontaneous mutation rates in humans are unlikely to be similar to those in mice have been discussed (Sankaranarayanan 1998). Briefly, these have to do with the differences in the number of cell divisions between the zygote and the mature germ cell in the two species. Vogel and Motulsky (1997) estimate that in human females, the number of cell divisions from zygote to the mature egg (N_f) is of the order of about 24. For the mouse female, estimates of Drost and Lee (1995) suggest that N_f is of the same order. So, from the standpoint of N_f, human and mouse females are similar.

In human males, however, the comparable number of cell divisions is much higher; it is about 30 until the age of puberty (taken to be 15 years), ~23 per year thereafter, and 6 for proliferation and meiosis. Thus, the number of cell divisions prior to sperm production (N_m) in a 20-year-old male can be estimated to be $30 + (5 \times 23) + 6 = 151$, increasing to 381 at age 30 years, 611 at age 40 years, and 841 at age 50 years (Crow 1999). The N_m/N_f thus increases with paternal age, being 6.3 at age 20, 15.9 at age 30, 25.5 at age 40, and 35.0 at age 50. In the male mouse, the number of cell divisions from zygote to sperm is of the order of about 62 at age 9 months, assuming a 9-month generation (Chang and others 1994; Drost and Lee 1995; Li and others 1996). The N_m/N_f ratio in the mouse is therefore 2.5 (*i.e.*, 62/25), which is much lower than in humans. The committee notes that in most mouse experiments, the parental animals were used at a rather uniform age (usually about 12 weeks), and the question of paternal age effects has not been specifically addressed.

Since most spontaneous mutations arise as a result of errors in DNA replication, one would expect that the mutation rate in human males would be higher than that in females and that there would be an increase in the likelihood of spontaneous germinal mutations with the age of the male (so-called paternal age effect). By and large, these expectations have been fulfilled. The literature on this subject and the recent evidence from molecular studies have been reviewed (Crow and Denniston 1985; Crow 1993, 1997, 1999; Vogel and Motulsky 1997; Sankaranarayanan 1998; Green and others 1999).

When one considers the large differences in life span between humans and mice and the paternal age effect for spontaneous mutations in humans, it is clear that extrapolation from short-lived mice to humans is unlikely to provide a reliable average spontaneous rate in a heterogeneous human population of all ages. This is one reason to abandon the use of the mouse data on spontaneous mutation rates in DD calculations and to use human spontaneous mutation rates instead. The following arguments support this: (1) estimates of spontaneous mutation rates in humans are unweighted averages of the rates in the two sexes (and therefore automatically incorporate sex differences and paternal age effects), and (2) the sex-averaged rate is relevant in the context of DD calculations (Sankaranarayanan 1998).

A second reason for not using the mouse spontaneous mutation rates for DD calculations is that the whole question of spontaneous mutation rates in mice has now assumed an unexpected complexity due to the noninclusion, until recently, of mutations that originated as germinal mosaics (resulting in progeny carrying the same mutation ["clusters"] in the following generation) in estimates of spontaneous mutation rates in the specific locus experiments (Russell and Russell 1996; Selby 1998a, 1998b; Russell 1999). According to Russell and Russell (1996), if mosaic data are included, the total spontaneous rate becomes twice that of 6.6 × 10^{-6} per gene based on mutations that arose singly. However, Selby (1998a, 1998b) has argued that (1) the data on clusters should be included in calculating the total spontaneous mutation rate; (2) his computer simulation studies (which incorporate clusters in his model) suggest an increase of the rate by a factor of about 5 compared to that based on mutations that arose singly; (3) the fivefold higher total spontaneous rate is the appropriate numerator in DD calculations; and (4) if paternal age effects are extrapolated from humans to mice, the estimate of spontaneous rate is even higher. In the view of this committee, the above argument cannot be sustained for humans for the following reasons:

First, while there is no doubt that a proportion of spontaneous mutations in human genes arise as germinal mosaics (and can potentially result in clusters in the following gen-eration), the limited data available on mosaics and clusters at present preclude a quantitative assessment of their contribution to spontaneous mutation rates. The main relevance of germinal mosaicism in the human context is this: the parent who carries a mosaic mutation for an autosomal dominant or X-linked trait does not have a mutant phenotype and therefore would not be considered as having a risk of producing affected children. However, because his or her gonads contain mutant and normal cells, he or she may run the risk of having more than one progeny who carries the mutant gene (mutational "clusters").

Second, if a substantial proportion of human mutations arise as germinal mosaics in one generation and result in clusters in the following generation, the frequencies of at least autosomal dominant and X-linked diseases also have to be corrected upwards to account for this possibility; there is no reliable way of doing this at present. The published estimates of human spontaneous mutation rates do not provide sufficient grounds for assuming that substantial proportions of mutations in the germ cells first arose as mosaics and subsequently resulted in clusters of mutations; if this had been the case, major increases in the frequencies of affected individuals from one generation to the next would have been observed, but this does not appear to be true. Further, family sizes in present-day human populations are limited (in fact, they are so small that there is almost never more than one affected offspring from a mating, in contrast to the situation in mice where large numbers of progeny are obtained from a single male). Both of these arguments support the view that mutational clusters are much less relevant in humans than in mice.

The advantages of using human spontaneous mutation rates for DD calculations are (1) they pertain to human disease-causing genes; (2) as mentioned earlier, the mutation rate estimates in humans, because they are averaged over both sexes, automatically include sex differences and paternal age effects; and (3) in mutation rate calculations, human geneticists count all mutants that arise anew irrespective of whether they were part of a cluster or not; if clusters had occurred, they would have been included. The committee therefore accepts the view that the use of human spontaneous rates and mouse induced rates for DD calculations (*i.e.*, the procedure used in BEIR I; NRC 1972) is more logical, and it has assessed published data on spontaneous mutation rate in humans and induced rates of mutations in mice.

Doubling Dose Estimation Using Spontaneous Mutation Rates of Human Genes and Induced Rates of Mouse Genes

Estimation of the Average Spontaneous Mutation Rate of Human Genes

To calculate a representative average spontaneous mutation rate of human genes, the available estimates for individual autosomal dominant diseases published by Childs (1981) and Vogel and Motulsky (1997) were used, irrespective of whether these diseases have high or low prevalence or high or low mutation rates. However, the analysis took into account the numbers of genes thus far known or estimated to underlie each of these disease phenotypes (Vogel and Motulsky 1997; Sankaranarayanan 1998; McKusick 2000). This represents an important departure from earlier estimates based on disease phenotypes alone, which generally assumed a one-to-one relationship between mutation and disease. Details of these diseases, estimates of mutation rates, and selection coefficients are given in Table 4-2. The (unweighted) average mutation rate derived from these data (for some 26 autosomal dominant phenotypes with an esti-

mated 135 loci) is $(2.95 \pm 0.64) \times 10^{-6}$ per locus per generation. This figure is within the range of 0.5×10^{-5} to 0.5×10^{-6} per locus used in the 1972 BEIR I report (NRC 1972).

The list of autosomal dominant diseases used to provide the basis for the prevalence estimate (*P* in Equation (4-3)) encompasses many more than the 26 diseases used in the above calculations (Sankaranarayanan 1998); these other diseases could not be included in the present analysis because of lack of information on mutation rates. Further, the mutation rate estimates for X-linked phenotypes have not been included in these calculations; instead, it has been assumed that the average spontaneous mutation rate for autosomal dominant genes calculated above can also be used for X-linked genes. The justification for this assumption rests on the following lines of reasoning: (1) among Mendelian diseases, autosomal dominants constitute the most important group from the standpoint of genetic risks, and (2) although X-linked recessive diseases are also expected to respond directly to an increase in mutation rate, since their prevalence is an order of magnitude lower than that of autosomal dominants (*i.e.*, 0.15% versus 1.5%) the assumption of similar spontaneous rates of mutations for autosomal dominants and X-linked recessives is unlikely to result in any significant underestimation of the total risk. In fact, for this reason, these two classes of diseases are considered together in risk estimation.

The Average Rate of Induced Mutations in Mice

To calculate the average rate of induced mutations in mice, the committee used all available data on rates of induced mutations in defined genes in mice; these relate to recessive specific locus mutations at 12 loci, biochemical mutations (null enzyme mutations, also recessive at a large number of loci), and autosomal dominant mutations at 4 loci incidentally detected in the course of the specific locus experiments. The data on these autosomal dominant mutations are all from studies carried out in Harwell; comparable data from Oak Ridge studies were unavailable. Inclusion of the data on dominant mutations in mutation rate calculations was dictated by the consideration that although the underlying genes were not well defined at the time these experiments were performed (but mutations were "frequently" observed and recorded, indicating that they were among the more radiation-mutable loci), we now know not only their identity (and the molecular nature of the mutations) but also their human counterparts (the mouse *Sl*, *W*, *Sp*, and *T* correspond to, respectively, the *MGF*, *KIT*, *PAX3*, and *T* genes in humans; see McKusick 2000). All of the data considered here come from experiments involving stem cell spermatogonia.

The data from female mice have not been used because there is uncertainty about whether mouse immature oocytes are a good model for assessing the mutational radiosensitivity of human immature oocytes (UNSCEAR 1988). The arguments rest on (1) the strikingly higher sensitivity of mouse

TABLE 4-2 Database for Estimating Average Spontaneous Mutation Rate of Human Autosomal Genes Associated with Autosomal Dominant Diseases and Their Selection Coefficients(s)

Disease Phenotype	Estimated No. of Loci	Mutation Rate $(\times 10^6)^a$	Selection Coefficient(s)^b
Achondroplasia	1	11.0	0.8
Amelogenesis imperfecta	1	1.0	0
Aniridia	2	3.8	0.1
Apert's syndrome	1	3.5	0
Blindness	9	10.0	0.7
Cataracts (early onset)	30	6.0	0.3
Cleft lip	1	1.0	0.2
Deaf mutism	15	24.0	0.7
Dentinogenesis imperfecta	2	1.0	0
Huntington disease	1	5.0	0.2
Hypercholesterolemia	1	20.0	0
Marfan syndrome	1	5.0	0.3
Multiple exostoses	3	7.7	0.3
Myotonic dystrophy	1	18.0	0.3
Neurofibromatosis	2	70.0	0.5
Osteogenesis imperfecta	2	10.0	0.4
Osteopetrosis	1	1.0	0.2
Otosclerosis	1	20.0	0
Polyposis of intestine	1	10.0	0.2
Polycystic kidney disease	2	87.5	0.2
Porphyria	2	1.0	0.05
Primary basilar impression	1	10.0	0.2
Rare diseases (early onset)	50	30.0	0.5
Retinoblastoma	1	8.7	0.5
Spherocytosis	1	22.0	0.2
Tuberous sclerosis	2	8.0	0.8
Total	135		
Average		(2.95 ± 0.64)	0.294

^{a}For some entries, mutation rate estimates are uncertain (see Childs 1981 for details).

^{b}Estimated from reproductive fitness.

SOURCE: Childs (1981); Vogel and Motulsky (1997).

immature oocytes to radiation-induced killing (the majority are destroyed by 0.5 Gy; Oakberg and Clark 1964) in contrast to those of human and rhesus monkey immature oocytes, for which the dose required is at least 100 times higher (Baker 1971) and (2) the observations that no mutations were recovered from oocytes sampled 7 weeks after irradiation in contrast to the situation with mature and maturing oocytes (Russell 1965). In view of this uncertainty and in order not to underestimate the risk, the committee has used the assumption that the rate estimated for males will also be applicable to females.

Details of the data used are summarized in Tables 4-3A to 4-3C and are from experiments involving acute X-irradiation or from high-dose fractionated X-irradiation (usually two fractions separated by 24 h) appropriately normalized to acute X-irradiation conditions (see Table 4-3A, footnote *d*; and Table 4-3B, footnotes *a* and *b*) to permit easy comparisons. Table 4-3A shows that the average rate of induced

mutations is highest at the original seven specific loci (3.03×10^{-5} per locus per gray) and is about one-third of the above at the six loci used in the experiments of Lyon and Morris (1969; *i.e.*, 0.78×10^{-5} per locus per gray; one locus, *a*, is common to both sets). For various sets of biochemical loci at which null mutations have been scored, the estimates vary over a range from 0.24×10^{-5} to 1.64×10^{-5} per locus per gray. The average rate for dominant visible mutations is within the above range. The unweighted average of the induced mutation rates is 1.09×10^{-5} per locus per gray for acute irradiation. The use of this rate for DD calculations, however, is somewhat problematic since (1) there is overlap of one or more loci in different data sets; (2) in some studies (see footnote *e*, Table 4-3A), all of the loci involved could not be ascertained; and (3) there is no simple way of taking into account the interlocus variation and sampling variance of induced rates from the derived average estimate of 1.09×10^{-5} per locus per gray.

TABLE 4-3A Database for Calculating Rates of Induced Mutations in Mice

System	No. of Loci	Average Rate/ Locus/Gy ($\times 10^5$)	Reference
1. The 7-locus system (Lyon and others 1964) (3 and 6 Gy; acute X- or γ-irradiation or 3 + 3 Gy, 24 h interval)	7[a]	3.03	Phillips (1961); Russell (1965, 1968); Lyon and others (1972); Cattanach and Rasberry (1994); Pretsch and others (1994)
2. The 6-locus system (Lyon and others 1964) (6 Gy; acute X-irradiation)	6[b]	0.78	Lyon and Morris (1969)
3. Biochemical loci (recessive, null enzyme) (3 + 3 Gy, 24 h interval; X-rays)	12[c]	0.70[d]	Charles and Pretsch (1986); Pretsch and others (1994)
4. Biochemical loci (recessive, null enzyme) (3 Gy, 3 + 3 Gy, 24 h interval and 6 Gy; X-rays)	32[e] 32 32	1.64 0.67[d] 0.24	Unpublished data of S.E. Lewis, cited in Neel and Lewis (1990)
5. Biochemical loci (recessive, null enzyme) (3 + 3 Gy, 24 h interval; X-rays)	4[f]	1.24[d]	Unpublished data of J. Peters, cited in Neel and Lewis (1990)
6. *Dominant* visibles (*Sl*, *W*, *Sp* and *T*)[g] (X rays)	4	0.44	See Table 4-3B

Unweighted average: $8.74/8 = 1.09 \times 10^{-5}$ per locus per gray

NOTE: Data are from experiments involving irradiation of males (stem cell spermatogonia) and all rates are normalized to single acute X-irradiation conditions.

[a]*a:* non-agouti; *b:* brown; *c:* chinchilla; *d:* dilute; *p:* pink-eyed dilution; *s:* piebald; *se:* short ear; in the work of Pretsch and others (1994), with some strains, mutations at four or five of these loci were scored.

[b]*a:* non-agouti; *bp:* brachypodism; *fz:* fuzzy; *ln:* leaden; *pa:* pallid; *pe:* pearl.

[c]*Ldh1, Tpi, Gpi1, Pgk, G6pd1, G6pd2, Pk, Gr, Mod1, Pgam, Gapdh, Ldr.*

[d]Normalized assuming additivity of the effect of dose fractionation.

[e]*Acy1, Car2, G6pd1, Ggc, Es1, Es3, G6pd1, Gpi1, Hba, Hbb, Idh1, Ldh1, Ldh2, Mod1, Mod2, Np1, Pep2, Pep3, Pep7, Pgm1, Pgm2, Pgm3, Pk3, Trf* (the identity of the other 8 loci could not be ascertained).

[f]*Hba, Hbb, Es3, Gpi1.*

[g]*Sl:* steel; *W:* dominant spotting; *Sp:* splotch; *T:* brachyury.

TABLE 4-3B Dominant Visible Mutations Recovered in the Course of Mouse Specific Locus Experiments (Spermatogonial Irradiation)

| Expt No. | X-ray Dose (Gy) | Number of Progeny | Number of mutations at | | | | | Mutations per Locus per Gray ($\times 10^5$) | Reference |
			Sl	*W*	*Sp*	*T*	Total		
1	6 + 6 (8-week interval)	3,612	1	—	—	—	1	0.58[a]	Lyon and others (1964)
2	6	16,735	—	1	—	—	1	0.25	Lyon and Morris (1969)
3	5 + 5	7,168	1	—	—	—	1	0.35[a]	Cattanach and Moseley (1974) Cattanach and others (1985)
4	3 + 3	7,645	2	—	—	—	2	1.09[a]	Cattanach and Rasberry (1994) Cattanach and others (1985)
5	3 + 3	15,849	1	1	1	3	6	0.35[b]	Cattanach and Rasberry (1994) Cattanach and others (1985)
6	6	10,897	1	—	—	—	1	0.38	Cattanach and Rasberry (1994)
7	6	19,285	1	—	—	—	1	0.22	Cattanach and Rasberry (1994)
8	1 + 9	10,318	1	—	—	1	2	0.24[a]	Cattanach and others (1985)
9	1 + 9	14,980	—	—	—	3	3	0.50[a]	Cattanach and others (1985)
								Unweighted average: 3.96/9 = 0.44 per locus per gray	

NOTE: Experiments were carried out during 1964–1994 in Harwell, England. All rates are normalized to single acute X-irradiation conditions.

[a]Normalized to single unfractionated irradiation conditions under the assumption of additivity of yields.

[b]Normalized to single unfractionated irradiation (by dividing the rate by 3) on the basis of observations of the enhancement of specific locus mutation frequency (in the same experiment by a factor of 3 [3H1 strain of mice]).

The committee therefore used the following approach to derive the average induced rate of mutations. All experimental data were first grouped by loci, so that an unweighted estimate of the locus-specific induced rates could be derived from the average of the estimates from all experiments involving each of the loci. Subsequently, these locus-specific rates were averaged across loci to arrive at the average induced mutation rate. This procedure permitted calculation of the standard error of the estimated rate that incorporated the sampling variability across loci as well as the variability of the rates in individual experiments. In this approach, unpublished data of Neel and Lewis (1990) were excluded since details of the identity of all the loci and the loci at which mutations were recovered were unavailable. Although fewer data were used (the total number of loci became 34), this approach was considered preferable since (1) no locus is double-counted while averaging over all loci, (2) the loci and the corresponding mutant phenotypes are clear, and (3) an estimate of the standard error of the mean (which takes into

account both intra- and interlocus variability) can be given. These data permit an overall average estimate of $(1.08 \pm 0.30) \times 10^{-5}$ per locus per gray (Table 4-3C). With a dose-rate reduction factor of 3 traditionally used[1] (Russel 1965;

---

[1]In the mouse, the dose-rate reduction factor of 3 for spermatogonial irradiations comes not only from the 6 Gy data of Dr. William Russell but also from the analysis of Dr. Tony Searle published in the *Proceedings of the Cortina International Radiation Reseach Conference* in 1967. Dr. Searle analyzed all of the chronic radiation data in the range from 37.5 to 861 R statistically and showed that the exposure-frequency relationship is linear and that the straight line of best fit could be described by

$$Y = 8.34 \times 10^{-6} + 6.59 \times 10^{-8} X,$$

where Y is the yield of mutations and X is the exposure in roentgens. The slope is one-third of that for acute X-irradiation (300 and 600 R).

Further, the following statement from BEIR V (NRC 1990, p. 110) provides additional substantiation for the dose-rate reduction factor of 3: "The other important baseline value for spermatogonia is for the response to low dose-rate, low-LET irradiations . . . the rate is $(7.3 \pm 0.8)10^{-8}$/locus/ rad for total doses between 35 and 900 rad (Ru82a). The dose-rate factor is 3.0 ± 0.4."

TABLE 4-3C Locus-Specific Rates for Radiation-Induced Mutations in Mice Estimated from Data Tables 4-3A and 4-3B

Locus[a]	Rate per Gray ($\times 10^5$)	SE ($\times 10^5$)
pa	0	0
pe	0	0
G6pd1	0	0
G6pd1	0	0
Ldh2	0	0
Ldr	0	0
Pgk1	0	0
Tpi	0	0
Hba2	0	0
Hbb1	0	0
Hbb2	0	0
Gapdh	0	0
Pk	0	0
Mod1	0	0
Sp	0.04	0.04
W	0.15	0.12
Gpi	0.33	0.33
a	0.45	0.24
T	0.45	0.18
ln	0.67	0.67
Ldh1	0.97	0.69
se	0.97	0.33
Sl	1.31	0.51
bp	1.34	0.95
Es3	1.67	1.67
Hba1	1.67	1.67
c	1.90	0.48
Gr	2.19	1.40
b	2.35	0.52
fz	2.68	1.34
p	2.93	0.56
d	3.14	0.62
Pgam	3.91	1.93
s	7.59	0.89
Average rate (acute irradiation)	1.08	0.30[b]
Chronic irradiation	0.36	0.10[b]

NOTE: For raw data and their analysis, see Sankaranarayanan and Chakraborty (2000a).

[a]In these calculations, two additional loci (*Ldh2* in the experiments of Pretsch and others 1994; *Hba2* in the experiments of Peters) have been included based on current evidence (Lewis and Johnson 1986).

[b]The standard error of the average rate was calculated taking into account variation of the rates among loci as well as sampling variation of the experimental data for each locus.

Searle 1967), the rate for chronic low-LET radiation conditions becomes $(0.36 \pm 0.10) \times 10^{-5}$ per locus per gray.

It is worth reiterating here that this is the first time an attempt has been made to use the mutation data coming not only from the 7 specific loci but also from all loci for which there are published data (a total of 34 loci; see Table 4-3C) taking into account interlaboratory and interexperimental variations in induced rates. Unfortunately, all of the data from biochemical loci and for dominant visibles were from

experiments involving acute X- or fractionated X-irradiation experiments. In trying to put together all of these data, there was no alternative but to use the correction factors suggested by the authors of the respective papers to estimate the rate for chronic radiation conditions from the available data. The committee feels that the procedures adopted in estimating an induced rate of $(0.36 \pm 0.10) \times 10^{-5}$ per gray are sound and that it is justifiable to use a single estimate for the induced rate of mutations.

THE DOUBLING DOSE ESTIMATE

With the estimates of $(2.95 \pm 0.64) \times 10^{-6}$ per locus for the rate of origin of spontaneous mutations in humans and $(0.36 \pm 0.10) \times 10^{-5}$ per locus per gray for induced mutations in mice, the DD becomes 0.82 ± 0.29 Gy. This new estimate is not very different from 1 Gy that has been used thus far and was based entirely on mouse data. The conceptual basis and the database used for estimating the average spontaneous and induced rates of mutations, however, are now different. The committee suggests retaining the use of 1 Gy for the DD estimate.

MUTATION COMPONENT OF GENETIC DISEASES

Background

As noted earlier, the MC is one of the quantities in the equation used to estimate risk of genetic disease using the doubling dose method (*i.e.*, risk per unit dose = $P \times [1/\text{DD}] \times$ MC, where P = baseline disease prevalence, 1/DD = the relative mutation risk per unit dose, and MC = the mutation component). The rationale for including MC in the risk equation is that the relationship between mutation and disease varies between different classes of genetic diseases—simple for autosomal dominant and X-linked diseases, slightly complex for autosomal recessive diseases, and very complex for multifactorial diseases—and the use of disease class-specific MC makes it possible to predict the impact of an increase in mutation rate on the frequencies of all classes of genetic diseases (Chakraborty and others 1998b; Denniston and others 1998; ICRP 1999).

General Definition

Let P be the disease prevalence before an increase in mutation rate and ΔP its change due to a Δm change in spontaneous mutation rate, m. The mathematical identity

$$\frac{\Delta P}{P} = \frac{\Delta m}{m} \cdot \frac{\Delta P / P}{\Delta m / m} \qquad (4\text{-}4)$$

formalizes the definition of MC. In this equation, since $\Delta P/P$ is the relative change in disease prevalence and $\Delta m/m$ is the relative change in mutation rate, the formal definition of MC becomes

$$MC = \frac{\Delta P / P}{\Delta m / m} \qquad (4\text{-}5)$$

In other words, MC is the relative change in disease prevalence per unit relative change in mutation rate. Because of the paucity of human data, until recently, estimates of $\Delta m/m$ have been obtained from mouse data and assumed to be applicable to the human situation.

It should be stressed that (1) the MC concept is applicable only when there is a change in mutation rate; (2) MC is *not* the same as the genetic component of the disease; rather, MC quantifies the responsiveness of the genetic component of the disease to increases in mutation rate; (3) if the disease is only partly genetic, since only the genetic component will respond to an increase in mutation rate, the MC for such a disease will be lower than that for a fully genetic disease; and (4) if the disease is entirely environmental in origin, the MC concept does not apply.

Note that despite the different notations used, Equation (4-4) is the same as the Equation (4-3), the basic risk equation (*i.e.*, risk per unit dose = $P \times [1/DD] \times MC$). The latter can be rewritten as risk per unit dose $\div P = (1/DD) \times MC$ in which risk per unit dose $\div P = \Delta P/P$ in Equation (4-4) and $(1/DD) = \Delta m/m$ (since $DD = m/\Delta m$). Therefore, if m increases to $m(1 + k)$ the disease incidence increases from P to $P(1 + kMC)$ showing that the MC concept is relevant only in the context of a change in mutation rate.

MC for Autosomal Dominant Diseases

The MC concept and its application are more easily illustrated with respect to autosomal dominant diseases for which the relationship between mutation and disease is straightforward. Two population genetic concepts are used in estimating MC, namely, the Hardy-Weinberg equilibrium and mutation-selection balance. The first of these relates the frequencies of mutant alleles to those of genotypes in large randomly mating populations, and the second describes the dynamics of mutant genes in populations.

Hardy-Weinberg Equilibrium

For a single locus with two alleles, the Hardy-Weinberg equilibrium concept is an application of the binomial expansion $(p + q)^2 = p^2 + 2pq + q^2$, where p and q are the proportions of alleles A and a (and $p + q = 1$), and $= p^2$, $2pq$, and q^2 are the proportions of the three genotypes AA, Aa, and aa. If the parents mate at random, which is equivalent to combining genes at random from a large pool to which each parent has contributed equally, the zygotes are in Hardy-Weinberg proportions. The larger the population, the closer the num-

bers agree with these binomial expectations. The Hardy-Weinberg concept thus summarizes the basic characteristic of stability of allele frequencies (and therefore of genotype frequencies) over time in large, randomly breeding populations in the absence of differences in viability or fertility among the genotypes, migration, mutation, and geographical subdivision of the population. In the case of genetic diseases, this is reflected as their stable prevalences in the population. With more than two alleles, the extension is straightforward: the binomial expansion becomes multinomial (Crow 2001).

Mutation-Selection Balance

Spontaneous mutations arise in each generation at a finite rate, and most are eliminated sooner or later by natural selection. At equilibrium, the rate of origin of new alleles by spontaneous mutation will be equal to the rate at which they are eliminated by selection and is called the mutation-selection equilibrium. The equilibrium frequency of the mutant allele depends on whether that allele is recessive or dominant.

Consider a one-locus, two-allele situation in a large, randomly mating population in Hardy-Weinberg equilibrium and assume that the fitness of the three genotypes (AA, Aa, and aa) can be represented by $1 - s$, $1 - s$ and 1, respectively. The zygotic frequencies, counted before selection, will be p^2, $2pq$, and q^2, respectively, for the three genotypes, where p denotes the frequency of the dominant allele A, and $q = 1 - p$, that of the normal allele a. In a stable gene pool, with the allele a mutating to A at a rate of m per generation, ignoring back mutations, there will be an mq amount of new disease-causing mutant alleles per generation; this will be counterbalanced by an elimination of these alleles by selection, which amounts to $pqs + p^2 s$. At equilibrium, these two quantities should be equal, yielding an equilibrium allele frequency of A (e.g., $\hat{p}$) that satisfies the equation

$$m\hat{q} = \hat{p}s, \qquad (4\text{-}6)$$

or

$$\hat{p} = \frac{m}{m + s} \approx \frac{m}{s} \qquad (4\text{-}7)$$

because the mutation rate (m) is generally smaller than the selection coefficient (s). At low mutant allele frequencies, the frequency of dominant diseases at equilibrium is then predicted to be $\hat{p}^2 + 2\hat{p}\hat{q} \approx 2\hat{p}$. For example, if for an autosomal dominant disease the spontaneous mutation rate is $m = 1 \times 10^{-5}$ and the selection coefficient $s = 0.5$, the equilibrium frequency of the mutant allele $p \approx m/s \approx 2 \times 10^{-5}$ and the disease frequency $2p \approx 4 \times 10^{-5}$ (since q is very nearly 1).

Estimation of MC

In estimating MC for autosomal dominant diseases, it is important to take into account the fact that some of these diseases (*e.g.*, Apert's syndrome, Crouzon's syndrome, osteogenesis imperfecta) are due entirely to germline mutations, whereas with some other diseases (*e.g.*, retinoblastoma, and breast cancers), only a proportion is due to germline mutations, the remainder being due to somatic mutations. As discussed later, for diseases of the latter type (referred to as those with a "sporadic" component), the predicted MCs will be less than those for the former in view of the fact that MC is related to the germline genetic component. In what follows, only the most relevant equations are given for MC estimations for two scenarios of radiation exposure, namely, exposure in one generation only or in every generation with and without the sporadic components. For details of the derivations of the equations, see the International Commission on Radiological Protection (ICRP 1999) Task Group report and Chakraborty and colleagues (1998b).

The starting assumption in these computations is that the population is in mutation-selection equilibrium prior to radiation exposure. When the population sustains radiation exposure, the mutation rate is increased, which in turn will impact disease frequency. As shown below, if the exposure occurs in one generation only, MC and ΔP are maximal in the first postradiation generation, progressively diminishing in subsequent generations until the population returns to the old equilibrium. When this occurs, MC becomes zero.

If, on the other hand, the population is exposed to radiation generation after generation (*i.e.*, the mutation rate is permanently changed from m to $[m + \Delta m]$), the *MC* and ΔP will continue increase with time (in generations) until the population reaches a new equilibrium between mutation and selection. At equilibrium, MC = 1 if the disease is entirely due to germinal mutations. Note that MC = 1 signifies that if the mutation rate is increased by $x\%$, the disease frequency at the new equilibrium (under conditions of radiation in every generation) will be increased by $x\%$. The magnitude of the increase in MC and the increase in disease frequency in intermediate generations will depend on Δm and the number of generations following radiation exposure.

MC Estimation for a Hypothetical Autosomal Dominant Disease Having No Sporadic Component in Its Etiology

For a one-time increase in mutation rate ("burst," indicated by the subscript b in MC_b below), the dynamics of change in MC with time, t, at any generation is given by

$$MC_b(t) = s(1 - s)^{t-1}. \qquad (4\text{-}8)$$

For example, if one assumes that $s = 0.5$, then MC_b at the first postradiation generation becomes $0.5 (1 - 0.5)^0 = 0.5$.

For a permanent increase in mutation rate (indicated by the subscript p), the equation is

$$MC_p(t) = [1 - (1 - s)^t]. \qquad (4\text{-}9)$$

Again assuming that $s = 0.5$, the MC_p at the first postradiation generation becomes $[1 - (1 - 0.5)^1] = 0.5$. Equations (4-8) and (4-9) thus show an interesting property of the effects on MC of a one-time or permanent increase in mutation rate in the first generation, namely, $MC_b = MC_p$. With no irradiation in subsequent generations, MC gradually decays to zero at a rate of $(1 - s)$ per generation, whereas under conditions of permanent increase in mutation rate, MC gradually increases in subsequent generations to attain a value of 1 at the new equilibrium.

The patterns of changes in MC and disease frequency with time, following a one-time or a permanent increase in mutation rate, are shown in Figure 4-1 and Table 4-4. In these illustrations, it is assumed that the mutation rate is increased from 1×10^{-5} to 2×10^{-5}, either in one generation only (broken line) or in every generation (solid line), and that the initial disease frequency (which corresponds to the baseline mutation rate of 1×10^{-5} and a selection coefficient of 0.5) is 4×10^{-5}.

It is clear that following a single-generation doubling of the mutation rate, both the disease frequency and MC show a transitory increase in the first postradiation generation. In subsequent generations, the disease frequency progressively declines to the old equilibrium value and MC declines to zero. With a permanent doubling of the mutation rate, for the selection coefficient of 0.5 used in these calculations, the disease frequency becomes twice that at the old equilibrium value by about the fifth postradiation generation by which time the mutation component becomes nearly 1.0.

MC Estimation for a Hypothetical Autosomal Dominant Disease with a Sporadic Component in Its Etiology

As mentioned earlier, some autosomal dominant diseases have a sporadic component in their etiology. For example, about 40% of retinoblastoma cases are due to germline mutations and the remaining ones are sporadic (Vogel 1979). For such diseases, the disease frequency at equilibrium can be assumed to take the form $P = A + Bm$. With A (sporadic component) and B (germinal component) as constants, only the second term will be responsive to an increase in mutation rate. If the dose dependence of induced mutations is linear, namely, $m = \alpha + \beta D$ and this form of m is substituted for P in the above equation,

$$P = A + B\alpha + \beta BD, \qquad (4\text{-}10)$$

so the relative increase in disease frequency $\Delta P/P = \beta BD/(A + B\alpha)$ and the relative increase in mutation rate $\Delta m/m = BD/\alpha$. Consequently,

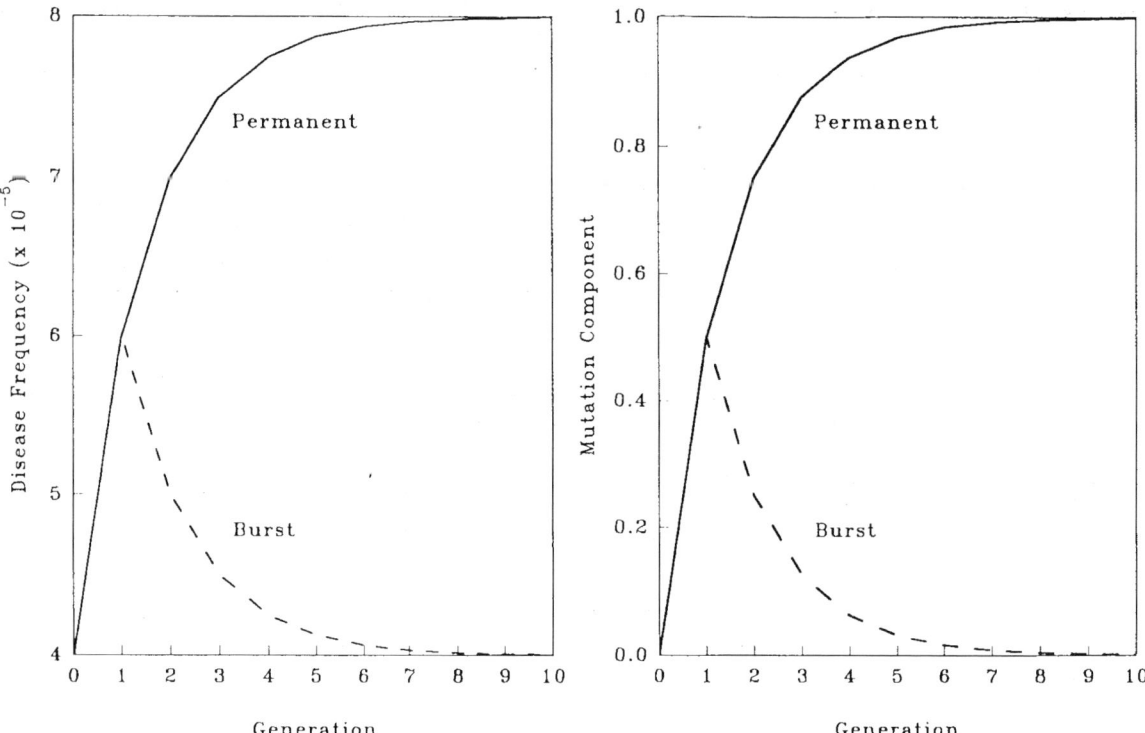

FIGURE 4-1 Changes in disease frequency (*y*-axis, left panel) and mutation component (*y*-axis, right panel) following a one-time (broken line) or a permanent (solid line) increase in mutation rate from 1×10^{-5} to 2×10^{-5} for an autosomal dominant disease. The disease frequency, before the doubling of the mutation rate, is 4×10^{-5} and the selection coefficient is 0.5. SOURCE: Figure reproduced with permission from Chakraborty and others (1998b).

$$MC = B\alpha/(A + B\alpha). \qquad (4\text{-}11)$$

It is clear that the larger the sporadic component, the smaller is the MC. When $A = 0$ and $B = 1$, as they are for most autosomal dominant diseases, MC at equilibrium will be 1. For diseases with a sporadic component in their etiology, MC at equilibrium will be less than 1 (*i.e.*, an *x*% increase in mutation rate will result in a <*x*% increase in disease frequency).

MC Estimation for X-Linked and Autosomal Recessive Diseases

The mathematical procedures for estimating MC for X-linked and autosomal recessive diseases are more complex than those for autosomal dominant diseases and are detailed in Chakraborty and colleagues (1998b). The relevant conclusions are the following:

For a one-time increase in mutation rate the response of X-linked diseases is similar to that of autosomal dominants (*i.e.*, MC in the first postradiation generation is equal to the selection coefficient, *s*). However, since only one-third of the X chromosomes are in males, *s* must be adjusted to take this into account. In other words, in Equation (4-7), *s* should be replaced by *s*/3. For example, if *s* = 0.6, the first-genera-

tion MC will be 0.2. For autosomal recessives, MC will be much smaller than for autosomal dominants, and it is close to zero in the first (as well as several successive) generations. This is due to the fact that when a recessive mutation first occurs (or is induced), it is present in heterozygotes and does not precipitate disease until the mutant allele frequency becomes sufficiently high in the population to produce homozygous individuals who will be affected by the disease.

For a permanent increase in mutation rate, the MC for both X-linked and autosomal recessive diseases progressively increases to reach a value of 1.0 at the new equilibrium. The rates of approach to the new equilibrium, however, are different and are dictated by selection coefficients and time (in generations) following radiation exposure. The effect of an increase in mutation rate on *MC* is most pronounced for autosomal dominants followed by that for X-linked and autosomal recessives, in that order.

Numerical Estimates of MC for Autosomal Dominant, X-Linked, and Autosomal Recessive Diseases Used in This Report

In Table 4-2, estimates of selection coefficients from published data for different autosomal dominant diseases are

TABLE 4-4 Effects of a One-Time or Permanent Doubling of the Mutation Rate on Mutant Gene Frequency (p), Disease Frequency (P), and Mutation Component (MC) for a Hypothetical Autosomal Dominant Disease

	Permanent Doubling			One-Time Doubling		
Generation	p	P	MC	p	P	MC
Initial	0.000020	0.000040	0.0000	0.000020	0.000040	0.0000
1	0.000030	0.000060	0.5000	0.000030	0.000060	0.5000
2	0.000035	0.000070	0.7500	0.000025	0.000060	0.2500
3	0.000038	0.000075	0.8750	0.000023	0.000045	0.1250
4	0.000039	0.000078	0.9375	0.000021	0.000043	0.0625
5	0.000039	0.000079	0.9688	0.000021	0.000041	0.0313
New equilibrium	0.000040	0.000080	1.0000	0.000020	0.000040	0.0000

NOTE: Values used in the computation are the following: mutation rate (m) = 1×10^{-5}; selection coefficient (s) = 0.5; initial mutant gene frequency (p) = $m/s = 2 \times 10^{-5}$, and initial disease frequency (P) = $2p = 4 \times 10^{-5}$.

General formulas for calculating the effects of an increase in mutation rate from m to $m(1 + k)$ on mutant gene frequency, disease frequency, and mutation component follow.

At Generation t	For a Permanent Increase	For a One-Time Increase
Mutant gene frequency, p_t	$p_0\{1 + k\,[1 - (1 - s)^t]\}$	$p_0[1 + ks\,(1 - s)^{t-1}]$
Disease frequency, P_t	$2p_t$	$2p_t$
Mutation component, MC	$[1 - (1 - s)^t]$	$s(1 - s)^{t-1}$

presented. The average of these values is $s = 0.29$. Similar estimates for X-linked diseases are not available. For estimating the risk to first-generation progeny, the committee uses a rounded value of MC = s = 0.3 for both autosomal dominant and X-linked diseases. The rationale for which rests on the following considerations: (1) the baseline incidence of X-linked diseases is an order of magnitude lower than that of autosomal dominant diseases (0.15% versus 1.5%; Table 4-1), (2) the net effect of selection for X-linked diseases is lower (*i.e.*, $s/3$ versus s for autosomal dominants), and (3) the use of the same MC value for both autosomal dominant and X-linked diseases therefore does not underestimate risk. The committee is cognizant of the fact that selection intensities in present-day human populations are probably lower. For autosomal recessives, the first-generation MC is close to zero.

MC ESTIMATION FOR CHRONIC MULTIFACTORIAL DISEASE

Introduction

As mentioned earlier, for most multifactorial diseases, knowledge of the number of genes involved, the types of mutational alterations, and the nature of environmental factors remains limited, and there is no simple relationship between mutation and disease. Further, unlike the situation for Mendelian diseases, no models have been proposed to explain the stable prevalences of multifactorial diseases in the population. Models such as the multifactorial threshold model of disease liability (see Annex 4B) are essentially *descriptive* models. They permit one to explain the transmission patterns of these diseases and make reasonable predictions of recurrence risks in families, but they are not, as such, suitable for the estimation of MC. There is, however, a wealth of literature about evolutionary population genetic models on the maintenance of quantitative variability (and traits) in populations, and these incorporate mechanisms (reviewed by Sankaranarayanan and others 1994). Although there are differences in detail between them and the applicability of these models to multifactorial diseases, all of them are based on equilibrium theory (*i.e.*, they invoke mutation and selection as opposing forces in the evolution and maintenance of variability for these traits). They are therefore similar to the models used to explain the dynamics of single mutant genes underlying Mendelian diseases in populations.

As a first approach to the problem of estimating MC for multifactorial diseases, an ICRP (1999) Task Group developed a "hybrid" model in which some concepts of the MTM and models for the evolution of quantitative traits in populations were incorporated. This "hybrid model" is henceforth referred to as the finite-locus threshold model (FLTM; ICRP 1999; see also Denniston and others 1998). The original aim was to use FLTM to estimate MC for both congenital abnormalities and chronic diseases. However, as discussed later, advances in human molecular biology and radiation genetics during the past few years suggest that it is not biologically meaningful to use the FLTM to estimate MC for congenital abnormalities, and therefore its use is limited to chronic diseases.

Finite-Locus Threshold Model

Rationale

As mentioned above, the FLTM uses the concepts of liability and threshold of the MTM (appropriately redefined for a finite number of loci) and that of mutation-selection equilibrium from evolutionary population genetic models on the maintenance of variability of quantitative traits. The choice of a finite number of loci rests on three main considerations: (1) although precise knowledge of the genetic basis is not yet available for most chronic diseases, for well-studied ones such as coronary heart disease, it is now clear that the number of underlying genes is probably small, and their mutant alleles have small to moderate effects; (2) estimates of the heritability of liability (h^2), a statistic that provides a measure of the relative contribution of genetic factors to *the* overall phenotypic variability for various chronic diseases, have been published in the literature; and (3) unlike the MTM, the FLTM permits quantitative analysis of the joint effects of mutation and selection. As emphasized in BEIR V (NRC 1990), the heritability of liability mentioned above should not be confused with heritability of the *trait*, which is very different (and much smaller than heritability of liability). This distinction is important since MC is related more to the heritability of liability than to the heritability of trait (see NRC 1990, Table 2-3, footnote *c*, for a mathematical formulation of the approximate relationship between heritability of liability and heritability of trait).

Assumptions and Predictions of the FLTM

Details of the assumptions and predictions of the FLTM are discussed in the report of the ICRP (1999) Task Group and by Denniston and colleagues (1998) and are summarized briefly in Annex 4C. In general terms, the FLTM assumes that the liability underlying a chronic disease, which is made up of both genetic and environmental factors, is a continuous variable and that the environmental contribution has a normal (Gaussian) distribution. Although the standard MTM assumes numerous (essentially an infinite number of) genetic factors (*i.e.*, mutant alleles), the FLTM assumes that the genetic component of liability is discrete (*i.e.*, it comes from mutant alleles of a finite number of gene loci). The latter is also true of the threshold. The FLTM incorporates mutation and selection (*s*) coefficients as additional parameters (the MTM does not include these). The effects of specified increases in mutation rate are evaluated in terms of changes in h^2 and MC.

Effects at Equilibrium Following a Permanent Increase in Mutation Rate

The predicted effects discussed below (and shown in Figure 4-2) are for the five-locus model when the spontaneous mutation rate per gene per generation (assumed to be 10^{-6}) is increased permanently to 1.15×10^{-6} (*i.e.*, a 15% increase) as a result of radiation exposures in every generation. The selection coefficients (*s* values) used were 0.2 to 0.8. The data points shown in Figure 4-2 are from different computer runs using different combinations of parameter values (selection coefficients, threshold, and environmental standard deviation). As can be seen, for h^2 values greater than about 0.1, MC > 0.8 at equilibrium, and for $h^2 > 0.4$, MC is essentially equal to 1.0. In other words, a 15% increase in mutation rate will result in a 15% increase in disease frequency at the new equilibrium.

Although the estimates discussed above are for the five-locus model ($n = 5$), these conclusions remain qualitatively unaltered for $n = 3, 4,$ and 6, which were also examined (data not shown).

Effects in Early Postradiation Generations Under Conditions of a Permanent Increase in Mutation Rate

The effects expected in early postradiation generations (*i.e.*, generations 1, 5, and 10) under the same radiation conditions as above are diagrammed in Figure 4-3. By noting the difference in the *y*-axis scales (compared to Figure 4-2), it is evident that the MC in early generations is very small, often being much less than 2% for the conditions specified for the model.

Comparison of the Effects at Equilibrium with Those in Early Generations Under Conditions of a Permanent Increase in Mutation Rate

Figure 4-4 compares the h^2 versus MC relationship at equilibrium with that at generation 10 (shaded areas in the figure are the ones of interest in MC estimation for chronic diseases). The conclusions from Figure 4-2 and Figure 4-3 are reinforced (*i.e.*, over a broad range of h^2 values from about 0.3 to 0.8, for the specified increase in mutation rate the MC at equilibrium is close to 1.0, whereas over the same h^2 range and the same increase in mutation rate, even after 10 postradiation generations the corresponding MCs are very small).

Effects on MC After an Increase in Mutation Rate in One Generation Only

The numerical algorithms used for the calculations above have also been used to examine the effects of a one-time increase in mutation rate (*i.e.*, the mutation rate was increased by 15% for one generation and then brought back to

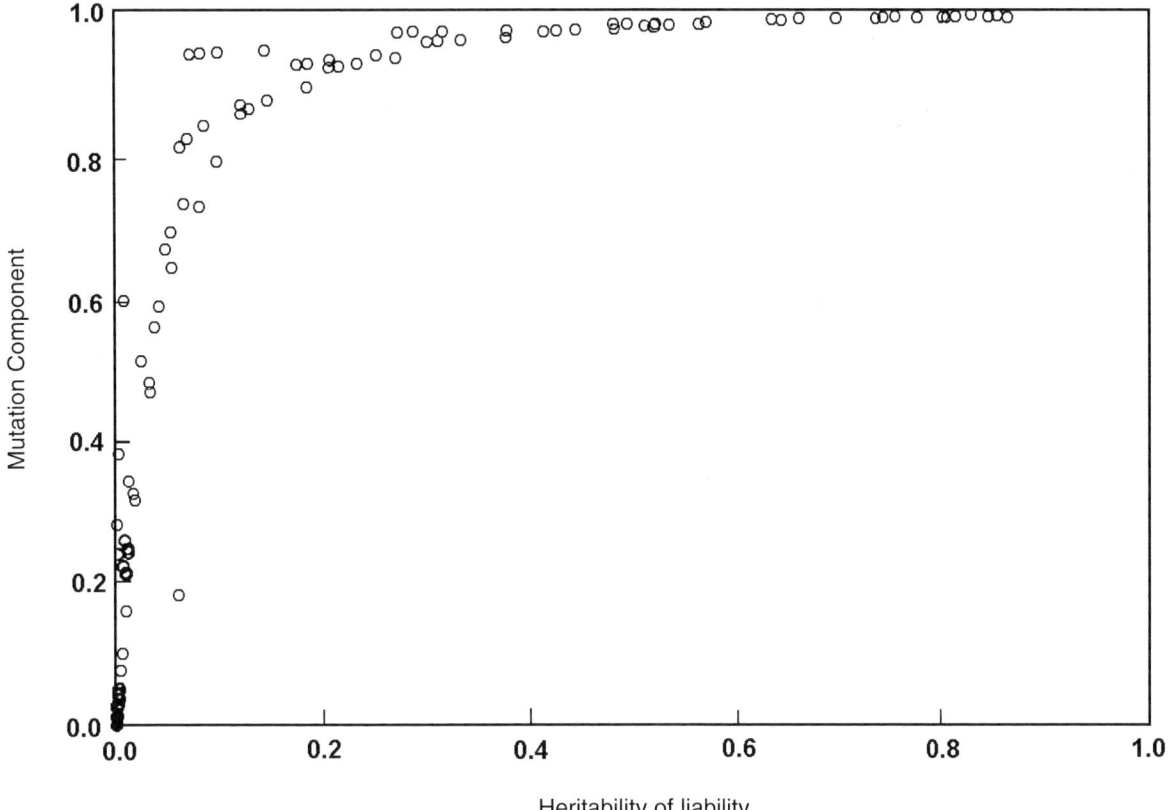

FIGURE 4-2 Relationship between heritability of liability (h^2) (x-axis) and mutation component (MC) (y-axis) at the new equilibrium between mutation and selection under conditions of radiation exposure in every generation. Results are for the five-locus FLTM when the assumed baseline mutation rate of 10^{-6} per gene is increased to 1.15×10^{-6} (*i.e.*, by 15%). Note that for h^2 estimates >30%, MC = 1.

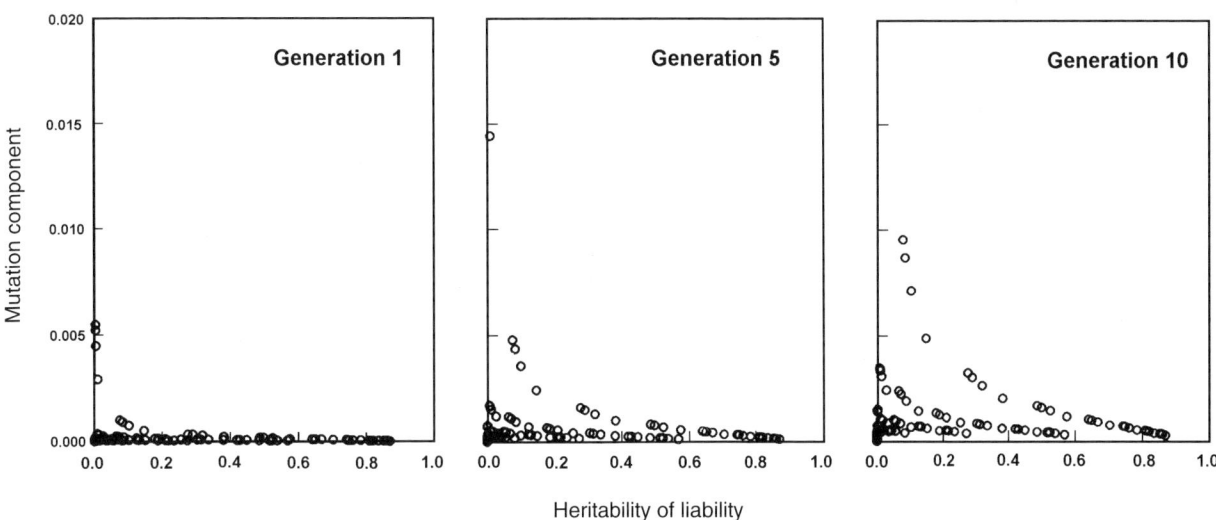

FIGURE 4-3 Relationship between heritability of liability (h^2) and mutation component for the first, fifth, and tenth postradiation generations, under the same conditions as specified in Figure 4-2. Note that the scale of the y-axis differs from Figure 4-2. MC values are very small, being <0.02 over a broad range of h^2.

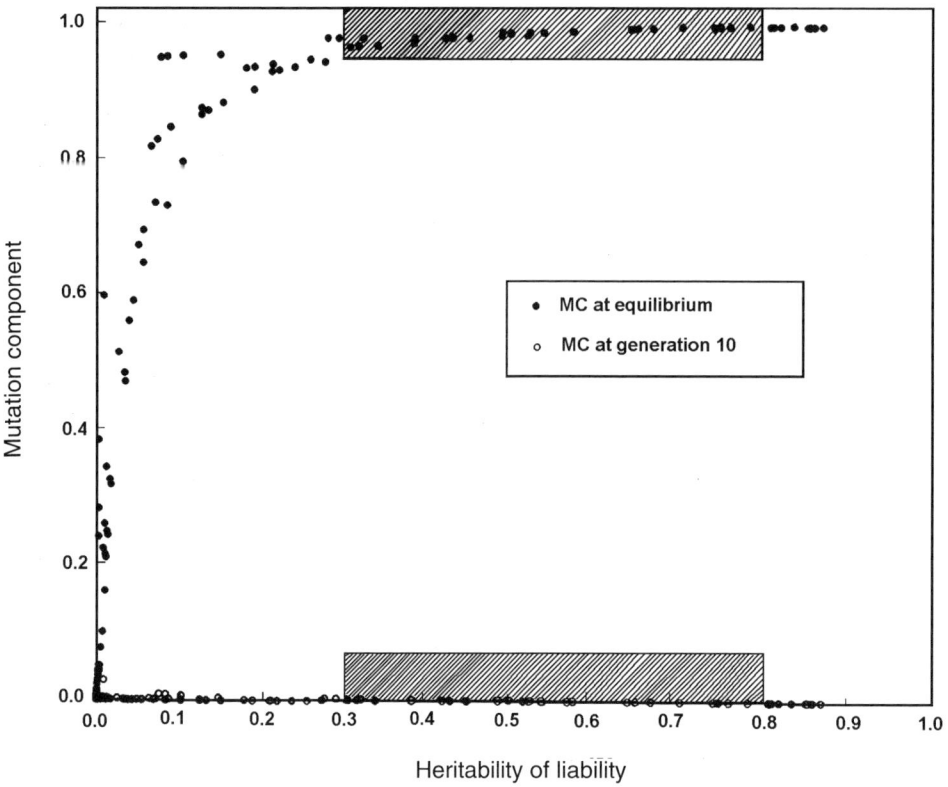

FIGURE 4-4 Comparison of the relationship between heritability of liability and mutation component for the tenth postradiation generation with that at the new equilibrium, under the same conditions as those specified for Figure 4-2. The shaded areas of h^2 (range: 0.3–0.8) are those of interest in the context of chronic diseases. Note that in this range, the MC at generation 10 is very small, whereas at the new equilibrium, it is equal to 1.

the original value for all subsequent generations). As expected, the first-generation MC is the same (*i.e.*, very small) as that shown in Figure 4-3, and this is followed by a gradual decline back to zero in subsequent generations (data not shown).

Effects of Gene Interactions (Epistasis) and Sporadics

The effects of gene-gene interactions on quantitative phenotypes at risk of complex diseases are varied and do not lend themselves readily to modeling. However, when some assumptions about these interactions were incorporated in the FLTM, it was found that the results (at the new equilibrium as well as for the early postradiation generations) were basically the same as those under conditions of no interactions (data not shown but discussed in Denniston and others 1998; ICRP 1999).

In all of the model predictions discussed thus far, the possibility that some individuals may be affected by the disease for reasons unrelated to their genotypes (sporadic cases) was not considered. When these were taken into account, as expected the magnitude of MC was lower, both at the new

equilibrium and in the early generations. The factor by which the numerical estimates of MC will change can be estimated to be $[1 - (a/P_T)]$, where (a/P_T) represents the proportion of sporadic cases among the total number of affected individuals (a = number of sporadic cases; P_T = total number of cases).

General Conclusions

The most important conclusion from computer simulation studies is that when the population is exposed to small doses of radiation in every generation, the MC for chronic diseases is very small, being of the order of 1 to 2% in the first several postradiation generations including the first. Since one of the assumptions of the model is the simultaneous increase in mutation rate of all of the genes underlying a given chronic disease, which is unlikely to occur at low radiation doses, the effective MC in the early generations is likely to be much less than 1–2%. One would therefore predict that the expected increases in the frequency of chronic diseases (relative to the baseline frequency) will be even smaller in the first few postradiation generations.

A second conclusion, again under conditions of a permanent increase in mutation rate, is that at the new equilibrium between mutation and selection (which will be achieved several tens—if not hundreds—of generations later, depending on the amount of increase in mutation rate and selection coefficients), the MC will become 1.0. In other words, for a sustained increase of $x\%$ in mutation rate, there will be an $x\%$ increase in the frequency of chronic diseases at the new equilibrium. This conclusion holds for several different combinations of assumed parameter values (selection coefficients, thresholds, numbers of loci, environmental variances, interactions among genes) and consequently can be considered robust.

Finally, if the population sustains radiation exposure in one generation only, the increase in MC will be transient and small, followed by a progressive decline to zero. The result will be a transient small increase in disease frequency followed by a decline toward the baseline frequency in subsequent generations.

This committee uses the 2% value in its calculations as the best MC estimate for the first postradiation generation, which was also the case for the ICRP (1999) Task Group and UNSCEAR (2001).

Bridging the Gap Between Rates of Radiation-Induced Mutations in Mice and Risk of Inducible Genetic Diseases in Humans

Introduction

Mouse data on rates of induced mutations (incorporated in the DD estimate) provide the basis for genetic risk prediction in humans. In predicting the risk as a product of P, 1/DD, and MC (*i.e.*, Equation (4-3) noted in the section on mutation component), an important assumption is implicit: mutations will be induced in those genes at which spontaneous mutations in humans cause disease (*i.e.*, the quantity P), the average rate of induced mutations in mice is applicable to induced human germline mutations, and such induced mutations will be compatible with viability and hence recoverable in the offspring of irradiated individuals. However, thus far, no radiation-induced genetic diseases have been found in the offspring of those who have sustained radiation exposures (*e.g.*, Byrne and others 1998; Meistrich and Byrne 2002; MGSC 2002).

Advances in human molecular genetics and radiation genetics during the last decade support the view that there are several fundamental differences (in mechanisms, nature, etc.) between spontaneous mutations that cause disease and radiation-induced mutations studied in experimental systems such as the mouse. More specifically, they suggest that a major proportion of human genes of relevance from the disease point of view may not yield "recoverable" induced mutations. Stated differently, the rate at which induced disease-causing mutations are seen in human live births following parental radiation exposures may be much lower than that of induced mutations in mice.

Concept of Potential Recoverability Correction Factor and Revision of the Risk Equation

Since there is no alternative to the use of mouse data on radiation-induced mutations for risk predictions in humans, methods have to be devised to bridge the gap between induced mutation rates in mice and the risk of genetic disease in humans. One such method has been developed recently and is based on the incorporation of a correction factor, termed the potential recoverability correction factor (PRCF), in the risk equation (Sankaranarayanan and Chakraborty 2000a). As a consequence, the risk now becomes a product of four quantities instead of the original three:

$$\text{Risk per unit dose} = P \times (1/DD) \times MC \times PRCF, \qquad (4\text{-}12)$$

where P, 1/DD, and MC are as defined earlier and PRCF is the disease-class-specific potential recoverability correction factor. Since PRCF is less than one, the estimate of predicted risk will be smaller when PRCF is incorporated than when it is not.

The differences between spontaneous disease-causing mutations in humans and radiation-induced mutations studied in experimental systems, which constitute the basis for the development of the PRCF concept, are discussed in detail by Sankaranarayanan (1999) and Sankaranarayanan and Chakraborty (2000b) and summarized in Annex 4D.

To assess PRCF, it was necessary first to define criteria on the basis of information available from molecular studies of radiation-induced mutations, to apply these to human genes of interest on a gene-by-gene basis, and to examine which among them can be considered candidates for *potentially recoverable* induced mutations. The operative words are the italicized ones, since there is as yet no evidence for a radiation-induced germ cell mutation in humans, our understanding of the structural and functional genomics of the genome is incomplete, and the criteria will undoubtedly change with advances in knowledge.

Among the attributes considered in defining the criteria are gene size, location, normal function, known mutational mechanisms, spectrum of spontaneous mutations, "gene-richness" or "gene-poorness" of the region, whether intragenic (including whole-gene) deletions and multigene deletions are known, and whether disruption of the gene or genomic region by rearrangements is associated with a mutant phenotype. Under the assumption that a deletion is induced in a genomic region containing the gene of interest, the question asked was, Given the structural and functional attributes of the gene or genomic region, can this deletion be considered potentially recoverable?

The criteria developed and how the genes examined are assigned to one of three groups—namely, unlikely to be recovered (group 1), recoverability uncertain (group 2), and potentially recoverable (group 3)—summarized in Annex 4D. Since the starting assumption is that the genomic region containing the gene of interest has sustained a multigene deletion, the assessments only tell us which disease-causing mutations, if induced, may be recovered in live births within the framework of the criteria used; they do not shed light on the absolute radiation risk of a given genetic disease. Also worth mentioning here is that assignment to group 1 (unlikely to be recovered) is somewhat less subjective, and therefore more reliable, than that to the other two groups. This aspect is taken into account in defining PRCF (*i.e.*, by lumping groups 2 and 3 and considering that the mutations in the genes included may be potentially recoverable).

In general terms, if one analyzes a total of *N* genes and if *n* among them can be excluded as unlikely to be recovered, the remainder, made up of the other two groups, constitutes $(N - n)$ and the fraction $(N - n)/N$ provides a crude measure of genes at which induced mutations may be recoverable. This fraction is referred to as the unweighted PRCF.

The PRCF as estimated above, however, does not take into account differences in the prevalence of diseases assigned to different groups. For example, if a disease with a high prevalence is assigned to group 1, societal concern about radiation effects will be far less than when it is assigned to the other two groups. Consequently, some weighting for disease prevalence is required.

If *P* is the total prevalence of diseases due to mutations in *N* genes and *p* is the prevalence of $(N - n)$ genes (in groups 2 + 3), then the weighted PRCF becomes $[p(N - n)/(PN)]$. For the purpose of risk estimation however, it is preferable to use a range provided by the unweighted and weighted PRCF estimates to avoid the impression of undue precision.

PRCF Estimates for Autosomal Dominant and X-Linked Diseases

A total of 67 genes involved in autosomal dominant (59) or X-linked (8) recessive diseases was included in the analysis. The results, given in Table 4-5, show that the unweighted and weighted PRCFs for autosomal dominants are 0.29 and 0.16, respectively; when X-linked diseases are included, the corresponding values become 0.36 and 0.20. Since the overall estimated prevalence of autosomal dominants is an order of magnitude higher than that of X-linked diseases (*i.e.*, 1.5% versus 0.15%), the use of the range of (rounded) PRCF values of 0.15 to 0.30 (encompassing the prevalence-weighted and unweighted estimates) for autosomal dominant and X-linked diseases seems reasonable.

PRCFs for Autosomal Recessive Diseases

The recoverability of induced recessive mutations is also subject to constraints imposed by the structure, function, and genomic contexts of the underlying genes. However, since induced recessive mutations are first present in the heterozygous condition (and 50% of the gene product is sufficient for normal functioning), one can assume that even large deletions may be recoverable in heterozygotes (unless the induced deletion encompasses neighboring essential structural genes, resulting in inviability of heterozygotes). Additionally, induced recessive mutations, at least in the first several generations, do not result in recessive diseases, and as discussed earlier, the MC for recessive diseases is close to zero

TABLE 4-5 Assessment of Potential Recoverability of Radiation-Induced Mutations in Autosomal Dominant and X-Linked Diseases and Calculation of PRCFs

Group	No. of Genes	Unweighted PRCF[a]	Prevalence ($\times 10^4$)	Weighted PRCF[b]
Autosomal dominants				
1 (unlikely to be recovered)	42	—	46. 5	—
2 + 3 (uncertain + potentially recoverable)	17	0.29	55.9	0.16
Subtotal	59	102.4		
Autosomal dominants + X-linked				
1 (unlikely to be recovered)	43	—	49.0	—
2 + 3 (uncertain + potentially recoverable)	24	0.36	60.9	0.20
Total	67	109.9		

[a]Unweighted PRCF, autosomal dominants: 17/59 = 0.29; autosomal dominants + X-linked: 24/67 = 0.36.

[b]Weighted PRCF, autosomal dominants: $(55.9 \times 17)/(102.4 \times 59) = 0.16$; autosomal dominants + X-linked: $(60.9 \times 24)/(109.9 \times 67) = 0.20$.

in the first few generations. In view of all these factors, it does not seem necessary to estimate PRCF for this class of diseases.

PRCFs for Chronic Diseases

In the FLTM used to estimate MC for chronic diseases, it is assumed that (1) the genetic component of liability is due to mutations in a finite number of gene loci, (2) the affected individuals are those whose genetic component of liability exceeds a certain threshold, and (3) radiation exposure can cause a simultaneous increase in mutation rate in all of the underlying genes, which in turn causes the liability to exceed the threshold. Consequently, the requirement for potential recoverability also applies to induced mutations in the underlying genes. A crude approximation of potential recoverability for each chronic disease is the xth power of that for mutation at a single locus, where x is the number of gene loci, assumed to be independent of each other. Since the PRCF for autosomal dominant and X-linked mutations has been estimated to be in the range from 0.15 to 0.30, for chronic diseases, these figures become $(0.15)^x$ to $(0.30)^x$. With the assumption of just two loci as a minimum, the PRCF estimate becomes 0.02 to 0.09, and with more loci, it will be much smaller. Intuitively, these conclusions are not unexpected given that one is estimating the simultaneous recoverability of induced mutations in two or more independent genes.

PRCFs for Congenital Abnormalities

Currently available data do not permit the estimation of PRCFs for congenital abnormalities. However, as discussed later, this does not pose any serious problem since at least a provisional estimate of risk for this class of diseases can now be made without recourse to the DD method.

Strengths and Weaknesses of the Use of PRCF Estimates

Development of the PRCF concept represents an example of how advances in human molecular biology and radiation genetics can be integrated for the purpose of genetic risk assessment. In principle, three ways of incorporating PRCFs into the risk equation (*i.e.*, Equation (4-3)) can be envisaged: (1) suitably *increase* the DD (*i.e.*, *reduce* the induced rate of mutations so that DD becomes higher and 1/DD becomes a smaller fraction); (2) *decrease* the MCs for the different classes of genetic disease; and (3) introduce disease class-specific PRCFs as an independent quantity into the risk equation. Of these, the last possibility has been preferred for two reasons. First, the original definition of the DD (a ratio of spontaneous and induced rates of mutations in a set of defined genes) and of MC (a quantity that *predicts* the relative increase in disease incidence per unit relative increase in mutation rate, both compared to the baseline) can be retained

without modifications. Second, with further advances in structural and functional genomics of the human genome and in the molecular analysis of radiation-induced mutations, there is the real prospect of defining PRCFs with greater precision.

In developing the PRCF concept, it has been assumed that the recoverability of an induced deletion is governed more by whether a given genomic region can tolerate large changes and yet be compatible with viability than by genomic organization per se. Considerable amounts of data exist that strongly support the view that in the case of deletion-associated naturally occurring Mendelian diseases, the deletions do not occur at random (*i.e.*, there are specificities of breakpoints dictated by the nucleotide sequence organization (reviewed in Sankaranarayanan 1999). *A priori*, therefore, one would not expect that radiation would be able to reproduce such specificities that nature has perfected over millennia, at least not in all genomic regions. Should this be the case, even the weighted PRCFs would be overestimates. However, until newer methods are developed to bridge the gap between induced mutation rates in mice and the risk of genetic diseases in humans, the PRCF range of 0.15 to 0.30 for autosomal dominant and X-linked diseases represents the best estimate that can be made at present.

The PRCF estimate range of 0.02 to 0.09 (*i.e.*, $[0.15]^2$ to $[0.30]^2$) for chronic diseases merits some comment since (1) it is based on the PRCFs for single-gene mutations and (2) it assumes just two loci (the minimum number required to call the disease multifactorial) underlying a chronic disease sustaining induced mutations simultaneously. On the first point, it is obvious that if the PRCFs for single-gene mutations change, the PRCFs for chronic diseases will also change. Secondly, the data on well-studied chronic diseases such as coronary heart disease (CHD), essential hypertension, and diabetes mellitus suggest that more than two loci may be involved. The implication is that the PRCFs for chronic diseases are likely to be smaller than cited above. For example, if there are three loci, the range becomes 0.003 to 0.03, with four loci, 0.0005 to 0.008, and so on. All this means is that the PRCF values for chronic diseases may turn out to be lower than 0.02 to 0.09.

The committee uses the PRCF ranges 0.15–0.30 for autosomal dominants and X-linked diseases and 0.02–0.09 for chronic diseases, as did UNSCEAR (2001).

Potential "Disease Phenotypes" of Radiation-Induced Genetic Damage in Humans

Introduction

For historical reasons, over the past four decades or so, the focus in the assessment of adverse genetic effects of radiation has been on the risk of inducible genetic diseases. The rationale for this rested on the premise that if spontaneous mutations can cause specific genetic diseases, so can radiation-induced mutations. This rationale gained support from experimental studies demonstrating that radiation-induced mutations in specific marker genes could be recovered in a number of biological systems, including the mouse. Consequently, efforts at risk estimation proceeded to use the mouse data on rates of induced recessive specific locus mutations as a basis for estimating the risk of genetic diseases due to mutations in single genes and assumed that the mouse rates can be used for this purpose.

Now, one can approach the question of adverse genetic effects of radiation from the perspective provided by our current understanding of the mechanism of radiation action, the molecular nature of radiation-induced mutations, increasing knowledge of human genetic diseases, and the mechanisms of their origin. One important outcome of this approach, discussed in the preceding section, is that it is now possible to conclude that the risk of single-gene diseases is probably much smaller than expected from the rates of induced mutations in mice. A second important outcome is the concept discussed in the present section, namely, that the adverse effects of gonadal irradiation in humans are more likely to be manifest as multisystem developmental abnormalities than as single-gene diseases.

Multisystem Developmental Abnormalities May Constitute the Major "Phenotypes" of Radiation-Induced Genetic Damage

The argument and findings that provide the basis for the above concept come from studies of the mechanism of induction of genetic damage by radiation, the nature of radiation-induced mutations, and the common phenotypic features of naturally occurring multigene deletions in humans. Some of these are discussed in the preceding section, and these studies and others are briefly considered below (see Sankaranarayanan 1999 for a detailed review).

Ionizing radiation produces genetic damage by random deposition of energy; the predominant type of radiation-induced genetic change is a DNA deletion, often encompassing more than one gene.

The whole genome is the target for radiation action, and deletions (and other gross changes) can be induced in any genomic region; however, since the recoverability of an induced deletion in a live birth is subject to structural and functional constraints, only a subset of these deletions that is compatible with viability may be recovered. Further, not all the recoverable deletions may have phenotypes that are recognizable from knowledge gained from naturally occurring genetic diseases.

Studies of naturally occurring human microdeletion syndromes, also termed "contiguous gene deletion syndromes" (Schmickel 1986) or segmental aneusomy[2] syndromes

(Budarf and Emanuel 1997), are instructive in the context of delineating phenotypes of multigene deletions. These syndromes result from deletions of multiple, functionally unrelated, yet physically contiguous genes that are compatible with viability in the heterozygous condition. Many have been reported in the human genetics literature, and they have been found in nearly all human chromosomes, but their distribution in different chromosomal regions seems to be nonrandom. This is not unexpected in the light of differences in gene density in different chromosomes and chromosomal regions. However, despite their occurrence in different chromosomes, the common features of the phenotypes of many of these deletions include mental deficiency, a specific pattern of dysmorphic features, serious malformations, and growth retardation (Schinzel 1988; Epstein 1995; Brewer and others 1998).

In considering all of these together, the concept was put forth that multisystem developmental abnormalities are likely to be among the principal phenotypes of deletions and other gross changes induced in different parts of the human genome. Because the underlying genetic change is a deletion, generally one would expect that these phenotypes would show autosomal dominant patterns of inheritance.

Experimental Data in Support of the Concept

Mouse data supporting the above concept come from studies on radiation-induced skeletal abnormalities (Ehling 1965, 1966; Selby and Selby 1977, 1978), cataracts (Kratochvilova and Ehling 1979; Ehling 1985; Favor 1989), congenital abnormalities ascertained *in utero* (Kirk and Lyon 1982, 1984; Nomura 1982, 1988, 1989, 1994; Lyon and Renshaw 1984; Rutledge and others 1986) and growth retardation (Searle and Beechey 1986; Cattanach and others 1993, 1996). The cases analyzed (*e.g.*, skeletal abnormalities, growth retardation) show that the underlying induced genetic changes are multigene deletions. It is worth mentioning here that the data on skeletal and cataract mutations were used earlier by both UNSCEAR and the BEIR committees to provide alternative estimates of the risk of dominant effects using what was referred to as the direct method. This method was not used by UNSCEAR (2001) or by the BEIR V committee (NRC 1990). The basic data from these studies, however, have now been used by UNSCEAR (2001) to obtain a provisional estimate of the risk of developmental defects without recourse to the DD method. This aspect is considered in the section on risk estimation.

There is no conceptual contradiction between naturally occurring and radiation-induced developmental abnormalities. As discussed earlier, naturally occurring human congenital abnormalities are classified as a subgroup of multifactorial diseases, whereas radiation-induced ones generally are predicted to show autosomal dominant patterns of inheritance. It may therefore seem that there is a conceptual contradiction. In reality, this contradiction is only apparent when

---

[2]Aberration in the number of chromosomes.

one considers the fact that the primary reasons for considering naturally occurring developmental abnormalities as multifactorial are their etiological heterogeneity (as a consequence of which their transmission patterns are inconsistent with Mendelian patterns of inheritance), the lack of knowledge of the genetic factors involved, and the nature of environmental factors. The concept that is emerging is that human developmental abnormalities may be treated as inborn errors in development or morphogenesis in obvious analogy with, and as an extension of, the classical concept of inborn errors of metabolism (Epstein 1995). Therefore, diverse dysmorphogenetic causes (including those "driven" by multigene deletions) can produce similar malformations.

OTHER POTENTIALLY RELEVANT DATA

Induction of Mutations at Expanded Simple Tandem Repeat Loci in the Mouse and Minisatellite Loci in Human Germ Cells

Introduction

Since the mid-1990s, several studies have been carried out on the induction of germ cell mutations at expanded simple tandem repeat (ESTR) loci in mice (formerly called minisatellites) and at minisatellite loci in humans. These are regions of the genome that do not code for any proteins but are highly unstable (mutable), both spontaneously and under the influence of radiation. These attributes have facilitated detection of increases in mutation rates at radiation doses and sample sizes substantially smaller than those used in conventional mutation studies with germ cells. Although these loci do not code for proteins and most spontaneous and radiation-induced mutational changes in them are not associated with adverse health effects, some limited evidence is suggestive of a possible role of minisatellites in human disease (reviewed in Bridges 2001). For example, there are data suggesting that minisatellites can affect transcription of the insulin gene (*IDDM2*) and *HRAS1* genes (Trepicchio and Krontiris 1992; Kennedy and others 1995) Further, it has been found that certain polymorphisms of the minisatellite at the 5′-flanking region of the *IDDM2* gene may be associated with predisposition to insulin-dependent diabetes mellitus (Bell and others 1984; Bennett and others 1995). Additionally, there is suggestive evidence of an association between the risk of cancer and mutations in the *HRAS1* gene (Krontiris and others 1993; Phelan and others 1996). Although it is not possible at present to use data from these studies for radiation risk estimation, they are considered in this report because some of the findings have exposed interesting aspects of the radiation response at these loci that have parallelisms to the genomic instability phenomenon recorded in irradiated somatic cell systems and therefore relevant for ongoing debates in radiobiology. Most of these studies have been reviewed recently (Bridges 2001; UNSCEAR 2001).

The principal conclusions are summarized here; and details are presented in Annex 4F.

Mouse Studies

Mutations at the ESTR loci can be induced by both low- and high-LET (neutrons from californium-252 [^{252}Cf]) irradiation of mouse germ cells (Dubrova and others 1993, 1998a, 1998b, 2000a, 2000b; Sadamoto and others 1994; Fan and others 1995; Niwa and others 1996). For both types of radiations, the dose-effect relationship for mutations induced in spermatogonial stem cells is consistent with linearity. The high frequency of induced mutations strongly supports the view that they are unlikely to result from direct radiation damage to these small genomic loci themselves (*i.e.*, they are nontargeted events arising indirectly as a result of genomic instability; Niwa and others 1996; Dubrova and others 1998b; Niwa and Kominami 2001). There is evidence that this instability is not the result of a general genome-wide increase in meiotic recombination rate (Barber and others 2000).

This genomic instability is transmissible to at least two generations resulting in increased frequencies of mutations (Dubrova and others 2000b; Barber and others 2002). These findings add further support to observations on genomic instability recorded in somatic cells—the occurrence of genetic changes in the progeny of irradiated cells at delayed times (in terms of cell generations) after irradiation.

Data on ESTR mutations obtained in experiments involving irradiated spermatogonial stem cells permit an estimate of the DD of about 0.33 Gy for acute X-irradiation, similar to that known for specific locus mutations in mice (Dubrova and others 1998b). It should be noted, however, that both the average spontaneous rate (0.111 per band) and the induction rate (0.338 Gy^{-1}) are orders of magnitude higher than those of specific locus mutations.

There are some discrepancies between the findings of Dubrova and colleagues and those of Niwa and colleagues: (1) In the work of Dubrova and colleagues, post-meiotic germ cells are not sensitive to mutation induction at the ESTR loci, whereas in the work of Niwa and colleagues, all germ cells are sensitive, albeit to different degrees; it is not yet clear whether these differences are due to differences in the mouse strains used or to some other reasons. (2) In the work of Niwa and colleagues, F_1 tests showed increased frequencies of mutations in the unirradiated maternal allele, suggesting the occurrence of destabilization in the zygote; however this occurs only after spermatozoal but not after gonial irradiation of the males; in the work of Dubrova and colleagues, the data imply that destabilization occurs in the F_1 zygote when the spermatozoa used for fertilization received irradiation either at the postmeiotic or premeiotic stages in spite of observations that postmeiotic germ cells were not sensitive to mutation induction.

Human Studies

The results of Dubrova and colleagues from the three post-Chernobyl studies (two in Belarus and one in Ukraine) and from a study conducted on the population in the vicinity of the nuclear test site in Semipalatinsk (Kazakhstan) provide evidence that mutations at minisatellite loci can be induced by radiation in human germ cells (Dubrova and others 1996, 1997, 2002b). The dose-response relationships, however, remain uncertain because of considerable difficulties in the estimation of parental gonadal doses. For example, in the first Belarus study (Dubrova and others 1996) the level of surface contamination by ^{137}Cs was used as a broad dose measure, and the children of parents inhabiting heavily contaminated areas (>250 kBq m^{-2}) were found to have twice the frequency of mutations compared to those of parents from less contaminated areas (<250 kBq m^{-2}). In the second Belarus study (with more exposed families and more loci sampled), based on estimates of individual doses, two groups were defined: <20 mSv and >20 mSv (Dubrova and others 1997). The mutation frequency in children from the latter group was 1.35 times that in the former and the frequency in both groups was about twofold higher than in unexposed UK controls.

In the Ukraine study (Dubrova and others 2002b), a 1.6-fold increase in mutation rate in the exposed fathers but not exposed mothers (both relative to unexposed controls) was found, but again the dose-response relationship is uncertain. The authors noted that the doses from external chronic irradiation and internal exposures together were of the order of ~100 mSv (excluding short-lived isotopes). In the Semipalatinsk study (Dubrova and others 2002a), again there was a 1.8-fold increase in the first-generation progeny of parents receiving relatively high doses of radiation (cited as >1.0 Sv, but could have been higher or lower). In this study, through the use of three-generation families, the authors obtained evidence for a decline in mutation frequency as population doses decreased. Although this is what one normally would expect, it becomes a puzzling observation in view of the earlier evidence from the authors on ESTR loci on transgenerational mutagenesis in mice (*i.e.*, the persistence of high mutation rate for at least two generations after the initial radiation exposure).

It is intriguing that in all studies discussed above, there is roughly a twofold increase in mutation rate (often less) despite the fact that the estimates of doses range from about 20 mSv to 1 Sv. Also noteworthy is that studies of the children of Chernobyl cleanup workers (estimated dose: <0.25 Sv; Livshits and others 2001) and of children of A-bomb survivors (estimated dose: 1.9 Sv; Kodaira and others 1995; Satoh and Kodaira 1996) do not show any increase in minisatellite mutation frequency. The same is true also of studies of cancer patients who had sustained chemo- and/or radiotherapy (Armour and others 1999; May and others 2000; Zheng and others 2000). The question of whether the induced mutation frequencies reach a plateau at low doses (unlike in the case of ESTR loci in mice) remains open. In a more recent study of the children born to Estonian Chernobyl cleanup workers, Kiuru and colleagues (2003) found that the minisatellite mutation rate was slightly but not significantly increased among children born after the accident relative to that in their siblings born before the accident; the recorded dose levels at which such an effect was seen were 200 mSv. At lower doses, there was no effect. It is obvious that much work is needed to validate the potential applications of minisatellite loci for monitoring mutation rate in human populations.

As discussed in Annex 4F, ESTR loci in mice and minisatellite loci in humans differ in a number of ways: the composition and size of the arrays, their distribution (apparently random in the case of ESTRs and subtelomeric in the case of minisatellites), the manifestation of instability (in both somatic cells and germline in the case of ESTRs, but almost completely restricted to the germline in the case of minisatellites, although the end result is the change in the number of repeat cores with both ESTRs and minisatellites), and mechanisms (ESTR instability appears to be a replication- or repair-based process involving polymerase slippage during replication, whereas minisatellite instability is due to gene conversion-like events involving recombinational exchanges). To what extent these differences may help explain the differences in response between mouse ESTR loci and human minisatellite loci remains to be determined. As pointed out by Yauk and others (2002), ". . . the use of mouse ESTR loci as models for human minisatellite instability should be treated with considerable caution." Apart from the evidence that the mutational events represent nontargeted ones, no real insights have emerged thus far on the mechanisms of instability or radiation mutagenesis at these loci. In view of this and the fact that "mutational events" at the mouse ESTR and human minisatellite loci do not pertain to protein-coding genes, these data are not used in risk estimation.

RISK ESTIMATION

Introduction

In this section, advances in knowledge reviewed in earlier sections are recapitulated briefly and used to revise the estimates of genetic risks presented in BEIR V (NRC 1990). Additionally the consistency of the main finding of the genetic studies carried out on atomic bomb survivors in Japan (*i.e.*, lack of demonstrable adverse genetic effects of radiation) with the present estimates and the strengths and weaknesses of the latter are discussed. Risks are estimated using the doubling dose method for Mendelian and chronic multifactorial diseases. For congenital abnormalities, mouse data on developmental abnormalities are used without recourse to the doubling dose method. No separate risks are estimated for chromosomal diseases since they are assumed to be subsumed in part under the risks for autosomal dominant + X-linked diseases and in part under those for congenital abnormalities. The estimates presented are for a population sustaining low-LET, low-dose or chronic radiation exposures at a finite rate in every generation and are applicable to the progeny of the first two postradiation generations.

The equation now used for risk estimation is:

$$\text{Risk per unit dose} = P \times (1/\text{DD}) \times \text{MC} \times \text{PRCF}, \qquad (4\text{-}13)$$

where P is the baseline frequency of the disease class under consideration, 1/DD is the relative mutation risk per unit dose, MC is the disease class-specific mutation component, and PRCF is the disease class-specific potential recoverability correction factor.

Summary of Advances Since the 1990 BEIR V Report

1. **Baseline frequencies of genetic diseases.** The baseline frequencies of Mendelian diseases have now been revised upwards. The revised estimates are the following: autosomal dominant diseases, 15,000 per million live births; X-linked diseases, 1500 per million live births; and autosomal recessive diseases, 7500 per million live births. For chromosomal diseases, the estimate remains unchanged at 4000 per million live births. For congenital abnormalities and chronic multifactorial diseases, the current estimates (respectively, 60,000 per million live births and 650,000 per million individuals in the population) are the same as those used in the UNSCEAR (1993, 2001) reports. BEIR V (NRC, 1990) used lower estimates of 20,000 to 30,000 for congenital abnormalities and did not provide any comparable estimate for chronic multifactorial diseases (see Table 4-1).

2. **Conceptual change in calculating the doubling dose.** Human data on spontaneous mutation rates and mouse data on induced mutation rates are now used to calculate the doubling dose, which was also the case in the NRC (1972) report. Although the conceptual basis for calculating the DD is now different (and the estimate itself is based on more data than has been the case thus far), its magnitude (*i.e.*, 1 Gy for chronic low-LET radiation conditions) is the same as that used in the BEIR V.

3. **Mutation component.** Methods to estimate the mutation component (the relative increase in disease frequency per unit relative increase in mutation rate) have now been elaborated for both Mendelian and chronic multifactorial diseases. For *autosomal dominant diseases*, the first postradiation generation MC = s = 0.3, where s is the selection coefficient. For the second postradiation generation, MC = 0.51 as given by the equation $\text{MC}_p = [1 - (1 - s)^t]$, where s = 0.3 and t = 2. For *X-linked diseases* (which are considered together with autosomal dominant diseases), the same values are used. For *autosomal recessive diseases*, MC

in the first few generations is close to zero. For *chronic multifactorial diseases*, MC in the first as well as the second postradiation generations is assumed to be about 0.02. For congenital abnormalities, it is not possible to calculate MC, but this does not pose any problem since the risk estimate for these does not use the doubling dose method.

4. **Potential recoverability correction factor.** A new disease class-specific factor, the PRCF, has been introduced in the risk equation to bridge the gap between radiation-induced mutations in mice and the risk of radiation-inducible genetic disease in human live births. The risk now becomes a product of four quantities (see Equation (4-13) above) instead of three, which was the case until the early 1990s (NRC 1990; UNSCEAR 1993). For *autosomal dominant and X-linked diseases*, the PRCF estimate is in the range 0.15 to 0.30; the lower value represents the "weighted PRCF" (*i.e.*, weighted by disease prevalence), and the higher value, the unweighted one (*i.e.*, the proportion of human genes at which induced disease-causing mutations are potentially recoverable in live births). For *autosomal recessive diseases*, no PRCF is necessary (since induced recessive mutations do not precipitate disease in the first few generations). For *chronic diseases*, PRCFs are estimated to be in the range between about 0.02 and 0.09 under the assumption that the number of genes underlying a given multifactorial disease is equal to 2 (the minimum number) and that the PRCF is the nth power of that for an autosomal dominant disease (*i.e.*, $[0.15]^2$ to $[0.3]^2$). It is not possible to calculate PRCF for congenital abnormalities.

5. The concept that the adverse effects of radiation-induced genetic damage in humans are likely to manifest predominantly as multisystem developmental abnormalities in the progeny of irradiated individuals has now been introduced in the field of genetic risk estimation.

The mouse data used to obtain a provisional estimate of the risk of developmental abnormalities (considered here under the risk of congenital abnormalities) pertain to those on radiation-induced dominant skeletal abnormalities, dominant cataract mutations, and congenital abnormalities ascertained *in utero* (see Table 4-3D). Details of these abnormalities are discussed in Sankaranarayanan and Chakraborty (2000b) and in UNSCEAR (2001).

Briefly, the data on skeletal abnormalities (Ehling 1965, 1966; Selby and Selby 1977) permit an overall estimate of about 6.5×10^{-4} per gamete per gray for acute X- or γ-irradiation of males (spermatogonial stem cells). This estimate takes into account the proportion of skeletal abnormalities in mice, which—if they occur in humans—are likely to impose a serious handicap. The comparable rate for dominant cataracts (Favor 1989) is lower, being ~0.33×10^{-4} per gamete per gray. The rate for congenital abnormalities (corrected for compatibility with live births) is 19×10^{-4} per gamete per gray based on two sets of data (Kirk and Lyon 1984; Nomura 1988). When these three estimates are combined, the resultant figure is about 26×10^{-4} per gamete per gray and this has been rounded upwards to 30×10^{-4} per gamete per gray. This estimate summarizes the overall risk of congenital abnormalities for acute X-irradiation of males. With a dose-rate reduction factor of 3, the rate applicable for chronic or low-dose irradiation conditions is about 10×10^{-4} per gamete per gray. Under the assumption that the rate in females will be the same, the rate applicable for irradiation of both sexes is about 20×10^{-4} per gamete per gray.

Current Risk Estimates

Estimates of risk for all classes of disease except congenital abnormalities have been obtained using the equation: Risk = $P \times 1/DD \times MC \times PRCF$. The values used for estimating the first-generation risk are the following:

Autosomal Dominant + X-Linked

$P = 16,500/10^6$; $1/DD = 1$; $MC = 0.3$; $PRCF = 0.15–0.30$
$= 16,500/10^6 \times 0.3 \times 0.15–0.30 = \sim750–1500$ per 10^6.

Chronic Diseases

$P = 650,000/10^6$; $1/DD = 1$; $MC = 0.02$; $PRCF = 0.02–0.09$
$= 650,000/10^6 \times 0.02 \times 0.02–0.09 = \sim250–1200$.

For the second postradiation generation, the MC value is 0.51 for autosomal dominant and X-linked diseases; the values for all others remain the same. Estimates for congenital abnormalities have been obtained using mouse data on developmental abnormalities (see Table 4-3D); the DD method was not used.

Table 4-6 (top part) presents the current estimates of genetic risks of radiation and compares them with those in BEIR V (NRC 1990; bottom part). All estimates are per million progeny per gray.

Risk to Progeny of the First Postradiation Generation

As can be seen, the risk is of the order of about 750 to 1500 cases for autosomal dominant and X-linked diseases (versus 16,500 cases of naturally occurring ones) and zero for autosomal recessive diseases (versus 7500 cases of naturally occurring ones). For congenital abnormalities, the estimate is about 2000 cases (versus 60,000 cases of naturally occurring ones), and for chronic diseases, it is about 250 to 1200 cases (versus 650,000 cases of naturally occurring ones). Overall, the predicted risks per gray represent 0.4 to 0.6% of the baseline frequency (738,000 per million).

Risk to Second Postradiation Generation Progeny

Under conditions of continuous radiation exposure in every generation, the risk to the second postradiation genera-

TABLE 4-3D Mouse Database Used for Estimating the Rate of Induction of Dominant Heritable Developmental Effects Listed as Congenital Abnormalities in Table 4-6

End Point	Dose (Gy)	Frequency of Affected Progeny		Rate per Gray ($\times 10^4$)[a]
References: Ehling (1965, 1966); Selby and Selby (1977)				
1. Skeletal abnormalities	6.0	5/754		11
2. Skeletal abnormalities	1+ 5	5/277	(30)[a]	15
3. Skeletal abnormalities	1+ 5	37/2646	(23)[a]	12
Overall average induction rate				~13
Rate applicable to humans				~6.5
Reference: Favor (1989)				
4. Cataracts[b]	1.5	2/23,157		0.28
	3.0	3/22,712		0.29
	5.3	3/10,212		0.47
	6.0	3/11,095		0.38
	6.0	3/17,599		0.21
Overall average induction rate				0.33
Reference: Nomura (1988)				
5. Congenital anomalies (detected *in utero*; ICR strain)	0.36	1/163[c]		56[d]
	1.08	3/234[c]		83
	2.16	9/496[c]		65
Reference: Kirk and Lyon (1984)				
6. Congenital anomalies (detected *in utero*; [(C3H/HeH) × (101/H) strain]	5.00	22/1014[c]		30[e]
Unweighted average induction rate				48
Corrected for viability in human live births				19[f]
Overall rate for developmental abnormalities				30[g]

NOTE: All these studies involved spermatogonial irradiation.

[a]Estimates in parentheses: observed rate per gray for fractionated radiation conditions (24 h interval between fractions); estimates without parentheses are standardized to acute radiation conditions by dividing the above by 2, the factor by which specific locus mutation frequencies are known to be enhanced under fractionated radiation conditions.

[b]Rates have been corrected for controls in which the frequency was 1/22,594.

[c]Denominator refers to the number of live fetuses screened.

[d]Rates corrected for controls (8/1967).

[e]Rate corrected for controls (5/720).

[f]Under the assumption that about 40% of the abnormalities may be compatible with live births in humans (see Nomura 1988).

[g]$(6.5 + 0.3 + 19)10^{-4} = 26 \times 10^{-4}$ rounded to 30×10^{-4}.

tion progeny is slightly higher for autosomal dominant and X-linked diseases and for congenital abnormalities. The overall increase in risk (all classes of disease) relative to the baseline is small (0.53%–0.91% of 738,000 per million progeny).

Comparisons of Present Estimates with those in BEIR V

The bottom part of Table 4-6 shows the risk estimates arrived at in BEIR V (NRC 1990). As evident, in the 1990 report (1) the estimates of baseline frequency of Mendelian diseases were lower and (2) no risk estimate was provided for chronic multifactorial diseases. It is worth mentioning that the differences between the current and the 1990 estimates stem from differences in the assumptions used (see NRC 1990 for details). For example, for congenital abnormalities, in the 1990 report it was assumed that the MC concept could be applied to these and that the first-generation

MC could be as high as 35% ("worst-case" assumption); in the present report, the MC concept has not been used for this class of diseases for reasons discussed earlier.

The genetic theory of equilibrium between mutation and selection that underlies the use of the doubling dose method predicts that when a population sustains radiation exposure in every generation, a new equilibrium between mutation and selection will eventually be reached, albeit after tens or hundreds of generations into the distant future. In principle, therefore, one can project risks at the new equilibrium. However, in the present report (in contrast to NRC 1990), this has not been done and calculations have been restricted to the

TABLE 4-6 Estimates of Current Genetic Risks from Continuing Exposure to Low-LET, Low-Dose, or Chronic Radiation and Comparisons with Estimates in BEIR V Not Estimated

Disease Class	Baseline Frequency per Million Live Births	Risk per Gray per Million Progeny at First Generation	Second Generation[a]
Current Estimates			
Mendelian			
Autosomal dominant and X-linked	16,500	~750 to 1500	~1300 to 2500
Autosomal recessive	7,500	0	0
Chromosomal	4,000	*b*	*b*
Multifactorial			
Chronic multifactorial	650,000[c]	~250 to 1200	~250 to 1200
Congenital abnormalities	60,000	~2000[d]	~2400 to 3000
Total	738,000	~3000 to 4700	~3950 to 6700[e]
Total risk of baseline per gray as percent		0.41 to 0.64	~0.53 to 0.91
ESTIMATES IN BEIR V REPORT (1990)			
Mendelian			
Autosomal dominant	10,000	~600 to 3500	NE
X-linked	400	<100	NE
Autosomal recessive	2,500	<100	NE
Chromosomal	4,400	<600	NE
Multifactorial			
Congenital abnormalities	20,000–30,000	1000	NE
Other disorders			
Heart disease	600,000	NE	
Cancer	300,000	NE	
Selected others	300,000	NE	

NOTE: The doubling dose used for both sets of estimates is 1 Gy. NE = N.

[a]Risk to the second generation includes that of the first generation.

[b]Assumed to be subsumed in part under the risk of autosomal dominant and X-linked diseases and in part under congenital abnormalities.

[c]Frequency in the population.

[d]Calculated using mouse data on developmental abnormalities without using the doubling dose method.

[e]Assumes that between 20 and 50% of the abnormal progeny in the first postradiation generation may transmit the damage to the second (*i.e.*, resulting in 400 to 1000 affected cases); this is in addition to the newly induced damage in the first postradiation generation and manifest in the second (2000 cases).

first two generations for the following reasons: (1) people are generally interested in genetic risks in the foreseeable future (*i.e.*, to children and grandchildren), and (2) embarking on prediction of risk tens or hundreds of human generations from now involves the unrealistic and untestable assumptions that circumstances (*e.g.*, demographic and health care possibilities) will remain constant over very long periods of time and that the various assumptions and quantities used will remain unchanged over tens or hundreds of human generations.

Reconciliation of Present Estimates with Main Conclusions of the Genetic Studies on A-Bomb Survivors in Japan

Introduction

The genetic studies of atomic bomb survivors carried out in Japan represent the largest and most comprehensive of the long-term human studies ever carried out on adverse hereditary effects of radiation. The various papers published over the past four decades on this research program have been compiled by Neel and Schull (1991). Since the beginning of these studies, their focus has always been on a direct assessment of adverse hereditary effects in the first-generation progeny of survivors, using indicators of genetic damage that were practicable at the time the studies were initiated in the early 1950s. They were not aimed at expressing risks in terms of genetic diseases. As the research progressed, it became clear that no statistically significant adverse effects could be demonstrated in the children of survivors, and this conclusion was found to hold when all of the available data until 1990 were analyzed (Neel and others 1990). The indicators used were: untoward pregnancy outcome (UPO), deaths among live-born infants through a period of about 26 years (exclusive of those resulting from malignancies), malignancies in children, frequency of balanced structural rearrangements of chromosomes, frequency of sex chromosomal aneuploidy, frequency of mutations affecting protein charge or function (electrophoretic mutations), sex ratio among children of exposed mothers, and growth and development of children.

DD Estimates from Japanese Data

Since the mid-1970s, several different DDs consistent with the emerging data had been estimated, and these are summarized in Annex 4A. The most recent DD estimates were those published by Neel and colleagues (1990) using five of the indicators mentioned above (*i.e.*, UPO, F_1 mortality, F_1 cancers, sex chromosomal aneuploidy, mutations affecting protein charge or function). Details of how these DDs were calculated are presented in Annex 4G. The important point here is that all of the past as well as the 1990 DD

estimates based on the Japanese data were higher by factors of 3 or more compared to the DD estimate of 1 Gy that has been used by UNSCEAR and the BEIR committees over the years. Since, in the public mind, the notion remains that the magnitude of DD defines the magnitude of risk (*i.e.*, a low DD is indicative of high risk and high DD of a low risk), the above discrepancy between the DDs has given rise to the concern that UNSCEAR and the BEIR V committees might have overestimated the risks.

The BEIR VII committee stresses that such comparison between DDs estimates is inappropriate for the following reasons: (1) the DDs used by the UNSCEAR and BEIR committees are based on spontaneous and induced rates of mutations at defined human and mouse genes, respectively (or mouse genes in the past); (2) DD enters the risk equation as 1/DD, and the latter is only one of the four quantities used to predict the risk of genetic disease prospectively; and (3) in the Japanese studies, DD is estimated retrospectively from empirical data on indicators of genetic damage that are totally different and not readily equatable to genetic diseases; further, based on current knowledge, most of the indicators would not have been expected to show a significant increase in frequency for mechanistic or other reasons (see Annex 4G).

Consistency of Findings from Japanese Studies with Present Risk Estimates

Notwithstanding these differences in end points, estimates of DDs, and the approaches used for risk estimation, the principal messages from the Japanese studies (no significant adverse effects in more than 30,000 progeny from parents with estimated conjoint gonadal dose of the order of about 0.4 Sv or less) and from estimates discussed in this document (3000 to 4700 cases per gray per million children of the first postradiation generation; Table 4-6) are basically the same—namely, that at low doses the genetic risks are small compared to the baseline risks of genetic diseases.

Strengths and Weaknesses of the Risk Estimates Presented in This Report

For the first time in genetic risk estimation, it has been possible to present risk estimates for all classes of genetic disease. In part, this is due to the incorporation of advances in human molecular biology within the conceptual framework of risk estimation. It is important to realize however that human data that bear on hereditary effects of radiation remain limited, and estimates of risk still have to be obtained indirectly using several assumptions. While the risk estimates presented in this document represent what is achievable at the present state of knowledge, it is instructive to examine the assumptions (and consequent uncertainties) and, more importantly, the overlap of the estimates made.

Equal Sensitivity of Both Sexes to Radiation-Induced Mutations

The first of these assumptions—namely, the sensitivity of human immature oocytes to radiation-induced mutations is equal to that of stem cell spermatogonia—was dictated by the view that mouse immature oocytes may not constitute a suitable model for assessing the response of human immature oocytes. If indeed human immature oocytes turn out to be less sensitive than stem cell spermatogonia, then the sex-averaged rate of induced mutations would be lower (*i.e.*, the DD would be higher, which means lower relative mutation risk). At present, it is not possible to address this issue.

The Doubling Dose

The average spontaneous mutation rate (the numerator in DD calculations) is based on 26 human disorders encompassing some 135 genes. Although it would have been ideal to use the average rate based on all of the genes contributing to diseases included under *P* (the baseline frequency of diseases), this was not possible because of lack of data. When full annotations of all of the genes in the human genome and knowledge of their disease potential and mutation rates become available, it is likely that the estimate of average rate of mutations will change. Likewise, the average rate of induced mutations (the denominator in DD calculations) is now based on induced mutations in 34 mouse genes with widely different locus-specific rates. Again, knowledge of the radiation response of mouse genes is expected to increase when the mouse genome, which has now been sequenced (FCRGERG 2002; MGSC 2002), becomes fully annotated and enables radiation mutagenesis studies with additional genes. At present, one can only speculate about how the induced mutation rate will change and how it will impact the DD estimate.

Mutation Component

In this report, it is assumed that the first-generation MC = *s* = 0.3 for autosomal dominant and X-linked diseases. The estimate for *s* has been obtained from an analysis of only a subset of naturally occurring autosomal dominant diseases for which such information was available and is therefore not applicable to all autosomal dominant and X-linked diseases included under *P*. If one were able to include these, the average *s* value might change. Further, it may be necessary to revisit the assumption that the *s* value for induced mutations that cause disease is similar to those of spontaneous disease-causing mutations.

The estimate that for chronic multifactorial diseases, MC = 0.02 for the first few generations has been obtained from computer simulation studies, which showed that the MC values were in the range of 0.01 to 0.02, often closer to the former than to the latter. If the actual MCs are in fact closer to 0.01, the currently used MC value will overestimate the risk by a factor of 2.

Potential Recoverability Correction Factors

For autosomal dominant and X-linked diseases, a range of PRCFs from 0.15 to 0.30 was used, the lower limit of the range being a weighted average (*i.e.*, weighted by disease incidence) and the upper limit, the unweighted average (*i.e.*, proportion of genes at which induced mutations are potentially recoverable in live births). However, the criteria developed for potential recoverability of induced deletions (the predominant type of radiation-induced DNA damage) do not include breakpoint specificities that are undoubtedly important in the case of deletion-associated naturally occurring Mendelian diseases. It seems unlikely that radiation-induced deletions would share these specificities, certainly not in all genomic regions. Should these specificities also be important for recovering induced deletions (with a disease phenotype similar to that associated with a naturally occurring deletion), even the weighted PRCF may be an overestimate.

For chronic multifactorial diseases, the assumption has been that the PRCF may simply be the *x*th power of that for single-gene diseases, with *x* = the number of genes that have to be simultaneously mutated to cause disease. The values of 0.02 to 0.09 have assumed *x* = 2, the minimum number. Although statistically such a calculation can be defended, the implicit biological assumption that at low doses of radiation, two independent mutations underlying a chronic disease may be induced simultaneously and recovered seems unrealistic.

There is one further point to be made, namely that the PRCF for chronic diseases is very sensitive to *x*. For example, if *x* = 3, the PRCF range becomes 0.003 to 0.03 (*i.e.*, $[0.15]^3$ to $[0.30]^3$). Since for many chronic diseases, mutations in more than two genes seem to be involved, the argument is that the PRCF range of 0.02–0.09 used in the present calculations may overestimate the risk.

Overlap in Estimates of Risk

It should be recalled that (1) the estimates of risk for autosomal dominant and X-linked diseases have been obtained using the DD method; (2) the risks of congenital abnormalities that are also adverse dominant effects have been obtained independently using mouse data without recourse to the DD method; and (3) the risk of "chromosomal diseases" has been assumed to be subsumed under the above two items. The important point is that since all of these represent dominant effects (and spontaneous mutations in many developmental genes are known to cause Mendelian diseases), there must be overlap between the classes of risk grouped under the headings of autosomal dominant + X-linked diseases and of congenital abnormalities, although at present it is difficult to assess its magnitude. The consequence is that the sum may overestimate the actual risk of dominant effects.

Overall Conclusions

The committee has evaluated Table 4-6 and expresses the view that these estimates of risk are the best that are possible at the present time.

ANNEX 4A: MODELS OF INHERITANCE OF MULTIFACTORIAL DISEASES IN THE POPULATION

Multifactorial Threshold Model of Disease Liability and the Concept of Heritability

Assessment of the relative importance of genetic and environmental factors in the etiology of multifactorial diseases is essential to explain their transmission patterns and predict their risks of recurrence in families. Conceptualized this way, it is a problem of quantitative genetics, the theoretical foundations for which were laid by Fisher (1918). Multifactorial diseases *per se*, however, are not quantitative traits, but qualitative ones (*i.e.*, all-or-none traits [presence or absence of disease]), although some of the surrogate clinical measures used are quantitative (*e.g.*, serum cholesterol levels in the case of CHD, systolic and diastolic blood pressure in the case of essential hypertension, fasting glucose concentrations in the case of diabetes mellitus). Consequently, methods originally developed for studies of quantitative traits and their inheritance were adapted to deal with these diseases.

Carter (1961) proposed the concepts of a hypothetical variable called *disease liability* that underlies multifactorial diseases and of *threshold*. The concept of disease liability enables one to envisage a graded scale of the degree of being affected or being normal. Likewise, the concept of threshold enables one to envision a certain value in the liability scale that, when exceeded, will cause the disease. Below the threshold, the individual would not be affected. Subsequently, Falconer (1965) formalized these concepts quantitatively by advancing what has come to be known as the MTM of disease liability.

Details of the MTM have been discussed extensively in a number of publications (*e.g.*, Falconer 1965, 1967; Smith 1975; Carter 1976a; Bishop 1990). The basic assumptions of the simple or standard version of the MTM are the following: (1) all environmental and genetic causes can be combined into a single continuous variable called liability, which is not measurable as such; (2) liability is determined by numerous genetic and environmental factors that act additively, each contributing a small amount of liability; (3) the liability in the population has a normal (Gaussian) distribution; and (4) affected individuals are those whose liability exceeds a certain threshold (see Figure 4A-1).

The MTM permits a number of predictions: First, when the population frequency of the disease is high, the relative risk to relatives of an index case (compared to the general population) would be expected to be greater, but proportionately less. This situation occurs because, when the popula-

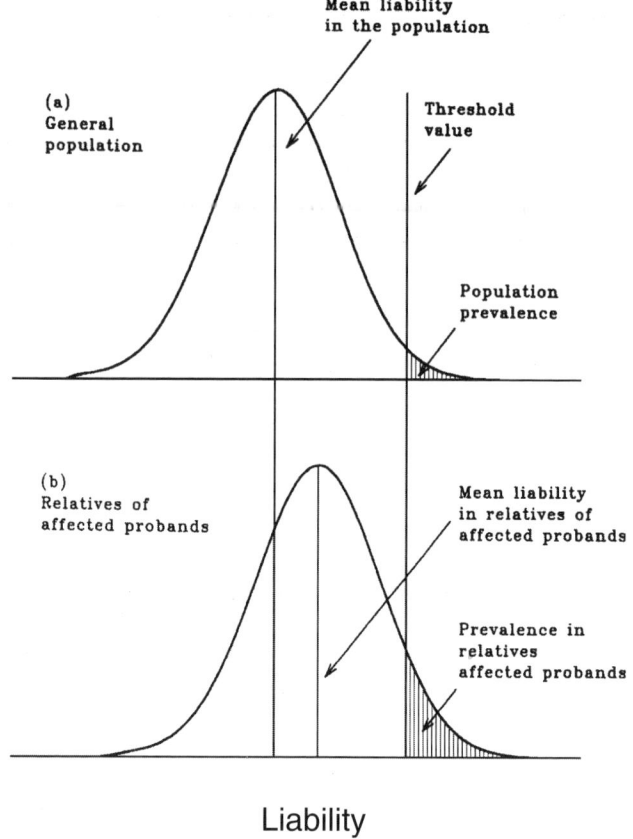

FIGURE 4A-1 Distribution of liability in the general population and in relatives of affected individuals according to the multifactorial threshold model.

tion frequency is high, the predisposing mutations for the condition are distributed throughout the population, so the likelihood of exceeding the threshold is high. When the population frequency of the disease is low, only relatives have a significant risk.

Second, for diseases that show marked differences in incidence between the sexes, the MTM—with the added assumption of different thresholds in the two sexes—would predict higher relative risks to relatives of the less frequently affected sex. For example, in Hungary, congenital pyloric stenosis is about three times more common in males than in females (0.22% versus 0.07%). The risk to brothers of affected females is about 20%, which is much higher than the value of 4% for the brothers of affected males (Czeizel and Tusnady 1984). On the assumption that the threshold is farther from the mean in females than in males (*i.e.*, more to the right upper tail of the distribution), one would expect that affected females would have more disease-predisposing mutations, on average, than affected males. Relatives of female patients would therefore receive more of these (thus being at correspondingly higher risk) than relatives of male patients (see Figure 4A-2).

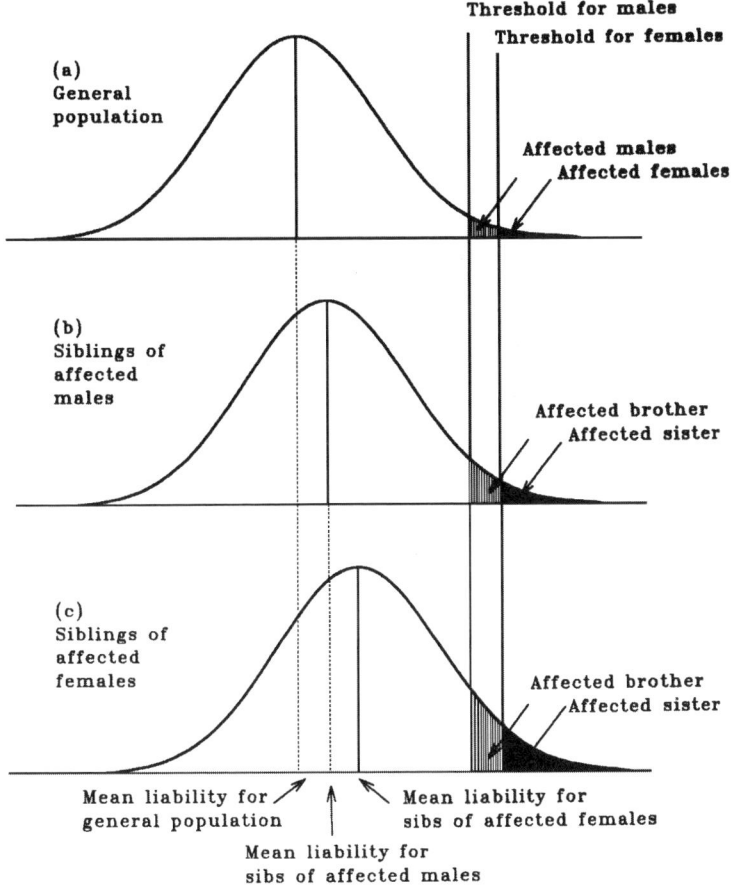

FIGURE 4A-2 Comparisons of the distribution liability in the general population with those in relatives of affected individuals when there are differences in the prevalence of multifactorial disease, according to the multifactorial threshold model with the additional assumption of different thresholds for disease liability in the two sexes.

Based on the properties of the normal distribution of liability (made up of both genetic and environmental components) that underlies the MTM, methods have been developed to use data on the population frequency of a given multifactorial disease to predict the risk to relatives of those affected and to estimate, on the basis of correlation in liabilities between relatives, the relative contribution of genetic factors to the overall phenotypic variability summarized in the statistic called "heritability of liability" (h^2).

Concept of Heritability

In quantitative genetics, the relative contributions of genetic and environmental factors to the overall phenotypic variation is assessed by analysis of variance (*i.e.*, by estimating the total phenotypic variance, V_P, and apportioning it into variance due to genetic factors, V_G, and variance due to environmental factors, V_E). Under the assumption that the genetic and environmental effects are independent of each other (*i.e.*, they are not correlated), $V_P = V_G + V_E$. The ratio

V_G/V_P is called "broad-sense heritability of liability," or "degree of genetic determination," and is symbolized by h_B^2. It provides a measure of the relative importance of genotype as a determinant of phenotypic value (Smith 1975).

The genotypic variance V_G can be subdivided into an additive component (V_A) and a component to deviations from additivity. Additive genetic variance is the component attributable to the average effect of genes considered singly, as transmitted in the gametes. The ratio V_A/V_P is called "narrow-sense heritability," or h_N^2, and expresses the extent to which the phenotypes exhibited by parents are transmitted to offspring, and it determines the magnitude of correlation between relatives. The nonadditive genetic variance is due to the additional effects of these genes when combined in diploid genotypes and arises from dominance (V_D), interaction (epistasis, V_I) between genes at different loci, and assortative mating (V_{AM}). In the absence of these sources of genetic variance, $h_N^2 = h_B^2$. It is important to note that most of the heritability estimates for chronic diseases published in the literature are broad-sense heritability of liability estimates and are in the range of about 0.3–0.8.

Other Models of Inheritance of Multifactorial Diseases

An important assumption of the MTM as discussed above is that a large number of factors, each with small effects, contributes to liability. However, the assumption of fewer contributing factors is also consistent with data from familial aggregation studies, and for this reason, it is not a good analytical tool for discriminating between different modes of inheritance. Consequently, attempts to fit the familial data to Mendelian models (with appropriate choice of assumptions on the numbers of loci, penetrance, dominance, etc.) or to a combination of major locus and polygenic models have been made, (*e.g.*, Elston and Stewart 1971; Morton and MacLean 1974; Kendler and Kidd 1986); although these models are of interest in catalyzing the search for the genes involved, they are now largely superseded by molecular approaches that hold the potential for direct identification of the genes.

ANNEX 4B: THE DOUBLING DOSE

Table 4B-1 provides a broad overview of the data used during the past four decades for estimating doubling doses. It is worth noting that although the present unit for expressing absorbed radiation dose is gray (or sievert when considering radiations of different qualities), in reviewing the earlier estimates in this section the DDs are expressed in the same units employed in the original publications, namely, roentgens (R), rads, roentgen-equivalent-man (rem), grays, and sieverts. Note that for low-LET radiation (*e.g.*, X-rays and γ-rays), 1 Gy = 100 rads ~ 100 R; 1 rem = 1 rad; and 1 Sv = 100 rem.

Briefly, the notion that the DD for genetic damage induced in human males at low-dose or chronic low-LET radiation conditions is likely to be of the order of about 100 R was already entertained in the early 1960s (UNSCEAR 1962). This estimate was guided by the findings (from mouse studies on recessive specific locus mutations) that chronic X-irradiation would be only about one-third as effective as acute X-irradiation in males and much less effective in females (Russell and others 1958, 1959). Consequently, it was suggested that the DD for chronic X-irradiation exposure conditions was probably at least three times that for acute X-irradiation (*i.e.*, three times that of about 30 R suggested in the 1958 UNSCEAR report for acute X-irradiation or about 100 R).

In 1971, Lüning and Searle broadened the original concept of the DD to include not only mutations at defined gene loci, but also four other end points of genetic damage (semisterility, dominant visible mutations recovered in the course of studies on recessive specific locus mutations, autosomal recessive lethals, and skeletal abnormalities, all from

experiments involving irradiation of male mice [spermatogonial stem cell irradiations]). They found that for acute X-irradiation of males, although individual estimates varied from 16 to 51 R (with wide confidence limits, except for specific locus mutations), the overall average was about 30 R. For low-dose or chronic low-LET radiation exposure, the suggestion was that it would be between three and four times that for acute X-irradiation (*i.e.*, about 100 R). UNSCEAR, however, did not use the DD method in its 1972 report, but in all reports published until 1993, the mouse data-based estimate of 1 Gy has been used.

The BEIR I report (NRC 1972) introduced the concept that DD estimates must be based on the average spontaneous mutation rate of human genes and the average induced rate of mutations in mouse genes. In that report it was assumed that (1) the spontaneous mutation rate of human genes might be in the range of 0.5×10^{-6} to 0.5×10^{-5} per gene and (2) the sex-averaged rate of induced recessive mutations in mouse was about 0.25×10^{-7} per locus per rem for low-LET radiation conditions. With these estimates, a range of DDs from 20 to 200 rem was calculated.

The induced rate of 0.25×10^{-7} per locus per rem mentioned above was the unweighted average of the rate of 0.5×10^{-7} per locus per rem for males (at 12 loci, including 7 of the specific loci have been used in most mouse experiments and the additional 5 used in the studies of Lyon and Morris 1969) and that of zero assumed for females. It was noted, however, that the estimate of 0.25×10^{-7} per locus per rem might be too high for at least two reasons: (1) "the gene loci at which these studies were made, were to some extent preselected for mutability" and (2) "the rate of induction of dominant visible mutations in mice is lower than for recessives by at least an order of magnitude and dominant mutations constitute a substantial part of the human genetic risk." This procedure of using human data on spontaneous mutation rates was driven by one of the principles stated by the committee—namely, that emphasis should be placed on human data when feasible—the implicit idea being that if the induced rate was extrapolated from mouse to humans, there would be one extrapolation uncertainty and if both spontaneous and induced rates were extrapolated to humans, there would be two such uncertainties.

When UNSCEAR (1977) first used the mouse data-based DD of 100 rads, it did not actually specify the induced rates. This was because the estimate of 100 rads was arrived at by assuming that the DD for low-LET chronic radiation conditions would be three times that of ~30 rads for high-dose-rate acute X-irradiation conditions (for five different end points; see Lüning and Searle 1971).

In BEIR III (NRC 1980), however, the committee abandoned the method that was used in BEIR I, namely, using human data on spontaneous mutation rates and mouse data on induced mutation rates in defined genes. The stated objection to the BEIR I method was that it mixed the induced rate of a set of mouse genes preselected for high mutability

TABLE 4B-1 Doubling Dose Estimates Used in Risk Estimation from the 1950s to the Early 1990s

Reference	DD	Radiation Conditions	Comments
1956 BEAR report (NRC 1956)	50–80 R 40 R	High dose rate (acute)	Guided more by general radiation genetic principles (established mostly from *Drosophila* studies) than by knowledge of mouse or human mutation rates and, therefore, nothing more than educated guesses; among the principles were (1) linear dose-effect relationship for induced mutations and (2) effect independent of dose rate or dose fractionation. The general philosophy and "best" estimates of the Medical Research Council (MRC 1958) and UNSCEAR (1958) were roughly similar
UNSCEAR (1962)	100 R	Chronic	Based on *mouse data* on the reduced effectiveness of chronic γ-irradiation for the induction of specific locus mutations (Russell and others 1958); assumed that DD for males will be about 3 times that of 30 R assumed in UNSCEAR (1958) for acute X-irradiation conditions; noted that DD for females may be higher
Lüning and Searle (1971)	16–51 rads ~100 rads	Acute Chronic	Based *on mouse data for 5 different end points* for males; no DD estimate provided for females
1972 BEIR report (NRC 1972)	20–200 rem	Chronic	Based on a range of spontaneous rates *in humans* (0.5×10^{-6} to 0.5×10^{-5}) and a sex-averaged rate of induction of specific locus mutations of 0.25×10^{-7} per locus per rem in *mice*
Neel and others (1974)	46 rem (Petersen and others 1990) 125 rem (females)	Acute	Based on data on *mortality of children born to A-bomb survivors* through the first 17 years of life; assumed that for chronic irradiation, the DD for males might be 3 to 4 times 46 rem and as much as 1000 rem for females
Sankaranarayanan (1976); Searle (1976)	80–240 rads	Chronic	Based on *mouse data* for specific locus mutations induced in spermatogonia and in mature + maturing oocytes and dominant visibles and translocations induced in spermatogonia
UNSCEAR (1977)	100 rads	Chronic	Rationale stated as follows: "Examination of available evidence in the mouse suggests that the use of a 100-rad DD will not underestimate the risk. The ICRP Task Group has also this figure in its calculations . . ."
1980 BEIR report (NRC 1980)	50–250 rem	Chronic	Based on the "best substantiated" estimate of DD of 114 rem for spermatogonial irradiation of *male mice* and approximately halving and doubling the above estimate to arrive at the range of 50–250 rem
UNSCEAR (1982)	100 rads	Chronic	No change from the 1977 report
Neel and others (1982); Schull and others (1982)	60 ± 93 rem 135 ± 388 rem 535 ± 2416 rem 135 ± 156 rem	Acute	The first three estimates are based, respectively, on data on UPOs, survival through childhood, and sex chromosomal aneuploids in the Japanese studies; the authors considered that the weighted average of 135 ± 156 rem (last entry) should be multiplied by a factor of 3 to make it applicable to chronic radiation conditions
UNSCEAR (1986)	1 Gy	Chronic	No change from the 1977 report
UNSCEAR (1988)	1 Gy	Chronic	No change from the 1977 report
1990 BEIR report (NRC 1990)	100 rads	Chronic	Overall estimate based on *mouse data* (both sexes) on several different end points; most estimates given as ranges that vary by factors between about 2 and 30 (a reflection of differences in estimated spontaneous and induction rates); multiplication factors between 5 and 10 used when necessary to convert DD estimates for high-dose-rate irradiation to those for chronic irradiation
Neel and others (1990)	1.69–2.23 Sv	Acute	Composite estimates of "minimal DDs" (DDs at 95% lower confidence limits) compatible with Japanese results on UPOs, F_1 mortality, F_1 cancer, sex chromosomal aneuploids, and mutations altering protein charge or function; on the assumption of a dose-rate reduction factor of 2, the authors suggest that for chronic low-LET, low-level radiation, the figures are likely to be twice those estimated (*i.e.*, about 3.4 to 4.5 Sv)
Neel and Lewis (1990)	1.35 Gy	Acute	Based on an analysis of *mouse data* on 7 mutational end points (spermatogonial irradiation experiments); the authors suggest that with the use of a dose-rate factor of 3, the DD will be about 3 Gy
UNSCEAR (Rabes and others 2000)	1 Gy	Chronic	No change from the 1977 report

SOURCE: Sankaranarayanan and Chakraborty (2000a).

with estimates of human spontaneous rates for more typical genes. The BEIR III committee adopted the view that it was preferable to use a DD estimate obtained from spontaneous and induced mutations in the same set of loci in the same species and used exclusively the data on the seven specific loci obtained in experiments with male mice. The figures used were 7.5×10^{-6} per locus for spontaneous rates and 6.6×10^{-8} per locus per rem for induced rates from which "the best substantiated" DD estimate of 114 R was calculated. To derive DDs for risk predictions, it approximately halved and doubled the above estimate of 114 R to obtain a range of 50 to 250 rem.

In BEIR V (NRC 1990), the committee again used primarily mouse data but included several additional end points in both sexes (dominant lethals, recessive lethals, dominant visibles, recessive visibles, reciprocal translocations, congenital malformations, and aneuploidy). On the basis of all these data, it concluded that "considering all endpoints together, the direct estimates of doubling dose for low dose rate radiation have a median value of 70–80 rad, indirect estimates based on high dose-rate experiments have a median value of 150 rad, and the overall median lies in the range of 100 to 114 rad. These estimates support the view that the doubling dose for low dose-rate, low-LET radiation in mice is approximately 100 rad for various genetic endpoints."

Table 4B-1 also shows that the DD estimates made over the years based on genetic data from A-bomb survivors (Neel and others 1974, 1982, 1990; Schull and others 1981, 1982; Otake and others 1990; Neel 1998) were at least some three to four times that of 1 Gy used by UNSCEAR and the BEIR committee; the so-called Japanese DD estimates, however, were never used by the above committees. For the first time, the BEIR V (NRC 1990) report gave a formal "status" to the Japanese results by noting that "a doubling dose of 100 rem approximates the lower 95% confidence limit for the data from atomic bomb survivors in Japan and it is also consistent with the range of doubling doses in mice."

ANNEX 4C: ASSUMPTIONS AND SPECIFICATIONS OF THE FINITE-LOCUS THRESHOLD MODEL

The assumptions and specifications of the FLTM have been discussed in detail by Denniston and colleagues (1998) and in the ICRP (1999) Task Group report. Briefly, the FLTM assumes that (1) the genetic component of liability of a chronic multifactorial disease is discrete and is determined by mutant alleles at a finite number (n) of autosomal gene loci; the total number of mutant alleles at these n loci in a given genotype is a random variable g; (2) the environmental component is continuous and represented by a random variable e, which has a Gaussian distribution with mean of zero and variance of V_e; (3) the total liability $x = f(g) + e$, where $f(g)$ is a function of the number of mutant alleles in the n-locus genotype of the individual and e is the environmen-

tal effect; (4) individuals with liability exceeding the threshold T (*i.e.*, $x > T$) are affected by the disease, and those for whom $x < T$ are unaffected; and (5) unaffected individuals have a fitness of 1 and unaffected ones of $(1 - s)$. The impact of an increase in total mutation rate as a result of radiation exposures—from m to $m(1 + k)$, with k measuring the increase relative to the baseline—is assessed in terms of changes in heritability of liability (h_x^2), and consequent changes in the MC. This assessment was carried out by assuming that the effects of the mutant alleles are either additive or synergistic.

Unlike the case of Mendelian diseases, the algebraic formulations of the FLTM do not permit expressing the effects in the form of a single equation. However, the predictions of the model can be evaluated iteratively using the computer program that was developed for this purpose. The program is first run using a specified set of parameter values (mutation rate, selection coefficients, threshold, etc.) until the population reaches equilibrium between mutation and selection. Once this occurs, the mutation rate is increased either once or permanently corresponding to radiation exposure in one generation only or in every generation, and the computer run is resumed with the new mutation rate while the other parameters remain the same. The changes in mutation component and its relationship to heritability of liability are then examined in desired generations and at equilibrium. It is worth mentioning that the h^2 estimates are not inputs but outputs of the program obtained using different combinations of s values, environmental standard deviation, and threshold.

ANNEX 4D: DIFFERENCES BETWEEN SPONTANEOUS DISEASE-CAUSING MUTATIONS IN HUMANS AND RADIATION-INDUCED MUTATIONS IN EXPERIMENTAL SYSTEMS

The molecular alterations recorded in spontaneous disease-causing mutations in humans include a wide variety ranging from base-pair changes to whole-gene deletions and some multigene deletions. Radiation-induced mutations studied in experimental systems (including the mouse), however, are often multigene deletions, although scored through the phenotype of the marker loci. The extent of the deletion varies with the locus and the genomic region in which it is located.

Spontaneous mutations arise through a number of different mechanisms, and most are dependent on the DNA sequence organization of the genes and their genomic context. In contrast, radiation-induced mutations originate through random deposition of energy in the cell. One can, therefore, assume that the initial probability of radiation inducing a deletion may not differ between different genomic regions. However, their recoverability in live-born offspring seems dependent on whether the loss of the gene or genomic region is compatible with viability in heterozygotes.

Spontaneous mutations can cause either loss or gain of function of the normal gene through different mechanisms. For example, loss-of-function mutations in genes that code for structural or regulatory proteins may result in dominant phenotypes through haploinsufficiency (*i.e.*, a single normal gene is not sufficient for normal functioning) or through dominant negative effects (*i.e.*, the mutant product interferes with the function of the normal gene in the heterozygote). While loss of function of a gene can result from a variety of molecular alterations including deletions, gain-of-function mutations are likely only when specific changes in the gene cause a given disease phenotype. Radiation-induced mutations, because they are often multigene deletions, cause loss of function through haploinsufficiency.

Despite the existence of a number of differences between spontaneous and radiation-induced mutations as outlined above, radiation mutagenesis studies with a variety of experimental systems have been very successful. The possible reasons for this are now becoming evident: although the choices of marker genes in early studies of induced mutations were dictated more by practical considerations (*e.g.*, obtaining sufficient numbers of mutants, unambiguous identification through their respective phenotypes) than by their relevance to human genetic diseases, in retrospect it is clear that the "successful" mutation test systems have been those in which most of these marker genes, and the genomic regions in which they are located, are nonessential for the viability of heterozygotes (*in vivo*) or of the cell carrying the induced genetic change (*in vitro*). Consequently, induced mutations—predominantly deletions—could be recovered and studied. Most human genes, however, do not appear to be of this type.

ANNEX 4E: CRITERIA USED TO ASSIGN HUMAN GENES TO ONE OF THREE GROUPS FROM THE STANDPOINT OF THE RECOVERABILITY OF INDUCED MUTATIONS IN LIVE BIRTHS

The genes included in the analysis are a subset of those in which mutations cause autosomal dominant and X-linked diseases, which have provided the basis for the overall incidence estimates for these diseases discussed earlier (Sankaranarayanan 1998). Since not all of them fulfilled the requirements for inclusion (because of insufficient information about one or more of the following: gene size, structure, function, genomic context, etc.), only a subset could be used. The "gene-richness" or "gene poorness" of given genomic regions was assessed using the MIM (Medelian Inheritance in Man) gene maps that present the cytogenetic location of "disease genes" and other expressed genes in given cytogenetic bands (McKusick 2000.).

A gene is assigned to group 1 (induced deletions unlikely to be recovered and/or unlikely to cause the phenotype of the disease under study) when the phenotype of the naturally occurring disease is due to specific (1) gain-of-function mutations (*e.g.*, the *FGFR3* gene involved in achondroplasia); (2) trinucleotide repeat expansions (*e.g.*, Huntington's disease); (3) dominant negative mutations (*e.g.*, the *COL1A1* gene involved in osteogenesis imperfecta); and (4) restricted array of point mutations (*e.g.*, mutations in the *APOB* gene involved in one form of familial hypercholesterolemia). Also included in this group are genes that are relatively small in size and located in putative gene-rich regions (*e.g.*, the *VMD2* gene in Best's macular dystrophy).

The gene is assigned to group 2 (uncertain recoverability) when (1) it is large, it codes for an essential structural protein, and the known genetic changes are missense or nonsense mutations; (2) whole-gene deletions are rare; (3) whole-gene deletions are not rare, but the gene is located in a putative gene-rich region; and (4) information on these other genes and their function is insufficient (*e.g.*, *BRCA2*; *VHL* [von Hippel-Lindau syndrome]).

Group 3 (potentially recoverable) includes genes that are generally large and constitutional deletions, some extending beyond the confines of genes, and translocations or inversions with breakpoints in the gene causing the disease phenotype are known despite the putative gene-rich nature of the genomic region (*e.g.*, *EXT1* [multiple exotoses]; *RB1* [retinoblastoma]).

For X-linked genes, the assessment is based on whether the induced deletion will be compatible with viability in males and cause disease (since the loss of the whole X chromosome is compatible with viability but results in 45,X females).

ANNEX 4F: RADIATION STUDIES WITH EXPANDED SIMPLE TANDEM REPEAT LOCI IN THE MOUSE AND MINISATELLITE LOCI IN HUMAN GERM CELLS

Introduction

The mouse and human nuclear genomes, like those of other complex eukaryotes, contain a large amount of highly repeated DNA sequence families most of which are transcriptionally inactive (Singer 1982). Among these are the simple sequence repeats that are perfect or slightly imperfect tandem repeats of one or a few base pairs (bp). In the mouse genome, the tandem repeat loci are represented by (1) relatively short microsatellites (<500 bp) with a repeat size of 1 to 4 bp; (2) long expanded simple tandem repeats (0.5 to 16 kilobases, repeat size 4 to 6 bp); and (3) true minisatellites (0.5 to 10 kb) with repeat size of 14 to 47 bp (Gibbs and others 1993; Bois and others 1998a, 1998b; Blake and others 2000).

Mouse ESTRs

The ESTRs were originally called minisatellites but have recently been renamed to distinguish them from the much more stable true minisatellites in the mouse genome (Bois and others 1998a, 1998b). The ESTRs are highly unstable (*i.e.*, they manifest high spontaneous mutation rates) in both somatic and germ cells. The mutational changes are manifest as changes in the number of tandem repeat cores and, hence, allele length. The available data suggest that the ESTR instability is a replication- or repair-based process involving polymerase slippage similar to mechanisms suggested for microsatellite instability (Ellegren 2000).

Human Minisatellites

In contrast to mouse ESTRs, the minisatellites in humans consist of longer repeats (10 to 60 bp) that may span from about 0.5 kb to several kilobases and show considerable sequence variation along the array (Jeffreys and others 1991; 1994; May and others 1996; Buard and others 1998; Tamaki and others 1999; Stead and Jeffreys 2000; Vergnaud and Denoeud 2000). The majority of the classical minisatellites are GC rich. The fact that some of the human minisatellite loci studied are highly unstable and have very high spontaneous mutation rates of the order of a few percent is now well documented (Jeffreys and others 1985, 1988, 1995; Smith and others 1990; Vergnaud and Denoeud 2000). Mutation at these loci is almost completely restricted to the germline and is attributed to complex gene conversion-like events involving recombinational exchanges of repeat units between alleles (Jeffreys and others 1994; May and others 1996; Jeffreys and Neumann 1997; Tamaki and others 1999; Buard and others 2000; Stead and Jeffreys 2000; Vergnaud and Denoeud 2000).

Radiation Studies with Mouse ESTR Loci

The Loci Used

Two ESTR loci have been used thus far in mouse mutation studies, namely, the *Ms6-hm*, and *Hm-2*, both of which show multiallelism and heterozygosity within inbred strains. The *Ms6-hm* is <10 kb in size (varying greatly between different mouse strains) and consists of tandem repeats of the motif GGGCA. Linkage analysis localized *Ms6-hm* near the brown (*b*) coat color gene on chromosome 4. The germline mutation rate is about 2.5% per gamete (Kelly and others 1989). The *Hm-2* locus is located on chromosome 9 and consists of GGCA tetranucleotide repeats with alleles containing up to 5000 repeat units (*i.e.*, up to 5 kb). The germline mutation rate of this locus is estimated to be of the order of at least 3.6% (Gibbs and others 1993). As discussed below, Dubrova and colleagues studied mutation induction at both of the above loci, whereas the Japanese workers focused their attention only on the *Ms6-hm* locus.

Low-LET Radiation Studies

In the studies of Dubrova and colleagues (1993) involving irradiation of spermatagonial stem cells (0.5 and 1 Gy of γ-rays; CBA/H strain), significant increases in the frequencies of mutations at the *Ms6-hm* and *Hm-2* loci were found. Subsequent work with X-irradiation doses of 0.5 and 1 Gy established that for mutations induced in the above cell stage, the dose-effect relationship was consistent with linearity ($y = 0.111 + 0.338D$), where *D* is the dose in grays (Dubrova and others 1998a, 1998b). From these data, the authors estimated that the DD for ESTR mutations induced in spermatogonia was 0.33 Gy for acute X-irradiation, similar to that reported for specific locus mutations in mice.

In the above work, spermatids were found to be insensitive to mutation induction, a finding at variance with those of Sadamoto and colleagues (1994) and Fan and coworkers (1995) with the C3H/HeN mouse strain. These authors showed that for *Ms6-hm* locus mutations, all male germ cell stages were sensitive (3 Gy of γ-irradiation). Nonetheless, both sets of studies demonstrated that increases in mutation frequencies could be detected at radiation doses and sample sizes substantially smaller than those used in conventional genetic studies with specific locus mutations.

High-LET Radiations Studies

Niwa and collegues (1996) found that acute neutrons from a ^{252}Cf source (65% neutrons + 35% γ-rays) were 5.9, 2.6, and 6.5 times more effective, respectively, in spermatozoa, spermatids, and spermatogonia, than acute γ-irradiation in inducing mutations at the *Ms6-hm* locus. In similar studies, Dubrova and colleagues (2000a) noted that in spermatogonial cells, chronic neutrons also from a ^{252}Cf source had a relative biological effectiveness of about 3 relative to chronic γ-irradiation (regression equations: $y = 0.136 + 1.135D$, neutrons; doses of 0.125, 0.25, and 0.5 Gy; $y = 0.110 + 0.373D$, γ-rays; doses of 0.5 and 1 Gy). Additionally (and not unexpectedly), they found that at the above γ-ray doses of 0.5 and 1 Gy, there was no dose-rate effect. It should be remembered that the lower effectiveness of chronic γ-irradiation recorded in earlier specific locus mutation studies (Russell and others 1958) occured at total doses of 3 and 6 Gy. This observation is in contrast to earlier results with specific locus mutations (Russell and others 1958) at 3 and 6 Gy showing that chronic γ-irradiation was only one-third as effective as acute X-irradiation in inducing specific locus mutations.

Mutation Induction at the ESTR Loci—An Untargeted Process Arising as a Result of Radiation-Induced Genomic Instability

One important conclusion that emerges from these studies is that mutation frequencies in the progeny of irradiated animals are too high to be accounted for by the direct induction of mutations at the loci studied (*i.e.*, radiation induction of germline mutations at ESTR loci is an untargeted process). Dubrova and colleagues (1998a, 1998b) concluded that there might be two associated processes: structural damage elsewhere in the genome or in other sensor molecules and, subsequently, indirect mutation at ESTR loci. This nontargeted origin of radiation-induced mutations at the ESTR loci is reminiscent of the phenomenon of delayed radiation-induced genomic instability in somatic cells (discussed in Chapters 2 and 3). The experiments of Barber and colleagues (2000) showed that the ESTR mutations in unirradiated or irradiated mice are not associated with a general genome-wide increase in meiotic recombination rate.

Further support for the concept of the nontargeted origin of induced ESTR mutations comes from the work of Niwa and Kominami (2001). In their study, male mice received 6 Gy of γ-irradiation and were mated to unirradiated females to produce F_1 progeny from irradiated spermatozoa and stem cell spermatogonia. As in their earlier studies, mutations at the *Ms6-hm* locus were studied. The mutant frequencies for the paternally derived allele increased to 22% and 19% in the F_1 progeny from irradiated spermatozoa and spermatogonia, respectively (about a twofold increase over the control rate). The surprising finding was that the mutation frequency also was higher (20%) in the maternally derived allele in progeny descended from irradiated spermatozoa, but not from spermatogonia. The authors' interpretation is that the introduction of damage into the egg by irradiated spermatozoa triggers genomic instability in zygotes and in embryos of subsequent developmental stages, and that this genomic instability induces untargeted mutation in *cis* (in the paternally derived allele) and in *trans* (in the unirradiated maternally derived allele).

Transgenerational Instability

Dubrova and colleagues (2000a) and Barber and coworkers (2002) provided additional evidence for the involvement of radiation-induced germline genomic instability in the origin of induced ESTR mutations. In these experiments involving chronic neutron irradiation (0.5 Gy) of spermatogonial stem cells, the mutation frequency in the F_1 progeny was about sixfold higher than in the control. Breeding from the unirradiated F_1 mice revealed that the mutation rate remained high in transmissions from both F_1 males (6×) and F_1 females (3.5×; scored in F_2). A part of this increase is due to germline mosaicism in F_1 animals, suggesting that paternal exposure to radiation results in a destabilization of ESTR loci in the germline of offspring and that some of the mutations occur sufficiently early in germline development for significant levels of mosaicism to arise. More importantly, this instability is transmissible through meiosis and mitosis to the F_2 generation and appears to operate in *trans* in the F_1

germline (*i.e.*, affecting alleles not only from the exposed F_0 male but also from the unexposed F_0 female). The latter finding is similar to that of Niwa and Kominami (2001).

In subsequent experiments, Barber and colleagues (2002) confirmed the transgenerational effects of chronic neutron irradiation and extended the observations to acute X-irradiation. Additionally, the response of two other inbred mouse strains (C57BL/6 and BALB/c) was compared with that of the CBA/H strain used in their studies. The rationale for the comparisons rests on earlier findings that BALB/c and CBA/H mice show higher levels of radiation-induced genomic instability in somatic cells than C57BL/6 mice and that this difference can be attributed to the strain-specific polymorphism at the *Cdkn2a* (cyclin-dependent kinase inhibitor) and *Prkdc* (DNA-dependent protein kinase catalytic subunit) genes (Zhang and others 1998; Yu and others 2001).

In these experiments, (1) spermatogonial neutron (0.4 Gy) or X-irradiation (2.0 Gy) of CBA/H mice resulted in an increase in the mutation rate in both the F_1 and the F_2 generations (derived from unirradiated F_1 males and females); however, although spermatid irradiation did not cause an increase in mutation rate in the F_1 generation (which was also the case in their earlier work), there was a clear increase in mutation rate in the F_2 progeny, suggesting that destabilization of the F_1 germline occurs after fertilization, regardless of the stage of spermatogenesis exposed to radiation, and that the radiation-induced signal also persists and destabilizes the F_2 germline; (2) transgenerational effects were also observed in neutron-irradiated (0.4 Gy) C57BL/6 and X-irradiated (1 Gy) BALB/c mice; and (3) there were clear differences in the levels of spontaneous and transgenerational instability in the order BALB/c > CBA/H > C57BL/6. In summary, these data permit the conclusion that the instability associated with radiation-induced germ cell mutations at the ESTR loci persist for at least two generations.

Direct Studies of ESTR Mutations in Mouse Sperm

In a recent paper, Yauk and colleagues (2002) have reported on mouse experiments involving single molecular polymerase chain reaction (PCR) analysis of genomic DNA for studying spontaneous and radiation-induced mutations at the *Ms6-hm* locus. These X-irradiated male mice (1 Gy) were killed 10 weeks postirradiation, and spermatozoa collected from caudal epididymis from the mice were screened for mutations. The findings were that (1) significant increases in mutation frequency could be detected, with the magnitude being similar to that established by conventional pedigree analysis, and (2) the majority of mutations resulted from small gains or losses of three to five repeat units.

Radiation-Induced Mutations at Human Minisatellite Loci

Studies After the Chernobyl Accident and Around the Semipalatinsk Nuclear Test Site

Dubrova and colleagues (1996) first reported on radiation-induced minisatellite mutations among children born between February and September 1994 to parents who were continuously resident in the heavily polluted rural areas of the Mogilev district of Belarus following the Chernobyl accident. Blood samples were collected from 79 families (father, mother, and child) for DNA analysis. The control sample consisted of 105 nonirradiated Caucasian families from the United Kingdom, sex-matched to the offspring of the exposed group. DNA fingerprints were produced from all families by using the multilocus minisatellite probe 33.15 and two hypervariable single-locus probes, MS1 and MS31. Additionally, most families were profiled with the minisatellite probes MS32 and CEB1. For the Mogilev families, the level of ^{137}Cs contamination was used as a dose measure, and the families were divided according to the median ^{137}Cs contamination levels into those inhabiting less contaminated areas (<250 kBq m^{-2}) and those inhabiting more contaminated areas (>250 kBq m^{-2}).

The data showed that the frequency of mutations (1) was higher by a factor of about 2 in the children of exposed families relative to control families and (2) showed a correlation with ^{137}Cs contamination levels as demarcated above. The authors suggested that these findings were consistent with radiation induction of germline mutations but also noted that other nonradioactive contaminants from Chernobyl, such as heavy metals, could be responsible. These results have been subject to criticism on the grounds that the U.K. control population was ethnically and environmentally different and therefore inappropriate for comparisons (UNSCEAR 2001). Furthermore, from the data presented, it would seem that the estimated germline doses in the whole region remain sufficiently uncertain to question the true significance of an approximately twofold difference in mutation frequencies.

In a subsequent extension of the above study, Dubrova and colleagues (1997) recruited 48 additional families and used five additional probes and found that the data confirmed the approximately twofold higher mutation rate in exposed families compared to nonirradiated families from the United Kingdom. In these studies, (1) approximate individual doses for chronic γ-ray exposures were computed for 126 families in the exposed group using published data on the annual external and internal exposure to ^{137}Cs in soil, milk, and vegetables and family histories after the Chernobyl accident; (2) the parental dose for each family was taken as the mean value of the paternal and maternal doses up to conception of the child; (3) families within the exposed group could be divided according to the median of the distribution, into less exposed (<20 mSv) and more exposed (>20 mSv); and (4) the mutation rate in the latter was significantly higher than in the former, and both were higher than in the unexposed UK controls.

Further evidence showing an increase in minisatellite mutation frequencies has also been obtained from two studies, one in the Kiev and Zhitomir regions of Ukraine that sustained heavy radioactive contamination after the Chernobyl accident (Dubrova and others 2002b) and another at the Semipalatinsk nuclear test site in Kazakhstan (Dubrova and others 2002a). In the Ukraine investigation, the control and exposed groups were composed of families containing children conceived before ($n = 98$) and after ($n = 240$) the Chernobyl accident. Eight hypervariable minisatellite probes (CEB1, CEB15, CEB25, CEB36, MS1, MS31, MS32, and B6.7) were used.

A statistically significant 1.6-fold increase in mutation rate was found in the germline of exposed fathers, whereas the maternal germline mutation rate was not elevated. More than 90% of the children in the exposed cohort came from the most heavily radioactively contaminated areas of Ukraine, with a level of surface contamination from ^{137}Cs of >2 Ci/km^2. According to gamma spectrometric measurements of radionuclide concentration in soil and measurements of external exposures (γ-exposure rate in air), the whole-body doses from external exposures did not exceed 50 mSv, and similar doses from the ingestion of ^{137}Cs and ^{134}Cs for the Ukrainian population were also reported. The authors note that that all of these doses are well below all known estimates of the DD for mammalian germline mutation of 1 Sv (Sankaranarayanan and Chakraborty 2000b; UNSCEAR 2001) and, therefore, cannot explain the 1.6-fold increase in mutation rate found in exposed families

Between 1949 and 1989, the Semipalatinsk site was the former Soviet Union's premier test site for 456 nuclear tests; it was closed in 1991. The surrounding population was exposed mainly to the fresh radioactive fallout from four surface explosions conducted in 1949, 1951, 1953, and 1956, and the radioactive contamination outside the test zone currently is assessed to be low. A total of 40 three-generation families around the test site (characterized by the highest effective dose >1 Sv) along with 28 three-generation nonirradiated families from a geographically similar noncontaminated rural area of Kazakhstan were included in the study (Dubrova and others 2002a). Note that the above dose estimate cited in the paper is from Gusev and colleagues (1997; based mostly on external radiation), and the World Health Organization (WHO 1998) states that the estimates range from <0.5 Sv to 4.5 Sv. All parents and offspring were profiled with the eight hypervariable minisatellite probes previously used in the Belarus and Ukraine studies. The mutation rates in the P_0 and F_1 generations were established from the observed frequencies, respectively, in the F_1 and F_2 generations (controls and exposed progeny).

The findings were (1) in the controls, the spontaneous mutation rates in the P_0 and F_1 generations were similar;

(2) in the irradiated groups, the P_0 rate was significantly higher (1.8-fold) and the F_1 rate was nonsignificantly (1.5-fold) higher compared to controls; and (3) plotted against the parental year of birth (1950–1960, 1961–1965, and 1966–1974), the mutation rate in the exposed F_1 generation showed a negative correlation (*i.e.*, decreased) with the parental year at birth, with the highest rate in the 1950–1960 cohort (similar to that in the P_0 families) and much lower in the later two time periods (similar to that in the control cohorts).

The authors have interpreted these findings as follows: (1) all P_0 parents born between 1926 and 1948 would have been directly exposed to relatively high levels of radiation from the nuclear tests, and this would explain the 1.8-fold increase in mutation rate; (2) F_1 parents born between 1950 and 1956 would be heterogeneous with respect to the doses received: some would also have been exposed to high radiation doses, while those born later would have received considerably lower doses, and this heterogeneity in the parental doses could explain the 1.5-fold increase in mutation rate; and (3) the negative correlation with the year of birth may reflect the decreased exposure after the decay of radioisotopes in the late 1950s and after the cessation of surface and atmospheric nuclear tests.

Other Population Studies

In the mid-1990s, subsequent to publication of the radiation studies with mouse ESTR loci discussed earlier, Kodaira and colleagues (1995) conducted a pilot feasibility study on germline instability in cell lines established from the children of atomic bomb survivors in Japan. The cell lines were from 64 children from the 50 most heavily exposed families (combined gonadal equivalent dose of 1.9 Sv) and 50 children from control families. Mutations at six minisatellite loci were studied using the following six probes: Pc-1, 8TM-18, ChdTC15, p8g3, 8MS1, and CEB1. A total of 28 mutations were found, but these were at the p8g-3, 8MS-1, and CEB-1 loci only, and there were no mutations at the other three loci. Twenty-two of these were in the controls (of 1098 alleles tested; 2%), and six were in children from irradiated parents (among 390 alleles; 1.5%). Thus, there was no significant difference in mutation frequencies between the control and the exposed groups. The use of probes 33.16 and 33.15 in subsequent work did not alter the above conclusion (Satoh and Kodaira 1996; Satoh and others 1996).

The discrepancy between the results of Kodaira and colleagues, on the one hand, and those of Dubrova and colleagues (1996, 1998b, 2000a, 2000b) in the Belarus and other cohorts discussed earlier appears real. To what extent this might be due to differences in type and duration of radiation exposure remains unclear. For instance, the A-bomb survivors were externally exposed to considerable acute doses of radiation, whereas in the Belarus, Ukraine, and Semipalatinsk studies the exposures were chronic (both in-

ternal and external). Secondly, in the case of A-bomb survivors, most of their children were born more than 10 years after the single, acute parental exposure; in Belarus and Ukraine, however, the affected areas have been irradiated constantly since the Chernobyl accident. Finally, the Japanese data are derived from families in which most of the children were born to parents of whom only one had sustained radiation; in the work of Dubrova and colleagues, the data pertain to children for whom both parents had been exposed to chronic irradiation.

Livshits and colleagues (2001) found that the children of Chernobyl cleanup workers (liquidators) did not show an elevated rate of minisatellite mutations compared to a Ukrainian control group. The dose estimate for the liquidators was <0.25 Gy but is subject to uncertainty (Pitkevich and others 1997), and the main exposure was from external γ-irradiation (with a relatively minor contribution from the intake of radionuclides) received as repeated small daily doses. Interestingly, children conceived within 2 months of the fathers' employment had a higher mutation rate than those conceived more than 4 months after the fathers stopped working there. This would be consistent with an effect on cells undergoing spermatogenesis, but not on spermatogonial stem cells. However, none of these differences was statistically significant.

More recently, Kiuru and colleagues (2003) compared the frequencies of minisatellite mutations among children of 147 Estonian Chernobyl cleanup workers. The comparisons were within families (*i.e.*, between children born before and after their fathers were exposed to radiation). The post-Chernobyl children ($n = 155$) were conceived within 33 months of their fathers' return from Chernobyl; the "control" children were siblings ($n = 148$) born prior to the accident. Mutations were studied at eight minisatellite loci (*CEB1*, *CEB15*, *CEB25*, *CEB36*, *MS1*, *MS31*, *MS32*, and *B6.7*). The estimated mean dose to the workers was 100 ± 60 mSv, with fewer than 1.4% of the cohort receiving more than 250 mSv.

A total of 94 mutations (42 in the pre-Chernobyl group and 52 in the post-Chernobyl group) were found at the eight tested loci. Within-family (*i.e.*, pre- and post-Chernobyl) comparisons of mutation rates showed that the post-Chernobyl children had a slightly but not significantly higher mutation rate (0.042 per band) than the pre-Chernobyl children (0.035 per band) with an odds ratio of 1.33 (95% CI: 0.80, 2.20). The available data do not permit an assessment of the extent to which differences in paternal age might have contributed to this difference. When the cleanup workers were subdivided according to their radiation doses, the mutation rate in children born to fathers with recorded doses of 200 mSv, showed a nonsignificant increase relative to their siblings; at lower doses there was no difference.

Weinberg and colleagues (2001) screened children born in families of cleanup workers (currently either in Ukraine or Israel) for new DNA fragments ('mutations') using "multisite DNA fingerprinting." In contrast to the results of Livshits and colleagues (2001), they reported a sevenfold increase in mutation rate in these children compared to those conceived before the Chernobyl accident and external controls. However, the mutants were detected using random amplified polymorphic DNA-PCR, an unreliable technology. These mutants were not validated and had no obvious molecular basis (Jeffreys and Dubrova 2001)

Studies of Cancer Patients

There are some limited data on minisatellite mutations detected directly in sperm sampled from cancer patients who have sustained radiotherapy and/or chemotherapy (Armour and others 1999; May and others 2000; Zheng and others 2000). All of these studies used the so-called small-pool PCR approach (SP-PCR) originally developed for the analysis of spontaneous mutations at human minisatellite loci (Jeffreys and others 1994). While this method can overcome the small sample size limitations encountered in pedigree analysis, a major shortcoming of the SP-PCR approach, compared to the pedigree approach, is the very large variation in spontaneous mutation rates of individual alleles at a single locus. Although SP-PCR can be used to evaluate the mutation rate in the same male before and after mutagenic treatment, it does not allow amplification of very large minisatellite alleles (longer than 5 kb), thus restricting mutation scoring to a subset of relatively small minisatellite sizes.

In the first of these studies (Armour and others 1999), sperm DNA of two men exposed to the anticancer drugs cyclophosphamide, etoposide, and vincristine, plus 2.2 Gy of X-rays (scattered radiation from mediastinal radiotherapy), were analyzed for mutations at the MS205 locus known to have a high germline mutation rate (~0.4–0.7% per gamete). There were no significant differences in mutation frequencies in the pretherapy and posttherapy samples (11 and 16 months, respectively, in the two individuals). Mutation rates were 0.38% versus 0.47% in the former and 0.10% versus 0.11% in the latter. It should be noted, however, that in mouse experiments, cyclophosphamide is mutagenic only in postmeiotic germ cells, etoposide (a topoisomerase II inhibitor) is mutagenic only in meiotic cells, and vincristine is not mutagenic, although it is known to prevent the assembly of tubulin into spindle fibers (Witt and Bishop 1996; Russell and others 1998).

In the second study (Zheng and others 2000), sperm DNA from 10 men treated for Hodgkin's disease (with different combinations of chemotherapeutic agents plus 2.5 Gy of abdominal X-rays) were analyzed using the MS205 locus. Nine patients treated with either vinblastine or adriamycin and bleomycin did not show any increases in mutation frequency. Vinblastine binds to tubulin and, in mice, results in aneuploidy but not chromosome breakage or mutations. Adriamycin is an intercalating agent and an inhibitor of topoisomerase-II, and in mice, this compound is toxic to germ cells but does not cause mutations (Witt and Bishop 1996). Bleomycin, a radiomimetic agent, selectively targets mouse oocytes, but no mutation induction in male germ cells has been observed. The only patient treated with procarbazine + oncovin + prednisone (for six cycles with 3–4 week intervals between cycles) showed a slight increase in mutation frequency (1.14% versus 0.79%). Procarbazine is known to be mutagenic to mouse spermatogonia.

In the work of May and colleagues (2000), sperm DNA samples from three seminoma patients who underwent orchiectomy and external beam radiotherapy were used to study induction of mutations at the B6.7 and CEB1 loci. These men received 15 fractions of acute X-irradiation, with a total testicular dose (from scattered radiation) ranging between 0.4 and 0.8 Gy. No induced mutations were found.

ANNEX 4G: DOUBLING DOSES ESTIMATED FROM GENETIC DATA OF CHILDREN OF A-BOMB SURVIVORS

The most recent DD estimates consistent with the Japanese data are those of Neel and colleagues (1990). These were expressed as "end-point-specific minimal DDs" excluded by the data at specified probability levels and "most probable gametic DD" (note that all of these are for the acute radiation conditions obtained during the bombings). For example, the minimal DDs at the 95% probability level were the following: 0.05 to 0.11 Sv (F_1 cancers); 0.18 to 0.29 Sv (UPO); 0.68 to 1.10 Sv (F_1 mortality); 1.60 Sv (sex-chromosomal aneuploidy), and 2.27 Sv (electrophoretic mutations). When only UPO, F_1 cancers, and F_1 mortality were considered together, the estimated DD at the 95% probability level was 0.63 to 1.04 Sv. The comparable estimate for sex chromosomal aneuploidy and electrophoretic mutations considered together was 2.71 Sv.

The oft-quoted DD range of 1.69 to 2.23 Sv, called the "most probable gametic DD" by Neel and colleagues, was obtained by calculating overall spontaneous and induced "mutation rates" for the above-mentioned five end-points and obtaining a ratio of these two. The former was estimated by summing the five individual estimates of spontaneous rates (which yielded 0.00632 to 0.00835 per gamete) and the latter, likewise, by summing the individual rates of induction (which yielded 0.00375 per gamete per parental Sv). The ratio 0.00632-0.00835/0.00375 is the DD range which is 1.69 to 2.23 Sv. The overall DDs thus calculated were found to be between 1.69 Sv (i.e., 0.00632/0.00375) and 2.23 Sv (i.e., 0.00835/0.00375) for the acute radiation conditions during the bombings. In these estimates, the limits reflect biological uncertainties about the parameters, but do not take into account the additional error inherent in the estimation process itself, which must be relatively large (Neel and others 1990). With a dose-rate reduction factor of 2 (which was used) for chronic low-LET radiation conditions, the relevant DD becomes about 3.4 to 4.5 Sv. Note, however, that the dose-rate reduction factor traditionally used by UNSCEAR and the BEIR committees is 3, based on specific locus mutation experiments with male mice.

For reasons discussed in the main text, the DDs estimated from these data cannot readily be compared with those used by UNSCEAR and the BEIR committees. However, the results with one indicator of damage used in the Japanese studies, namely, untoward pregnancy outcome, which includes stillbirths, congenital abnormalities, and early neonatal deaths, permit a crude comparison with the risk of congenital abnormalities estimated in this report. The rate of induction defined by the regression coefficient for UPO is $(26.4 \pm 27.7) \times 10^{-4}$ per parental sievert, compared to the background risk of 500×10^{-4} assumed in the calculations. The risk of congenital abnormalities (estimated from mouse data in this document) is 60×10^{-4} per Gy^{-1} for acute X-irradiation, compared to the background risk (human data) of 600×10^{-4}. Considering the uncertainties involved in both of these estimates, one can conclude that they are of the same order.

The other end points—namely, F_1 mortality, F_1 cancers, sex chromosomal aneuploidy, and electrophoretic mobility or activity mutations—that have been used in the Japanese studies have not been used in this report and so do not lend themselves to comparisons. It should be noted that the first two of the above (*i.e.*, F_1 mortality, F_1 cancers) are multifactorial traits (similar to UPO), and their responsiveness to an increase in mutation rate will depend on the magnitude of the mutation-responsive component, which is quite small, as Neel and colleagues point out. Consequently, the rates of induced genetic damage underlying these traits are expected to be small, and increases will be undetectable with the available sample sizes at the relatively low radiation doses (about 0.4 Sv) sustained by most of the survivors.

The reasons for the lack of significant effects on sex chromosomal aneuploidy and electrophoretic mutations are different. There is no evidence from mouse studies that radiation is capable of inducing chromosomal nondisjunction (the principal basis for the origin of sex chromosomal aneuploidy). Since radiation is a poor inducer of point mutations, *a priori* one would not expect electrophoretic mutations to be induced by radiation to any great extent as they are known to be due to base-pair changes. Null enzyme mutations would be expected to be induced, but they are unlikely to be found at the low dose levels experienced by most survivors. Consequently, it is not surprising that the DD estimates of Neel and colleagues for these end points (1.60 Sv for sex-chromosomal aneuploids and 2.27 Sv for electrophoretic mutations) are higher than those for the other end points.

5

Background for Epidemiologic Methods

INTRODUCTION

Epidemiology is the study of the distribution and determinants of disease prevalence in man (MacMahon and others 1960). Epidemiologists seek to describe the populations at risk and to discover the causes of diseases. This entails quantification of the risk of disease and its relationship to known or suspected causal factors. In radiation epidemiology, exposure to radiation is the factor of primary interest, and epidemiologists seek to relate risk of disease (primarily cancer) to different levels and patterns of radiation exposure. Epidemiologic studies have been of particular importance in assessing the potential human health risks associated with radiation exposure.[1]

As part of the study of the causes of disease, epidemiologists measure factors that are suspected of leading to its development. A basic comparison used in radiation epidemiology is to measure the rate of a specific disease among persons who have been exposed to radiation and among persons who have not. The two rates are compared to assess whether they are similar or are different. A logical extension of this basic mode of comparison is to stratify the exposed subjects on the basis of amount (dose) of radiation in order to assess whether disease rates vary with dose, that is, whether there is a dose-response relationship.

If the rates of a disease are essentially the same in the exposed and unexposed groups, there is said to be no association between radiation exposure and disease. This does not necessarily mean that in all populations at all times, radiation is not related to the disease, but it does mean that in this population at this time, sufficient evidence does not exist for an association between radiation and disease. If the disease rate is higher among those exposed to radiation, there is a positive association. If the disease rate is higher among the unexposed group, there is a negative (inverse) association between radiation exposure and disease.

Epidemiologists use the term "risk" in two different ways to describe the associations that are noted in data. *Relative risk* is the ratio of the rate of disease among groups having some risk factor, such as radiation, divided by the rate among a group not having that factor. Relative risk has no units (*e.g.*, 75 deaths per 100,000 population per year ÷ 25 deaths per 100,000 per year = 3.0). Excess relative risk (ERR) is the relative risk minus 1.0 (*e.g.*, 3.0 − 1.0 = 2.0). *Absolute risk* is the simple rate of disease among a population (*e.g.*, 75 per 100,000 population per year among the exposed or 25 per 100,000 per year among the nonexposed). Absolute risk has the units of the rates being compared. Excess absolute risk (EAR) is the difference between two absolute risks (*e.g.*, (75 per 100,000 per year) − (25 per 100,000 per year) = 50 per 100,000 per year). If the rates of disease differ in the exposed and unexposed groups, there is said to be an association between exposure and disease. None of these measures of risk is sufficient to infer causation. A second step in data analysis is necessary to assess whether or not the risk factor is simply a covariate of a more likely cause.

In modeling the relation between radiation exposure and disease, either the ERR or the EAR may be used. In addition, the estimated dose of radiation exposure is integrated into the models, so that estimation is made of the ERR or EAR as a function of dose. Relative risk and ERR have certain mathematical and statistical advantages and may be easier to understand for small risks, but absolute risk and EAR are more closely related to the burden of disease and to its impact on the population. Thus, each type of measure has its advantages, and each is used in this report.

Having assessed whether or not there is evidence of an association between radiation exposure and a disease in the population of interest, the next task of the epidemiologist is to assess whether noncausal factors may have contributed to the association. An association might not represent a causal link between radiation and disease, but rather could be due to chance, bias, or error. It should be noted that chance can never be ruled out as one possible explanation for an asso-

---

[1]See Glossary for definition of specific epidemiologic terms.

132

ciation that is observed in epidemiologic data, although the probability may be extremely small.

Having judged that an association in a population under study cannot be demonstrated to have occurred because of error or bias, an investigator computes a measure of association that takes into account any relevant differences between the exposed and the unexposed group. Also it is usual to quantify the uncertainty in a measured association by calculating an interval of possible values for the true measure of association. This *confidence interval* describes the range of values most likely to include the true measure of association if the statistical model is correct. It always is possible that the true association lies outside the confidence interval either because the model is incomplete or otherwise in error or because a rare event has occurred (with rare defined by the probability level, commonly 5%).

Another step in assessing whether radiation exposure may be the cause of some disease is to compare the results of a number of studies that have been conducted on populations that have been exposed to radiation. If a general pattern of a positive association between radiation exposure and a disease can be demonstrated in several populations and if these associations are judged not to be due to confounding, bias, chance, or error, a conclusion of a causal association is strengthened. However, if studies in several populations provide inconsistent results and no reason for the inconsistency is apparent, the data must be interpreted with caution. No general conclusion can be made that the exposure is a cause of the disease.

An important exercise is assessing the relation between the dose of exposure and the risk of disease. There is no question that radiation exposure at relatively high doses has caused disease and death (NRC 1990; UNSCEAR 2000b). However, at relatively low doses, there is still uncertainty as to whether there is an association between radiation and disease, and if there is an association, there is uncertainty about whether it is causal or not.

Following is a discussion of the basic elements of how epidemiologists collect, analyze, and interpret data. The essential feature of data collection, analysis, and interpretation in any science is *comparability*. The subpopulations under study must be comparable, the methods used to measure exposure to radiation and to measure disease must be comparable, the analytic techniques must ensure comparability, and the interpretation of the results of several studies must be based on comparable data.

COLLECTION OF EPIDEMIOLOGIC DATA

Types of Epidemiologic Studies

Research studies are often classified as experimental or observational depending on the manner in which the levels of the explanatory factors are determined. When the levels of at least one explanatory factor are under the control of the

investigator, the study is said to be experimental. An example is a clinical trial designed to assess the utility of some treatment (*e.g.*, radiation therapy). When the levels of all explanatory factors are determined by observation only, the study is observational. If treatment is assigned by a random process, the study is experimental. The majority of studies relevant to the evaluation of radiation risks in human populations are observational. For example, in the study of atomic bomb survivors, neither the conditions of exposure nor the levels of exposure to radiation were determined by design.

Two basic strategies are used to select participants in an observational epidemiologic study that assesses the association between exposure to radiation and disease: select exposed persons and look at subsequent occurrence of disease, or select diseased persons and look at their history of exposures. A study comparing disease rates among exposed and unexposed persons, in which exposure is not determined by design, is termed a "cohort" or a "follow-up" study. A study comparing exposure among persons with a disease of interest and persons without the disease of interest is termed a "case-control" or "case-referent" study.

Randomized Intervention Trials

Intervention trials are always prospective—for example, subjects with some disease are enrolled into the study, and assignment is made to some form of treatment according to a process that is not related to the basic characteristics of the individual patient (Fisher and others 1985). In essence, this assignment is made randomly so that the two groups being studied are comparable except for the treatment being evaluated. Random is not the same as haphazard; a randomizing device must be used, such as a table of random numbers, a coin toss, or a randomizing computer program. However, random assignment does not guarantee comparability. The randomization process is a powerful means of minimizing systematic differences between two groups ("confounding bias") that may be related to possible differences in the outcome of interest such as a specific disease. Further, blinded assessment of health outcome will tend to minimize bias in assessing the utility of alternative methods of treatment. Another important aspect of randomization is that it permits the assessment of uncertainty in the data, generally as *p*-values or confidence intervals. Intervention trials related to radiation exposure are conducted with the expectation that the radiation will assist in curing some disease. However, there may be the unintended side effect of increasing the risk of some other disease.

Although a randomized study is generally regarded as the ideal design to assess the possible causal relationship between radiation and some disease in a human population, there are clearly ethical and practical limitations in its conduct. There must be the expectation that in the population under study, radiation will lead to an improvement in health status relative to any alternative treatment. Such studies are usually conducted with patients who need therapeutic intervention; randomly selected patients may be treated with radiation and some other form of treatment or with different types or doses of radiation. In these trials the sample size is relatively small and the follow-up time is relatively short. Therefore, most studies to assess the long-term adverse outcomes of exposure to therapeutic radiation, are, of necessity cohort studies.

Cohort Studies

Cohort studies may be retrospective or prospective. In a retrospective cohort study of a population exposed to radiation, participants are selected on the basis of existing records such as those maintained by a company or a hospital (*e.g.*, radiation badge records). These records were made out at the time an individual was working or treated and thus may be used as the historical basis for classification as a member of the exposed cohort. In a prospective cohort study, participants are selected on the basis of current and expected future exposure to radiation, and exposure information is measured and recorded as time passes. In both types of cohort study, the members of the study population are followed in time for a period of years, and the occurrence of new disease is measured. In a retrospective cohort study, the follow-up has already occurred, while in a prospective cohort study, the follow-up extends into the future. Many studies that are initiated as retrospective cohort studies become prospective as time passes and follow-up is extended.

The information available in a retrospective cohort study is usually limited to what is available from the written record. In general, members of the cohort are not contacted directly, and information on radiation exposure and disease must come from other sources. Typically, information on exposure comes from records that indicate the nature and amount of exposure that was accumulated by a worker or by a patient. On occasion, all that is available is the fact of exposure, and the actual dose may be estimated based on knowledge of items such as the X-ray equipment used (Boice and others 1978).

Information on disease also must come from records such as medical records, insurance records, or vital statistics. Cancer mortality is readily evaluated by retrospective cohort studies, because cancer registries exist in a number of countries or states and death from cancer is fairly reliably recorded.

Most studies that have followed patients treated with therapeutic radiation are retrospective cohort studies. Series of patients are assembled from medical and radiotherapy records, and initial follow-up is done from the date of therapy until some arbitrary end of follow-up. Patients treated as long ago as the 1910s have been studied to assess the long-term effects of radiation therapy (Pettersson and others 1985; Wong and others 1997a).

The information available in a prospective cohort study is potentially much greater than that available in a retrospective cohort study. Exposure is contemporaneous and may be measured forward in time, and members of the cohort may be contacted periodically to assess the development of any new disease. Direct evaluation of both exposure and disease may be done on an individual basis, with less likelihood of missing or incomplete information due to abstracting records compiled for a different purpose.

The follow-up of survivors of the Japanese atomic bomb explosions is largely prospective, although follow-up did not begin until 1950 (Pierce and others 1996). Exposure assessment was retrospective and was not based on any actual measurement of radiation exposure to individuals. Reconstruction of the dose of radiation exposure is an important characteristic of this study, and improvements in dose estimation continue to the present with a major revision of the dosimetry published in early 2005 (DS02).

The primary advantage of a retrospective cohort study is that time is compressed. If one wishes to evaluate whether radiation causes some disease 20–40 years after exposure, a retrospective study can be completed in several years rather than in several decades. The primary disadvantage of a retrospective cohort study is that limited information is available on both radiation exposure and disease. The primary advantage of a prospective cohort study is that radiation exposure and disease can be measured directly. The primary disadvantage is that time must pass for disease to develop. This leads to delay and expense. Most studies in radiation epidemiology are retrospective cohort studies.

Case-Control Studies

Case-control studies may be prospective or retrospective. The cases are those individuals with the disease being studied. Cases in a retrospective case-control study are usually selected on the basis of existing hospital or clinic records (*i.e.*, the cases are "prevalent"). In a prospective case-control study, the cases are "incident," that is, they are selected at the time their disease was first diagnosed. Controls are usually nondiseased members of the general population, although they can be persons with other diseases, family members, neighbors, or others.

After the cases and controls have been identified, it is necessary to determine which members of the study population have been exposed to radiation. Usually, this information is obtained from interviewing the cases and the controls. However, if the case or control is deceased or unable to respond, exposure information may come from a relative or from another proxy.

The information available in case-control studies usually is less reliable than that collected in cohort studies. For example, consider the accuracy of dietary history for the past year *versus* that of a year from several decades in the past. Exposure information may be available only from interview of the study subjects and therefore be less reliable than reliance on contemporary records. There may be differential recall of exposure to radiation depending on case or control status, which leads to a lack of comparability in the information available. It is rare to be able to quantify the amount of past exposure in a case-control study. However, in some situations related to radiation exposure, only data from case-control studies are available.

The critical differences between a retrospective cohort study and a case-control study are that subjects in the former are selected on the basis of exposure category at the start of the follow-up period and exposure measures are concurrent with the actual exposure. Conversely, in a case-control study, subjects and controls are selected on the basis of disease outcome, and past exposures must be reconstructed.

On occasion in epidemiology, a hybrid study is performed: the "nested" case-control study. A cohort study is conducted, and subsequently, additional information on exposure is collected for persons with disease and for a sample of persons without disease. For example, radiation exposure among persons with a second cancer may be compared to that among a sample of those without a second cancer. Nested case-control studies are best thought of as a form of retrospective cohort study, in that the study population is initially defined on the basis of exposure rather than of disease.

In evaluation of the possible health effects of exposure to ionizing radiation, many of the informative case-control studies have been nested within cohorts. Exposure measures in these studies are generally not based on interview data, but rather on review of available records, sometimes supplemented by extensive modeling and calculations. In some nested studies, the objective is to obtain information on dose or other factors that would be too expensive to obtain for the entire cohort. Examples are a case-control study of selected cancers in women irradiated for cervical cancer to obtain individual dose estimates (Boice and others 1985); a breast cancer study of A-bomb survivors to obtain data on reproductive factors through interview (Land and others 1994b); and a study of lung cancer in Hanford workers to extract smoking histories from medical records (Petersen and others 1990).

Comparability in Study Design

The design of an epidemiologic study must assume comparability in the selection of study participants, comparability in the collection of exposure and disease information relevant to each study subject, and comparability of the basic characteristics of the study subjects. Any lack of comparability may undermine inferences about an association between exposure and disease, so that interpretation is ambiguous or impossible.

Comparability in a clinical trial ordinarily is straightforward, because study subjects are assigned randomly to the various forms of treatment being evaluated. Random assignment prevents selection on the basis of outcome and provides the optimum strategy for minimizing differences between the two groups being studied. Comparability in a cohort study means that subjects exposed to radiation and unexposed subjects are enrolled without knowledge of disease status, that information on disease is obtained without knowledge of exposure status, and that other factors related to disease occurrence are not related to exposure status.

Lack of comparability in any of these epidemiologic study designs may lead to one or another form of bias, which in turn may minimize or invalidate any information contained in the data from the study. Three common and potentially serious forms of bias are *selection bias*, when enrollment into a study is dependent on both radiation exposure and disease status; *information bias*, when information on disease or on radiation exposure is obtained differentially from exposed or from diseased persons; and *confounding bias*, when a third factor exists that is related to both radiation exposure and disease effects.

Selection bias is generally a minor issue in clinical trials and cohort studies, including retrospective cohort studies. In a prospective cohort study, disease has not yet occurred, so there is little possibility of selecting exposed persons on the basis of their future disease status. Exceptions are rare and limited to situations in which some preclinical sign or symptom affects selection—for example, when persons volunteer for one or another intervention because they know that they are at special risk.

By contrast, *selection bias* can be a major issue in case-control studies, because both exposure and disease already have occurred when the study subjects are enrolled; there is the danger that persons who are both exposed and diseased will be overselected to participate in the study. If this occurs, the data contain invalid information on the true relation between exposure and disease. Self-selection (volunteering) for a nonexperimental study can be a particularly potent source of bias.

An example of selection bias occurred in a study of leukemia among workers at the Portsmouth, New Hampshire, Naval Shipyard (Najarian and Colton 1978). In an initial case-control study, persons with leukemia who had been occupationally exposed to radiation were widely known and hence more likely to be located and enrolled than were unexposed workers with leukemia, and a positive association between radiation and leukemia was reported. Subsequently, after an extensive follow-up of all members of the workforce, no association between radiation exposure and leukemia was found (Greenberg and others 1985). The initial *preferential* selection of diseased workers who were exposed to radiation led to an erroneous appearance of a positive association between radiation and leukemia.

Information bias may occur in a clinical trial or a cohort study if knowledge of exposure is available when information on disease is being obtained; there is the possibility that disease will be diagnosed more among exposed persons than among nonexposed persons. For this reason, in obtaining information on disease among participants, information on exposure is kept hidden (blinded), so that any error in disease ascertainment occurs equally among exposed and unexposed persons.

Information bias is a major threat in a case-control study if knowledge of disease is available when information on exposure is being obtained; there is a possibility that exposure will be ascertained more among diseased persons than among nondiseased persons. For this reason, in obtaining information on exposure among participants, information on disease is kept hidden from the interviewer and, if possible, from the respondent (blinded), so that any error in exposure ascertainment occurs equally among diseased and nondiseased persons. Further protection against information bias may come from blinding subjects and/or interviewers to the hypothesis under study.

Information bias as well as selection bias affected the Portsmouth Shipyard Study (Najarian and Colton 1978). In the initial case-control study, information on radiation exposure was obtained by interview of relatives of workers with and without leukemia. Subsequently, it was found that relatives of those with leukemia tended to overreport radiation exposure, whereas relatives of those without leukemia tended to underreport exposure (Greenberg and others 1985).

Confounding bias is a basic issue in all epidemiologic studies where no random assignment of exposure has occurred; this is the usual situation except for randomized clinical trials. No one type of nonexperimental epidemiologic study is inherently more subject to confounding bias. If information is available on each factor that is suspected of being a confounder, confounding bias may be minimized in a study design by matching on the relevant factors or in data analysis by stratification or statistical adjustment. However, if some confounding factor has not been measured, the data may be wrong. Thus, interpretation of the data must take into account the possible influence of potential confounding. Confounding bias is especially troublesome when the association under investigation is weak. In this case, a confounder has the potential to mask an association completely or to create an apparent effect. Because the risks associated with low levels of ionizing radiation are small, confounding bias is potentially important in low-level radiation studies.

A third factor (other than exposure and disease) can be confounding *only* when it is associated with both the exposure and the disease. Association only with exposure or only with disease is not sufficient for a factor to be confounding.

The so-called healthy worker effect is an example of confounding in studies of mortality among occupational groups, including those employed in the nuclear industry (Monson 1990). Ordinarily, persons who enter the workforce are healthy, and if mortality among workers is compared to that among the general population, the workers are found to be at a relatively low risk. If all members of the workforce were exposed to radiation, one interpretation would be that radiation reduces the risk of death.

In a clinical trial, assignment to a type of specific exposure is ordinarily a random process so that, on average, the two groups being compared are comparable with respect to possible confounding factors. Thus, in a randomized trial, confounding—although possible—is less of a concern than in a cohort or a case-control study.

Statistical Power

An important part of any epidemiologic study is its statistical power (*i.e.*, the probability that under the assumptions and conditions implicit in the model, it will detect a given level of elevated risk with a specific degree of significance). The power of a cohort study will depend on the size of the cohort, the length of follow-up, the baseline rates for the disease under investigation, and the distribution of doses within the cohort, as well as the magnitude of the elevated risk. Similarly, statistical power in a case-control study depends on the number of cases, the number of controls per case, the frequency and level of exposure, and the magnitude of the exposure effect. Statistical power is generally evaluated before a study is conducted. Afterwards it is more useful to refer to statistical precision, which is reflected in the width of the confidence intervals for risk estimates (UNSCEAR 2000b).

ANALYSIS OF EPIDEMIOLOGIC DATA

The basic data collected in an epidemiologic study are data on exposure and data on disease. In the simplest form, an individual may be exposed or not and may be diseased or not. Thus, there are four possibilities: exposed and diseased, exposed and not diseased, not exposed and diseased, or not exposed and not diseased. Typically, these data are entered into a "fourfold table" (Table 5-1).

It can be seen that in a study of N individuals, $a + b$ are exposed, $a + c$ are diseased, and a are both exposed and diseased. Interest is generally focused on whether a is larger than expected in relation to the other entries. Mathematically this is the same as asking whether d is larger than expected, or whether b or c are smaller than expected. Accurate counts in all four cells are necessary for valid inferences

TABLE 5-1 The Fourfold Table

Exposure	Disease		
	Yes	No	Total
Yes	a	b	$a + b$
No	c	d	$c + d$
Total	$a + c$	$b + d$	N

about whether the disease is associated with the exposure. The rate of disease among the exposed subjects (R_e) is equal to $a/(a + b)$, and the rate of disease among the unexposed subjects (R_n) is equal to $c/(c + d)$.

Measures of Association

Two measures are commonly used to compare the disease rates between exposed and unexposed subjects. The relative risk (RR) is the ratio of the two rates; that is, $RR = R_e/R_n$. The ERR is given by $ERR = RR - 1 = R_e/R_n - 1 = (R_e - R_n)/R_n$. These ratios are dimensionless. The rates can also be subtracted rather than divided. The difference between R_e and R_n, that is, $R_e - R_n$, is termed the "attributable risk," or "risk difference." It is also referred to as the excess risk (ER) or the EAR, with the latter terminology commonly used in radiation epidemiology. The ER and EAR are often expressed as the number of excess cases or deaths per person-year (PY) or, for convenience, per 1000 PY.

In radiation studies, information on radiation dose is often available. Either of the measures, ERR or EAR, can be expressed per unit of radiation dose. In the simplest situation, one has exposed and unexposed groups and information on the average dose D received by exposed subjects. The ERR coefficient is then defined as

$$ERR = (R_e - R_n)/(R_n D),$$

and absolute risk coefficient is defined as

$$EAR = (R_e - R_n)/PY \cdot D,$$

where PY is the number of person-years of follow-up.

Both measures may depend on variables such as sex, age at exposure, time since exposure, and age at risk (attained age). The ERR expresses risk and its dependencies relative to risk in the unexposed, whereas the EAR expresses risk and its dependencies independent of risk in the unexposed. The RR (or ERR) has certain statistical advantages and is the more commonly used measure for epidemiologic studies, especially etiologic studies. The EAR is a useful measure for estimating the burden of risk in a population, including the dependence of this burden on various factors. Both measures can be used to estimate absolute lifetime risk as discussed in Chapters 11 and 12.

In some of the more informative radiation studies, dose estimates for individual subjects are available. In this case, more complex statistical regression methods are used to estimate the ERR and EAR per unit of radiation dose based on the assumption of a linear dose-response. These methods have been used in analyses of data on Japanese A-bomb survivors and on some medically exposed populations. The reader should consult Chapters 6 and 7 for further discussion of this approach.

Instead of categorizing persons with radiation exposure as simply being exposed or not, subjects may be categorized as having high, medium, or no exposure. In this case, there would be a sixfold table—three rows and two columns. Such data are of value in assessing whether or not there is a dose-response relationship between radiation exposure and disease. If the rate of disease is highest among the most exposed, intermediate in the middle exposure group, and lowest among those with no exposure, a dose-response relationship exists. In this report, only data that are of utility to a quantitative assessment of a dose-response relationship between radiation exposure and disease are included.

For radiation, we are generally interested in going beyond just deciding if there is a causal relationship. An important strength of radiation epidemiology is the availability of quantitative information on dose. Only by relating effects to dose can results be compared across studies or used to predict risks from exposures in other settings.

Tools of Statistical Inference

The second task in data analysis is assessing the statistical precision of an ERR or other measure of association calculated from data. Statistical estimates calculated from data are imprecise, or variable, in the sense that replication of the study (with identical conditions of exposure and levels of exposure, but with a different random sample of subjects) would likely result in a different estimate of risk. Thus, it is important to determine whether the actual observed association (*e.g.*, an RR different from 1.0) can be explained by chance (random variation) alone. In epidemiologic studies the assessment of precision is usually accomplished via the calculation of *p*-values or confidence intervals.

The validity of both *p*-values and confidence limits rests on many assumptions about the study design and the data. Statistical results are often most correct when deviations from the assumptions are small, that is, the procedures are "robust." It is the task of the investigator and any subsequent analyst to know the assumptions and to ensure that they are sufficiently close to reality.

Consider a hypothetical replication of the study in which the true RR is 1.0 (*i.e.*, disease outcome is not related to exposure). The ERR from the hypothetical replication will not equal 1.0 exactly, but will vary randomly around the true value of 1.0. The *p*-value of the actual study is the probability that the RR estimated from the hypothetical data is more extreme in its difference from 1.0 (in either direction) than the RR estimated from the actual sample. A small *p*-value means that it is unlikely that the actual RR was calculated from data having a true RR of 1.0. In other words, a small *p*-value provides evidence that the true RR is different from 1.0; the smaller the *p*-value, the stronger is the evidence.

The confidence interval and *p*-value are based on the same theory; they use the theory in slightly different ways to answer slightly different questions. A *p*-value is appropriate for answering a confirmatory question such as, Is 1.0 a believable value of RR? A confidence interval is appropriate for answering an exploratory question, such as, What are the believable values of RR? Obviously, a confidence interval lends partial information to the confirmatory question since values not in the 95% confidence interval are "rejected" at the significance level of 0.05. The p-value does add additional information, however, since it provides a degree of evidence. For example, p-values of .049 and .00000049 provide quite different measures of the believability of the hypothesis (of RR equal to 1.0, say), even though the 95% confidence interval excludes 1.0 in both cases.

Statistical precision is determined largely by study size (number of subjects). Larger studies generally result in more precise estimates. Small effects (RRs near 1.0) are generally more difficult to detect than large effects, because a confidence interval centered close to 1.0 is likely to include 1.0 unless the sampling variance is small. One consequence is that very large studies are required to estimate small effects precisely. This explains in part why risk models cannot be based exclusively on low-dose studies. The RRs associated with low doses are close to 1.0 and thus can be estimated precisely only in very large studies.

Control of Confounding

The third task in data analysis is to assess whether or not the crude association that is observed in a study is due to confounding by one or more other factors. For example, in assessing the relation between radiation and lung cancer, one should consider whether cigarette smoking is a confounding factor. Cigarette smoking is a recognized cause of lung cancer, and thus there is an association between smoking and lung cancer. If persons who are exposed to radiation, such as uranium miners, smoke more than persons who are not exposed, they may have an increased risk of lung cancer just from the smoking. Thus, unless the analysis deals with smoking as well as radiation, it is possible that an association between radiation and lung cancer seen in data only reflects the confounding influence of cigarette smoking.

In data analysis, the simplest way to assess whether or not confounding is present is to stratify on the confounding factor. That is, two fourfold tables are set up that relate the exposure (radiation) to the disease (lung cancer). If it is assumed that all smokers smoke the same, one table contains data only for smokers and a second table contains data only for nonsmokers. Within each of these two tables, no confounding by smoking is possible.

If it is necessary to control more than one confounding factor in the analysis of epidemiologic data, it is usual to construct a multivariate model relating exposure to disease and controlling for the potential confounding effect of a number of other factors. For example, sex and age are two factors that are commonly included in multivariate models. Such modeling is similar to stratification on a number of

confounders and summarizing results in a standardized RR with associated confidence interval.

Linear Relative Risk Model

A model that plays a prominent role in radiation epidemiology studies is one in which the RR is a linear function of dose. In its simplest form,

$$RR(D) = 1 + \beta D,$$

where D is dose, $RR(D)$ is the relative risk at dose D, and β is the ERR per unit of dose, which is usually expressed in grays or sieverts. In more complex forms, β is allowed to depend on gender, age at diagnosis, and other variables.

This linear RR model has been used extensively in radiation epidemiology, including studies of A-bomb survivors (Chapter 6), persons exposed for medical reasons (Chapter 7), and nuclear workers (Chapter 8). The model has served as the basis of cancer risk estimation by three BEIR committees (NRC 1988, 1990, 1999), by the 2000 UNSCEAR committee (2000b), and by the National Institutes of Health (NIH 2003). It also plays an important role in developing the BEIR VII committee's cancer risk estimates (Chapter 12). The linear model has been chosen because it is supported by radiobiological models (Chapter 2) and because it fits the data from most studies (although in many studies, statistical power is inadequate to distinguish among different dose-response functions).

In the simplest situation, in which one has exposed and unexposed groups and information on the average dose D received by exposed subjects, β is estimated by $(R_e - R_n)/(R_n D)$ as discussed earlier. In many radiation studies, however, doses for individual subjects are available and more complex estimation procedures are required to make use of this information. Preston and colleagues (1991) have developed the EPICURE software that allows for flexible modeling of both relative and absolute risks, including the fitting of linear RR models.

Prentice and Mason (1986) and Moolgavkar and Venzon (1987) discuss inferences based on the linear RR model and note that the distribution of the maximum likelihood estimate of β may be highly skewed, and that confidence intervals based on the estimates of the asymptotic standard error (Wald method) can be seriously misleading. Re-parameterizing the model as $\beta = \exp(\alpha)$ is sometimes helpful but does not allow for the possibility that β or its lower confidence bound may be negative. Another difficulty is that, to ensure that the RR is nonnegative, it is necessary to constrain the parameter β to be larger than $-1/D_{MAX}$, where D_{MAX} is the maximum dose in the study. These problems may be particularly severe in studies of nuclear workers, where dose distributions are highly skewed and estimates of β are often very imprecise. For this reason, tests and confidence intervals in nuclear worker studies have sometimes been based on the likelihood ratio, or on score statistic approximations, or on computer simulations (Gilbert 1989), which can lead to intervals that are not symmetric on either a linear or a logarithmic scale. In some situations, especially in studies with sparse data, the estimate and/or the lower confidence bound for β may be negative; some investigators report such findings simply as <0.

INTERPRETATION OF EPIDEMIOLOGIC DATA

Assessment of Associations

After epidemiologic data have been collected and analyzed, the associations noted in the data must be interpreted. The measures of association and of statistical precision that have been computed have no inherent meaning; they reflect only the data that have been accumulated in the study. It is possible that these data have resulted from bias, error, or chance and thus have no interpretive meaning. A formal evaluation of the study design and of the methods used to collect and analyze the data is needed to assess the meaning of the data.

The first step in the interpretation of data is to assess the methods used in the study itself. The following questions must be considered:

• Is there evidence that selection bias has been avoided in enrolling the study subjects?
• Is there evidence that information bias has been minimized in assessing exposure or disease?
• Is there evidence that the potential confounding influence of other factors has been addressed?
• Is there evidence for sufficient precision in the measure of exposure or of disease to permit a reasonable basis for interpretation?

The possible occurrence of *selection bias* or of *information bias* may be assessed only by evaluation of the methods used in data collection. If either of these biases is judged to have an appreciable likelihood of being important, no analyses can be conducted to adjust for the error that may have been introduced. The data must be regarded as unsuitable for the purpose at hand. In contrast, potential confounding bias can be assessed and usually controlled by analytic strategies for factors on which information has been collected. There will always remain factors that have the potential for confounding but for which no information is available, including factors that are not even suspected of being confounders. This does not mean that no interpretation is possible, but it does mean that some degree of caution is needed in interpreting any association between radiation exposure and disease.

Chance is always a possible explanation for any association (or lack of association) in a scientific study, no matter how strong or how statistically significant the association.

The *p*-value or confidence interval that is computed estimates only the likelihood that chance alone could have accounted for the observed association. The *p*-value does not distinguish between a true association and one that is due to bias or error. Also, interpretation of the likely range of an association based on its confidence interval reflects only the play of chance, not of error or bias. In addition, rare events do happen. Each *p*-value of the confidence interval should be examined with some care to determine whether a rare event is a plausible explanation for the statistical findings. Interpretation of the results of statistical analysis is as much an art as a science.

In all epidemiologic studies, measures of exposure and measures of disease are imprecise. This *imprecision* is not considered an error in methodology, but rather an inevitable occurrence associated with the assessment of observational data. When errors in measuring disease or exposure are random, unrelated to true disease and exposure, and independent among subjects, it is usually the case that measures of association are attenuated. That is, RRs are biased toward 1.0, the case of no association. In radiation epidemiology, errors in measuring disease (*e.g.*, misdiagnosing cancer) are not different from disease misclassification problems in other epidemiology studies. Thus, the effect of disease misclassification is reasonably well understood. However, exposure measurement error problems in radiation epidemiology are often unique to radiation studies, and the effect of such errors generally is less well understood.

For most radiation epidemiology studies, measurements of exposure were not made at the time of exposure, but rather have been reconstructed some time after exposure using available information. For example, exposures for A-bomb survivors are calculated using sophisticated models for the spatial intensity of radiation and information about a subject's location and local shielding at the time of exposure. It is likely that such measurements contain both random and nonrandom components. The effects of random errors in exposure measurements are reasonably well understood and include, in general, attenuation of estimated associations, underestimation of linear risk coefficients, and possible distortion of the shape of the dose-response relationship. The severity of these effects generally depends on the magnitude of the measurement errors (as measured by their variance) relative to the variability in true exposures. The effects of nonrandom errors in exposure measurements are specific to the nature of the error. For example, if a dosimetry system systematically overestimated exposures by 10%, the dose-response relationship would erroneously be stretched over a greater range of doses, the slope of the fitted line would be reduced, and linear risk coefficients would be underestimated by approximately 10%.

A second step in evaluating whether some exposure causes some disease is to assemble all of the relevant literature and to display all of the data that are regarded as relevant and of adequate quality. On occasion, a so-called meta-analysis is conducted in which there is a quantitative summarization of the data. Such an analysis is not a necessary step and in fact may not be indicated. Only data from valid studies may be included in a meta-analysis, and among valid studies, all studies must contain similar information. In essence, a meta-analysis is a formal rather than an informal summarization of the epidemiologic literature.

A pooled analysis of data from similar studies is not the same procedure as a meta-analysis, but rather a useful extension of basic data analysis. An important tool for obtaining a broad assessment of the evidence from several studies is to conduct combined analyses of data from groups of similar studies. Analyses based on combined data provide tighter confidence limits on risk estimates than analyses based on data from any single study population. To the extent that biases found in individual studies tend to cancel out, combined analyses may help to reduce bias that results from confounding and other potential sources of bias. Such analyses also help to determine if differences in findings among studies are truly inconsistent or are simply the result of chance fluctuations. The application of similar methodology to data from all populations, in addition to the presentation of results in a comparable format, facilitates comparison of results from different studies.

A third step in interpretating epidemiologic data is to compare the results of an individual study with those of similar studies. The goal of such an exercise is to reach a judgment about whether, in general, it may be concluded that under certain conditions, an exposure causes a disease.

The so-called Bradford Hill criteria are the standard criteria used to assess whether the general epidemiologic literature on some exposure or some disease provides sufficient information to judge causality (Hill 1966). These criteria have been expanded, reduced, revised, and reinterpreted by countless authors to meet their special needs, but the core idea remains—use rational operational criteria to judge evidence from observational studies. A revised version of the Hill criteria follows:

- *Consistency*—An association is seen in a variety of settings.
- *Specificity*—The association is well defined rather than general.
- *Strength*—The association is high or low rather than close to 1.0.
- *Dose-response*—The higher the exposure, the higher is the rate of disease.
- *Temporal relationship*—The exposure occurs before the disease.

- *Coherence*—The association is believable based on information from other scientific disciplines.
- *Statistical significance*—The association is statistically significant or not.

Each of these criteria should be considered in assessing whether an association between exposure and disease can be judged to be causal. Except for temporal relationship, there need not be evidence for each of these criteria.

With respect to the use of the Hill criteria in assessing the association between exposure to ionizing radiation and health outcome, they are of limited current value for human cancer. Ionizing radiation at high doses is acknowledged to be a cause of most relatively common human cancers (IARC 2000). The presence of a dose-response relationship for many cancers is considered strong evidence for a causal relationship. For less common cancers and for diseases other than cancer, there are not sufficient data to apply the Hill criteria. IARC (2000) notes: "A number of cancers, such as chronic lymphocytic leukaemia, have not been linked to exposure to x or γ rays."

Assessment of Dose-Response Relationships

As noted above, evaluation of a dose-response relationship is one of the Hill criteria to be applied in assessing whether or not an association is judged to be causal. With respect to providing a risk estimate for low-dose, low-linear energy transfer radiation in human subjects, other information is necessary. Specifically, one needs relatively accurate information for individuals on dose from ionizing radiation, as well as a relatively complete measure of the incidence of or mortality from diseases. To date, the data from the survivors of the atomic bomb in 1945 in Hiroshima and Nagasaki have been the primary source of such information. The Radiation Effects Research Foundation has been responsible for estimating the exposure of individuals and for measuring the incidence and mortality of cancer and other diseases.

One of the primary tasks of this committee has been to evaluate the data that are available from studies of populations exposed to medical radiation, occupational radiation, and environmental radiation so as to assess whether information on dose-response associations from these data sources can be assembled and to evaluate whether such information can be compared to that obtained from the populations exposed to radiation from the atomic bombs. Chapters 7, 8, and 9 address these studies.

6

Atomic Bomb Survivor Studies

INTRODUCTION

The Life Span Study (LSS) cohort consists of about 120,000 survivors of the atomic bombings in Hiroshima and Nagasaki, Japan, in 1945 who have been studied by the Radiation Effects Research Foundation (RERF) and its predecessor, the Atomic Bomb Casualty Commission. The cohort includes both a large proportion of survivors who were within 2.5 km of the hypocenters at the time of the bombings and a similar-sized sample of survivors who were between 3 and 10 km from the hypocenters and whose radiation doses were negligible. The LSS cohort has several features that make it uniquely important as a source of data for developing quantitative estimates of risk from exposure to ionizing radiation. The population is large, not selected because of disease or occupation, has a long follow-up period (1950–2000), and includes both sexes and all ages at exposure, allowing a direct comparison of risks by these factors.

Doses are reasonably well characterized and cover a useful range. Doses are lower than those usually involved in medical therapeutic exposures, but many survivors were exposed at doses that are sufficiently large to estimate risks with reasonable statistical precision. In addition, the cohort includes a large number of survivors exposed at low doses, allowing some direct assessment of effects at these levels. The exposure is a whole-body exposure, which makes it possible to assess risks for specific cancer sites and to compare risks among sites. Because of the use of the Japanese family registration system, mortality data are virtually complete for survivors who remained in Japan. High-quality tumor registries in both Hiroshima and Nagasaki allow the study of site-specific cancer incidence with reasonably reliable diagnostic data. In addition, the LSS cohort is probably less subject to potential bias from confounding than many other exposed cohorts because a primary determinant of dose is distance from the hypocenter, with a steep gradient of dose as a function of distance. Finally, special studies involving subgroups of the LSS have provided clinical data, biological measurements, and information on potential confounders or effect modifiers.

The LSS also has limitations, which are important to consider in using and interpreting results based on this cohort. The subjects were Japanese and exposed under wartime conditions and, in this sense, differ from various populations for which risk estimates are desired. To be included in the study, subjects had to survive the initial effects of the bombings, including the acute effects of radiation exposure, and it is possible that this might have biased the findings. Dose estimates are subject to uncertainty, especially that due to survivor location and shielding. The cohort provides no information on dose-rate effects since all exposure is at high dose rates. Estimates of linear risk coefficients tend to be driven by doses that exceed 0.5 Gy; although estimates based only on survivors with lower doses can be made, their statistical uncertainty is considerably greater than those that include survivors with higher doses. Even at higher doses, data are often inadequate for evaluating risks of cancers at specific sites, especially those that are not common (although, for many site-specific cancers, the LSS provides more information than any other study).

Because of its many advantages, the LSS cohort of A-bomb survivors serves as the single most important source of data for evaluating risks of low-linear energy transfer radiation at low and moderate doses. This chapter describes the LSS cohort and presents findings for leukemia and for solid cancers as a group. The most recent major publications on cancer mortality (Preston and others 2003) and incidence (Preston and others 1994; Thompson and others 1994) are emphasized, but papers addressing special issues such as the shape of the dose-response function are also considered. Results for cancers of specific sites, including results from the three publications just noted, are discussed along with material from various special studies. Risks from *in utero* exposure are discussed separately. Although cancer is the main late effect that has been demonstrated in the A-bomb survivor studies, several studies have addressed the effects of radiation exposure on other health outcomes including benign tumors and mortality from causes of death other than cancer. These are discussed at the end of the chapter. In general, the committee has summarized papers on cancer incidence, cancer mortality, and noncancer mortality in the LSS cohort that have been published since BEIR V (NRC 1990).

This chapter is based on published material and does not include results of analyses conducted by the committee, which are described in Chapter 12. At the time of this writing, detailed analyses of mortality data covering the period 1950–1997 and of incidence data covering the period 1958–1987 had been published. The committee's analyses were based on the most recent DS02 dosimetry system, whereas most of the published analyses described in this chapter were based on the earlier DS86 dosimetry system (see discussion of dosimetry below for further comment). Preston and colleagues (2004) recently evaluated the impact of changes in dosimetry on cancer mortality risk estimates using mortality data through 2000; these results are summarized in the discussion of dosimetry.

DESCRIPTION OF THE COHORT

The full LSS cohort consists of approximately 120,000 persons who were identified at the time of the 1950 census. It includes 93,000 persons who were in Hiroshima or Nagasaki at the time of the bombings and 27,000 subjects who were in the cities at the time of the census but not at the time of the bombings. This latter group has been excluded from most analyses since the early 1970s because of inconsistencies between their mortality rates and those for the remainder of the cohort.

Health End Point Data

Data on health end points are obtained from several sources. Vital status is updated in 3-year cycles through the legally mandated Japanese family registration system in which deaths, births, marriages, and divorces are routinely recorded. This ensures virtually complete ascertainment of death regardless of where individual subjects reside in Japan. Death certificates provide data on the cause of death. The Leukemia Registry has served as a resource for leukemia and related hematological disease (Brill and others 1962; Ichimaru and others 1978). In the 1990s, it became possible to link data from both the Hiroshima and the Nagasaki tumor registries to the LSS cohort, which allows the evaluation of cancer incidence (Mabuchi and others 1994). An advantage of the registry data, in addition to the inclusion of nonfatal cancers, is that diagnostic information is of higher quality than that based on death certificates. Both tumor registries employ active approaches for case ascertainment and provide high-quality data from 1958 onward. Published analyses based on these data cover the period 1958–1987 (Thompson and others 1994). Limitations of the incidence data are that

they are not available before 1958 and do not include subjects who have migrated from Hiroshima or Nagasaki.[1]

The Adult Health Study (AHS) is a resource for data on health end points that require clinical data. The AHS cohort is a 20% subsample of the LSS, oversampled to provide greater representation of subjects in high-dose categories. Since 1958, AHS subjects have been invited to participate in biennial comprehensive health examinations at RERF. The level of participation has been between 70 and 85% for those living in the Hiroshima and Nagasaki areas (Ron and others 1995a).

Dosimetry

Most results presented in this chapter were based on the dosimetry system adopted in 1986 (DS86). The committee's analyses, described in Chapter 12, are based on the revised DS02 system, adopted in 2004. The DS02 system is the result of a major international effort to reassess and improve survivor dose estimates. This effort was initiated because reports in the early 1990s on thermal neutron activation measured in exposed material (*e.g.*, Straume and others 1992; Shizuma and others 1993) were interpreted as suggesting that the then-current survivor dosimetry system (DS86) might systematically underestimate neutron doses for Hiroshima survivors who were more than about 1 km from the hypocenter. However, the revised estimates of neutron dose do not differ greatly from the DS86 estimates. The new dosimetry system also introduces improved methods for the computation of γ-radiation doses and better adjustments for the effects of external shielding by factory buildings and local terrain features.

Preston and colleagues (2004) analyzed mortality data on solid cancer and on leukemia using both DS86 and DS02 dose estimates. They found that both the risk per sievert for solid cancer and the curvilinear dose-response for leukemia were decreased by about 10% by the dosimetry revision. They also found that parameters quantifying the modifying effects of gender, age at exposure, attained age, and time since exposure were changed very little by the revision.

Table 6-1, based on Preston and colleagues (2003), shows the distribution of survivors in the LSS cohort by their estimated DS86 doses to the colon. The dose to the colon is taken to be the γ-ray absorbed dose to the colon plus the neutron absorbed dose to the colon times a weighting factor 10. This weighted dose is denoted by d, and its unit sieverts;[2] such estimates were available for 86,572 survivors. The

---

[1]Analyses of cancer incidence data have included an adjustment of person-years to account for migration (Sposto and Preston 1992).

[2]Use of the symbol Sv for the unit of d is an extension of the convention to use sievert as a special name of the unit joules per kilogram (J/kg) with regard to the effective dose or the equivalent organ doses (*i.e.*, the dose quantities that contain the radiation weighting factor recommended by ICRP 1991).

TABLE 6-1 Number of Subjects, Solid Cancer Deaths, and Noncancer Disease Deaths by Radiation Dose

	DS86 Weighted Colon Dose (Sv)[a]							
	Total	0 (<0.005)	0.005–0.1	0.1–0.2	0.2–0.5	0.5–1.0	1.0–2.0	2.0
Number of subjects	86,572	37,458	31,650	5,732	6,332	3,299	1,613	488
Solid cancer deaths (1950–1997)	9,335	3,833	3,277	668	763	438	274	82
Noncancer disease deaths (1950–1997)	31,881	13,832	11,633	2163	2,423	1,161	506	163

[a]These categories are defined using the estimated dose to the colon, obtained as the sum of the γ-ray dose to the colon plus 10 times the neutron dose to the colon.

SOURCE: Based on data from Preston and others (2003).

37,458 survivors (43%) with doses less than 0.005 Sv were primarily survivors who were located more than 2.5 km from the hypocenter. Only 2101 (2.4%) had doses exceeding 1 Sv. Table 6-1 also shows the number of solid cancer deaths and noncancer disease deaths in the period 1950–1997.

STATISTICAL METHODS

The material in the sections that follow draws heavily on results presented by Thompson and colleagues (1994) and Preston and colleagues (1994, 2003). Here, features of the statistical methods that were used for most analyses in these papers are described. Readers should consult the source papers for details. In nearly all cases, analyses were based on Poisson regression using the AMFIT module of the computer software EPICURE (Preston and others 1991).

Most recent analyses have been based on either excess relative risk (ERR)[3] models, in which the excess risk is expressed relative to the background risk, or excess absolute risk (EAR)[4] models, in which the excess risk is expressed as the difference in the total risk and the background risk. The age-specific instantaneous risk is given either by

$$\lambda(c,s,a,b) \, [1 + \mathrm{ERR}(s,e,a,t,d)] \qquad (6\text{-}1)$$

or

$$\lambda(c,s,a,b) + \mathrm{EAR}(s,e,a,t,d) \qquad (6\text{-}2)$$

where λ denotes the background rate at zero dose and depends on city (c), sex (s), attained age (a), and birth year (b), and the excess may depend on sex (s), age at exposure (e),

attained age (a), and time since exposure (t). Not all variables are included in all models; in fact, any two of the variables e, t, and a determine the third. Parametric models are used for the ERR and EAR. The most recent analyses of solid cancer mortality (Preston and others 2003) have been based on models of the form

$$\mathrm{ERR} \text{ or } \mathrm{EAR} = \rho(d)_{\beta_s} \exp(\gamma e) \, a^{\eta}. \qquad (6\text{-}3)$$

Earlier analyses (Thompson and others 1994; Pierce and others 1996) were based primarily on ERR models of the form

$$\mathrm{ERR} = \rho(d)_{\beta_s} \exp(\gamma e). \qquad (6\text{-}4)$$

The function $\rho(d)$ is usually taken to be a linear or linear-quadratic function of dose, although threshold and categorical (nonparametric) models have also been evaluated. With the linear function, $\rho(d) = \beta_s d$, and β_s is the excess relative risk per sievert (ERR/Sv), which provides a convenient summary statistic. The parameters γ and η measure the dependence of the ERR/Sv on age at exposure and attained age.

Preston and colleagues (2003) and Thompson and colleagues (1994) used parametric models for the background risks. Some past analyses, such as those by Pierce and co-workers (1996) treated the background risk in ERR models by including a separate parameter for each category defined by city, sex, age at risk, and year. Thompson and colleagues did not fit EAR models; however, average EARs were estimated by dividing the estimated number of excess cancers by the total person-year-Sv.

Analyses of leukemia are based on bone marrow dose; analyses of the combined category of all solid cancers are based on colon dose; and analyses of site-specific cancers are based on specific organ doses. Dose is expressed in sieverts and is a weighted dose obtained as the sum of the dose of γ-radiation and 10 times the neutron dose. This approach is based on the assumption of a constant relative biological effectiveness (RBE) of 10 for neutrons. In most

---

[3]The ERR is the rate of disease in an exposed population divided by the rate of disease in an unexposed population minus 1.0.

[4]The EAR is the rate of disease in an exposed population minus the rate of disease in an unexposed population.

analyses, the kerma[5] doses are truncated at 4 Gy, equivalent to truncating organ doses at 3 Gy. Analyses by Preston and colleagues (2003) and by Pierce and colleagues (1996) were adjusted for random errors in doses using an approach described by Pierce and colleagues (1990) and based on the assumption of a coefficient of variation of 35% for the error in individual dose estimates. This adjustment generally increases estimated risk coefficients by about 10%. Earlier papers, such as analyses by Thompson and coworkers (1994) and by Preston and coworkers (1994), did not include this adjustment.

For analyses based on tumor registry data, adjustments were necessary to account for migration from the two cities. These are described briefly by Thompson and colleagues (1994) and Preston and colleagues (1994) and in more detail by Sposto and Preston (1992).

Leukemia

This section reviews analyses of mortality data for the period 1950–1990 (Pierce and others 1996) and of incidence data for the period 1958–1987 (Preston and others 1994). Leukemia mortality data for the period 1950–2000 were analyzed by Preston and colleagues (2004) and used to develop the committee's models for estimating leukemia risks; these analyses are described in Chapter 12.

Leukemia was the first cancer to be linked with radiation exposure in A-bomb survivors (Folley and others 1952) and has the highest relative risk of any cancer. Pierce and colleagues estimated that 78 of 176 (44%) leukemia deaths among survivors with doses exceeding 0.005 Sv were due to radiation exposure. Leukemia risks increased with dose up to about 3 Sv, with evidence of upward curvature; that is, a linear-quadratic function fitted the data significantly better than a linear function. With this linear-quadratic function, the excess risk per unit of dose at 1 Sv was about three times that at 0.1 Sv.

For those exposed under about age 30, nearly all of the excess deaths occurred before 1975, but for those exposed at older ages, the excess risk appeared to persist throughout the follow-up period. The temporal trends also differed by sex, with evidence of a steeper decline in risk for males than for females. Both the nonlinear dose-response and the complex patterns by age and time since exposure mean that simple models cannot adequately summarize leukemia risks.

Preston and colleagues (1994) analyzed data from the leukemia registry. An important recent development in studies of leukemia is the reclassification of leukemia cases by new systems and criteria (Matsuo and others 1988; Tomonaga

and others 1991), which allows meaningful analyses of specific types of leukemia. Preston and colleagues evaluated patterns of risk by sex, age at exposure, and time since exposure for four major subtypes of leukemia: acute lymphocytic leukemia (32 cases), acute myelogenous leukemia (103 cases), chronic myelogenous leukemia (57 cases), and adult T-cell leukemia (39 cases). Dose-response relationships were seen for the first three but not for adult T-cell leukemia. The estimated numbers of cases in excess of background were 17.1 for acute lymphocytic leukemia, 29.9 for acute myelogenous leukemia, and 25.9 for chronic myelogenous leukemia. The other major type of leukemia, chronic lymphocytic leukemia, showed no excess, but it is infrequent in Japan.

Results of analyses of all types of leukemia showed dependencies on sex, age at exposure, and time since exposure similar to those for the mortality data and led to a model similar to that based on mortality data. Preston and colleagues note that allowing overall modification by sex and age at exposure in an EAR model did not significantly improve the fit once time since exposure was included in the model, but that these factors significantly modified the time since exposure effects. Specifically, risks for those exposed early in life decreased more rapidly than for those exposed later, and the decrease was less rapid for women than for men. Analyses of specific leukemia types indicated that there were significant differences in the effects of age at exposure and sex and in the temporal pattern of risks. The shape of the dose-response did not show statistically significant differences among the subtypes.

ALL SOLID CANCERS

Analyses of cancers in this category, which excludes leukemia and other hematopoietic cancers, are useful for providing summary information and models based on larger numbers than are available for cancers of specific sites (discussed below). The discussion in this section is based on both mortality (Preston and others 2003) and incidence data (Thompson and others 1994). Mortality analyses were based on 9335 solid cancer deaths that occurred during 1950–1997, whereas incidence analyses included 8613 incidence cases occurring during 1958–1987.[6] The incidence data do not include cases of subjects who migrated and were diagnosed with cancer outside of Hiroshima and Nagasaki; as noted above, analyses were adjusted for migration.

Preston and collegues estimate that 8% of the 5502 solid cancer deaths among those with doses exceeding 0.005 Sv were due to radiation, much lower than the corresponding percentage of 44% for leukemia. This percentage was

---

[5]Kinetic energy released in material. A dosimetric quantity, expressed in grays, that equals the kinetic energy transferred to charged particles per unit mass of irradiated medium when indirectly ionizing (uncharged) particles, such as neutrons, traverse the medium. If all of the kinetic energy is absorbed "locally," the kerma is equal to the absorbed dose.

[6]These numbers contrast with 10,127 solid cancer deaths occurring in 1950–2000 and 12,778 incident cases of solid cancer excluding thyroid and nonmelanoma skin cancer occurring in 1958–1998, the periods covered by analyses conducted by the committee and described in Chapter 12.

slightly higher for the incidence data, where 11% of 4327 cancers in the exposed were estimated to result from radiation exposure (Thompson and others 1994). For both the mortality and the incidence data, risks of solid cancer increased with dose up to about 3 Sv, with little evidence of nonlinearity in the dose-response for doses in the 0–3 Sv range. For mortality data, this is illustrated by Figure 6-1, taken from Preston and colleagues (2003). Estimates based on only the low-dose portion of the mortality data are similar to those based on the range from 0 to 2 Sv. For example, there was a statistically significant dose-response ($p = .025$) based on analyses restricted to the 0–0.125 Sv dose range, with the ERR/Sv estimated to be 0.74 (SE = 0.38). This estimate did not differ significantly ($p > .5$) from the estimate of 0.54 Sv^{-1} (SE = 0.07) based on the 0–2 Sv range (Preston and others 2003, Table 4).

Figure 6-2 shows plots of the ERR and EAR for solid cancer mortality by age at exposure and attained age. The ERR for females was about twice that for males, but the EARs were similar for the two sexes since baseline risks for females are about half those for males. Both the ERR and the EAR were found to decrease with increasing age at exposure. The EAR increased with increasing attained age within age-at-exposure groups, while the ERR decreased with increasing attained age, especially for those exposed in childhood. Preston and colleagues emphasize results based on a model that allows the ERR to vary with both age at exposure and attained age, but also pay attention to a model in which the ERR varies only with age at exposure since the evidence for this effect was stronger.

Similar plots based on the committee's analyses of cancer incidence data are presented in Figures 12-1 and 12-2. These data show similar patterns to those for mortality except that the evidence for modification of the ERR by attained age was stronger with the updated incidence data than with the mortality data.

Preston and colleagues (2003) also present lifetime risk estimates for an LSS cohort member exposed to 1 Sv. These estimates were 18–22% for a person exposed at age 10, 9% for a person exposed at age 30, and 3% for a person exposed at age 50. These estimates did not differ greatly from those based on earlier mortality data (Pierce and others 1996).

Additional Analyses Addressing the Shape of the Dose-Response Function

Several additional papers address the shape of the dose-response function and evidence for risk at the lower end of the dose distribution; these include analyses by Kellerer and Nekolla (1997), Little and Muirhead (1997), Hoel and Li (1998), and Pierce and Preston (2000). These analyses take advantage of the large number of survivors with lower doses and investigate the possibility of a threshold, departures from linearity, and the degree to which effects might be overesti-

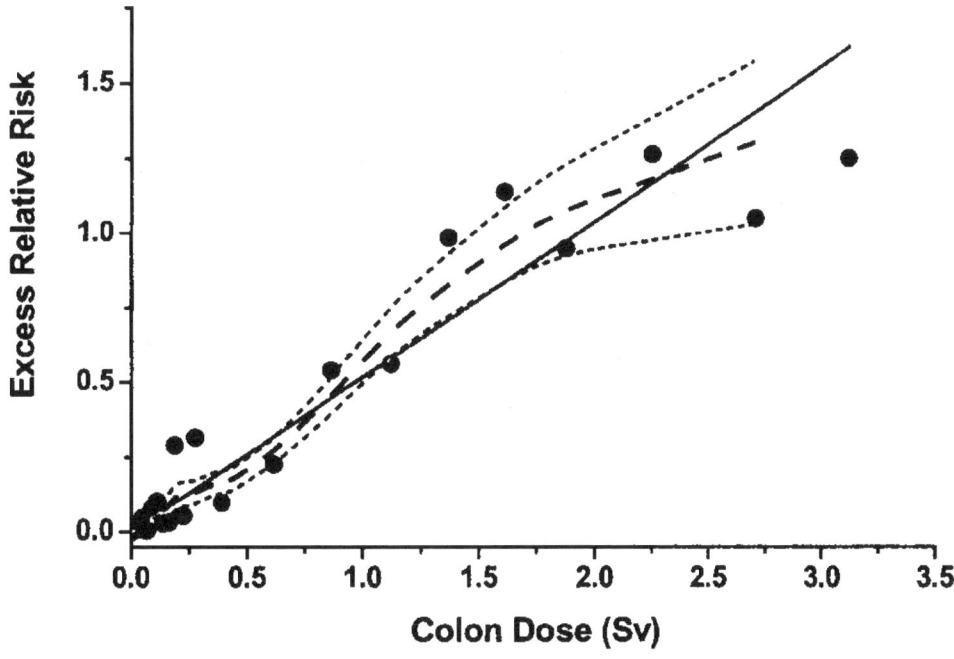

FIGURE 6-1 Solid cancer mortality dose-response function averaged over sex for attained age 70 after exposure at age 30. The solid straight line is the linear slope estimate, the points are dose-category-specific ERR estimates, the dashed curve is a smoothed estimate derived from the points. Dotted curves indicate upper and lower one-standard-error bounds on the smoothed estimate. SOURCE: Reproduced with permission from Preston and others (2003).

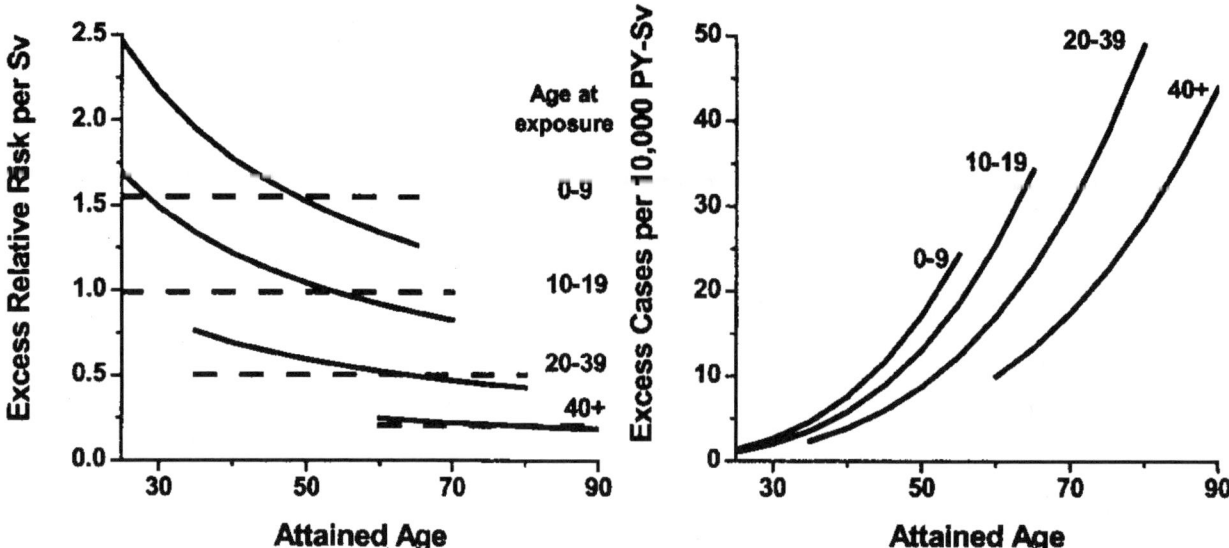

FIGURE 6-2 Primary descriptions of the excess risk of solid cancer mortality. Left panel: fitted sex-averaged ERR estimates using both attained-age-declining (solid black line) and attained-age-constant (dashed lines) forms, for age-at-exposure groups 0–9, 10–19, 20–39, and 40+. ERR estimates for women are about 25% greater, and ERR estimates for men 25% lower, than the values shown. Right panel: fitted EAR estimates for the same dose groups. There is no evidence of significant sex differences in the fitted EAR. SOURCE: Reproduced with permission from Preston and others (2003).

mated based on linear extrapolation from high to low doses. The committee discusses the analyses by Pierce and Preston (2000) because these are the only analyses that include updated cancer incidence data.

Pierce and Preston (2000) investigated solid cancer risks at low doses using cancer incidence data for 1958–1994, thus adding 7 years of data to that available in previously published incidence data analyses. Because experimental data have indicated that the RBE of neutrons decreases with increasing dose, the RBE was assumed to be a function of dose, with a value of 40 at very low doses that decreased to about 8 when the neutron dose reached 0.02 Gy (where the gamma dose was about 2 Gy). Because of evidence that survivors located more than 3000 m from the bombings had higher cancer rates than other survivors estimated to have zero doses, these distally located survivors were omitted from the analyses described below. This exclusion had little effect on analyses based on the full dose range, but did affect analyses directed specifically at low-dose effects.

In analyses based on the range 0–2 Sv, Pierce and Preston (2000) found little evidence of nonlinearity in the dose-response except for a small elevation in risk over linearity in the 0.15–0.3 Sv range. They estimated a curvature parameter θ, defined as the ratio of the quadratic and linear coefficients for gamma dose, and found that the upper 95% confidence limit for θ was 0.75 Gy^{-1}. At this value, the linear coefficient was estimated to be a factor of 1.9 smaller than that obtained from a strictly linear model, and the factor 1.9 (*i.e.*, the dose rate effectiveness factor, DREF) was termed

the "overestimation factor." This result might be interpreted as indicating that the maximum DREF that is reasonably compatible with the A-bomb survivor data is unlikely to be greater than 2. In addition, Pierce and Preston (2000) evaluated threshold models in which the risk was zero up to a given threshold and then increased linearly. They estimated the threshold to be 0 Sv with an upper confidence limit of 0.06 Sv. Evidence of a statistically significant dose-response was found in the dose range 0–0.10 Sv.

Pierce and Preston (2000) warn against overinterpretation of the minimum dose at which evidence of a significant dose-response is found, indicating that "in the presence of available data, it is neither sound statistical interpretation nor prudent risk evaluation to take the view that the risk should be considered as zero in some low-dose range due to lack of statistical significance when restricting attention to that range." They further call attention to the large potential for bias due to confounding in analyses based on low doses, noting particularly that A-bomb survivor results in the low-dose range are influenced by whether or not distally located survivors are included.

Other Analyses

The A-bomb survivor data have been combined with data from cohorts of persons exposed for medical reasons, primarily for the purpose of further exploration of the modifying effects of age at exposure, attained age, and time since exposure (Little and others 1998, 1999a, 1999c; Little, 1999).

Although these analyses provide valuable information on the comparability of risks and of modifying factors in different cohorts, the results for the A-bomb survivor cohort itself generally confirm the findings reported earlier in the chapter, and they are not discussed further here. Biologically based models have also been applied to the A-bomb survivor data (Kai and others 1997; Pierce and Mendelsohn 1999).

SITE-SPECIFIC CANCERS

Because the exposure of A-bomb survivors was whole-body exposure, studies of the LSS cohort afford the opportunity to compare cancer risks by site. Inferences for site-specific cancers are based on smaller numbers than those for all solid cancers and involve smaller ERRs than leukemia. This often means that there is considerable uncertainty in quantifying risk, in evaluating modifying factors, and even in determining whether or not there is a dose-response relationship. Although it is likely that radiosensitivity varies across sites, it is often not possible to separate true differences from chance fluctuations. Cancers at some sites may fail to exhibit associations because of small numbers of cases and diagnostic misclassification, which is more problematic for mortality data than for incidence data.

Preston and colleagues (2003) used common models for expressing risks for cancers at different sites. Specifically, 15 sites were analyzed with parameters expressing the modifying effects of age at exposure and attained age set equal to those for all solid cancers. Results of these analyses are summarized in Figure 6-3, which shows the ERR/Sv for exposure at age 30 and attained age 70. Except for sex-specific cancers, estimates are averaged for the two sexes. Preston and colleagues (2003) note that the variability in this plot is generally consistent with what would be expected if the true site-specific ERRs were all equal to that for all solid cancers. More detailed analyses of the five most common types of solid cancer (stomach, colon, liver, lung, and female breast) were conducted. With ERR models, the age-time patterns were similar for these sites, although the decrease in risk with attained age was more rapid for colon cancer. With EAR models, statistically significant departures from the solid cancer temporal model were found for lung cancer, which increased more rapidly with attained age than other solid cancers, and breast cancer, which decreased more rapidly with age at exposure than other solid cancers.

Data from the Hiroshima and Nagasaki tumor registries are preferable to mortality data for evaluating site-specific risks. These data have the major advantages of including

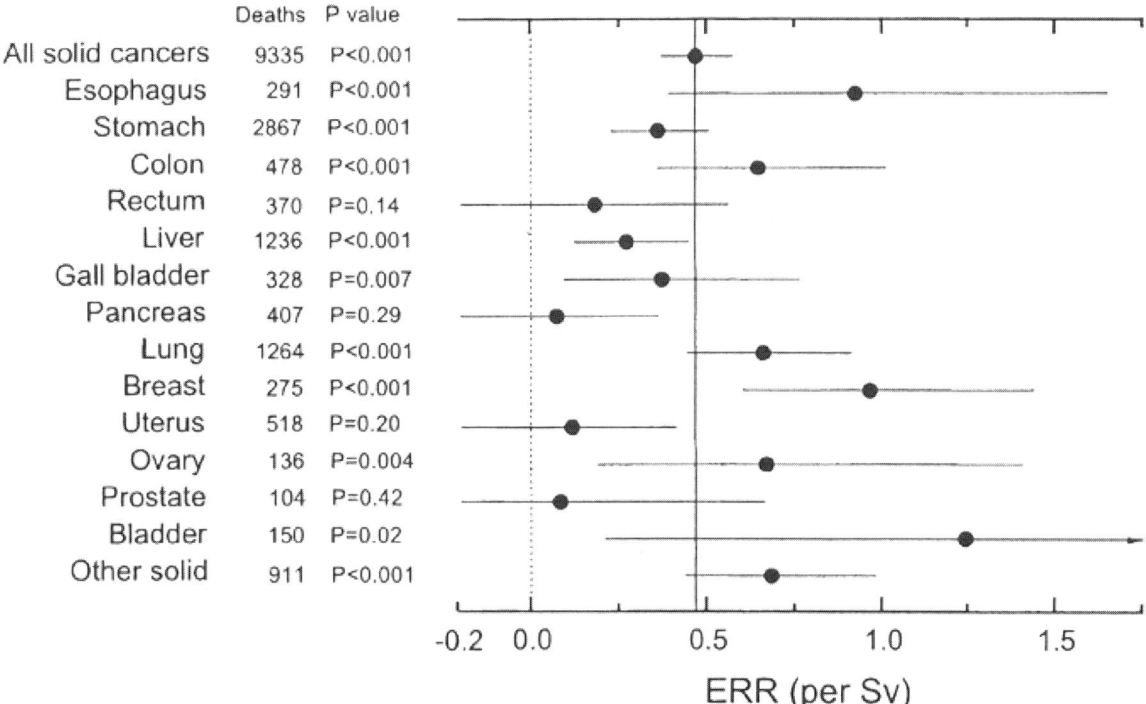

FIGURE 6-3 Estimates of the site-specific solid cancer mortality ERR with 90% confidence intervals and one-sided *p*-values for testing the hypothesis of no dose-response. Except for sex-specific cancers (breast, ovary, uterus, and prostate) the estimates are averaged over sex. All estimates and *p*-values are based on a model in which the age-at-exposure and attained-age effects were fixed at the estimates for all solid cancers as a group. The light dotted vertical line at 0.0 corresponds to no excess risk; the dark solid vertical line indicates the sex-averaged risk for all solid cancers. SOURCE: Reproduced with permission from Preston and others (2003).

nonfatal cancers and of more accurate diagnostic information with data on histological types of cancer. Results based on analyses by the committee of updated incidence data (1958–1997) are discussed in Chapter 12.

Thompson and colleagues evaluated cancer incidence data from 1958 to 1987 for the cancer sites shown in Figure 6-4 and Table 6-2. For each site, they evaluated whether there was a significant association with dose, whether there were departures from linearity, and whether risks were modified by city, sex, age at exposure, attained age, or time since exposure.

Of the cancer sites shown in Figure 6-4 and Table 6-2, the largest ERR/Sv was for breast cancer. Relatively large values were also seen for nonmelanoma skin cancer and for cancers of the ovary, urinary bladder, and thyroid. In addition to these sites, the 95% confidence intervals excluded zero for cancers of the stomach, colon, liver, and lung. It should be noted that the size of the ERR/Sv may be affected by the size of the baseline risk. These ERRs/Sv were obtained from a model with no modifying factors and are not strictly comparable to those based on mortality data and shown in Figure 6-3, which included modifying factors and were intended to be applicable to a person exposed at age 30 at attained age 70.

In addition to analyses by Thompson and colleagues (1994), several papers provide further analyses that, in some cases, give more attention to histological type and, in other cases, are based on case-control studies that include data on possible modifying factors that were not available for the full cohort. These results are summarized below for selected cancer sites.

Female Breast Cancer

In a case-control interview study nested within the LSS cohort and including cases occurring in 1950–1985, Land and colleagues (1994b) investigated known risk factors for breast cancer: age at the time of a first full-term pregnancy, number of children, and cumulative period of breast-feeding. The influence of these factors on breast cancer risks in women in the LSS cohort was similar to that found in other studies. The relationship of these factors and radiation exposure was reasonably well described by a multiplicative model (in which known risk factors for breast cancer do not modify the ERR/Sv), whereas an additive model could be rejected.

Preston and colleagues (2002a) conducted pooled analyses of breast cancer incidence in eight cohorts. Analyses from this paper based on the LSS cohort alone that included cases

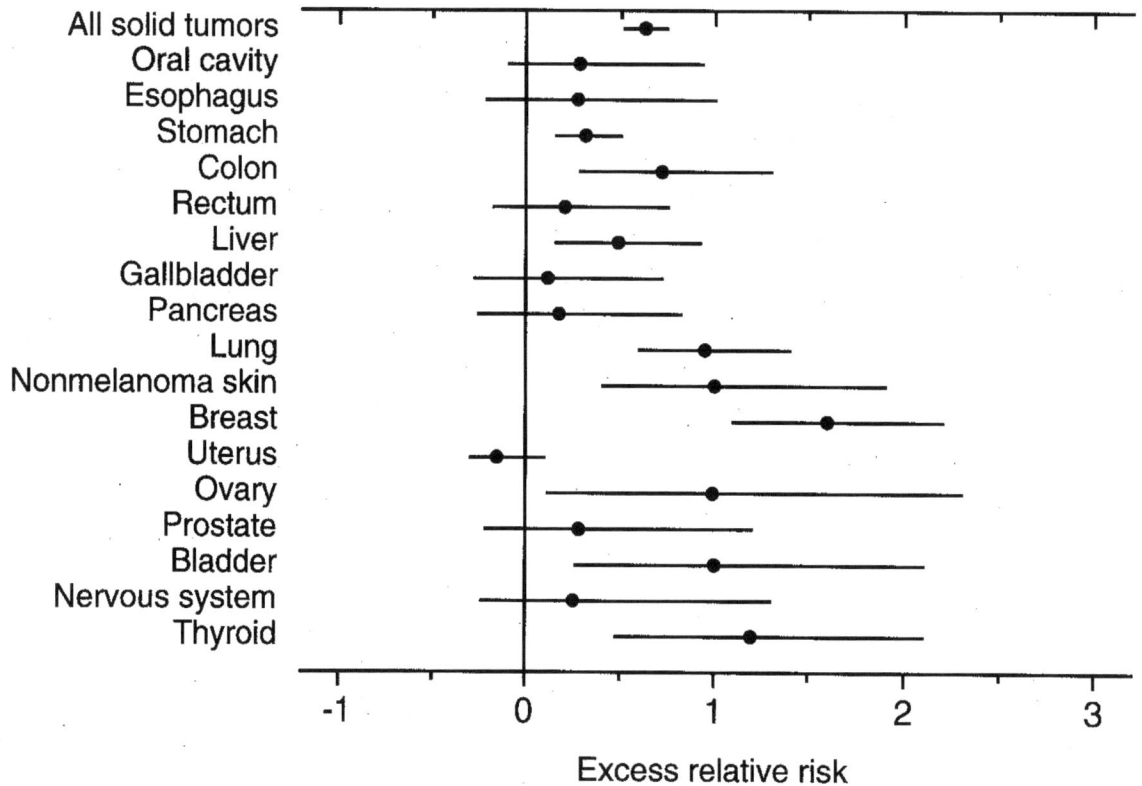

FIGURE 6-4 Excess relative risk at 1.0 Sv (RBE 10) for solid cancer incidence and 95% confidence interval, 1958–1987. SOURCE: Reproduced with permission from Thompson and others (1994).

TABLE 6-2 Summary of Risk Estimates for Solid Cancer Incidence by Cancer Site or Organ System

Cancer Site or Organ System	Percentage of Total Cases[a]	ERR$_{1Sv}$	EAR per 10,000 PY-Sv	AR,[b] %
Total solid tumors	100.0	0.63 (0.52, 0.74)[c]	29.7 (24.7, 34.8)	11.6 (10.2, 14.3)
Oral cavity and pharynx	1.5	0.29 (−0.09, 0.93)	0.23 (−0.08, 0.65)	9.1 (−3.0, 25.9)
Digestive system	55.7	0.38 (0.25, 0.52)	10.4 (7.0, 14.0)	7.8 (5.3, 10.6)
Esophagus	2.1	0.28 (−0.21, 1.0)	0.30 (−0.23, 1.0)	6.5 (−5.0, 22.5)
Stomach	30.9	0.32 (0.16, 0.50)	4.8 (2.5, 7.4)	6.5 (3.5, 10.5)
Colon	5.3	0.72 (0.29, 1.3)	1.8 (0.74, 3.0)	14.2 (5.9, 23.9)
Rectum	4.1	0.21 (−0.17, 0.75)	0.43 (−0.35, 1.5)	4.4 (−3.6, 14.6)
Liver	6.8	0.49 (0.16, 0.92)	1.6 (0.54, 2.9)	10.9 (3.6, 19.4)
Gallbladder	3.4	0.12 (−0.27, 0.72)	0.18 (−0.41, 1.1)	2.2 (−5.1, 13.1)
Pancreas	2.8	0.18 (−0.25, 0.82)	0.24 (−0.36, 1.1)	3.5 (−5.2, 15.3)
Respiratory system	11.9	0.80 (0.50, 1.2)	4.4 (2.9, 6.1)	16.3 (10.6, 22.6)
Trachea, bronchus, and lung	10.1	0.95 (0.60, 1.4)	4.4 (2.9, 6.0)	18.9 (12.5, 26.0)
Nonmelanoma skin	2.0	1.0 (0.41, 1.9)	0.84 (0.40, 1.4)	24.1 (11.5, 38.6)
Female breast	6.1	1.6 (1.1, 2.2)	6.7 (4.9, 8.7)	31.9 (23.2, 41.1)
Uterus	8.4	−0.15 (−0.29, 0.10)	−1.1 (−2.1, 0.68)	−3.3 (−6.4, 2.1)
Ovary	1.5	0.99 (0.12, 2.3)	1.1 (0.15, 2.3)	17.7 (2.4, 37.3)
Prostate	1.6	0.29 (−0.21, 1.2)	0.61 (−0.46, 2.2)	7.0 (−5.3, 25.5)
Urinary organs and kidney	3.8	1.2 (0.62, 2.1)	2.1 (1.1, 3.2)	22.3 (11.8, 34.2)
Urinary bladder	2.4	1.0 (0.27, 2.1)	1.2 (0.34, 2.1)	16.3 (4.8, 30.1)
Kidney	0.8	0.71 (−0.11, 2.2)	0.29 (−0.50, 0.79)	15.2 (−2.6, 41.3)
Nervous system	1.5	0.26 (−0.23, 1.3)	0.19 (−0.17, 0.81)	5.7 (−5.3, 24.5)
Thyroid	2.6	1.2 (0.48, 2.1)	1.6 (0.78, 2.5)	25.9 (12.4, 40.7)

[a]254 solid cancers of other and ill-defined sites are included in the total solid tumors category.

[b]AR is the attributable risk, which in this case is the percentage of cases in exposed survivors attributed to radiation exposure.

[c]Values in parentheses are the 95% confidence limits.

SOURCE: Thompson and others (1994).

occurring in the period 1958–1993 showed a clear decline in the ERR/Sv with either age at exposure or attained age when evaluated separately. The EAR was also found to decrease with age at exposure, but to increase with attained age at least up to age 50. These analyses, as well as earlier analyses by Tokunaga and colleagues (1994) and by Thompson and coworkers (1994), found that the dose-response for breast cancer was well described by a linear function. Tokunaga and colleagues (1994) also report a strong attained age effect, with an ERR/Sv of 13 for breast cancer occurring before age 35 compared to an ERR/Sv of about 2 for breast cancer occurring after age 35.

Land and colleagues (2003) reported on an incidence survey of breast cancers diagnosed during 1950–1990. As in previous analyses a strong linear dose-response was found. A modified isotonic regression approach, which required only that the ERR/Sv be monotonic in age, was used to evaluate in detail the modification of the dose-response by age at exposure and attained age. The abstract notes that "exposure before age 20 was associated with higher ERR$_{1Sv}$ compared to exposure at older ages, with no evidence of consistent variation by exposure age under 20. ERR$_{1Sv}$ was observed to decline with increasing attained age, with by far the largest drop around age 35."

Thyroid Cancer

Like breast cancer, thyroid cancer risks are described well by a linear dose-response function and also show a strong dependence on age at exposure. In fact, there is little evidence of a dose-response for persons exposed in adulthood (Thompson and others 1994; Ron and others 1995a), while the ERRs/Sv for those exposed as children were large (9.5 for persons exposed under age 10, and 3.0 for those exposed at ages 10–19; Thompson and others 1994). Although several other cohorts provide data on thyroid cancer risks from external radiation exposure in childhood (Ron and others 1995a), the LSS cohort is the only cohort providing much information on thyroid cancer risk from external radiation exposure in adulthood.

Salivary Gland Cancer

Because some types of salivary gland tumors are not readily identified by the conventional disease classification codes used by tumor registries, a special evaluation that included pathology reviews of both benign and malignant salivary gland tumors was undertaken by Land and colleagues (1996). This resulted in an estimated ERR/Sv of 3.5 (95% CI

1.5, 7.5) for malignant tumors, higher than any of the ERR/
Sv shown in Table 6-2, although very uncertainly estimated.
The ERRs/Sv was 0.7 (0.1, 1.7) for benign tumors. Most of
the dose-response for malignant tumors resulted from mu-
coepidermoid carcinoma with an ERR/Sv of 8.3 (2.5, 29.6),
whereas most of the dose-response for benign tumors resulted from Warthin's tumor with an ERR/Sv of 3.1 (0.6,
10.3).

Stomach Cancer

This site merits special comment primarily because stomach cancer is the most common type of cancer in Japan and,
specifically, in the LSS cohort. Based on cancer incidence
data evaluated by Thompson and colleagues (1994), stomach cancer had a relatively small but precisely estimated
ERR/Sv of 0.32 (0.16, 0.50). The ERR/Sv for females was
about three times that for males, and the ERR/Sv decreased
with increasing age at exposure. Nearly one-third (31%) of
the solid cancer cases included in the incidence data were
stomach cancers, so this cancer potentially has a strong impact on overall solid cancer results. However, analyses of
solid cancer mortality data with stomach cancer excluded
resulted in parameter estimates that were similar to those
obtained for all solid cancers (Preston and others 2003).

Liver Cancer

Liver cancer is one of the most frequently occurring cancers in Japan and the third most common cancer (after stomach and lung) in the LSS. Liver cancers reported on death
certificates might in fact be cancers originating in other organs because the liver is a frequent site for metastatic cancer.
This can be a problem even for tumor registry data, since
some cases were based only on death certificate information.
For this reason, Cologne and colleagues (1999) conducted a
study of primary liver cancer based on extensive pathology
review of known or suspected cases of liver cancer. This
study showed a clear dose-response with an estimated ERR/
Sv (with 95% CI) of 0.81 (0.32, 1.43). The ERRs/Sv for
males and females were very similar (0.81 and 0.78, respectively), in contrast to findings for many other cancers, and
somewhat remarkable given that background rates for males
were about three times those for females. The modifying
effect of age at exposure was also different from that for
other cancers, with excess risk peaking for those exposed in
their twenties, but little evidence of excess risk for those
exposed under age 10 or over age 45.

Of the 364 cases analyzed, there were 307 hepatocellular
carcinomas (HCCs), 53 cholangiocarcinomas, two mixed
hepatocellular-cholangiocarcinomas, and one each of he-
patoblastoma and hemangiosarcoma. This is in contrast to
liver cancers associated with Thorotrast exposure, which are
dominated by cholangiocarcinomas and hemangiosarcomas.
Cologne and colleagues found no difference in the dose-response for HCC compared to cholangiocarcinoma, although

this may have been because the number of cancers of the
latter type was small.

It has been estimated that more than 60–75% of HCC
cases in Japan are related to chronic hepatitis C infection and
that 20–25% are positive for hepatitis B surface antigen
(Fujiwara and others 2000). Neriishi and others (1995) reported a radiation dose related increase in the prevalence of
hepatitis B surface antigen in atomic bomb survivors.
Fujiwara and colleagues (2000) did not find such a relationship for hepatitis C infection, but their data suggest that the
radiation dose-response for chronic liver disease was greater
for survivors who were positive for hepatitis C antibody than
for survivors who were negative.

Lung Cancer

Next to stomach cancer, lung cancer was the most common cancer in the LSS cohort. This cancer showed a strong
sex association with the ERR/Sv for females about four times
as large as that for males based on the incidence data evaluated by Thompson and colleagues (1994), which probably
reflects at least in part the larger baseline risks for males.
Lung cancer also deviated from the usual pattern of decreasing risk with increasing age at exposure. Instead, lung cancer
risks appeared, if anything, to increase with increasing age at
exposure, although, based on the incidence data, this trend
was not statistically significant.

Recently, Pierce and coworkers (2003) evaluated the joint
effects of smoking and radiation on lung cancer incidence
through 1994 in a subset of about 45,000 members of the
LSS cohort for whom both radiation dose and smoking data
were available. The smoking data were obtained from mail
surveys of the LSS cohort and clinical interviews of members of the AHS conducted during 1963–1993. Pierce and
colleagues (2003) found that the effects of smoking and
radiation were significantly submultiplicative and consistent
with an additive model. They note that the aging of the cohort
and higher smoking levels among more recent birth cohorts
resulted in a stronger basis for evaluating the joint effects of
smoking and radiation than in previous analyses by Kopecky
and colleagues (1986), Prentice and colleagues (1983), and
the National Research Council (NRC 1988); these earlier
investigations were unable to distinguish between additive
and multiplicative effects. Pierce and colleagues (2003) also
found that adjustment for smoking substantially reduced the
female-to-male ERR/Sv ratio; about 85% of the men and
16% of the women were smokers. With adjustment for smoking, there was evidence of a decline in the ERR/Sv with increasing attained age (comparable to other solid cancer sites),
but no evidence of modification by age at exposure.

Skin Cancer

Ron and colleagues (1998b) conducted a detailed study
of skin cancer that included pathologic review of cases. Basal
cell carcinoma (80 cases) was found to be associated with

radiation dose with some evidence of nonlinearity in the dose-response, but with no evidence of an interaction with ultraviolet radiation. No dose-response association was found for squamous cell carcinoma (69 cases). The relationships with dose for melanoma (10 cases) and Bowen's disease (26 cases) were not statistically significant, but estimates of the ERR/Sv were large.

Central Nervous System Cancers

See discussion of central nervous system tumors at the end of the section "Benign Neoplasms."

Lymphoma

Analyses of mortality data by Pierce and colleagues (1996) showed no evidence of an association for lymphoma; with the mortality data, it was not possible to distinguish between Hodgkin's and non-Hodgkin's cases. Lymphoma was not included in more recent mortality analyses. The incidence data included 210 lymphoma cases, of which 22 were Hodgkin's and 188 were non-Hodgkin's. A statistically significant dose-response was found for males, but not for females, for whom the estimated ERR/Sv was negative (Preston and others 1994).

Multiple Myeloma

Multiple myeloma exhibited a statistically significant dose-response based on the mortality data (Pierce and others 1996), but incidence data showed little evidence of such an association (Preston and others 1994). The discrepancy in these findings appears to be due to deaths with questionable diagnoses and second primary tumors that were included in the mortality analyses, but not the incidence analyses.

CANCERS RESULTING FROM EXPOSURE *IN UTERO*

Delongchamp and colleagues (1997) analyzed data on cancer mortality among atomic bomb survivors who were exposed either *in utero* or, for comparison, during the first 5 years of life. These analyses covered the period 1950–1992, adding an additional 8 years of follow-up to data available to the BEIR V committee (Yoshimoto and others 1988). Analyses were restricted to cancers occurring between the ages of 17 and 45. Ten cancers were observed in the cohort exposed *in utero*, and a significant dose-response was observed with an estimated ERR/Sv of 2.1 (90% CI 0.2, 6.0). This estimate did not differ significantly from that observed for survivors exposed during the first 5 years of life. An unusual aspect of the finding was that 9 of the 10 cancers occurred in females, and significant differences between the sexes persisted even when the three female cancer sites (breast, ovary, and uterus) were excluded.

BENIGN NEOPLASMS

Studies addressing benign neoplasms have generally been based on either the AHS or the tumor registries. Fujiwara and colleagues (1992) used the AHS to investigate hyperparathyroidism in Hiroshima survivors. About 4000 individuals with DS86 doses were tested for hyperparathyroidism, and a dose-response relationship was found ($p < .001$). The estimated relative risk at 1 Gy was 4.1 (95% CI 1.7, 14.0), and a decrease in relative risk with increasing age at exposure was suggested. The authors concluded that doses lower than those used in radiotherapy might induce this disorder. Nagataki and colleagues (1994) used Nagasaki AHS data to investigate thyroid diseases in 2587 subjects with diagnoses based on uniform procedures including ultrasonic scanning. Significant dose-response relationships were observed for all solid nodules (females), adenoma, and nodules without histological diagnosis (females). An association was also found for autoimmune hypothyroidism, one of the nonneoplastic end points investigated. However, the dose-response for hypothyroidism was not monotonic; risk increased to about 0.7 Sv and then decreased.

Ron and colleagues (1995b) used data from the Hiroshima and Nagasaki tumor and tissue registries to evaluate benign tumors of the stomach, colon, and rectum for 1958–1989. A total of 470 cases with histologically confirmed benign gastrointestinal tumors (163 stomach, 215 colon, and 92 rectum) were identified. A positive dose-response relationship was observed for stomach tumors, with an estimated ERR/Sv of 0.52 (95% CI 0.01–1.43), similar to that for stomach cancer. There was little evidence of dose-response for either colon or rectal tumors.

Tokunaga and colleagues (1993) investigated proliferative and nonproliferative breast disease using breast tissue samples from 88 high-dose and 225 low-dose autopsy cases of members of the LSS cohort. Both proliferative disease in general and atypical hyperplasia in particular were found to be positively associated with radiation dose, with the strongest association for subjects who were 40–49 years of age at exposure. The authors hypothesized that this finding might be "related to the age dependence of radiation-induced breast cancer, in that potential cancer induced in this age group by radiation exposure may receive too little hormonal promotion to progress to frank cancers."

Kawamura and colleagues (1997) conducted a study of uterine myoma based on ultrasound examination of 1190 female AHS participants in Hiroshima. The reason for conducting this study was concern that the previously identified dose-response associations (Wong and others 1993), discussed below, might have resulted from bias in case detection. This study resulted in an estimated ERR/Sv of 0.61 (95% CI 0.12, 1.31). It was judged unlikely that bias could explain the association. In earlier analyses by Wong and colleagues, time since exposure was found to be a significant modifier for uterine myoma, with younger survivors showing a decrease with time and older survivors showing an increase with time.

Preston and colleagues (2002b) investigated tumors of the nervous system and pituitary gland based on cases ascertained through the Hiroshima and Nagasaki Tumor and Tissue Registries and through medical records from RERF and major medical institutions in Hiroshima and Nagasaki. Histologic diagnoses were obtained by having four pathologists independently review slides and medical records. The majority of the 228 central nervous system tumors included in the study were benign. A statistically significant dose-response association was observed for all nervous system tumors with an estimated ERR/Sv of 1.2 (95% CI 0.6, 2.1). The ERR/Sv was highest for schwannomas (4.5; 95% CI 1.9, 9.2), but the dose-response for all other central nervous system tumors evaluated as a group was also statistically significant. The dose-responses for all nervous system tumors and for schwannomas were both statistically significant when limited to subjects with doses of less than 1 Sv, and there was no evidence that the slope for this low-dose range was different from that for the full range. Modification of risk by sex, age at exposure, and attained age was also investigated.

NONNEOPLASTIC DISEASE

Findings Based on Mortality Data

A statistically significant dose-response relationship with mortality from nonneoplastic disease in A-bomb survivors was demonstrated by Shimizu and colleagues (1992) based on mortality data for 1950–1985. The addition of five years of mortality data (through 90) strengthened the evidence for this effect and allowed a more detailed evaluation (Shimizu and others 1999). In these analyses, statistically significant associations were seen for the categories of heart disease, stroke, and diseases of the digestive, respiratory, and hematopoietic systems.

Preston and colleagues (2003) updated these results and present analyses of deaths from all causes excluding neoplasms, blood diseases, and external causes such as accidents or suicide. They give considerable attention to the fact that for a few years after the atomic bomb explosions, baseline risks for noncancers in proximal survivors (within 3000 m of the hypocenter) were markedly lower than those in distal survivors. They refer to this as the "healthy survivor effect" and note that it could lead to distortion of the dose-response, particularly in the early years of follow-up. They also note that a small difference (2%) in baseline risks for proximal and distal survivors persisted in later years, which they consider likely to be due to demographic factors such as urban-rural differences. They address this potential source of bias by conducting analyses restricted to the period 1968–1997 and by including an adjustment for differences in proxi-

mal and distal survivors (although results without the adjustment are also presented).

The estimated ERR/Sv for noncancers based on a linear model with no dependence on age at exposure or sex was 0.14, generally lower than that for all solid cancers (where the ERR/Sv depends on age and sex). There was no evidence of a statistically significant dependence on either age at exposure or sex, but the data were compatible with effects similar to those estimated for solid cancers. A linear dose-response function fitted the data well, but it was not possible to rule out a pure quadratic model or a model with a threshold as high as 0.5 Sv. Similar to Shimizu and colleagues (1999), significant dose-response relationships were found for heart disease, stroke, respiratory disease, and digestive disease. There was no evidence of radiation effects for infectious diseases or all other noncancer diseases in the group evaluated. Lifetime noncancer risks for people exposed to 1 Sv were estimated to be similar to those for solid cancer for those exposed as adults, and about half those for solid cancer for those exposed as children. Because baseline risks for the noncancer category evaluated are greater than those for all solid cancers, even the relatively small ERR/Sv leads to a fairly large absolute lifetime risk.

Because small ERRs can easily arise from bias, Shimizu and colleagues (1999) evaluated several potential sources of bias, including misclassification of cause of death, confounding, and cohort selection effects. Although Preston and coworkers (2003) discuss cohort selection effects in detail, they did not reevaluate other sources of bias. The committee summarizes the discussion provided by Shimizu and colleagues in the remainder of this section.

With regard to misclassification, they note that Sposto and coworkers (1992) investigated the possibility of bias from this source using mortality data through 1985. These investigators used estimated age-dependent misclassification probabilities obtained from RERF autopsy data to conduct analyses that corrected for misclassification and found that estimates for noncancer mortality were reduced by 20%, but remained highly statistically significant. Shimizu and colleagues (1999) used mail survey and interview data to examine the possible effect of several potential confounders including educational history and smoking. Although most of the factors evaluated were found to affect noncancer mortality, they were not found to be associated strongly with dose. Analyses adjusted for various confounders, based on survivors with available data, resulted in ERRs/Sv that were very similar to the unadjusted values.

Shimizu and colleagues (1999) also evaluated noncancer diseases of the blood, benign neoplasms, and deaths from external causes. Because these categories were not reevaluated by Preston and coworkers (2003), the committee summarizes these findings. The ERR/Sv for the 191 deaths from noncancer diseases of the blood was estimated to be 1.9 (90% CI 1.2, 2.9), larger than the estimated values for most solid cancers. The accuracy of death certificate diagnosis is known to be poor for this category and likely to include many misclassified leukemias and malignant lymphoma deaths. Among 128 deaths for which additional diagnostic information was available, there were 57 nonneoplastic disease deaths. When these deaths were analyzed separately, the resulting ERR/Sv was 2.0 (90% CI 0.6, 4.4), nearly identical to that based on the full 191 deaths. Analyses suggested that the effect was limited to nonaplastic anemias (29 cases), since the estimate for aplastic anemias (31 cases) was essentially zero. There was also a suggestion of a strong dose-response based on 13 deaths from myelodysplastic syndrome, a neoplastic disease thought to be a precursor of acute myelogenous leukemia.

Although the data evaluated by Shimizu and colleagues (1999) included 379 deaths attributed to benign neoplasms or neoplasms of unspecified nature, only 31 deaths were specifically indicated on the death certificate as being due to benign neoplasms. There was no convincing evidence of a dose-response for these 31 deaths.

With regard to deaths from external causes, suicide rates showed a statistically significant decline with increasing dose, whereas no evidence of a dose-response relationship was found for deaths from other external causes.

Findings Based on the Adult Health Study (AHS) or on Autopsy Data

Wong and colleagues (1993) evaluated the relationship between exposure to radiation and the incidence of 19 nonmalignant disorders using data from the AHS cohort for 1958–1986. They found statistically significant positive dose-response relationships ($p < .05$) for thyroid disease ($p < .001$), chronic liver disease and cirrhosis ($p = .007$), and uterine myoma ($p < .001$). In addition, myocardial infarction showed a significant dose-response for 1968–1986 among those who were under 40 years of age at exposure ($p = .03$). Statistically significant relationships were not detected for hypertension, hypertensive heart disease, ischemic heart disease, occlusion and stenosis of precerebral and cerebral arteries, aortic aneurysm, stroke, cataract, gastric ulcer, duodenal ulcer, viral hepatitis, calculus of kidney and ureter, cervical polyp, hyperplasia of prostate, dementia, and Parkinson's disease. Modification of the ERR/Sv by sex, city, age at exposure, and time since exposure was also investigated for those end points that showed overall associations. Age at exposure was found to be a significant modifier of risk for thyroid disease (decreasing ERR/Sv with increasing age); modifying effects for uterine myoma are discussed above ("Benign Neoplasms").

Kodama and colleagues (1996) reviewed results of studies addressing noncancer diseases and their relationship to radiation exposure in A-bomb survivors. They also update some of the analyses by Wong and colleagues (1993) to include data through 1990, but do not present nearly as much detail as the latter. They found a statistically significant as-

sociation for myocardial infarction based on all of the data ($p = .02$), with an estimated ERR/Sv of 0.17 (95% CI 0.01, 0.36). The association remained significant when analyses were adjusted for various risk factors including blood pressure and cholesterol. Positive dose-response relationships were also found for several other end points of atherosclerosis, which the authors interpreted as supporting a real association between radiation exposure and atherosclerosis. Kodama and colleagues (1996) confirmed previously identified radiation associations for uterine myoma, hyperparathyroidism, and chronic liver disease with an ERR/Gy of 0.46 (0.27, 0.70), 3.1 (0.7, 13), and 0.14 (0.04, 0.27) for the three respective end points.

Wong and colleagues (1999) used AHS data to examine long-term trends in total serum cholesterol levels over the 28 years from 1958 to 1986. Dose-response relationships for the increase in cholesterol levels over time were demonstrated for women in general but only in the youngest birth cohort (1935–1945) for men. Age, body mass index, city, and birth year were considered in the analyses, and some analyses were adjusted for cigarette smoking. These results may partially explain the dose-response relationship for coronary heart disease that has been observed in other studies of atomic bomb survivors.

LIFE SHORTENING

Cologne and Preston (2000) investigated life shortening in the LSS cohort using mortality data through 1995. Although dose-related increases in both cancer and noncancer mortality imply that longevity is also related to dose, earlier papers addressing these effects (Pierce and others 1996; Shimizu and others 1999) did not specifically attempt to quantify the degree of radiation-induced life shortening, an end point that reflects the effects of both cancer and noncancer mortality. The investigation of longevity was undertaken in part because of earlier reports in both the scientific literature and the press that certain atomic bomb survivors had greater-than-average life expectancy.

A clear decrease in median life expectancy with increasing radiation dose was found. Among cohort members with estimated doses between 0.005 and 1.0 Gy, the median loss of life was estimated to be about 2 months, while among cohort members with estimated doses of 1 Gy or more, the median loss of life was estimated to be about 2.6 years. The median loss of life among all cohort members with doses estimated to be greater than zero was about 4 months.

Cologne and Preston (2000) present estimates of life expectancy for groups defined by dose. For those with zero dose, separate estimates are presented for groups defined by distance from the hypocenter, including estimates for those who were not in the city (>10 km from the hypocenter). Although the relative mortality for all nonzero-dose groups compared to the combined in-city, zero-dose group was 1.0 or greater, results for those in the lowest-dose category

(0.005–0.25 Gy) were somewhat dependent on the choice of comparison group. Cohort members in this low-dose category had a median life expectancy that was shorter than that of zero-dose survivors who were within 3 km of the hypocenter (229 d), shorter than the not-in-city group (365 d), but slightly longer (52 d) than survivors located 3 km or more from the hypocenter. These results do not support the hypothesis that life expectancy for atomic bomb survivors exposed at low doses is greater than that for comparable unexposed persons.

SUMMARY

The LSS cohort of survivors of the atomic bombings in Hiroshima and Nagasaki continues to serve as a major source of information for evaluating health risks from exposure to radiation, and particularly for developing quantitative estimates of risk from exposure to ionizing radiation. Its advantages include its large size, the inclusion of both sexes and all ages, a wide range of doses that have been estimated for individual subjects, and high-quality mortality and cancer incidence data. In addition, the whole-body exposures received by this cohort offer the opportunity to assess risks for cancers of a large number of specific sites and to evaluate the comparability of site-specific risks. The full LSS cohort consists of approximately 120,000 persons who were identified at the time of the 1950 census. However, most recent analyses have been restricted to approximately 87,000 survivors who were in the city at the time of the bombings and for whom it is possible to estimate doses. Special studies of subgroups of the LSS have provided clinical data, biological measurements, and information on potential confounders or modifiers.

Mortality data for the period 1950–1997 have been evaluated in detail, adding 12 years to the follow-up period available at the time BEIR V (NRC 1990) was published. The longer follow-up period not only increases statistical precision, but also allows more reliable assessment of the long-term effects of radiation exposure, including modification or risk by attained age and time since exposure. Importantly, cancer incidence data from both the Hiroshima and the Nagasaki tumor registries became available for the first time in the 1990s. These data not only include nonfatal cancers, but also offer diagnostic information that is of higher quality that that based on death certificates, which is especially important for evaluating site-specific cancers. Although published evaluations described in Chapter 6 are based on DS86 dosimetry, a revised DS02 system—the result of a major international effort to reassess and improve survivor dose estimates—has recently become available and was used to develop BEIR VII risk models. An initial evaluation indicates that this revision will slightly reduce risk estimates.

The more extensive data on solid cancer that are now available have allowed more detailed evaluation of several issues pertinent to radiation risk assessment. Several investigators have evaluated the shape of the dose-response, focusing on the large number of survivors with relatively low doses. These analyses have generally confirmed the appropriateness of linear functions to describe the data. The modifying effects of sex, age at exposure, and attained age have also been explored in detail using both ERR and EAR models. The ERR/Sv has been found to decrease with both increasing age at exposure and increasing attained age, and it now appears that both variables may be necessary to provide an adequate description of the data. By contrast, the EAR shows a sharp increase with increasing attained age and a decrease with increasing age at exposure.

The availability of high-quality cancer incidence data has resulted in several analyses and publications addressing specific cancer sites. These analyses often include special pathological review of the cases and sometimes include data on additional variables (*e.g.*, smoking for evaluation of lung cancer risks). Papers focusing on the following cancer sites have been published in the last decade: female breast cancer, thyroid cancer, salivary gland cancer, liver cancer, lung cancer, skin cancer, and central nervous system tumors. Special analyses have also been conducted of cancer mortality in survivors who were exposed either *in utero* or during the first 5 years of life.

Health end points other than cancer have been linked to radiation exposure in the LSS cohort. Of particular note, a dose-response relationship with mortality from nonneoplastic disease was demonstrated in 1992, and subsequent analyses in 1999 and 2003 have strengthened the evidence for this association. Statistically significant associations were seen for the categories of heart disease, stroke, and diseases of the digestive, respiratory, and hematopoietic systems. The data were inadequate to distinguish between a linear dose-response, a pure quadratic response, or a dose-response with a threshold as high as 0.5 Sv.

7

Medical Radiation Studies

INTRODUCTION

Studies of patients irradiated for the treatment or diagnosis of diseases have provided considerable information for the understanding of radiation risks, particularly for specific cancer types, including thyroid and breast cancer (IARC 2000; UNSCEAR 2000b). Today, approximately 50% of cancer patients are treated using radiation (Ron 1998), and several million cancer survivors are alive in the United States, emphasizing the importance of investigating the long-term consequences of radiotherapy and examining the features of epidemiologic studies of medical radiation.

Large cohorts of radiation-treated patients who have been followed for long periods are available, allowing evaluation of cancer and other late effects. Population-based cancer registries in many countries have been used to identify these patients and to facilitate patient enrollment, thus allowing investigators to determine the risks of a second primary cancer after treatment with radiation for a primary cancer (Boice and others 1985). The characteristically detailed radiotherapy records for cancer patients and patients treated for nonmalignant conditions allow precise quantification of the doses to the organs of individuals, which in turn facilitates the evaluation of dose-response relationships. Frequently, patients with the same initial condition that receive treatments other than radiation are available for comparison, although the clinical indications for treatment may differ.

In most cases, patients received high doses of radiation on the order of 40–60 Gy to the targeted region, aimed at producing cell killing. These "high" doses would decrease with distance from the target tissue, and some tissues might receive doses that are referred to in this report as "low dose" (100 mGy or less). The use of such studies to estimate the effect of low-dose exposures raises a number of questions. The exposures were generally only partial-body exposures in persons who were ill, possibly resulting in a different risk than an equivalent whole-body uniform exposure. Because of their disease, patients may have a different sensitivity to radiation-induced disease than persons who do not have the disease. However, these studies are valuable and will likely become more important in the next decade, both for radiation protection of patients and for radiation protection in general because they provide a unique opportunity to address the following issues:

- Effects of different radiation types
- Risk of specific tumor types
- Effects of potential risk modifiers, including sex, age, and exposure fractionation
- Possible genetic susceptibility to radiation-induced cancer

In addition to studies of cancer survivors, long-term studies of patients who received radiation therapy for benign conditions such as enlarged tonsils and tinea capitis have also provided important information about radiation-induced cancer risk (UNSCEAR 2000b). These patients are particularly important in the evaluation of radiation risks in the absence of the possibly confounding effects of the malignant disease being treated and/or of concomitant therapy for cancer. Diagnostic radiation procedures, in contrast, generally result in small doses to target organs, and most studies of such exposure provide little information about radiation risks. A number of procedures, however, in particular repeated examinations of air collapse therapy for tuberculosis and of spine curvature for scoliosis, have resulted in sizable doses to specific tissues, and studies of patients who have undergone these examinations provide valuable information on radiation risks (UNSCEAR 2000b). It is noted that, although no informative studies are available, the recent use of computed tomography (CT) can deliver sizable doses, typically of the order of tens of millisieverts per examination (Brenner and Elliston 2004); UNSCEAR (2000b) reports cumulative doses of the order of 100 mSv for children.

As in the other review chapters in this report, studies were judged to be informative for the purpose of radiation risk estimation if (1) the study design was adequate (see Chapter 5 concerning informative study designs and limitations); (2) individual quantitative estimates of radiation dose to the organ of interest were available for the study subjects; (3) if so, the details of the dose reconstruction approach were evaluated, and (4) a quantitative estimate of disease risk in relation to radiation dose—in the form of an estimated relative risk (ERR) or excess absolute risk (EAR) per gray—was provided.

Overall, more than 100 studies of patients receiving diagnostic or therapeutic radiation have evaluated the association between exposure to radiation and risk of cancer at multiple sites (IARC 2000; UNSCEAR 2000b). Studies that provide information about the size of radiation cancer risks are reviewed in detail in this chapter. Articles included in this chapter were identified principally from searching the PubMed database of published articles from 1990 through December 2004. Searches were restricted to human studies and were broadly defined: key words included radiation; neoplasms; cancers; radiation-induced; medical exposures; radiotherapy; diagnostic radiation; and iodine-131. Articles were also identified from UNSCEAR (2000b), from the references cited in papers reviewed, and from direct contacts with some of the main scientists who have been involved with studies of medical exposures in recent years. The data and confidence intervals are those given in the cited papers.

MEDICAL USES OF RADIATION

Medical use of radiation usually occurs under three circumstances: (1) treatment of benign disease, (2) diagnostic examination, and (3) treatment of malignant disease (Table 7-1). Diagnostic imaging using X-rays goes back to the time of Roentgen's discovery in 1896. Diagnostic procedures, particularly the widespread use of X-rays, continue to be the most common application of radiation in medicine, even as non-ionizing radiation methods—ultrasound and magnetic resonance imaging—have become more generally accepted. Approximately 400 million diagnostic medical examinations and 150 million dental X-ray examinations are performed annually in the United States (Mettler and others 1996). On average, each person receives at least two examinations per year. The annual individual and collective effective doses from diagnostic medical X-rays have been estimated as 0.5 mSv and 130,000 person-Sv (UNSCEAR 2000b).

The range of X-ray techniques used includes radiography, fluoroscopy, CT, interventional radiology, and bone densitometry. These procedures are intended to provide diagnostic information and in principle are conducted with the lowest practicable levels of patient dose to meet clinical objectives. Ranges of typical doses from various medical diagnostic exposures are shown in Table 7-1.

TABLE 7-1 Estimated Range of Effective Doses from Diagnostic Radiation Exposures

Procedure	Type of Examination	Range of Doses
Conventional simple X-rays	Chest films X-rays of bones and skull X-ray of abdomen	0.02–10 mGy
Conventional complex X-rays	GI series Barium enema Intravenous urogram	3–10 mGy
Computed tomography (CT)	Head injuries Whole-body examinations	5–15 mGy
Spiral CT	Head injuries Whole-body examinations	10–20 mGy
Angiography	Coronary, aortic, peripheral, carotid, abdominal	10–200 mGy
Interventional procedures	Angioplasties with stent placement Percutaneous dilatations, closures, biopsy procedures	10–300 mGy
Internal emitters	Radioisotope studies	3–14 mSv

Although doses of single procedures are typically low, there is concern that populations of pediatric patients who may need repeated exams over time to evaluate their pulmonary, cardiac, urinary, or orthopedic conditions may receive relatively high cumulative doses. Similarly, adult patients may also require repeated examinations to evaluate fracture healing, or progression of pulmonary disease, or the regression or progression of neoplastic lesions.

In contrast, therapeutic exposures are less frequent, and the dose levels are higher in view of the different purpose. Currently, radiotherapy is used mainly for the treatment of cancer, where the intention is to deliver a lethal dose to malignant tissue within a well-defined target volume, while minimizing the irradiation of surrounding healthy tissue. In the past, high doses of radiation have also been used for the treatment of a number of benign conditions, such as enlarged thymus and ringworm of the scalp (tinea capitis). Doses from radiotherapy to the target organs are generally above 1 Gy (and typically in the range of 50–60 Gy for the treatment of malignant diseases). Radiotherapy involves mainly partial-body irradiation, however; hence very different doses are delivered to different organs or tissues of the body. Doses to distant organs are generally considerably lower (of the order of fractions of a gray), and studies of cancer risk in these organs are therefore potentially informative for the assessment of risks associated with low-level exposure. Further, many of the patients treated with radiotherapy received fractionated doses, and studies of these patients provide the potential to study the effects of exposure fractionation and protraction.

Radiotherapy for Malignant Disease

Studies of second cancer following radiotherapy have generally focused on patients treated for cervical cancer, breast cancer, Hodgkin's disease (HD), and childhood cancers (*i.e.*, patients that generally have a favorable long-term prognosis). Survivors of these cancers may live long enough to develop a second, treatment-related malignancy. It should be noted that chemotherapy and/or hormonal therapy used in the treatment of cancers is a potential confounding factor in investigations of the risk of a second primary cancer.

Cervical Cancer

The treatment of cervical cancer involves external beam radiotherapy or radium or cesium in applicators to deliver high local doses of X-rays and gamma rays to the cervix uteri and adjacent organs in the abdomen and pelvic area. Treatment is usually successful, and patients survive for years after radiotherapy. Although doses to the cervix are very high (typically 40–150 Gy), doses to distant organs are significantly lower: of the order of 0.1 Gy to the thyroid, 0.3 Gy to the breast and the lung, 2 Gy to the stomach, and 7 Gy to the active bone marrow (Kleinerman and others 1995).

Most of the information on second cancers following radiotherapy for cervical cancer comes from an international cohort study of approximately 200,000 women treated for cervical cancer. The study involved the follow-up, based on 15 cancer registries in eight countries (Canada, Denmark, Finland, Norway, Sweden, the United Kingdom, the United States, and Yugoslavia [Slovenia]), of a multinational cohort of nearly 200,000 women patients treated for cancer of the cervix after 1960. In 1985, Boice and colleagues reported on 5146 second cancers that were diagnosed in this cohort up to 1980 and showed an increased risk of cancer following radiotherapy at a number of sites (Boice and others 1985). Kleinerman and coworkers (1995) extended the follow-up of this cohort, adding an additional 10 years of incident cases. Several registries from the original study were retained, and other registries were added to increase the number of nonexposed comparison subjects. A total of 7543 cases were included. This study confirmed earlier findings of increased risk of malignancies following radiotherapy and the persistence of increased risk over time.

Case-control studies of specific cancer types, nested within this cohort, allowed the reconstruction of individual doses to specific organs and the estimation of site-specific cancer risks (Boice and others 1987, 1988, 1989). These studies are based on incidence data; the numbers of exposed and unexposed patients were large; there was long and com-

plete follow-up (hundreds of cases and controls, with follow-up of 10–20 years or more); chemotherapy was rarely used; and the existence of radiotherapy records facilitated the development of a comprehensive dose reconstruction system to estimate individual doses.

In an expanded case-control study nested within this international cohort (Boice and others 1988), radiation doses for selected organs were reconstructed from original radiotherapy records. Very high doses, of the order of several hundred grays delivered to the cervix, significantly increased the risks for cancers of the bladder, rectum, and vagina and possibly bone, uterine corpus, cecum, and non-Hodgkin's lymphoma (NHL). Doses of several grays increased the risks for stomach cancer and for leukemia. The ERR[1] for stomach cancer was 0.54 Gy^{-1} (90% CI 0.05, 1.5), with an excess attributable risk of 3.16 per 10^4 person-years (PY) per gray (0.05, 10.4), based on 348 cases and 658 controls. A nonsignificant twofold increase in the risk of thyroid cancer was observed, with an average dose of 0.11 Gy (43 cases and 81 controls).

More detailed dose-response investigations were carried out for leukemia and breast cancer after treatment for cervical cancer. The case-control study of leukemia risk (Boice and others 1987) included 195 cases and 745 controls, of whom 181 and 672, respectively, had received radiotherapy. Radiation dose to the active bone marrow was estimated from detailed radiotherapy records of the subjects. Radiation exposure did not affect the risk of chronic lymphocytic leukemia (CLL; 52 cases). For other forms of leukemia taken together (143 cases), there was a significant twofold increase in risk associated with radiotherapy; the risk increased with increasing dose up to about 4 Gy and then decreased at higher doses and was modeled adequately by a linear-exponential function. The linear term of this model for leukemia other than CLL provides an estimate of the ERR per gray in the low-dose range, where cell killing is negligible; this estimate is 0.88 Gy^{-1} (standard error = 0.69).

The case-control study of breast cancer included 953 cases and 1806 controls (Boice and others 1989). Radiation doses to the breast (average 0.31 Gy) and ovaries (average 32 Gy) were reconstructed from original radiotherapy records. Overall, there was no association between radiotherapy and risk of breast cancer. Among women with intact ovaries (561 cases), radiotherapy was associated with a significant reduction of risk, probably attributable to cessation of ovarian function. Among women with no ovaries, there was a slight increase in breast cancer risk and a suggestion of a dose-response with a relative risk (RR)[2] of 1.0, 0.7, 1.5, and 3.1, respectively, for the dose groups 0, 0.01–0.24, 0.25–0.49, and 0.5 + Gy. From these data, UNSCEAR (2000b)

---

[1]ERR is the rate of disease in an exposed population divided by the rate of disease in an unexposed population minus 1.0.

[2]RR is the rate of disease in an exposed population divided by the rate of disease in an unexposed population.

estimated an ERR per gray of 0.33 (< –0.2, 5.8) for women with no ovaries and of –0.2 (< –0.2, 0.3) overall.

A cohort study of second cancer risk following radiation therapy for cancer of the uterine cervix was also carried out in Japan among 11,855 patients (Arai and others 1991). Significant excesses of leukemia and of cancers of the rectum, bladder, and lung were observed. No estimation of organ dose is available.

Hodgkin's Disease

The large radiation therapy fields used to treat HD and the young age and long survival of patients provide an opportunity to study the risk of second cancer after exposure to ionizing radiation. Most patients, however, in the past 20 years, have been treated with a combination of radiotherapy and chemotherapy.

Following a first report by Arseneau and collaborators (1972), a number of authors have studied the risk of second cancer following treatment for HD (Boivin and others 1984). Initial reports focused mainly on the risk of leukemia following this treatment, but as longer follow-up periods were considered, an excess risk of a number of solid cancers (in particular breast and lung) became apparent.

The results of the first multinational study were published in 1987 by Kaldor and collaborators. The study involved the follow-up (based on 11 cancer registries in seven countries: Denmark, Finland, Norway, Sweden, Slovenia, Canada, and the United Kingdom) of a cohort of 28,462 patients treated for HD between 1950 (in the earliest countries) and 1984. Increases in the risk of NHL, leukemia, lung, bladder, and breast cancer were reported in this cohort. No treatment information was available in this study, and no information is provided on radiation risks. Nested case-control studies of leukemia and of lung cancer were carried out, allowing reconstruction of individual doses for the subjects and estimation of site-specific cancer risks (Kaldor and others 1990a, 1992).

The case-control study of leukemia included 163 cases and 455 controls. Radiation dose to the active bone marrow was estimated for subjects who had undergone radiotherapy, and doses were classified into three categories (<10, 10–20, and 20+ Gy). Among patients who did not receive chemotherapy, a significant increase in the risk of leukemia was seen at doses of more than 20 Gy (Kaldor and others 1990a).

Another case-control study from the same collaborative group involved 98 cases of lung cancer occurring between 1960 and 1987 and 259 matched controls (Kaldor and others 1992). Radiation dose to the lung as a whole was estimated for the 60 cases and 275 controls who had undergone radiotherapy, and doses were classified into three categories (<1, 1–2.5, and 2.5+ Gy). Among patients treated with radiotherapy alone, there was a nonsignificant increase in risk in relation to radiation dose level. It is noted that the follow-up was short in this study, with three-quarters of the lung cancer cases having been diagnosed within 10 years of their initial disease (Kaldor and others 1992).

In 1995, Boivin and collaborators published results of a joint Canada-U.S. study of second cancer risk among 10,472 patients treated for HD between 1940 and 1987. A total of 122 leukemia and 438 solid tumors were found, and nested case-control studies were carried out. Significant increases in the risk of cancers of the respiratory system, intrathoracic organs, and female genital system were observed among patients followed for 10 years or more after surgery. Estimates of organ doses were not available, and analyses by level of radiation dose are not shown.

Van Leeuwen and collaborators (1995) conducted a case-control study of lung cancer nested in a cohort of 1939 patients treated for HD between 1966 and 1986 in the Netherlands. Radiation dose to the parenchyma, bronchi, and trachea were estimated for patients who had received radiotherapy (30 cases and 82 controls). A statistically significant increase in the risk of lung cancer was observed, with an RR of 9.6 (95% CI 0.93, 98.0) for patients who had received 9 Gy or more compared to patients who had received less than 1 Gy. The increase was greater among those who either continued smoking or started smoking after diagnosis, and a multiplicative interaction was observed between radiation dose and tobacco smoking.

Swerdlow and collaborators (2001) carried out a nested case-control study of lung cancer in a cohort of 5519 patients with HD treated in Britain between 1963 and 1993. The study included 88 cases and 176 controls for whom treatment and other risk factor information was abstracted from medical records. An increased risk of lung cancer following radiotherapy was observed. No individual reconstruction of dose to the lung was carried out.

Travis and colleagues (2002) carried out a case-control study of lung cancer nested within a multinational cohort of 19,046 HD patients diagnosed between 1965 and 1994 and reported to population-based cancer registries in Connecticut, Iowa, Denmark, Finland, the Netherlands, Sweden, and Canada (Ontario). The study included 222 cases and 444 matched controls. Nineteen of the cases were included in the previous case-control study by Kaldor and coworkers (1992). Dose to the specific location of the lung where the tumor had developed (and to a comparable location for matched controls) was calculated from radiotherapy records. The mean dose was 27.2 Gy in cases and 21.8 Gy in controls. In subjects who had not undergone chemotherapy, a significantly increased risk of lung cancer was observed (odds ratio [OR][3] 5.9; 95% CI 2.7, 13.5) for a dose of 5 Gy or more. A significant trend in risk was observed with increasing dose.

In a follow-up to this study, Gilbert and colleagues (2003) analyzed radiation effects among 227 lung cancer cases and

---

[3]OR is the odds of being exposed among diseased persons divided by the odds of being exposed among nondiseased persons.

455 controls (the 199 cases and 393 controls from the Travis 2002 study who had adequate radiation dose information and 28 cases and 62 controls from the Dutch study of van Leeuwen and others 1995). Doses to the lung ranged from zero to more than 60 Gy; the distribution of doses was bimodal, with most subjects having received doses of less than 5 or more than 30 Gy. To account for a possible latent period between radiation exposure and lung cancer resulting from that radiation exposure, only doses received more than 5 years in the past were considered. Among the 146 cases and 271 controls who had received radiotherapy more than 5 years in the past, a significant association was seen between radiation dose and risk of lung cancer, with an ERR per gray of 0.15 (95% CI 0.06, 0.39). There was little evidence for nonlinearity of the dose-response, despite the fact that the majority of patients received doses to the lung in excess of 30 Gy. Information about smoking and radiotherapy was available for the study subjects. A multiplicative interaction was seen between radiation dose and tobacco smoking and an additive interaction with chemotherapy. The ERRs for men and women were respectively 0.18 (0.063, 0.52) and 0.044 (–0.009, 0.53); the difference between the sexes was not statistically significant.

Breast cancer following treatment for Hodgkin's disease has also been studied in a number of cohorts. Travis and collegues (2003) carried out a nested case-control study of breast cancer in a cohort of 3817 one-year survivors of HD diagnosed at age 30 years or younger between 1965 and 1994 and included in cancer registries in Iowa, Denmark, Finland, Sweden, the Netherlands, and Ontario, Canada. Individual doses to the area of the breast from which the tumor arose were reconstructed using detailed radiotherapy records and results of experiments with phantoms. Mean dose delivered to the location of the breast where cancer developed was 25.1 Gy (range: 12.0–61.3 Gy) in cases and 21.1 Gy (range 0–56.0 Gy) in controls. The study included 105 cases and 266 controls. A significant increase in the risk of breast cancer was seen following doses of 4 Gy or more (OR 3.2; 95% CI 1.4, 8.2); the increase remained significant even following very high doses (OR 8, 95%; CI 1.6, 26.4, of 40 Gy or more). No significant association between age at exposure or reproductive history was seen in this study, but the risk was lowered among women who received 5 Gy or more to the ovaries or who were also treated with alkylating agents. The estimated ERR per gray for women who did not receive alkylating agent chemotherapy or high radiation doses to their ovaries was 0.15 (95% CI 0.04, 0.74).

Van Leeuwen and colleagues (2003) also studied the risk of breast cancer among female survivors of HD treated in the Netherlands. The study included 48 cases who developed breast cancer 5 years or more after HD diagnosis and 175 matched controls. It should be noted that 40 of the 48 cases in the study of van Leeuwen and colleagues were also included in the study by Travis and coworkers (2003). The object of the study was to evaluate the joint roles of radiation

dose, chemotherapy, and hormonal factors in breast cancer following HD. As in the Travis study, the risk of breast cancer increased with radiation dose up to at least 40 Gy. A substantial risk reduction was associated with chemotherapy, which affects menopausal age, suggesting that ovarian hormones promote tumorigenesis after radiation-induced initiation. No estimate of ERR or EAR[4] per gray is given. Little if any increased risk was seen for patients treated after age 30.

Most recently, Dores and colleagues (2002) studied the risk of second cancers in general among 32,581 HD patients (including 1111 25-year survivors of HD) registered in 16 population-based cancer registries in North America and Europe. A total of 2153 second cancers were observed between 1935 and 1994. As before, significant increases in the risk of a number of second malignancies were observed. Although the elevated risks of cancers of the stomach, breast, and uterine cervix appeared to persist for 25 years or more, an apparent decrease in the risk of other solid tumors is suggested. These cohort studies, although they provide important information concerning treatment-related second neoplasms and their patterns of risk over time, do not provide quantitative information on the risk of radiation-induced cancer because of the absence of individual dose estimates.

The risks of breast, thyroid, and gastrointestinal cancers were also investigated in patients treated for HD at Stanford University Medical Center (Hancock and others 1991, 1993a; Birdwell and others 1997). Increases in these diseases were observed, but no dose estimates were available. Hancock and colleagues (1993b) also investigated mortality from heart disease following treatment for HD in a cohort of 2232 patients treated from 1960 to 1991 with an average follow-up of 9.5 years. The RR for mortality due to heart diseases was 3.5 (95% CI 2.7, 4.3) among those who received mediastinal radiation doses of more than 30 Gy. The increased risk was highest for exposures that occurred before the age of 20 and increased with time since exposure. No increased risk was observed among subjects who received doses lower than 30 Gy. In a separate study, Heidenreich and coworkers (2003) found a high prevalence of asymptomatic heart disease—specifically aortic valvular disease—following mediastinal irradiation.

Breast Cancer

Leukemia, lung cancer, soft tissue sarcoma, and contralateral breast cancer have been studied in patients receiving radiotherapy for breast cancer.

A case-control study of leukemia (excluding CLL) was carried out nested within a cohort of 82,700 women with breast cancer in the United States. A total of 90 cases and 264 controls were included with individual estimates of dose

---

[4]EAR is the rate of disease in an exposed population minus the rate of disease in an unexposed population.

to the active bone marrow. A significant dose-response was seen for acute nonlymphocytic leukemia after adjustment for the amount of chemotherapy, with an RR of 2.4 among those who received radiotherapy alone (Curtis and others 1992). No information was provided on the magnitude of the risk per gray or on the risk of other forms of leukemia.

A case-control study of contralateral breast cancer was carried out nested within a cohort of 41,109 women in Connecticut diagnosed with breast cancer between 1935 and 1982. A total of 655 cases and 1189 controls were included. The average dose to the contralateral breast was 2.8 Gy. A significant increased risk was seen only among women who received radiotherapy before age 45 (RR 1.59; 95% CI 1.07, 2.36, based on 78 exposed cases); a significant dose-response was observed in this group (Boice and others 1992).

No excess risk of contralateral breast cancer was seen in a cohort of 14,000 women treated between 1946 and 1982 in Denmark (Basco and others 1985). The study included 194 cases with individual dose estimates (mean doses ranging from 1.4 to 3.3 depending on the type of radiotherapy and the field considered). The RR per 100 cGy was 0.99 (95% CI 0.76, 1.30); little difference was seen for those diagnosed 5–10 years or more after their first tumor.

A case-control study of contralateral breast cancer was conducted among women with primary breast cancer entered in the Danish Cancer Registry from 1943 to 1978 (Storm and others 1992). A total of 529 cases and 529 controls were included, and individual doses to the contralateral breast were estimated from detailed radiotherapy records for all subjects who had received radiotherapy. The mean dose to the contralateral breast was estimated to be 2.5 Gy. There was no significantly increased risk of contralateral breast cancer in this study (RR = 1.04; 95% CI 0.74, 1.46).

A case-control study of lung cancer was conducted based on the Connecticut Tumor Registry (Inskip and others 1994). The study included 61 cases of lung cancer and 120 controls. Cases were diagnosed between 1945 and 1981 among women who had been treated for invasive breast cancer between 1935 and 1971 and survived at least 10 years. Individual radiation dose to different segments of the lung was estimated from detailed radiotherapy records. Average dose to the lung was 15.2 Gy to the ipsilateral lung and 4.6 Gy to the contralateral lung. Patients who received radiotherapy had a 1.8 times higher risk of developing lung cancer than those who did not (95% CI 0.8, 3.8). The risk increased with time since exposure and appeared to be higher among women exposed under age of 45, although this was not significant. The risk was highest for the ipsilateral lung. The ERR was estimated to be 0.2 Gy^{-1} to the affected lung (95% CI –0.62, 1.03), based on 15 exposed cases.

A nested case-control study of second malignant neoplasms was carried out for a cohort of 7771 women initially treated for breast cancer between 1954 and 1983 at the Institut Gustave Roussy near Paris, France (Rubino and others 2003). Individual doses to the location of the second tumor were estimated from detailed radiotherapy records. More than 40% of the irradiated patients received a local dose of less than 1 Gy. A significant quadratic dose-response was found in this study, with an excess risk of all second malignant neoplasms combined of 0.2% (95% CI 0.05, 0.5%) at 1 Gy.

Darby and coworkers (2003) studied cardiovascular mortality in a cohort of 89,407 Swedish women identified from the Swedish cancer registry as having had unilateral breast cancer at the ages of 18 to 79 years between 1970 and 1996. Mortality from cardiovascular disease was higher in women who had left-sided tumors (OR 1.10; 95% CI 1.03, 1.18) 10 years or more after the diagnosis of breast cancer; for ischemic heart disease, the OR was 1.13 (95% CI 1.03, 1.25). No dose estimates were available in this record linkage study, but the fact that the increase was restricted to women with tumors in the left breast and that no increase in mortality from other causes (except breast cancer) was seen in this population lends plausibility to the hypothesis of a radiation effect on the risk of heart disease.

Ovarian Cancer

A case-control study of leukemia within an international cohort of 99,113 survivors of ovarian cancer showed no significant excess risk for leukemia associated with radiotherapy alone (Kaldor and others 1990b). A more recent case-control study was carried out, nested within an international cohort of 28,971 patients in whom ovarian cancer was diagnosed between 1980 and 1993 (Travis and others 1999). The study included 96 leukemia cases and 272 controls. Individual dose to the active bone marrow was estimated for the 26 cases and 79 controls who had received radiotherapy. The median dose to the bone marrow was 18.4 Gy. Radiotherapy increased the risk of leukemia following platinum-based chemotherapy. No increased risk of leukemia was observed in subjects who had radiotherapy alone; the data are sparse: one exposed case and 36 exposed controls.

Testicular Cancer

Travis and colleages (1997) studied second cancer incidence in a multinational cohort of 28,843 men who had been diagnosed with testicular cancer between 1935 and 1993 in the United States, Denmark, Finland, the Netherlands, Sweden, and Canada (Ontario). Cases of second cancers occurring between 1965 and 1994 in this cohort were identified from population-based cancer registries in these countries. Significantly increased risks of second cancers in general, as well as of leukemia (64 cases) and stomach cancer (93 cases), were observed among patients who had received radiotherapy 5 years or more in the past. No individual doses were available.

Travis and colleagues (2000) conducted a case-control study of leukemia nested within a multinational cohort of 18,567 patients diagnosed with testicular cancer between

1970 and 1993 and registered in cancer registries in Iowa, Connecticut, New Jersey, Canada (Ontario), Denmark, Finland, the Netherlands, and Sweden. The study included 36 cases and 106 matched controls. Individual radiation dose to the active bone marrow was estimated from detailed radiotherapy records. In men who did not receive chemotherapy (mean radiation dose to 12.6 Gy), a 3.1-fold elevation of leukemia risk was observed (95% CI 0.7, 22). The risk increased with radiation dose to the active bone marrow, with an OR of 19.7 (95% CI 1.5, 59) for doses of 20 Gy or more (based on four exposed cases). No estimate of ERR or EAR per gray is given.

Thyroid Cancer

A cohort of 834 thyroid cancer patients treated with iodine-131 and of 1121 thyroid cancer patients treated by other means in Sweden between 1950 and 1975 was followed for cancer occurrence (Hall and others 1991). The average ^{131}I cumulative activity administered was 4.55 GBq. The average duration of follow-up was 14 years. A total of 99 second cancers were found 2 years or more after ^{131}I therapy among those treated with this modality and 122 among those treated by other means. The incidence of second malignancy was higher among those treated with [131]I. Among women, the overall standardized incidence ratio (SIR)[5] was 1.45 (95% CI 1.14, 1.83), and significantly elevated SIRs were found for tumors of the salivary glands, genital organs, kidney, and adrenal gland. A significant trend was seen with increasing ^{131}I activity, with a SIR of 1.80 (95% CI 1.20, 2.58) for administered activities of 3.66 GBq and above.

A cohort of 1771 patients treated with ^{131}I for thyroid cancer was followed up for incidence of second cancers (de Vathaire and others 1997). The average ^{131}I cumulative activity administered was 7.2 GBq, resulting in an estimated average dose of 0.34 Sv to the bone marrow and 0.80 Sv to the whole body. After a mean follow-up of 10 years, no case of leukemia was seen. Eighty patients developed a secondary solid cancer, including 13 colorectal cancers. The risk of colorectal cancer was related to the total activity administered (ERR = 0.47 GBq^{-1}; 95% CI 0.1, 1.6). The overall ERR for solid tumors in this study was 0.38 per estimated effective sievert (95% CI –0.22, 1.2); when tumors of the digestive track were excluded, the ERR was reduced to –0.15 Sv^{-1} (95% CI –0.35, 0.22).

Childhood Cancers

The treatment for childhood cancers, often a combination of both radiotherapy and chemotherapy, has prolonged the

life expectancy of children with cancer and increased the chance of development of second cancers. Since childhood cancer is rare, national and international groups such as the Late Effects Study Group (Tucker and others 1987a, 1987b, 1991) and several groups in the United Kingdom (Hawkins and others 1987) and France (de Vathaire and others 1989, 1999) have combined their data to evaluate risks. Results from these cohort studies have indicated that the risk for developing a second cancer in the 25 years after the diagnosis of the first cancer was as high as 12% (Tucker and others 1991). Further, genetic predisposition appears to have a substantial impact on risk of subsequent cancers. Among patients treated for hereditary retinoblastoma, the risk of developing a second cancer in the 50 years after the initial diagnosis was as high as 51% (Wong and others 1997b).

Three nested case-control studies including 64 cases of bone cancer and 209 controls (Tucker and others 1987a), 23 cases of thyroid cancer and 89 controls (Tucker and others 1991), and 25 cases of leukemia and 90 controls (Tucker and others 1987b) were conducted from the Late Effects Study Group cohort of 9170 children who developed a second malignant tumor at least 2 years after diagnosis of the first tumor. A significant increased risk of bone cancer was found among patients who received radiation therapy (RR 2.7; 95% CI 1.0, 7.7), with a sharp dose-response gradient reaching a fortyfold risk following doses to the bone of more than 60 Gy. A significant increased risk of thyroid cancer was also found among patients who had received radiation therapy; most of the increase was among those who had received doses of 2 Gy or more. There was no evidence of a dose-response relationship for leukemia.

In a U.K. cohort of 10,106 3-year survivors of childhood cancer, Hawkins and colleages (1987; Hawkins 1990) reported an excess of second tumors among subjects who had received radiotherapy in comparison with the general population. In addition, two nested case-control studies of 59 cases of second bone cancer and 220 controls (Hawkins and others 1996) and 26 cases of second leukemia and 96 controls (Hawkins and others 1992) were conducted within this cohort, with individual dose reconstruction to the organs of interest. The risk of bone cancer increased substantially with increased cumulative radiation dose to the bone ($p < .001$), although a decline in risk was seen at doses equal to or greater than 50 Gy. A nonsignificant increased risk of leukemia was observed among those who had received radiotherapy (RR 8.4; 95% CI 0.9, 81.0 based on seven exposed cases). A significant dose-response relationship was observed.

In a cohort study of 634 children treated for childhood cancer from 1942 and 1969 in the Institut Gustave Roussy in Paris, a twofold increase in the risk of second malignancy was seen after doses from radiotherapy of more than 25 Gy, based on two exposed cases (de Vathaire and others 1989). A nonsignificant dose-response was seen based on 13 cases who had received radiotherapy alone.

---

[5]SIR is the ratio of the incidence rate of a disease in the population being studied divided by the comparable rate in a standard population. The ratio is similar to an RR times 100.

In a French-British cohort study (de Vathaire and others 1999) that overlapped partially with the Late Effects Study, the French study, and British studies, described above, an excess of second cancers was seen among 1045 children who received radiotherapy alone (based on 31 second malignant neoplasms, including 8 brain cancers). Fourteen cases of thyroid carcinoma were identified in the entire cohort of 4096 3-year survivors of childhood cancers. All 14 had received radiotherapy. The average dose to the thyroid in this study was 7 Gy. A significant dose-response was observed for thyroid cancer in this study—RRs were 4.0 (90% CI 0.7, 44), 11.0 (90% CI 2.3, 123), 13.0 (90% CI 2.2, 141), and 26.0 (90% CI 3.4, 308) for doses within the ranges of 0.25 to <1 Gy (3 cases), 1.0 to <10 Gy (5 cases), 10 to <30 Gy (3 cases), and 30+ Gy (2 cases), respectively.

In a joint analysis of data from childhood cancer survivor cohorts from France, Britain, and Nordic countries, a nested case-control study of melanoma was carried out. Radiotherapy appeared to increase the risk of melanoma for local doses greater than 15 Gy (OR 13.0; 95% CI 0.94, 174.0), based on three exposed cases; the ORs for doses less than 1 Gy and of 1–15 Gy were 1.4 (95% CI 0.28, 7.0) and 3.2 (95% CI 0.37, 27) based on very small numbers of exposed cases—five and two, respectively (Guerin and others 2003).

A partially nested case-control study of soft tissue sarcoma (STS) was carried within the French-U.K. cohort of 4400 3-year survivors of childhood cancer survivors; 25 cases and 121 controls were included. Individual dose to the site of STS development was calculated. A significant increase in the risk of STS was seen among those who received radiotherapy (OR 19.0; 95% CI 3.0, 60.0). The risk increased with the square of the radiation dose and was independent of chemotherapy (Menu-Branthomme and others 2004).

Other Cancers

The health effects of radiotherapy for a number of other cancer types have also been considered in single studies. Travis and colleagues (1991) studied the risk of second cancers among 29,153 patients diagnosed with NHL between 1973 and 1987 in nine areas of the United States. Radiation therapy appeared to increase the risk of acute nonlymphocytic leukemia and possibly of cancers of the lung, bladder, and bone. No estimate of radiation dose was available.

Curtis and coworkers (1994) studied the risk of leukemia following cancer of the uterine corpus in a cohort of 110,000 women assembled from nine population-based cancer registries in the United States, Canada, Denmark, Finland, and Norway. Radiation doses were computed to 17 sections of the active bone marrow for 218 women who developed leukemia and 775 matched controls. There was no association between radiation dose and risk of CLL (RR 0.90; 95% CI 0.4, 1.9). For all leukemia excluding CLL, however, the RR was 1.92 (95% CI 1.3, 2.9). There appeared to be no associa-

tion with age at exposure in this study. A complex dose-response was observed, with a relative risk of 1.8 (95% CI 1.1, 2.8) following continuous exposures from brachytherapy[6] at comparatively low doses and low dose rates (mean total dose 1.7 Gy). The risk was of the same order (RR 2.3; 95% CI 1.4, 3.7) after fractionated exposures at much higher doses and dose rates from external beam treatment (mean total dose 9.9 Gy).

Summary

Studies of second cancer following radiotherapy have generally focused on patients treated for malignant diseases with a favorable long-term prognosis, such as cervical cancer, breast cancer, HD, and childhood cancers. Because many survivors of these cancers live long enough to develop a second, treatment-related malignancy, these studies have provided valuable information on the magnitude of risk following radiation exposure. The cohort studies generally do not provide quantitative information on the risk of radiation-induced cancer because of the absence of individual dose estimates.

Case-control studies of specific cancer types have been carried out, nested within cohorts of cancer survivors. In allowing the reconstruction of individual doses to specific organs for the subjects, they have provided important information for the estimation of site-specific cancer, even if the average doses to the target organs have generally been high. Studies of patients treated for HD have provided quantitative estimates of the risk of cancers of the lung and breast—organs that generally received fairly high doses (of the order of 20 Gy on average) from the radiotherapy. Studies of patients treated for cancer of the cervix have provided estimates of the risk of breast cancer, leukemia, and stomach cancer (at average doses of 0.2, 7, and 2 Gy, respectively). Studies of women treated for a first breast cancer have provided quantitative estimates of the risk of lung cancer, at average doses of the order of 5–15 Gy. These estimates are reviewed in detail, and compared with risk estimates derived from other medical exposure studies, in the section "Evaluation of Risk for Specific Cancer Sites."

Radiotherapy for Benign Disease Among Adults

In the past, radiotherapy has been used in different countries for the treatment of a number of benign conditions in children (skin hemangioma, tinea capitis, enlarged thymus) and adults (*e.g.*, benign breast and gynecological disease, ankylosing spondylitis, peptic ulcer). Studies of patients treated with radiation (X-rays and gamma rays) for benign disease provide valuable information about the carcinogenicity of low-LET (linear energy transfer) radiation. Doses

---

[6]Radiation therapy in which a radioactive material sealed in needles, or wires, or other small delivery devices is placed directly into or near a tumor.

used in the treatment of benign conditions were generally not as high as those used to treat malignant disease so that cell-killing effects do not predominate, survival after treatment is good because the conditions treated were generally not life-threatening, and there is minimal confounding from concomitant treatment.

Benign Breast Disease

A U.S. cohort of 601 women treated with radiotherapy for acute postpartum mastitis and 1239 women treated by other means between 1940 and 1957 was followed for 29 years. The average dose to the breasts was 3.8 Gy. A significant increase in the risk of breast cancer was seen among women who had received radiotherapy, based on 51 exposed breasts with cancer. Using a linear multiplicative model the risk increased by 0.4% per rad (ERR per Gy 0.4; 90% CI 0.2, 0.7). A dose-response curve that appeared to be essentially linear up to about 7 Gy was demonstrated, and an increased risk for breast cancer was observed based on 56 cases (Shore and others 1986).

A Swedish cohort of 1216 women treated for benign breast disease with radiotherapy and 1874 women treated by other means from 1925 to 1954 was followed for an average of 27 years for development of a subsequent cancer. Mean absorbed doses to the breast were determined from detailed radiotherapy records and experiments with phantoms. The average dose to the breast was 5.8 Gy (range 0–50): 278 cases of breast cancer were diagnosed; 183 of these cases had received radiotherapy. A significant linear dose-response relationship was seen, with a downturn at approximately 10 Gy and higher. The estimated ERR for breast cancer was 1.63 Gy^{-1} (95% CI 0.77, 2.89, based on 47 exposed cases) among subjects with less than 3 Gy and 1.31 Gy^{-1} (95% CI 0.79, 2.04, based on 75 exposed cases) among subjects with less than 5 Gy (Mattsson and others 1995).

Mattsson and colleagues (1997) also studied the risk of malignancies other than breast cancer. Average doses were estimated to 14 organs. A significant increase in the risk of all cancers combined (excluding breast) was observed. A significant linear dose-response was seen for stomach cancer: ERR per Gy 1.3 (95% CI 0.0, 4.4), based on 14 exposed cases and a mean dose to the stomach of 0.66 Gy (range 0–5.4). No significantly increased risk was seen for any other cancer site, including leukemia, based on a small number of exposed cases (Mattsson and others 1997). The estimated ERR for lung cancer was 0.38 (95% CI <0, 0.6), based on 10 exposed cases and a mean lung dose of 0.75 (range 0–9.0).

Peptic Ulcer

Cancer mortality up to 1985 was studied in a U.S. cohort consisting of 1831 patients irradiated between 1937 and 1965 for the treatment of peptic ulcer and 1778 who were not (Griem and others 1994). An elevated risk of circulatory disease mortality was observed among those who received radiotherapy compared to those who did not. Overall, a 50% increase in the risk of all cancers combined was observed. Significant increases were seen for cancers of the stomach, pancreas, and lung; the average doses to the organs were estimated to be 15, 13, and 1.7 Gy, respectively. For stomach cancer, a threefold increase in risk was observed in this study; the RR at 1 Gy was estimated to be 1.15, and the absolute risk was 4.19 per 10^4 PY per gray. The estimated RR of lung cancer was 1.66 at 1 Gy.

In an updated follow-up of this cohort up until 1997 (average follow-up 25 years), Carr and colleagues (2002) also reported significant exposure-related increases in the risk of cancers of the stomach, pancreas, and lung among 1859 patients treated with radiotherapy. For stomach cancer, the ERR was estimated to be 0.20 Gy^{-1} (95% CI 0, 0.73), based on analyses restricted to subjects who had received doses to the stomach of 10 Gy or less (mean dose to the stomach 8.9 Gy; number of exposed cases 11). The corresponding estimate for cancer of the pancreas was 0.34 Gy^{-1} (95% CI 0.09, 0.89), with a mean dose of 8.2 Gy and 14 exposed cancer cases. For lung cancer, the ERR was estimated to be 0.43 Gy^{-1} (95% CI 0.12, 1.35) among subjects in the lowest-dose quartile (<1.4 Gy—mean dose 1.1 Gy), based on 21 deaths from lung cancer. Although the risk of pancreatic cancer decreased with increasing age at exposure, no association with age at exposure was observed for stomach and lung cancer.

Benign Gynecological Diseases

A U.S. cohort of 4153 women treated with intrauterine ^{226}Ra between 1925 and 1965 for uterine bleeding disorders was followed for an average of 27 years up to 1983 (Inskip and others 1990b). Individual organ doses were estimated based on detailed radiotherapy records and simulation of pelvic irradiation treatments on phantoms. A significantly increased standardized mortality ratio (SMR)[7] for death from all cancers was seen in this population compared to the general population. In addition, significant increases were observed for deaths from colon and uterine cancer, cancers of the female genital organs, and leukemia. Estimated ERR per Gy were 0.006 (90% CI –0.01, 0.05) for cancer of the uterus, 0.41 (90% CI –0.69, 1.51) for other genital organs, 0.51 (90% CI –0.08, 5.61) for colon cancer, and 0.20 (90% CI 0.08, 0.35) for bladder cancer.

Inskip and colleagues (1990a) studied the risk of leukemia in relation to radiation dose among 4483 of these women. Individual doses to various sections of the red bone marrow were calculated from detailed radiotherapy records. The median dose to red bone marrow was 0.53 Gy. A significant excess of leukemia was observed; the risk was highest 2–5 years after treatment (SMR 8.1; 95% CI 2.6, 18.8,

---

[7]SMR is the ratio of the mortality rate from a disease in the population being studied divided by the comparable rate in a standard population. Often the ratio is multiplied by 100.

compared to the general population) and among women over 55 years at irradiation (SMR 5.8; 95% CI 2.5, 11.3). The average ERR in this study was 0.19 Gy^{-1} (95% CI 0.08, 0.32) for intrauterine ^{226}Ra exposure, and the average absolute excess mortality from leukemia was 2.6 per 10^4 PY per gray.

The risk of leukemia, lymphoma, and multiple myeloma was studied in an expanded cohort of 12,955 women treated for benign gynecological disorders at one of 17 hospitals in Massachusetts, Connecticut, Rhode Island, or New York State between 1925 and 1965 (Inskip and others 1993). Of these women, 9770 were treated with radiation (either intracavitary ^{226}Ra or external beam X-rays), while the rest were treated by other methods. The average age at treatment was 46.5 years, and the average dose to active bone marrow in exposed women was 1.2 Gy. The RR for all cancers of hematopoietic and lymphatic tissue was 1.3 (95% CI 1.2, 1.5) for irradiated women, compared to nonirradiated. The risk of lymphomas, multiple myeloma, and nonacute lymphocytic leukemia was similar between irradiated and nonirradiated women. The RRs for acute lymphocytic leukemia and for myeloid leukemia were elevated, however: RR 3.7 (95% CI 1.3, 16) and 3.7 (95% CI 0.9, 36), respectively. For acute lymphocytic and nonlymphocytic leukemia, the SMRs were similar for women treated with radium only and with both modalities, and were smallest for X-ray patients (difference not statistically significant). The ERR per Gy was 0.29 overall; 0.37 Gy^{-1} (95% CI <0, 1.5) for treatment with radium only; 0.05 per Gy (95% CI –0.06, 0.33) for X-rays only; and ERR 0.21 per Gy (95% CI 0.05, 0.83) for the combination of both modalities. Average doses for the different treatment types were 0.6, 2.3, and 2.0 Gy, respectively, indicating a complex dose-response relationship.

A cohort of 2067 women who received radiotherapy for metropathia hemorrhagica (uterine bleeding disorders) in Scotland between 1940 and 1960 was followed until the end of 1990 (Darby and others 1994). The average follow-up was 28 years. Absorbed doses to the active bone marrow and to 20 solid organs or anatomical sites were estimated from treatment records. Overall, 331 deaths from cancer were observed, and significantly elevated SMRs were observed for cancers at heavily irradiated sites (average local dose >1 Gy): cancer of pelvic sites, particularly urinary bladder cancer (mean dose 5.2 Gy); colon cancer (mean dose 3.2 Gy); leukemia, and multiple myeloma (mean total active bone marrow dose 1.3 Gy). A deficit of breast cancer mortality was also observed in this cohort, due mainly to a large deficit in women who had received doses to the ovary of 5 Gy or more. No estimate of risk per unit dose was presented.

A Swedish cohort study included 2007 women treated for metropathia hemorrhagica between 1912 and 1977. Of these, 788 received radiotherapy for this condition. The population was followed up for cancer mortality and incidence from 1958 to 1982, with a mean follow-up period of 28 years (Ryberg and others 1990). A total of 107 cancers were ob-

served among irradiated women. The SIR for cancer was 1.22 among irradiated women and 1.09 among nonirradiated. A significant increase in the SIR for cancers at heavily irradiated sites in the pelvic area was only observed 30 years or more after irradiation. A decreased risk for breast cancer was also observed in this cohort, except for women treated at the age of 50 or more. No estimate of risk per unit dose was presented.

Hormonal Infertility

A U.S. cohort of 816 women who received X-ray therapy to the ovaries and/or pituitary gland for refractory hormonal infertility and amenorrhea between 1925 and 1961 was followed up until the end of 1990 (Ron and others 1994). The average duration of follow-up was 35 years. Individual organ doses were estimated from radiotherapy records. Average doses were 0.01 Gy to the breast, 0.9 Gy to the ovary, and 1.0 Gy to the sigmoid colon. Seventy-eight deaths from cancer occurred in this cohort. No increase in mortality rates was found for leukemia or sites directly exposed to radiation, such as the ovary or brain, based on a very small number of deaths (two leukemia, three ovary, and one brain cancer death). The SMRs were significantly elevated, however, for cancer of the colon (15 deaths) and for NHL (6 deaths). No estimate of risk per unit dose was presented.

Ankylosing Spondylitis

A U.K. cohort consisting of 15,577 patients diagnosed with ankylosing spondylitis between 1935 and 1957 was followed for mortality up to the end of 1991 (Weiss and others 1994). The average duration of follow-up was 25 years. Of these subjects, 14,566 had received X-ray treatment for their disease. Radiation doses to various organs were calculated for a sample of patients, and average estimated doses from all treatment courses occurring within 5 years of the initial treatment courses were attributed to all patients. The mean total body dose was estimated to be 2.6 Gy. Irradiated patients had a significantly greater mortality rate from cancer than expected from the national rates for England and Wales, and significant increases were seen for leukemia, NHL, multiple myeloma, and cancers of the esophagus, colon, pancreas, lung, bones, connective and soft tissue, prostate, bladder, and kidney. A linear dose-response model for all cancers except leukemia gave an ERR of 0.18 Gy^{-1} (95% CI 0.10, 0.27) 5–24 years after treatment (based on 741 deaths), decreasing significantly to 0.11 Gy^{-1} 25 years or more (based on 845 deaths) after treatment. For lung cancer, the ERR was 0.09 Gy^{-1} (95% CI 0.03, 0.15) 5–24 years after treatment, based on 282 deaths and an average dose to the bronchi of 8.88 Gy. For stomach cancer, the ERR was –0.004 Gy^{-1} (95% CI –0.05, 0.05) 5–24 years after treatment, based on 127 deaths and an average stomach dose of 3.21 Gy. There was no increased risk in breast cancer in this population, based on 84 deaths (ERR 0.08 Gy^{-1}; 95% CI – 0.30, 0.65); this may result from the fact that the average dose the ovaries was high—5.5 Gy).

The risk of leukemia mortality in this cohort was studied further by Weiss and colleagues (1995), using a case-subcohort approach. A total of 60 leukemia deaths were observed during the follow-up period. Radiotherapy records were obtained for all but six of the deaths from leukemia excluding CLL, and individual doses to the red bone marrow were estimated as in the previous study; estimated doses were also available for the subcohort, as described in Weiss and colleagues (1994). The average dose to the total red bone marrow was estimated to be 4.44 Gy, but doses were non-uniform, with the heaviest dose to the lower spine. A linear-exponential model (in which the exponential term allows for cell sterilization in heavily exposed parts of the bone marrow), varying with time since exposure, provided a good description of the risk for non-CLL. The estimated ERR per Gy was 12.4 (95% CI 2.3, 52.1) 10 years after exposure; 1–25 years after exposure, the average ERR per Gy was estimated to be 7.0, based on 35 cases.

A Swedish cohort of 20,024 patients who received X-ray therapy between 1950 and 1964 for painful benign conditions of the locomotor system (including arthrosis and spondylosis) was followed for cancer incidence and mortality until the end of 1988 (Damber and others 1995). The average length of follow-up was 25 years. Average conversion factors between surface dose and mean absorbed dose in the red bone marrow were estimated by treatment site (for six sites), based on the treatment records of random samples of 30 subjects drawn from the cohort (Damber and others 1995). The conversion factors were applied to the entire cohort and used for stratification of subjects in different levels of exposure. The average absorbed dose to the red bone marrow was estimated to be 0.39 Gy. A total of 116 leukemia cases (115 deaths) were observed during the study period. The SIR and SMR for subjects with mean absorbed doses of 0.5 Gy or more were 1.40 (95% CI 1.00, 1.92) and 1.50 (95% CI 1.05, 2.04), respectively. No estimate of risk per unit dose was presented.

Thyroid Diseases

Iodine-131 is currently the treatment of choice for hyperthyroidism, largely because no serious side effects are known. Concerns remain, however, about the subsequent risk of cancer. Several studies of patients treated with ^{131}I for hyperthyroidism have been carried out in the United States, Sweden, and the United Kingdom.

The occurrence of leukemia and of thyroid neoplasms (both benign and malignant) was studied among 36,050 patients treated for hyperthyroidism between 1946 and 1968 and included in the Cooperative Thyrotoxicosis Therapy Follow-up Study (Saenger and others 1968; Dobyns and others 1974). Approximately 20,000 subjects had been treated

with ^{131}I. The follow-up was active, with an average duration of 8 years. No excess of leukemia or thyroid cancer was observed among patients treated with ^{131}I.

In a follow-up to this study, Hoffman (1984) studied cancer risk up to 1979 in the subgroup of 3696 women who had been treated at the Mayo Clinic, one of the original participating centers. Among these, 1005 had received ^{131}I therapy alone and 2141 had been treated with surgery alone. A total of 527 cancer cases were identified in these two study groups; 175 were excluded because they occurred within a year of treatment. The mean observation period was 15 years for patients treated with ^{131}I. The average whole-body dose is estimated to be of the order of 0.06–0.4 Gy in this cohort. There was no increased cancer risk among those treated with ^{131}I and no indication of a relation with ^{131}I activity delivered. Nonsignificant increased risks were seen for cancers in the two most exposed organs (thyroid and salivary glands, based on three and two cases, respectively).

Goldman and colleagues (1988) reported on an extended follow-up of 1762 women, included in the Cooperative Thyrotoxicosis Therapy Follow-up Study, who were treated at the Massachusetts General Hospital between 1946 and 1964. A total of 1406 had been treated with ^{131}I. No dose estimation was conducted. The average follow-up duration was 17.2 years. An elevated SMR was noted in this cohort (SMR 1.3; 95% CI 1.2, 1.4) for all causes of death but not for all cancers (SMR 0.9; 95% CI 0.7, 1.1). A nonsignificantly increased SMR was noted for breast cancer (SMR 1.2; 95% CI 0.9, 1.5); no association with ^{131}I activity was found.

Ron and colleagues (1998a) reported on mortality to the end of 1990 in the Cooperative Thyrotoxicosis Therapy Follow-up Study. The cohort included 35,593 hyperthyroid patients, 91% of whom had been diagnosed with Grave's disease. Fewer than 500 subjects were less than 15 years of age at the time of treatment. The mean length of follow-up was 21 years, and 51% of the subjects had died during the study period. Doses from ^{131}I to 17 organs (other than the thyroid) were estimated for each study subject by multiplying the amount of administered activity by the age-specific dose factor and 24-h thyroid uptake provided for each organ by the International Commission on Radiological Protection (ICRP 1988). Treatment with ^{131}I was not related to all cancer mortality (SMR 1.02; 95% CI 0.98, 1.07) or to mortality from any specific cancer, with the exception of thyroid cancer (SMR 3.94; 95% CI 2.52, 5.86, based on 27 cases). A nonsignificant increase in mortality from thyroid cancer was seen with increasing ^{131}I administered activity—when deaths occurring in the first 5 years after treatment were excluded, there was no evidence of a relationship with total activity; it is therefore likely that the underlying thyroid disease played a role in the observed cancer increase.

Cancer incidence was also studied in 4557 patients who received ^{131}I therapy for hyperthyroidism in Sweden between 1950 and 1975 at Radiumhemmet, Sweden (Holm 1984). Information on thyroid disease and treatment was abstracted from medical records. Cancer diagnoses in these patients were identified from the Swedish Cancer Registry for 1958 to 1976. The average length of follow-up was 9.5 years and 398 cases were identified. There was no increased risk of cancer as a whole or of leukemia in this population. Only for central nervous system (CNS) tumors among women was an increased risk seen (RR 1.89, based on 17 cases); the RR was higher among those who received total ^{131}I activities of 370 MBq or more (RR 2.30, based on 9 cases).

The risk of cancer was studied in 10,552 patients (including the 4557 in the previous study) treated for hyperthyroidism with ^{131}I in seven hospitals in Sweden between 1950 and 1975 (Holm and others 1991). The mean age at treatment was 57 years. Information on thyroid disease and treatment was abstracted from medical records. The mean total activity administered was 506 MBq (360 MBq to patients with Grave's disease and 700 MBq to those with toxic nodular goiter). The dose to various organs was estimated using conversion factors from ICRP (1988) tables and other sources. Dose to the thyroid was in the range of 60–100 Gy. Doses to other organs were lower: 0.25 Gy to the stomach, 0.07 Gy to the lung, and 0.06 Gy to the red bone marrow. A total of 1543 cancer cases were identified during 1958–1985. The mean follow-up time of subjects who survived more than a year after treatment was 15 years, with a maximum of 28 years. The SIR for all cancers in this population was 1.06 (95% CI 1.01, 1.11) compared to the Swedish population. Significant increases were seen for cancers of the lung and kidney and, among 10-year survivors, for cancers of the stomach, kidney, and brain. Only the risk for stomach cancer increased with the level of administered ^{131}I dose and this increase was not statistically significant; the estimated RR at 1 Gy for stomach cancer was 2.32 and the absolute risk was 9.6 per 10^4 PY per gray.

A population-based study of cancer incidence in a cohort of 7417 patients treated with ^{131}I in the West Midlands region of the United Kingdom between 1950 and 1991 was carried out (Franklyn and others 1999). The subjects were followed up for cancer incidence and mortality from 1971 to 1991. No estimation of dose from ^{131}I is presented. Significant decreases in all cancer incidence (634 cases: SIR 0.83; 95% CI 0.77, 0.90) and mortality (448 cases: SMR 0.90; 95% CI 0.82, 0.98) were observed in this cohort. Significant increases in incidence were seen for cancers of the small bowel (six cases: SIR 4.8; 95% CI 2.2, 10.7) and thyroid (nine cases: SIR 3.3; 95% CI 1.7, 6.3).

Summary

Studies of patients treated with radiation (X-rays and gamma rays) for benign disease provide valuable information about the carcinogenicity of low-LET radiation. Doses used in the treatment of benign conditions were generally not as high as those used to treat malignant disease, so that cell-killing effects do not predominate, survival after treat-

ment is good since the conditions treated were generally not life-threatening, and there is minimal confounding from concomitant treatment.

Studies of patients treated for ankylosing spondylitis, benign breast disease, benign gynecological disease, and peptic ulcer have provided valuable information for the quantification of radiation risk estimates for cancers of the lung, breast, and stomach and for leukemia. These estimates are reviewed in detail, and compared with risk estimates derived from other medical exposure studies, in section "Evaluation of Risk for Specific Cancer Sites."

Studies of patients treated with ^{131}I for thyroid diseases provide little quantitative information about radiation risks.

Radiotherapy for Benign Disease Among Children

Tinea Capitis

Between 1948 and 1960 nearly 20,000 children, primarily immigrants to Israel or children of immigrants from North Africa and the Middle East, were treated with radiation for tinea capitis (ringworm of the scalp) in Israel (Ron and others 1988b). This treatment modality was used in other countries as well, and a study also was carried out in New York (Shore and others 1984).

In Israel, mortality in a cohort of 10,834 irradiated children, 10,834 matched comparison subjects, and 5392 sibling controls was studied by Ron and colleagues (1989). Crude dose estimates were derived from treatment information (dosage, area), age of the child, and the use of filtration. Between 1950 and 1982, 609 subjects died. Radiotherapy in childhood was associated with an increased risk of mortality from tumors of the head and neck (particularly brain and thyroid tumors) and leukemia. For the latter (leukemia), the EAR was estimated to be 0.9 per 10^4 PY per gray (the mean average dose to the bone marrow was 0.3 Gy in this cohort).

The relation between radiation dose and risk of tumors of the brain and CNS in this cohort was examined further by Ron and colleagues (1988a). The dose reconstruction method used was improved compared to the above paper, relying heavily on dosimetric studies and measurements in a simulated phantom model of a 6-year-old child. The average dose to the brain in irradiated patients was 1.5 Gy (range 1–6 Gy), and the average minimal and maximal doses to specific areas of the brain were 0.8 and 1.8 Gy, respectively. Sixty neural tumors developed in irradiated subjects. The RR was 6.9 (95% CI 4.1, 11.6) overall, compared to the nonirradiated control groups; for neural tumors of the head and neck it was 8.4 (95% CI 4.8, 14.8). Increased risks were apparent for meningioma (RR 9.5, 19 deaths), gliomas (RR 2.6, 7 deaths), nerve sheath tumors (RR 18.8, 25 deaths), and other neural tumors. A strong dose-response relation was found, with the RR approaching 20 after doses of the order of 2.5 Gy. No estimate of risk per dose is presented in this study. Since then, a descriptive study of 253 meningioma cases diagnosed in the above cohort has been published (Sadetzki and others 2002), but there were no risk estimates. A more recent study (Sadetski and others 2005) conducted a survival analysis using Poisson regression to estimate the excess relative and absolute risks for brain tumors. After a median follow-up of 40 years, ERRs/Gy of 4.63 and 1.98 (95% CI 2.43, 9.12 and 0.73, 4.69) and EARs/Gy per 10^4 PY of 0.48 and 0.31 (95% CI 0.28, 0.73 and 0.12, 0.53) were observed for benign meningiomas and malignant brain tumors, respectively. The risk of both types of tumors was positively associated with dose. The estimated ERR/Gy for malignant brain tumors decreased with increasing age at irradiation from 3.6 for exposures below the age of 5 to 0.5 for exposures at ages 10 or above ($p = .04$), while no trend with age was seen for benign meningiomas. The ERR for both types of tumor remains elevated 30-plus years after exposure.

Modan and colleagues (1989) reported on an additional 5-year follow-up (until 1986) of the Israeli tinea capitis cohort. While the previously observed increases in the incidence of head and neck tumors (mainly brain, CNS, and thyroid tumors) continued, an elevated risk of breast cancer was observed for the first time in this cohort, based on 13 new cases in 1982–1986. The estimated average dose to the breast was low—0.016 Gy. The increase was seen only among women who were 5–9 years of age at the time of radiation exposure (10 cases). No estimate of risk per dose is presented.

Ron and colleagues (1989) reported on the risk of thyroid cancer following irradiation in childhood for tinea capitis, based on an extended follow-up (until 1986). The dose reconstruction method is based on the approach described above for the brain and CNS study. To adjust the dose for possible head movement during treatment, individual dose estimates were multiplied by a factor of 1.5 as suggested by results of dosimetric studies. Average doses to the thyroid were 0.13, 0.09, and 0.06 Gy, respectively, for children aged less than 5, 5–10, and 10–15 years at the time of exposure. Overall, 98 thyroid tumors were identified among the exposed and 57 among the two control populations. An estimated dose of 0.09 Gy was related to a fourfold increase (95% CI 2.3, 7.9) in the risk of thyroid cancer and a twofold increase in benign tumors. The dose-response was consistent with linearity. The risk diminished with age at exposure, and the RR appeared to be constant over time. The ERR was estimated to be 30 Gy^{-1} and the EAR was 13 per 10^4 PY per gray.

Modan and coworkers (1998) also reported a 4.5-fold increase in the incidence of malignant salivary gland tumors ($p < .01$) and a 2.6-fold increase of benign tumors in subjects irradiated for tinea capitis. A clear dose-response association with both cancer and benign tumors was demonstrated. No estimate of risk per dose level was presented.

In New York, about 2200 children who received X-ray treatment for tinea capitis during the 1940s and 1950s and a comparable group of 1400 treated without X-rays were fol-lowed by mail questionnaire to evaluate the incidence of skin cancer (Shore and others 1984). The average length of follow-up was 26 years. Delivered doses ranged from 3 to 6 Gy depending on the portion of the scalp, with lower doses to the skin of the face and neck (0.1–0.5 Gy). In the irradiated group, 41 subjects had a diagnosis of basal carcinoma of the scalp or face, compared to 3 in the control group. The prevalence of multiple lesions was high in the exposed group. The minimum latent period was long (about 20 years); skin cancers were more pronounced on the face, where the potential for exposure to ultraviolet is higher, and were restricted to Caucasians although one-quarter of the study population was African American. No estimate of risk per dose is presented.

Enlarged Thymus Gland

Patients in Rochester, New York who received X-ray treatment between 1926 and 1957 in infancy (before 6 months of age) for an enlarged thymus gland and their nonirradiated siblings have been followed up periodically through the use of a mail questionnaire (Shore and others 1985, 1993a, 1993b; Hildreth and others 1985). Information on X-ray treatment factors was extracted from medical records and supplemented by interviews with the treating physicians. These, along with anatomic measurements for infants, allowed estimation of doses to various nearby organs. The thyroid doses were estimated by irradiating a radiological phantom of an infant. The irradiated group had a statistically significant increase of both benign and malignant thyroid tumors (Shore and others 1985) and extrathyroid tumors (Hildreth and others 1985), particularly benign tumors of the bone, nervous system, salivary glands, skin, and breast (women only) and malignant tumors of the skin and breast.

In the most recent paper on thyroid cancer, which reports on follow-up to 1986, the cohort included 2657 exposed subjects and 4833 unexposed siblings with at least 5 years of follow-up (Shore and others 1993a). The average duration of follow-up was 37 years. Thyroid doses could be estimated for 91% of the subjects. The thyroid dose distribution was skewed, ranging from 0.03 to more than 10 Gy, with a mean of 1.4 Gy and a median of 0.3 Gy. There were 37 pathologically confirmed thyroid cancers among the irradiated group and 5 among the sibling controls. A linear dose-response was found in this cohort with an ERR of 9.0 at 1 Gy (90% CI 4.0, 24.0). An increased risk was seen even at low doses, with a significant positive slope in the dose range 0–0.3 Gy, based on four exposed cases. The risk ratio decreased over time, but was still highly elevated 45 years after exposure. There was no evidence of a decrease in the absolute excess risk over time (EAR varying from 2.1 per 10^4 PY per gray 5–14 years postirradiation to 6.0 per 10^4 PY per gray after 45 years or more). Analyses of interactions suggested that all Jewish subjects and women with older ages at menarche or at first childbirth were at greater risk of radiation-induced thyroid cancer.

The risk of benign thyroid adenomas was also studied in more detail (Shore and others 1993b). There were 86 pathologically confirmed thyroid adenomas among the irradiated group and 11 in the sibling controls. The estimated ERR was 6.3 Gy^{-1} (90% CI 3.7, 11.2) overall and 7.8 Gy^{-1} when restricted to subjects with doses less than 6 Gy. Adenoma rates were elevated even at lower doses, with a significant increase in the lowest-dose group (<0.25 Gy). The risk continued to be elevated to the end of follow-up.

Analyses of the risk of breast cancer in relation to radiation dose were also carried out in this population. Hildreth and colleagues (1989) reported on the follow-up to 1985 of 1200 women who received X-ray treatment and their 2469 nonirradiated sisters. Twenty-two breast cancer cases were diagnosed in the irradiated group and twelve in the control group. The estimated average dose to the breast was 0.69 Gy. A linear dose-response was observed, with an ERR of 2.48 Gy^{-1} (95% CI 1.1, 5.2) and an EAR of 5.7 per 10^4 PY per gray (95% CI 2.9, 9.5).

Skin Hemangioma

Two Swedish cohort studies have been performed of patients treated for skin hemangioma in infancy. In the first study (Lundell and others 1994), the cohort consisted of 14,351 infants (less than 18 months of age) treated between 1920 and 1959 at Radiumhemmet, Stockholm, who were followed up for cancer incidence over the period 1958–1986. Radiotherapy was given with β-particles, X- and/or γ-rays, and usually, with some type of ^{226}Ra applicator. Individual organ doses were calculated using treatment information and, for ^{226}Ra needles and tubes, phantom simulations. Seventeen thyroid cancers were registered in this cohort during the follow-up period. The mean dose to the thyroid was 1.07 Gy (range <0.01, 4.34 Gy). A significant excess thyroid cancer incidence was seen in this cohort, starting 19 years after treatment and persisting at least 40 years after irradiation. A significant dose-relationship was observed, with an ERR of 4.92 Gy^{-1} (95% CI 1.26, 10.2) and an EAR of 0.90 per 10^4 PY per gray.

Lundell and Holm (1995) also studied the risk of other solid tumors in this cohort. Statistically significantly increased SIRs were seen for cancer of the pancreas and tumors of the endocrine glands, based on small numbers of cases (9 and 16, respectively). For lung cancer (mean dose 0.12 Gy), a nonsignificant ERR of 1.4 Gy^{-1} was reported (confidence interval not given) and an EAR of 0.33 per 10^4 PY per gray, based on 11 cases. For stomach cancer (mean dose 0.09 Gy), both the ERR and the EAR were negative (values not reported), based on five cases.

Lundell and colleagues (1996) reported more specifically on the risk of breast cancer among women from this cohort. The mean absorbed dose to the breast was 0.39 Gy (range <0.01, 35.8 Gy). During the follow-up period, 75 breast cancer cases were found in the cohort. A significant linear dose-response relationship was observed, with an ERR of 0.38 Gy^{-1} (95% CI 0.09, 0.85) and an EAR of 0.41 per 10^4 PY per gray. This was not modified by age at exposure or by dose to the ovaries. The ERR increased significantly with time since exposure, however, with an ERR at 1 Gy of 2.25 (95% CI 0.59, 5.62) 50 years or more after exposure. The EAR was 22.9 per 10^4 PY per gray.

In an analysis of leukemia mortality in the same cohort, 20 deaths from leukemia were observed (11 in childhood and 9 among adults). The weighted bone marrow dose was 0.13 Gy on average (range <0.01–4.6 Gy). There was no association between radiation dose and leukemia (childhood or adult) in this cohort. Among those who received more than 0.1 Gy, the ERR was estimated to be 5.1 Gy^{-1} (95% CI 0.1, 15) for childhood leukemia, –0.02 Gy^{-1} (95% CI –0.8, 1.9) for adult leukemia, and 1.6 Gy^{-1} (95% CI –0.6, 5.5) overall.

The second Swedish hemangioma study included 11,807 patients treated with ^{226}Ra between 1930 and 1965 at Sahlgrenska University Hospital in Göteborg (Lindberg and others 1995). The cohort was followed up for cancer incidence over the period 1958–1989. Doses to 11 organs were calculated on the basis of ^{226}Ra activity, location of the hemangioma, and estimated absorbed dose rate in these organs per unit activity in a phantom the size of a 5–6-month-old child. No correction was made for different body sizes according to the age of the child at the time of treatment. A total of 248 malignancies were observed during the study period. A significantly increased risk of cancer was seen overall, as well as tumors of the CNS (34 cases), thyroid (15 cases), and other endocrine glands (23 cases). The mean absorbed dose to the thyroid in this cohort was 0.12 Gy; the ERR for thyroid cancer was estimated to be 7.5 Gy^{-1} (95% CI 0.4, 18.1) and the EAR 1.6 per 10^4 PY per gray.

Karlsson and others (1997) studied the risk of intracranial tumors in this cohort further in a cohort and a case-control study. Dose estimation was similar to that described above for subjects in the cohort study, although a correction was made for different age groups (0–4, 5–11, 12–18, and 18+ months). Activity was considered to be uniformly distributed over the treatment area. In the case-control study, the dose at the exact tumor site was calculated by considering the exact distance between the treatment location, according to the record, and the site of the tumor. For the controls, the dose was calculated at the location of the tumor in the corresponding case. In the cohort, 47 intracranial tumors developed in 46 individuals. An excess was found for many histopathological subgroups but was significant only for gliomas and meningiomas. The mean dose to the brain was 0.072 Gy (median 0.023 Gy; range <0.001–2.4 Gy). There was an excess of brain tumors in all dose categories, but no clear dose-response relationship. When analyses were restricted to subjects treated before the age of 7 months, both a linear and a linear-quadratic model described the data. The estimated ERR for the entire cohort was 1.05 Gy^{-1}; the EAR was 1.20 per 10^4 PY per gray. In the case-control study, the mean absorbed dose at the site of the tumor was 0.031 Gy for cases and 0.09 for controls. The estimated OR was 1.65 Gy^{-1} (95% CI 0.63, 4.32).

Pooled analyses of the data on breast cancer and intracranial tumors from the two Swedish hemangioma cohorts were also carried out. In the pooled breast cancer analyses (Lundell and others 1999), 245 breast cancer cases diagnosed between 1958 and 1993 were available. The ERR was estimated to be 0.35 Gy^{-1} (95% CI 0.18, 0.59) and the EAR 0.72 per 10^4 PY per gray (95% CI 0.39, 1.14). There was no evidence of an effect of time since exposure on the ERR; the EAR, however, increased with time since exposure. Neither age at exposure, dose rate, nor ovarian dose appeared to have an effect on the ERR.

In the pooled analysis of intracranial tumors (Karlsson and others 1998), 88 tumors were found in 86 individuals between 1958 and 1993. There was a significant dose-response relationship, and increasing age at exposure decreased the magnitude of the risk. The ERR was 2.7 Gy^{-1} (95% CI 1.0, 5.6) overall; 4.5 Gy^{-1} for exposure before the age of 5 months; 1.5 Gy^{-1} for exposures between 5 and 7 months; and 0.4 Gy^{-1} for exposures at older ages. The overall EAR was 2.12 per 10^4 PY per gray (95% CI 0.27, 4.38). There was no effect of time since exposure on the ERR, while the EAR increased with time since exposure.

Cancer mortality was studied in a cohort of 7037 patients less than 15 years of age treated for a skin hemangioma between 1940 and 1973 at the Institut Gustave Roussy, near Paris, France (Dondon and others 2004). Among them, 4940 had received radiotherapy. The cohort was followed up from 1969 to 1997, during which time 16 patients died of cancer; 14 of these had received radiotherapy. A nonsignificant excess of cancer-related mortality was observed for irradiated patients compared to the general population (SMR 1.53; 95% CI 0.86, 2.48). The excess was highest among those treated with ^{226}Ra (RR 2.53; 95% CI 0.84, 7.07), in comparison to those who did not receive radiotherapy. No estimate of risk per dose is presented.

Enlarged Tonsils and Other Benign Conditions

In 1974, a prospective follow-up program was set up at the Michael Reese Hospital in Chicago to screen patients who had received X-ray treatment between 1939 and 1962 during childhood for benign head and neck conditions (primarily enlarged tonsils). During the screening, more than 35% of the subjects were found to have thyroid nodules (Schneider and others 1985). Analyses of dose-response relationships for thyroid cancer and thyroid nodules in this cohort were conducted by Schneider and colleagues (1993) with a mean follow-up of 33 years. Individual doses were estimated for study subjects on the basis of treatment records

and experiments with an anthropomorphic phantom of a 6-year-old child, together with conversion factors for children of different ages. The average dose to the thyroid was estimated to be 0.6 Gy. Overall uncertainty in thyroid dose estimates for an individual subject is of the order of 50% and is related to the child's movements during treatment and deviations in height and weight. The cohort included 4296 patients, of whom 3843 had estimated dose to the thyroid and 2634 could be followed up. A total of 1043 nodules and 309 thyroid cancers were diagnosed in the cohort. The ERR/Gy was 3.0 overall; it decreased with increasing age at exposure (from 3.6 for exposures below the age of 1 year to 1.4 for exposures between ages 5 and 15). There was no apparent difference between men and women. The slope of the dose-response relationship appeared to reach a maximum 25–29 years after exposure, but response continued to be elevated 40 years after exposure. The ERR appeared to be greater for cases diagnosed before 1974, when the screening program started (ERRs 9.2 and 1.8, respectively, for the period before 1974 and for 1974 and later, based on 109 and 200 cases), but this difference was not significant ($p = .4$).

From 1975 through 1982, a follow-up study was performed in Sweden for patients treated with X-rays for cervical tuberculous adenitis (Fjalling and others 1986). Of these patients, 444 underwent thyroid examination on average 43 years after their initial treatment. 101 had undergone surgery for thyroid nodules, including 25 for thyroid cancer. The absorbed dose to the thyroid was estimated to range from 0.4 to 51.0 Gy. A dose-response relationship was seen both for thyroid cancer and for nodules. No estimate of risk per dose is presented.

Thyroid Diseases

No study has focused specifically on populations exposed to ^{131}I in childhood or adolescence for the treatment of hyperthyroidism. As indicated in the earlier section on treatment of adult benign thyroid diseases, the number of subjects under age 20 at diagnosis in the hyperthyroidism cohorts is very small. In a review paper, Shore (1992) carried out an analysis of risk in those exposed below age 20 in the Swedish and U.S. studies. The total population was estimated to be 602, with an approximate average follow-up of 10 years and a mean dose to the thyroid of about 88 Gy. Two cases of thyroid cancer were reported compared to about 0.1 expected. The estimated ERR was 0.3 Gy^{-1} (90% CI 0.0, 0.9) and the EAR 0.1 per 10^4 PY per gray (90% CI 0.0, 0.2).

Summary

Studies of children treated with radiation (X-rays and γ-rays) for benign disease also provide valuable information about the carcinogenicity of low-LET radiation. Studies of patients treated for tinea capitis, enlarged thymus, and benign head and neck diseases have provided much of the quantitative information on the risk of thyroid cancer related to external radiation in children. Studies of children treated for hemangioma (with average doses ranging from 0.09 to 0.4 Gy, depending on the target organ) have provided radiation risk estimates for cancers of the lung, breast, thyroid, and stomach, as well as for leukemia. These estimates are reviewed in detail, and compared with risk estimates derived from other medical exposure studies, in the section of this chapter "Evaluation of Risk for Specific Cancer Sites."

No study has focused specifically on populations exposed to ^{131}I in childhood or adolescence for the treatment of hyperthyroidism.

Diagnostic Radiation Among Adults

Chest Fluoroscopy for Follow-up of Pulmonary Tuberculosis

A cohort study of 64,172 tuberculosis patients was carried out in Canada to assess the risk of cancer associated with multiple fluoroscopies (Miller and others 1989). In this cohort, 25,007 patients were exposed to highly fractionated radiation from repeated fluoroscopic examinations used to monitor lung collapse from pneumothorax treatment. Howe (1995) studied the risk of lung cancer in this cohort. Absorbed lung doses from fluoroscopy were estimated for each patient for each year since admission for tuberculosis. This involved obtaining counts from medical records of the number of fluoroscopies each year. This number was combined with information on dose per fluoroscopy obtained from the output of typical fluoroscopes used during the relevant period, the estimated organ dose per unit of surface exposure based on human phantom experiments, and interviews with physicians who administered pneumothorax during the relevant period. The average lung dose per fluoroscopy session was estimated to be 11 mGy. The mean total dose to the lung was 1.02 Gy (range 0–24.2 Gy), and the mean number of fractions was 92. During the study period (1950–1987), 1178 lung cancer deaths occurred. There was no evidence of an association between risk of lung cancer and dose: the ERR at 1 Gy was 0.00 (95% CI –0.06, 0.07). The authors conclude that their study supports the hypothesis of a substantial reduction in risk related to fractionation for low-LET radiation.

Howe and McLaughlin (1996) also reported analyses of breast cancer mortality in relation to radiation dose in this cohort. Estimates of dose to the breast were derived as above. The mean dose to the breast varied across provinces: it was 2.1 Gy (range 0–18.4) in Nova Scotia and 0.79 Gy (range 0–14.4) elsewhere. This difference was related to a difference in practices, with a much larger proportion of the examinations in Nova Scotia being carried out in an anterior-posterior orientation, and hence resulting in a higher breast dose than elsewhere. A total of 681 (103 in Nova Scotia and 578 elsewhere) deaths from breast cancer were identified during the study period. A strong dose-response association was seen in this study, with the ERR decreasing significantly with increasing age at exposure. The ERR appeared to be constant from 5 to 39 years after exposures, with a suggestion of a decrease after that. The ERR for exposure at age 15 was estimated to be 0.90 Gy[-1] overall but differed significantly between regions: it was 3.56 Gy[-1] (95% CI 1.85, 6.82) in Nova Scotia and 0.40 Gy[-1] (95% CI 0.13, 0.77) elsewhere. The EAR 20 years later for exposure at age 15 was estimated to be 10.3 per 10[4] PY per gray (95% CI 6.37, 16.2) in Nova Scotia and 1.22 per 10[4] PY per gray (95% CI 0.42, 2.34) elsewhere; the overall estimate is 3.16 per 10[4] PY per gray. The difference in risk between Nova Scotia and elsewhere does not have an obvious explanation. Doses in Nova Scotia were generally less fractionated than elsewhere, and the dose per fraction was higher; hence this could be a dose-rate or fractionation effect. Comparisons of results with estimates from the study of atomic bomb survivors, however, show that the ERRs/Gy are fairly consistent and do not therefore suggest a major effect of fractionation.

The mortality experience of 6285 women patients who received repeated fluoroscopic examinations to monitor lung collapse for treatment of tuberculosis between 1925 and 1954 in Massachusetts was studied by Davis and colleagues (1989). A control cohort of 7100 nonirradiated tuberculosis patients was also studied. No significant increase in the mortality from all cancers, from lung cancer, or from leukemia was seen in the exposed cohort. Increases in mortality from breast and esophageal cancers were observed based on 62 and 14 deaths, respectively.

The incidence of breast cancer was studied further among 4940 women from the above cohort who could be followed from 1970 or 1980 (depending on the subcohort) until 1989 (Boice and others 1991b). Estimates of absorbed dose to the breast were derived by taking into account the number of lung collapse treatments, calendar year of exposure, age at exposure, and exposure settings of the fluoroscopy machines in use at the time. Physicians who conducted the examinations—and patients themselves—were interviewed about the fluoroscopy procedures during the lung collapse sessions. Among exposed women, the mean dose to the glandular tissue of the breast was estimated to be 0.79 Gy and the mean number of fluoroscopies was 88. The average length of follow-up of the cohort was 22 years. Among the 2573 women examined with fluoroscopy, 147 breast cancers developed during the follow-up period (SIR 1.29; 95% CI 1.1, 1.5). No increased risk of breast cancer was seen among the 2367 women treated by other means. Breast cancer risk increased significantly with increasing radiation dose. The overall ERR was estimated to be 0.61 Gy[-1] (95% CI 0.3, 1.01) and the EAR 10.7 per 10[4] PY per gray (95% CI 6.0, 15.8). The risk decreased significantly with increasing age at exposure; the ERRs were estimated to be 1.0, 0.7, 0.1, and 0.1 Gy[-1], respectively, for exposures at age 15, 20, 35, and 45 years. The excess breast cancer risk was not apparent until 15 years after exposure and remained high for the period of observation (over 50 years).

Other Uses of Diagnostic X-Rays in Adults

Preston-Martin and coworkers (1988) carried out two population-based, case-control studies of cancer risk in relation to prior exposure to diagnostic X-rays, one of tumors of the parotid gland and the other of chronic myeloid and monocytic leukemia. Significant associations between reported numbers of X-rays (and estimated doses) were seen for both diseases. Results from these studies should be interpreted with caution, however, because the information on past exposures was self-reported, obtained by questionnaire, and therefore subject to recall bias and uncertainty.

In a U.S. case-control study of 565 leukemia patients, 318 NHL patients, 208 multiple myeloma patients, and 1390 matched controls, the history of diagnostic X-ray exposure was ascertained from medical records held by two health plans (Boice and others 1991a) and, hence, was not subject to recall bias. Each diagnostic X-ray procedure was assigned a probable dose to the active bone marrow (averaged over the whole body) based on an extensive literature review. After excluding examinations within 2 years of diagnosis, no association was found between dose of radiation from diagnostic X-rays and the risk of leukemia or NHL. The risk of multiple myeloma, however, was increased among those patients who were frequently exposed to X-rays. No estimate of risk per dose is presented.

Inskip and colleagues (1995) carried out a case-control study of thyroid cancer among residents of the Uppsala Health Care Region in Sweden to assess the relationship between diagnostic X-ray exposure and the risk of thyroid cancer. The study included 484 cases diagnosed between 1980 and 1992 and an equal number of age-, sex-, and country of residence-matched controls. Lifetime residential history of study subjects was compiled and radiological records were searched at all hospitals serving regions where study subjects had lived. Approximate radiation doses to the thyroid gland were estimated for different types of X-ray examinations based on historical measurements made in Sweden and the United States. No association was seen between estimated radiation dose and the risk of thyroid cancer.

Diagnostic Iodine-131 Exposures

Holm and colleagues (1988) studied the incidence of thyroid cancer in a cohort of 35,074 patients who had received diagnostic ^{131}I exposures for suspected thyroid disorders in Sweden between 1951 and 1969; 50 thyroid cancer cases were observed between 1958 and 1985, compared to 39.4 expected in the general population (SIR 1.27; 95% CI 0.94, 1.67). The risk of thyroid cancer increased with increasing ^{131}I activity. Those who were examined for a suspected thyroid tumor tended to have received higher ^{131}I activities, however, and there was no increased incidence among those examined for other reasons (SIR 0.62; 95% CI 0.35, 1.00,

based on 16 cases). The increased risk was highest 5–9 years after examination, and there was no evidence of a dose-response 10 years or more after exposure.

In a follow-up paper, Holm and colleagues (1989) reported on cancer risk in this population over the same period. For each patient, information on delivered ^{131}I activity, date of administration, 24-h thyroid uptake, and reason for examination was abstracted from medical records. Dose estimates were derived from this information using ICRP (1988) conversion factors. The average activity delivered to the thyroid was 1.9 MBq, which would result in radiation doses of less than 10 mGy to organs other than the thyroid. A total of 3746 cancers occurred in this cohort in the first 5 years after examination. No significant increase was seen in the incidence of cancer overall (SIR 1.01; 95% CI 0.98, 1.04). Significantly increased SIRs were noted, however, for endocrine tumors other than the thyroid, lymphoma, leukemia, and nervous system tumors. Cancer risk appeared to increase with increasing dose of ^{131}I in years 5–9 after examination only and was significantly elevated only for subjects who received more than 2.7 MBq. No dose-related increase was seen 10 or more years after follow-up.

Hall and colleagues (1992) reported on combined analyses of leukemia incidence among 46,998 Swedish patients who had received ^{131}I for either diagnostic or therapeutic purposes (these include most of the 35,074 subjects from Holm and others (1988) and the 10,552 from Holm and others (1991). The mean absorbed dose to the bone marrow was estimated to be 14 mGy (range 0.01–2.23). The average duration of follow-up was 21 years: 195 leukemias occurred more than 2 years after exposure and the SIR was 1.09 (95% CI 0.94, 1.25). Similar but not significantly increased risks were seen for CLL and non-CLL. A significant excess of CLL was observed among those who received more than 100 mGy to the red bone marrow. The leukemia risk did not vary with level of ^{131}I radiation dose.

Hall and colleagues (1996) also reported on follow-up up to the end of 1990 of thyroid cancer incidence among 34,104 of the patients who had received ^{131}I diagnostic exposures for suspected thyroid disorders in Sweden between 1951 and 1969 (Holm and others 1988). Dose to the thyroid was estimated as in Holm and coworkers (1989), taking into account information on thyroid mass from patient records and scintigrams. The mean dose to the thyroid was estimated to be 1.1 Gy, and 67 thyroid cancers were identified during the study period. The SIR was 1.35 (95% CI 1.05, 1.71). The excess thyroid cancer risk was restricted to those patients who had been examined for suspicion of thyroid tumor. Risk was not related to radiation dose to the thyroid, time since exposure, or age at exposure. Among the 2408 patients under age 20 at the time of the examination, a small excess risk was seen (3 cases observed versus 1.8 expected).

Summary

Studies of populations with diagnostic radiation exposures are, in principle, more suited than studies of therapeutic exposures to the evaluation of health risk following low doses of radiation. However, most of the studies on diagnostic X-rays reviewed do not provide risk estimates and hence are not informative for the purpose of this report. The exception is the studies of patients who received repeated chest fluoroscopies to monitor lung collapse in the treatment of tuberculosis. Careful dose reconstruction to the lung was carried out in a study in Canada and to the breast in studies in Canada and Massachusetts, allowing the quantification of risk to these organs. It is noted that average doses in studies of patients receiving fluoroscopies are high, however, of the order of 1–2 Gy to the lung and breast because of the very large number of such procedures the patients had to undergo. Estimates from these studies are reviewed in detail, and compared with risk estimates derived from other medical exposure studies, in the section "Evaluation of Risk for Specific Cancer Sites."

Studies of diagnostic ^{131}I exposures reviewed did not provide estimates of risk per unit dose.

Diagnostic Radiation Among Children

Scoliosis

In 1989, Hoffman and colleagues reported a doubling in the incidence of breast cancer in a pilot study of 1030 women who had received multiple diagnostic X-rays between 1935 and 1965 for evaluation of scoliosis during childhood and adolescence. These results were explored further in the U.S. Scoliosis Cohort Study (Doody and others 2000), which included 5573 women patients with scoliosis who had been referred to one of 14 orthopedic centers in the United States. The cohort included only cases of scoliosis diagnosed before age 20 between 1912 and 1965. Information on personal characteristics and scoliosis history was abstracted from medical records of participating institutions, together with radiology reports, radiograph jackets, and radiology logbooks to determine for each examination the date, field, view, position, size of the radiograph, and other factors necessary to determine dose to the breast.

Manufacturers of the radiograph machines that had been used in the study centers completed a questionnaire concerning machines and parameters during the study period. Dose to the breast was estimated for each examination for which the breast was in the beam. For each examination, the breast was classified as preteen (<13 years old) or teen and adult combined (13 years) depending on the age of the patient at the time; dose to the breast was estimated at a depth of 1 cm for preteen breasts and at 2.5 cm otherwise. The average number of examinations per patient was 24.7 (range 0–618), and the mean cumulated dose to the breast was 0.11 Gy

(range 0–1.7 Gy); 631 women had no radiographic examination. The mean age at diagnosis of scoliosis in the cohort was 10.6 years, and the average length of follow-up was 40.1 years. A total of 77 breast cancer deaths were observed in the cohort compared to 45.6 expected based on U.S. national rates. The risk increased significantly with cumulative radiation dose; the unadjusted ERR was 5.4 Gy^{-1} (95% CI 1.2, 14.1); adjustment for type of treatment or age at first radiographic examination reduced the risk, as did restriction of analysis to women who had at least one radiographic examination (ERR 2.7; 95% CI –0.2, 9.3). Findings from this study must be interpreted with caution because dose to the breast could be underestimated (records were available only of radiographs from participating institutions and did not include those made before referral to these institutions). Further, a number of factors may confound the association between radiation dose and risk of breast cancer, such as the severity of disease, which may affect reproductive history and hence breast cancer risk.

CT scans

As indicated in the introduction to this chapter, there is concern about the potential health effect of repeated CT scan exposures, particularly in childhood. No epidemiologic study of populations exposed to CT was available to the committee.

Brenner and colleagues (2001) and Brenner and Elliston (2004) have evaluated the possible consequences of CT exposures based on estimated doses to specific organs. They conclude that lifetime risks of cancer are not negligible.

Exposure In Utero

Prenatal X-rays were first associated with increased risk of childhood leukemia and cancer in the 1950s in the Oxford Survey of Childhood Cancers (OSCC), a U.K.-wide study begun in 1955 (Stewart and others 1958). Results were based on a case-control study of 1416 childhood cancer deaths and the same number of controls, in which mothers of the study subjects were asked about their child's history of radiographic examinations (*in utero* and after birth). This association was confirmed by MacMahon (1962) in a study of a cohort of 734,243 children born in the northeastern United States between 1947 and 1954, in which 584 subjects had died of cancer in childhood and information about prenatal X-rays was obtained from medical records, thus eliminating the possibility of recall bias.

The OSCC is the largest study of childhood cancer after prenatal exposure to X-rays. It has continued and been expanded to cover all children dying from malignant disease in the United Kingdom under the age of 16 (Bithell and Stewart 1975; Knox and others 1987; Gilman and others 1989); in 1981, it included 15,276 matched case-control pairs. The magnitude of the association appears to have diminished over time (Muirhead and Kneale 1989), but so has the dose of radiation to which pregnant women have been exposed during examinations (Doll and Wakeford 1997). A decrease over time also was reported in the northeastern U.S. study (Monson and MacMahon 1984).

The possible effect of prenatal exposure has been studied in a number of other populations in the United States and Europe. Results of the case-control studies have been combined in meta-analyses by Bithell (1989, 1990). Although dominated by the OSCC, results of these studies show a significant RR of 1.4 for *in utero* radiation in association with childhood cancer (Doll and Wakeford 1997).

Controversy continues, however, about the existence and size of the risk following prenatal exposure. Boice and Miller (1999) noted that the increases were restricted to case-control studies and were not seen in cohort studies; they also commented on the similarity of relative risks for leukemia and solid cancers, suggesting an underlying bias in the case-control studies. In their review, Doll and Wakeford (1997) discuss these arguments. In regard to cohort studies, they combine the results of cohort studies for which relative risks can be calculated reliably and note that, when the atomic bomb survivors are excluded, an increased risk is obtained that is consistent with the combined results of case-control studies. They note further that the incomplete follow-up of the Japanese atomic bomb survivor cohort in the years after the bombings may be partially responsible for the apparent inconsistency of results concerning the effects of prenatal exposures. The argument that radiation risks for leukemia and solid cancers differ is based on observations of exposure in childhood and later years. Doll and Wakeford (1997) note that the carcinogenic effects of radiation exposure *in utero* and in childhood are not expected to be the same because the cells that give risk to most of the typical childhood cancers other than leukemia persist and are capable of dividing for only a short time, if at all, after birth. Doll and Wakeford further conclude that the idea of a causal relationship is supported by the increase in RR with increasing number of X-ray examinations conducted in the third trimester of pregnancy and the significant decline in RR with year of birth, paralleling the decline in fetal doses that occurred over the same period (UNSCEAR 1972).

Based on the results of the Oxford survey and other studies of the effects of maternal irradiation, UNSCEAR (1996) reported a statistically significant leukemia risk (up to age 15 years) and estimated a 40% increase in risk of childhood cancers (up to 15 years) at doses of 10–20 mGy (low LET). Risk estimates have been derived since then by a number of authors and committees (UNSCEAR 1996; Doll and Wakeford 1997; Wakeford and Little 2003). In the most recent analyses, Wakeford and Little (2003) derive an ERR for childhood cancer following prenatal exposure of about 50 Gy^{-1}, with an EAR of about 8% Gy^{-1}. They comment, however, that statistical, dosimetric, modeling, and other uncertainties associated with these risk estimates are appreciable. They also note that when these uncertainties and those

associated with equivalent risk coefficients from the Japanese atomic bomb survivor cohort exposed *in utero* are taken into account, the risk estimates for childhood cancer from these two sources of data are compatible and they conclude that "doses to the fetus *in utero* of the order of 10 mSv discernibly increase the risk of childhood cancer."

Diagnostic [131]I Exposures

The use of ^{131}I for diagnostic purposes in childhood is rare; hence information on risk is very sparse. In the cohort of 34,104 patients who had received ^{131}I diagnostic exposures for suspected thyroid disorders in Sweden between 1951 and 1969, reported by Hall and colleagues (1996), only 2408 patients were under age 20 at the time of the examination. Among these, a small excess risk was seen (3 cases observed vs. 1.8 expected).

Summary

Information on radiation risks following diagnostic radiation exposure in childhood comes from a study of women who received multiple diagnostic X-rays for the evaluation of scoliosis during childhood and adolescence. This study, in which important efforts were made to reconstruct dose to the breast, has provided an estimate of the risk of radiation-induced breast cancer. This estimate is reviewed, and compared with risk estimates derived from other medical exposure studies, in the following section.

Studies of prenatal exposure to diagnostic X-rays have, despite long-standing controversy, provided important information on the existence of a significantly increased risk of leukemia and childhood cancer following diagnostic doses of 10–20 mGy *in utero*.

Only one study has examined the effects of using ^{131}I for diagnostic purposes in childhood. A small excess of thyroid cancer risk was seen—based on very small numbers—and no risk estimate is provided.

EVALUATION OF RISK FOR SPECIFIC CANCER SITES

This section reviews radiation risk estimates for five types of malignancies (lung cancer, female breast cancer, thyroid cancer, leukemia, and stomach cancer). The results of analyses of the risk of heart disease following medical radiation exposures are also reviewed.

Lung cancer was chosen because it is the most common malignancy among humans. Breast cancer was selected because breast tissue in young women is responsive to low levels of low-LET radiation, and thyroid cancer was chosen because of the inherent radiosensitivity of the thyroid gland. Leukemia was chosen because bone marrow is sensitive to low levels of low-LET radiation, and stomach cancer was selected because of its high incidence in many parts of the world, including Japan. Breast cancer, thyroid cancer, and leukemia were evaluated in the BEIR V (NRC 1990) report.

As far as feasible, the effects of both external and internal (^{131}I) low-LET radiation exposures are considered separately. The focus on low-LET irradiation is related to the goal of BEIR VII to investigate the magnitude of risk from these radiation types. Although the focus of this report is the effects of low doses (*i.e.*, doses less than 100 mGy), results of medium- and high-dose studies are also reviewed because they provide important insights into modifiers of radiation risks that cannot, at present, be studied in populations with lower-dose medical exposures.

The information presented in the previous section of this chapter was used to identify the studies that are informative for radiation risk estimation and have provided estimates of risk per gray in a comparable fashion (either as ERR or as EAR). The estimates have been taken from the original publications. When such estimates were not available in the original study reports, these studies have not been included in this section, with the exception of the study of breast cancer in cervical cancer survivors (for which risk estimates were taken as derived by UNSCEAR 2000). No estimation was attempted by the committee because there is consider-

able uncertainty in deriving risk estimates based on only an average dose in populations with dose distributions that are often skewed and include subjects with very high, cell-killing doses.

The estimates from these studies are summarized in Tables 7-2 to 7-6 and Figures 7-1 to 7-6. In the following sections, differences and similarities among risk estimates are discussed, and conclusions are drawn, where possible, about radiation risks for each of the cancer sites of interest. Since the conditions of exposure, the characteristics of the study populations, and the extent and quality of the dosimetry and follow-up differ widely, the risk estimates derived for individual studies are not strictly comparable. They do, however, illustrate the range and significance of estimates obtained and provide some indication of the influence of the study-specific factors involved.

Lung Cancer

Lung cancer is the leading cause of cancer mortality in industrialized countries, and its incidence is rising in many developing countries. Cigarette smoking is accepted to be the primary cause of lung cancer. Also, ionizing radiation

TABLE 7-2 Risk Estimates for Cancer Incidence and Mortality from Studies of Radiation Exposure: Lung Cancer

Reference	Study	Radiation Type	Average Dose (Gy)	Dose Range	Cases	Controls/ Population	ERR/ Gy	LB	UB	EAR/ 10⁴ PY/ Gy	LB	UB	Comments
Incidence													
Inskip and others (1994)	Breast cancer	External	Cont. 4.6 Ipsi. 15.2		61	120	0.20	−0.62,	1.03				
Lundell and Holm (1995)	Hemangioma	Mostly Ra	0.12		11	14,351	1.40	ns	0.3				
Mattson and others (1997)	Benign breast disease	External	0.75	0–8.98	10	1216	0.38	<0,	0.6				
Gilbert and others (2003)	HD treatment	External	20	0–>60	146	271	0.15	0.06,	0.39				Overall
					107	200	0.18	0.06,	0.52				Men
					39	71	0.04	−0.01,	0.53				Women
Mortality													
Weiss and others (1994)	Ankylosing spondylitis	External X-ray	8.88	0.78–16.3	282		0.09	0.03,	0.15				Decrease afterwards
Howe (1995)	Fluoroscopy	External	1.02	0–24.2	1178	25,007	0.00	−0.06,	0.07				
Carr and others (2002)	Peptic ulcer	External	1.1	NA	21		0.43	−0.12,	1.35				Among subjects with lung dose <1.4 Gy

NOTE: The number of cases and controls (or population size in cohort studies), as well as the mean dose and range, relate only to exposed persons. Empty cells indicate data not available from publication. LB = lower bound; UB = upper bound of CI (usually 95%). EAR/10⁴ PY/Gy an all tables.

has been implicated as a lung carcinogen (UNSCEAR 2000b).

Of all the studies reviewed above, only seven provide dose-specific estimates of ERR and only one provides an estimate of the EAR. Table 7-2 and Figure 7-1 summarize the results from these studies. In the figure, results are shown for all studies as well as those restricted to an average dose to less than 1 Gy.

In the incidence studies, the ERR/Gy ranges from 0.15 Gy^{-1} in survivors of HD to 1.4 Gy^{-1} in patients treated for hemangioma in infancy. In mortality studies, estimates range from 0.00 per gray among tuberculosis patients exposed to fluoroscopy to 0.43 Gy^{-1} in patients treated with

radiation for peptic ulcer. Although risk estimates from these studies vary, confidence intervals are very large and the estimates shown are therefore statistically compatible.

In interpreting the results of these studies, differences in study populations and exposure patterns must be taken into account: the hemangioma study (which had 11 cases of lung cancer) included only patients who were exposed in infancy, while the average age in other cohorts ranged from 28 in tuberculosis patients exposed to fluoroscopic X-rays to 50 years among breast cancer survivors. Exposure rate and exposure pattern also vary across studies, with hemangioma patients having received a low-dose-rate, protracted exposure (over a period of 1 d to more than 2 years) and tubercu-

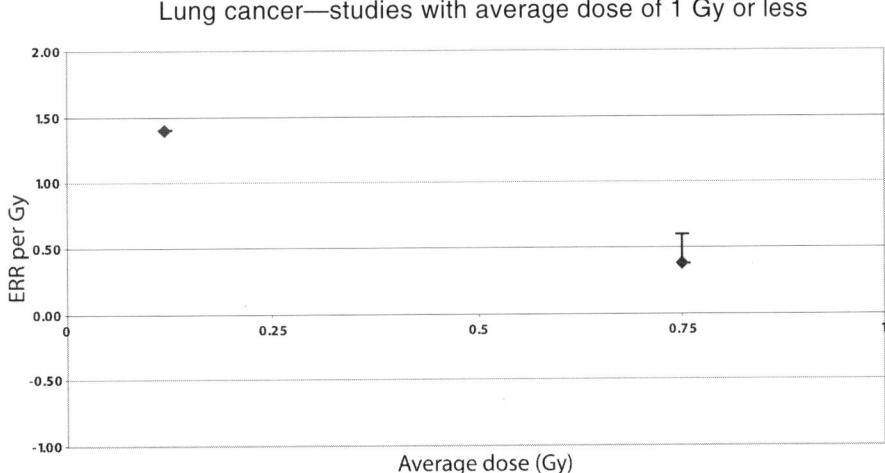

FIGURE 7-1 Distribution of study-specific estimates of ERR per gray for lung cancer according to average dose to the lung. Results are shown for all studies as well as studies in which the average dose to the lung was less than 1 Gy.

losis patients having received a highly fractionated dose (with 92 fluoroscopy sessions on average) at a low dose rate.

The ERR/Gy from the studies of acute high-dose-rate exposures are compatible and in the range 0.1–0.4 Gy^{-1}. The study of HD survivors showed little evidence for nonlinearity of the dose-response, despite the fact that the majority of patients received very high doses to the lung (in excess of 30 Gy). A multiplicative interaction was seen in this study between radiation dose and tobacco smoking, with a smaller ERR/Gy among women than men (difference not statistically significant).

It is difficult to evaluate the effects of age at exposure or of exposure protraction based on these studies because only one study (the hemangioma cohort) is available in which exposure occurred at very young ages and protracted low-dose-rate exposures were received. Risk estimates from that study are higher than those seen in other studies, but the difference is not statistically significant.

The study of tuberculosis patients, based on a very large number of lung cancer deaths, appears to indicate that substantial fractionation of exposure leads to a reduction in risk.

Female Breast Cancer

Breast cancer is the most commonly diagnosed cancer and cause of cancer mortality among women in North America and Western Europe. Incidence rates are lower in Asian countries. Ionizing radiation is well documented as a cause of breast cancer in women, especially when exposures occur in childhood and around puberty (UNSCEAR 2000b).

Of all the studies reviewed in the medical uses of radiation section above, only 11 provide dose-specific estimates of ERR and/or EAR. Table 7-3 and Figure 7-2 summarize the results from these studies. In the figure, results are shown for all studies as well as restricted to studies in which the average dose to the breast was less than 1 Gy.

In the incidence studies, the ERR/Gy ranges from 0.15 Gy^{-1} in women who received very high doses for HD radiotherapy (mean among cases = 25 Gy) to 2.5 Gy^{-1} in populations irradiated for enlarged thymus in infancy. The range in mortality studies is similar: from 0.08 Gy^{-1} among patients irradiated for treatment of ankylosing spondylitis to 2.7 Gy^{-1} in women repeatedly exposed to X-rays to monitor scoliosis. As indicated previously, in the international cervical cancer follow-up study (Boice and others 1988) the significant reduction of risk seen among women with intact ovaries was probably attributable to the cessation of ovarian function related to radiotherapy; only the risk estimates in women with no ovaries are considered here. Similarly, the results from Travis and colleagues are restricted to women who had chest radiotherapy only (and hence exclude women with high doses to the ovaries).

Although the risk estimates from these studies vary considerably, confidence intervals are very large and the estimates shown are therefore statistically compatible, except

for the study of HD patients where the doses to the breast (up to more than 60 Gy) will have led to cell killing and hence a reduction of risk per gray.

The situation is somewhat different for EAR. Few studies have reported risk estimates in terms of EAR. The estimates shown in Table 7-3 and Figure 7-2 are quite variable, and several of the confidence intervals do not overlap, indicating heterogeneity in risk estimates across these studies. In reviewing these results, differences in study populations and exposure patterns must be taken into account. These include the following:

- The thymus and hemangioma studies relate mainly to patients who were irradiated in infancy; in the scoliosis study, the mean age at first exposure was 10.1 years, while in the other studies, the majority of subjects were adults at the time of radiation exposure. Mean ages ranged from 25 to 52 years, respectively, in the Massachusetts fluoroscopy study and the cervical cancer survivor study.
- Exposure patterns ranged from very protracted low-dose-rate or fractionated exposures in the scoliosis and tuberculosis studies, where diagnostic radiation was used to monitor the evolution of the disease, to high-dose-rate, acute or much less fractionated exposures received for treatment of disease.
- Exposures in childhood and adolescence, particularly in the time around puberty (Doody and others 2000), have been shown to be associated with higher risks of radiation-induced breast cancer than exposures later in life. Figure 7-3 shows the relation between ERR/Gy and average age at exposure. The ERR appears to decrease with increasing age at exposure.

Exposure in infancy led to an ERR of 2.5 Gy^{-1} in the Rochester thymic irradiation study, based on 22 exposed cases, but the ERR in the pooled analysis of data on hemangioma patients was much lower, 0.35 Gy^{-1}. A previous report of a much higher ERR in the Stockholm hemangioma cohort 50 or more years after exposure was not confirmed in the pooled analysis.

Exposure at age 15 in the Massachusetts fluoroscopy study was estimated to result in an ERR/Gy of 1.0, and the ERR for breast cancer mortality was 2.7 Gy^{-1} in the scoliosis study (with average age at first exposure of 10.1 years) and 0.9 Gy^{-1} overall in the Canadian fluoroscopy study for exposures between 15 and 25 years of age. Exposures at older ages tended to result in lower risk estimates, ranging from 0.33 among cervical cancer patients with no ovaries to about 0.61 in women exposed to fluoroscopy, and to about 1.63 following high-dose-rate irradiation for benign breast disease. The ERR was lower, but not statistically incompatible, in the ankylosing spondylitis study, probably related to the fact that the dose to the ovary was high among these women (ERR not shown in Figure 7-3).

Exposure fractionation does not appear to be an important determinant of risk per gray in the fluoroscopy studies.

TABLE 7-3 Risk Estimates for Cancer Incidence and Mortality from Studies of Radiation Exposure: Female Breast Cancer

Reference	Study	Radiation Type	Average Dose (Gy)	Dose Range	Cases	Controls/ Population	ERR per Gy	LB	UB	EAR/ 10⁴ PY/ Gy	LB	UB	Comments
Incidence													
Boice and others (1988)	Cervix	External + ²²⁶Ra	0.3		953	1,806	−0.20	<−0.2,	0.3				Overall
					145	284	0.33	<−0.3,	5.8				Women with no ovaries
Hildreth and others (1989)	Thymus		0.69	[0.02–7.5]ᵃ	22	1,201	2.48	1.1	5.2	5.7	2.9,	9.5	
Boice and others (1991b)	Fluoroscopy	External	0.79	[0.02–5]ᵃ	147	2,573	0.61	0.3	1.01	10.7	6.0	15.8	Overall
							1.00						Exposure at age 15
							0.70						Exposure at age 25
							0.10						Exposure at age 35
Mattsson and others (1995)	Benign breast disease	External	5.8	0–50	47		1.63	0.77	2.89				
Shore and others (1986)	Postpartum mastitis	External	3.8	[0.6–14]ᵃ	51	601	0.40	0.2	0.7				
Lundell and others (1996)	Hemangioma	Mostly ²²⁶Ra	0.39 0.22	<0.01–5.8	75 16	9,675	0.38 2.25	0.09 0.59	0.85 5.62	0.4 22.9			Overall 50+ years after exposure
Lundell and others (1999)	Hemangioma	Mostly ²²⁶Ra	0.29	<0.01–35.8	236	17,202	0.35	0.18	0.59	0.7	0.4	1.1	
Travis and others (2002)	HD survivors	External	22	<0.1–61.3	67	122	0.15	0.04	0.73				Women with chest radio-therapy only
Mortality													
Weiss and others (1994)	Ankylosing spondylitis	External	0.59	0.07–1.27	42		0.08	−0.3,	0.65	NA		NA	Note—high dose to the ovary
Howe and others (1996)	Fluoroscopy	External	2.13	0–18.4	103		3.56	1.85	6.82	10.3	6.4	16.2	Nova Scotia: for exposures at age 15— EAR 20 years after exposures
			0.79	0–14.4	578		0.40	0.13	0.77	1.2	0.4	2.3	Other provinces: for exposures at age 15— EAR 20 years after exposures
					681		0.90	NA	3.2				Overall: for exposures at age 15— EAR 20 years after exposures
Doody and others (2000)	Scoliosis	External	0.11	0–1.7	70	4,942	2.7	−0.2	9.3				Among women with at least one radiographic examination

NOTE: The number of cases and controls (or population size in cohort studies) as well as the mean dose and range relate only to exposed persons. Empty cells indicate data not available from publication. LB = lower bound; UB = upper bound of CI.

ᵃFrom Preston and others 2002.

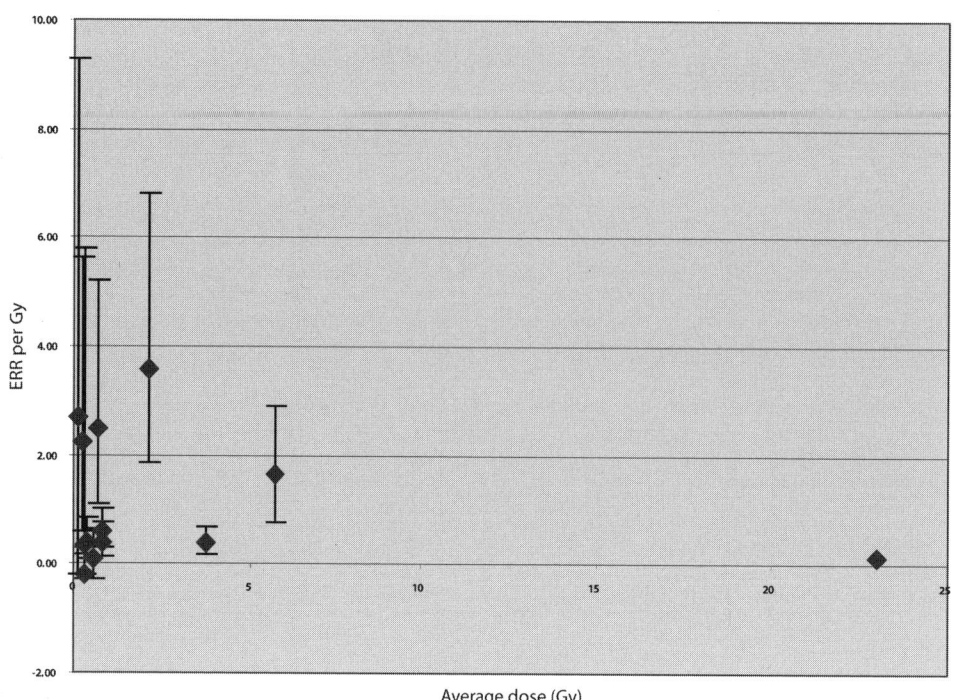

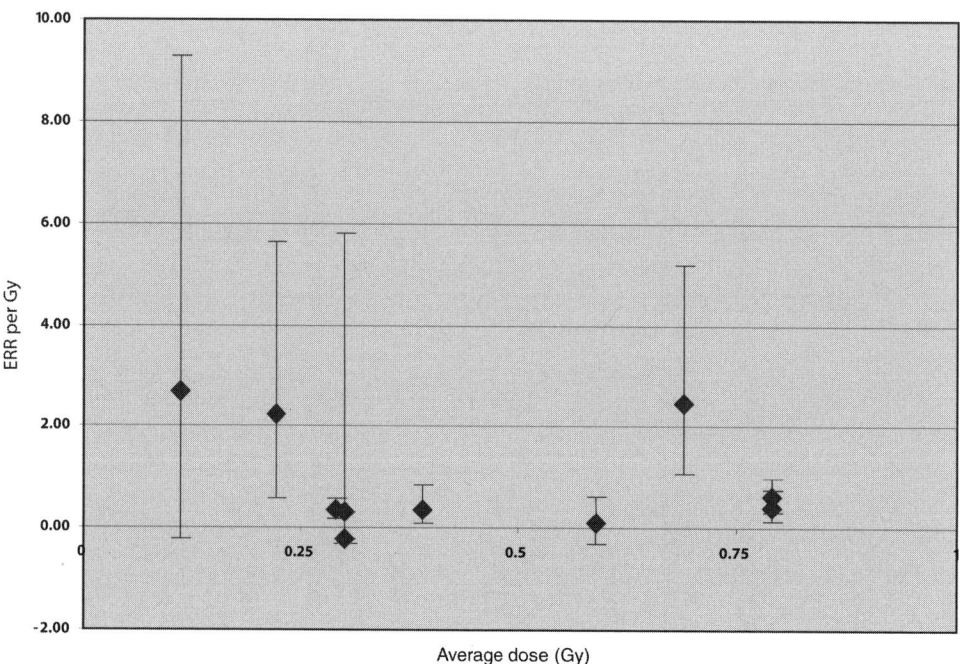

FIGURE 7-2 Distributions of study-specific estimates of ERR and EAR for breast cancer according to level of average dose to the breast.

EAR per 10⁴ PY per Gy—breast

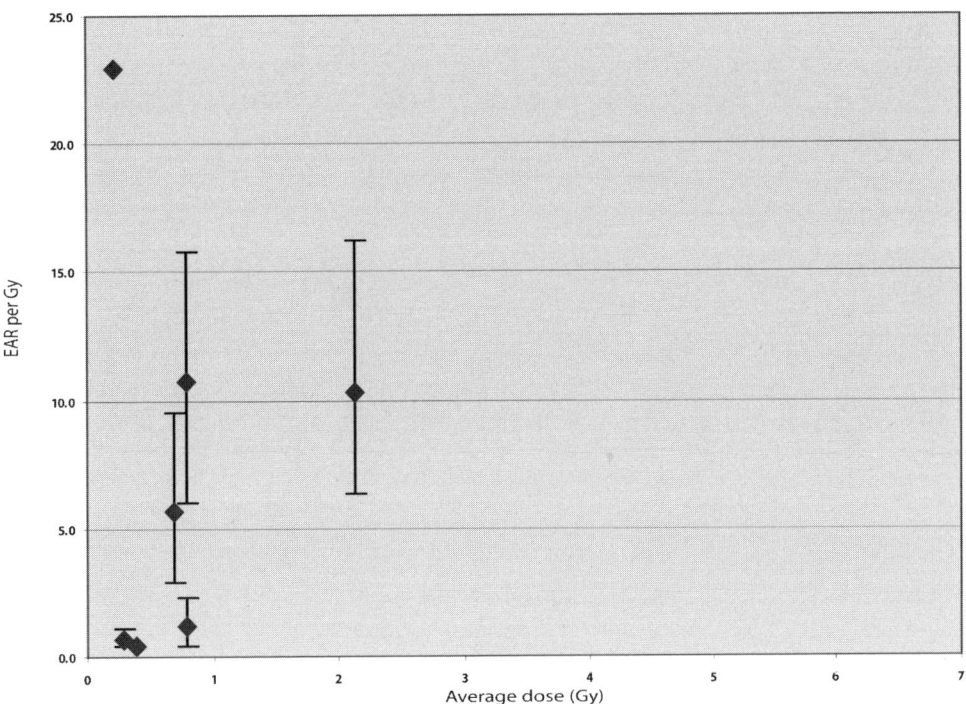

EAR per 10⁴ PY per Gy—breast—studies with average dose of 1 Gy or less

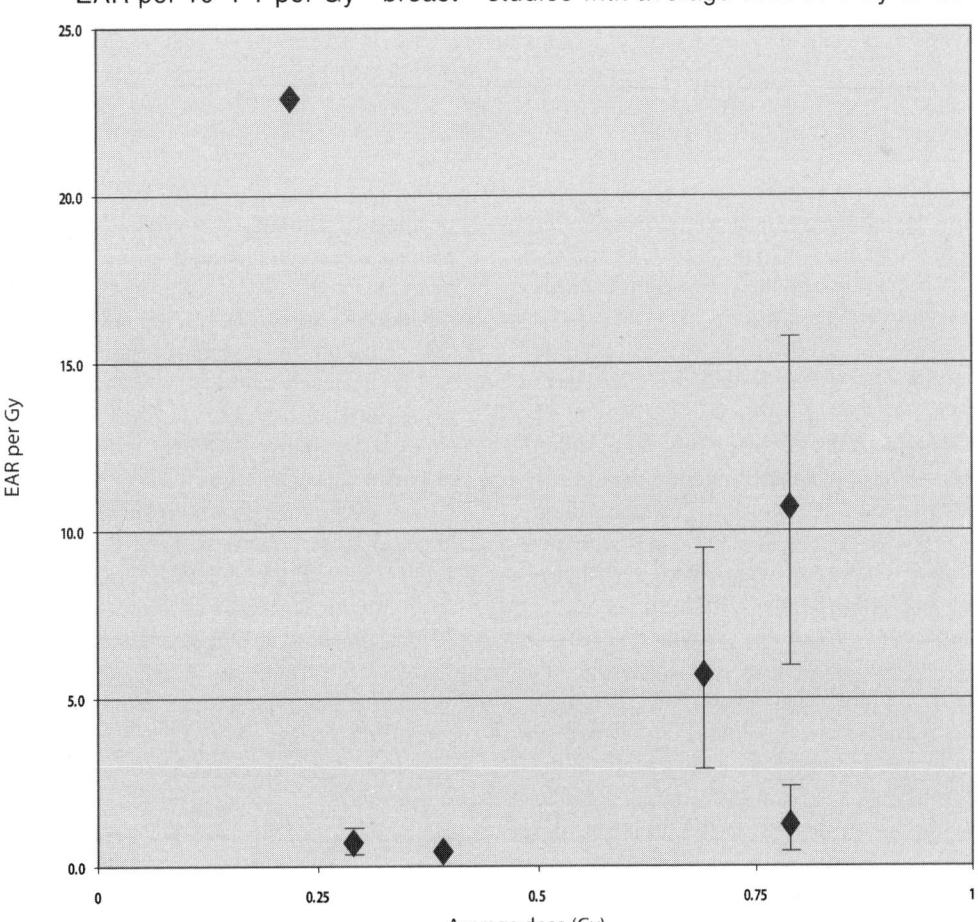

Breast cancer ERR by age at exposure

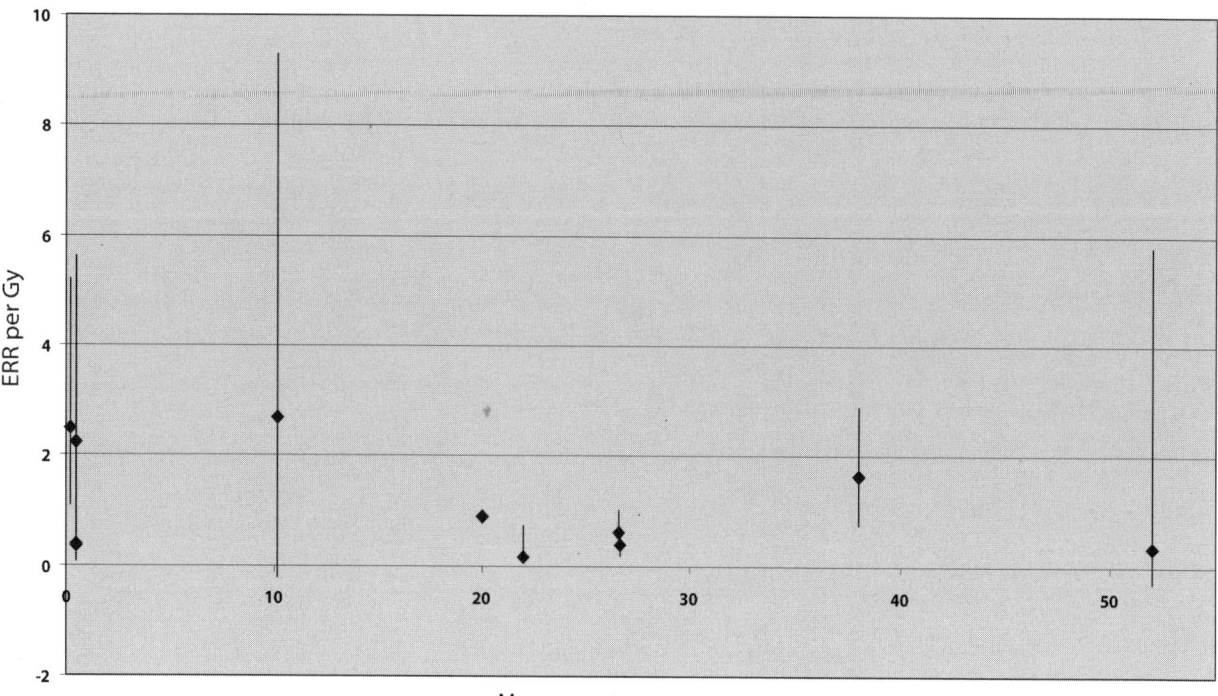

FIGURE 7-3 Distribution of study-specific estimates of ERR/Gy for breast cancer according to average age at exposure.

Protraction of low-dose-rate exposure in the hemangioma cohort may account for the reduced risk following exposures in infancy in this cohort, although analyses within this study do not indicate a significant association.

Preston and colleagues (2002b) carried out a pooled analysis of eight cohorts to estimate radiation-induced breast cancer risk and evaluate the role of modifying factors. The analyses included studies of the following populations: Japanese atomic bomb survivors (Thompson and others 1994), the original and extended Massachusetts TB fluoroscopy cohorts (Boice and others 1991b), the New York acute post partum mastitis cohort (Shore and others 1986), the Rochester infant thymic irradiation cohort (Hildreth and others 1989), the Swedish benign breast disease cohort (Mattsson and others 1993), and the Gothenburg and Stockholm skin hemangioma cohorts (Lindberg and others 1995). The analyses included 1502 breast cancer cases among 77,527 women, about half of whom were exposed to radiation, with 1.8 million person-years of follow-up. No simple unified summary model adequately described the excess risk in all of these studies.

The excess risks for the thymus, tuberculosis, and atomic bomb survivor cohorts showed similar temporal trends, de-

pending on attained age in the ERR model and on both age at exposure and attained age in the EAR model. The excess rates appeared to be similar in these cohorts, with a combined EAR estimate of 9.9 per 10^4 PY per gray (95% CI 7.1, 1.4) at age 50, suggesting similarity of risks following acute and fractionated low-dose-rate exposure. The ERR/Gy was greater among Japanese atomic bomb survivors; this difference may be partly attributed to the lower background rates of breast cancer in Japan.

The excess rates were higher for the mastitis and benign breast disease cohorts with EAR estimates of 15 (95% CI 7.7, 24) and 32 (95% CI 21, 47) per 10^4 PY per gray, respectively, suggesting that women with some benign breast conditions may be at an elevated risk of radiation-induced breast cancer.

The hemangioma cohorts showed lower risks (EAR: 5.1 per 10^4 PY per gray; 95% CI 1.3, 11), suggesting a reduction of risks following protracted low-dose-rate exposures.

Thyroid Cancer

Thyroid cancer is one of the less common forms of cancer. Its incidence is relatively high before age 40, it increases comparatively slowly with age, and it is about three times higher in women than men. Ionizing radiation is a well-documented cause of thyroid cancer (UNSCEAR 2000b).

Of all the studies reviewed in the medical uses of radiation section above, only six provide dose-specific estimates of ERR and/or EAR. Table 7-4 and Figure 7-4 summarize the results from these studies. In Figure 7-4, results are shown for all studies as well as restricted to studies in which the average dose to the thyroid was less than 1 Gy.

All of the studies shown are studies of children who received radiotherapy for benign conditions. All results relate to thyroid cancer incidence. Because of the relatively good prognosis of most papillary thyroid cancers, studies of thyroid cancer mortality add little information about radiation risks.

In studies of external radiation exposure, the ERR/Gy ranges from 3 Gy^{-1} in children exposed for enlarged tonsils and other benign head and neck disorders to 30 Gy^{-1} among those exposed in Israel for the treatment of tinea capitis. Similarly, the estimates of EAR vary from 0.9 per 10^4 PY per gray in the hemangioma study to 13 in the tinea capitis study. Although risk estimates from these studies vary considerably, the confidence intervals tend to be large; it is likely that the estimates shown are statistically compatible.

Ron and colleagues (1995a) conducted combined analyses of data from seven studies including five cohort studies—the atomic bomb survivors study, the Rochester thymic irradiation study (Shore and others 1993a), the Israeli tinea capitis study (Ron and others 1989), and the Michael Reese and Boston enlarged tonsil studies (Pottern and others 1990; Schneider and others 1993)—as well as two case-control studies of thyroid cancer nested respectively within the International Cervical Cancer Survivor Study (Boice and others 1988) and the International Childhood Cancer Survivor Study (Tucker and others 1991). The analyses included a total of 707 cases, the majority of which (apart from the A-bomb and cervical cancer survivors) were below age 15 at time of exposure.

For subjects exposed below age 15, a linear dose-response was seen with a leveling or decrease in risk at the higher doses used for cancer therapy. The pooled ERR was 7.7 Gy^{-1} (95% CI 2.1, 28.7), based on a random effects model that took into account the heterogeneity of risk across studies. The EAR was estimated to be 4.4 per 10^4 PY per gray (95% CI 1.9, 10.1). Both of these estimates were significantly affected by age at exposure, with a strong decrease in risk with increasing age at exposure and little apparent risk for exposures after age 20. The ERR appeared to decline over time

TABLE 7-4 Risk Estimates for Cancer Incidence and Mortality from Studies of Radiation Exposure: Thyroid Cancer

Reference	Study	Radiation Type	Average Dose (Gy)	Dose Range	Cases	Controls/ Population	ERR/ Gy	95% CI	EAR/ 10^4 PY/ Gy	LB	UB	Comments
Incidence												
Ron and others (1989)	Tinea capitis	External X-ray	0.09		98	10,834	30.0		13.0			
Shore (1992)	Meta-analysis of hyperthyroidism studies	^{131}I	88		2	602	0.30	(0, 0.9)[a]	0.1	(0.0, 0.2)[a]		
Schneider and others (1993)	Benign head and neck	External X-ray	0.6		309	234	3.00					Overall
			0.6		109		9.20					Before 1974 and screening program
			0.6		200		1.80					After 1974
Shore and others (1993a)	Enlarged thymus	External X-ray	1.36	0.03–10	37	2,657	9.00	(4, 24)[a]	2.9	(2.1, 3.9)[a]		
Lundell and others (1994)	Hemangioma	Mostly Ra	1.07	<0.01–4.34	17	14,351	4.92	(1.26, 10.2)	0.9			
Lindberg and others (1995)	Hemangioma	^{226}Ra	0.12		15	11,807	7.50	(0.4, 18.1)	1.6			
Ron and others (1995a)	Pooled analysis of 7 studies	External X-ray			700	58,000	7.7	(2.1, 28.7)	4.4	(1.0, 10.1)		

NOTE: The number of cases and controls (or population size in cohort studies), as well as the mean dose and range, relate to exposed persons only. Empty cells indicate data not available from publication. LB = lower bound; UB = upper bound of CI.

[a]90% CI.

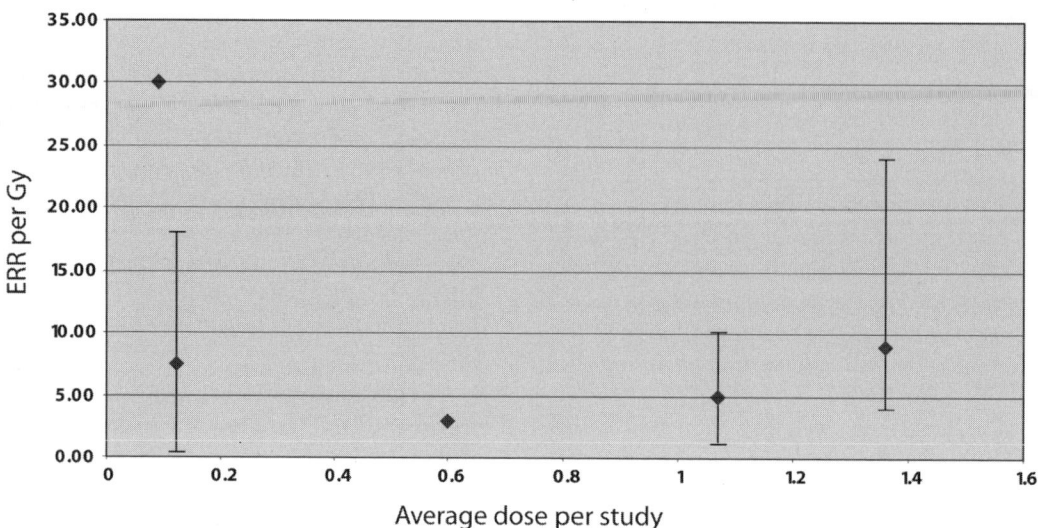

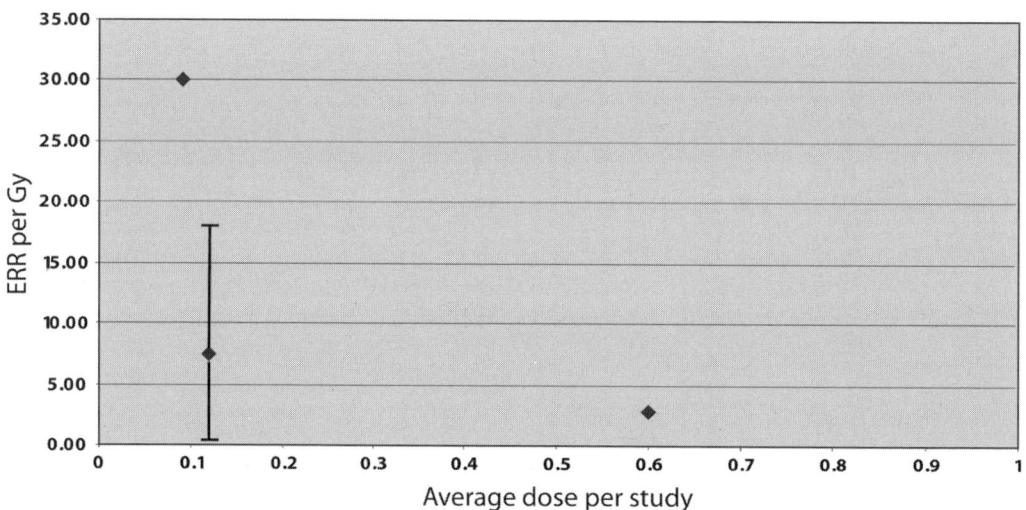

FIGURE 7-4 Distribution of study-specific estimates of ERR/Gy for thyroid cancer according to level of average thyroid dose. NOTE: The estimate from populations exposed to ^{131}I is excluded.

about 30 years after exposure but was still elevated at 40 years.

Three studies provided data on exposure protraction or fractionation (thymus, tinea capitis, and Michael Reese); analyses indicate that a small nonsignificant decrease in risk may be related to exposure fractionation in these studies.

A meta-analysis of hyperthyroidism studies provides a risk estimate of thyroid cancer in relation to ^{131}I exposure in childhood (Shore 1992). The ERR estimate from that study

is 0.3 Gy^{-1}, lower than that from studies of external exposures, but based on only two exposed cases. This study therefore provides little information about the risk of thyroid cancer in relation to exposure to this nuclide. Studies of the effects of ^{131}I exposure later in life are reviewed in the preceding section, although no dose-related estimate of risks have been provided. These studies, taken together, provide little evidence of an increased risk of thyroid cancer following ^{131}I exposure after childhood.

Leukemia

Leukemia is one of the less common malignancies, but substantial epidemiologic and experimental information exists on the leukemogenic effects of ionizing radiation (UNSCEAR 2000b).

Of all the studies reviewed in the "Medical Uses of Radiation" section, six provide dose-specific estimates of ERR and/or EAR. Table 7-5 and Figure 7-5 summarize the results from these studies. In the figure, results are shown for all studies as well as restricted to studies in which the average dose to the active bone marrow was less than 1 Gy. Results shown are for leukemia excluding CLL in all studies except the tinea capitis and uterine bleeding studies.

The ERRs/Gy shown in Table 7-5 range from 0.88 Gy^{-1} in women who received an average dose to the active bone marrow of 7 Gy from radiotherapy for cervical cancer to 12.4 Gy^{-1} in subjects treated for ankylosing spondylitis (average dose 4.4 Gy). All other estimates, from studies with average doses ranging from 0.1 to 2 Gy, are relatively close, in the range 1.9 to 5 Gy^{-1}, and are statistically compatible.

Three studies have provided estimates of EAR per 10^4 PY per gray. Risk estimates between these studies are relatively close, ranging from 1 to 2.6.

In most of the studies included here, the majority of subjects were adults at the time of exposure (with average ages at exposure between 45 and 52 years in the uterine bleeding, benign breast disease, and cervical cancer survivor studies). Only the tinea capitis and hemangioma studies provide information about exposures in childhood. In the hemangioma study—where all subjects were irradiated in infancy—the overall ERR/Gy is similar to that seen in other studies; it is notable, however, that this is driven mainly by a higher ERR for childhood leukemia; the ERR for adult leukemia in this study was very close to zero. In the tinea capitis study, in which all exposures were below age 15, no ERR is shown; the EAR is similar to that seen in the other studies.

In one study (Inskip and others 1993), an effort was made to estimate separately the effects of external exposures, ^{226}Ra, and the combination of the two. Estimates of risk from ^{226}Ra alone or in combination with external radiation are

TABLE 7-5 Risk Estimates for Cancer Incidence and Mortality from Studies of Radiation Exposure: Leukemia Excluding CLL

Reference	Study	Radiation Type	Average Dose (Gy)	Dose Range	Cases	Controls/ Population	ERR per Gy	95% CI	EAR/ 10^4 PY/ Gy	LB	UB	Comments
Incidence												
Boice and others (1985)	Cervix	External X-rays + intracavitary ^{226}Ra	7		143	745	0.88	(SE: 0.69)				
Ron and others (1988b)	Tinea capitis	External X-ray	0.3		14[a]	10,834			0.9			
Inskip and others (1990b)	Uterine bleeding	^{226}Ra	0.53		34[a]	4,483	1.90	(0.8, 3.2)	2.6			
Inskip and others (1993)	Benign gynecological disease	Overall			39	8,352	2.90		1.2			
		Rad + ext	2.03		9	1,437	2.10	(0.5, 8.3)	1.0	(0.3, 1.9)		
		^{226}Ra	2.31		26	5,508	3.70	(−1, 15)	1.5	(0.3, 2.9)		
		External	0.59		4	1,407	0.50	(−0.6, 3.3)	0.1	(−0.2, 0.6)		
Lundell and Holm (1996)	Hemangioma	Mostly Ra	0.13	<0.01–4.6	20	14,624	1.60	(−0.6, 5.5)				Overall
			0.13	<0.01–4.6	9	14,624	5.01	(0.1, 15)				Childhood leukemia only
			0.13	<0.01–4.6	11	14,624	−0.02	(−0.8, 1.9)				Adult leukemia only
Mortality												
Weiss and others (1995)	Ankylosing spondylitis	External X-ray	4.38	1.27–6.99[b]	35	1,745[c]	12.4	(2.25, 52.1)				

NOTE: The number of cases and controls (or population size in cohort studies) as well as the mean dose and range relate to exposed persons only. Empty cells indicate data not available from publication. LB = lower bound; UB = upper bound of CI.

[a]All forms of leukemia combined.

[b]10–90% range.

[c]Subcohort with reconstructed doses.

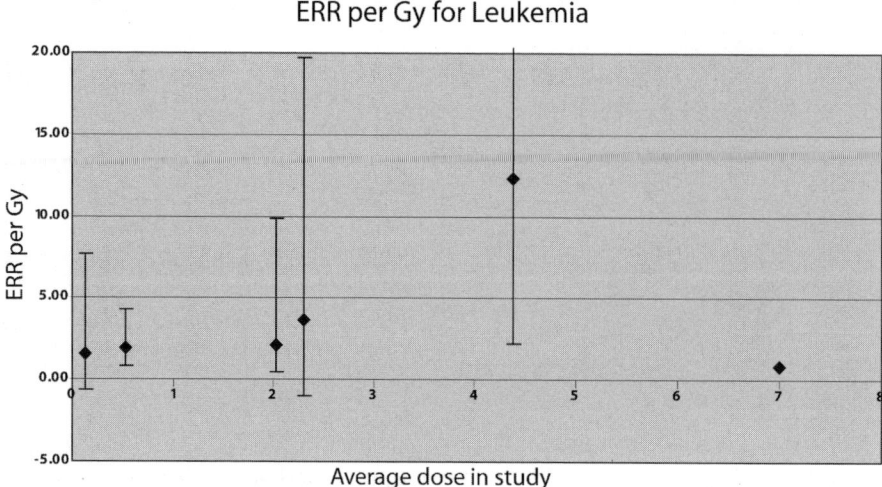

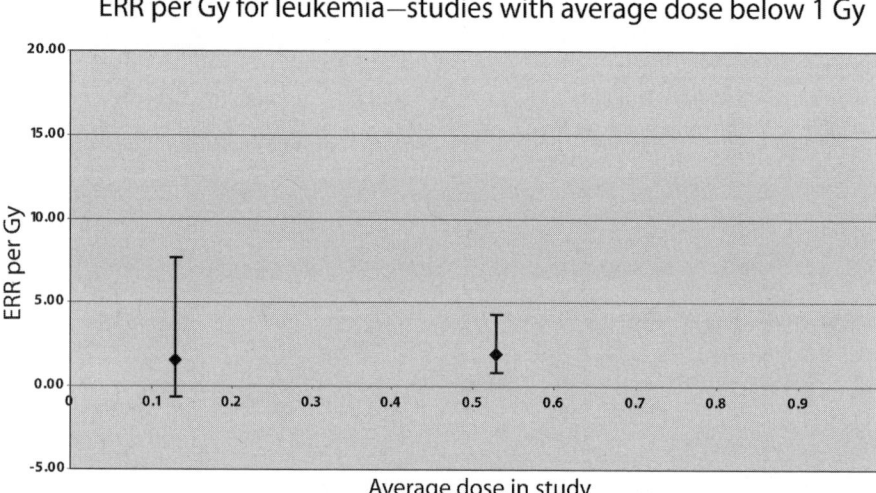

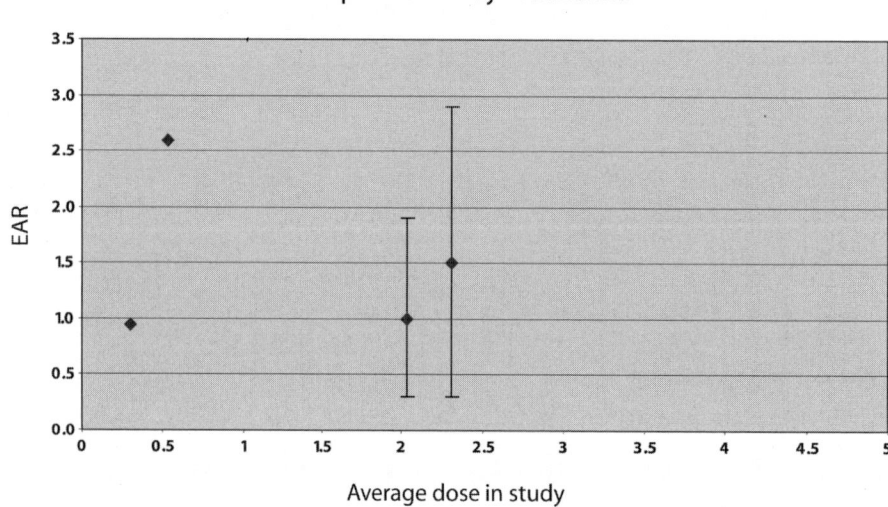

FIGURE 7-5 Distribution of study-specific estimates of ERR/Gy for leukemia according to level of average dose to the active bone marrow.

similar; the ERR/Gy for external exposures appears to be lower, but this result is based on only four cases exposed.

Stomach Cancer

Incidence rates for stomach cancer vary considerably throughout the world, with particularly high rates in Japan. Many countries have seen decreases in incidence and mortality over the past 50 years or so, believed in large part to be due to healthier diets with increased fruits and vegetables and less salt.

Of all the studies reviewed on medical uses of radiation, five provide dose-specific estimates of ERR and/or EAR. Table 7-6 and Figure 7-6 summarize the results from these studies. In the figure, results are shown for all studies as well as restricted to studies in which the average dose to the active bone marrow was less than 1 Gy.

Among the studies of populations with external radiation exposure and/or ^{226}Ra, the estimates of ERR/Gy range from negative (in the hemangioma study) to 1.3 Gy^{-1} in the study of benign breast disease. The confidence intervals are wide, and they all overlap, indicating that these estimates are statistically compatible. An ERR of 1.32 Gy^{-1} (not significantly different from zero) was seen among patients treated for hyperthyroidism with ^{131}I.

Radiation and Circulatory Diseases

Although radiation exposure is well established as a risk factor for cancer, a clear understanding of the relationship between radiation exposure and other diseases is lacking. It has been postulated that the cardiovascular system is resistant to radiation-induced injury (Stewart and others 1995). However, it appears that tissue damage may occur as a result of both therapeutic (Stewart and Fajardo 1984) and A-bomb radiation exposure (Villeneuve and Morrison 1997; Shimizu and others 1999). Capillaries represent the most radiosensitive component of the cardiovascular system, with characteristic changes including detachment of endothelial cells and thrombosis. Arterial changes resulting from radiation exposure depend on vessel size, with small and medium-sized arteries undergoing changes in all vessel layers, and large arteries appearing to be relatively radioresistant, although radiation exposure may predispose larger vessels to the development of atherosclerosis (Louis and others 1974).

Radiation exposure has also been implicated in the development of cerebrovascular injury (O'Connor and Mayberg 2000). Specific conditions postulated to arise from irradiation include vasculopathy, intracranial aneurysm formation, cerebral radiation necrosis, intracranial atherosclerosis, and stroke (Trivedi and Hannan 2004).

Both animal and human studies have identified intimal thickening, lipid deposition, and adventitial fibroses of the

TABLE 7-6 Risk Estimates for Cancer Incidence and Mortality from Studies of Radiation Exposure: Stomach Cancer

Reference	Study	Radiation Type	Average Dose (Gy)	Dose Range	Cases	Controls/ Population	ERR/ Gy	95% CI	EAR/ 10^4 PY/ Gy	LB	UB	Comments
Incidence												
Boice and others (1989)	Cervix	External X-rays + intracavitary ^{226}Ra	2	0.5–3.5	348	658	0.54	(0.05, 1.5)[a]	3.2	(0.1, 10.4)		
Holm and others (1991)	Hyperthyroidism	^{131}I	0.07		29		1.3		9.6			
Lundell and Holm (1995)	Hemangioma	Mostly Ra	0.09		5	14,351	<0		<0			
Mattsson and others (1997)	Benign breast disease	External	0.66	0–5.4	14	1,216	1.3	(0, 4.4)				
Weiss and others (1994)	Ankylosing spondylitis	External X-ray	3.2	0.52–5.8	127	1,745[b]	–0.004	(–0.05, 0.05)				
Carr and others (2002)	Peptic ulcer	External	8.9		11	1,859	0.20	(0, 0.73)				Among subjects with stomach dose 10 Gy

NOTE: The number of cases and controls (or population size in cohort studies) as well as the mean dose and range relate to exposed persons only. Empty cells indicate data not available from publication. LB = lower bound; UB = upper bound of CI.

[a]90% confidence interval.

[b]Subcohort with reconstructed doses.

ERR per Gy—Stomach cancer

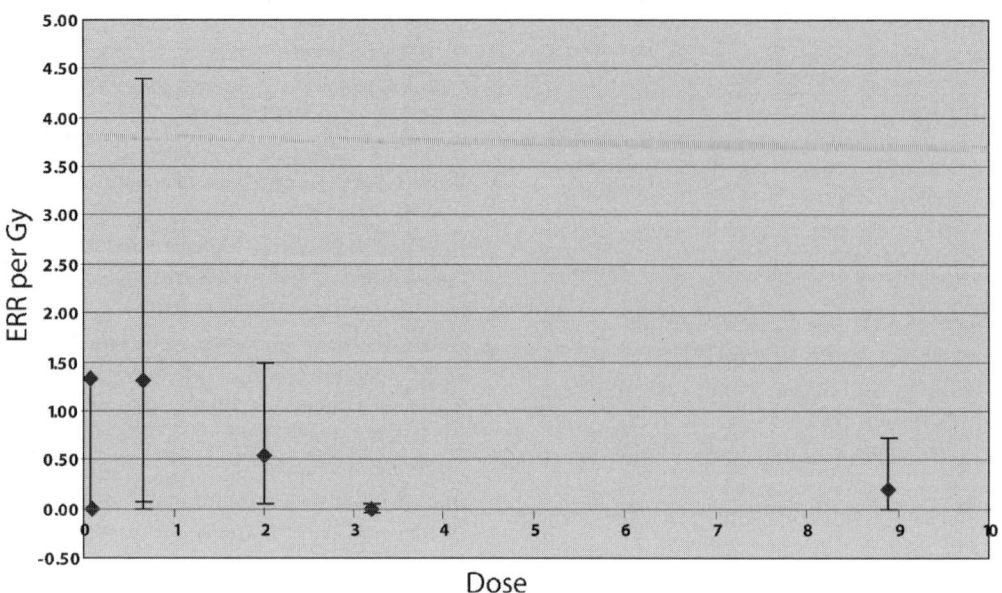

ERR per Gy—Stomach cancer—Studies with doses below 1 Gy

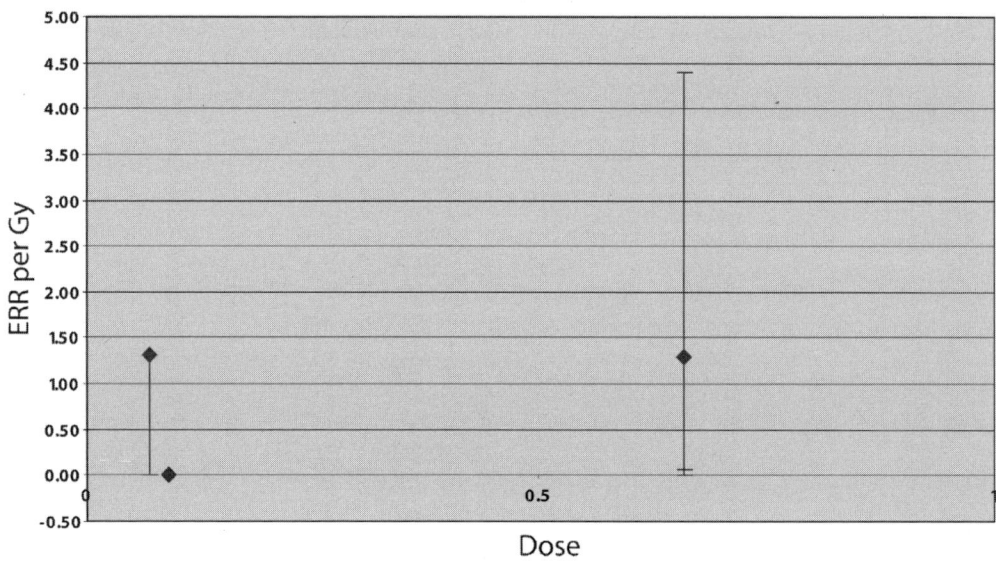

FIGURE 7-6 Distribution of study-specific estimates of ERR/Gy for stomach cancer according to level of average dose to the stomach.

vascular system following irradiation. These changes are associated with atherosclerosis and the normal aging process, although irradiation may accelerate the development of these conditions (Trivedi and Hannan 2004).

Although the dose required to produce specific conditions or vascular effects is uncertain, it appears that over extended periods, the nature of the changes induced are similar for low doses (on the order of 5 Gy) and for high doses (in the region of 40 Gy). There is a broad spectrum and severity of cardiovascular diseases, with radiation being only one of many possible risk factors that may act directly or indirectly on the vasculature. To clarify the role of radiation in the etiology of cardiovascular diseases, further studies involving long-term, low-level exposures are needed, taking into account all of the known risk factors for cardiovascular outcomes.

Excess heart disease mortality has been observed among women with breast cancer who were irradiated with cobalt-

60 (Host and Loeb 1986, not reviewed here) and among persons with HD who received mediastinal irradiation (Boivin and Hutchison 1982; Hancock and others 1993b, 1993c). Most affected patients had received at least 30 Gy to the mediastinum, although some had received less (Trivedi and Hannan 2004).

DISCUSSION

Since the publication of BEIR V (NRC 1990), new information concerning health effects of radiation exposures has become available from epidemiologic studies of populations exposed to medical uses of ionizing radiation. The longer follow-up periods in recent reports have increased the statistical power in examining dose-response relationships at the doses used for medical purposes.

Available studies of the effects of radiotherapy for malignant or benign diseases confirm the presence of a heightened risk of development of a number of primary or second primary cancers on follow-up. Because the doses in most series far exceed 100 mGy to the site of interest, they provide limited direct quantitative information on the risk of low-level radiation, particularly when they involve large doses where cell killing may lead to underestimation of the risk per unit dose. These studies provided valuable information for the study of risk modifiers, including age at exposure, attained age, and possible differences in patterns of risk across countries.

Analyses that are restricted to populations with low doses are complicated by the limitations of statistical variability as well as by limitations of sample size and study design, including dose reconstruction. Limitations also include chance, small undetected biases, and the consequences of doing multiple tests of statistical significance. Indeed, among diagnostic radiation studies, only studies of repeated chest fluoroscopy and scoliosis examinations are informative concerning the magnitude of ERR and EAR as a function of dose. It must be noted that although the dose rates in these studies are low, the cumulative doses received by tuberculosis patients are high, and even scoliosis patients followed radiologically for spine curvature received average cumulative doses of the order of 100 mGy or more.

Most of the information on radiation risks therefore still comes from studies of populations with medium to high doses, with the notable exceptions of childhood cancer risk following *in utero* exposures and thyroid cancer risk following childhood exposures, for which significant increases have been shown consistently in the low- to medium-dose range.

SUMMARY

In this chapter on medical radiation, particular attention has been paid to estimating the risk of cancer at specific sites—namely, the lung, breast, thyroid, and stomach, as well

as leukemia risk. Information that has become available since 1989 has contributed to the examination of risks for these malignancies.

A large number of studies involving radiation exposure for medical reasons have been described and discussed. Although these studies of medically exposed cohorts have increased our general knowledge of radiation risks, not all of them contribute substantially to quantitative risk assessment. Many studies lack the sample size and high-quality dosimetry that are necessary for precise estimation of risk as a function of dose, a point that is illustrated by the large confidence intervals for many of the risk estimates shown in Tables 7-2 to 7-6 and by the limited number of studies for which risk estimates per gray are available.

Nevertheless, studies of populations exposed to therapeutic and diagnostic radiation provide information on issues that cannot be addressed with atomic bomb survivor data alone. Some examples are the evaluation of risk in Caucasian populations where baseline cancer and other disease risks may be very different from those in a Japanese cohort. Also, studies of medically exposed cohorts allow for the evaluation of risk from protracted exposures. In addition, studies of medical uses of radiation have been important in establishing the lack of radiation risk for CLL, since this cancer is very rare in Japan.

Often there is interest in comparing results from different studies to gain information on the modifying effects of factors such as baseline risks and protraction of exposure that may differ among the studies. It should be kept in mind that such comparisons can be difficult to interpret since there are nearly always several differences among the cohorts being compared. As an illustration, the ERR/Gy for breast cancer in the Life Span Study cohort has been found to be higher than the ERR/Gy in tuberculosis fluoroscopy patients (Howe and McLaughlin 1996; Little and Boice 1999; Preston and others 2002a). However, it is not clear whether this difference occurs because of the higher baseline risks in the Caucasian fluoroscopy cohorts, the lower dose rate in these patients, the lower energy of the X-ray exposure used in fluoroscopy (Brenner 1999), or some combination of these factors.

For lung cancer, the ERR/Gy from the studies of acute, high-dose-rate exposures are statistically compatible and in the range 0.1–0.4 Gy^{-1}. It is difficult to evaluate the effects of age at exposure or of exposure protraction based on these studies because only one study (the hemangioma cohort) is available in which exposure occurred at very young ages and in which protracted low-dose-rate exposures were received. The study of tuberculosis patients appears to indicate that substantial fractionation of exposure leads to a reduction of risk.

For breast cancer, EARs appears to be similar (of the order of 9.9 per 10^4 PY per gray at age 50) following acute and fractionated exposures to moderate- to high-dose-rate radiation. Effects of attained age and age at exposure are important modifiers of risk. The excess risks appear to be higher in populations of women treated for benign breast conditions, suggesting that these women may be at an elevated risk of radiation-induced breast cancer. The hemangioma cohorts showed lower risks, suggesting a possible reduction of risks following protracted low-dose-rate exposures.

For thyroid cancer, all of the studies providing quantitative information about risks are studies of children who received radiotherapy for benign conditions. A combined analysis of data from some of these cohorts with data from the atomic bomb survivors and from two case-control studies of thyroid cancer nested within the International Cervical Cancer Survivor Study and the International Childhood Cancer Survivor Study provides the most comprehensive information about thyroid cancer risks. For subjects exposed below the age of 15, a linear dose-response was seen, with a leveling or decrease in risk at the higher doses used for cancer therapy. The pooled ERR was 7.7 Gy^{-1} and the EAR 4.4 per 10^4 PY per gray. Both estimates were significantly affected by age at exposure, with a strong decrease in risk with increasing age at exposure and little apparent risk for exposures after age 20. The ERR appeared to decline over time about 30 years after exposure but was still elevated at 40 years.

Little information on thyroid cancer risk in relation to ^{131}I exposure in childhood was available. Studies of the effects of ^{131}I exposure later in life provide little evidence of an increased risk of thyroid cancer following ^{131}I exposure after childhood.

For leukemia, ERR estimates from studies with average doses ranging from 0.1 to 2 Gy are relatively close, in the range 1.9 to 5 Gy^{-1}, and are statistically compatible. Estimates of EAR are also similar across studies, ranging from 1 to 2.6 per 10^4 PY per gray. Little information is available on the effects of age at exposure or of exposure protraction.

For stomach cancer, the estimates of ERR/Gy range from negative to 1.3 Gy^{-1}. The confidence intervals are wide and they all overlap, indicating that these estimates are statistically compatible.

Finally, results of two studies of patients having undergone radiotherapy for HD or breast cancer suggest that there may be some risk of cardiovascular morbidity and mortality for very high doses and dose-rate exposures. The magnitude of the radiation risk and the shape of the dose-response curve for these outcomes, if an effect exists, are uncertain.

In conclusion, studies of medically irradiated populations provide information on the magnitude of risk estimates (mainly in the medium- to high-dose range) and on the effects of factors, such as exposure pattern and age at exposure, that may modify risk. Further studies of medically exposed populations are needed to study possible gene-radiation interactions that may render parts of the population more sensitive to radiation-induced health effects. Studies of populations (particularly children and infants) with lower- to medium-dose diagnostic exposures also are needed because of the increasing use of procedures such as CT and radiological monitoring of infants.

8

Occupational Radiation Studies

INTRODUCTION

The risk of cancer among physicians and other persons exposed to ionizing radiation in the workplace has been a subject of study since the 1940s, when increased mortality from leukemia was reported among radiologists compared to mortality among other medical specialists (March 1944; Dublin and Spiegelman 1948). An extensive retrospective cohort study (Court Brown and Doll 1958) confirmed the earlier reports and also noted excess mortality from other cancers. Since then, numerous studies have considered the mortality and cancer incidence of various occupationally exposed groups, in medicine (radiologists and radiological technicians), nuclear medicine, specialists (dentists and hygienists), industry (nuclear and radiochemical industries, as well as other industries where industrial radiography is used to assess the soundness of materials and structures), defense, research, and even transportation (airline crews as well as workers involved in the maintenance or operation of nuclear-powered vessels). The type of ionizing radiation exposure varies among occupations, with differing contributions from photons, neutrons, and α- and β-particles.

Studies of populations with occupational radiation exposure are of relevance for radiation protection in that most workers have received protracted low-level exposures (a type of exposure of considerable importance for radiation protection of the public and of workers). Further, studies of some occupationally exposed groups, particularly in the nuclear industry, are well suited for direct estimation of the effects of low doses and low-dose rates of ionizing radiation (Cardis and others 2000) for the following reason: large numbers of workers have been employed in this industry since its beginning in the early to mid-1940s (more than 1 million workers worldwide); these populations are relatively stable; and by law, individual real-time monitoring of potentially exposed personnel has been carried out in most countries with the use of personal dosimeters (at least for external higher-energy exposures) and the measurements have been kept.

Individual epidemiologic studies of occupational exposure to ionizing radiation, however, face a number of obstacles with respect to assessment of the dose-response relationship in the low-dose region (*e.g.*, NRC 1990; Ron 1998; Boice and others 2000):

• The statistical power necessary to detect an adverse health effect from the low doses encountered in occupational settings requires a large number of exposed workers and sufficiently long follow-up to account for the latency periods. Thus, follow-ups of individual cohorts of workers ordinarily have insufficient statistical power. A number of large, combined multinational studies and analyses of mortality among nuclear industry workers have been carried out in order to address these issues (Cardis and others 2000).

• In some studies, such as those of radiologists and other medical personnel, the lack of individual dose estimates is a major limitation, as is the lack of a suitable comparison group.

• The usefulness of analyses involving external comparison groups is limited due to the "healthy worker effect" often found in many occupational cohorts (Howe and others 1988; Carpenter and others 1990).

Articles included in this chapter were identified principally from searching the PubMed database of published articles from 1990 through December 2004. Searches were restricted to human studies and were broadly defined: key words included radiation; neoplasms; cancers; radiation-induced; occupational radiation; nuclear industry; nuclear workers; radiation workers; Mayak; Chernobyl; accident recovery workers; liquidators; radiologists; radiological technologists; radiotherapists; radiotherapy technicians; dentists; dental technicians; pilots; airline crew; airline personnel; and flight attendants. Articles were also identified from UNSCEAR (2000b), from references cited in papers reviewed, and from direct contacts with some of the main scientists who have been involved with studies of occupational exposures in recent years.

Studies of occupationally exposed persons have been reviewed in BEIR V (NRC 1990) and in more detail in UNSCEAR (2000b). Because of the large number of studies of radiation workers, they are not described exhaustively in this chapter, which focuses mainly on the most informative studies for the current BEIR VII evaluation (*i.e.*, studies in which the sample size is sufficiently large and the historical individual dosimetric information is sufficiently complete for radiation risk estimation). As in the other review chapters in this report, studies were judged to be informative for the purpose of radiation risk estimation if (1) the study design was adequate and no major bias could be identified (see Chapter 5 concerning informative study designs and limitations); (2) individual quantitative estimates of radiation dose to the organ of interest were available for study subjects; (3) if so, the details of the dose reconstruction or estimation approach were evaluated; and (4) a quantitative estimate of disease risk in relation to radiation dose—in the form of an excess relative risk (ERR) or excess absolute risk (EAR) per gray—was provided. The data and confidence intervals are those given in the cited papers.

NUCLEAR INDUSTRY WORKERS

A direct assessment of the carcinogenic effects of protracted, generally low-level radiation exposure can be made from studies of cancer risk among workers in the nuclear industry, many of whom have been exposed to above-background levels of ionizing radiation over several decades and whose exposures have been monitored through the use of personal dosimeters. Throughout this report, the term "nuclear industry" will be used to refer to facilities engaged in the production of nuclear power, the manufacture of nuclear weapons, the enrichment and reprocessing of nuclear fuel, or reactor research. Uranium mining is not included.

Principal References

Many studies of mortality—and, in some instances, cancer incidence—among nuclear industry workers have been carried out over the past 20 years. Published studies have covered workers in Canada, Finland, France, India, Japan, Russia, Spain, the United Kingdom, and the United States. Most have been cohort studies. The main studies in which mortality or morbidity has been examined by level of individual radiation dose are listed in Table 8-1. The characteristics of the cohorts and results are summarized briefly in Table 8-2. A number of published studies are not described in Table 8-2, for the following reasons:

• The studies of Mayak workers in the former USSR are described in the next section of this chapter. Many of these workers received mixed exposures to low- and high-LET (linear energy transfer) ionizing radiation, including considerable doses from internal contamination with plutonium-239.

• Studies of nuclear industry workers in which analyses were not reported in relation to individual external dose estimates are not discussed further in this chapter. These are studies of the employees of the U.S. Department of Energy (DOE) facilities of Linde (Dupree and others 1987), Oak Ridge Y-12 plant (workers employed between 1943 and 1947; Polednak and Frome 1981), Pantex (Acquavella and others 1985), Savannah River (Cragle and others 1988), and United Nuclear Corporation (Hadjimichael and others 1983); studies of mortality of nuclear industry workers in Slovakia (Gulis 2003) and at the French Atomic Energy Commission (Telle-Lamberton and others, 2004); and the proportional mortality studies of workers in nuclear installations in India (Nambi and Soman 1990; Nambi and others 1991, 1992).

• Nested case-control studies of specific cancers in the cohort studies including melanoma (Austin and Reynolds 1997; Moore and others 1997); leukemia (Stern and others 1986); prostate cancer (Rooney and others 1993); and lung cancer (Rinsky and others 1988; Petersen and others 1990) are not included.

Studies of combined cohorts comprising many of the workers included in individual studies have been carried out in the United Kingdom and the United States, as well as studies of all workers included in the national dose registries in Canada, Japan, and the United Kingdom. In the USA, combined analyses of the data on workers from Hanford, Rocky Flats, and Oak Ridge National Laboratory (ORNL) have been reported by Gilbert and collaborators (1989, 1993a). The latest analysis included 35,933 workers, followed until the end of 1986 (Gilbert and others 1993a). A study of workers employed in one of 15 commercial nuclear power facilities was also conducted (Howe and others 2004). The study included 53,698 workers followed up for mortality from 1979 to 1997.

The British study of the National Registry of Radiation Workers (NRRW; Kendall and others 1992a, 1992b; Little and others 1993; Muirhead and others 1999) includes 124,743 monitored workers in the above-mentioned U.K. cohorts as well as employees of Nuclear Electric, the Defense Radiological Protection Service, and a number of other nuclear facilities. The latest publication covers follow-up for mortality until the end of December 1992. Combined analyses of three U.K. nuclear industry workforces (the Atomic Energy Authority [AEA], Atomic Weapons Establishment [AWE] and Sellafield) with follow-up extended to the end of 1988 have also been carried out (Carpenter and others 1994, 1998).

In Canada, the study of the National Dose Registry (NDR) covered 206,620 workers in the industrial, medical, and dental fields, as well as nuclear power, followed for mortality through 1987 (Ashmore and others 1998) and cancer incidence through 1988 (Sont and others 2001). About 25% of these were nuclear industry workers, but detailed results were not presented for this group. The average dose of the entire cohort is low (6.6 mSv). The average length of follow-

TABLE 8-1 Cohort Studies of Nuclear Workers in Which Mortality or Morbidity Has Been Studied by Level of Individual Radiation Dose

Country	Cohort	Reference
Canada	Chalk River plant of Atomic Energy of Canada Ltd.	Howe and others (1987); Gribbin and others (1993)
France	Electricité de France	Rogel and others (2005)
Finland	Finnish power plants and research reactor	Auvinen and others (2002)
Spain	Spanish Nuclear Energy Board	Artalejo and others (1997)
United Kingdom	Atomic Energy Authority	Duncan and Howell (1970); Beral and others (1985); Fraser and others (1993)
	Atomic Weapons Establishment	Beral and others (1988); Atkinson and others (2004)
	Sellafield	Smith and Douglas (1986); Douglas and others (1994); Omar and others (1999)
	Chapelcross	Binks and others (1989)
	Capenhurst	McGeoghegan and Binks (2000b)
	Springfields	McGeoghegan and Binks (2000a)
United States	Fernald	Ritz (1999)
	Hanford Site	Kneale and others (1981); Gilbert and others (1989); Gilbert and others (1993b); Kneale and Stewart (1993)
	Mound Facility	Wiggs and others (1991a, 1991b)
	Oak Ridge National Laboratory	Checkoway and others (1985); Wing and others (1991); Richardson and Wing (1999b)
	Oak Ridge Y-12 Plant	Checkoway and others (1988); Loomis and Wolfe (1996, 1997)
	Oak Ridge X-10 Plant	Frome and others (1997)
	Rocketdyne/Atomics International	Ritz and others (1999a)
	Rocky Flats	Wilkinson and others (1987); Voelz and others (1997)
	Portsmouth Naval Shipyard	Rinsky and others (1981)

up was slightly less than 10 years in the incidence study, which covered a total of 191,333 person-years of follow-up. A study of mortality in the subgroup of nuclear power industry workers registered in the NDR has recently been published (Zablotska and others 2004). The study included 45,468 workers monitored for more than 1 year between 1957 and 1994. The average cumulative dose was 1.5 mSv. The average length of follow-up was 13.4 years (607,979 person-years of follow-up).

In Japan, the study (ESGNWJ 1997) covered a large cohort of 114,900 Japanese nuclear workers. The follow-up time was short (average 4.6 years), and the cumulative dose was relatively low (average 13.9 mSv). Consequently the study had little power to assess possible health effects of occupational ionizing radiation exposure; in particular, the test for trend for all cancers had a one-sided p-value of 0.65, and the test for trend for leukemia had a one-sided p-value of 0.22 (ESGNWJ 1997).

In addition to the national combined analyses, a multinational combined analysis was carried out to maximize the information from studies of nuclear industry workers (IARC 1994, 1995; Cardis and others 1995). Individual data from seven of the cohorts are included in Table 8-2 (Hanford, ORNL, Rocky Flats, AEA, AWE, Sellafield, and the Chalk River plant of Atomic Energy of Canada Ltd. [AECL]) and from the U.S. Rocky Flats facility (Wilkinson and others 1987). Overall, 95,673 workers employed between 1943 and 1988 in one of the participating facilities were included. They contributed 2,124,526 person-years of follow-up (an average follow-up of 22.2 years). The collective dose was 3843 Sv, most of which (98%) was received by men.

Characteristics of Studies of Nuclear Industry Workers

In the majority of the studies listed above, study subjects are defined as workers employed in the nuclear industry for whom detailed individual external dose estimates were available. Exceptions include the Canadian NDR study (Ashmore and others 1998), which included many other types of radiation workers, and a number of cohorts (Hanford, ORNL, Sellafield, AEA, and AWE) in which both monitored and nonmonitored workers are included. In the latter studies, estimates of risk per unit dose are restricted to monitored workers, except in the study of ORNL (Wing and others 1991; Richardson and Wing 1999b), where doses were estimated for a number of workers who had not been monitored.

TABLE 8-2 Main Characteristics of Principal Studies of Nuclear Industry Workers

Study Population	References	Dates of Exposure	Dates of Follow-up	No. of Subjects	Person-Years	Average Radiation Dose (Sv)	Collective Dose (Sv)	Comments
United States								
Hanford Site	Gilbert and others (1993b)	1944–1978	1944–1986	32,643[a]		0.026	854	Workers employed 6 months or more
Oak Ridge National Laboratory	Richardson and Wing (1999b)	1943–1985	1943–1990	14,095	425,486	NA		Cohort includes white males
Oak Ridge (X-10, Y-12)	Frome and others (1997)	1943–1985	1943–1984	28,347	603,365	NA		
Rocky Flats	Gilbert and others (1993a)	1951–1979	1952–1983	5,952	81,237[b]	0.041	241	White males
Los Alamos	Wiggs and others (1994)	1943–1977	1943–1990	15,727	456,637	NA		White males
Mound Facility	Wiggs and others (1991b)	1947–1979	1947–1979	3,229	54,151	0.030	1625	Monitored white males
Savannah River Site	Cragle and others (1994)	1943–1986	1952–1986	9,860	NA	0.041		
Rocketdyne/AI	Ritz and others (1999a;1999b)	1950–1993	1950–1994	4,563	118,749	0.012		
Portsmouth Naval Shipyard	Rinsky and others (1981)	1952–1977	1952–1977	7615[c]	98,223	0.028	212	White males
United Kingdom								
Sellafield	Douglas and others (1994)	1947–1975	1947–1988 (1971–86—incidence)	10,276[d]	370,329[e]	0.128	1317	Mortality and morbidity study
	Omar and others (1999)	1947–1975	1947–1992 (1971–86—incidence)	10,382[f]	415,431	0.130	1352	All workers—mortality and morbidity study
AEA	Fraser and others (1993)	1946–1979	1946–1986	39,718	873,796	0.022		Mortality and morbidity study
AWE	Beral and others (1988)	1951–1982	1951–1982	22,552	419,467	0.003		
Springfields	McGeoghegan and Binks (2000a)	1946–1995	1946–1995	13,960	479,146[g]	0.020–0.023		
Capenhurst	McGeoghegan and Binks (2000b)	1946–1995	1946–1995	3,244	334,473[g]	0.010	32	
Canada								
AECL	Gribbin and others (1993)	1956–1980	1956–1985	8,977	157,101	0.015		Males
France								
Electricité de France (EDF)	Rogel and others (2005)	1961–1994	1961–1994	22,395		5.5 (median)	402	EDF
Combined Cohorts								
Canadian NDR	Ashmore and others (1998)	1951–1983	1951–1983	206,620	2,861,093	0.063		
Canadian nuclear workers	Zablotska and others (2004)	1957–1994	1957–1994	45,468	607,979	13.5		Canadian nuclear workers

continues

TABLE 8-2 Continued

Study Population	References	Dates of Exposure	Dates of Follow-up	No. of Subjects	Person-Years	Average Radiation Dose (Sv)	Collective Dose (Sv)	Comments
	Sont and others (2001)	1951–1983	1969–1988	191,333	2,667,903	0.066		Morbidity study
Combined analyses of U.K. nuclear workers	Carpenter and others (1994)	Varied 1946–1988	1946–1988	75,006	1,800,144	0.037		
NRRW	Muirhead and others (1999)	<1976–1992	<1976–1992	124,743	2,063,300	0.031	3810	
Combined analyses of U.S. workers (Hanford, ORNL, Rocky Flats)	Gilbert and others (1993a)	Varied 1944–1979	Varied 1944–1986	44,943	835,070	0.027	1237	
U.S. nuclear facility workers	Howe and others (2004)	1945–1997	1979–1997	53,698	698,051	25.7		
Three-country combined analyses (Canada, U.K., U.S.)	Cardis and others (1995)	Varied 1943–1982	Varied 1943–1988	95,673	2,124,526	0.04	3843	

NOTE: NA = not available.

[a]Monitored workers only.
[b]Excludes first 5 years of follow-up.
[c]Includes only workers with doses >0.001 rem.
[d]Radiation workers.
[e]Includes nonradiation workers.
[f]Including 5203 plutonium workers.
[g]Includes nonradiation workers (more than 5000 at Springfields and more than 9000 at Capenhurst).

The number of workers and person-years of follow-up in the major studies are listed in Table 8-2. In general, exposure in most of these cohorts was predominantly to low levels of external radiation (X- and γ-rays and some neutrons). Internal contamination (through inhalation, ingestion, skin absorption, or wounds) by tritium, plutonium, uranium, and other radionuclides occurred in some subgroups of workers.

Assessment of Exposure to Radiation

Control of radiation dose to workers in occupational settings is achieved by demarcating radiation levels in work areas, conducting routine radiation monitoring (*e.g.*, by air sampling and the use of *in situ* radiation monitors), and by individual monitoring of workers. The studies of nuclear industry workers considered here are based on workers for whom individual monitoring of dose from external "higher"-energy (300–3000 keV) photon radiation was carried out routinely.

Individual monitoring at its simplest consists of assigning radiation-sensitive dosimeters to each worker. Dosimeters, which consist of one or more of ionization chambers, photographic film, luminescent phosphors, or electronic devices, are worn by workers while they are present in designated radiation areas. Dosimeters are normally placed on the chest, and it is usually assumed that the measured radiation dose is representative of the whole-body dose (*i.e.*, estimates "whole-body equivalent dose"); the dose to different parts of the body is assumed to be uniform.

In nearly all cases, dosimeters are sensitive to the penetrating photon radiation of intermediate (>100 keV) to higher photon (*i.e.*, X- and γ-rays) energies typical of radiation fields in the respective facilities. Specialized dosimeters and calibration methods are generally needed to measure accurately the dose from low-energy photons, beta, or neutron radiation present in some occupational environments. Monitoring for the intake of radioactive material is performed by bioassay, by whole-body *in vivo* counting, or by wearing personal air samplers. In most of the facilities that have been the object of the epidemiologic studies described above, measurements of dose to individuals have generally been recorded on a routine basis using the available dosimetry technology.

Occupational radiation dose data constitute the most complete and detailed information currently available to researchers for studying the carcinogenic effects of low-dose, protracted exposures to ionizing radiation. They are generally presented in the form of annual summaries of doses from different types of radiation (penetrating photons, beta, and where appropriate and measured, tritium and neutrons).

These data were, however, compiled to monitor worker exposure for compliance with radiation protection guidelines, which have changed over time, and not specifically for epidemiologic purposes. Overall, the accuracy and precision of recorded individual doses and their comparability will therefore depend on:

- the dosimetry technology, which includes the physical capabilities of the dosimetry system, such as the response to different types and energies of radiation, in particular in mixed radiation fields;
- the radiation fields in the work environment, which may include mixed types of radiation, variations in exposure geometries, and environmental conditions; and
- the administrative practices adopted by facilities to calculate and record personnel dose based on technical, administrative, and statutory compliance considerations.

Consequently, detailed examination of dosimetry practices, including sources and magnitude of errors, is important in considering whether sufficiently accurate and precise estimates of dose can be obtained for use in an epidemiologic study.

Information on internal contamination with radionuclides other than tritium is generally sparse, particularly in early years, and consists of information on the fact of monitoring or on a percentage of the annual limit of intake. Very few studies have attempted to reconstruct individual doses from nuclides other than tritium. One exception is the study of Sellafield workers in the United Kingdom, where efforts have been made to reconstruct plutonium exposures (Omar and others 1999).

In high-dose studies, the majority of excess deaths from cancer have been demonstrated in subjects exposed to doses of at least 1 Sv. There were approximately 3000 such subjects among atomic bomb survivors. Doses received by employees of nuclear industry facilities are considerably lower. In the Sellafield cohort (Douglas and others 1994), in which the highest doses among the nuclear industry worker studies have been reported, only about 60 out of more than 10,000 individuals monitored for external radiation exposure had received doses of 1 Sv or more, and these doses were accumulated over the course of a working life. The mean cumulative radiation dose in the three-country combined analyses was 40.2 mSv per worker and the collective dose was 3843 Sv (IARC 1995). Women comprised fewer than 15% of the workers, and their mean cumulative dose was low (6.2 mSv) compared to that of men (46.0 mSv). Overall, the distribution of doses was very skewed; almost 60% of subjects had cumulative doses less than 10 mSv, 80% were less than 50 mSv, and less than 2% had doses greater than 400 mSv.

The majority of cohort studies collected only information that could readily be obtained from employment and dosimetry records. This consists, in addition to information on individual annual radiation dose from different types of radiation, date of birth, date and cause of death, sex, socioeconomic status based on occupational group or education, and dates of beginning and end of employment. Nested case-control studies have allowed the exploration of additional factors including tobacco smoking and other occupational exposures.

Results

In most of the nuclear industry workers studies, death rates among worker populations were compared with national or regional rates. In most cases, rates for all causes and all cancer mortality in the workers were substantially lower than in the reference populations. Possible explanations include the healthy worker effect and unknown differences between nuclear industry workers and the general population.

In most studies where external radiation dose estimates were available, death rates were also compared in relation to levels of radiation exposure within the study population. For all cancer mortality (excluding leukemia), the estimates of radiation-induced excess risk varied from negative to several times greater than those derived from linear extrapolation from high-dose studies (Table 8-3). Moreover, because of the large degree of uncertainty, many of these estimates were consistent with an even wider range of possibilities, from negative risks to excess risks at least an order of magnitude greater than those on which the current radiation protection recommendations have been based.

In most of the large studies of nuclear industry workers, estimates of ERR[1] per gray (ERR/Gy) have been derived, mostly using Poisson regression. Estimates of excess death rate per 10^6 person-years (PY) per gray have also been presented in some studies. Results of such analyses are shown in Tables 8-3 and 8-4 for all cancers excluding leukemia and for leukemia, respectively. Table 8-5 is a listing of the results from other studies of nuclear workers that could not be used in computation of ERRs or EARs.[2]

Cancer mortality was observed to increase significantly with increasing level of exposure in four studies: AWE (Beral and others 1988), ORNL (Wing and others 1991; Richardson and Wing 1998), Canadian NDR (Ashmore and others 1998), and Rocketdyne (Ritz and others 1999a). The ERR estimate based on the three-country combined analysis was close to zero, but was compatible with a range of possi-

---

[1]ERR is the rate of disease in an exposed population divided by the rate of disease in an unexposed population minus 1.0.

[2]EAR is the rate of disease in an exposed population minus the rate of disease in an unexposed population.

TABLE 8-3 Radiation Risk Estimates in Studies of Nuclear Industry Workers—Summary of Risk Estimates per Gray for Mortality from All Cancers Excluding Leukemia

Study Population	References	Number of Cancer Deaths	ERR/Sv (90% CI)	EAR/10^4 PY/Sv (90% CI)	Comments
United States					
Hanford	Gilbert and others (1993b)	1413	–0.0 (<0, 1.9)		
ORNL	Richardson and Wing (1999a)	879	—	1.21%/10 mGy (SE 0.65) 4.98%/10 mGy (SE 1.48)	Leukemias included Restricted to dose received after age 45
Oak Ridge Y-10, X-12	Frome and others (1997)	1134	1.45 (0.15, 3.48)[a]	—	Leukemias included
Rocky Flats	Gilbert and others (1993a)	114	<0 (<0, 0)	—	
United Kingdom					
AEA	Fraser and others (1993)	720	0.8 (–1.0, 3.1)[a]	20.3 (–26.0, 71.1)[a]	
AWE	Beral and others (1988)	275	7.6 (0.4, 15.3)[a]		
Capenhurst	McGeoghegan and Binks (2000b)	174	–1.3 (<0, 2.4)		Males only
Sellafield	Douglas and others (1994)	567	0.11 (–0.4, 0.8)	5.6 (90% CI 15.86, 27.15)[b]	
Springfields	McGeoghegan and Binks (2000a)	939	0.64 (–0.95, 2.7)		Males only
Canada					
AECL	Gribbin and others (1993)	221	0.049 (–0.68, 2.17)		
Combined cohorts					
Canadian NDR	Ashmore and others (1998)	1632	3.0 (1.1, 4.8)	—	
Canadian nuclear workers	Zablotska and others (2004)	531	2.80 (–0.038, 7.13)[a]		
Combined UK nuclear industry workforce	Carpenter and others (1994, 1998)	1824	–0.02 (–0.5, 0.6)[a]	–0.68 (–23.3, 20.9)[a]	
NRRW	Muirhead and others (1999)	3020	0.086 (–0.28, 0.52)[a]	—	
Hanford, ORNL, Rocky Flats	Gilbert and others (1993a)	1789	0.0 (<0 .8)[a]	—	
U.S. nuclear facility workers	Howe and others (2004)	368[c]	0.506 (–2.01, 4.64)[a]		
Three-country combined analyses (Canada, U.K., U.S.)	Cardis and others (1995)	3830	–0.07 (–0.39, 0.30)	—	

NOTE: Doses are lagged by 10 years unless otherwise indicated.

[a]95% confidence interval.
[b]Doses are lagged by 15 years.
[c]All solid cancers only.

TABLE 8-4 Radiation Risk Estimates in Main Studies of Nuclear Industry Workers—Summary of Risk Estimates per Gray for Mortality from Leukemia Excluding Chronic Lymphocytic Leukemia

Study Population	References	Observed Cases	ERR/Sv (90% CI)	EAR/10^4 PY/Sv	Comments
United States					
Hanford	Gilbert and others (1993b)	44	−1.1 (<0, 3.0)	—	
ORNL	Wing and others (1991)	28	6.4 (-11.2, 24.0)		
Hanford, ORNL, Rocky Flats	Gilbert and others (1993a)	67	−1.0 (<0, 2.2)[a]	—	
United Kingdom					
UKAEA	Fraser and others (1993)	31	−4.2 (-5.7, 2.6)		
UKAWE	Beral and others (1988)	4			
Capenhurst	McGeoghegan and Binks (2000b)	4	−1.27 (<0, 2.75)		Males only
Sellafield	Douglas and others (1994)	12	13.92 (90% CI 1.94, 70.52)	2.47 (90% CI 1.21, NE)	Upper bound for EAR could not be estimated (NE)
Springfields	McGeoghegan and Binks (2000a)	23	−1.89 (< −1.97, 13.1)		Males only
Canada					
AECL	Gribbin and others (1993)	4	19.0 (0.14, 113)	—	
Combined Cohorts					
Canadian NDR	Ashmore and others (1998)	46	0.4 (−4.9, 5.7)	—	Males
Canadian nuclear workers	Zablotska and others (2004)	18	52.5 (0.21, 291)[a]		
Combined U.K. nuclear industry workforce	Carpenter and others (1994, 1998)	49	4.18 (0.4, 13.4)[a]	2.10 (0.4, 3.6)[b]	2-year lag; adjusted for age, sex, calendar period, social class, and facility
NRRW	Muirhead and others (1999)	91	2.55 (−0.032, 7.16)	—	2-year lag
Hanford, ORNL, Rocky Flats	Gilbert and others (1993a)	67	−1.0 (<0, 2.2)[a]		
U.S. nuclear facility workers	Howe and others (2004)	26	5.67 (−2.56, 30.4)[a]		
Three-country combined analyses (Canada, U.K., U.S.)	Cardis and others (1995)	119	2.18 (0.13, 5.7)	—	2-year lag; adjusted for age, socioeconomic status, facility, and calendar time

NOTE: Doses are lagged by 2 years unless otherwise specified.

[a]95% confidence interval.
[b]Absolute risk estimate is number of deaths per person-year per sievert.

TABLE 8-5 Results of Studies of Nuclear Industry Workers with Individual External Dosimetry that Did Not Provide ERRs or EARs

Country	Facility	No. of Subjects	All Cancers		Leukemia	
			No. of Deaths	Results (90% CI)	No. of Deaths	Results (90% CI)
United States	Mound (Wiggs and others 1991a, 1991b)	3,229	66	No association with radiation dose	4	Significant ($p < .01$) positive trend with radiation dose
	Los Alamos (Wiggs and others 1994)	15,727	732	No association with radiation dose	44	No association with radiation dose
	Portsmouth Naval Shipyard (Rinsky and others 1981)	7,615	201	No association with radiation dose	7	No association with radiation dose
	Rocky Flats (Wilkinson and others 1987)	5,413	50	Slope = −3.65/10 mSv (−12.02, 4.71)	4	RR = 1.0 (0.8, 9.1) for 10mSv vs. <10 mSv
	Rocketdyne (Ritz and others 1999a)	4,563	258	Significant ($p = .036$) trend	28[a]	Significant ($p = .003$) trend
United Kingdom	BNFL (McGeoghegan and Binks 1999)	2,467	—[b]	Significant ($p < .01$) positive trend when doses are lagged by 15 years	—[b]	No association with radiation dose
	AWE (Atkinson and others 2004)	26,395	1560	No association with radiation dose	38	No association with radiation dose
Slovakia	Jaslovske Bohunice power plant (Gulis 2003)	2,776	14	No association with radiation dose	0	
France	Electricité de France (Rogel and others 2005)	22,395	116	No association with radiation dose	5	No association with radiation dose

[a]Hemato- and lymphopoietic cancers.
[b]Not specified.

bilities, from a reduction of risk at low doses to risks twice those on which current radiation protection recommendations are based.

In most studies, analyses of mortality in relation to cumulative external radiation dose were conducted for many specific types of cancer. These studies have generally not shown significant increases in risk among exposed workers for most cancer types examined, although a few positive associations have been found (Table 8-3).

For leukemia, risk estimates varied considerably from study to study (Table 8-4). In the pooled study of workers in the United States (Gilbert and others 1993a), the estimate of ERR per gray based on the combined data was negative, although the upper confidence bound was slightly larger than the estimate currently recommended by the International Commission on Radiological Protection (ICRP 1991). By contrast, significant positive associations were observed in AECL and nuclear worker studies in Canada (Gribbin and others 1993; Zablotska and others 2004) and in the U.K. study of Sellafield workers (Douglas and others 1994), as

well as in the NRRW cohort (Muirhead and others 1999) and the three-country combined analyses (Cardis and others 1995). The confidence intervals in these studies were wide, and the estimates of risk were consistent with those on which current radiation protection recommendations are based.

Statistically significant ($p < .05$, one-sided) positive associations between cumulative external radiation dose and mortality from multiple myeloma were found in the Hanford (Gilbert and others 1989) and Sellafield (Douglas and others 1994) studies. A similar association was also found in the NRRW (Muirhead and others 1999) and three-country analyses (Cardis and others 1995), largely reflecting the previously reported associations in individual cohorts. The association in the Hanford study was not significant when follow-up was extended to 1986 (Gilbert and others 1993b).

An association between radiation dose and mortality from cancer of the prostate was found in two studies, the AEA (Beral and others 1985; Fraser and others 1993) and the AWE (Beral and others 1988); in AWE workers it was statistically significant only among workers who had been monitored for exposure to radionuclides (Rooney and others 1993) in the period 1946–1979. No such increase was observed in the NRRW (Muirhead and others 1999), which included all of the monitored workers in these two studies.

A significant positive association with lung cancer was observed in the AWE and ORNL studies (Beral and others 1988; Wing and others 1991), particularly among those exposed to radionuclides in the AWE and in nonmonthly workers at ORNL. Information on tobacco smoking was available systematically in these studies. A few other significant associations were reported in single studies (Table 8-4). Given the number of associations examined, some of the significant results observed may have been due to chance.

Several points must be kept in mind when making comparisons of these worker-based risk estimates and confidence intervals with those based on high-dose-rate studies. The most important are possible biases and uncertainties in dose estimates, errors in outcome data, and inadequate adjustment for confounders.

Design Issues

Among the very large and potentially most informative cohort studies reviewed in this chapter, two present a number of problems that limit their informativeness. In the Canadian NDR study (Ashmore and others 1998), the very low standardized mortality ratio (SMR)[3] for all-cause mortality (61) suggests that record linkage procedures between the Canadian National Dose Registry and the Canadian Mortality Data Base may have been imperfect. There could have been some confounding of the dose-response because of associations between the probability of successful linkage and factors (*e.g.*, socioeconomic status [SES]) associated with occupational radiation dose. This is the only study in which associations have been observed between radiation dose and all-cause mortality, all cancer mortality (without any clear relation to specific cancers), mortality due to cardiovascular diseases (males and females), and fatal accidents (males only). Moreover, no information is available on SES—a factor that has been shown in a number of previous cohorts to be a confounder of the association between radiation dose and cancer risk. Almost three-quarters of the cohort consists of radiation workers employed in different settings (dentistry, medicine, industrial radiography), where radiation control may be very different (possibly less uniform and systematic due to the much smaller numbers of persons monitored in individual workplaces) than in the nuclear industry.

In the Japanese NDR study (ESGNWJ 1997), SES information is also not available. Further, because of difficulties in carrying out vital status follow-up in Japan, the very large

cohort had to be restricted to those who were employed in the previous 5 years; hence the follow-up time of this cohort is very short, and older workers as well as workers with higher doses (who were employed in early years and left employment more than 5 years in the past) have been excluded from the follow-up. Consequently the study has little power to estimate possible health risks associated with occupational radiation exposure.

Adequacy of the Dose Estimates

High-Energy Photon Doses

The accuracy and precision of individual dose estimates in the nuclear industry is a function of time, place, radiation energy and quality, the geometry of the radiation exposure, and the location of the dosimeter on the body of the worker.

Efforts were made in some of the studies (AEA, Sellafield, ORNL, U.S. DOE combined analyses, three-country study, Saclay site in France) to assess the importance of dosimetric errors due to administrative practices adopted by facilities to calculate and record personnel dose based on technical, administrative, and statutory compliance considerations (Adams and Langmead 1962; Smith and Inskip 1985; Taylor 1991; Telle 1995; Tankersley and others 1996; Mitchell and others 1997; Watkins and others 1997; Telle-Lamberton and others 1998). Results of reanalyses of data using different approaches to estimate doses from missing dosimeters or below-threshold readings have yielded similar results to the analyses based on original data (Inskip and others 1987; Little and others 1993).

In the three-country combined analyses, a retrospective dosimetry study was carried out to identify the various sources of biases and random errors in dosimetry for workers in each of the facilities included and to estimate the magnitude of these errors. As a result, it was concluded that for the majority of workers with predominant high-energy (300–3000 keV) photon exposures at levels greater than the detection threshold of the dosimeter, there is no strong reason to believe that available dose estimates substantially underestimate or overestimate deep dose. The dose estimates were judged to be compatible across facilities and over time. However, available dose estimates may have overestimated dose to the bone marrow by up to 20%. Estimation of leukemia risk adjusting for this overestimation yielded an ERR of 2.6 Sv^{-1} instead of 2.2. For deep organs, the factor is likely to be smaller, of the order of several percent. Random errors in dose estimates are likely to bias the risk estimates downwards, compared to estimates from high-dose studies, which have been based on organ doses. At lower exposure levels however, practices for recording subthreshold doses have resulted in a slight underestimation of doses from predominant higher-energy photon exposure (Fix and others 1997).

At the Hanford plant in the United States, based on experiments and expert assessments, efforts were made to

---

[3]SMR is the ratio (multiplied by 100) of the mortality rate from a disease in the population being studied divided by the comparable rate in a standard population.

quantify systematic errors related to the dosimetry technology and radiation fields as well as errors related to laboratory practices (Fix and others 1994, 1997; Gilbert 1998). When these errors were taken into account in the risk estimation process, it resulted in a widening of the confidence intervals around the ERR (Gilbert and Fix 1995) as shown in Table 8-6.

Doses from Neutrons, Low- and Very-High-Energy Photons, and Internal Contamination

In the three-country study, efforts were also made to identify workers with substantial doses from radiations other than high-energy photons (mainly from neutrons, low-energy radiation, and contamination with radionuclides, particularly plutonium), for whom recorded dose estimates may be in error. Although it was not possible to identify all such workers, risk estimates based on restricted dosimetry analyses, which excluded all such workers who could be identified, did not differ greatly from those based on the standard approach (-0.04 and 2.05 Sv^{-1} respectively, for all cancers excluding leukemia and for leukemia excluding chronic lymphocytic leukemia (CLL) compared to -0.07 and 2.18 Sv^{-1} in the standard population). In addition, the estimate of risk for all cancers excluding leukemia and lung cancer (the organ that would receive the majority of the dose from plutonium contamination) was identical to that of all cancers excluding leukemia (-0.07 Sv^{-1}; 90% CI -0.39, 0.30). It is therefore unlikely that the risk estimates in this study are substantially biased by inclusion in the analyses of a minority of workers with dose from neutrons, low-energy photons, and internal contamination (Cardis and others 1995).

Possible Confounding and Modifying Factors

Tobacco Smoke

As in most occupational cohort studies, information on life-style factors such as smoking habits, diet, and other oc-

cupational exposures could not be obtained retrospectively for all members of the cohort. In the three-country combined analyses, there was little indirect evidence for an association between cumulative dose and mortality from smoking-related cancers, respiratory diseases, or liver cirrhosis; thus, it is unlikely that smoking or alcohol consumption are strongly correlated with radiation dose (Cardis and others 1995). This is supported by the observation that the risk estimates for all cancers excluding leukemia and all cancers excluding both leukemia and lung cancer were nearly identical (Cardis and others 1995): also, the results of two studies, carried out within the Hanford (Petersen and others 1990) and AEA (Carpenter and others 1989) cohorts, respectively, showed little evidence for an association between smoking and occupational radiation dose. A positive association between smoking and occupational radiation dose was found in the AECL cohort in Canada (Howe and others 1987).

Socioeconomic Status

A positive association between radiation dose and mortality from circulatory disease was observed in the four cohorts included in the three-country study in which information on SES was least detailed (Rocky Flats, Sellafield, AECL, Canadian NDR). It may reflect residual confounding by life-style factors for which the SES variable is an inadequate proxy.

Radionuclides in the Working Environment

At uranium fuel production facilities, inhalation of airborne uranium dust may represent an important potential source of radiation exposure. Workers in these facilities have two main possible sources of radiological exposure to tissues of the whole body: external γ-ray exposure and internal depositions that deliver radiation doses (mainly from α-particles) primarily to the lung and lymphatic system. If the uranium dust is soluble, exposure of other tissues may also occur such as liver, kidney, and bone, although organ doses would be expected to be small. Low-LET radiation risk estimates for tumors in these organs are possibly confounded by high-LET radiation exposure for workers at uranium production facilities, since workers with a significant dose from internal contamination are often persons with substantial external exposure. A number of studies of such workers have been reviewed (Cardis and Richardson 2000; NRC 2000).

Comparison of findings among uranium-processing facilities is complicated by the fact that processes and historical periods of operation have differed among facilities, leading to differences in exposure conditions and follow-up among cohorts. Further, assessment of past internal uranium exposure of nuclear workers is complicated by the methodological difficulties of internal dosimetry, as well as by inadequate historical information with which to quantify internal radiation doses accurately. These exposure measurement

TABLE 8-6 Estimates of the ERR per Sievert with 90% CIs for the Hanford Worker Study Based on Recorded Doses and Based on Estimated Organ Doses

	All Cancers Excluding Leukemia	Leukemia Excluding CLL
Recorded doses	0.23 (90% CI <0, 1.5)	–0.9 (90% CI <0, 2.7)
Organ doses (corrected for systematic errors related to radiation fields)	0.20 (90%CI <0, 1.7)	–1.3 (90% CI <0, 3.6)

NOTE: CLL = chronic lymphocytic leukemia.

problems pose significant difficulties for epidemiology: the inability to classify workers accurately by level of internal radiation exposure may lead to confounding of the analyses of association between external low-LET radiation dose and cancer risk.

Lung cancer has been the primary outcome of interest in studies of workers in fuel enrichment and production facilities. Lung cancer mortality was found to be significantly elevated, compared to national rates, among workers in nuclear fuel processing facilities in three reports (Loomis and Wolf 1997; Checkoway and others 1988; Frome and others 1990), but not in others (Brown and Bloom 1987; Dupree and others 1987, 1995; Ritz and others 1999b). An association between external low-LET radiation dose and lung cancer mortality was observed in two cohorts in the United States (Fernald and Y-12; Checkoway and others 1988; Ritz and others 1999a), and an association with lung cancer incidence (using a 20-year lag) was observed in one study in the United Kingdom (McGeoghegan and Binks 2000a). No association was found in other papers on the U.S. (Hadjimichael and others 1983; Ritz and others 2000) and U.K. (McGeoghegan and Binks 2000b) cohorts. No information on dose to the lung from internal contamination was available for analysis in these studies. In studies where estimation of dose to the lung from internal contamination was carried out, an association was observed at Y-12, but not at Rocketdyne (Ritz and others 1999a). In contrast, a U.S. multifacility case-control study of lung cancer among workers exposed to uranium dust at TEC, Y-12, Fernald, and Mallinckrodt found no such association; there was a suggestion, however, of positive associations among workers hired over age 45 (Dupree and others 1995). Therefore, risk estimates for low-LET radiation-induced lung cancer risk in these cohorts should be treated with caution.

Following the observation of increased prostate cancer mortality related to cumulative external radiation dose in the AEA (Beral and others 1988; Rooney and others 1993) a nested case-control study was conducted of prostate cancer risk among employees of that facility. The study showed that exposure to five radionuclides (tritium, chromium-51, iron-59, cobalt-60, and zinc-65), evaluated separately, was associated with an increased risk of prostate cancer. Analyses of the association between external radiation dose and prostate cancer risk were carried out both for workers with probable exposure to these radionuclides and for those who had no such exposure. The association between external dose and prostate cancer was restricted to those with radionuclide exposure.

In the Combined UK Industrial Workforce study, Carpenter and colleagues (1998) carried out analyses of cancer mortality in relation to external radiation dose in two groups—those who had been monitored for internal radionuclide contamination and those who had not. A positive association was seen in both groups of workers, although it was statistically significant only among those who had been monitored for internal contamination.

Cancer mortality and incidence was studied among Sellafield workers in relation to exposure to plutonium and to external low-LET radiation (Omar and others 1999). A significant association between mortality from leukemia excluding CLL (13 deaths) was seen in relation to external radiation dose using a 2-year lag, as had been seen in the previous follow-up of this cohort (Douglas and others 1994). When analyses were restricted to plutonium workers and took into account both external low-LET radiation dose and estimated plutonium dose, the association was no longer statistically significant, based on six deaths.

Other Occupational Exposures

Wing and colleagues (1993) evaluated the effect of potential exposure to beryllium, lead, and mercury in the ORNL cohort by identifying workers with potential for these exposures from employment records. Adjustment for these potential exposures had little effect on the radiation risk estimates. The interpretation of these results is limited by the absence of individual exposure estimates for the chemicals considered.

Rinsky and colleagues (1981) considered exposure to a number of workplace carcinogens in a case-control study of lung cancer among civilian employees of the Portsmouth naval shipyard. Asbestos and welding by-products were found to confound the association between radiation exposure and lung cancer risk in this population, where radiation workers appear to be more heavily exposed to asbestos and welding fumes than other workers. The unadjusted lung cancer odds ratio for workers with a cumulative dose of 10–49.99 mSv was 1.8 (95% CI 1.1, 3.1) compared to workers with no history of radiation exposure; adjustment for asbestos and welding fumes reduced it slightly to 1.7 (95% CI 1.0, 2.9).

Modifiers of Radiation Risk

Several authors have reported an association between age at exposure and/or attained age and the risk of radiation-induced cancer. This has been reported in the Hanford, ORNL, and Rocketdyne cohorts (Gilbert and others 1993a; Stewart and Kneale 1996; Richardson and Wing 1999a; Ritz and others 1999b), but not in five other cohorts in which it was considered—Rocky Flats, AECL, AEA, AWE, and Sellafield (Cardis and others 1995; IARC 1995).

The three-country and the NRRW studies (Cardis and others 1995; Muirhead and others 1999) of nuclear industry workers currently provide the most comprehensive and precise direct estimates of the effects of protracted exposures to low levels of low-LET radiation. Although the estimates are lower than the linear estimates obtained from studies of atomic bomb survivors, as seen in Table 8-7, they are compatible with a range of possibilities, from a reduction of risk at low doses, to risks twice those on which current radiation

TABLE 8-7 Comparison of Estimates of ERR/Gy Between Major Nuclear Industry Workers Combined Analyses and the Atomic Bomb Survivors

Study Population	All Cancers but Leukemia	Leukemia, Excluding CLL
Atomic bomb survivors[a]	0.24 (0.12, 0.4)	2.2 (0.4, 4.7)
Nuclear workers		
Three-country study	−0.07 (−0.39, 0.30)	2.2 (0.1, 5.7)
NRRW	0.09 (−0.28, 0.52)	2.6 (−0.03, 7.2)

[a]Based on male atomic bomb survivors, aged 20–60 years at exposure, as presented by Muirhead and others (1999).

protection recommendations are based. Overall, they do not suggest that current radiation risk estimates for cancer at low levels of exposure are appreciably in error. Uncertainty concerning the exact size of this risk, remains, however, as indicated by the width of the confidence intervals presented.

WORKERS FROM THE MAYAK FACILITY

A cohort of about 21,000 Russian nuclear workers who worked at the Mayak plutonium production complex between 1948 and 1972 is under study. The Mayak complex, which is located in the Chelyabinsk region of the Russian Federation, includes three main plants: a reactor complex, a radiochemical separation plant, and a plutonium production plant. Workers at all three plants had the potential for exposure to external radiation, and workers at the radiochemical and plutonium production plants also had the potential for exposure to plutonium. Recently, data on workers at two auxiliary plants, who had much less potential for exposure, have been added to the cohort under study to expand the comparison group. As for other nuclear worker cohorts, estimates of annual external doses are available from individual film badge monitoring data. Some workers were also monitored for plutonium exposure; however, since routine testing based on large urine samples did not begin until about 1970, only about 40% of workers with the potential for such exposure have been monitored.

External exposures and exposures of Mayak workers to plutonium far exceed those of other nuclear worker cohorts discussed previously in this chapter. For example, for the nearly 11,000 monitored workers hired before 1959, the mean cumulative external dose was 1.2 Gy, more than an order of magnitude higher than any of the cohorts described in Table 8-2. Thus, the Mayak cohort offers a unique opportunity to obtain reasonably precise estimates of risks from medium- to high-dose *protracted* external exposure that can then be compared to estimates based on acute exposure, such as those obtained from A-bomb survivors.

The first estimates of risk from external exposure were reported by Shilnikova and colleagues (2003). Analyses focused on leukemia (excluding CLL); cancers of the lung, liver, and bone (analyzed as a group); and solid cancers excluding lung, liver, and bone (also analyzed as a group). Lung, liver, and bone are the organs that receive the largest doses from plutonium, and excess cancers in all three organs have been linked clearly to plutonium exposure among Mayak workers (Gilbert and others 2000; Koshurnikova and others 2000; Kreisheimer and others 2000). Analyses were adjusted for internal exposure to plutonium by using the estimated body burden for workers who had plutonium-monitoring data and by using a plutonium surrogate variable for workers who were not monitored for plutonium. The plutonium surrogate variable was developed recently from detailed work histories.

For leukemia, the estimated ERR/Gy was 6.9 (90% CI 2.9, 15) for the period 3–5 years after exposure and 0.5 (90% CI 0.1, 1.1) for the period 5 or more years after exposure. The estimate based on the entire period was 1.0 (90% CI 0.5, 2.0). There was no statistically significant departure from linearity and no evidence of modification by sex or age at hire.

Estimates and confidence intervals for the solid cancer end points are shown in Table 8-8. For these end points, linear-quadratic functions provided significantly better fits than linear functions with a "downturn" in the dose-response at high doses. This may have resulted from overestimation of doses of certain workers in early years due to inadequacies in early film dosimeters. If this is the case, estimates of the linear term from the fitted linear-quadratic function may be more reliable. The estimates for cancers of the lung, liver, and bone were higher than those for other organs, possibly because the adjustment for plutonium exposure was less adequate for these cancers. There was no evidence of modification of the dose-response by sex, age at hire, or time since exposure.

TABLE 8-8 Estimated ERR/Gy for Solid Cancers Among Mayak Workers

Model	ERR/Sv (90% CI)		
	Lung, Liver, or Bone	Other Solid Cancers	All Solid Cancers
Linear	0.30 (0.18, 0.46)	0.08 (0.03, 0.14)	0.15 (0.09, 0.20)
Linear quadratic[a]	0.54 (0.27, 0.89)	0.21 (0.06, 0.37)	0.30 (0.18, 0.43)

[a]Estimates are for the linear coefficient of a fitted linear-quadratic function.

Summary

Studies of workers employed at the Mayak complex in the Russian Federation offer a unique opportunity, because of the magnitude of the doses received (mean cumulative external dose of 1.2 Gy among monitored workers hired before 1959), to obtain reasonably precise estimates of risk from medium- to high-dose protracted external exposures. Substantial doses from plutonium have also been received by a number of these workers. Estimates of the radiation-related risks of leukemia; solid cancers; and lung, liver, and bone cancer have been derived from this cohort. Uncertainties in external dose estimates and in plutonium doses to specific organs must be considered in the interpretation of these results. Further studies of this population will be important to understand the effects of protracted exposure.

CHERNOBYL CLEANUP WORKERS

The Chernobyl accident resulted in widespread radioactive contamination of areas populated by millions of people in the three most affected countries of Belarus, the Russian Federation, and Ukraine. The populations at risk can be separated into the following groups (see Table 8-9):

1. the "liquidators," also referred to as "cleanup workers," include persons who participated in the cleanup of the accident (cleanup of the reactor; construction of the sarcophagus; decontamination; building of roads; destruction and burial of contaminated buildings, forests, and equipment), as well as many others, including physicians, teachers, cooks, and interpreters who worked in the contaminated territories;

2. the "evacuees" who were evacuated from the town of Pripyat and the 30 km zone around the Chernobyl reactor in April–May 1986;

3. the residents of the "strict control zones"—those members of the general population who have continued to live in the more heavily contaminated areas (with levels of ^{137}Cs deposition greater than 555 kBq m^{-2}), typically within a few hundred kilometers of the Chernobyl Nuclear Power Plant (NPP). Within these areas, radiation monitoring and preventive measures have been taken to maintain doses within permissible levels; and

4. the general population of the contaminated territories in the three countries.

The "liquidation" of the consequences of the Chernobyl accident lasted for about 3 years (1986–1989). During that time, different tasks were carried out, including the initial localization of the catastrophe (firefighting; closing down unaffected units of the power plant); evacuation of Pripyat and the population in the 30 km zone; decontamination of the inside of the Chernobyl NPP buildings, as well as the roofs of nearby buildings and nearby territories; renovation and maintenance of the other blocks of the power plant; construction of the sarcophagus; actions to decrease the spread of radioactive materials in the environment; safeguard of the 30 km zone and settlements and miscellaneous activities in the 30 km zone (health care, ecological monitoring, bringing in food, water, etc., for the liquidators). Different groups of liquidators were involved in these tasks; they worked under differing conditions of radiation monitoring and safety and were exposed to various types and levels of radiation. From 600,000 to 800,000 persons took part in the cleanup activities to liquidate the consequences of the Chernobyl accident. The exposure level was highest for those (approximately 200,000 liquidators) who worked in the 30 km zone in 1986–1987.

Follow-up

In 1987, an "All-Union Distributed Registry" was established following a directive of the Ministry of Public Health of the USSR (Tsyb and others 1989). The objective was to set up a comprehensive registration and active follow-up system for the persons most affected by the Chernobyl accident, including the liquidators. This system foresees an annual medical examination in which individuals are examined systematically by a general practitioner and a number of different specialists. All data on diseases diagnosed during the annual medical examination, as well as any other time during the year, are sent to the Chernobyl Registry for inclusion in the registry database. A study in Russia (Cardis and Okeanov 1996) indicates that the diagnostic information in the Chernobyl Registry is not always completely accurate. The lack of verification and quality control is actively being remedied but must be kept in mind when interpreting results

TABLE 8-9 Estimates of Collective Effective Doses for Chernobyl Population Groups of Interest

Population	Number	Collective Effective Dose (Sv)
Evacuees	135,000	1,300
Liquidators (1986–1987)	200,000	20,000
Persons living in contaminated areas[a]		
Deposition density of ^{137}Cs >15 Ci km^{-2}	270,000	10,000–20,000
Deposition density of ^{137}Cs >1 to 15 Ci km^{-2}	3,700,000	20,000–60,000

[a]Doses are for 1986–1995; over the longer term (1996–2056) the collective dose will increase by approximately 50%.

SOURCE: Cardis and others (1996).

of studies of cancer frequency among exposed populations in these countries. Results from this follow-up may also be biased because participation in the annual examination may be related to illness and/or to level of exposure.

Means also exist in the affected countries to carry out "passive" follow-up of exposed persons and of the general population with the use of population registries—of mortality, cancer, and other diseases. In each country of the former USSR, population registration is carried out at the local level in the address bureaus (where the addresses of current residents are kept) and the ZAGS (*buro zapicii akta grazhdanskovo sostoyania*), which compiles all information about birth, marriage, divorce, and death of persons living in the administrative area. No centralized registry exists, however, and results of a pilot study (Cardis and Okeanov 1996) indicate that considerable time and effort may be needed to trace subjects who have moved from one area to another.

A computerized national Cancer Registry has been functioning in Belarus since the 1970s and registers all cases of malignant neoplasms. A comprehensive registry of hematological diseases also exists in Belarus, in the Institute of Haematology and Blood Transfusology. In Russia and the Ukraine, no centralized cancer registration system was in place at the time of the accident. Work has been carried out in both countries to set one up—at least in contaminated areas in Russia (Okeanov and others 1996; Storm and others 1996)—and quality control activities are continuing.

Information is also available systematically on the general (*i.e.*, not only cancer) morbidity of the population of the three countries. In the countries of the former USSR, regional outpatient clinics systematically collect information on disease diagnoses on all the residents of the region they cover (not only those included in the Chernobyl Registry). This information is summarized locally and is sent on special statistical reporting forms at yearly intervals to the Ministry of Health. These forms contain information about the number of cases of acute and chronic diseases diagnosed in a given year in the population in all areas of the country. This information is not broken down by age or sex. No verification of completeness or duplicates is possible. This passive system of collecting morbidity data on the population contrasts with the active follow-up carried out, as described above, for persons included in the Chernobyl Registry. Comparisons of morbidity based on these sources must therefore be interpreted with caution.

Radiation Doses to Different Groups: Dose Levels and Available Estimates

The dosimetric information available for liquidators is subject to controversy because personal dosimeters in use in the early days after the accident were too few and generally too sensitive. A reasonable estimate of the average dose received by the group of 200,000 people who worked in 1986–1996 is 100 mSv (Ivanov and others 1996). Thus, the collec-

tive effective dose would be approximately 20,000 Sv. Some workers received their dose in a few minutes—for example working on the roof of the reactor—while others received it over months or even years, and the predominant radiation type and route of exposure varied according to the time and activity of liquidators.

Dose estimates have generally been derived in one of three ways:

1. individual dosimetry: the liquidator was given a personal dosimeter;
2. group dosimetry: an individual dosimeter was assigned to one member of a group of liquidators; or
3. itineraries: measurements of γ-ray levels were made at various points where liquidators worked, and an individual's dose was estimated as a function of the points where he or she worked and the time spent in these places.

Liquidators are in principle included in the State Chernobyl Registries of Belarus, Russia, and Ukraine. Doses for a substantial proportion of them are missing from these registries. Liquidators who worked in the first year generally had higher recorded doses than those who worked in subsequent years. The level of dosimetric control and the adequacy of dose estimates vary between civilian liquidators (construction workers, logistic support), military liquidators (soldiers and officers who worked in decontamination, dosimetric control, and evacuation), and radiation specialists.

Results

Increases (doubling or tripling) in the incidence of leukemia and thyroid cancer have been observed in most of the studies of liquidators from Belarus, Russia, and Ukraine. Increases in leukemia risk are not unexpected since predictions from risk estimates in atomic bomb survivors have shown that if the experience of the A-bomb survivors is applicable to the Chernobyl situation, a tripling of leukemia mortality could be expected in the first 10–12 years following exposure (Cardis and others 1996).

These results are difficult to interpret since, as indicated above, the follow-up of liquidators is much more active than that of the general population in the three countries. There are questions about the adequacy and completeness of the diagnostic information on liquidators in the Chernobyl Registry (Cardis and others 1996). For thyroid cancer in adults, the depth of screening to which the liquidators are subjected may greatly influence the observed incidence.

In a case-control study based on the limited dosimetric data of the Chernobyl Registry in Russia, no significant association was seen between the risk of leukemia and radiation dose (Ivanov and others 1997a, 1997b). A recent cohort study of Russian liquidators showed no association between external radiation dose and risk of thyroid cancer among 72,000 liquidators from six regions (Ivanov and others

2002); no information on internal dose from iodine isotope was available in this study.

It is noteworthy that no increase in the incidence of leukemia or thyroid cancer has been reported to date among Baltic country liquidators (Kesminiene and others 1997; Rahu and others 1997). These findings do not contradict the findings reported in Belarus, Russia, and Ukraine in that the number of liquidators in the Baltic countries is small, and the results are also consistent with a radiation-related increase.

At this time, no conclusion can be drawn concerning the presence or absence of a radiation-related excess of cancer—particularly leukemia—among Chernobyl accident recovery workers. There is a pressing need for well-designed, sound analytical studies of recovery workers from Belarus, Russia, Ukraine, and the Baltic countries, in which special attention is given to individual dose reconstruction and the effect of screening and other possible confounding factors.

Summary

Studies of Chernobyl cleanup workers offer an important opportunity to evaluate the effects of protracted exposure in the low- to medium-dose range. No reliable risk estimates can be drawn at present from studies of these workers, however, because of the difficulties of follow-up and lack of validated individual dose estimates.

AIRLINE AND AEROSPACE EMPLOYEES

Airline pilots and flight attendants are exposed to increased cosmic radiation during flights. In 1991, the ICRP recommended that exposures to natural cosmic radiation should be considered occupational exposures for aircrews (ICRP 1991). Although aircrew members are not thought to exceed the National Council on Radiation Protection and Measurements (NCRP 1995) recommendation for occupationally exposed workers of 20 mSv per year averaged over 5 years, they do exceed the safety level set for the general public (1 mSv per year). The exposure varies with altitude, latitude, and solar flare activity. Solar activity varies on an 11-year cycle; however, prediction of short-term intense periods of activity is not possible. At 41,000 feet over the poles, the equivalent dose may vary from a norm of about 12 μSv to an extreme of 100 μSv (Friedberg and others 1989). The mean annual dose from galactic cosmic radiation can be modeled using knowledge of altitude, latitude, solar activity, and the Earth's geomagnetic field. Friedberg and colleagues (1989) estimated the annual equivalent doses that would be received on 32 U.S. domestic and international flight routes as 0.2–9.1 mSv, considerably less than recommended annual adult occupational exposures.

Several review articles have been published recently on epidemiologic studies of the occupational cancer risk for pilots and flight attendants (Blettner and others 1998; Blettner and Zeeb 1999; Boice and others 2000). The ability of stud-

ies to detect an association with ionizing radiation has been limited by several factors. Few studies have included internal comparisons, basing results instead on proportional mortality ratios, SMRs, or standardized incidence ratios. As a group, pilots and flight attendants differ appreciably from the general population. Pilots and other aircrew members are required to be very healthy and undergo frequent medical checkups, leading to the possibility of enhanced early detection of cancers in this occupational group. Disrupted circadian rhythms and, in females, relatively late age of first parity are other characteristics that complicate the choice of a suitable comparison group. Increased sun exposure, exposure to elevated ozone levels, fuel exhaust fumes, and electromagnetic fields are factors that may also confound any relationship observed between adverse health effects and cosmic radiation. Moreover, small study group sizes and the relatively low exposure levels of restricted range are further obstacles to the precise quantification of any risk.

Whether epidemiologic studies of airline personnel can have sufficient power and precision to detect so small an association has been questioned. Based on published values of annual radiation exposure of aircrew flying at high altitudes, Boice and colleagues (1992) estimated that a flying career of 20–30 years duration would result in only an 80–180 mSv cumulative dose, corresponding to a relative risk (RR) of only about 1.06, if causal. The cosmic radiation to which aircrews are exposed is predominantly in the form of high-LET neutrons and low-LET γ-radiation, the former of which can contribute as much as half of the total equivalent dose at typical flight altitudes (Boice and others 1992; Hammer and others 2000). The choice of an appropriate weighting factor for the conversion of neutron dose estimates to equivalent doses is thus crucial for dosimetry in this occupational group and for assessment of the contribution of low-LET γ-radiation to any adverse health effects. At present, the evidence for an adverse health effect in aircrews due to ionizing radiation is inconclusive.

Summary

Studies of airline and aerospace employees do not currently provide estimates of radiation-related risks because dose estimates have not been used in the studies to derive quantitative risk estimates.

MEDICAL AND DENTAL OCCUPATIONAL EXPOSURES

Early studies of patterns of mortality among radiologists and other physician specialists produced a suggestion of an excess risk of specific cancers. Excess mortality from leukemia and lymphoma, especially multiple myeloma, and also from skin, lung, pancreatic, and prostate cancer (*e.g.*, Matanoski and others 1975a, 1975b; Smith and Doll 1981; Logue and others 1986; Wang and others 1988) have been suggested, although findings were not consistent across studies for all cancers. Matanoski and colleagues (1987) reported higher overall mortality and higher cancer mortality in radiologists compared to other specialists with lower expected exposures.

A survey of the health of radiologic technologists (Boice and others 1992) gathered information on risk factors including smoking status, reproductive history, use of oral contraceptives, personal exposure to radiographs, height, weight, use of hair dye, and postmenopausal estrogens, and family and personal medical history of cancer. Members of the study population ($n = 143,517$, registered for more than 2 years with the American Registry of Radiologic Technologists, ARRT) were predominantly female and white. Personal dosimetric information was available for 64% of all the registered technologists, but only 34% of the breast cancer cases and 35% of the controls. Cases and controls were generally older and more likely to have stopped work before computerized records of dosimetry information were begun in 1979. Occupational exposure was estimated through the number of years worked as a technologist obtained from questionnaire data.

A cohort study using the ARRT database (Doody and others 1998) reported SMRs and RRs adjusted for age, calendar year of follow-up, and gender. No significant excess mortality among radiological technologists was observed for lung cancer, breast cancer, or leukemia. The SMR for all malignant neoplasms exhibited a significant trend with the number of years certified ($p < .001$), as it did for breast cancer. In the absence of complete personal dosimetry information, accurate estimates of risk due to exposures to ionizing radiation are not possible.

Yoshinaga and colleagues (1999) reported results from a retrospective cohort study of radiological technologists in Japan. External comparisons were also made with all workers and with professional and technical workers to address the issue of the healthy worker effect. The study used all Japanese men as the external comparison group; the SMR for all cancers in this study was 0.81 (95% CI 0.73, 0.95). Although elevated SMRs were observed for cancers of the colon, skin, lymphoma, multiple myeloma, and leukemia, none was statistically significant. The SMR for leukemia was significant in comparison to the total workforce as the reference group (SMR = 1.99; 95% CI 1.09, 3.33) and also for professional and technical workers as the reference group (SMR = 1.82; 95% CI 1.00, 3.06). No quantitative information on dosimetry was given in the report, nor was there an internal comparison, thus limiting the usefulness of the report for the estimation of risk.

Since 1990, a number of studies of radiologists have been published that utilized measurements of individual exposure (Andersson and others 1991). Andersson and colleagues (1991) studied the cancer risk among staff at two radiotherapy departments in Denmark. The average cumulative radiation dose was 18.4 mSv, although 63% of the persons had doses <5 mSv. The expected number of cancers was estimated using cancer incidence rates from the Danish Cancer Registry. The overall relative risk was 1.07 (95% CI 0.91, 1.25) for all cancers, and no significant dose-response was observed. The risks for cancers that are considered radiation sensitive were not elevated.

Berrington and colleagues (2001) reported the results of 100 years of follow-up of British radiologists who registered with a radiological society between 1897 and 1979 and who were followed until January 1, 1997. A progressive increase was observed in the SMRs for cancer with number of years since first registration. It appears that excess risk of cancer mortality in the period more than 40 years after first registration is likely a long-term effect of radiation exposure for radiologists registering between 1921 and 1954. Radiologists whose first registration was after 1954 demonstrated no increase in cancer mortality, possibly because of their lower overall radiation exposure.

SUMMARY

Epidemiologic studies of radiation workers and other persons exposed to ionizing radiation in the workplace started in the late 1950s with the study of British radiologists. Since then, numerous studies have considered the mortality and cancer incidence of various occupationally exposed groups in medicine, industry, defense, research, and aviation.

Studies of occupationally exposed groups are, in principle, well suited for the direct estimation of the effects of low doses and low dose rates of ionizing radiation. Potentially, the most informative studies at present are those of nuclear industry workers (including the workers of Mayak in the former USSR), for whom individual real-time estimates of doses have been collected since the 1940s with the use of personal dosimeters. More than 1 million workers have been employed in this industry since its beginning. However, studies of individual worker cohorts are limited in their ability to estimate precisely the potentially small risks associated with low levels of exposure. Risk estimates from these studies are variable, ranging from no risk to risks an order of magnitude or more than those seen in atomic bomb survivors.

Combined analyses of data from multiple cohorts offer an opportunity to increase the sensitivity of such studies and provide direct estimates of the effects of long-term, low-dose, low-LET radiation. The most comprehensive and precise estimates to date are those derived from the U.K. National Registry of Radiation Workers and the three-country study (Canada-United Kingdom-United States), which have provided estimates of leukemia and all cancer risks. Although the estimates are lower than the linear estimates obtained from studies of atomic bomb survivors, they are compatible with a range of possibilities, from a reduction of risk at low doses to risks twice those upon which current radiation protection recommendations are based. Overall, there is no suggestion that the current radiation risk estimates for cancer at low levels of exposure are appreciably in error. Uncertainty regarding the size of this risk remains as indicated by the width of the confidence intervals.

Because of the absence of individual dose estimates in most of the cohorts, studies of occupational exposures in medicine and aviation provide minimal information useful for the quantification of these risks.

Because of the uncertainty in occupational risk estimates and the fact that errors in doses have not formally been taken into account in these studies, the committee has concluded that the occupational studies are currently not suitable for the projection of population-based risks. These studies, however, provide a comparison to the risk estimates derived from atomic bomb survivors. As with survivors of the atomic bomb explosions, persons exposed to radiation at Mayak and at Chernobyl should continue to be followed for the indefinite future.

Summary

Studies of medical and dental occupational exposures do not currently provide quantitative estimates of radiation-related risks, due to the absence of radiation dose estimates.

9

Environmental Radiation Studies

INTRODUCTION

A considerable number of epidemiologic studies have been reported that have attempted to determine whether persons exposed, or potentially exposed, to ionizing radiation from environmental sources are at an increased risk of developing cancer. All epidemiologic studies are inherently uncertain, because they are observational in nature rather than experimental. Nevertheless, not all study designs are equally informative regarding the estimation of radiation risk to humans, and not all epidemiologic studies are of the same quality. Therefore, in evaluating the evidence regarding the risk of exposure to environmental sources of radiation, it is important to consider carefully the specific methodological features of the study designs employed.

Studies of environmental radiation exposure are of three basic designs: (1) descriptive studies, often referred to as ecologic; (2) case-control studies; and (3) cohort or follow-up studies. The existing published literature consists primarily of reports that are descriptive in nature and ecologic in design. The preponderance of this type of study is due to the fact that they are relatively easy to carry out and are usually based on existing data. Such investigations have utilized incidence, mortality, and prevalence data to estimate disease rates and, typically, to evaluate whether rates of disease vary in a manner that might be related to radiation exposure. If these analyses are based on large numbers of cases or large population groups, such studies may give the appearance of very precise results. Most often, geopolitical boundaries or distance from a source of radiation are used as surrogate means to define radiation exposure. For example, cancer incidence rates might be evaluated as a function of distance from a nuclear facility, or specialized statistical techniques might be employed to determine whether cases of cancer cluster or aggregate in a particular region or time period characterized by potential radiation exposure more than would be expected to occur by chance (*i.e.*, in the absence of any exposure).

Weaknesses associated with studies of this type make them of limited value in assessing risk. The primary limitation is that the unit of analysis is not the individual; thus, generally little or no information is available that is specific to the individual circumstances of the people under study. Of most concern in this regard is the definition of radiation exposure. Ecologic studies generally do not include estimates of individual exposure or radiation dose. Either aggregate population estimates are used to define population dose for groups of people, or surrogate indicators such as distance or geographic location are used to define the likelihood or potential for exposure or, in some cases, an approximate magnitude or level of exposure. This approach has serious limitations. It implies, for example, that residents who live within a fixed distance from a facility are assumed to have received higher radiation doses than those who live at greater distances or than individuals in the larger population as a whole who do not live in the vicinity of the facility. Further, it assumes that everyone within the boundary that defines exposure (or a given level of exposure) is equally exposed or has the same opportunity for exposure. In most situations, such assumptions are unlikely to be accurate, and variability in exposure of individuals within the population may be substantially greater than the exposure attributed on a population basis. The resulting almost certain misclassification of exposure can lead to a substantial overestimation or underestimation of the association of the exposure with the disease under study.

Similarly, there is usually no information available in ecologic studies regarding other factors that might influence the risk of developing the disease(s) under study (*i.e.*, other risk factors). Thus, there is no way to evaluate the impact of such factors in relation to the potential effect of radiation exposure. This inability to evaluate or account for the potential confounding effect of other important factors, or the modifying effect of such factors on risk, makes the ecologic approach of limited use in deriving quantitative estimates of radiation risk.

A third limitation of the ecologic design is that disease outcome usually is not confirmed at the individual level. Most studies rely on routine reporting, either of mortality through death certificates or of cancer incidence through cancer registration and surveillance systems. Such sources of information vary in their degree of accuracy and completeness, and they can sometimes vary in relation to the surrogate measures being used to define exposure (*e.g.*, geographic area). This can lead to the identification of spurious associations.

Fourth, ecologic studies seldom estimate or account for population migration or movement. This, too, can result in the appearance of spurious associations if aggregate or population measures of radiation exposure actually reflect underlying changes in population mobility with factors such as time, age, or geographic area.

Finally, descriptive studies are often based on a small number of cases of disease. Such studies have low statistical power to detect an association if it truly exists, and they are very sensitive to random fluctuations in the spatial and/or temporal distribution(s) of the disease(s) under study. This is especially true for diseases such as cancer, particularly childhood cancer, which are relatively uncommon on a population basis.

There have also been attempts to evaluate the effect of environmental radiation exposures using the two most common analytical study designs employed in epidemiology: the case-control and the cohort study. Such studies are almost always based on individual-level data and thus are not subject to many of the limitations summarized above for ecologic studies. Nevertheless, each of these study designs is subject to specific weaknesses and limitations. Of most concern in case-control studies is the potential bias that can result in relation to the selection of cases and controls, such that the two groups are differentially representative of the same underlying population. A second important source of bias can be differential recall of information about exposure for cases relative to controls. In cohort studies, a common limitation is the relatively small number of cases for uncommon disease outcomes and the resultant low statistical power. A second concern is the completeness of follow-up of the cohort under study, and equal follow-up and determination of disease status according to exposure. Such limitations of both types of analytic epidemiologic studies may be particularly problematic in investigations of low doses and relatively small increases in disease risk. Under such circumstances, the magnitude of the impact on risk estimates of small or modest biases may be as great or greater than the magnitude of the true disease risk.

In summary, most existing published studies of environmental radiation exposure are ecologic in design. Such studies are limited in their usefulness in defining the risk of disease in relation to radiation exposure or dose. They can sometimes be informative in generating new hypotheses or suggesting directions of study but seldom, if ever, are of value in testing specific hypotheses or providing quantitative estimates of risk in relation to specific sources of environmental radiation. Epidemiologic studies, in general, have limited ability to define the shape of the radiation dose-response curve and to provide quantitative estimates of risk in relation to radiation dose, especially for relatively low doses. To even attempt to do so, a study should (1) be based on accurate, individual dose estimates, preferably to the organ of interest; (2) contain substantial numbers of people in the dose range of interest; (3) have long enough follow-up to include adequate numbers of cases of the disease under study; and (4) have complete and unbiased follow-up. Unfortunately, the published literature on environmental radiation exposures is not characterized by studies with such features.

The accompanying tables provide a summary of the principal studies of environmental radiation exposure published since the BEIR V report (NRC 1990). Articles included in this summary were identified principally from searching the PubMed database of published articles from 1990 through July 2004. Searches were restricted to human studies and were broadly defined: key words included radiation; neoplasms; radiation-induced; radioactive fallout; and environmental radiation. Searches specific to the Chernobyl accident included Chernobyl, Russia, Ukraine, and Belarus as key words. Articles were also identified from UNSCEAR (2000b) and from the usual scientific interactions with other investigators. The tables are organized according to the type of exposure situation under study as follows: (1) populations living around nuclear facilities; (2) populations exposed from atmospheric testing, fallout, or other environmental releases of radiation; (3) populations exposed from the Chernobyl accident; (4) populations exposed from natural background; and (5) children of adults exposed to radiation. Within each type of exposure situation, the tables are further grouped according to study design: ecologic studies, case-control studies, and cohort studies. Each table contains a brief description of the principal design features and results of each study. The principal criteria used to assess the utility of each study in evaluating the risk of disease in relation to radiation exposure were the following: (1) Was there a quantitative estimate of radiation dose; (2) if so, was the estimate for individuals in the study (*i.e.*, individual-level estimates of radiation dose received); and (3) was there a quantitative estimate of disease risk in relation to radiation dose?

POPULATIONS LIVING AROUND NUCLEAR FACILITIES

Table 9-1A lists 16 ecologic studies of populations living around nuclear facilities, 13 of the locations being outside the United States. Most define exposure, or potential for exposure, based on a measure of distance from the facility, although the two studies of exposures at Three Mile Island by Hatch (1992) utilized some information on measurements

TABLE 9-1A Populations Living Around Nuclear Facilities—Ecologic Studies

Reference	Incidence/Mortality	Population Studied	Type of Exposure	Dates of Accrual	Type of Dosimetry	Outcomes Studied	Number of Cases	Summary of Results
Hatch and Susser (1990)	Incidence and mortality	Residents (ages 0–24) within 10 miles of Three Mile Island	Background gamma	1975–1985	Outdoor measurements taken in 1976	All cancer; leukemia	49 (0–14) 104 (0–24)	Increased risk for highest vs. lowest quartile; childhood cancer and leukemia
Hatch and others (1990)	Incidence	Residents within 10 miles of Three Mile Island	Xenon and iodine	1975–1985	Dispersion modeling, based on monitoring data	All cancer; childhood cancer (ages 0–14, 0–24); leukemia (ages 0–14, 0–24, 25); and all lymphoma	5493 total	No evidence of an effect on cancer incidence
Jablon and others (1991)	Mortality	Residents of 107 counties in U.S. with or near nuclear installations	Unspecified	1950–1984	County with a nuclear facility that began operation before 1982, or an adjacent county if at least 20% of the county was within a 16 km radius	15 cancer sites; benign and unspecified neoplasms	900,000 deaths in 107 counties	No evidence of excess mortality in study counties
Sofer and others (1991)	Incidence	Children and young adults living near nuclear plant in Israel	Unspecified	1960–1985	Distance from Negev nuclear plant	Leukemia	192	No overall increase; some increase with time among 0–9 in Western Negev; increase in girls 0–4 from 1970 to 1979
Michaelis and others (1992)	Incidence	Children living near nuclear installations in Germany	Unspecified	1980–1990	Distance from nuclear facility	Childhood cancer; acute leukemia	81 within 5 km	No increase for all cancer, acute leukemia; suggested increases in subgroups of early ages or close proximity
McLaughlin and others (1993b)	Incidence and mortality	Children born to mothers residing near nuclear installations in Ontario, Canada	Unspecified	1950–1987	Distance from nuclear facility	Leukemia in children	Range by facility: 2–72	Suggestion of some excess over expected for some analyses; none significant
Bithell and others (1994)	Incidence	Children in England and Wales	Unspecified	1966–1987	Distance from nuclear facilities based on ward	Leukemia and non-Hodgkin's lymphoma (NHL)	Range for 25 km zones: 7–570	Linear risk score significantly elevated in Sellafield and Burghfield
Black and others (1994a)	Incidence	Residents of Dalgety Bay, Scotland	Particles of radium-226	1975–1990	Routine monitoring measurements	All cancer; 18 specific sites	211 (total)	No evidence of increase over expected
Black and others (1994a)	Incidence	Children and young adults in Dounreay, Scotland	Contamination from nuclear reprocessing plant	1968–1991	Distance from Dounreay	Leukemia and NHL	12 in nearest zone	Evidence of increase over expected in nearest zone

209

continues

210

TABLE 9-1A Continued

Reference	Incidence/ Mortality	Population Studied	Type of Exposure	Dates of Accrual	Type of Dosimetry	Outcomes Studied	Number of Cases	Summary of Results
Zaridze and others (1994)	Incidence	Children in Kazakhstan	Unspecified	1981–1990	Distance from nuclear testing sites	All cancer; six specific sites	Total: 1408; leukemia: 512	Increase in leukemia in areas closest to testing sites; some evidence of increase in brain tumors
Viel and others (1995)	Incidence	People under age 25 living around La Hague reprocessing plant in France	Unspecified	1978–1992	Distance from the La Hague plant	Leukemia	25	Cluster of cases located close to La Hague plant
Waller and others (1995)	Incidence	Children in 2594 parishes of Sweden	Unspecified	1980–1990	Distance from nuclear facility	Acute lymphocytic leukemia (ALL)	656	No significant clustering of cases found
Gulis and Fitz (1998)	Incidence	Residents of Trnava, Slovakia	Unspecified	1986–1995	Distance from nuclear power plant	13 cancer sites	Range for zones: 0–323	Suggestion of increasing incidence closer to the site; nonsignificant
Kaatsch and others (1998)	Incidence	Children living near nuclear facilities in Germany	Unspecified	1991–1995	Distance from a nuclear facility	All cancer; leukemia; lymphoma; selected sites	Total 550; leukemia 182	No evidence of an increase in incidence
Guizard and others (2001)	Incidence	Residents under age 25 in areas around the La Hague plant in France	Unspecified	1978–1998	Distance from the La Hague plant	Leukemia	38	Increase over expected in area less than 10 km from site
Boutou and others (2002)	Incidence	Nord Cotentin, France	Population mixing—near nuclear power plant and reprocessing unit	1979–1998	Population mixing index per geographic unit (commune), based on number of workers born outside department of La Manche	Childhood leukemia in persons under age 25		Incidence rate ratio 2.7 in rural communes in highest tertile of mixing, relative to urban communes. Positive trend in leukemia with increasing mixing index. Risk stronger for ALL in children 1–5

taken around the site after the accident. All but one (Jablon and others 1991) are based on incidence data, and one study in Canada (McLaughlin and others 1993a) uses mortality data as well as incidence data. The focus of most of these investigations is leukemia and/or childhood cancer, although a few include all cancers as an outcome. The size of the studies, in terms of numbers of cases, ranges from very small (Black and others, 1994a; 12 cases in the most highly exposed zone) to extremely large (Jablon and others 1991). Notably, most of the studies do not specify the nature of the radiation exposure, and none of the 16 contains individual estimates of radiation dose. Although some of these studies report an increased occurrence of cancer that could potentially be related to environmental radiation exposures, none provides a direct quantitative estimate of risk in relation to radiation dose.

Table 9-1B summarizes three case-control studies of persons living around a nuclear facility. Two studies are of leu-

kemia, one in children under age 15 (Urquhart and others 1991) and the other in people under age 25 (Pobel and Viel 1997). Both studies are based on a small number of cases and focus primarily on parental radiation exposure and X-ray exposure of the child. Neither study found an increased risk associated with these types of radiation exposure. Both, however, did find an increased risk associated with playing on beaches near the nuclear facility. The third study (Shields and others 1992) focuses on congenital and perinatal conditions, stillbirths, and infant deaths in relation to exposures from uranium mines. Exposures include environmental exposures from living near a mine or mine dumps or tailings, or living in a home made from mine rock, as well as from working in a uranium mine. This study does not provide an estimate of radiation risk associated with any of the indicators of exposure.

In summary, most of the studies of populations living around nuclear facilities have not included individual esti-

TABLE 9-1B Populations Living Around Nuclear Facilities—Case-Control Studies

Reference	Population Studied		Number of Subjects		Dates of Accrual	Type of Exposure	Type of Dosimetry	Summary of Results
	Cases	Controls	Cases	Controls				
Urquhart (1991)	Leukemia and NHL in children under age 15 resident in Caithness	Selected from birth register; matched by zone of residence at birth, date of birth, sex	14	55	Diagnosis 1970–1986	Paternal preconception whole-body dose; antenatal X-ray	Employment at Dounreay; recorded dose from employment records; questionnaire for X-ray	No increased risk with employment at Dounreay, recorded radiation dose, antenatal X-ray; evidence of increased risk from playing on beaches within 50 km of Dounreay
Shields and others (1992)	Congenital and perinatal conditions, stillbirths, infant deaths	Chronologically nearest normal single birth; matched by sex, mother's age within 5 years, gravidity	266	266	1964–1981	Environmental exposure from working or living near, or working in uranium mines	Environment: time prior to child's birth worked in uranium mine; residence within 0.5 mile of mine, dumps, or tailings; living in home made with mine rock. Workers: recorded WLM, estimated gonadal dose	Only significant association with mother living near tailings or mine dumps. Overall, associations with measures of radiation exposure were weak
Pobel and Viel (1997)	Leukemia diagnosed in people <25 years of age living within 35 km of La Hague nuclear plant	Sample of children cared for by general practitioners of the cases; matched to cases on sex, age, place of birth; and residence at diagnosis of case	27	192	1978–1993	Antenatal and postnatal X-ray exposure; parental occupational exposures (including radiation); viral infections, life-style	For parents employed in nuclear facility, whole-body external dose (mSv) was obtained from company records. Other information obtained by questionnaire	No association with occupational radiation exposure of parents; increased risk for use of local beaches, consumption of local fish, length of residence in granitic area or house

TABLE 9-2A Populations Exposed from Atmospheric Testing, Fallout, or Other Environmental Release of Radiation—Ecologic Studies

Reference	Incidence/ Mortality	Population Studied	Type of Exposure	Dates of Accrual	Type of Dosimetry	Outcomes Studied	Number of Cases	Summary of Results
Darby and others (1992)	Incidence	Children under age 15 in Nordic countries	Fallout from nuclear weapons tests	Denmark (1948), Finland, Norway, Iceland (1958), Sweden (1961–1987)	Estimates of bone marrow dose to fetus, 1-year-old, testes, received during fallout period: low, medium, high	Leukemia	Not given	Little increase in high-fallout years; slightly elevated in high vs. medium group
Gilbert (1998)	Incidence and mortality	United States	Fallout from nuclear weapons tests in Nevada	Deaths: 1957–1994; incident cases: 1973–1994	Mean thyroid dose by county, derived from measurements and environmental modeling	Thyroid cancer	4602 deaths; 12,657 incident cases	No increased risk with cumulative dose or dose received at ages 1–15; suggested increase for those exposed under age 1 and those in 1950–1959 birth cohort

mates of radiation dose and have therefore not provided an estimate of disease risk. The three case-control studies described above found no increased risk of disease associated with radiation exposure.

POPULATIONS EXPOSED FROM ATMOSPHERIC TESTING, FALLOUT, OR OTHER ENVIRONMENTAL RELEASE OF RADIATION

Table 9-2A describes two ecologic studies of populations exposed to fallout from atmospheric nuclear testing, fallout, or other sources of environmental release of radiation. The nature of the exposure is not specified beyond "fallout." These studies utilize population-based measures of exposure rather than individual estimates of radiation dose. They address two separate outcomes (leukemia and thyroid cancer), but provide no quantitative estimates of risk associated with the exposure.

Table 9-2B summarizes two cohort studies of persons who participated in U.K. atmospheric nuclear weapons tests. The study by Darby and colleagues (1993) is an extension of an earlier analysis from this cohort and uses doses from film badges to characterize individual external whole-body radiation dose. It investigates all causes of mortality as well as all major forms of cancer. Overall, the study found no increased risk of developing cancer or other fatal diseases as a function of estimated dose received, based on follow-up through 1991 and relatively large numbers of cases. There was some evidence of an increase in leukemia, based on only 29 cases. The most recent update of this cohort (Muirhead and others 2003) found little increase in overall mortality or cancer incidence and no increase in other types of cancer, but continuing evidence of a small increased risk of non-chronic lymphocytic leukemia (CLL).

In contrast, a recent study of U.S. veterans (Dalager and others 2000) who participated in atmospheric nuclear weapons tests reported a significant increase in death from all causes, and for all lymphopoietic cancers combined, although the number of cases in the latter group was very small. This study focused on veterans whose external γ-radiation dose, as recorded on film badges, was 5 rem, and compared mortality in this group to veterans who participated in one nuclear test and whose dose was 0.25 rem. The mean dose among the 5 rem group was 7.8 rem and among the controls was 0.08.

Also included in Table 9-2B are several studies of the population of residents living near the Techa River in the southern Urals of the Russian Federation. More than 25,000 residents were exposed to external γ-radiation as well as internally from fission products (primarily cesium-137, strontium-90, ruthenium-106, and zirconium-95) released into the Techa River from the nearby Mayak plutonium production facility, predominately in the early 1950s. Studies have been conducted of cancer mortality in residents and their offspring, as well as pregnancy outcomes. Initial dose estimates were based on average doses reconstructed for settlements. Efforts to estimate individual doses for members of this resident cohort continue. To date, there is no evidence of a decrease in birth rate or fertility in the exposed population, and there is no increased incidence of spontaneous abortions or stillbirths (Kossenko and others 1994). There is some evidence of a statistically significant increase in total cancer mortality (Kossenko 1996). Current estimates of the excess absolute risk (EAR)[1] of leukemia in this cohort is 0.85 per 10,000 person-years (PY) per gray (95% CI 0.2, 1.5), and for

---

[1]EAR is the rate of disease in an exposed population minus the rate of disease in an unexposed population.

TABLE 9-2B Populations Exposed from Atmospheric Testing, Fallout, or Other Environmental Release of Radiation—Cohort Studies

Reference	Incidence/Mortality	Cohort Definition	Comparison Group	Dates of Accrual	Type of Exposure	Type of Dosimetry	Outcomes Studied	Number of Cases	Summary of Results
Darby and others (1993)	Incidence and mortality	Persons who participated in U.K. atmospheric nuclear weapons tests	Men identified from Ministry of Defense archives who did not participate	1950s–1991	External whole-body dose	Recorded on film badges obtained from Ministry of Defense	Broad causes of death; 27 specific cancer sites	All causes: 2753 (control group—2939)	No effect on risk of developing cancer or other fatal diseases; some evidence of an increase over expected for leukemia, based on 29 cases
Kossenko and others (1994)	Pregnancy outcome and mortality	Children born to 28,100 residents exposed to discharges of radioactive waste into Techa River	Unexposed populations living in the same area	1953–1974	External and internal dose: primarily from ^{137}Cs, ^{90}Sr, ^{106}Ru, ^{95}Zr	Gonadal doses estimated as average for each settlement	Birth rate, fertility, fetal loss, infant mortality	56 cancer deaths	No decrease in birth rate or fertility in exposed population; no increased incidence of spontaneous abortions or stillbirths; no change in cancer mortality
Kossenko and others (1994)	Mortality	28,000 residents exposed to discharges of radioactive waste into Techa River, 1950–1953	Unexposed populations living in the same area	1950–1982	External and internal dose: primarily from ^{137}Cs, ^{90}Sr, ^{106}Ru, ^{95}Zr	Average absorbed dose to bone marrow estimated for each settlement	All cancer and 13 major site categories	163 cancers in exposed population	Increase in total cancer mortality. Leukemia: absolute risk 0.85 per 10,000 PY per gray; relative risk for esophagus, stomach, and lung similar to atomic bomb survivors
Kossenko (1996)	Mortality	28,000 residents exposed to discharges of radioactive waste into Techa River	Matched control group from unexposed area	33-year period from 1949 through 1982	External and internal dose: primarily from ^{137}Cs, ^{90}Sr, ^{106}Ru, ^{95}Zr	Average absorbed dose to bone marrow estimated for each settlement	Leukemia and solid cancer in residents; cancer in offspring		Leukemia: absolute risk 0.85 per 10,000 PY per gray; solid cancer: relative risk 0.65 Gy^{-1}. No increase in offspring of exposed residents
Davis and others (2001)	Cumulative incidence	Persons born to mothers resident in one of 7 counties surrounding Hanford Site from 1940 to 1946	Internal control according to estimated individual thyroid radiation dose	Birth through date of exam in 1992–1997	Primarily ^{131}I	Estimated individual absorbed dose to thyroid	Thyroid cancer and 12 categories of noncancer thyroid diseases	19 thyroid cancer cases	No increase in thyroid cancer or any noncancer thyroid disease outcome associated with increasing radiation dose to the thyroid
Dalager and others (2000)	Mortality	Persons who participated in U.S. atmospheric nuclear weapons tests and received highest doses	Navy veterans who participated in HARDTACK and received minimal radiation dose	Date of first exposure through 1996	External gamma dose	Film badges	All deaths; lymphopoietic, leukemia, digestive, respiratory, other cancer	300 deaths in veterans with 5 rem; 11 cases of lymphopoietic cancer	All-cause mortality: relative risk (RR) 1.22 (95% CI 1.04–1.44); lymphopoietic cancer 3.72 (95% CI 1.28–10.83)

continues

TABLE 9-2B Continued

Reference	Incidence/ Mortality	Cohort Definition	Comparison Group	Dates of Accrual	Type of Exposure	Type of Dosimetry	Outcomes Studied	Number of Cases	Summary of Results
Kossenko and others (2000)	Mortality	10,459 offspring of parents exposed to discharges of radioactive waste into Techa River	None	1950–1992	External and internal dose: primarily from ^{137}Cs, ^{90}Sr, ^{106}Ru, ^{95}Zr	None	Cancer	25 cancer deaths	Descriptive analyses only—no estimates of risk
Koshurnikova and others (2002)	Mortality and incidence	72,185 persons living in Ozyorsk for at least 1 year under age 15 and born 1948–1988; or born elsewhere 1934–1988 but moved to Ozyorsk before age 15	Ozyorsk Population	1948–1988	Fallout from Mayak facility	None	Deaths, cancer deaths, leukemia, thyroid cancer	4636 deaths; 371 cancer deaths; 53 leukemia deaths; 31 thyroid cancer cases	Thyroid cancer 3–4 times expected relative to Russia; 1.5–2-fold higher based on Chelyabinsk Oblast rates
Muirhead and others (2003)	Mortality and incidence	21,357 persons who participated in the U.K. atmospheric nuclear weapons tests	22,333 men who did not participate in tests identified from Ministry of Defense records, matched on a number of characteristics	1952–1998	External gamma	Film badge readings and potential for exposure based on duties	All deaths, 27 types of cancer	2089 deaths; 785 cancer deaths; 16 leukemia deaths; 2641 cases of cancer; 67 cases of leukemia	Little difference in overall mortality or cancer incidence between exposed and controls; no increase in multiple myeloma; evidence of a small risk of non-CLL leukemia
Takahashi and others (2003)	Prevalence	3709 Marshall Island Residents born before the Castle BRAVO atmospheric nuclear weapons tests on March 1, 1954	Internal control according to estimated dose level	1993–1997	Fallout from Castle BRAVO test	Surrogate estimates of dose based on ^{137}Cs soil deposition levels	Thyroid cancer	57 cases	Prevalence increased with quartile of estimated dose, but was not significant

solid tumors the relative risk estimate is 0.65 Gy^{-1} (95% CI –0.3, 1.0). Median dose estimates for soft tissue in this cohort are 7 mSv (maximum 456 mSv) and for bone marrow 253 mSv (maximum 2021 mSv). Estimates of the relative risk for cancer of the esophagus, stomach, and lung are similar to those reported for atomic bomb survivors. There is no evidence of an increase in cancer mortality in the offspring of exposed residents (Kossenko 1996). There has also been one study (Koshurnikova and others 2002) of persons living in the town of Ozyorsk exposed to fallout from the nearby Mayak nuclear facility. This study reported an excess of thyroid cancer three to four times that expected relative to rates for all of Russia and a somewhat lower excess (1.5 to twofold higher) based on a comparison with Chelyabinsk Oblast rates. No estimates of radiation dose were included in this study.

Two other cohort studies of persons exposed to atmospheric releases of radioactive materials are also summarized in Table 9-2B. One is a follow-up study of 3440 persons exposed as young children to atmospheric releases of primarily ^{131}I from the Hanford nuclear facility in eastern Washington State (Davis and others 2001, 2004a). No increased risk of thyroid cancer was found associated with individual radiation dose to the thyroid. The other (Takahashi and others 2003) is a prevalence study of thyroid cancer conducted through screening of 3,709 Marshall Island residents born before the Castle BRAVO atmospheric nuclear weapons test on March 1, 1954. Radiation dose was based on a surrogate constructed from age-specific doses estimated for the Utirik atoll and ^{137}Cs deposition levels on atolls where the participants resided. There was some indication that the prevalence of thyroid cancer increased with quartile of estimated dose, but the increase was not statistically significant.

In summary, some but not all studies of persons exposed to fallout or other environmental releases of radiation have found increased risks of specific disease outcomes. Most notable are findings of a significant increase in death from all causes and for all lymphopoietic cancers combined in a recent study of U.S. veterans who participated in atmospheric nuclear weapons tests, and evidence of an increase in total cancer mortality and thyroid cancer incidence among residents living near the Techa River in the southern Urals of the Russian Federation.

POPULATIONS EXPOSED FROM THE CHERNOBYL ACCIDENT

The explosion at the Chernobyl Power Station Unit 4 in Ukraine on April 26, 1986, released large quantities of radionuclides into the atmosphere, resulting in the contamination of a large geographic area. Initially exposures were due principally to radioisotopes of iodine, primarily iodine-131 (^{131}I), and subsequently to radiocesium, primarily cesium-137 (^{137}Cs), from both external exposure and the consumption of contaminated milk and other foods. Numerous epidemiologic studies have been carried out since the Chernobyl accident to investigate the potential late health consequences of exposure to ionizing radiation from the accident. These studies have focused largely on thyroid cancer in children, but have also included investigations of recovery operation workers and residents of contaminated areas, and have investigated the occurrence of leukemia and solid tumors other than thyroid cancer among exposed individuals.

Overwhelmingly, the published findings are from studies that are ecologic in design and therefore do not provide quantitative estimates of disease risk based on individual exposure circumstances or individual estimates of radiation dose. Most reports are descriptive incidence and prevalence studies that utilize population or aggregate estimates of radiation dose. The principal studies are summarized in Table 9-3A. Only four analytical studies are published that report dose-response results based on individual dose estimates (Table 9-3B). In the sections that follow, current evidence is summarized separately regarding the risk of thyroid cancer, leukemia, and other solid tumors associated with radiation exposure from the Chernobyl accident. Studies of recovery operations workers are considered in Chapter 8 on occupational exposures.

Thyroid Cancer

An increase in the incidence of thyroid cancer first began to appear in Belarus and Ukraine in 1990. After the initial few reports, there was immediate skepticism that such increases were related directly to radiation exposure from Chernobyl. The very early onset of disease after exposure (only 4 years) was unexpected based on existing knowledge of the latent period for radiation-related thyroid cancer; there was doubt about the certainty of the pathologic diagnoses; and there was speculation that the apparent increases were largely the result of widespread population screening.

Numerous reports have continued to describe an increasing number of cases of thyroid cancer, particularly in the most heavily contaminated regions of Ukraine and Belarus, and also in Russia. Collectively, findings reported to date have demonstrated an association between radiation exposure from the Chernobyl accident and an increase in thyroid cancer incidence. Among those under age 18 at the time of the accident, it has been estimated that approximately 2000 thyroid cancers were diagnosed from 1990 to 1998 in Ukraine, Belarus, and Russia. The increase in all three countries for this period was approximately fourfold, with the highest increase observed in the Gomel region in Belarus. More recent data indicate that excess thyroid cancer continues to occur among people in Belarus, Ukraine, and the contaminated regions of Russia. This increase cannot be explained only by the aging of the cohort and the improvement in case detection and reporting. Although there is now little doubt that an excess of thyroid cancer has occurred in highly contaminated areas, there is still very little information re-

TABLE 9-3A Populations Exposed from the Chernobyl Accident—Ecologic Studies

Reference	Incidence/Mortality	Population Studied	Type of Exposure	Dates of Accrual	Type of Dosimetry	Outcomes Studied	Number of Cases	Summary of Results
Prisyazhiuk and others (1991)	Incidence	Three contaminated districts in Ukraine: Polesskoye, Naroditchy, Ovrutch	Fallout from Chernobyl	1981–1990	Calendar year (before and after accident) and district (contaminated areas)	Leukemia, thyroid cancer, all other cancer	Leukemia: 105; thyroid: 25; all other: 3804	Overall, incidence rates were not different before and after the accident. Leukemia in age 65+ group increased in 1987 and remained 2–3 times higher; three cases of thyroid cancer diagnosed in 1990 in <14 age group (none 1981–1989); all others increased in 1987 by a third
Ramsay and others (1991)	Incidence	Population of Lothian, Scotland	Fallout from Chernobyl	1978–1989	Calendar year (i.e., from Chernobyl, before and after accident)	Down's syndrome	Ave.: 12.4 cases per year; range 7 (1989)–26 (1987)	Significant increase in 1986–1987
Baverstock and others (1992)	Incidence	Belarus	Fallout from Chernobyl	1986–1992	Calendar year and region	Thyroid cancer	104	Marked increase beginning in 1990; highest rates in Gomel
Kazakov and others (1992)	Incidence	Six regions of Belarus and Minsk city	Fallout from Chernobyl	1986–1992	Calendar year and region	Thyroid cancer	131	Average of 4 cases per year 1986–1989 55 in 1991; projected 60 in 1992. Most increase in Gomel
Ivanov and others (1993)	Incidence	Belarus: children ages 0–14	Fallout from Chernobyl	1979–1991	Two time periods: 1979–1985; 1986–1991. Three levels of contamination by region or city	Childhood leukemia	Not given	No change in incidence after Chernobyl accident, and no increase after accident in areas with higher contamination levels
Parkin and others (1993)	Incidence	20 European countries: children ages 0–14	Fallout from Chernobyl	1980–1988	Estimated dose (effective equivalent dose) in 30 countries or regions, obtained from UNSCEAR	Childhood leukemia	3679	Risk of leukemia 1987–1988 relative to before 1986 was not related to radiation exposure
Auvinen and others (1994)	Incidence	Finland: children 0–14 in 1976–1992	Fallout from Chernobyl	1976–1992	Estimated cumulative dose in 2 years after the accident. Based on measurements of dose rate in 455 municipalities. Internal dose estimated from whole-body measurements on sample of 81 children. Municipalities divided into fifths of exposure	Childhood leukemia	Not given	Incidence did not increase in 1976–1992. Relative excess in 1989–1992 was not significantly different from zero

Reference	Type	Population	Exposure	Time period	Comparison	Endpoint	Number	Findings
Hjalmars and others (1994)	Incidence	Sweden: children 0–15	Fallout from Chernobyl	1980–1992	^{137}Cs contamination by geographic area	Childhood acute leukemia	888	No significant increase in childhood acute leukemia in contaminated areas
Petridou and others (1994)	Incidence	Greece: children 0–14	Fallout from Chernobyl	1980–1991	Three time periods: 1980–June 1986; July 1986–June 1988; July 1988–June 1991. Mean fallout levels (based on ^{137}Cs measurements) grouped into 17 geographic regions	Childhood leukemia	968	No evidence of an increased incidence of childhood leukemia in periods after Chernobyl accident. No association between childhood leukemia and region by radiation fallout level
Likhtarev and others (1995)	Incidence	Ukraine: children ages 0–14	Fallout from Chernobyl	1986–1993	Calendar year; 7 geographic zones defined by estimated average thyroid dose to children	Thyroid cancer	418 cases in 0–14 year olds; 248 cases in those 15 and older	Increase beginning in 1989; rate in 1993 was 5 times higher than 1986; higher incidence in zones with higher contamination levels
Prisyazhiuk and others (1995)	Incidence	Four districts in Ukraine: Narodichy, Ovrutch, Ivankov, Polesskoye	Fallout from Chernobyl	1980–1993	Three time periods: 1980–1985 (before accident); 1986–1993 (after accident); 1980–1993	All cancer; leukemia and lymphoma; thyroid cancer	Not given	Statistically significant increase in thyroid cancer after the accident; no significant increase in all cancer, or leukemia and lymphoma
Stsjazhko and others (1995)	Incidence	Belarus, Russia, Ukraine	Fallout from Chernobyl	1981–1994	Three time periods: 1981–1985 (before accident); 1986–1990; 1991–1994; 6 geographic regions	Thyroid cancer	Since the accident: Belarus, 333; Russia, 23; Ukraine, 209	Increase in thyroid cancer incidence after the accident; most pronounced in most heavily contaminated areas
Sugenoya and others (1995)	Prevalence	Two cities in Belarus (Chechelsk and Bobruisk): children ages 10–15	Fallout from Chernobyl	October 1991–August 1992	Contamination levels (^{137}Cs): Chechelsk 5–>40 Ci/km^2; Bobruisk, control area	Thyroid abnormalities	888 screened in Chechelsk; 521 screened in Bobruisk	Significantly higher prevalence of multiple micronodular lesions in diffuse goiter in contaminated city
Gunay and others (1996)	Incidence	Bursa, Turkey: pediatric cases of malignancy	Fallout from Chernobyl	1986–1995	Calendar year, 1986–1995	Acute leukemia, lymphoma, solid tumors	Acute leukemia: 101; lymphoma: 44; solid tumor: 31	Significant increase in acute leukemia after 1986; no significant increase in lymphoma or solid tumor
Ivanov and others (1996)	Incidence	Seven regions of Belarus: children 0–15	Fallout from Chernobyl	1982–1994	Calendar year; 7 geographic regions	Childhood leukemia	Not given	No increase associated with calendar; no difference in rates by geographic region
Kumpusalo and others (1996)	Prevalence	Two villages in Bryansk region of Russia (Mirnyi and Krasnyi): residents ages 3–34	Fallout from Chernobyl	1993	Contaminated area (Mirnyi) and control area (Krasnyi)	Thyroid ultrasound findings	302 screened in Mirnyi; 200 screened in Krasnyi	No pathological U.S. findings in either city. Prevalence of thyroid abnormalities higher in contaminated area: ages 0–9, 8.1% in Mirnyi; 1.6% in Krasnyi

continues

TABLE 9-3A Continued

Reference	Incidence/ Mortality	Population Studied	Type of Exposure	Dates of Accrual	Type of Dosimetry	Outcomes Studied	Number of Cases	Summary of Results
Parkin and others (1996)	Incidence	34 regions of 23 countries in Europe: children 0–15	Fallout from Chernobyl	1980–1991	Estimated dose (effective equivalent dose) in 34 countries or regions, obtained from UNSCEAR for first year, 0–4 years and 0–70 years after the accident	Childhood leukemia	25,820	Small increase in incidence over time, but no association in the period 1987–1991 with estimated dose
Petridou and others (1996)	Incidence	Greece: children ages 0–4	Fallout from Chernobyl	1980–1994	Three geographic areas based on ^{137}Cs measurements. Children born during second half of 1986 and all of 1987 considered exposed in utero; remaining births considered unexposed	Infant leukemia	Exposed in utero: 55 unexposed in utero: 297	Incidence in those exposed in utero 2.6 times higher than unexposed (95% CI 1.4, 5.1) Children born to mothers resident in high-contamination areas were at significantly higher risk of leukemia. No increase associated with preconception exposure of parents
Remennik and others (1996)	Incidence	Bryansk, Kaluga, Tula, Orel, Ryazan, Kursk regions of Russia	Fallout from Chernobyl	1981–1994	Calendar year, geographic region	All malignancies; thyroid cancer	Not given	Increase in thyroid cancer incidence in children in the Bryansk region after the accident. Incidence of all cancer higher in 6 study regions than all of Russia since 1987
Tondel and others (1996)	Incidence	Six most contaminated counties of Sweden: persons age 0–19	Fallout from Chernobyl	1978–1992	Three levels of contamination based on ^{137}Cs measurements; two time periods: 1978–1986 and 1987–1992	Brain cancer, acute lymphatic leukemia, other neoplasms	746	No clear associations between cancer incidence and level of radiation contamination; or increases over time since the accident
Ashizawa and others (1997)	Prevalence	119,178 children examined at 5 centers in Ukraine, Belarus, and Russia	Fallout from Chernobyl	Examined May 15, 1991– April 30, 1996	Five geographic areas	Goiter	42,470	Variation in prevalence by region: highest in Kiev (54%); lowest in Gomel (18%). Significant inverse association between goiter and median urine iodine level
Bleuer and others (1997)	Incidence	Belarus: individuals born 1963 or later	Fallout from Chernobyl	1986–1995	^{131}I contamination levels by raion (geographic unit comparable to a county in the U.S.)	Thyroid cancer	528 in persons <15 in 1986	Feasibility study to assess the use of existing data to evaluate geographic and time trends—no conclusions risk of thyroid cancer

219

continues

Reference	Measure	Population	Exposure	Period	Design	Outcomes	Numbers	Results
Ivanov and others (1997a)	Incidence and mortality	Kaluga Oblast, Russia	Fallout from Chernobyl	1981–1995	Contaminated areas of oblast, based on ^{137}Cs measurements and estimates of whole-body doses by settlement; calendar time (by year and grouped to reflect before and after accident	All cancer, 8 specific cancer sites; grouped as GI, respiratory, leukemia	All cancer: 2052; GI cancer: 808; respiratory cancer: 446; leukemia: 35	Time trends similar before and after the accident. Significant increase in thyroid cancer incidence in women after the accident
Kasatkina and others (1997)	Prevalence	Two rural areas of the Orel region of Russsia: Uritzky and Kolpnyansky regions. Two samples of children with enlarged thyroids by palpation: (1) 2–3 trimester gestation—1 year at exposure; (2) ages 8–9 at exposure	Fallout from Chernobyl	Not given	Two regions based on ^{137}Cs contamination	Endemic goiter; thyroid volume; cytology; thyroid autoantibodies	Goiter: 88 in contaminated area; 20 in control area	Five times the prevalence of thyroid enlargement in contaminated vs. control area. No evidence of thyroid dysfunction. Higher prevalence of autoantibodies and greater cellular proliferation in contaminated areas
Lazjuk and others (1997)	Incidence	Three regions of Belarus: Gomel and Mogilev (contaminated, based on ^{137}Cs measurements) and Minsk city (control)	Fallout from Chernobyl	1982–1994	Two contaminated and one control region; 1982–1985 (before), 1987–1994 (after)	Congenital anomalies: total and 9 specific types	Total: contaminated areas: before 1201, after 2561; control before 255, after 649	Increase in congenital and fetal abnormalities in contaminated regions (1.6-fold increase based on examination of abortuses; 1.8-fold increase based on examination of tissues). Most increased were multiple congenital malformations, polydactyly, reduction limb defects
Lomat and others (1997)	Incidence	Belarus: children 0–14	Fallout from Chernobyl	1986–1995	Three groups: children evacuated from 30 km zone; children residing or moved into areas >15 Ci/km^2; children born to parents in 30 km zone or resettled from areas >15 Ci/km^2	All cancer; thyroid cancer; noncancer: digestive, endocrine, anemia, nervous system, respiratory diseases	Not given	Large increase in thyroid cancer from 1987 to 1995 ($0.2 = 4.0 \times 10^5$). Also increases in the incidence of endocrine and dermatologic diseases and mental disorders. Largest increases in those evacuated from 30 km zone
Michaelis and others (1997)	Incidence	Germany: children 0–15	Fallout from Chernobyl	Children born 1980–1990	Three levels of ground deposition of ^{137}Cs; three time periods: born 7/1/86–12/31/87 (exposed); 1/1/80–12/31/85 and 1/1/88–12/31/90 (unexposed)	Infant leukemia	Exposed: 35; unexposed: 143	Higher incidence in exposed than unexposed cohorts: RR 1.48 (95% CI 1.02, 2.15). Subgroup analyses not consistent with a relationship to exposure levels

TABLE 9-3A Continued

Reference	Incidence/ Mortality	Population Studied	Type of Exposure	Dates of Accrual	Type of Dosimetry	Outcomes Studied	Number of Cases	Summary of Results
Pacini and others (1997)	Incidence	Belarus: individuals diagnosed with thyroid cancer from May 1986–December 1995 under age 21	Fallout from Chernobyl	May 1986–December 1995	Geographic distribution of radioactive contamination	Thyroid cancer	472 (372 <14 age; 100 14–21)	Excess thyroid cancer in both children and adolescents: cases age 5 or under account for majority; youngest ages have greatest risk
Sobolev and others (1997)	Incidence	Ukraine: children born 1968–1986	Fallout from Chernobyl	1986–1995	Calendar year; four geographic regions based on estimated thyroid doses	Thyroid cancer	2077	Increase in incidence of thyroid cancer; especially in youngest age group and highest-dose area
Vykhuvanets and others (1997)	Prevalence	Ukraine: 53 children age 7–14 in 15 contaminated settlements in Chernigov and Kiev regions; 45 children age 6–14 in uncontaminated areas of Poltava region	Fallout from Chernobyl	Examined 1993–1994	Estimated thyroid doses based on information from the Ministry of Health regarding personal absorption doses, average absorption doses, and average summary doses	Thyroid ultrasound, autoantibodies, thyroid hormones, lymphocyte subsets	NA	Significant association between autoimmune thyroid disorders and radiation dose
Ivanov and others (1998)	Incidence	Belarus: children born 1982–1994	Fallout from Chernobyl	Born 1982–1994	Three time periods: 7/1/86–12/31/87 (exposed); 1/1/82–12/31/85 and 1/1/88–12/31/94 (unexposed); all Belarus, Mogilev, and Gomel combined	Infant leukemia	Exposed 17; unexposed 89	Slight increase in infant leukemia in exposed: Belarus RR 1.26 (95% CI 0.76, 2.1); Mogilev and Gomel RR 1.51 (0.63, 3.6)
Jacob and others (1998)	Incidence	Three regions of Ukraine: Kiev, Zhytomyr, and Chernigov; three regions in Belarus: Gomel/Mogilev, Minsk city, Gomel city; Bryansk region of Russia	131I from Chernobyl	1991–1995	Estimated average thyroid dose by region	Thyroid cancer	Ukraine 175; Belarus 201; Russia 31	EAR per 10^4 PY per gray ranges from 0.9 in Zhytomyr to 3.8 in Kiev city; in Belarus from 2.3 in Minsk city to 3.1 in Mogilev/Gomel; in Bryansk is 2.7
Pacini and others (1998)	Prevalence	Two areas of Belarus: Hoiniki and Braslav; children 12 at time of accident	Fallout from Chernobyl	Examined 1992–1994	Contaminated area (Hoiniki) and uncontaminated area (Braslav)	Thyroid autoantibodies, thyroid hormones	Hoiniki: 287 examined; Braslav: 208 examined	Significantly higher prevalence of autoantibodies in girls in Hoiniki; highest in ages 9 years or older. No increase in free T_4, free T_3, or TSH in contaminated area

Reference	Type	Population	Exposure	Years	Dose/Time periods	Endpoint	Number of cases	Results
Steiner and others (1998)	Incidence	West Germany, children age <15	Fallout from Chernobyl	1980–1990	Three time periods: 7/1/86–12/31/87 (exposed); 1/1/80–12/31/85 and 1/1/88–12/31/90 (unexposed)	Infant leukemia	Exposed 325; unexposed 1934	Slight increase in leukemia in exposed areas: Overall RR 1.48 (95% CI 1.21, 2.78). No clear trend in incidence associated with exposure
Ivanov and others (1999)	Incidence	Bryansk, Tula, Kaluga, Orel regions of Russia, population ages 0–60	Fallout from Chernobyl	1982–1996	Three time periods: 1982–1986; 1986–1990; 1991–1996. Thyroid doses estimated for those age 0–17 in Bryansk	Thyroid cancer	3082 cases	Highest risk in children up to age 4 at exposure (14 times risk in adults). EAR in those 0–17 in Bryansk in girls is 2.21 (95% CI 0.74, 3.68) per 10^4 PY per gray and for boys is 1.62 (−0.04, 3.23)
Jacob and others (1999)	Incidence	Belarus: 2 cities and 2122 settlements; Bryansk, Russia: 1 city and 607 settlements; persons born 1971–1985	^{131}I from Chernobyl	1991–1995	Thyroid dose from ^{131}I by settlement or city	Thyroid cancer	243 cases	EAR 2.1 per 10^4 PY per gray Excess relative risk (ERR) 23 per Gy^{-1}. No differences by countries or cities or rural areas
Kofler and others (1999)	Incidence	Belarus: children 0–15 at the time of the accident	Fallout from Chernobyl	1986–1997	Calendar year and geographic area (raion)	Thyroid cancer	805 cases	Shorter latency periods (4–5 years) in areas with higher exposure and higher incidence rates
Tronko and others (1999)	Incidence	Ukraine: 27 regions, children 0–18, cases treated at Institute of Endocrinology and Metabolism in Kiev	Fallout from Chernobyl	1986–1997	Calendar year and geographic area (raion). Estimated thyroid dose based on contamination levels	Thyroid cancer	577 cases	Significant increase in incidence after the accident. Most affected group was 5 years in 1986. Largest increase in those with estimated dose 0.5 Gy
Vermiglio and others (1999)	Prevalence	Tula region of Russia, 143 iodine-deficient children 5–15 years of age from moderately contaminated area and 40 sex- and age-matched children from nearby uncontaminated area	Fallout from Chernobyl	Not stated	^{137}Cs contamination by geographic area	Thyroid autoantibodies	183 children and adolescents examined	High prevalence of autoimmunity in persons from contaminated area, most notably in those born or in utero in 1986
Heidenreich and others (2000)	Incidence	Ukraine: individuals born 1968–1997	Fallout from Chernobyl	1986–1998	Age at exposure and time since exposure	Thyroid cancer	Not given	No increase for 3 years after accident; then linear excess absolute risk through 1998; decrease in excess risk with increasing age at exposure
Jacob and others (2000)	Incidence	Belarus: 2 cities and 2122 settlements; individuals born 1971–1986	^{131}I from Chernobyl	1991–1996	Thyroid dose from ^{131}I by settlement or city	Thyroid cancer	657 cases	Number of cases in first decade after accident is a small fraction of what is expected in the following four decades

continues

TABLE 9-3A Continued

Reference	Incidence/ Mortality	Population Studied	Type of Exposure	Dates of Accrual	Type of Dosimetry	Outcomes Studied	Number of Cases	Summary of Results
Romanenko and others (2000)	Prevalence	236 patients with renal cell carcinoma in Ukraine (Institute of Urology and Nephrology in Kiev) and 112 patients in Spain (University Hospital in Valencia)	Fallout from Chernobyl	Kiev: 1993–1999 Valencia: 1975–1989	^{137}Cs contamination level by geographic area	Renal cell carcinoma pathology and cell proliferation activity	Kiev 236 cases; Valencia 112 cases	Significant and strong increase of proliferative activity and aggressivity in Ukraine cases
Noshenko and others (2001)	Incidence	Zhitomir and Poltava regions of Ukraine: children born in 1986	Fallout from Chernobyl	1986–1996	Two geographic areas: Zhitomir—contaminated and Poltava—uncontaminated	Acute leukemia	Zhitomir 21 cases; Poltava 8	Incidence rates in contaminated region increased relative to uncontaminated region
Shibata and others (2001)	Prevalence	Four districts of the Gomel region of Belarus: children born 1983–1989	Fallout from Chernobyl	Screened 2/2/98–12/22/00	Three time periods: Group III 1/1/83–4/26/86; Group II 4/27/86–12/31/86; Group I 1/1/87–12/31/89	Thyroid cancer	32 cases	Significant increase relative to Group I in Group III (OR 121; 95% CI 9, 1,000) and Group II (OR 11; 3, 76). Increase was highest in those youngest in 1986
Romanenko and others (2002)	Prevalence	Two areas in Ukraine (not specified): males with benign prostatic hyperplasia and females with chronic cystitis, living more than 15 years in the geographic areas selected	Fallout from Chernobyl	1999–2000	Two geographic areas: contaminated and uncontaminated with ^{137}Cs; measurements of ^{137}Cs in urine for some patients	DNA damage repair indicators in bladder urothelium	156 males; 48 females	Significant activation of DNA damage repair in persons from contaminated area compared to those from uncontaminated area
Tronko and others (2002)	Incidence	Ukraine: persons 0–18 in 1986	Fallout from Chernobyl	1986–2000	Calendar year; two geographic areas: (1) 6 regions contaminated and (2) 21 remaining regions	Thyroid cancer	Ages 0–18: 1876 cases; ages 0–15: 1318 cases	Steady rise in incidence after the accident in 6 contaminated regions. Greatest increase in those ages 0–4 in 1986
Ivanov and others (2003)	Incidence	Bryansk region of Russia: residents age 15–69 at accident	Fallout from Chernobyl	1986–1998	Mean thyroid doses by raion	Thyroid cancer	1051 cases	Incidence relative to Russia: twofold higher. ERR at 1 Gy, based on external controls: -0.4 (males), -.3 (females); based on internal controls: 0.7 (males), -0.9 (females)

Romanenko and others (2003)	Prevalence	Two areas in Ukraine (not specified): males with benign prostatic hyperplasia and females with chronic cystitis, treated at the Institute of Urology in Kiev	Fallout from Chernobyl	Not specified	^{137}Cs measurements in urine; three geographic areas with different levels of soil contamination with ^{137}Cs	Urinary bladder lesions	Males: 159; females: 5	Strong relationship between ^{137}Cs exposure from Chernobyl and development of chronic proliferative atypical cystitis. Greatly elevated levels of p38, p50, and p65 expression in the urothelium in those with higher levels of ^{137}Cs in the urine
Shakhtarin and others (2003)	Incidence	3070 persons from 75 settlements in the most highly contaminated areas of the Bryansk region of Russia	Fallout from Chernobyl	1968–1986	Estimated thyroid dose for each settlement based on available thyroid ^{131}I measurements and ^{137}Cs contamination levels	Thyroid cancer	34 cases	ERR significantly associated with increasing thyroid dose and inversely associated with urinary iodine excretion levels. Joint effect of radiation and iodine deficiency: at 1 Gy the ERR in areas with severe iodine deficiency was two times that in areas of normal iodine intake
Tukiendorf and others (2003)	Incidence	Opole province, Poland	Fallout from Chernobyl	1994–1998	^{134}Cs and ^{137}Cs measurements in 1993 in 3224 settlements	Thyroid cancer	27 cases in males; 94 in females	Increased thyroid cancer in females in areas with elevated cesium measurements; areas 20–50 kBq/m^2, 1.5-fold excess; >50 kBq/m^2, 2 to 5-fold excess
Verger and others (2003)	Incidence	Eastern France, residents <15 at accident	Fallout from Chernobyl	1991–2000 (predicted)	Thyroid doses estimated by geographic area from contamination measurements	Thyroid cancer	Predicted No. of excess cases: 1.3–22	Predicted excess cases due to Chernobyl fallout is less than the uncertainty in the number of spontaneous cases predicted

NOTE: T$_3$ = triiodothyronine; T$_4$ = thyroxine; TSH = thyroid-stimulating hormone.

TABLE 9-3B Populations Exposed from the Chernobyl Accident—Case-Control Studies

Reference	Population Studied Cases	Controls	Number of Subjects Cases	Controls	Dates of Accrual	Type of Exposure	Type of Dosimetry	Summary of Results
Astakhova and others (1998)	Thyroid cancer in children in Belarus <15 at the time of the accident	Type I: Random sample of children in contaminated raions Type II: Sample of children with same opportunity for diagnosis as cases Both types matched on age, sex, rural or urban residence in 1986	107	Type I: 107 Type II: 107	1987–1992	Chernobyl fallout: major contributor to thyroid dose is ^{131}I. Lesser contributions from ^{132}I, and ^{133}I, and external radiation	Retrospective dose reconstruction. Thyroid dose estimated for individuals based on settlement doses for most cases and controls. For 12 cases (no controls) dose was estimated based on thyroid measurements	Significant differences between cases and both sets of controls regarding dose. Strong and significant dose-response relationship. Odds ratio (highest- vs. lowest-dose group) in Gomel, Type I controls: rural 10.4 (3.5, 31.2); urban 5.1 (1.3, 20.0)
Noshchenko and others (2002)	Leukemia in children age 0–20 at the time of the accident in Zhytomir and Rivno Oblasts in Ukraine	Two controls per case, randomly selected from the same oblast as the case but not the same raion, matched on age, sex, type of settlement	98	151	1987–1997	Chernobyl fallout: major contributor to bone marrow dose is external gamma from fallout and ingestion of ^{134}Cs and ^{137}Cs with food	Retrospective dose reconstruction. Individual accumulated dose to bone marrow estimated, based on settlement measurements and individual dosimetry interviews	Statistically significant risk (OR for >10 mSv 2.5; CI 1.1, 5.4). Higher risk in males. Risk highest 1993–1997 (OR 4.1; 1.5–11.3), especially for acute lymphoblastic type (OR 13.1; 2.6–65.0)
Davis and others (2004b)	Thyroid cancer in children 0–19 at the time of the accident residing in 7 most contaminated raions in the Bryansk Oblast of Russia	Two controls per case, randomly selected from the same raion as the case, matched on age, sex, type of settlement	26	52	April 26, 1986– October 1, 1997	Chernobyl fallout: major contributor to thyroid dose is ^{131}I. Lesser contributions from ^{132}I, and ^{133}I, and external radiation	Retrospective dose reconstruction. Individual accumulated dose to thyroid estimated, based on environmental measurements and individual dosimetry interviews	Significant dose response ($p < .009$). OR by dose quartile: 3–60 mGy, 1.0; 66–240 mGy, 1.65 (0.3, 8.5); 290–600 mGy, 3/05 (0.4, 22.1); 610–2730 mGy, 44.7 (3.3, 604)
Cardis and others (2005b)	Thyroid cancer in children age 0–14 at the time of the accident in Belarus (Gomel and Mogilev) and 0–18 in Russia (Kaluga, Tula, Orel, Bryansk)	Randomly selected from the same oblast as the case, matched on age and sex	276	1300	1992–1998	Chernobyl fallout: major contributor to thyroid dose is ^{131}I. Lesser contributions from short-lived isotopes of iodine and tellurium and external radiation from long-lived radionuclides	Retrospective dose reconstruction. Individual accumulated dose to thyroid estimated, based on environmental measurements and individual dosimetry interviews	Significant dose-response linear up to 1.5–2 Gy. RR at 1 Gy 5.5 (95% CI 3.1, 9.5). Significant effects of iodine deficiency and iodine supplementation as modifiers of RR per gray

garding the quantitative relationship between radiation dose to the thyroid from Chernobyl and the risk of thyroid cancer.

There are only three published population-based case-control studies of thyroid cancer in children that utilize individual estimates of radiation dose and provide quantitative information on thyroid cancer risk (Table 9-3B). The first is based on 107 cases diagnosed in Belarus (Astakhova and others 1998). Although a strong relationship between estimated radiation dose and thyroid cancer was found, thyroid doses were inferred for children from estimates for adults who lived in the same villages. The second is based on confirmed cases of thyroid cancer in children and adolescents aged 0–19 years at the time of the accident, residing in the more highly contaminated areas of the Bryansk Oblast of Russia (Davis and others 2004b).

Based on 26 cases and 52 controls and using a log-linear dose-response model treating estimated individual thyroid radiation dose as a continuous variable, the trend of increasing risk with increasing dose was statistically significant (one-sided $p = .009$). The third is a population-based, case-control study of thyroid cancer carried out in contaminated regions of Belarus and the Russian Federation (Cardis and others, 2005). The study included 276 cases and 1300 matched controls aged less than 15 years at the time of the accident. Individual doses were calculated for each subject. A very strong dose-response relationship was observed in this study ($p < .0001$). At 1 Gy, the odds ratio (OR) varied from 5.5 (95% CI 3.1, 9.5) to 8.4 (95% CI 4.1, 17.3) depending on the form of the risk model used. A clear linear dose-response relationship was observed up to about 1 Gy, followed by a marked flattening. The risk appeared to be related mainly to exposure to [131]I. Collectively, data from these studies suggest that exposure to radiation from Chernobyl is associated with an increased risk of thyroid cancer and that the relationship is dose dependent. These findings are consistent with descriptive reports from contaminated areas of Ukraine and Belarus, and the quantitative estimate of thyroid cancer risk is generally consistent with estimates from other radiation-exposed populations.

A number of the studies have also focused on the potentially modifying influence of a number of host and environmental factors. Results from studies of atomic bomb survivors and persons exposed to external irradiation have shown that exposure at the youngest ages is associated with the greatest risk of thyroid cancer. The available data on exposure from the Chernobyl accident are largely in agreement with this observation. For example, a recent paper (Tronko and others 2002) found the highest incidence of thyroid cancer among those exposed at ages 0–4, who also had the highest doses. There have been few studies in persons exposed at older ages, however. One study of thyroid cancer diagnosed in adolescents and adults in the Bryansk region of Russia reported a small excess of thyroid cancer among adults (Ivanov and others 2003), but the excess was not correlated

with the imputed doses, and larger studies with longer follow-up and greater statistical power are needed. It has also been postulated that the risk of thyroid cancer may be especially high among persons exposed *in utero*, because developing fetal thyroid tissue may be highly susceptible to thyroid cancer induction by ^{131}I exposure. At present there are no data available from Chernobyl regarding the risk of thyroid cancer from *in utero* exposure.

Fifteen years after the Chernobyl accident, thyroid cancer incidence is still highly elevated. Although based on studies of thyroid cancer in other radiation-exposed populations there is no reason to expect a decrease in the next several years; at the present time the follow-up of Chernobyl-exposed children is too short to determine long-term risks. An increase in thyroid cancer has been observed in both males and females. Most, but not all, of the Chernobyl studies have reported similar relative risks per unit dose for males and females.

Iodine deficiency may also be an important modifier of the risk of radiation-induced thyroid cancer. Some regions contaminated by the Chernobyl accident are areas of mild to moderate iodine deficiency. To date, only two published studies have investigated the relationship between iodine deficiency, radiation dose, and the risk of thyroid cancer in young people. In a study carried out in the Bryansk region of Russia, Shakhtarin and colleagues (2003) report a significantly increased risk of thyroid cancer with increasing radiation dose from Chernobyl that was inversely associated with urinary iodine excretion levels. At 1 Gy, the ERR in territories with severe iodine deficiency was approximately two times that in areas of normal iodine intake, thereby suggesting that iodine deficiency may enhance the risk of thyroid cancer following radiation exposure. The evidence is not conclusive because the study is ecologic and uses approximations for both radiation dose and iodine deficiency. In their case-control study in Belarus and Russia, Cardis and colleagues (2005) also investigated the effects of iodine deficiency and its interaction with radiation exposure in the risk of thyroid cancer. Subjects who resided in the areas of lowest soil iodine content had a 3.1 times (95% CI 1.7, 5.4) higher risk at 1 Gy than subjects residing in areas of higher soil iodine content. It is noted that administration of potassium iodide as a dietary supplement significantly reduced the risk of radiation-induced thyroid cancer.

Finally, relatively little has been published regarding thyroid outcomes other than thyroid cancer, although one study has reported an elevated risk of benign thyroid tumors (Ivanov and others 2003). There have been reports of increases in autoimmune disease and antithyroid antibodies following childhood exposure to Chernobyl (Lomat and others 1997; Vykhovanets and others, 1997; Pacini and others 1998; Vermiglio and others 1999). However, a study by the Sasakawa Foundation, which screened 114,000 children, found no association between a surrogate for thyroid dose

(^{137}Cs) and thyroid antibodies, hypothyroidism, hyperthyroidism, or goitre (Ashizawa and others 1997).

Leukemia

The evidence from epidemiologic studies regarding the risk of leukemia in populations exposed to radiation from Chernobyl comes from studies of recovery operation workers, some of whom were exposed at a high or moderate dose levels and dose rates (depending on when and where they worked), and the general population who have been subject to low-dose-rate exposure (primarily from ^{137}Cs) for a number of years and will continue to be exposed in the future. Worker populations were exposed as adults and are considered in Chapter 8. Resident populations were exposed at all ages, but studies of residents are primarily of persons exposed as children and/or *in utero*.

Several studies have investigated the risk of leukemia in children exposed to Chernobyl fallout *in utero*. All are ecologic in design, and results are inconsistent. The initial study compared rates for temporal cohorts born during "exposed" and "unexposed" periods in Greece and found a 2.6-fold increase in leukemia risk and elevated rates for those born in regions with higher levels of radioactive fallout (Petridou and others 1996). However, the numbers of cases in each exposure group were small, and the results could not be duplicated when a similar approach comparing areas with the same categories of contamination (<6 kBq m^{-2}, 6–10 kBq m^{-2}, >10 kBq m^{-2}) was applied to the analysis of data from the German Childhood Cancer Registry (Steiner and others 1998).

In a study in Belarus (Ivanov and others 1998), where levels of contamination are higher by a factor of 10 or more, the results were similar to the Greek study but the trend was weaker. Nevertheless, although the findings are based on small numbers and are not statistically significant, the highest annual incidence rate was in 1987, the year after the accident, and the largest rate ratio (RR = 1.51; 95% CI 0.63, 3.61) was in the two most contaminated regions: Gomel and Mogilev.

A more recent small study published by Noshchenko and colleagues (2001) compared leukemia incidence during 1986 to 1996 among children born in 1986 and thus exposed *in utero* in Zhitomir, a contaminated region, to children born in Poltava, a relatively uncontaminated region. The reported risk ratios based on cumulative incidence show significant increases for all leukemia (relative risk [RR]^2 =2.7; 95% CI 1.9, 3.8) and for the subtype of acute lymphoblastic leukemia (RR = 3.4; 95% CI 1.1, 10.4).

The ongoing European Childhood Leukemia-Lymphoma Incidence Study (ECLIS) has evaluated the risk of leukemia

by age using data from population-based cancer registries in Europe (including Belarus and Ukraine). Focusing on the risk of leukemia by age of diagnosis in 6-month intervals in relation to estimated doses from the Chernobyl fallout received *in utero*, preliminary results suggest a small increase in risk in infant leukemia and leukemia diagnosed between 24 and 29 months.

Thus, at present the available evidence from ecologic studies does not convincingly indicate an increased risk of leukemia among persons exposed *in utero* to radiation from Chernobyl. However, the statistical power of these studies is low for detecting moderate-sized associations, and the exposure measures are crude. There are no data from analytic epidemiologic studies in which individual dose estimates are available. Consequently, there is neither strong evidence for or against an association between *in utero* exposure to Chernobyl fallout and an increased risk of leukemia.

Several ecologic studies also have investigated the association between radiation exposure of children from Chernobyl and the occurrence of leukemia. The ECLIS utilized incidence data in children under age 15 from 36 cancer registries in 23 countries. Parkin and colleagues (1996) compared acute leukemia incidence rates before the Chernobyl accident (1980–1985) with those for 1987 and 1988. Although the number of leukemia cases for 1987–1988 significantly exceeded the number of cases expected on the basis of 1980–1985 data, there was no evidence that the excess in leukemia rates was more pronounced in areas that were most affected by Chernobyl-related ionizing radiation exposure. Similar results were observed in the 5-year ECLIS follow-up report.

Additional reports have focused on changes in childhood leukemia rates before and after the accident in individual European countries and elsewhere. Overall, there was little evidence for an increase in rates of childhood leukemia in Ukraine, Belarus, Russia, Finland, Sweden, Greece, or a number of other countries from Central, Eastern and Southern Europe after the Chernobyl accident. Furthermore, there was no association between the extent of contamination and the increase in risk in these countries. However, one Swedish study (Tondel and others 1996), reported a non-statistically significant increase of acute lymphocytic leukemia (ALL) after the accident in children younger than 5 (OR 1.5; 95% CI 0.8, 2.6). A small study in northern Turkey showed that in one pediatric cancer treatment center, more patients with ALL were seen after the accident than before, but no incidence rates were reported (Gunay and others 1996).

There has been only one analytic (case-control) study of childhood leukemia reported (Noshchenko and others 2002) based on cases identified among residents of the Rivno and Zhytomir Oblasts in Ukraine. Cases were under age 20 at the time of the accident and were diagnosed between 1987 and 1997. Data were collected on 272 cases; however the analysis was based on only 98 cases that were independently verified and interviewed. Controls were selected randomly from

---

^{2}RR is the rate of disease in an exposed population divided by the rate of disease in an unexposed population.

the same oblasts, excluding the raion of residence of the case, and matched according to age at the time of the accident, sex, and type of settlement. The mean estimated dose to the bone marrow among study subjects was 4.5 mSv and the maximum was 101 mSv. The study found a statistically significant increased risk of acute leukemia among males with cumulative doses greater than 10 mSv diagnosed from 1993 to 1997. A similar association was found for acute myeloid leukemia (AML) diagnosed in 1987–1992. These results should be interpreted cautiously, however, because they are based only on approximately one-third of the cases and a lesser proportion of controls, and it is not clear whether cases and controls were selected for dose estimation in an unbiased manner.

On balance, the existing evidence does not support the conclusion that rates of childhood leukemia have increased as a result of radiation exposures from the Chernobyl accident. However, ecologic studies are not particularly sensitive to detecting relatively small changes in the incidence of a disease as uncommon as childhood leukemia over time or by different geographic areas. Further, existing descriptive studies vary in several aspects of study design: methods of case ascertainment (cancer registries versus retrospective record review), methods of classifying radiation exposure, and length of follow-up after the accident (range 2–10 years). The single analytical study is insufficient to draw convincing conclusions regarding leukemia risk after Chernobyl exposure of children.

A few studies have investigated adult resident populations living in highly contaminated areas. Osechinsky and Martirosor (1995) investigated the incidence of leukemia and lymphoma in the general population of the Bryansk region of Russia for 1979–1993 using an ad hoc registry of hematological diseases established after the Chernobyl accident. The incidence rates in the six most contaminated districts (more than 37 kBq m^2 of ^{137}Cs deposition density) did not exceed the rates in the rest of the region or in Bryansk city, where the highest rates were observed. Comparisons of crude incidence rates before and after the accident (1979–1985 and 1986–1993) showed a significant increase in the incidence of all leukemia and non-Hodgkin's lymphoma (NHL), but this was due mainly to increases in the older age groups in rural areas. The incidence of childhood leukemia and NHL was not significantly different in the six most contaminated areas from the incidence in the rest of the region. Similarly, Ivanov and colleagues (1997a, 1997b) found no evidence of an increase in leukemia rates in the most contaminated areas of the Kaluga district of the Russian Federation after the Chernobyl accident.

In Ukraine, Bebeshko and colleagues (1997) examined incidence rates for leukemia and lymphoma in the most highly contaminated areas of the Zhytomir and Kiev districts before and after the Chernobyl accident. Total incidence in adults increased from 5.1 per 100,000 during 1980–1985 to 11 per 100,000 PY during 1992–1996, but there was no excess in contaminated areas of the regions. Similarly, Prisyazhniuk and colleagues (1995) investigated the incidence of leukemia and lymphoma in the three most contaminated regions of Ukraine. There was a steady increase in leukemia and lymphoma rates for both men and women between 1980 and 1993, but there was no evidence of a more pronounced increase after the accident.

Thus, on balance, there is no convincing evidence that the incidence of leukemia has increased in adult residents of the exposed populations that have been studied in Russia and Ukraine. However, few studies of the general adult population have been conducted to date, and they have employed ecologic designs that are relatively insensitive.

Solid Tumors Other Than Thyroid Cancer

There has been relatively little study of the incidence of or mortality from solid cancers other than thyroid cancer in populations exposed to radiation from the Chernobyl accident. Two studies have investigated solid cancer incidence in liquidation workers (Prisyazhnik and others 1996; Ivanov and others 2004a, 2004b) and are considered in Chapter 8. No descriptive or analytical epidemiologic studies of breast cancer risk in populations exposed to radiation from Chernobyl have been published in the peer-reviewed literature. However, one monograph report has cited elevated breast cancer incidence rates based on members of Ukrainian registries (Prysyazhnyuk and others 2002). These included 150,000 residents of contaminated areas close to Chernobyl; 90,000 liquidation workers in 1986 (with mean dose evaluated as 100–200 mSv) and 1987 (mean dose 50–100 mSv); and 50,000 evacuees from Pripyat (mean dose 10–12 mSv) and the 30 km zone (mean dose 20–32 mSv). For breast cancer among the women in these cohorts, the standardized incidence ratio (SIR), based on comparisons to Ukrainian female population rates, was reported as 1.50 (95% CI 1.27, 1.73) for 1993-1997 among residents of contaminated territories. For evacuees from the 30 km zone, the SIR during 1990-1997 was 1.38 (95% CI 1.06, 1.70), and for women who served as liquidation workers during 1986-1987, who comprised only about 5% of the liquidation worker cohort, the SIR for 1990-1997 was 1.51 (95% CI 1.06, 1.96). These registry-derived estimates must be interpreted with considerable caution because they were not subject to diagnostic confirmation and may be influenced by differences in screening intensity.

Similarly, although no descriptive or analytical epidemiologic studies of bladder or kidney cancer risk in relation to Chernobyl radiation have been published in the peer-reviewed literature, there has been a series of papers investigating aspects of possible radiation carcinogenesis in these organs. Romanenko and colleagues (2003) have continued to monitor the incidence of urinary bladder cancer in Ukraine, reporting that it increased from 26.2 to 43.3 per 100,000 PY between 1986 and 2001. In a study of 204 urothelial biopsies of Ukrainian patients, they concluded that activation of DNA damage repair was detected more frequently among residents of contaminated areas, compared to those of putatively uncontaminated areas (Romanenko and others 2002). Morimura and colleagues (2004) observing p53 gene mutations in 54.5% of 11 and 16.7% of 18 Ukrainian bladder cancers collected before and after the Chernobyl accident, respectively, suggesting the possibility of distinct molecular genetic pathways of bladder cancer induction before and after the accident. Romanenko and colleagues (2000) have also reported that renal carcinoma incidence has increased from 4.7 to 7.5 per 100,000 PY.

In summary, there is now little doubt that an excess of thyroid cancer has occurred in areas highly contaminated by radiation from the Chernobyl accident. Analytical studies further indicate that exposure to radiation from Chernobyl is associated with an increased risk of thyroid cancer and that the relationship is dose dependent. Quantitative estimates of risk from these studies are consistent with estimates from other radiation-exposed populations. There is evidence that young age at exposure and iodine deficiency may be important modifiers of the risk of radiation-induced thyroid cancer. There is no convincing evidence that the incidence of leukemia has increased in children or adult residents of the exposed populations; however, few studies of leukemia have been conducted to date and most have employed ecologic designs that are relatively insensitive. There have been very

few studies of the incidence of or mortality from solid cancers other than thyroid cancer in populations exposed to radiation from the Chernobyl accident, and there is no evidence of an increase in any solid cancer type to date.

POPULATIONS EXPOSED FROM NATURAL BACKGROUND

Table 9-4 summarizes four studies of populations exposed from natural background radiation. Two were conducted in China, one in Great Britain, and one in India. A number of different cancer outcomes were studied, based on incidence, mortality, and prevalence data. These studies did not find higher disease rates in geographic areas with high background levels of radiation exposure compared to areas with lower background levels. However, these studies were ecologic in design and utilized population-based measures of exposure rather than individual estimates of radiation dose. Thus, they cannot provide any quantitative estimates of disease risk associated with the exposure levels found in the areas studied.

CHILDREN OF ADULTS EXPOSED TO RADIATION

Table 9-5A lists three ecologic studies of children of adults exposed to radiation. The focus is on preconception parental exposure and the risk of leukemia and lymphoma in

TABLE 9-4 Populations Exposed from Natural Background Radiation—Ecologic Studies

Reference	Incidence/ Mortality	Population Studied	Type of Exposure	Dates of Accrual	Type of Dosimetry	Outcomes Studied	Number of Cases	Summary of Results
Wang and others (1990)	Prevalence	Women ages 50–65 living in Yangjiang, China, vs. nearby control areas	Natural background (mostly external whole-body gamma)	1986 (survey)	Measured external exposure (average annual dose in high-background area: 330 mR; in control area: 114 mR)	Thryoid nodularity, serum thyroid hormone levels, chromosome aberrations	Nodules in high areas (95); in control areas (93)	No difference in prevalence of nodules; no difference in thyroid hormone levels; increased frequency of unstable chromosome aberrations
Lu-xin (1994)	Mortality	Population of Yangjiang, China, vs. control area (not specified)	Natural background radiation	1970–1986	Measured annual external exposure (mR)	11 cancer sites	High-exposure area 914; control 1032	No increase in high-background areas except cervix
Richardson and others (1995)	Incidence	Children under age 15 in Great Britain	Natural background (gamma and radon)	1969–1983	Survey of radon and gamma concentrations in homes; gamma outside; 459 districts	Leukemia	6691	No association of childhood leukemia with indoor or outdoor gamma levels
Nair and others (1999)	Incidence	Population of Karunagappally tuluk in Kerala, India	Thorium deposited along coastal areas (gamma)	1990–1996	Gamma measurements made in each house	All cancers	Not given	No evidence of higher incidence of cancer in areas of higher natural gamma radiation exposure

TABLE 9-5A Children of Adults Exposed to Radiation—Ecologic Studies

Reference	Incidence/ Mortality	Population Studied	Type of Exposure	Dates of Accrual	Type of Dosimetry	Outcomes Studied	Number of Cases	Summary of Results
Kinlen (1993a)	Incidence	Residents of Seascale below age 25 in 1951–1991	Paternal preconception whole-body dose	1951– 1991	Lifetime preconception dose obtained from employment records (mSv)	Leukemia and NHL	Leukemia: 5 in Seascale; NHL: 3 in Seascale	Significant excess of leukemia and NHL in Seascale among those born in Seascale, and those born elsewhere
Parker and others (1993)	NA	Children born in Cumbria from 1950 to 1989 to fathers employed at Sellafield	Paternal preconception whole-body dose	NA	Total cumulative and 6-month preconception dose, obtained from employment records	Radiation doses (no disease outcomes)	9256 births to fathers exposed to radiation before conception	7% of collective preconception dose is associated with children born in Seascale; mean individual preconception doses consistently lower in Seascale
Wakeford and Parker (1996)	Incidence	Residents of West Cumbria under age 25	Paternal preconception whole-body dose	1968– 1985	Cumulative preconception dose obtained from worker records	Leukemia	41	Increased incidence in some groups defined by area and age; no increase associated with paternal preconception dose

the offspring of exposed parents. These studies followed the findings first published by Gardner and colleagues (Gardner and others 1990a, 1990b) suggesting that an excess incidence of leukemia in children in West Cumbria may be due to parental preconception exposure to ionizing radiation during employment at the nearby Sellafield nuclear fuel processing plant. All three studies were conducted in relation to exposures received by parents working at the Sellafield nuclear facility in Great Britain. One study (Parker and others 1993) is a radioecologic study, examining the distribution of possible doses received by fathers employed at Sellafield of children born in Cumbria from 1950 to 1989; it does not address disease outcome. Although there is some evidence of an increased risk associated with measures of individual dose in the other two studies, the findings are based on very small numbers of cases and the results across studies are not consistent.

A larger number of case-control studies have been conducted to investigate the possible relationship between radiation exposure of adults and subsequent cancer in their offspring. Table 9-5B summarizes the results of seven published case-control studies. Six of the seven studies included in the table are investigations that are related to findings first published by Gardner and colleagues (1990b). The six studies summarized here include investigations in England and Wales, Scotland, and Canada. All but one investigated leukemia and/or childhood cancer. The seventh study by Sever and colleagues (1988) is a study of congenital malformations. All but the study by Sorahan and Roberts (1993) used employment records and recorded doses to estimate individual preconception radiation dose. The study by Sorahan and Roberts (1993) used job histories to estimate paternal exposure to ionizing radiation and the potential for exposure

to radionuclides in the 6 months prior to the conception of 14,869 children dying of cancer. For all childhood cancers, the RR was 2.9 (95% CI 1.2, 7.1) for those potentially exposed to radionuclides. There was no evidence of an association between external ionizing radiation and cancer risk. The most recent study by Draper and colleagues (1997) found an increased risk of childhood leukemia and NHL among children whose fathers were radiation workers (RR 1.8; 95% CI 1.1, 3.0). The risk was also elevated for all other childhood cancers among offspring of mothers who were radiation workers (RR 5.0; 95% CI 1.4, 26.9). There was no evidence of a dose-response trend. In summary, none of the studies provides quantitative information from dose-response analyses or quantitative estimates of the risk of disease associated with exposure, and results across studies are inconsistent.

Table 9-5C describes cohort studies published regarding the risk of cancer and adverse reproductive outcomes in children of adults exposed to radiation. Two are studies by Gardner and colleagues (1987) that are not based on individual estimates of radiation dose but rather on proximity to the Sellafield nuclear plant at different ages (at birth and while attending school). A third (Roman and others 1999) is an attempt to confirm Gardner's findings of an increased risk of leukemia and lymphoma in children born to fathers with preconception radiation exposure. Individual paternal preconception exposure was estimated from employment records. Person-years at risk were accrued from date of birth for 39,557 children of male workers and 8883 children of female workers until age 25, cancer diagnosis, or death. A total of 111 cases of malignant cancer were found, but there was no evidence of increased risk relative to the general population. Rate ratios for all cancers (adjusted for calendar

TABLE 9-5B Children of Adults Exposed to Radiation—Case-Control Studies

Reference	Population Studied — Cases	Population Studied — Controls	Number of Subjects — Cases	Number of Subjects — Controls	Dates of Accrual	Type of Exposure	Type of Dosimetry	Summary of Results
Sever and others (1988)	Congenital malformations, identified from 3 hospitals in two counties near Hanford	Selected from hospital delivery room records, next live birth, matched by sex, mother's age (5 years), race	672	977	1957–1980	External whole-body radiation	Recorded doses obtained from Hanford records; estimates in millisieverts	Overall, no association with employment at Hanford; suggestion of increase with parental preconception dose; some increases evident in subgroups
Gardner and others (1990b)	Leukemia and lymphoma in people under 25 born in West Cumbria	From birth register, matched by date of birth and sex: local group, matched by parish; area group, unmatched	Leukemia (52); NHL (22); Hodgkin's disease (23)	1001	Diagnosis: 1950–1985	Total and 6-month external whole-body preconception exposure; antenatal X-ray	Doses from worker radiation records (British Nuclear Fuels)	Leukemia and NHL higher in children born near Sellafield, and with fathers employed at the plant especially those with high preconception doses
Kinlen (1993b)	Leukemia and lymphoma in people born in Scotland since 1958, diagnosed under age 25	Randomly selected from births, matched by county and sex	1024 leukemia; 237 NHL	3783	1958–1990	Total, 3-month, and 6-month preconception external whole-body dose	Doses from worker records (Scottish nuclear industry)	No significant excess in any subgroup; no association with preconception radiation dose
McLaughlin and others (1993a)	Children in Ontario, 0–14, died from leukemia 1950–1963 or diagnosed 1964–1988, born to mothers living near nuclear facility	Selected from births, matched to case by date of births (3 months) and region of mother's residence at child's birth	112	890	Deaths: 1950–1963; diagnosis: 1964–1988	Whole-body external dose, whole-body external tritium dose, radon dose	Recorded doses from National Dose Registry	No increased risk for any exposure period or exposure type
Roman and others (1993)	Leukemia or NHL, diagnosed ages 0–4, born in West Berkshire, Basingstoke, and North Hampshire	Two controls per case selected from birth registers; four per case from delivery registers in study area; matched by sex, date of birth (6 months), area of residence at birth, time of diagnosis	54	324	1972–1989	Exposure to radiation at work	Record of employment in nuclear industry; recorded film badge dose if monitored	Cases were more likely to have a parent employed in the nuclear industry; fathers of cases were more likely to be monitored for radiation; no dose-response evident for fathers monitored
Sorahan and Roberts (1993)	Children dying of cancer under age 16 in England, Wales, and Scotland	Selected from birth register, matched by local authority, sex, date of birth	15,279	15,279	1953–1981	6-month preconception; external whole-body dose; exposure to radionuclides (unsealed sources)	Expert assignment, based on job titles	No association with external exposure; increased risk with potential exposure to radionuclides
Draper and others (1997)	Childhood cancer in Great Britain and Scotland	Selected from birth register for same area, born within 6 months of case, same sex	35,949	38,323	Great Britain: 1952–1986; Scotland: 1987–1990	Total, 3-month and 6-month preconception external whole-body dose	Doses recorded by National Registry for Radiation Workers	Fathers of cases more likely to be radiation workers; no dose-response for any exposure periods for fathers or mothers

TABLE 9-5C Children of Adults Exposed to Radiation—Cohort Studies

Reference	Incidence/ Mortality	Cohort Definition	Comparison Group	Dates of Accrual	Type of Exposure	Type of Dosimetry	Outcomes Studied	Number of Cases	Summary of Results
Gardner and others (1987)	Mortality	Children attending school in Seascale up to 11/84, born since 1950	U.K. national rates	Beginning school–6/30/86	Presumed exposures from Sellafield	Attending school in community near Sellafield plant	Major categories of causes of death	Total deaths: 10	No increase in relation to national rates for all cancer, all causes, leukemia, or lymphoma
Gardner and others (1987)	Mortality	Children born to mothers resident in Seascale from 1950 to 1983	U.K. national rates	1950–6/30/86	Presumed exposures from Sellafield	Born in community near Sellafield plant	Major categories of causes of death	Total deaths 27; leukemia 5	Approximately tenfold excess of leukemia deaths vs. national rates; 2.5-fold excess for other cancer; no increase for other causes
Black and others (1992)	Incidence	Children born in Dounreay area 1969–1988; children attending local schools in the same period born elsewhere	Scottish national rates by tumor site, sex, and calendar year	1969–1988	Presumed exposures from Dounreay nuclear reprocessing plant	Born in or living in Dounreay area of Caithness, Scotland	Leukemia and NHL, Hodgkin's disease, other cancers	Total cancer cases in birth cohort 5; total cases in school cohort 3	Increased incidence of leukemia in both birth and school cohorts: birth cohort O/E — 2.3 (0.7, 5.4); schools cohort O/E — 6.7 (1.4, 19.5)
Dickinson and others (1996)	Incidence	260,060 singleton births to mothers resident in Cumbria, U.K.	Children of parents who worked at Sellafield anytime between 1947 and 1989	1950–1989	External dose from ionizing radiation to fathers prior to conception of the child	Recorded radiation dose obtained from Sellafield facility	Sex ratio	Live births to fathers with dose prior to conception: 10,272	Significantly higher sex ratio (1.09; CI 1.06, 1.13) for children of fathers exposed at Sellafield than other Cumbria children. Increased sex ratio (1.4; CI 1.13–1.73) for children of fathers with >10 mSv in 90 d prior to conception
Dummer and others (1998)	Mortality	256,066 live and 4034 stillbirths to mothers resident in Cumbria, U.K.	Observed and expected stillbirth rates by distance (in circles of 5, 10, 15, 20, and 25 km) and direction. Expected estimated from rates in remainder of Cumbria	1950–1989	Presumed exposures from Sellafield	Proximity to and direction from mother's residence	Stillbirths	Live births to mothers within 25 km of Sellafield: 54,746; stillbirths 888	No evidence that proximity to Sellafield increased risk of stillbirth. No significant increase in stillbirths with distance within any of six directional sectors
Parker and others (1999)	Mortality	248,097 live and 3715 stillbirths to mothers resident in Cumbria, U.K.	Children of father who worked at Sellafield anytime between 1947 and 1989	1950–1989	External and internal dose from ionizing radiation to fathers prior to conception of the child	Recorded radiation dose obtained from Sellafield facility	Stillbirths	Live births in fathers exposed prior to conception: 9078; stillbirths 130	Significant increase in stillbirth with father's external radiation dose prior to conception: OR per 100 mSV 1.24 (CI 1.04, 1.45). Risk higher for stillbirths with congenital anomaly and highest for neural tube defects

TABLE 9-5C Children of Adults Exposed to Radiation—Cohort Studies

Reference	Incidence/ Mortality	Cohort Definition	Comparison Group	Dates of Accrual	Type of Exposure	Type of Dosimetry	Outcomes Studied	Number of Cases	Summary of Results
Roman and others (1999)	Incidence	Children under age 25 of male employees of three nuclear authorities in Great Britain	External: national rates from England and Wales. Internal: within the cohort by radiation exposure levels	For external analyses: born 1965 or later; for internal, born 1985 or later	External whole-body preconception dose	Employment in nuclear industry; whether monitored for radiation exposure; dose estimates from records	All cancer, leukemia, and NHL	Total cancer 111 leukemia 28	No excess incidence over expected; leukemia in children whose fathers received >100 mSv preconception dose was 5.8 times that in children conceived prior to father's employment, based on 3 cases; no evidence of any dose-response for leukemia
Doyle and others (2000)	Incidence and mortality	Employees of AWE, AEA and BNF, and for AEA and BNF past employees <75 years old who were included in the pension database	Within the cohort by radiation exposure level	1993–1996	External and internal dose from ionizing radiation to fathers prior to conception of the child	Whether monitored for radiation exposure; if so, dose estimates from records of the nuclear facility	Fetal deaths and congenital malformations	Live births: women 3048; men 20,899 Fetal deaths: women 526; men 2723	Risk of fetal death and congenital malformations not related to whether father was monitored for radiation prior to conception or to the dose of radiation received. Risk of early miscarriage (<13 weeks) was higher if mother was monitored before conception (OR 1.3; CI 1.0, 1.6), but no trend with radiation dose. Risk of stillbirth was also higher (OR 2.2; CI 1.0, 4.6). Risk of any major malformation not associated with maternal monitoring or dose prior to conception

period, age and sex of child, and the number of children born to each parent) were significantly greater than 1.0 among offspring of fathers who received cumulative external doses of 100 mSv or 10 mSv in the 6 months prior to conception (4.1, 95% CI 1.4, 11.8, 5.1, 95% CI 1.6, 16.9), respectively. It should be noted that these results were based on very few cases (four and three, respectively). No trend of increasing risk with cumulative dose was apparent. None of the three studies provide quantitative estimates of risk based on dose-response analyses, and the results across studies are not consistent. Thus, there is little evidence from epidemiologic studies of a link between parental preconception exposure to ionizing radiation and childhood leukemia or other cancers.

Other possible indices of the occurrence of transmissible genetic damage from preconception exposures include spontaneous abortions, congenital malformations, neonatal mortality, stillbirths, and the sex ratio of offspring. Relatively few epidemiologic studies have been conducted to evaluate these outcomes in relation to preconception radiation exposure. Dickinson and colleagues (1996) examined the sex ratio among children born to fathers employed at Sellafield. Exposure was assessed using two methods: total cumulative radiation dose prior to conception and dose received in the 90 days prior to conception. Total cumulative dose did not account for a significant amount of variation in the sex ratio during the period 1950–1988. No significant trend was observed between sex ratio and exposure 90 d prior to conception, although the sex ratio was increased in children of fathers in the highest-dose category (>10 mSv). Chance could not be ruled out as the reason for this result.

A companion study investigated stillbirths in the offspring of men employed at Sellafield (Parker and others 1999). Individual film badge doses were available by record linkage with the British Nuclear Fuels (BNF) dosimetry database. Significant positive associations between both the total cumulative dose (OR per 100 mSv = 1.24; 95% CI 1.04, 1.45) and the dose during the 90 d prior to conception (OR per 100 mSv = 1.86; 95% CI 1.21, 2.76) and risk of stillbirth were observed.[3] A nested case-control study was conducted among radiation workers alone using live births matched on sex and date of birth. In contrast with the cohort analysis, the adjusted OR for exposure 90 d preconception was not significantly different from 1.00 (OR per 100 mSv = 1.08; 95% CI 0.68, 1.74). The total cumulative dose, however, did show a significant association with the occurrence of stillbirth (OR per 100 mSv = 1.24; 95% CI 1.04, 1.45). Although based on only a few exposed individuals, neither analysis indicated the presence of an association with internal exposure to radionuclides. Limitations of the study noted by the authors included the possibility of the existence of residual con-

founding by year of birth, a time-varying uncertainty (30%) in the recorded film badge doses, and the absence of information on concurrent exposures to organic chemicals in the workplace. An earlier study of stillbirth rates around Sellafield (Dummer and others 1998) found no increase in stillbirths in the resident population within 25 km of the facility.

The Nuclear Industry Family Study in the United Kingdom has also investigated possible links between occupational radiation exposures and reproductive health (Maconochie and others 1999). This study population includes all current employees of the Atomic Energy Authority, Atomic Weapons Establishment, and BNF, as well as past employees who were under age 75 and on record at the pension administration office. Information on reproductive health and health of children was obtained through a mailed questionnaire and linked with data from the employers on occupational exposure to ionizing radiation. The database consists of 53,672 pregnancies, 39,557 reported by men and 8,883 by women. Results of the analysis of fetal deaths and congenital malformations were reported by Doyle and colleagues (2000). The risk of neither fetal death nor major congenital malformation was related to paternal preconception radiation dose. Although early miscarriage was more common among mothers who had been monitored prior to conception (OR 1.3; 95% CI 1.0, 1.6), there was no evidence of a dose-response. Risk of fetal death was higher among mothers who had been monitored prior to conception (OR 2.2; 95% CI 1.0–4.6). ORs were adjusted for parental age, birth order, previous fetal loss, calendar year of the end of pregnancy, and manual versus nonmanual job status. No dose response was evident.

In summary, there have been a number of studies of children of adults exposed to radiation. Ecologic studies are based on very small numbers, and none provide quantitative information from dose-response analyses or quantitative estimates of the risk of disease associated with exposure. There is little conclusive evidence from epidemiologic studies of a link between parental preconception exposure to radiation and childhood leukemia or other cancers. Few studies have been conducted to evaluate other possible indices of the occurrence of transmissible genetic damage from preconception radiation exposures, such as spontaneous abortions, congenital malformations, neonatal mortality, stillbirths, and the sex ratio of offspring. Some but not all studies have found a significant positive association between total cumulative dose, as well as dose during the 90 d prior to conception, and the risk of stillbirth. The risk of neither fetal death nor major congenital malformation has been related to paternal preconception radiation dose.

EXPOSURE TO RADIOACTIVE IODINE 131

In evaluating the evidence regarding the risk of cancer associated with exposure to environmental sources of radia-

---

[3]OR represents the odds of being exposed among diseased persons divided by the odds of being exposed among nondiseased persons.

tion, internal exposure to ^{131}I is of particular concern regarding the risk of thyroid cancer. In contrast to the considerable amount of information that is available from numerous studies of external radiation exposure, there is relatively little information regarding the risk of thyroid cancer in humans exposed to ^{131}I. Existing evidence comes from studies of ^{131}I administered for therapeutic or diagnostic purposes and from various environmental exposure settings, most notably from recent studies of persons exposed to radiation from the Chernobyl accident (reviewed above).

Studies of therapeutic and diagnostic ^{131}I exposures are described in detail in Chapter 7. In brief, early studies of persons receiving therapeutic ^{131}I for hyperthyroidism found no convincing evidence that the risk of thyroid cancer was increased (Dobyns and others 1974; Safa and others 1975; Holm and others 1980a; Holm 1984); most of the participants were adults at the time of exposure, were followed for very short periods, had existing thyroid disease at the time of treatment, and were treated with radiation doses that were quite high (generally 20,000–100,000 mGy). Results from a follow-up (Ron and others 1998a) of one of these studies (Dobyns and others 1974) suggest an increased risk of death from thyroid cancer in patients previously treated with ^{131}I, but the numbers of excess deaths were small and it is likely that underlying thyroid disease might have contributed to these results. Similar results were obtained from a study of 7400 patients who were treated with radioiodine from 1950 to 1991 in England (Franklyn and others 1999). Studies have also evaluated persons exposed to much lower doses (generally 500–1000 mGy) through diagnostic procedures (Holm and others 1980a, 1980b; Hall and others 1996). Although there is some evidence of a small increase in thyroid cancer associated with such exposures, there is a lack of consistency and the small increases in thyroid cancer in some studies are likely due to the underlying thyroid condition. As for the therapeutic studies described above, these too are primarily of persons exposed as adults. The thyroid gland is more radiosensitive in children than adults, most likely because of more rapid growth in infants and children (Williams 2003) and because of differences in metabolism (Mettler and others 1996).

Only a few studies have evaluated the effects of environmental exposure to radioactive iodine. In contrast to the medical exposures summarized above, which were due exclusively to ^{131}I, environmental exposures have generally contained mixtures of ^{131}I, external radiation, and short-lived radioiodines. Initial studies of thyroid disease incidence in Utah schoolchildren exposed to fallout from atmospheric nuclear weapons testing at the Nevada Test Site appeared to show no difference in thyroid disease outcomes compared to children from unexposed areas (Rallison and others 1975). However, a follow-up study reported a slight excess risk of thyroid neoplasms associated with radioiodine exposure (Kerber and others 1993). Although positive dose-response trends were also noted for total nodules and thyroid cancer

(when analyzed separately), they were not statistically significant. The study was limited by small numbers of exposed individuals and a low incidence of thyroid neoplasms and by the fact that the examiners were not blinded to exposure. In contrast, a follow-up study of 3440 persons exposed as young children to atmospheric releases of primarily ^{131}I from the Hanford Site found no increased risk of thyroid cancer associated with individual radiation dose to the thyroid (Davis and others 2001, 2004a).

The explanation for the apparent difference in results in the Utah study and the Hanford study is not clear. One possibility is that the exposures were substantially different in terms of the mix of radionuclides and the dose rate. Thyroid dose at Hanford was due almost entirely to ^{131}I, whereas in Utah there was greater contribution from other radioiodines as well as external sources. Exposures in Utah were also more concentrated and episodic than at Hanford, corresponding to specific nuclear tests. This likely resulted in doses being delivered at substantially higher dose rates (although the total dose among 3545 study participants for whom thyroid doses could be estimated [mean 98 mGy] was similar to Hanford doses). A second possibility is that the Utah study's estimated dose-response could have been biased in the direction of finding an association because the collection of dietary consumption data took place after thyroid disease classification was known for each participant.

Extensive evaluation of the population of the Marshall Islands has shown an increase in benign and malignant thyroid nodules in residents of the northern atolls of Rongelap and Utirik (Conard 1980, 1984). In addition, a retrospective cohort study of more than 7000 Marshall Islanders showed that the prevalence of palpable thyroid nodularity (1.0 cm) decreased linearly with increased distance from the Bikini test site (Hamilton and others, 1987). More recently, there has been extensive investigation of populations exposed to radioactive fallout (including ^{131}I as a substantial component) after the Chernobyl accident. Findings from these studies are reviewed and summarized above.

In summary, studies of exposure to ^{131}I from therapeutic and diagnostic uses provide some evidence of a small increase in thyroid cancer associated with such exposures, but there is lack of consistency in the findings. Furthermore, the small increases in thyroid cancer observed in some studies are likely due to the underlying thyroid condition, not to radiation exposure. Results from environmental exposures have been inconsistent. Findings of an increase in thyroid neoplasia in persons exposed to fallout in the Marshall Islands are limited by the lack of individual dosimetry. No excess risk of thyroid cancer was found in residents exposed to radiation from Hanford, and the slight excess risk of thyroid neoplasms associated with radioiodine exposure of Utah residents from the Nevada Test Site was based on small numbers.

In contrast, substantial increases in thyroid cancer have been reported in areas contaminated with radioactive fallout from Chernobyl, primarily among children. Although much of the thyroid dose from Chernobyl is due to ^{131}I, exposure to a mix of other radionuclides and the lack of individual dose estimates in most of the studies to date have made it difficult to develop quantitative risk estimates for radiation dose from ^{131}I. However, there is now emerging evidence indicating that exposure to radiation from Chernobyl is associated with an increased risk of thyroid cancer and that the relationship is dose dependent. These findings are based on individual estimates of thyroid radiation dose and reveal strong and statistically significant dose-related increased risks that are consistent across studies. Thus, although the precise quantitative relationship between radiation dose from ^{131}I and the development of thyroid neoplasia remains uncertain at this time, recent findings from studies around Chernobyl and Hanford provide important quantitative estimates of risk as a function of dose.

DISCUSSION

A considerable number of papers have been published from studies that have attempted to determine whether persons exposed, or potentially exposed, to ionizing radiation from environmental sources are at an increased risk of developing cancer. The existing published literature consists primarily of reports that are descriptive in nature and ecologic in design. Such studies are limited in their usefulness in defining risk of disease in relation to radiation exposure or dose. They can sometimes be informative in generating new hypotheses or suggesting directions for study, but seldom, if ever, are they of value in testing specific hypotheses or providing quantitative estimates of risk in relation to specific sources of environmental radiation. Fewer attempts have been made to evaluate the effect of environmental radiation exposures using the two most common analytical study designs employed in epidemiology: the case-control study and the cohort study. Such studies are almost always based on individual-level data and thus are not subject to many of the limitations inherent in ecologic studies. They can potentially provide quantitative estimates of risk based on individual radiation dose.

Epidemiologic studies, in general, have limited ability to define the shape of the radiation dose-response curve and to provide quantitative estimates of risk in relation to radiation dose, especially for relatively low doses. To be informative in this regard a study should (1) be based on accurate, individual dose estimates, preferably to the organ of interest; (2) contain substantial numbers of people in the dose range of interest; (3) have long enough follow-up to include adequate numbers of cases of the disease under study; and (4) have complete and unbiased follow-up. Unfortunately, the published literature on environmental radiation exposures is not characterized by studies with such features.

Sixteen ecologic studies of populations living around nuclear facilities are summarized, thirteen of the locations

being outside the United States. Most define exposure, or potential for exposure, based on a measure of distance from the facility, and the focus of most of these investigations is leukemia and/or childhood cancer, although a few include all cancers as an outcome. Notably, most of the studies do not specify the nature of the radiation exposure, and none of the 16 contain individual estimates of radiation dose. Although some of these studies report an increased occurrence of cancer that could be related potentially to environmental radiation exposures, none provide a direct quantitative estimate of risk in relation to radiation dose. There have been three case-control studies of persons living around a nuclear facility. One focuses on congenital and perinatal conditions, stillbirths, and infant deaths in relation to exposure from uranium mines. This study does not provide an estimate of radiation risk associated with any of the indicators of exposure. The other two are of leukemia in children and young adults. Neither study found an increased risk associated with parental radiation exposure and X-ray exposure of the child, but both did find an increased risk associated with playing on beaches near the nuclear facility.

Several cohort studies have been reported of persons exposed to environmental radiation under various circumstances: participation in atmospheric nuclear weapons tests conducted by the United Kingdom and the United States; residents and their offspring living near the Techa River in the southern Urals of the Russian Federation and exposed from the nearby Mayak nuclear complex; residents living near the Hanford Site in eastern Washington State; and residents of the Marshall Islands. Overall, studies of persons who participated in U.K. atmospheric nuclear weapons tests found no increased risk of developing cancer or other fatal diseases as a function of estimated dose received, but there was some evidence of an increase in non-CLL leukemia. In contrast, a recent study of U.S. veterans who participated in atmospheric nuclear weapons tests reported a significant increase in death from all causes and for all lymphopoietic cancers combined.

Results from studies of residents living near the Techa River have found no evidence of a decrease in birth rate or fertility in the exposed population and no increased incidence of spontaneous abortions or stillbirths. There is some evidence of a statistically significant increase in total cancer mortality. Estimates of the relative risk for cancer of the esophagus, stomach, and lung are similar to those reported for atomic bomb survivors. There is no evidence of an increase in cancer mortality in the offspring of exposed residents. The one study of persons living in the town of Ozyorsk exposed to fallout from the nearby Mayak nuclear facility reported an excess of thyroid cancer three to four times that expected relative to rates for all of Russia and a somewhat lower excess (1.5 to twofold higher) based on a comparison with Chelyabinsk Oblast rates.

A follow-up study of persons exposed as young children to atmospheric releases primarily of ^{131}I from the Hanford

Site in eastern Washington State found no increased risk of thyroid cancer associated with individual radiation dose to the thyroid. A prevalence study of thyroid cancer conducted through screening of 3709 Marshall Island residents born before the Castle BRAVO atmospheric nuclear weapons test on March 1, 1954, found some indication that the prevalence of thyroid cancer increased with quartile of estimated dose, but the increase was not statistically significant.

Numerous epidemiologic studies have been carried out since the Chernobyl accident to investigate the potential late health consequences of exposure to ionizing radiation from the accident. These studies have focused largely on thyroid cancer in children, but have also included investigations of recovery operation workers and residents of contaminated areas, and have investigated the occurrence of leukemia and solid tumors other than thyroid cancer among exposed individuals. Overwhelmingly, the published findings are from studies that are ecologic in design and therefore do not provide quantitative estimates of disease risk based on individual exposure circumstances or individual estimates of radiation dose. Most reports are descriptive incidence and prevalence studies that utilize population or aggregate estimates of radiation dose. Only three analytical studies are published that report dose-response results based on individual dose estimates.

Numerous reports have continued to describe an increasing number of cases of thyroid cancer, particularly in the most heavily contaminated regions of Ukraine and Belarus, as well as in Russia. Collectively, findings reported to date have demonstrated an association between an increase in thyroid cancer incidence and radiation exposure from the Chernobyl accident. This increase cannot be explained only by the aging of the cohort and the improvement of case detection and reporting. Although there is now little doubt that an excess of thyroid cancer has occurred in highly contaminated areas, there is still very little information regarding the quantitative relationship between radiation dose to the thyroid from Chernobyl and the risk of thyroid cancer. Results from three analytical studies published indicate that exposure to radiation from Chernobyl is associated with an increased risk of thyroid cancer and that the relationship is dose dependent. The findings from these studies are consistent with descriptive reports from contaminated areas of Ukraine and Belarus, and the quantitative estimate of thyroid cancer risk is generally consistent with estimates from other radiation-exposed populations. Available data on exposure from the Chernobyl accident are largely in agreement with observations from other studies showing that exposure at the youngest ages is associated with the greatest risk of thyroid cancer. At present no data are available from Chernobyl regarding the risk of thyroid cancer from *in utero* exposure. Fifteen years after the Chernobyl accident, thyroid cancer incidence is still highly elevated. An increase in thyroid cancer has been observed in both males and females, and most of the Chernobyl studies have reported similar rela-

tive risks per unit dose for males and females. Iodine deficiency also appears to be an important modifier of the risk of radiation-induced thyroid cancer, and there is some evidence that iodine deficiency enhances the risk of thyroid cancer following radiation exposure. Finally, relatively little has been published regarding thyroid outcomes other than thyroid cancer, although one study has reported an elevated risk of benign thyroid tumors and there have been reports of increases in autoimmune disease and antithyroid antibodies following childhood exposure to Chernobyl.

Evidence from epidemiologic studies regarding the risk of leukemia in the general population reflects low-dose-rate exposure (primarily from ^{137}Cs), which has occurred for a number of years and will continue to occur in the future. These resident populations were exposed at all ages, but studies of residents are primarily of persons exposed as children and/or *in utero*.

At present, the available evidence from ecologic studies does not convincingly indicate an increased risk of leukemia among persons exposed *in utero* to radiation from Chernobyl. There are no data from analytic epidemiologic studies in which individual dose estimates are available. The existing evidence does not support the conclusion that the rates of childhood leukemia have increased as a result of radiation exposure from the Chernobyl accident. However, ecologic studies of the types conducted to date are not particularly sensitive to detecting relatively small changes in the incidence of a disease as uncommon as childhood leukemia over time or by different geographic areas. The single analytical study is insufficient to draw conclusions regarding leukemia risk after exposure of children to Chernobyl. There is also no convincing evidence that the incidence of leukemia has increased in adult residents of the exposed populations that have been studied in Russia and Ukraine. However, few studies of the general adult population have been conducted, and they have employed ecologic designs that are relatively insensitive.

There has been relatively little study of the incidence or mortality from solid cancers other than thyroid cancer in populations exposed to radiation from the Chernobyl accident. Two studies have investigated solid cancer incidence in liquidation workers. They reported increases of cancer incidence during the periods, but generally the excesses were relatively small and not statistically significant. No descriptive or analytical epidemiologic studies of breast cancer risk in populations exposed to radiation from Chernobyl have been published in the peer-reviewed literature; however, one monograph has cited elevated breast cancer incidence rates based on Ukrainian registries. Similarly, although no descriptive or analytical epidemiologic studies of bladder or kidney cancer risk in relation to Chernobyl have been published in the peer-reviewed literature, there has been a series of papers investigating aspects of possible radiation carcinogenesis in these organs.

Four ecologic studies of populations exposed to natural background radiation have been reported. Two were conducted in China, one in Great Britain, and one in India. These studies did not find any association between disease rates and indicators of high background levels of radiation, and they do not provide any quantitative estimates of disease risk.

Three ecologic studies of children of adults exposed to radiation have been published, with a focus on preconception parental exposure and the risk of leukemia and lymphoma in the offspring of exposed parents. All three studies were conducted in relation to exposures received by parents working at the Sellafield nuclear facility in Great Britain. Although there is some evidence of an increased risk associated with measures of individual dose, the findings are based on very small numbers of cases and the results across studies are not consistent. A larger number of case-control studies have been conducted to investigate the possible relationship between radiation exposure of adults and subsequent cancer in their offspring. In summary, none of the studies provide quantitative information from dose-response analyses or quantitative estimates of the risk of disease associated with exposure, and results across studies are inconsistent. There have been three cohort studies published regarding the risk of cancer in children of adults exposed to radiation. None of the three provide quantitative estimates of risk based on dose-response analyses, and the results across studies are not consistent. Thus, there is little conclusive evidence from epidemiologic studies of a link between parental preconception exposure to ionizing radiation and childhood leukemia or other cancers.

Other possible indices of the occurrence of transmissible genetic damage from preconception exposures include spontaneous abortions, congenital malformations, neonatal mortality, stillbirths, and the sex ratio of offspring. Relatively few epidemiologic studies have been conducted to evaluate these outcomes in relation to preconception radiation exposure, and there is no consistent evidence of an association of any such outcomes with exposure to environmental sources of radiation.

Studies of exposure to ^{131}I from therapeutic and diagnostic uses provide some evidence of a small increase in thyroid cancer, but the small increase observed is likely due to the underlying thyroid condition, not to radiation exposure. Findings of an increase in thyroid neoplasia in persons exposed to fallout in the Marshall Islands are limited by the lack of individual dosimetry. No excess risk of thyroid cancer was found in residents exposed to radiation from Hanford, and only a slight excess risk of thyroid neoplasms was found associated with radioiodine exposure of Utah residents from the Nevada Test Site. In contrast, substantial increases in thyroid cancer have been reported in areas contaminated with radioactive fallout from Chernobyl, primarily among children. Recent evidence from three population-based case-control studies indicates that exposure to radiation from Chernobyl is associated with an increased risk of thyroid cancer and that the relationship is dose dependent. These findings are based on individual estimates of thyroid

radiation dose and reveal strong and statistically significant dose-related increased risks that are consistent across studies. They provide important quantitative estimates of risk as a function of dose, primarily from ^{131}I.

SUMMARY

This chapter reviews the evidence from peer-reviewed articles published since BEIR V (NRC 1990) of the relationship between exposure to ionizing radiation from environmental sources and human health.

Ecologic studies of populations living around nuclear facilities neither contain individual estimates of radiation dose nor provide a direct quantitative estimate of risk in relation to radiation dose. Similarly, the one case-control study of congenital and perinatal conditions, stillbirths, and infant deaths in relation to exposures from uranium mines does not provide an estimate of the risk associated with any of the indicators of exposure, and two ecologic studies of populations exposed to fallout from atmospheric nuclear testing or other sources of environmental release of radiation provide no quantitative estimates of the risk associated with presumed exposure.

Several cohort studies have been reported of persons exposed to environmental radiation under various circumstances. No increased risk of developing cancer or other fatal diseases was found in persons who participated in U.K. atmospheric nuclear weapons tests, but there was some evidence of an increase in non-CLL leukemia. U.S. veterans who participated in atmospheric nuclear weapons tests reported a significant increase of death from all causes and for all lymphopoietic cancers combined. There is no evidence of a decrease in birth rate or fertility or an increased incidence of spontaneous abortions or stillbirths in residents living near the Techa River in the Russian Federation. There is some evidence of a statistically significant increase in total cancer mortality, but no evidence of an increase in cancer mortality in the offspring of exposed residents. Persons living in the town of Ozyorsk (Russia) exposed to fallout from the nearby Mayak nuclear facility reported an excess of thyroid cancer (1.5–4 times higher than expected). No increased risk of thyroid cancer was found associated with individual radiation dose to the thyroid in persons exposed as young children to atmospheric releases primarily of ^{131}I from the Hanford Site in eastern Washington State. There is some indication that the prevalence of thyroid cancer among Marshall Island residents born before the Castle BRAVO atmospheric nuclear weapons test increased with quartile of estimated dose, but the increase was not statistically significant.

There continues to be an increasing number of cases of thyroid cancer in populations exposed to radiation from the Chernobyl accident that cannot be explained only by the aging of the cohort and the improvement in case detection and reporting. Results from three analytical studies indicate that exposure to radiation from Chernobyl is strongly associated with an increased risk of thyroid cancer in a dose-dependent manner, and the quantitative estimate of thyroid cancer risk generally is consistent with estimates from other radiation-exposed populations and is observed in both males and females. At present, no data are available from Chernobyl regarding the risk of thyroid cancer from *in utero* exposure. Iodine deficiency appears to be an important modifier of risk, enhancing the risk of thyroid cancer following radiation exposure from Chernobyl. Relatively little has been published regarding thyroid outcomes other than thyroid cancer, although one study has reported an elevated risk of benign thyroid tumors and there have been reports of increases in autoimmune disease and antithyroid antibodies following childhood exposure to Chernobyl.

Evidence from ecologic studies does not indicate an increased risk of leukemia among persons exposed *in utero* to radiation from Chernobyl nor that rates of childhood leukemia have increased. A single analytical study is insufficient to draw conclusions regarding leukemia risk after exposure of children to Chernobyl. There is no convincing evidence that the incidence of leukemia has increased in adult residents of the exposed populations that have been studied in Russia and Ukraine. There has been very little study of the incidence or mortality from solid cancers other than thyroid cancer in populations exposed to radiation from the Chernobyl accident, and there is no evidence of significant excesses of any other solid cancer type.

Four ecologic studies of populations exposed from natural background radiation did not find any association between disease rates and indicators of high background levels of radiation exposure (for a general discussion of the limitations of ecologic studies see the introduction to this chapter and, more specifically in reference to studies of populations exposed from natural background radiation, see Appendix D, "Hormesis and Epidemiology").

Ecologic studies of children of adults exposed to radiation while working at the Sellafield nuclear facility in Great Britain have suggested some increased risk of leukemia and lymphoma associated with individual dose, but the findings are based on very small numbers of cases and the results across studies are not consistent. A larger number of case-control studies provides no quantitative estimates of the risk of disease in offspring of exposed parents, and results across studies are inconsistent. None of three published cohort studies provide quantitative estimates of risk based on dose-response analyses, and the results across studies are not consistent. Relatively few epidemiologic studies have been conducted to evaluate outcomes such as spontaneous abortions, congenital malformations, neonatal mortality, stillbirths, and the sex ratio in relation to preconception radiation exposure, and there is no consistent evidence of an association of any such outcomes with exposure to environmental sources of radiation.

In contrast to the considerable amount of information that is available from numerous studies of external radiation exposure, there is relatively little information regarding the risk of thyroid cancer in humans exposed internally to ^{131}I. There is some evidence of a small increase in thyroid cancer associated with exposure to ^{131}I from therapeutic and diagnostic uses, but the findings are inconsistent and the small increases in thyroid cancer observed in some studies are likely due to the underlying thyroid condition, not to radiation exposure. Results from environmental exposures have also been inconsistent. An increase in thyroid neoplasia has been observed in persons exposed to fallout in the Marshall Islands, but no excess risk of thyroid cancer was found in residents exposed to radiation from Hanford, and the slight excess risk of thyroid neoplasms associated with radioiodine exposure in Utah residents from the Nevada Test Site was based on very small numbers. In contrast, substantial increases in thyroid cancer have been reported in areas contaminated with radioactive fallout from Chernobyl, primarily among children. Recent evidence indicates that exposure to radiation from Chernobyl is associated with an increased risk of thyroid cancer and that the relationship is dose dependent. These findings are based on individual estimates of thyroid radiation dose and reveal strong and statistically significant dose-related increased risks that are consistent across studies.

10

Integration of Biology and Epidemiology

INTRODUCTION

Previous chapters of this report have reviewed major elements of experimental and epidemiologic studies relating to the tumorigenic and heritable effects of low-LET (linear energy transfer) ionizing radiation. The development of views on the risks to health from exposure to ionizing radiation depends increasingly upon the establishment of scientific coherence between judgments that stem from knowledge of the biological mechanisms underlying radiation-induced health effects and the direct epidemiologic quantification of such effects. The epidemiologic modeling of radiation-induced health effects for the purposes of risk estimation relies in many cases on biological concepts developed from experimental studies with cultured cells and laboratory animals. This chapter draws together the most important conclusions reached from the reviews of the data. The principal topics considered here are the intimate relationship between cellular responses to DNA damage and health effects; the possible implications of the knowledge of cancer mechanisms for projections of cancer risk over time and the transport of that risk between populations; the shape of the dose-response relationship for cancer risk at low doses; dose and dose-rate effects for cancer risk; the possible implications for cancer risk or other forms of cellular response to radiation; genetic factors in radiation cancer risk; and the heritable effects of radiation.

DNA DAMAGE RESPONSE AND CANCER RISK

Chapters 2 and 3 review the largely cellular and molecular data that strongly support the proposition that chromosomal DNA, the genetic material of the cell, represents the principal target for the deleterious effects of ionizing radiation. In brief, energy deposition from low-LET electron tracks intersecting DNA or its local environment can lead to radical-mediated disruption of covalent bonds in DNA. The cell responds to the presence of such DNA damage in a biochemically complex fashion, but the outcome of greatest importance is the repair or misrepair of critical DNA lesions. Depending on its location, misrepaired DNA damage can lead to the appearance of gene and chromosomal mutations in both somatic (Chapter 2) and germline cells (Chapter 4).

The establishment of an intimate relationship between DNA damage responses, somatic mutation, and cancer development represents one of the most important advances in cancer research during the last decade (Vogelstein and Kinzler 1993; UNSCEAR 2000b). There is good evidence that this relationship applies to a range of human tumor types arising spontaneously or induced by certain environmental carcinogens (UNSCEAR 1993, 2000b).

Epidemiologic studies (Chapters 5–9) show that exposure to low-LET radiation can lead to the age- and time-dependent development of a wide range of tumor types that, in general, are not distinguishable from those arising in nonirradiated populations; studies with experimental animals provide essentially the same message (Chapter 3). Therefore, an initial conclusion would be that the multistage process of cancer development after ionizing radiation is unlikely to be substantially different from that which applies generally (*i.e.*, that the DNA-damaging capacity of radiation is a crucial element in cancer risk). DNA or chromosomal mutations and the heritable effects of radiation are summarized in a subsequent section of this chapter. This initial conclusion receives much support from reviews of data on the links between radiosensitivity and DNA damage response deficiency in humans and mice (Chapters 2 and 3) and findings of candidate radiation-associated mutations in tumors of humans and experimental rodents (Chapter 3). This broad conclusion, while not excluding other mechanistic components of radiation cancer risk, particularly at high doses, underpins many of the judgments summarized below.

PROJECTION OF RISKS OVER TIME

Although the Life Span Study (LSS) cohort (Hiroshima and Nagasaki) has been followed for more than 50 years, most survivors who were young (under age 20) at the time of the bombings are still alive, and thus their risks at older ages, when baseline risks are greatest, have not yet been studied. This is also true of other exposed cohorts. For leukemia, risks in A-bomb survivors had dropped to negligible levels by the end of the follow-up period (Preston and others 1994; Pierce and others 1996). However, estimating lifetime risks of solid cancer for those who are young at exposure requires assumptions about the time-response patterns of disease. Approaches that have been used in past risk assessments include a multiplicative projection based on the assumption that the excess cancer rate increases in direct proportion to the baseline cancer rate and an additive projection based on the assumption that the excess rate is constant and independent of the background rate. Currently available data on A-bomb survivors and other cohorts make it clear that the additive projection method is not appropriate, and this method has not been used in recent years.

From a biological standpoint, a multiplicative projection of risk implies a mechanism whereby all host and environmental factors that modify the background cancer rate have an equivalent impact on radiation-induced disease. This would be the case if radiation were to act predominantly on an early stage in multistage tumorigenesis (i.e., as a tumor initiator). By contrast, additive projection of risk would apply if radiation acted independently as one of many cancer-modifying factors during postinitiation cellular development (e.g., as a tumor promoter). Cytogenetic and molecular studies on tumorigenic mechanisms in experimental animals (Chapter 3) suggest that acute doses of low-LET radiation act predominantly to initiate tumorigenesis rather than to promote its development. Thus, the monoclonal tumorigenic mechanism of initiation proposed for low-LET radiation is also most consistent with a multiplicative projection of cancer risk. In addition, epidemiologic studies of Japanese A-bomb survivors and of persons exposed for medical reasons indicate that exposure early in life results in greater risks than exposure later in life, which also argues against strong tumor-promoting activity and favors an initiation role.

Although multiplicative risk projection is clearly better supported than additive risk projection, current epidemiologic data indicate that relative risks may decrease with increasing attained age or time since exposure, especially for those who were young at exposure (Thompson and others 1994; Little and others 1998; Preston and others 2002b). Thus, it may not be appropriate to use the multiplicative projection method without modification. Risk assessments conducted by the United Nations Scientific Committee on the Effects of Atomic Radiation (UNSCEAR 2000), the National Institutes of Health (NIH 2003), and this committee have allowed for a decline in relative risk with attained age (see Chapter 12). Because experimental animal data seldom include detailed information on age-specific baseline and radiation-associated cancer, these data do not inform us about a decline in the relative risk with time since exposure or attained age.

Finally, because follow-up is now reasonably complete for all but the youngest A-bomb survivors, there is less uncertainty in projecting risks forward in time than in past risk assessments.

THE TRANSPORT OF CANCER RISK BETWEEN DIFFERENT POPULATIONS

Another important issue in risk assessment is applying risks estimated from studying a particular exposed population to another population that may have different genetic and life-style characteristics and different baseline cancer risks. Specifically, the application of risk estimates developed from Japanese atomic bomb survivors to a U.S. population is a concern. Two approaches that have been used are multiplicative or relative risk transport, in which it is assumed that the risks resulting from radiation exposure are proportional to baseline risks, and additive or absolute risk transport, in which it is assumed that radiation risks (on an absolute scale) do not depend on baseline risk and thus are the same for the United States and Japan. Estimates based on relative and absolute risk transport can differ substantially. For example, baseline risks for stomach cancer are much higher in Japan than in the United States, and for this reason, estimates of stomach cancer risks from radiation exposure from a recent report based on absolute risk transport are nearly an order of magnitude higher than those based on relative risk transport (UNSCEAR 2000).

In general, if the factors that account for the difference in baseline risks act multiplicatively with radiation, then relative risk transport would be appropriate, whereas if they act additively, then absolute risk transport would be appropriate. If some factors act multiplicatively and others additively, the correct estimate might be intermediate to those obtained with the relative or absolute transport models. Whether a factor acts multiplicatively or additively with radiation will depend on whether radiation and the factor of interest act principally as initiators of cancer or act at later stages in multistage cancer development as discussed below.

Two approaches based on epidemiologic data can inform us regarding the most appropriate transport method. The first is to compare risk estimates based on A-bomb survivors with those obtained from studies of non-Japanese populations, particularly predominantly Caucasian populations. If estimates of the excess relative risk (ERR)[1] per sievert are comparable, this suggests that relative risk transport may be appropriate, whereas if estimates of the excess

---

[1]ERR is the rate of disease in an exposed population divided by the rate of disease in an unexposed population minus 1.0.

absolute risk (EAR)[2] per 10[4] person-year (PY) per sievert are comparable, this suggests that absolute risk may be appropriate. However, other differences in the populations often confound such comparisons. Most of the relevant exposures occured for medical reasons, were generally protracted, and often were at higher doses than those received by atomic bomb survivors, making it difficult to interpret comparisons. Additional difficulties are dosimetry uncertainties and statistical variation, which is quite large in some studies. Furthermore, although many studies report estimates of the ERR per gray (ERR/Gy), few report estimates of the EAR per gray. Comparisons of estimates from the LSS and medical studies are also discussed in the material below on breast cancer and at the end of Chapter 12 after the BEIR VII risk estimates have been presented.

A second approach based on epidemiologic data is to investigate interactions of various risk factors with radiation. However, there are few studies with available data on both radiation and other risk factors and with sufficient power to distinguish multiplicative and additive interactions. Relevant data are reviewed below. A detailed discussion of interactions is given by UNSCEAR (2000b, Appendix H).

In the sections that follow, the committee first discusses the type of interaction that would be expected based on consideration of whether radiation and other risk factors act primarily as initiators or promoters. Because the correct transport model is not necessarily the same for all cancer sites, this is followed by a discussion of cancers of each of several specific sites. The etiology of each site-specific cancer is discussed briefly, including the role of various risk factors. A discussion of epidemiologic studies that address interactions of radiation and other risk factors then follows.

Although baseline risks for all solid cancers (as a single category) do not differ greatly between the United States and Japan, this occurs because of the canceling out of site-specific cancers that are higher in the United States (including breast, colon, lung, and prostate) and site-specific cancers that are higher in Japan (including stomach and liver). If the correct transport models differ by site, estimates of all solid cancers based on relative and absolute risk transport may not fully reflect the transport uncertainty.

Postirradiation Cancer Mechanisms and Choice of Transport Model

Animal studies (Chapter 3) suggest that low-LET radiation acts principally as an initiator of tumorigenesis and is at best a weak tumor promoter. In addition, for many tumor types, relative risks (ratio of radiation-associated and spon-

taneous risks) are more comparable across animal strains than are absolute risks (Storer and others 1988). Thus, quantitative animal tumorigenesis data are most consistent with a relative risk transport model, although there are exceptions.

Current knowledge implies the following: (1) at low doses, radiation acts principally as an initiator of cancer (Chapter 3), and (2) many of the known cancer risk factors such as hormonal or reproductive factors, particularly for breast cancer risk, and chronic inflammation associated with microbial infection, for stomach and liver cancers (discussed in this chapter), tend to act at later stages in multistage tumorigenesis. In these latter cases, cancer risk modification is believed to be associated largely with the postinitiation clonal expansion of preneoplastic or malignant cells (Chapter 3). Genetic factors acting throughout cancer development may also modify risk (Chapter 3).

Biologically based risk projection models provide a simplistic, but useful, intuitive framework to evaluate the possible role of radiation in populations with different distributions of risk factors for specific cancer types. An example of such modeling approaches is given in Annex 10A, which summarizes judgments that can be made on the transport of cancer risk using the Moolgavkar and Knudson two stage clonal expansion model, viewing low-LET radiation as a tumor initiator. In simple terms, the model predicts that in the case of a radiogenic tumor type with a strong influence of promoters, one would favor a relative risk transportation model, whereas in the case of a tumor type with a strong influence of initiators, one would favor an absolute risk transportation model.

Etiology of Cancer at Different Sites

As briefly illustrated in Annex 10A, knowledge of the mechanistic factors that underlie tumor etiology can provide an important input to judgments on the most appropriate methodology for transportation of radiation cancer risk between different populations. This section provides an overview of the etiology of a selection of radiogenic human tumors.

Stomach Cancer

Stomach cancer is a disease with a much higher background incidence in Japan than in the United States (IARC 2002). Risk factors for gastric cancer include the presence of conditions such as chronic atrophic gastritis, gastric ulcer, atrophic gastritis, and autoimmune gastritis associated with pernicious anemia. These cause an excessive rate of cell proliferation in the gastric epithelium and are therefore likely to act as promoters, increasing the chance of fixation of replication errors induced by radiation and dietary carcinogens (IARC 2003). *Helicobacter pylori* infection of the

---

[2]EAR is the rate of disease in an exposed population minus the rate of disease in an unexposed population.

stomach appears to be a strong risk factor for stomach cancer, and its effect is likely to be mainly through tumor promotion (although there is increasing evidence that it may also cause tumor initiation; Parsonnet and others 1994; Aromaa and others 1996; Goldstone and others 1996). Environmental risk factors include low consumption of fruit and vegetables; consumption of salted, smoked, or poorly preserved foods; and cigarette smoking (Fuchs and Mayer 1995). The majority of these agents are likely to influence the promotion of tumors.

The above considerations would therefore suggest that for stomach cancer, relative risk transport may be better supported than absolute risk transport. This is also supported by a study of predominantly male peptic ulcer patients, where the estimated ERR/Gy based on patients with doses to the stomach of less than 10 Gy (mean 8.2 Gy) was 0.20 (95% CI 0, 0.73), very similar to that based on male A-bomb survivors (Carr and others 2002; see Table 12-2).

Liver Cancer

The incidence of liver cancer (mainly hepatocellular carcinoma) is also much higher in Japan than in the United States (IARC 2002). The main risk factors for this disease are chronic infection with hepatitis B or C virus, dietary exposure to aflatoxins, and chronic alcohol consumption (IARC 2003). Tobacco smoking also plays a role in the etiology of liver cancer (IARC 2004).

Aflatoxins induce mutations in several genes involved in hepatocellular carcinoma and are thus likely to be involved in the early or initiating stages of carcinogenesis. Hepatitis B and C infections and alcohol consumption, on the other hand, are likely to be involved in the promotion of tumors. They are thought to increase the risk of liver cancer through inflammation that may result in liver cirrhosis. The latter is the major clinical determinant of hepatocellular carcinoma, with 70–90% of these tumors developing in patients with macronodular cirrhosis (IARC 2003).

Baseline risks for liver cancer are much higher in Japan than in the United States, and rates of infection with hepatitis B and C undoubtedly contribute to this difference. The mechanistic arguments above and the limited epidemiologic data tend to support the use of the multiplicative transportation model.

Lung Cancer

Lung cancer is the most common cancer worldwide and the major cause of death from cancer, particularly among men (IARC 2003). In the United States, based on SEER (Surveillance, Epidemiology, and End Results) registry data, the annual incidence rates, age-standardized to the world population, were 55.7 and 33.5 per 100,000, respectively, in men and women in 1993–1997. Comparable rates in Hir-

oshima and Nagasaki during the same period were lower (40–44 per 100,000 in men and 11.8–12.9 per 100,000 in women), particularly among women (IARC 2002).

The major risk factors for lung cancer are tobacco consumption, occupational exposure to a number of carcinogens, and air pollution (Pope and others 2002; IARC 2003). Geographic and temporal differences in lung cancer incidence are determined overwhelmingly by tobacco consumption (IARC 2003).

Tobacco smoke contains approximately 4000 specific chemicals, including nicotine, polycyclic aromatic hydrocarbons, *N*-nitroso compounds, aromatic amines, benzene, and heavy metals. Lung cancer is not thought to be attributable to any one chemical component, but rather to the effect of a complex mixture of chemicals in tobacco smoke, which may act at different stages of the carcinogenic process. Based on the mechanistic arguments above, this suggests that neither a pure absolute nor a pure relative risk transport model is appropriate.

The estimated ERRs/Gy for lung cancer in several studies involving medical exposures in predominately Caucasian patients are lower than those based on A-bomb survivors (Table 6-3), and this might be interpreted as indicating that absolute risks are more comparable than relative risks. However, the lower ERR estimates may also have resulted from other differences in the study populations, particularly the much higher doses in several of the medical studies.

Pierce and colleagues (2003) evaluated the joint effect of smoking and radiation exposure on lung cancer risks in A-bomb survivors and found that they were significantly submultiplicative and consistent with an additive model. They also demonstrate that inferences about the modifying effects of gender and age at exposure on the ERR/Gy can be distorted if analyses do not account for smoking; this is because smoking habits in the LSS cohort depend strongly on both factors.

By contrast, studies of lung cancer risks in underground miners exposed to radon (NRC 1999) or of Hodgkin's disease (HD) patients treated with high doses of radiation (Gilbert and others 2003) rejected additive interactions and found that multiplicative interactions were compatible with the data. However, these studies may be less relevant for estimating the risks of low doses of low-LET radiation than those of A-bomb survivors. Underground miners were exposed to α-emitting (high-LET) radon progeny. In addition, the evidence for a multiplicative relation of radiation and smoking comes primarily from analyses of data on miners in Colorado and China, where doses to the lung (in sieverts) were much higher than in the LSS cohort (NRC 1999). Although data on miners were compatible with a multiplicative effect and not with an additive one, the estimated interaction was submultiplicative. HD patients were also exposed to very high doses (mean dose to the lung 25 Gy) and, in addition, were subject to the immunodeficiency inherent to this lymphoma and associated with the chemotherapy that was also given to many of these patients.

In summary, the absolute risk transport model has greater support for lung cancer than for stomach or liver cancer. Mechanistic considerations suggest that the correct model may be intermediate between relative and absolute risk.

Breast Cancer

Breast cancer is the most common cancer and one of the leading causes of death from cancer among women worldwide, with nearly 1,000,000 new cases per year. Known risk factors for breast cancer include reproductive factors, postmenopausal increased weight, and history of proliferative benign breast disease (IARC 2003). Differences in cancer incidence between U.S. and Japanese populations have been attributed to the tumor promotion effects of hormonal factors (Moolgavkar and others 1980).

In addition, a strong genetic contribution to the risk of spontaneous breast cancer has been shown by the increased cancer incidence among women with a family history of breast cancer. A number of genes involved in DNA damage response pathways, including *BRCA1*, *BRCA2*, and less certainly *ATM*, have been found to confer genetic susceptibility to breast cancer. Alterations in the activity of *ATM*, *BRCA1*, and *BRCA2* proteins may have far-reaching consequences in the control of genetic stability and the risk of tumor development. The presence of sequence variants that alter either the expression or the function of these genes could therefore influence gene-environment interactions and enhance the increased breast cancer risk in women following radiation exposure (see Chapter 3).

There is no study published on *BRCA1* or *BRCA2* mutation frequency in the Japanese population. However, since the prevalence of these mutations in relatively large studies of breast and breast-ovarian cancer in Japanese families is similar to that in Europe and North America, it is likely that *BRCA1* and *BRCA2* mutation frequencies will be the same in Japanese and Caucasians. In Caucasians, the frequency of *BRCA1* was estimated to be 0.051% (95% CI 0.021, 0.125) and of *BRCA2* 0.068% (95% CI 0.033, 0.141; Antoniou and others 2002). Thus, slightly more than one individual in 1000 is a carrier of the *BRCA1* or *BRCA2* mutation. For *ATM* there is no information about the frequency of heterozygotes in the Japanese population. However, for *ATM* and other possible breast cancer genes, as a first approximation it is assumed that there are not major differences in gene frequencies among populations in Japan and Europe or North America.

Thus, in the absence of more detailed data on mutation and polymorphism frequencies in Japan and the United States, the main differences in breast cancer incidence between these two countries are judged to relate to reproductive history and, implicitly, to hormonal factors that would

be expected to act as tumor promoters. The above considerations would therefore suggest that the preferred transportation model for breast cancer should be based on a multiplicative model.

The female breast is one of the few cancer sites for which extensive epidemiologic data on predominantly Caucasian populations are available, and this makes it possible to base risk estimates directly on Caucasian data, avoiding the need to transport risks. Nevertheless, it is useful to evaluate what these data tell as about appropriate transportation models.

Land and colleagues (1980) conducted parallel analyses of cancer incidence data in Japanese A-bomb survivors, Massachusetts tuberculosis fluoroscopy patients, and New York women treated with radiation for mastitis, and found that absolute risks were comparable for the three cohorts whereas relative risks were much larger in the Japanese cohort. This was recently confirmed in a pooled analysis of breast cancer incidence in several cohorts by Preston and coworkers (2002a). In this study, models that were similar in form could be used to describe breast cancer incidence in A-bomb survivors and in U.S. women (Massachusetts fluoroscopy patients and the Rochester infant thymus irradiation cohort). The overall ERR/Gy was about three times as large in the Japanese cohort, whereas the EAR/Gy was similar for the LSS and the U.S. cohorts. However, since fluoroscopy exposure is protracted and involves lower-energy photons than A-bomb exposure, these differences in exposures might confound the comparison. Also, in a pooled analysis of breast cancer mortality in Canadian fluoroscopy patients and A-bomb survivors, neither the ERRs nor the EARs were found to differ significantly between the cohorts (Howe and McLaughlin 1996), although the ERR for the combined LSS women was nearly four times that for non-Nova Scotia Canadian women. Little and Boice (1999) and Brenner (1999) provide additional discussion of these issues with a commentary by Ullrich (1999).

In a case-control study of breast cancer among A-bomb survivors, Land and colleagues (1994a) evaluated the interaction of several risk factors for breast cancer with radiation and found that the relationship was better described by a multiplicative model than an additive one. This, together with the etiological and mechanistic considerations above, would seem to favor relative risk transport, in contradiction to the higher ERR/Gy observed in A-bomb survivors and noted in the preceding paragraph; these observations, however, might have come about because of other differences between the Japanese and U.S. cohorts.

In summary, mechanistic considerations and some epidemiologic data support relative risk transport. However, direct use of data on predominantly Caucasian populations results in estimates that are comparable to those based on A-bomb survivors on an absolute risk scale, but not on a relative risk scale.

Thyroid Cancer

Cancer of the thyroid is a rare disease, accounting for only about 1% of cancer cases in developed countries. The incidence is highest in Iceland and Hawaii (IARC 2002). Annual incidence rates in Japan and the United States are similar (of the order of 2–3 per 100,000 among men and 7 10 per 100,000 in women, for rates age-standardized to the world population), and incidence rates have been increasing worldwide in the last decades. Thyroid cancer in childhood is a very rare disease, with an annual incidence of less than one case per million per year in most developed countries.

Thyroid cancer is about three times as frequent in women as in men, suggesting that hormonal factors may play a role in its etiology, although results from epidemiologic studies of reproductive factors are inconsistent. Iodine deficiency is thought to be involved in the development of papillary thyroid cancer, as may the consumption of some cruciferous and goitrogenic vegetables (IARC 2003). Experimental studies have shown that excessive long-term stimulation of the thyroid gland by thyroid-stimulating hormone, such as results from iodine deficiency, can lead to tumor formation with or without addition of a mutagenic agent (Thomas and Williams 1991).

History of goiter and benign thyroid nodules is associated with papillary thyroid cancer risk, as is family history of thyroid cancer; the possible role of increased thyroid screening in these associations is unclear at present. Medullary thyroid carcinoma, a rare type of thyroid cancer, has a very strong genetic component (IARC 2003).

Because the majority of the risk factors listed above (hormones, iodine deficiency) are likely to influence the promotion of tumors, mechanistic considerations suggest that the preferred transportation model for thyroid cancer should be based on relative risk transport. It is noted that the BEIR model for thyroid cancer risk is based on a combined analysis of epidemiologic studies carried out in different countries (U.S., Japan, Israel). The multiplicative model developed by Ron and coworkers (1995b) was applied directly with the uncertainty that reflects international variation in thyroid cancer risk.

Leukemia

Leukemia comprises about 3% of all incident cancers worldwide; the age-standardized incidence in the United States (standardized to the world population), based on SEER registry data, was 10.8 and 6.7 per 100,000, respectively, in men and women during 1993–1997. Rates in Nagasaki prefecture during the same period were similar (9.4 and 6.2 per 100,000, respectively, in men and women), while they were lower in Hiroshima (6.1 and 4.7 per 100,000, respectively; IARC 2002). It should be noted that these rates include chronic lymphocytic leukemia, which is known to be rare in Japan but is more frequent in the United States.

The etiology of leukemia is not well established. Apart from ionizing radiation, occupational exposure to agents such as benzene can increase the risk of leukemia, as can exposure to some chemotherapeutic agents. Some risk factors such as Down's syndrome and exposure to extremely low frequency magnetic fields have been postulated as risk factors for childhood leukemia. Infection by the HTLV-1 virus is responsible for adult T-cell leukemia, a disease observed in Japan, but rarely in the United States (IARC 2003). Conversely, chronic lymphocytic leukemia, a neoplasm of B lymphocytes, is rare in Japan but more frequent in the United States.

Based on the above, it is not currently possible to draw conclusions about mechanisms of carcinogenesis, and therefore transport models, except to note that the different prevalence of infection with a number of viruses including HTLV-1 and viruses involved in B-cell lymphomas may account for a difference in the incidence of specific leukemia subtypes between Japan and the United States.

Conclusions

At present, neither knowledge of biological mechanisms nor data from epidemiologic studies are sufficient to allow definitive conclusions regarding the appropriate method for transporting risks, although mechanistic considerations suggest somewhat greater support for relative risk than for absolute risk transport. For this reason, the committee provides estimates based on both relative risk and absolute risk transport. When a single estimate is needed, a weighted mean of the two estimates can be used. For cancer sites other than breast, thyroid, and lung, the committee recommends a weight of 0.7 for the estimate obtained using relative risk transport and a weight of 0.3 of the estimate obtained using absolute risk transport with the weighting done on a logarithmic scale. This choice of weights, which clearly involves subjective judgment, was made because the mechanistic considerations discussed above suggest somewhat greater support for relative risk transport, particularly for cancer sites (such as stomach, liver, and female breast) for which known risk factors act mainly on the promotion or progression of tumors. The choice reflects uncertainty regarding which model is correct and also allows for the possibility that some factors interact additively with radiation, whereas others interact multiplicatively. The uncertainty involved in this choice is reflected in the subjective confidence intervals that are provided as discussed in Chapter 12.

Exceptions to the approach noted above are made for cancers of the breast, thyroid, and lung. For breast and thyroid cancer, extensive data on Caucasian populations are available and can be used directly to estimate risks. The committee's preferred models, which are described in Chapter 12, make use of these data. For lung cancer, analyses of the A-bomb survivor data by Pierce and colleagues (2003) support an additive interaction of smoking and radiation. Since differences in smoking habits undoubtedly contribute to the differences in baseline risks in Japan and the United States, this finding supports the use of absolute risk transport. Furthermore, lung cancer analyses of A-bomb survivor data based on EAR models may provide a more reliable evaluation of the dependence of radiation risk on factors such as gender and age at exposure than do ERR models, as discussed above. As indicated in Chapter 12, relative risk transport estimates are based on ERR models, whereas absolute risk transport estimates are based on EAR models. Thus, for lung cancer, the weighting scheme used for most other solid cancers is reversed, and a weight of 0.7 is used for the estimate obtained with absolute risk transport and a weight of 0.3 for the estimate obtained with relative risk transport.

For sites other than breast, thyroid, and lung, it is likely that the correct transport model varies by site. However, the committee judged that current knowledge was insufficient to provide separate approaches for other specific sites.

FORM OF THE DOSE-RESPONSE FOR RADIATION TUMORIGENESIS

Follow-up of cancer incidence and mortality in Japanese A-bomb survivors (the LSS study) continues to provide the most informative epidemiologic data on the shape of the dose-response for solid tumors and leukemia (Chapter 6), although studies of large-scale populations with low-dose chronic exposures are increasingly informative about the effects of low doses.

Atomic bomb survivor data for solid tumors combined provide statistical evidence of a radiation-associated excess at doses down to around 100 mSv; these combined data are well described by a linear no-threshold dose-response, although some low-dose nonlinearity is not excluded (Pierce and Preston 2000; Preston and others 2003). Indeed, dose-response relationships for individual tumor types in the LSS can differ, and for nonmelanoma skin cancer the dose response is highly curvilinear with an excess seen only at doses higher than around 500 mSv. The LSS dose-response for leukemia is also clearly curvilinear, with a statistically significant excess being evident at doses around 200 mSv.

The above human data well illustrate the problems of limited statistical power that surround epidemiologically based conclusions on the shape of the low dose-response for radiation cancer risk and how it might vary between tumor types. Similar difficulties surround judgments based on data obtained using experimental animals; many studies are broadly consistent with a linear no-threshold dose response, but there are a number of examples of highly curvilinear, threshold-like relationships (Chapter 3).

It is abundantly clear that direct epidemiologic and animal approaches to low-dose cancer risk are intrinsically limited in their capacity to define possible curvilinearity or dose thresholds for risk in the range 0–100 mSv. For this reason the present report has placed much emphasis on the mechanistic data that can underpin such judgments.

The following data and conclusions are given in Chapters 1, 2, and 3 and are most pertinent to radiation risks in the dose range 0–100 mSv where epidemiologic and animal data are inadequate.

First, there is evidence that most cancers are monoclonal in origin (i.e., they develop from progeny of a single abnormal cell; UNSCEAR 1993). Whatever molecular mechanism is envisaged for radiation, at very low doses (e.g., 0–5 mGy low LET), increases in dose simply increase the probability that a given single cell in the tissue will be intersected by an electron track which will have a nonzero probability of inducing a biological effect. Therefore, at these very low doses, a linearity of response is almost certain (Chapter 3).

Second, given the intimate relationship established between DNA damage response, gene or chromosomal mutations, and cancer development, the form of the dose-response for mutation induction in single cells should be broadly informative for cancer initiation. Data from a large-scale study noted in Chapter 2 suggest a linear relationship between low-LET dose and chromosomal mutation down to around 20 mGy.

A central question addressed in this report is the nature of critical DNA lesions after low-LET radiation and the extent to which they may be repaired by the cell without errors. This is a crucial judgment in radiation tumorigenesis since, at the level of cancer-associated gene or chromosomal mutation, the presence of a true dose threshold demands totally error-free DNA damage response and repair.

The detailed information available on the importance of a chemically complex DNA double-strand break (DSB) induced by a single ionization cluster for postirradiation biological effects (Chapter 1), together with the predominance of error-prone nonhomologous end joining (NHEJ) repair in postirradiation cellular response, argues strongly against a DNA repair-mediated low-dose threshold for cancer initiation (Chapters 1–3). The same data provide a strong counter to pro-threshold arguments based on the relative abundance of spontaneously arising and radiation-induced DNA damage. Those arguments fail to take account of the quality of the repair achievable for simple and complex forms of DNA damage.

In principle, complex DNA DSBs may be repaired with full fidelity by homologous recombination (HR) pathways. Since HR operates almost exclusively between sister chromatids in cells that have newly replicated their DNA (Chapters 1 and 2), the cell has a limited cell cycle window for such error-free repair. At any one time, only a small fraction of stem-like target cells in tissues are expected to reside within this postreplication window—many will be in a nonreplicative, quiescent state (e.g., Potten and Hendry 1997; Kountouras and others 2001; Young 2004). On this basis HR-mediated error-free repair is unlikely to be the dominant feature of *in vivo* cellular response and tumor induction.

Finally, evidence is emerging that the DNA deletions that are characteristic molecular footprints of NHEJ-mediated misrepair and gene loss in cultured cells are also seen as early events in radiation-induced tumors in rodents; there is also preliminary evidence pointing toward the involvement of NHEJ misrepair in the genesis of early arising *RET* gene rearrangements in post-Chernobyl childhood thyroid cancer (Chapter 3).

When considered together, these *in vitro* and *in vivo* data are seen to provide a scientifically coherent linkage between error-prone postirradiation repair of chemically complex DNA DSBs in target cells *in vivo* and tumor induction.

Mechanistic uncertainties remain, but the weight of available evidence would argue against the presence of a low dose threshold for tumor induction based on error-free repair of initial DNA damage. In summary, the committee judges that the balance of scientific evidence at low doses tends to weigh in favor of a simple proportionate relationship between radiation dose and cancer risk.

DOSE AND DOSE-RATE EFFECTS ON TUMOR INDUCTION

Since much of the informative epidemiologic data on low-LET radiation cancer risk derives from the study of acute exposures, it is necessary to make somewhat indirect judgments about the magnitude of the expected reduction in risk associated with low doses and dose protraction. This reduction in risk is conventionally described by the dose and dose-rate effectiveness factor (DDREF). As illustrated and discussed in Chapter 2 (see Figure 2-1), the reduction in risk for low doses (DEF) and the reduction in risk for dose protraction (*i.e.*, low dose rates; DREF) are assumed to be equal; therefore, the term DDREF is used for estimating effects for either low doses or low dose rates.

Information from cellular and molecular studies strongly suggests that dose and dose-rate effects of low-LET radiation are determined largely by the activity of DNA damage response process in cells. For the induction of gene and chromosomal mutations in cultured somatic cells, values for DDREF generally fall in the range of 2–4 (Lloyd and others 1992; Thacker 1992; UNSCEAR 1993, 2000b; Cornforth and others 2002), although higher and lower values have been recorded in some mutation systems. Together, these data are consistent with the view that the temporal abundance of radiation-induced DNA damage is a major factor in the efficiency or fidelity of DNA repair and hence the frequency of induced mutation (Chapter 2).

In vivo effects of dose protraction or fractionation are likely to be more complex, but available data on animal tumorigenesis show that the reduction of tumor yield with dose fractionation is determined by processes that operate on a time scale of up to 24 h. This time scale is more consistent with the activity of cellular DNA repair than with that of postirradiation whole-tissue remodeling, thus drawing together dose-rate effects at the cellular and whole-animal levels (Chapter 3).

Animal tumorigenesis data and related information from life-shortening studies (Chapter 3) may be used to provide judgments on DDREF that vary up to a value of 10 or more (UNSCEAR 1994). However, when those tumor types that, atypically, depend strongly on cell killing are excluded and analysis is restricted to doses up to a few grays, the DDREF values obtained are in the range of 2–3 (Chapter 3). These values are similar to those of gene mutation and, thereby, broadly consistent with the recurring theme of a close association between DNA damage response, mutation induction, and cancer.

The biological picture overall is that cellular and animal data relating to protracted radiation exposures provide a convincing argument for the inclusion of DDREF in judgments about cancer risk at low doses and low dose rates. The animal data showing reduction in carcinogenic effectiveness, including life shortening, following protracted exposure constitute the strongest element in this argument; the coherence of the mechanistic data adds additional weight.

An alternative approach is to estimate DDREF on the basis of the degree of curvature of the dose-response for excess cancer after acute irradiation. Conventional radiobiological theory holds that the initial linear (α) term of a linear-quadratic ($\alpha D + \beta D^2$) dose-response (where D is the dose) will represent the low-dose and low-dose-rate response. Accordingly, the α and β terms of the acute dose-response may be used to provide an estimate of DDREF. Note that the BEIR V committee did not apply a DREF (sic) in its analysis of solid tumor data and used a linear-quadratic model for leukemia (NRC 1990). Also, the UNSCEAR (2000) committee commented that the LSS data suggested a "value of about 1.5" for the DDREF. In its report, the International Commission on Radiological Protection (ICRP 1990) stated that "the Commission has decided to recommend that for radiation protection purposes the value of 2 be used for the DDREF, recognizing that the choice is somewhat arbitrary and may be conservative."

The committee has taken a computational approach to the estimation of DDREF that is based on a Bayesian analysis of combined dose-response data. The data sets considered were (1) solid cancer incidence in the LSS cohort of Japanese atomic bomb survivors; (2) cancer and life shortening in animals; and (3) chromosome aberrations in human somatic cells.

Derivation of the Dose and Dose Rate Reduction Factor by Bayesian Analysis

The BEIR VII cancer risk estimates are based on risk models derived primarily from analyses of data on the LSS cohort of Japanese atomic bomb survivors. Historically, and with the exception of leukemia, there has been little statistical evidence of a need for curvature in the LSS dose-response models and substantial reliance on models in which risk is simply a *linear* function of radiation dose (Pierce and Preston 2000; Preston and others 2003). There is stronger evidence of curvature from radiobiological considerations and experimental results. The DDREF has been used in the past as a device for allowing risk estimates to conform to this expected curvature but without abandoning the LSS linear models (ICRP 1991; NCRP 1993; EPA 1999; UNSCEAR 2000b; NIH 2003).

A rationale for DDREF is illustrated in Figure 10-1 for a setting that mimics a simple animal experiment on cancer induction by acute-dose low-LET radiation in which risks are observed only at two doses: zero and some particular "high dose." If the true dose-response relationship is concave up to that dose, as the radiobiological data tend to suggest, then a line connecting the excess risk at high dose to the origin would tend to have a larger slope than a line that approximates the dose-response curve at doses near zero. The DDREF in this case is the ratio of these two slopes (*i.e.*, the risk per unit of dose at high dose divided by the risk per unit of dose at low dose). If this ratio is known then it can be used

to convert a risk estimate from the high-dose linear approximation to the more appropriate low-dose linear approximation, as shown in the figure. The association between the form of the dose-response at acute doses and the effects of dose-*rate* is discussed in Chapter 2 and in Annex 10B.

This DDREF clearly must depend on what is meant by high dose and should not be mistakenly thought of as a universal low-dose correction factor. Furthermore, of particular interest here is what might more appropriately be called an LSS DDREF, where a curvature adjustment to risk estimates from LSS-estimated linear models is based on a wide range of doses. The line analogous to the "high-dose linear approximation" of Figure 10-1 is the one that results from linear model estimation with the LSS data. If a certain degree of curvature is presumed, then it is possible to define an LSS DDREF that correctly adjusts LSS linear risk in order to estimate cancer risk at low doses. Such a definition is provided after the discussion of a numerical characterization of dose-response curvature upon which it is based.

If, over some dose range of interest, the dose-response curve can be approximated by a linear-quadratic (LQ) function, $\alpha Dose + \beta Dose^2$, then the slope of the high-dose linear

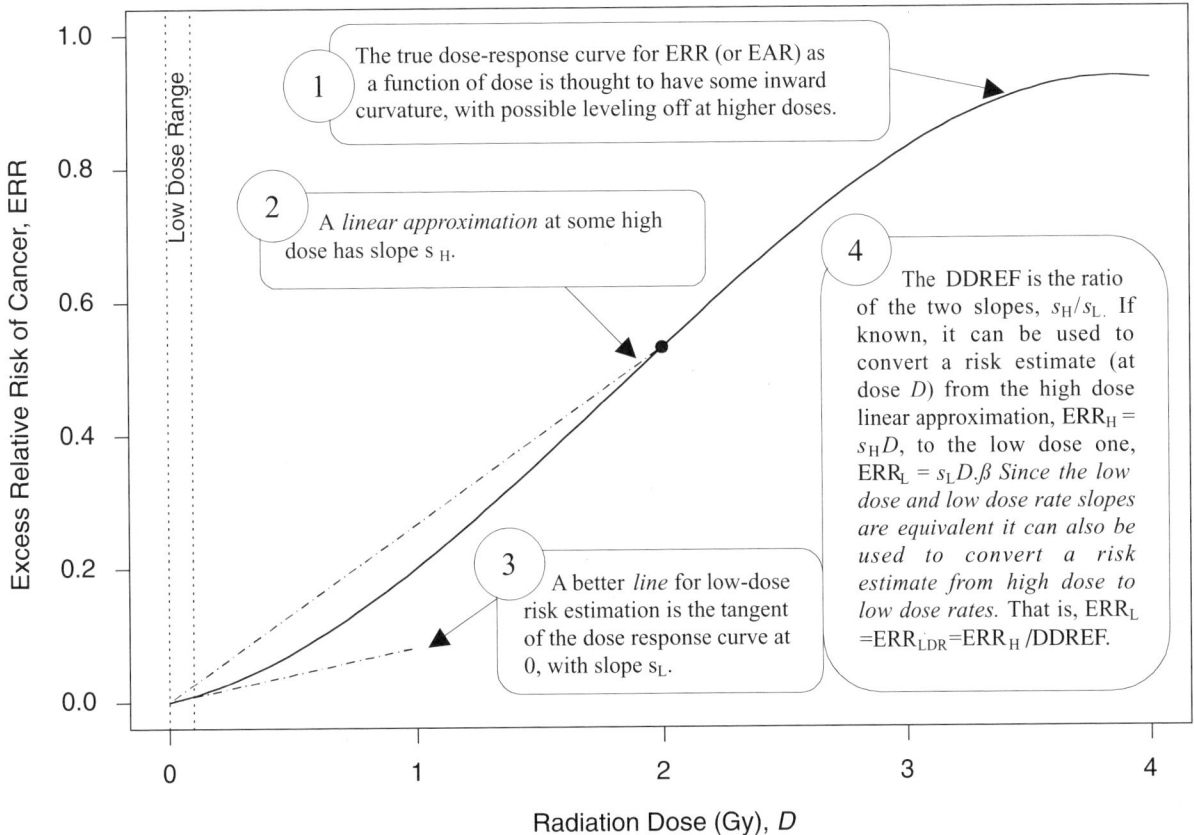

FIGURE 10-1 A hypothetical dose-response curve with a linear approximation for low doses (*i.e.*, the tangent of the curve at dose zero) and a linear approximation based on risk at one particular high dose (*i.e.*, the line that passes through the origin and the true dose-response curve at the high dose), when the high dose is taken to be 2 Gy. The DDREF at this high dose is the larger slope divided by the smaller slope. An explanation of why this low-dose effect also applies to low-dose-*rate* effects is provided in Chapter 3.

approximation (s_H in Figure 10-1) at a particular high dose D_H is $\alpha + \beta D_H$, the slope of the low-dose linear approximation (s_L in Figure 10-1) is α, and the DDREF corresponding to D_H is their ratio, $1 + (\beta / \alpha)D_H$ (UNSCEAR 1993). A natural numerical quantity for curvature characterization, therefore, is β / α, which is not tied to any particular high dose. This ratio is referred to here as the LQ "curvature" and is represented by the symbol θ (*i.e.*, the reciprocal of the so-called crossover dose).

If the correct curvature, θ, is known, then an LSS DDREF may be defined through the following steps: an LQ model for ERR or EAR is estimated from the LSS data in such a way that the curvature is constrained to be θ, that is, by fitting the relative risk model $\alpha_{LQ}(Dose + \theta Dose^2)$ for fixed θ and with unknown linear component α_{LQ}. A separate *linear* model is estimated from the same data: $\alpha_L Dose$, with linear component α_L. The LSS DDREF is the estimate of the ratio of the two linear components, α_L / α_{LQ}. The resulting DDREF can be used to convert a risk estimate based on the linear model projection to one based on the linear component of an estimated LQ model with curvature determined by a given choice of the value of θ. Figure 10-2 illustrates the definition for two possible choices of this value.

The two definitions of DDREF as a function of LQ curvature must be clearly distinguished: the fixed high-dose DDREF (or UNSCEAR definition), DDREF = $1 + \theta \times high$ *dose*, and the LSS DDREF defined by the estimation process in the preceding paragraph. The first is a function of θ and some particular *high dose*. The second is a function of θ and the LSS data. Their relationship, as illustrated in Table 10-1, indicates that the LSS DDREF based on A-bomb survivors with doses of 1.5 Sv or less is roughly equivalent to the fixed high-dose DDREF at an effective high dose of about 1 Sv. In other words, in terms of the familiar UNSCEAR single high-dose definition, one can act as if the nonzero LSS doses were concentrated at a dose of 1 Sv.

Table 10-1 may be used as an aid in interpreting radiobiological evidence for curvature. If, for example, radiobiology data indicate that a DDREF of 2 is appropriate for adjusting risks based on a linear model derived at the single *high dose* of 2 Sv, then the implicit curvature is 0.5 Sv^{-1} and the corresponding LSS DDREF is 1.5.

The committee estimates LSS DDREF in this report by combining radiobiological and LSS evidence concerning curvature via a Bayesian statistical analysis and applying the definition of LSS DDREF to the result. As detailed in Annex

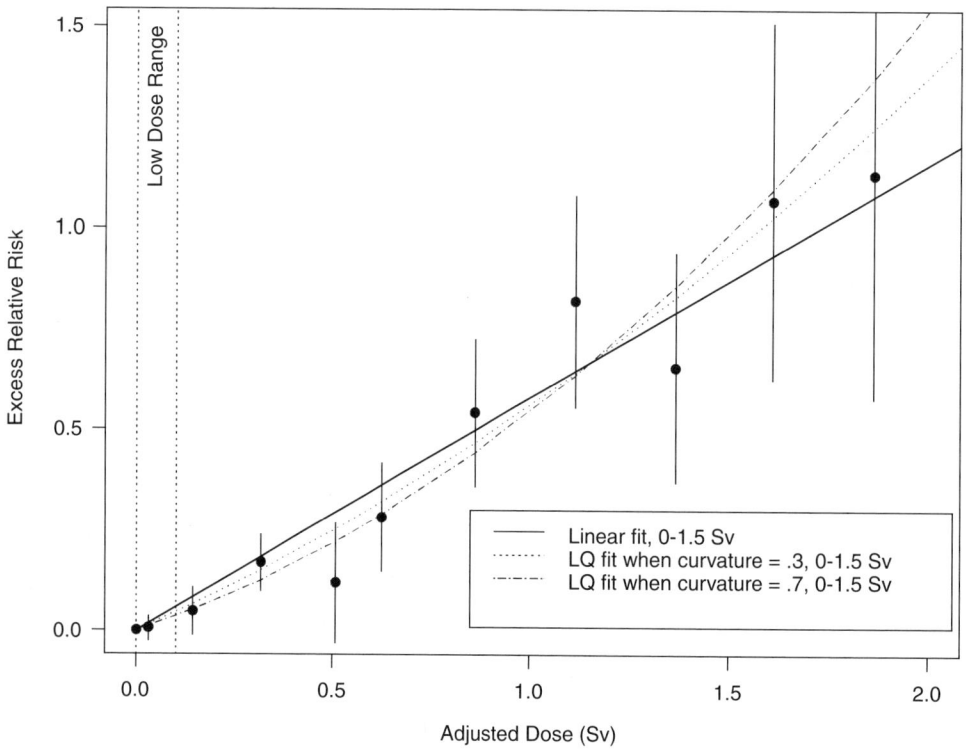

FIGURE 10-2 Illustration of LSS DDREF. Plotted points are the estimated ERRs for solid cancer incidence (averaged over sex, for individuals exposed at age 30 at attained age 60) from LSS subjects in each of 11 dose categories. The vertical lines extend two standard errors above and below the estimates. The solid line is a linear fit to the data for dose range 0–1.5 Sv, with slope $\alpha_L = 0.56$. The other two curves are estimated LQ models for the same dose range, when the curvature, θ, is constrained to be 0.3 Sv^{-1} (resulting in estimated linear coefficient $\alpha_{LQ} = 0.43$) and 0.7 Sv^{-1} (resulting in estimated linear coefficient $\alpha_{LQ} = 0.32$). The LSS DDREFs that result from these are 0.56 / 0.43 = 1.3 and 0.56 / 0.32 = 1.8, respectively.

TABLE 10-1 UNSCEAR Definition of DDREF and LSS DDREF[a] Corresponding to Three Values of Curvature

	UNSCEAR DDREF[b]			
Curvature (θ, Sv^{-1})	High Dose = 1 Sv	High Dose = 2 Sv	High Dose = 3 Sv	LSS DDREF[c]
0.5	1.5	2.0	2.5	1.5
1.0	2.0	3.0	4.0	2.1
2.0	3.0	5.0	7.0	3.1

[a]For incidence of solid cancers and based on doses between 0 and 1.5 Sv, as in Figure 10-2.

[b]DDREF = $1 + \theta \times high\ dose$.

[c]From estimating LQ models forced to have curvature θ.

10B, the radiobiological information comes from mouse experiments, via models estimated from direct cancer risk data and models estimated from cancer-associated life-shortening data. The resulting *posterior* distribution for possible values of LSS DDREF is displayed in Figure 10-3.

Table 10-2 summarizes the graphical results of Figure 10-3. A single estimate of curvature is the median of the posterior distribution: 0.5 Sv^{-1}, corresponding to an LSS DDREF of 1.5. On the basis of these analyses, there is little disagreement between the radiobiological and LSS estimates of LSS DDREF. While a quadratic term in an LSS LQ model is not significantly different from zero (twosided *p*-value = .2, for the 0–1.5 Sv dose range), the single best estimate of LSS DDREF from the LSS subset is 1.3. If the radiobiological estimate of 1.5 seems low, the committee believes that it is due not to a new interpretation of radiobiological curvature but rather to the use of an LSS DDREF that is specific to the needs of LSS linear model adjustment to account for the curvature. As evident in Table 10-1, a DDREF suitable for LSS adjustment is less than that expected for low-dose extrapolation of estimates based on high doses of 2 to 3 Sv.

The Bayesian approach formalizes the connection among the DDREF, the LQ curvature in radiobiology, and the LSS data. However, there are two reasons for the continuing uncertainty in the estimation of DDREF: (1) there is substantial inconsistency and imprecision in the data from animal experiments; and (2) the curvature estimates from radiobio-

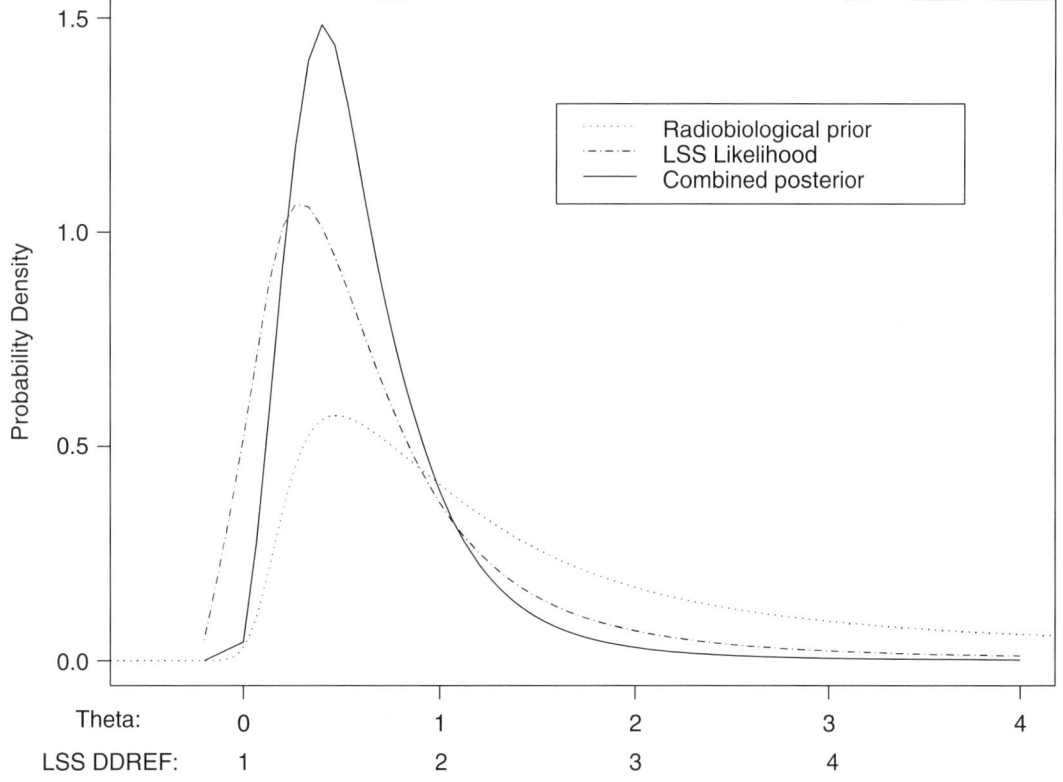

FIGURE 10-3 Results of a Bayesian statistical analysis of dose-response curvature and associated LSS DDREF values. The probability density labeled "radiobiological prior" expresses the belief about curvature deduced from animal data, as detailed in Annex 11B. Regions of high density correspond to more believable values of curvature. The LSS likelihood is the likelihood function of curvature θ from the data displayed in Figure 10-2. The "combined" density is the Bayesian posterior obtained by updating the radiobiological density to account for information from the LSS data. The scale below the plot shows the implied values of LSS DDREFs corresponding to the θ scale. NOTE: The committee judges it preferable to choose a cutoff dose that lies within the lower rather the higher portion of the possible range.

TABLE 10-2 Maximum Likelihood Estimates of Curvature and Corresponding Estimates of LSS DDREF[a] and the Posterior Median from the Bayesian Analysis that Combines the Two

Estimate of θ	(95 % interval)	LSS DDREF	(95% interval)
Radiobiology animal experiments	0.5 Sv	(0.1, 3.2)	1.5 (1.0, 4.4)
LSS data (0–1.5 Sv dose range)	0.3 Sv	(–0.1, 1.5)	1.3 (0.8, 2.6)
Combined (posterior)	0.5 Sv	(0.1, 1.2)	1.5 (1.1, 2.3)

NOTE: The 95% intervals are confidence intervals (likelihood ratio) in the first two rows and Bayesian posterior probability intervals in the last row.

[a]From radiobiological animal experiment results and LSS data.

logical data and from LSS data are sensitive to the range of doses used for estimation.

As shown in Annex 10B and evident in Figure 10B-1, there is a statistically significant difference in curvatures for the different mouse strains, sex, and cancer outcome combinations investigated (p-value < .001). Some results indicate large curvature, some no curvature, and some curvature in the opposite direction. The combined effect is weak evidence for small curvature. Because of the wide variability, the analysis is sensitive to the particular studies chosen and the approach for estimating a curvature that is presumed to be constant across studies.

The numerical results also are not robust because of the somewhat arbitrary choice in dose range subset for estimating linear-quadratic models, both from animal experiments and from LSS data. If the LQ model is fitted to a dose range that includes doses at which leveling off of the dose-response curve has occurred (as illustrated in Figure 10-1), the result will be biased for the intended purpose. If the dose range is too low, meaning it excludes doses for which the LQ approximation is still good, the estimates will be less precise than what is possible but will not lead to any bias. Given these consequences, it is judged preferable to choose a cutoff dose that is too low rather than one that is too high. The cutoffs of 1.5–2 Gy for animal experiments and 1.5 Sv for LSS data were chosen subjectively, based on the belief that these were sufficiently low that leveling off is not of great concern. Nevertheless, the fact remains that the relationships shown in Figure 10-3 would be quite different if different dose ranges were used.

The cutoff of 1.5 Sv for the LSS data is important for assessing curvature. The resulting LSS DDREF is appropriate for adjusting risks from linear models based on the same dose range. Since the LSS estimated linear model is insensitive to the choice of subset however, the particular choice of dose range upon which to estimate the linear model is not critical.

More generally, since a *linear* model fits the LSS data over the entire range (for cancers other than leukemia), it is important to question why the expected curvature fails to materialize and whether the absence of curvature necessarily implies that the LSS low-dose slope is too large. It could be that a linear relationship is the result of some cancelation of inward curvature and high-dose leveling-off. It is not obvious that the linear relationship resulting from such cancelation overestimates low-dose risk.

Given these unresolved issues, it is comforting that the estimate of LSS DDREF is consistent with the best-fitting LQ model based on LSS data alone; that is, low-dose risk estimates based on LSS linear models with DDREF adjustment will be essentially the same as risk estimates based on the best-fitting LQ model from LSS data over the range 0 to 1.5 Sv. In Figure 10-2, for example, it is clear that the linear component of an LQ curve with curvature 0.5 Sv^{-1} over the low-dose range is consistent with the data. The difference between that estimate and one based on the unadjusted linear model will be small relative to the size of the associated confidence interval.

The DDREF analysis has used LSS data on solid cancer *incidence*. A recent similar analysis on cancer *mortality* (Preston and others 2004) has provided the somewhat larger estimated curvature 0.94 Sv^{-1} (90% CI 0.16, 8.4) for the best-fitting LQ model over the range 0 to 2 Sv. Since there is considerable imprecision in the calculations, this result is not inconsistent with the committee's conclusions.

In summary, the approach used by the committee to make an analytical judgment about the value of DDREF has employed a combined Bayesian analysis of dose-response curvature for cancer risk using animal radiobiological data and human evidence from the LSS. The committee found a believable range of DDREF values for adjusting linear risk estimates from the LSS cohort to be 1.1–2.3. Based on this analysis, the committee elected to use the value of 1.5 for solid tumors; also, a linear-quadratic model was used for leukemia. The committee recognizes the limitations of the data and the uncertainties in estimating the DDREF.

OTHER FORMS OF CELLULAR AND ANIMAL RESPONSE TO RADIATION

This report has given much attention to biological responses to radiation that, although not well understood, may influence the development of views on tumorigenic mechanisms and the modeling of epidemiologic data.

Adaptive Responses

Adaptive responses to radiation are represented in a range of studies that purport to demonstrate that a low priming dose of radiation can influence the subsequent response of cells or experimental animals to subsequent challenge by a second higher dose. It is claimed by some that these adaptive responses will reduce low-dose cancer risk substantially, perhaps to zero, or even be beneficial to health (see Calabrese and Baldwin 2003 and references therein).

Cellular data and mechanistic considerations on adaptive responses are reviewed in Chapter 2. From this review it is concluded that adaptive responses are not expressed robustly in cells and that a mechanistic basis for the phenomena, particularly in the form of well-characterized DNA damage response, has yet to be established. This situation may be contrasted with the detailed knowledge that has accrued on many other aspects of DNA damage recognition or repair and cellular response (see Chapters 1 and 2). Accordingly, cellular and mechanistic data on adaptive responses are as yet insufficient to develop specific judgments on the fundamental aspects of low-dose cancer risk.

Recent animal studies on adaptive responses to radiation and cancer risk are considered in Chapter 3. These studies provide some evidence that under certain conditions, a low priming dose of radiation can modestly influence the rate of development of certain tumors. However this response is not accompanied by a reduction of overall lifetime cancer risk. Uncertainties remain about the specific conditions of irradiation under which this form of adaptive response is expressed, and its mechanistic basis is a matter of speculation. Accordingly, these animal data, although of considerable scientific interest, are not sufficiently well developed to influence the modeling and interpretation of epidemiologic data.

Induced Genomic Instability

Induced genomic instability is a term used to describe a set of cellular phenomena whereby radiation exposure alters the state of a cell in a way that generally leads to a persistent elevation of mutation rate over many cell generations. The cellular data reviewed in Chapter 2 highlight the inconsistent mode of expression of this phenomenon and the current lack of information on the cellular mechanisms that might be involved. It is notable that many of these data sets relate to cells established in culture for many years. Despite these problems of interpretation, there has been speculation that radiation-induced genomic instability might make a significant contribution to cancer induction *in vivo* and thereby confound the interpretation of epidemiologic data. Chapter 3 considers the *in vivo* expression of radiation-induced genomic instability, possible mechanistic links with cancer induction in animal models, and the expression of such instability in radiation-associated human tumors. Although some uncertainty remains, these *in vivo* data strongly question the proposition that radiation-induced, genome-wide instability plays a major role in radiation tumorigenesis. One possible exception to this is the instability of altered telomeric sequences at chromosome termini that may trigger ongoing cycles of chromosomal associations and rearrangement (Chapters 2 and 3). However, given the great uncertainty about the contribution of induced and persistent genomic instability to postirradiation tumor development, there is at present no meaningful way in which the phenomenon can be included in the general interpretation of epidemiologic data and, thereby, the derivation of new estimates of low-LET cancer risk.

Bystander Cellular Effects

Chapter 2 details the almost exclusively cellular data for high-LET radiation, indicating that cellular damage response signals may be passed from an irradiated cell to a non-irradiated neighbor. There are few consistent data sets for low-dose, low-LET radiation. The stress-related mechanisms that have been suggested to underlie postirradiation signal transfer via cellular gap junctions or cell culture medium are not well understood. In addition, the *in vivo* expression of bystander effects and their impact on tumor development have yet to be adequately addressed. For these reasons, the committee judges that current knowledge of these phenomena is insufficient for the purpose of interpreting epidemiologic data and developing judgments on cancer risks at low doses of low-LET radiation.

GENETIC SUSCEPTIBILITY TO CANCER

The data reviewed in Chapters 1 and 3 provide coherent evidence from cellular, animal, and clinical or epidemiologic studies that inheritance of certain germline gene mutations can predispose to radiation-induced cancer. The qualitative linkage between such epidemiologic or clinical and experimental data are particularly strong for rare, strongly expressing human mutations. However, with current knowledge, experimental data cannot quantitatively inform about the magnitude of the increased radiation risk in such genetic disorders. Accordingly, only broad judgments are possible—principally that strongly expressing human mutations of relevance to radiation cancer risk are too rare to an appreciably distort population-based estimates of low-dose risk as derived from epidemiologic data (Chapter 3).

The implication for population risk of weakly expressing but potentially common variant genes is a most difficult issue. Genetic studies with mice (Chapter 3) provide evidence of the potential complexity of germline gene-gene interactions in radiation tumorigenesis. However, human molecular epidemiologic studies in this area are at a very early stage of development, and no specific judgments are possible on the extent to which common genetic variation influences epidemiologic measures of radiation risk. The general judgment made in Chapter 3 is that the potential impact of such variant genes on radiation cancer risk in the population will depend on a complex interplay between their frequency in the population, their tissue specificity, and the strengths of the gene-gene and gene-radiation interactions that may apply.

HERITABLE EFFECTS OF RADIATION

As in the BEIR V report (NRC 1990), estimates of the risks of adverse heritable effects of radiation exposure are made indirectly through extrapolation from mouse data on rates of radiation-induced germ cell mutations using population genetic theory and a set of plausible assumptions (see Chapter 4). These estimates are expressed as increases in the frequencies of genetic diseases relative to their baseline frequencies in the population. The method that is used for this purpose is referred to as the "doubling dose method." Equation (10-1) below summarizes the method:

$$\text{Risk per unit dose} = P \times [1/DD] \times MC \times PRCF, \qquad (10\text{-}1)$$

where P is the baseline frequency of the disease class under consideration, DD is the doubling dose (*i.e.*, the dose of radiation required to double the rate of spontaneous mutation in a generation, estimated as a ratio of rates of spontaneous and induced mutations in defined genes), MC is the mutation component (a measure of the responsiveness of the disease class to an increase in mutation rate), and PRCF is the potential recoverability correction factor (the fraction of induced mutations that are compatible with live births and cause disease).

This report incorporates several important advances that have been made since the publication of BEIR V (NRC 1990), among which are those that relate to the four quantities mentioned above. It suffices to note that the estimates for P, MC, and PRCF are different for Mendelian and multifactorial diseases; however, the DD estimate of 1 Gy (for low-dose or chronic low-LET exposure) is common to both classes of disease.

The risk estimates provided in Chapter 4 are about 3000 to 4700 cases of excess genetic disease per million first-generation progeny per gray of radiation to the parental generation. Compared to the natural (*i.e.*, baseline) risk of genetic diseases of 738,000 per million live births in the population, the radiation risk (per gray) is very low (about 0.4 to 0.6% of the baseline).

As mentioned earlier, the results of the extensive genetic epidemiologic studies of A-bomb survivors in Japan have shown no adverse effects in the progeny that could be attributed to the radiation exposures (of the order of 0.4 Sv) sustained by most survivors. The indicators of adverse effects used in these studies were untoward pregnancy outcomes (UPOs), mortality of live born children through a period of about 26 years (exclusive of those resulting from malignant tumors), malignancies in the F_1 children, frequency of balanced structural rearrangements of chromosomes, frequency of sex chromosomal aneuploids, frequency of mutations affecting protein charge or function, sex ratio shift among children of exposed mothers, and growth and development of F_1 children. The important point here is that these indicators of adverse effects cannot be compared readily to what are formally called genetic diseases.

The total numbers of children included in the analyses to ascertain radiation effects were about 41,000 in the "unexposed" and 31,000 in the "exposed" groups, although the numbers were different for different indicators (*e.g.*, ~8000 children each in the exposed and control groups for balanced structural chromosomal rearrangements and sex chromosomal aneuploidy; ~41,000 in the exposed and ~31,000 in the exposed groups for malignancies in F_1).

Although no statistically significant effects of parental radiation exposures were found, Neel and colleagues (1990) estimated doubling doses on the basis of data for five of the indicators (*i.e.*, UPO, F_1 mortality, F_1 cancers, sex chromosomal aneuploids, mutations) that would be consistent with the findings. In order to do this, several assumptions had to be made (discussed in Annex 4G). The oft-quoted DD estimated from these data, corrected for low-dose or chronic, low-LET radiation conditions is 3.4 to 4.5 Sv.

The perception remains that the above estimate of the DD is indicative of far lower heritable risk than that implied by the DD of 1 Gy used by the present BEIR committee and UNSCEAR (2001; since 1/DD, the relative mutation risk per unit dose, is a smaller fraction with the Japanese DD). It should be stressed that comparison of the DDs alone does not present the correct picture of risks for the following reasons: (1) the Japanese DDs are estimated *retrospectively* from empirical observations using measures of genetic ill health that are totally different from those used by this committee; besides, these measures have not shown any significant differences between the control and radiated groups; and (2) the DD of 1 Gy used by the present committee (and by UNSCEAR 2001) is based on data on mutations in defined genes and is used *prospectively* as one of the four factors in predicting the risk of genetic diseases. Nonetheless, the principal message that emerges from the Japanese epidemiologic studies and the present risk estimates projected from mouse data on radiation-induced mutations is the same—namely, that at low or chronic doses of low-LET irradiation, the genetic risks are very small compared to the baseline risk in the population.

SUMMARY

The principal objective of this chapter is to highlight the ways in which cellular, molecular, and animal data considered in this report may be integrated with epidemiologic findings in order to develop coherent judgments on the health effects of low-LET radiation. Emphasis is placed on data integration for the purposes of modeling these health risks.

The principal conclusions from this work can be summarized as follows:

• Current knowledge of the cellular or molecular mechanisms of radiation tumorigenesis tends to support the application of models that incorporate the ERR projection over time.

• The choice of models for the transport of cancer risk from Japanese A-bomb survivors to the U.S. population is influenced by mechanistic knowledge and information on the etiology of different cancer types. Where specific epidemiologic evidence is lacking, the committee recommends that the weights attaching to relative and absolute risk transport should be 0.7 and 0.3, respectively.

• A combined Bayesian analysis of A-bomb epidemiologic information and experimental data has been employed to provide an estimate of the DDREF for cancer risk. The committee found a believable range of DDREF values to be 1.1 to 2.3 and uses a median value of 1.5 to estimate solid cancer risks.

• Knowledge of adaptive responses that may act to reduce radiation cancer risk was judged to be insufficient to be incorporated in a meaningful way into the modeling of epidemiologic data. The same judgment is made in respect of the possible contribution to cancer risk of postirradiation genomic instability and bystander signaling effects.

• Genetic variation in the population is a potentially important factor in the estimation of radiation cancer risk. Strongly expressing cancer-predisposing mutations are judged from modeling studies to be too rare to distort population-based estimates of risk appreciably but are a significant issue in some medical irradiation settings. The position regarding potentially more common variant genes that express only weakly remains uncertain.

• The estimation of the heritable effects of radiation by the committee takes advantage of new information on human genetic disease and on mechanisms of radiation-induced germline mutations. The application of a new approach to genetic risk estimation leads the committee to conclude that low-dose induced genetic risks are very small compared to baseline risks in the population.

• The committee judges that the balance of evidence from epidemiologic, animal, and mechanistic studies tends to favor a simple proportionate relationship at low doses between radiation dose and cancer risk. Uncertainties in this judgment are recognized and noted.

ANNEX 10A: APPLICATION OF THE MOOLGAVKAR AND KNUDSON TWO-STAGE CLONAL EXPANSION MODEL TO THE TRANSPORT OF RADIATION CANCER RISK

In the case of tumors whose background incidence is strongly influenced by initiating factors, one would expect the EAR to be directly transportable from one population to another. If one considers, for example, the Moolgavkar and Knudson two-stage clonal expansion model (Moolgavkar and Knudson 1981; Moolgavkar and Luebeck 1990) shown in Figure 10A-1, the hazard function $h(t)$ at time t is given approximately by the following formula:

$$h(t) = \mu(t)\int_0^t \{v(s)X(s)\exp\int_s^t [\alpha(u) - \beta(u)]du\}ds,$$

where $\mu(t)$ and $v(t)$ denote, respectively, the first and second mutation rates at time t; $\alpha(t)$ is the rate of division of intermediate (or initiated) cells; and $\beta(t)$ is the rate of death or differentiation of intermediate cells at time t.

If ionizing radiation and the other main risk factors for the tumor of interest are predominantly cancer initiators, their effect would be modeled additively on the first mutation rate μ, as follows:

$$\mu(t, \text{dose} \mid \text{risk factor}) = \mu(t) + \gamma \text{ dose} + \varepsilon \text{ risk factor}.$$

The resulting relative risk would then be of the form:

$$
\begin{aligned}
&RR(t, \text{dose} \mid \text{risk factor}) \\
&= h(t, \text{dose} \mid \text{risk factor}) / h(t \mid \text{risk factor}) \\
&= [\mu(t) + \gamma \text{ dose} + \varepsilon \text{ risk factor}] / [\mu(t) + \varepsilon \text{ risk factor}],
\end{aligned}
$$

while the absolute risk (AR) would be of the form:

$$
\begin{aligned}
&AR\ (\text{dose} \mid \text{risk factor})\ = h(t, \text{dose} \mid \text{risk factor}) - h(t) \\
&= [\mu(t) + \gamma \text{ dose} + \varepsilon \text{ risk factor}] \int_0^t \{v(s)X(s)\exp \\
&\quad \int_s^t [\alpha(u) - \beta(u)]du\}ds - [\mu(t) + \varepsilon \text{ risk factor}] \\
&\quad \int_0^t \{v(s)X(s)\exp\int_s^t [\alpha(u) - \beta(u)]du\}ds \\
&= \gamma \text{ dose}\int_0^t \{v(s)X(s)\exp\int_0^t [\alpha(u) - \beta(u)]du\}ds \\
&= \gamma \text{ dose }\, h(t) / \mu(t) \\
&= \delta(t) \text{ dose } h(t).
\end{aligned}
$$

According to this formulation, the effect of radiation would tend to be independent of the other risk factors on the AR scale. The AR per sievert could therefore be transported from one population to another.

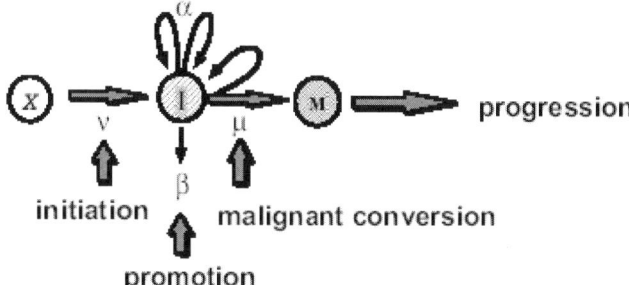

Two-stage clonal expansion model

FIGURE 10A-1 The two-stage clonal expansion model. SOURCE: Luebeck and others (1999).

If, on the contrary, the background incidence of a given tumor type is heavily influenced by host or environmental promoting factors (*e.g.*, breast cancer, stomach cancer), the effects of these factors can be thought to affect the expansion (increasing α to a value α_r, decreasing β to β_r, or both) of the clone of initiated or transformed cells and, thus, the expression of tumors. The resulting relative risk would then be of the form:

$$RR(t, \text{ dose} \mid \text{risk factor})$$

$$= h(t, \text{ dose} \mid \text{risk factor} / h(t \mid \text{risk factor})$$

$$= \frac{(\mu(t) + \gamma \text{ dose}) \int_0^t \{v(s) X(s) \exp \int_s^t [\alpha(u) - \beta(u)] du\} ds}{\mu(t) \int_0^t \{v(s) X(s) \exp \int_s^t [\alpha(u) - \beta(u)] du\} ds}$$

$$= (1 + \gamma / \mu(t) \text{ dose}).$$

This formulation is independent of the magnitude of the effect of promoting factors on the cell division and mortality rates α_r and β_r. Hence the ERR can be exported directly from one population to another.

Expressed in simple terms, low-LET radiation (viewed here as a tumor initiator) will tend to act additively with other tumor initiators and multiplicatively with tumor promoters. Thus, in the case of a radiogenic tumor type with a strong influence of promoters (*e.g.*, stomach cancer), one would favor an RR transportation model, while in the case of a tumor with strong influence of initiators, one would favor an AR transportation model.

The preceding formulations are consistent with more general analyses of the nature of risk relationships involving exposure to two carcinogens (Kodell and others 1991; Zielinski and others 2001).

ANNEX 10B: EVIDENCE FOR THE CONNECTION BETWEEN DOSE EFFECTS AND DOSE-RATE EFFECTS IN ANIMAL EXPERIMENTS

First consider fractionated acute exposures. If the relative risk due to the sum of K acute exposures of equal dose, D / K, administered at separate times, is the sum of the individual relative risks, and if an LQ dose-response model describes the effects at each fraction, then the total relative risk due to all K exposures is

$$RR_{\text{Total}} = K \{\alpha(D / K) + \beta(D / K)^2] = \alpha D + \beta D^2 / K.$$

Thus, for a given total dose D, the importance of the quadratic term diminishes with increasing number of fractions of exposure. The RR due to a protracted exposure may be thought of, at least approximately, as the limit as K approaches infinity. In this way, the total RR due to a protracted exposure is simply αD, where α is the linear coefficient in the LQ model. Therefore, if a risk estimate corresponding to a protracted exposure D is based on an LSS linear model, it

should be adjusted to correspond to the linear component of the estimated LQ model, which is exactly what the DDREF presented in this chapter is designed to do.

Figure 10B-1 shows data from mouse experiments that fitted to the model above (data from Table 6 of Edwards 1992). These data show that the slope in the linear dose-response for chronic exposure approximates the linear component of the LQ model for acute exposure.

Details of DDREF Estimation

An LQ model for ERR or EAR, with curvature constrained to be θ, may be written as $\alpha_{LQ}[Dose + \theta Dose^2]$. A Bayesian statistical analysis is used to update information about dose-response curvature from animal carcinogenesis studies with the information concerning curvature from the LSS cohort of Japanese A-bomb survivors (over the dose range 0–1.5 Sv). A posterior distribution for LSS DDREF follows directly from this, via its definition as a function of θ. The LSS DDREF is essentially $1 + \theta$ for the 0–1.5 Sv dose range and for values of θ of interest here. Pierce and Vaeth (1991) provide a more detailed discussion of this relationship over different dose ranges.

Two forms of animal experiment data were used to estimate curvature: estimated cancer risks and mean survival times (referred to as life-shortening data). These are two different summarizing results from the same experiments, so they are not independent but address the curvature in different ways. LQ models for risk as a function of dose can be estimated for each separate cancer and combined to form a single estimate of curvature, θ. On the other hand, the life-shortening studies ignore cause of death and therefore represent a cumulative effect of all radiation-induced deaths, the majority of which are cancer related. By using the relationship between survival rate and risk, the curvature of interest can be estimated from these, as detailed below.

The estimated risks of relevant cancers, plotted versus radiation dose in Figure 10B-2, were extracted from the summary tables of Edwards (1992), but exclude (1) the results in Tables 1 and 2 because those risk estimates were not adjusted for competing causes of death; (2) results for doses greater than 2 Gy; and (3) results on lymphomas, ovarian cancer, reticulum cell carcinoma, and nonmyeloid leukemias, because these are thought to arise via atypical biological mechanisms, as discussed in Chapter 3, or to reflect an ill-defined combination of cancer types. The risks presented here are based on acute exposures only.

There is substantial evidence that the curvature, θ, is not the same in all 11 situations (*p*-value < .0001, from a likelihood ratio test). Despite this evidence, the model with common curvature explains 97% of the variability and the model with different curvatures explains 98% of the variability in estimated risks, so the practical significance of the different curvatures may not be too important. Note in Figure 10B-2 that although the LQ curves seem to be highly divergent, the

Risks of Lung Adenocarinoma; BALB-c Female Mice

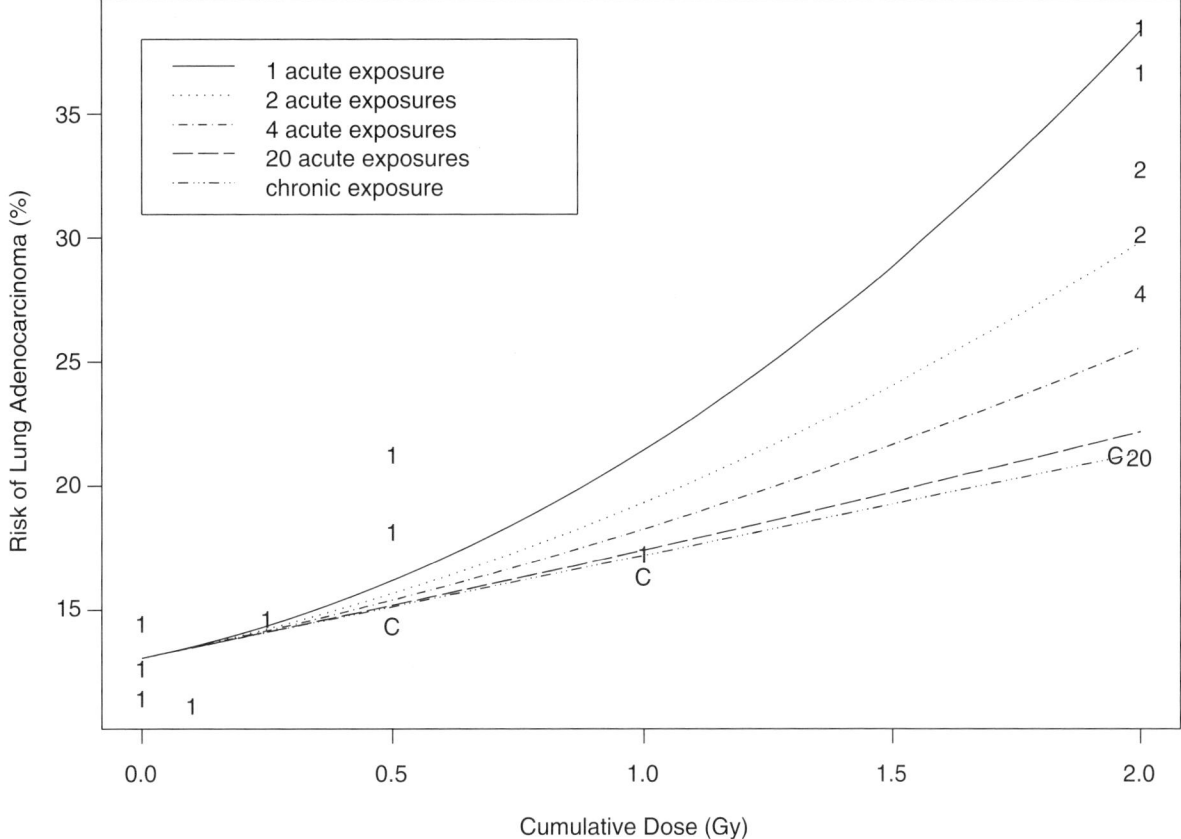

FIGURE 10B-1 Risks of lung cancer versus dose from experiments in which doses were administered fractionally or chronically. Each plotted point corresponds to an estimated risk from one experiment. The plotting symbol shows the number of fractions (number of separate acute exposures) or "C" if the administration was chronic. The curves show an estimated LQ model for risk from dose D administered in K fractions, $\alpha D + \beta D^2 / K$, for five different values of K (where K is taken to be infinity for chronic exposure).

dotted lines tend to intersect most of the error bars—the evidence for different curvatures is not as extreme as it might appear from simple visual inspection.

By acting as if there is a single value of θ, the evidence for it is summarized by the likelihood function labeled "Animal Experiments" in Figure 10B-4. This is a scaled profile likelihood function for θ from a model in which the risk estimates of Figure 10B-2 are normally distributed with variances that are proportional to the reciprocal of the squared standard errors. The means are modeled to depend on the particular condition—corresponding to each of the 11 graphs in Figure 10B-2—with linear-in-dose coefficients that depend on the particular condition and with a quadratic coefficient that is θ divided by the linear coefficient. Thus, the different conditions have different linear and quadratic terms, but the ratio of the quadratic to linear term is held constant.

The "life-shortening" data used here are the mean survival times of mice exposed acutely and chronically to γ-rays

at various doses (Storer and others 1979). Indications of a dose-rate effect from these data stem from the observation that the mean survival times are longer for chronically exposed mice than for acutely exposed mice given the same total dose. However, to extract specific information about curvature, it is necessary to understand the connection between age-specific failure rate and survival time. The human risk models estimated with the LSS data are for age-specific failure rates, also known as hazard functions. Interest here therefore concerns LQ models for hazard functions. Data on mouse survival times may, in principle, be used directly to estimate the hazard function, by employing standard statistical tools of survival analysis, but the unavailability of the raw data precluded this approach by the committee. If the survival times are assumed to follow an exponential probability distribution, then the hazard function is the reciprocal of the mean survival time. By using this exponential assumption (which is probably incorrect but useful nonetheless for extracting information about curvature, at least roughly), the

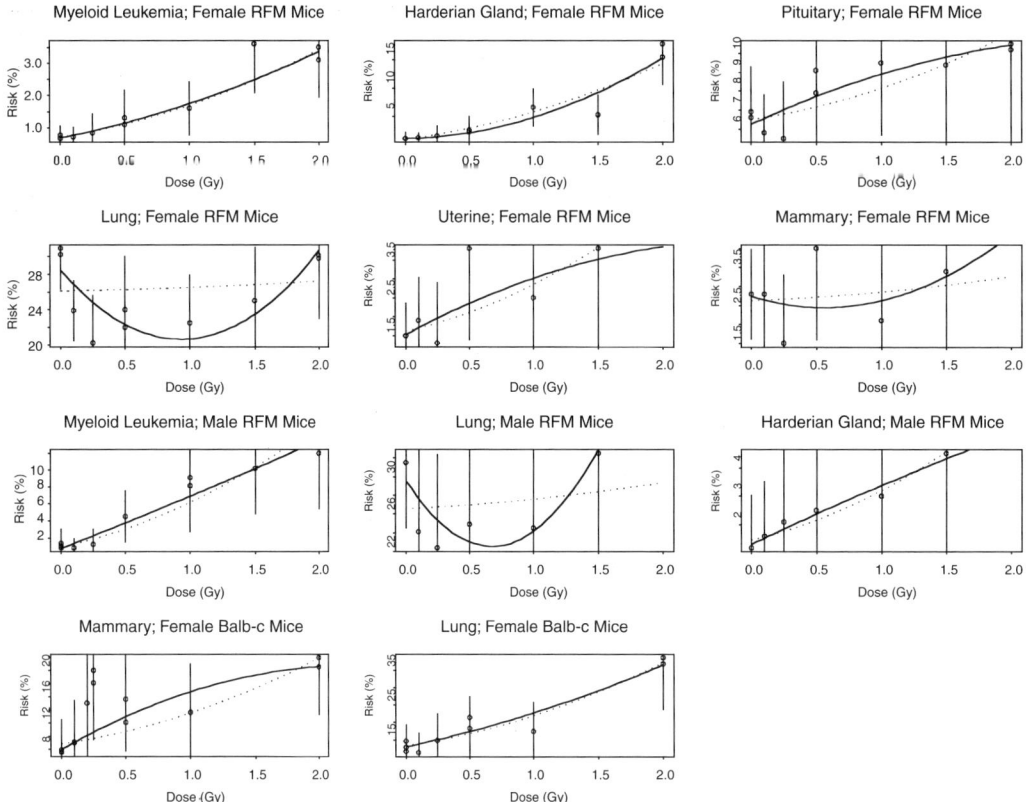

FIGURE 10B-2 Estimated risk of cancer versus radiation dose from various mouse experiments. SOURCE: Data from A.A. Edwards (1992) for cancer site, mouse strain, and sex combinations. Vertical bars extend two standard errors above and below each estimate. Solid curves are estimated LQ models based on each condition individually. Dotted curves are the best-fitting LQ models when curvature is constrained to be the same for all 11 conditions.

curvature of interest can be ascertained by fitting an LQ model to the reciprocal of the mean survival.

Figure 10B-3 shows the reciprocal mean survival times plotted versus dose, with different plotting characters for means based on acutely and chronically exposed mice. Also shown on the plot are the fits to the model that has the age-specific death rate equal to a constant plus $\alpha Dose$ for chronically exposed mice and the same constant plus $\alpha(Dose + \theta Dose^2)$ for acutely exposed mice, following the reasoning in the first section of this Annex. (The estimates are maximum likelihood estimates based on normality of the reciprocal means, which are estimates from a large number of mice.) The estimates depend highly on the dose range considered. This presents a difficulty since leveling off of the dose-response is expected (as shown in Figure 10-1), but the dose at which leveling off occurs is difficult to determine, both theoretically and empirically. The decision was made by the committee to use the 0–1.5 Gy dose range, but this is subjective and open to debate.

The (profile) likelihood function for θ is shown in Figure 10B-4. It is evident that the life-shortening data indicate slightly more curvature than the direct cancer risk results. While it is appropriate to multiply two likelihoods from in-

dependent data sets, these data sets are not independent. Instead, an average of the two is obtained, shown as the solid curve in Figure 10B-4, to represent an average effect based on the two ways of dealing with the data. The maximum likelihood estimate of θ from the average likelihood is 0.5, corresponding to an LSS DDREF of 1.5.

Evidence of curvature at the cellular level comes primarily from studies of chromosomal aberrations in human cells. Table 10B-1 shows estimated LQ models for the regression of chromosome aberration induction on dose. These results may be included weakly, by specifying a probability distribution with mean and variance equal to the sample mean and sample variance of the three curvatures in the table. The result of including such a distribution in the averaging of Figure 10B-4 is to increase the width of the resulting average likelihood, with little effect on the center of the distribution. Since they do not alter the results and because of the extra theoretical demand in incorporating cellular data into models for human cancer rates, chromosome aberration data were not included in the analysis.

The final step in the LSS DDREF estimation involves combining animal radiobiological information with LSS information about curvature. The average likelihood in Fig-

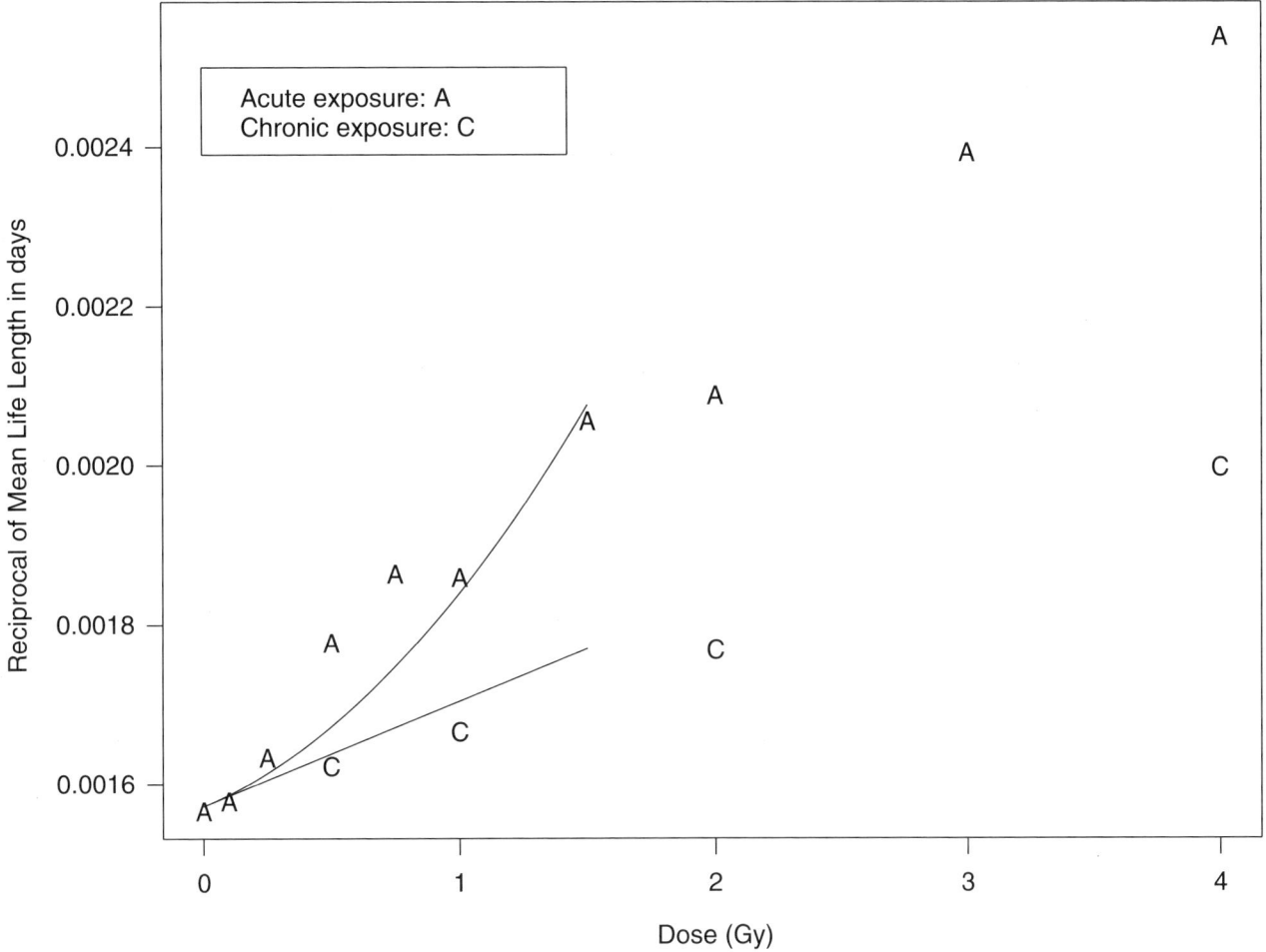

FIGURE 10B-3 Life-shortening data from Tables 1, 2, and 3 of Storer and others (1979). Plotted are the reciprocals of the mean life lengths of RFM female mice versus the dose of exposure, with different plotting symbols for acute (A) and chronically (C) exposed groups. The curves are the result of estimation of an LQ model for age-specific mortality rate fit to the 0–1.5 Gy dose range.

10B-4 is used as a "Bayesian prior distribution" in a Bayesian analysis. The resulting "posterior distribution" that results when this prior is multiplied by the LSS likelihood function is shown in Figure 10-3. The posterior density in the graph indicates the likely values of curvature and corresponding values of the LSS DDREF.

It should be evident that further study beyond that accomplished here could possibly lead to a better summarization of radiobiological information about curvature than provided in Figure 10B-4. Even a more thorough study, however, would similarly be obstructed by the subjectivity involved in the choice of dose range upon which LQ models are fit, by the inconsistency of animal experiment results, and by the difficulty in translating mouse results to human cancer rates.

The posterior density can be used directly to describe the uncertainty in LSS DDREF for the uncertainty analysis in Chapter 12. This distribution probably understates the uncertainty in knowledge of LSS DDREF, however, because of the various subjective choices involved. In an attempt to be more realistic about the state of knowledge of LSS DDREF, an inflated variance (of the distribution of the logarithm of LSS DDREF) is used in the uncertainty analysis.

TABLE 10B-1 Estimates of Linear and Quadratic Coefficients from Chromosome Aberration Induction Studies and the Implicit Curvatures

Human Cell Type	Radiation	α	β	θ
Lymphocytes[a]	Cobalt-60	0.015	0.06	4.0
Lymphocytes[a]	250 kV X-rays	0.04	0.06	1.5
Primary human fibroblasts[b]	Cesium-137 (acute)	0.059	0.029	0.5
Primary human fibroblasts[b]	Cesium-137 (chronic)	0.019		

[a]From Lloyd and others (1992).
[b]From Cornforth and others (2002).

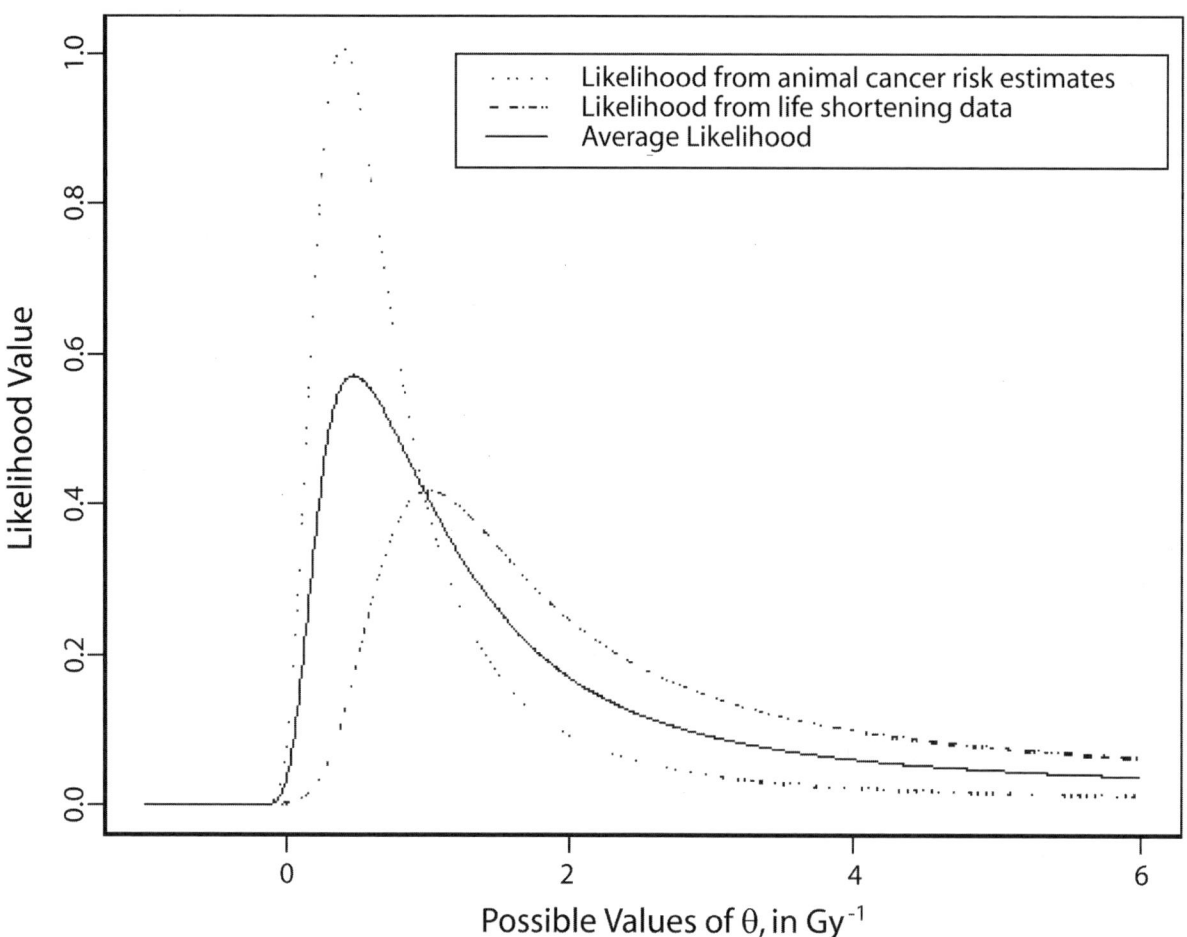

FIGURE 10B-4 A summary of radiobiological evidence for curvature: the (profile) likelihood for curvature from the risk estimate data in Figure 10B-2, the (profile) likelihood from the life shortening data in Figure 10B-3, and their average.

11

Risk Assessment Models and Methods

RISK ASSESSMENT METHODOLOGY

The occurrence of cancers is known to be related to a number of factors, including age, sex, time, and ethnicity, as well as exposure to environmental agents such as ionizing radiation. Understanding the role of exposure in the occurrence of cancer in the presence of modifying effects is a difficult problem. Contributing to the difficulty are the stochastic nature of cancer occurrence, both background and exposure related, and the fact that radiogenic cancers are indistinguishable from nonradiogenic cancers.

This section summarizes the theory, principles, and methods of risk assessment epidemiology for studying exposure-disease relationships. The two essential components of risk assessment are a measure of exposure and a measure of disease occurrence. Measuring exposure to radiation is a challenging problem, and dosimetry issues are discussed in detail elsewhere in this report; the common epidemiologic measures of disease occurrence are reviewed in this section. Evaluation of the association between exposure and disease occurrence is aided by the use of statistical models, and the types of models commonly used in radiation epidemiology are described below, as are the methods for fitting the models to data. This section ends with a description of the use of fitted models for estimating probabilities of causation and certain measures of lifetime detriment associated with exposure to ionizing radiation.

Rates, Risks, and Probability Models

Some individuals exposed to environmental carcinogens (*e.g.*, ionizing radiation) develop cancer and some do not; the same is true of unexposed individuals. Thus, cancer is not a necessary consequence of exposure, and exposure is not necessary for cancer. However, the greater incidence of cancer in individuals exposed to known carcinogens indicates that the probability or risk of developing cancer is in-

creased by exposure. Compared to unexposed individuals, the elevated risks of exposed individuals are manifest by increased cancer rates in the latter group. Risks and rates are the basic measures used to compare disease occurrence in exposed and unexposed individuals. This section describes rates and risks and their relationship to one another as a prelude to the sections on modeling and model fitting.

Incidence Rate

A common measure of disease occurrence used in cancer epidemiology is the *incidence rate*. Incidence refers to new cases of disease occurring among previously unaffected individuals. The population incidence rate is the number of new cases of the disease occurring in the population in a specified time interval divided by the sum of observation times, in that interval, on all individuals who were disease free at the beginning of the time interval. In general an incidence rate is time dependent and depends on both the starting point and the length of the interval.

With data from studies in which subjects are followed over time, incidence rates can be estimated by partitioning the following period into intervals of lengths L_j having midpoints t_j for $j = 1,...,J$, and estimating a rate for each interval. Let n_j denote the number of individuals who are disease free and still under observation at time t_j, and d_j the number of new diagnoses during the jth interval. An estimate of the incidence rate at time t_j is obtained by dividing d_j by the product of n_j and L_j:

$$\hat{\lambda}(t_j) = \frac{d_j}{n_j L_j}.$$

The denominator in $\hat{\lambda}(t_j)$ is an approximation to the sum of observation times on the n_j population members in the jth interval and in practice is usually replaced by the actual observation time, which accounts for the fact that the d_j diagnoses of disease did not occur exactly at time t_j.

Risk

Risk is defined as the probability that an individual develops a specified disease over a specified interval of time, given that the individual is alive and disease free at the start of the time period. As with the incidence rate, risk is time dependent and depends on both the starting point and the length of the interval. In a longitudinal follow-up study as described above, the proportion of new occurrences d_j among n_j disease-free individuals still under observation at time t_j,

$$\hat{p}(t_j) = \frac{d_j}{n_j},$$

is an estimate of the risk or probability of disease occurrence in the jth time interval.

Incidence rates and risks are related via the general formula, risk = rate × time. For the longitudinal follow-up study estimates defined above, the relationship is manifest by the equation

$$\hat{p}(t_j) = \hat{\lambda}(t_j)L_j.$$

Probability Models

The description of rates and risks in terms of estimates from a longitudinal follow-up study is informative and clearly indicates the relevance of these numerical quantities to the study of disease. However, the development of a general theory of risk and risk estimation requires definitions of rates and risks that are not tied to particular types of studies or methods of estimation. Probability models provide a mathematical framework for studying incidence rates and risks and also are used in defining statistical methods of estimation depending on the type of study and the data available.

Models for studying the relationship between disease and exposure are usually formulated in terms of the *instantaneous incidence rate*, which is the theoretical counterpart of the incidence rate estimate defined below. The instantaneous incident rate is defined in terms of the probability distribution function $F(t)$ of the time to disease occurrence. That is, $F(t)$ represents the probability that an individual develops the disease of interest in the interval of time $(0, t)$. Two functions derived from $F(t)$ are used to define the instantaneous incidence rate. One is the survivor function, which is the probability of being disease free throughout the interval $(0, t)$ and is equal to $1 - F(t)$. The second is the probability density function, which is the derivative of $F(t)$ with respect to t, that is, $f(t) = (d / dt)F(t)$, and measures the rate of increase in $F(t)$. The instantaneous incidence rate, also known as the hazard function, is the ratio

$$\lambda(t) = \frac{f(t)}{1 - F(t)}.$$

Integrating the instantaneous incident rate yields the *cumulative incidence rate*

$$\Lambda(t) = \int_0^t \lambda(u)du.$$

The cumulative incidence rate and the distribution function satisfy the relationship

$$F(t) = 1 - \exp\{-\Lambda(t)\}, \qquad (11\text{-}1)$$

from which it follows that the instantaneous incidence rate completely determines the first-occurrence distribution $F(t)$.

The risk of first disease occurrence in the interval $(t, t + h)$, given no previous occurrence, is the conditional probability

$$p(t, t + h) = \frac{F(t + h) - F(t)}{1 - F(t)}.$$

When h is not too large, so that the difference quotient $\{F(t + h) - F(t)\} / h$ approximates $f(t) = dF(t) / dt$,

$$p(t, t + h) = \frac{F(t + h) - F(t)}{h} \frac{h}{1 - F(t)} \approx \frac{f(t)}{1 - F(t)} h = \lambda(t)h.$$

Thus, among individuals who are disease free at time t, the risk of disease in the interval $(t, t + h)$ is approximately $\lambda(t)h$. This approximation is the theoretical counterpart of the relationship between risks and rates described in the discussion of risk. In the remainder of this chapter, incidence rate means instantaneous incidence rate unless explicitly noted otherwise.

Incidence Rates and Excess Risks

It is clear that the incidence rate plays an important role in the stochastic modeling of disease occurrence. Consequently, models and methods for studying the dependence of disease occurrence on exposure are generally formulated in terms of incidence rates. In the following it is assumed that individuals have been stratified on the basis of age, sex, calendar time, and possibly other factors related to disease occurrence, and that incidence rates are stratum specific. In the simple case of two exposure categories, exposed and unexposed, let $\lambda_E(t)$ and $\lambda_U(t)$ denote the incidence rates of the exposed and unexposed groups, respectively. If disease occurrence is unrelated to exposure, one expects that $\lambda_E(t) = \lambda_U(t)$, whereas lack of equality between these two incidence rates indicates an association between disease occurrence and exposure.

A common measure of discrepancy between incidence rates is the difference

$$\text{EAR}(t) = \lambda_E(t) - \lambda_U(t),$$

which by convention is called the *excess absolute risk* (EAR) even though it is, technically, a difference in rates. Rearranging terms results in

$$\lambda_E(t) = \lambda_U(t) + \text{EAR}(t),$$

showing that EAR(t) describes the additive increase in incidence rate associated with exposure. For example, if the EAR is constant, EAR(t) = b, then the effect of exposure is to increase the incidence rate by the constant amount b for all time periods. Note that $b = 0$ corresponds to the case of no association.

A second common measure of discrepancy is the *relative risk* (RR), defined as

$$RR(t) = \frac{\lambda_E(t)}{\lambda_U(t)}.$$

Rearranging terms shows that

$$\lambda_E(t) = RR(t)\lambda_U(t),$$

so that RR(t) describes the multiplicative increase in incidence rate associated with exposure. When the RR is constant, RR(t) = r, the effect of exposure is to alter incidence rate by the factor r. If exposure increases risk, then $r > 1$; if exposure decreases risk, then $r < 1$, and $r = 1$ corresponds to the case of no association. The *excess relative risk* ERR(t) is

$$ERR(t) = RR(t) - 1.$$

The ERR of the exposed and unexposed incidence rates are related via the equation

$$\lambda_E(t) = \lambda_U(t) \{1 + ERR(t)\}.$$

RISK MODELS

Direct Estimates of Risk

The previous section defined the fundamental quantities used in risk estimation: risks, rates, EAR, RR, and ERR, and established their relevance to the study of environmental carcinogens. These measures enable the study of differences in disease occurrence in relationship to time, by studying either EAR(t) or ERR(t) between unexposed and exposed groups. For most carcinogens, exposure is not a simple dichotomy (unexposed, exposed) but occurs on a continuum. That is, the exposure or dose d can vary from no exposure ($d = 0$) upward. In such cases the relationship between risk—or EAR(t) or ERR(t)—and dose is of fundamental importance. For all carcinogens it is generally agreed that sufficiently large doses increase the risk of cancer. By definition there is no increase in risk in the absence of exposure ($d = 0$). That is, when $d = 0$, both EAR(t) = 0 and ERR(t) = 0. Thus, for many carcinogens the only open or unresolved issue is the dependence of risk on small or low doses. Low-dose ranges are often the most relevant in terms of numbers of exposed individuals. They are also the most difficult ranges for which to obtain unequivocal evidence of increased risk. These difficulties result from the fact that small increases in risk associated with low levels of exposure are difficult to

detect (using statistical methods) in the presence of background risks.

The difficulties can be seen by considering the estimates of risk from the longitudinal follow-up study described in "Rates, Risks, and Probability Models." For a time period L_j, let $n_{j,E}$, $d_{j,E}$ and $n_{j,U}$, $d_{j,U}$ denote the number of individuals at risk at the start of the interval and the number of occurrences of disease during the interval for the exposed and unexposed subgroups, respectively. A direct estimate of the excess risk for the jth time period is the difference between two proportions $(d_{j,E} / n_{j,E}) - (d_{j,U} / n_{j,U})$. Even in the favorable situation in which the baseline risk is relatively well estimated compared to the risk of the exposed group (when $n_{j,U}$ is large relative to $n_{j,E}$), the ability to reliably detect small increases in risk associated with exposure requires a large number of exposed individuals at risk. For example, using the usual criterion for statistical testing in order to detect with probability .80 a 5% increase in risk when the baseline risk is 0.10, the number of individuals at risk in the exposed group would have to be approximately $n_{j,E}$ = 30,000.

A key objective of this report is the calculation of quantitative estimates of human health risks (*e.g.*, cancer) associated with exposure to ionizing radiation for specific subpopulations defined by stratification on variables such as sex, age, exposure profile, and smoking history. In theory, such estimates could be derived by identifying a large group of individuals having common exposure profiles within each stratum and following the groups over a long period of time. As described above, the proportion of individuals in each group who develop cancer in specific time periods provides the desired estimates of risk. However, this approach is not feasible because sufficient data are not available. At low levels of exposure, cancer risks associated with exposure are small relative to baseline or background risks. The increases in observed cancer rates associated with exposure are small relative to the natural random fluctuations in baseline cancer rates. Thus, very large groups of individuals would have to be followed for very long periods of time to provide sufficiently precise estimates of risk associated with exposure. Consequently, direct estimates of risk are not possible for stratified subpopulations. The alternative is to use mathematical models for risk as functions of dose and stratifying variables such as sex and age.

Estimation via Mathematical Models for Risk

Model-based estimation provides a feasible alternative to direct estimation. Model-based estimates efficiently exploit the information in the available data and provide a means of deriving estimates for strata and dose profile combinations for which data are sparse. This is accomplished by exploiting assumptions about the functional form of a risk model. Of course, the validity of estimates derived from models depends on the appropriateness of the model; thus model choice is important. The accepted approach in radiation epidemiology is to base models on radiobiological principles and theories of carcinogenesis to the fullest extent possible, keeping in mind statistical limitations imposed by the quantity and quality of data available for model fitting. Biologically based and empirically derived mathematical models for risk are discussed in the next two sections.

Biologically Based Risk Models

Biologically based risk models are designed to describe the fundamental biological processes involved in the transformation of somatic cells into malignant cancer cells. The use of biologically based risk models in epidemiologic analyses can result in a greater understanding of the mechanisms of carcinogenesis. These models can also help to expose the complex interrelationships between different time- and age-dependent exposure patterns and cancer risk. Biologically based risk models provide an analytical method that is complementary to the traditional, well-established, empirical approaches.

Armitage and Doll (1954) observed that for many human cancers the log-log plot of age-specific incidence rates versus age is nearly linear, up to moderately old ages. This observation has led to the development of models for carcinogenesis. In brief, Armitage and Doll's theory postulates that malignant transformation occurs following the *k*th stage of a series of spontaneous and irreversible changes (Armitage 1985). The corresponding hazard function is of the form $\lambda(t) = at^{k-1}$, where *t* denotes time and *a* is a constant reflecting the dependence of the hazard on the number of stages, *k*. These models have been fit to various data sets, leading to the observation that most cancers arise after the occurrence of five to seven stages. Comprehensive reviews of the mathematical theory of carcinogenesis have been given by Armitage and Doll (1961), Whittemore (1978), and Armitage (1985).

In response to the multiplicity of parameters produced by their earlier models, Armitage and Doll proposed a simpler two-stage model designed to avoid parameters not readily estimable from available data. A major limitation of these early two-stage models is their failure to address the multiplication and death of normal cells, which was known to occur in tissues undergoing malignant change (Moolgavkar and Knudson 1981). A revised two-stage model was later proposed by Moolgavkar and colleagues, which allowed for the growth of normal tissue and the clonal expansion of intermediate cells (Moolgavkar and Knudson 1981). Numerous two-stage models have since been described in the literature (Fisher 1985; Moolgavkar 1991; Sielken and others 1994; Luebeck and others 1996; Heidenreich and others 1999, 2002a, 2002b; Moolgavkar and others 1999; Heidenreich and Paretzke 2001; Moolgavkar and Luebeck 2003).

The two-stage clonal expansion (TSCE) model assumes a normal stem cell population of fixed size *X* and a rate of first mutation of *v(d)*, depending on the dose *d* of the carcinogen.

The number of initiated cells arising from the normal cell pool is described by a Poisson process with a rate of *vX*. The initiated cells then divide either symmetrically or non-symmetrically. Symmetrical division results in two initiated cells, while nonsymmetrical division results in an initiated cell and a differentiated cell. The rate of symmetrical division is designated by $\alpha(t)$, and the death differentiation rate by $\beta(t)$. The difference $\alpha - \beta$ is the net proliferation rate for initiated cells. The rate of division into one initiated cell and one malignant cell is designated by $\mu(t)$ (Hazleton and others 2001).

TSCE models for radiation carcinogenesis have now been applied successfully to a number of important data sets, including atomic bomb survivors (Kai and others 1997) and occupational groups such as nuclear power plant workers and miners (Moolgavkar and others 1993; Luebeck and others 1999; Sont and others 2001). A study of atomic bomb survivors illustrates the usefulness of the two-stage model in radiation epidemiology (Kai and others 1997). Findings from this analysis include the observation of a high excess risk among children that may not be explained by enhanced tissue sensitivity to radiation exposure. The temporal patterns in cancer risk can be explained in part by a radiation-induced increase in the pool of initiated cells, resulting in a direct dose-rate effect (Kai and others 1997). Exact solutions of the two-stage model (Heidenreich and others 1997) and multi-stage models (Heidenreich and others 2002b) have been applied to atomic bomb survivors' data.

Another data set to which application of the TSCE has been useful is the National Dose Registry (NDR) of Canada. This database contains personal dosimetry records for workers exposed to ionizing radiation since 1951, with current records for more than 500,000 Canadians (Ashmore and others 1998). Application of the TSCE model to the NDR suggests an explanation of the apparently high excess relative risk observed, relative to the A-bomb data (Sont and others 2001). The TCSE model reveals that the dose-response for the NDR cohort is consistent with the lung cancer incidence in the A-bomb survivors' cohort, provided that proper adjustments are made for the duration of exposure and differences in the background rate parameters.

In addition to the TSCE model, the Armitage-Doll model of carcinogenesis has evolved into several other analytic methods, including the general mutagen model (Pierce 2002). The basic assumption of this model is that a malignant cell results from the accumulation of mutations, with *k* mutations required for malignancy. The effect of exposure is that an increment of dose at age *a*, at rate *d(a)*, results in a multiplicative increase $\lambda_r[1 + \beta d(a)]$ in the rate of all *k* mutations. Although this model applies to both recessive and dominant mutations, it does not explicitly allow for selective proliferation of cells having only some of the required mutations. The general mutagen model has been applied successfully to A-bomb survivor data (Pierce and Mendelsohn 1999; Pierce and Preston 2000) and to underground miners exposed to radon (Lubin and others 1995).

Whereas empirical approaches to risk modeling rely on statistical models to describe data, biologically based models depend on fundamental assumptions regarding the mechanisms of radiation carcinogenesis. The parameters created by modern biologically based risk models have direct biological interpretation, provide insight into cancer mechanisms, and generate substantive questions about the pathways by which exposure to ionizing radiation can increase cancer risk. These models also provide a way of describing temporal patterns of exposure and risk.

Although biologically based risk models have many strengths, some general limitations are associated with their use. Such models can only approximate biological reality and require an understanding of the complex mechanisms of radiation carcinogenesis for interpretation. In addition, it is difficult to distinguish among alternative models that yield similar dose-response curves without direct information on the fundamental biological processes represented by the model, which are often unknown. Biologically based risk models are generally more complex than empirical models and may require richer databases to develop properly. Despite these limitations, biologically based models have found many applications for important epidemiologic data sets, and the successes achieved to date afford support for the continual development of such models for future analyses that will directly inform the association between radiation exposure and human cancer risk.

Biologically based models have not been employed as the primary method of analysis in this report for several reasons. The mechanisms of radiation carcinogenesis are not fully understood, which makes the development of a fully biologically based model difficult. The data required for a biologically based model, such as rates of cell proliferation and mutation, are also generally not available. The availability of empirical risk models that provide a good description of the available data on radiation and cancer permits the preparation of useful risk projection.

Empirically Based Risk Models

The following symbols are used to describe the variables that enter into risk models based on the Japanese A-bomb survivor data:

- a: attained age of an individual
- e: age at exposure to radiation
- d: dose of radiation received
- s: code for sex (1 if the individual is a female and 0 if male)
- p: study population-specific factors

Models also sometimes include time since exposure (t). Since $t = a - e$, models that include a and e implicitly include t.

Models for the incidence rate for individuals of age a, exposed to dose d, at age e, generally depend on sex s (1 for females, 0 for males) and other study population-specific factors generically represented by p. For example, the study population-specific parameters for A-bomb survivor data models are city c and calendar year y, that is, $p = (c, y)$. The incidence rate is, in general, a function $\lambda(a, e, d, s, p)$ of all of these factors. By definition, the background incidence rate does not depend on either d or e, so the EAR formulation of the exposed incidence rate has the form

$$\lambda(a, e, d, s, p) = \lambda(a, s, p) + \text{EAR}(a, e, d, s, p),$$

and the ERR formulation is

$$\lambda(a, e, d, s, p) = \lambda(a, s, p) \{1 + \text{ERR} (a, e, d, s, p)\},$$

where EAR (a, e, d, s, p) and ERR (a, e, d, s, p) are the EAR and ERR, respectively. When the excess risk functions are dependent on the study population—that is, when they depend on the factor p—estimates of risk derived from the models are specific to the study population and therefore of limited utility for estimating risks in other populations. Thus, it is desirable to find suitable models in which either the excess risk or the excess relative risk does not depend on population-specific parameters. Consequently, models used in radiation risk estimation are often of the form

$$\lambda(a, e, d, s, p) = \lambda(a, s, p) + \text{EAR}(a, e, d, s)$$

or

$$\lambda(a, e, d, s, p) = \lambda(a, s, p) \{1 + \text{ERR} (a, e, d, s)\}.$$

That is, the excess risk functions depend only on a, e, d, and s, but not p. Note that if t represents time after exposure, then because $t = a - e$, any two of the variables t, a, and e determine the third, so at the current level of generality, the excess risk functions could also be written as functions of t, e, d, and s. Also, because there is no excess risk at ages prior to exposure ($a < e$), ER$(a, e, d, s) = 0$ ($a < e$), EAR$(a, e, d, s) = 0$ and ERR$(a, e, d, s) = 0$ for $a < e$ and thus, $\lambda(a, e, d, s, p) = \lambda(a, s, p)$ for $a < e$. The formulas and equations in the remainder of this chapter are described only for the relevant case $a \geq e$.

Radiobiological considerations suggest that for low-dose, low-LET (linear energy transfer) radiation, the risk of disease for an individual exposed to dose d depends on a linear or quadratic function of d. That is, risk depends on dose d through a function of the form

$$f(d) = \alpha_1 d + \alpha_2 d^2,$$

where α_1 and α_2 are parameters to be estimated from the data. At higher doses of radiation, cell sterilization and cell death compete with the process of malignant transformation, thereby attenuating the risk of cancer at higher doses. A more general model applicable to a broader dose range and used extensively in radiation research is

$$f(d) = \alpha_1 d + a_2 d^2) \exp(-\alpha_3 d - \alpha_4 d^2).$$

The models for dependence on dose are generally incorporated into risk models by assuming that the excess risk functions are proportional to $f(d)$, where the multiplicative constant (in dose) depends on a, e, and s.

VARIABLES THAT MODIFY THE DOSE-RESPONSE RELATIONSHIP

In general, cancer rates vary considerably as functions of attained age, and there is strong evidence indicating that cancer risks associated with radiation exposure also vary as functions of attained age and age at exposure. For example, it has been observed that after instantaneous exposure to radiation, leukemia and bone cancer rates rise for a short period of time ($\approx$ years) and then decrease to baseline rates over a longer period of time ($\approx$ years). In contrast, the available evidence suggests, and it is generally believed, that rates for most other cancers increase after exposure to radiation and possibly remain at elevated levels at all ages.

Models for the dependence of risk on variables such as age at exposure, attained age, and time since exposure are often empirical and are justified more by epidemiologic and statistical principles than by radiobiological theory. A useful class of models that includes the modifying effects on radiation dose-response of attained age, age at exposure, and gender has the form

$$\lambda(a, e, d, s, p) = \lambda(a, s, p) + f(d)g(a, e, s);$$

for EAR models, and

$$\lambda(a, e, d, s, p) = \lambda(a, s, p)\{1 + f(d)g(a, e, s)\};$$

for ERR models, where $g(a, e, s)$ is a function of attained age, age at exposure, and gender. Because time since exposure is equal to the difference $t = a - e$, this class of models includes models defined as functions of time since exposure. Often g depends on e and t via exponential and power functions.

For example, the committee's preferred model for solid cancer uses

$$g(a, e, s) = \exp(\gamma \tilde{e} + \eta \ln(a) + \theta s),$$

where $\tilde{e}$ is $e - 30$ years if e is less than 30, and 0 if e is greater than or equal to 30; and γ, η, and θ are unknown parameters, which must be estimated from the data.

Model Parameter Estimation

Models describe the mathematical form of a risk function, but the parameters in the model must be estimated from data. For example, a linear dose model presupposes that risk increases linearly with dose but the slope of the line, which measures the increase in risk for a unit increase in dose, must be estimated from data. Similarly, models for the effect of

modifying factors depend on parameters that must be estimated from data. The most common method of fitting risk model data (*i.e.*, estimating the unknown parameters in the model) is the method of maximum likelihood reference. Given a model for the probability density of the observed data, a likelihood is obtained by evaluating the density at the observed data. The likelihood is a function of the data and the unknown parameters in the probability density model. The parameters are estimated by those values in the parameter space (the set of all allowable parameter values) that maximize the likelihood for the given data values.

There are several approaches for the numerical calculations of likelihood analysis. Estimation based on grouped data using a Poisson form of the likelihood (Clayton and Hills 1993) has been used for the analyses of atomic bomb survivors and other major epidemiologic studies of radiation health risks.

This analysis is facilitated by forming a table so that individuals contributing information to each cell of the table have equal, or approximately equal, background rates. In particular, the table is formed by the cross-classification of individuals into categories of age at exposure, time period, exposure dose, and all other variables that appear in the model. The key summary variables required for each cell are the total person-years (PY) of observation in the cell, the number of new cases of cancer, the mean dose, the mean age at exposure, and the mean age or mean time since exposure.

For an RR model, the contribution to the likelihood from the data in each cell of the table has the same form as a Poisson likelihood (thus permitting well-understood and straightforward computations), with the mean equal to the product of PY; a parameter for the common, cell-specific background rate; and the RR $1 + fg$, where f and g are functions of dose and of age, age at exposure, and sex, described previously.

The full likelihood is the product of the cell-specific Poisson likelihoods. Numerical optimization is required to maximize the likelihood, and statistical inference generally is based on large-sample approximations for maximum likelihood estimation.

Using the Estimated Model

The models developed as described above can be used to estimate both lifetime risks and probabilities of causation, both of which are discussed below. Following this, several limitations in the use of these models, which lead to uncertainties in estimated risks, are discussed. Further discussion of uncertainties and the committee's approach to quantifying them can be found in Chapter 12.

Estimating Lifetime Risks

To calculate the lifetime risk for a particular age at exposure and a particular gender, one essentially follows a subject forward in time and calculates the risk of developing a radiation-induced cancer at each age subsequent to age at exposure. This requires probabilities of survival to each subsequent age, which are obtained from life tables for the population of interest. ERR models are expressed in terms of a relative increase in the sex- and age-specific background rates for the cancer of interest; these rates are usually obtained from cancer mortality vital statistics for the population of interest (or incidence rates if cancer incidence is to be estimated).

An important issue in estimating lifetime risks is the extrapolation of risks beyond the period for which follow-up data are available. No population has been followed for more than 40 or 50 years; thus, it is not possible to model the EAR or ERR directly for the period after follow-up has ended, a limitation that is primarily important for those exposed early in life. Estimating lifetime risks for this group thus requires assumptions that are usually based on the observed pattern of risk over the period for which data are available. For example, if the ERR appears to be a constant function of time since exposure, it may be reasonable to assume that it remains constant. Alternatively, if the EAR or ERR has declined to nearly zero by the end of the follow-up period, it may be reasonable to assume that the risk remains at zero.

Another important issue is how to apply risks estimated from studying a particular exposed population to another population that may have different characteristics and different background risks. Specifically, the application of estimates based on Japanese atomic bomb survivors to a U.S. population is a concern, since background rates for some specific cancers (including stomach, colon, liver, lung, and breast) differ substantially between the two populations. The BEIR V (NRC 1990) committee calculations were based on the assumption that relative risks (ERR) were comparable for different populations; however, the BEIR III (NRC 1980) committee modified its ERR models based on the assumption that absolute risks were comparable. Some recent efforts have used intermediate approaches with allowance for considerable uncertainty (NIH 1985, 2003).

Estimating Probabilities of Causation

The *probability of causation* (PC; NIH 1985, 2003) is defined as the ratio of ERR to RR:

$$PC = \frac{ERR}{1 + ERR}$$

where for brevity the dependence of ERR on dose, time variables, and possibly other individual characteristics is suppressed. For the RR models described previously, ERR $= fg$, where $f = f(d)$ and $g = g(a, e, s)$, in which case

$$PC = \frac{fg}{1 + fg}.$$

Thus, the ERR model provides immediate PC estimates.

Modeling Caveats

The theory of risk assessment, modeling, and estimation and the computational software for deriving statistically sound parameter estimates from data provide a powerful set of tools for calculating risk estimates. Risk models provide the general form of the dependence of risk on dose and risk-modifying factors. Specific risk estimates are obtained by fitting the models (estimating unknown parameters) to data. The role of data in the process of risk estimation cannot be overemphasized. Neither theory, models, nor model-fitting software can overcome limitations in the data from which risk estimates are derived. In human epidemiologic studies of radiation, both the quality and the quantity of the data available for risk modeling are limiting factors in the estimation of human cancer risks. The quality of data, or lack thereof, and its impact on risk modeling are discussed below under three broad headings. The primary consequence of less-than-ideal data is uncertainty in estimates derived from such data.

Incomplete Covariate Information

The specificity of risk models is limited by the information available in the data. Even the most extensive data sets contain, in addition to measurements of exposure, information on only a handful of predictor variables such as dose, age, age at exposure, and sex. Consequently, models fit to such data predict the same risk of cancer for individuals having the same values of these predictor variables, regardless of other differences between the two individuals. For example, two individuals who differ with respect to overall health status, family history of cancer (genetic disposition to cancer), exposure to other carcinogens, and so on, will be assigned the same estimated risk provided they were exposed to the same dose of radiation, are of the same age, and have the same age at exposure and the same gender.

Consequently, among a group of individuals having the same values of the predictor variables in the model, some will have a higher personal risk than that predicted by the model and some will have a lower personal risk. However, on average, the group risk will be predicted reasonably well by the model. The situation is similar to the assessment of insurance risk. Not all teenage males have the same personal risk of having an automobile accident (some are better drivers than others), yet as a group they are recognized as having a greater-than-average risk of accidents, and premiums are set accordingly. From the insurance company's perspective, the premiums are set fairly in the sense that their risk models adequately predict the claims experience of the group.

Radiation risk models are similar in that they adequately predict the disease experience of a group of individuals sharing common values of predictor variables in the model. However, such estimated risks need not be representative of individual personal risks.

Estimated Doses

The standard theory and methods of risk modeling and estimation are appropriate under the assumption that dose is measured accurately. Estimated radiation dose is a common characteristic of human epidemiologic data, and questions naturally arise regarding the adequacy of dose esti mates for the estimation of risk parameters and the calculation of risk estimates. These are different problems and are discussed separately.

First, consider the problem of calculating risk estimates from a given risk equation. Suppose that the risk equation has been estimated without bias and with sufficient precision to justify its use in the calculation of risks. Assume also that risk increases with dose: that is, the risk equation yields higher risks for higher doses. Suppose that an estimate of lifetime risk is desired for an individual whose dose is estimated to be d. If d overestimates the individual's true dose, the lifetime risk will be overestimated; if d underestimates the true dose, the risk will be underestimated. This is intuitive and is a consequence of the fact that risk is an increasing function of dose.

The problem of estimating risk equation parameters from data with estimated doses is a little more complicated. Errors in estimated doses can arise in a number of different ways, not all of which have the same impact on risk parameter estimation. For example, flaws in a dosimetry system have the potential to affect all (or many) dose estimates in the same manner, leading to systematic errors for which all (or many) dose estimates are too high or too low. Errors or incomplete records in data from which dose estimates are constructed (*e.g.*, badge data from nuclear industry workers) are likely to result in more or less random errors in dose estimates (*i.e.*, some individuals will have dose estimates that are too high and others will have estimates that are too low). Systematic errors can result in biased estimates of risk equation parameters. The type of bias depends on the nature of the systematic error. For example, risk equations derived from data with doses that are overestimated by a constant factor (>1) will result in an underestimation of risk at a particular given dose d; doses that are underestimated by a constant factor (<1) will result in an overestimation of risk. Random errors in dose estimates also have the potential to bias estimated risk equations. Random error-induced bias generally results in the underestimation of risk. That is, random errors tend to have the same qualitative effect as systematic overestimation of doses.

The estimation of risk models from atomic bomb survivors has been carried out with a statistical technique that accounts for the random uncertainties in nominal doses (Pierce and others 1990). To the extent that it is based on correct assumptions about the forms and sizes of dose uncertainties, it removes the bias due to random dose measurement errors.

Data from Select Populations

Ideally, risk models would be developed from data gathered on individuals selected at random from the population for which risk estimates are desired. For example, in estimating risks for medical workers exposed to radiation on the job, the ideal data set would consist of exposure and health information from a random sample of the population of such workers. However, data on specific populations of interest are generally not available in sufficient quantity or with exposures over a wide enough range to support meaningful statistical modeling. Radiation epidemiology is by necessity opportunistic with regard to the availability of data capable of supporting risk modeling, as indicated by the intense study of A-bomb survivors and victims of the Chernobyl accident.

A consequence of much significance and concern is the fact that risk models are often estimated using data from one population (often not even a random sample) for the purpose of estimating risks in some other population(s). Cross-population extrapolation of this type is referred to as "transporting" the model from one population to another. The potential problem it creates is the obvious one—namely, that a risk equation valid for one population need not be appropriate for another. Just as there are differences in the risk of cancer among males and females and among different age groups, there are differences in cancer risks among different populations. For example, the disparity between baseline rates for certain cancers (*e.g.*, stomach cancer) in Japanese and U.S. populations suggests the possibility of differences in the risks due to radiation exposure.

Transporting models is generally regarded as a necessity, and much thought and effort are expended to ensure that problems of model transportation are minimized. The decision to use EAR models or ERR models is sometimes influenced by concerns of model transport. Problems of transporting models from one population to another can never be eliminated completely. However, to avoid doing so would mean that risk estimates would have to be based on data so sparse as to render estimated risks statistically unreliable.

12

Estimating Cancer Risk

INTRODUCTION

This chapter presents models that allow one to estimate the lifetime risk of cancer resulting from any specified dose of ionizing radiation and applies these models to example exposure scenarios for the U.S. population. Models are developed for estimating lifetime risks of cancer incidence and mortality and take account of sex, age at exposure, dose rate, and other factors. Estimates are given for all solid cancers, leukemia, and cancers of several specific sites. Like previous BEIR reports addressing low-LET (linear energy transfer) radiation, risk models are based primarily from data on Japanese atomic bomb survivors. However, the vast literature on both medically exposed persons and nuclear workers exposed at relatively low doses has been reviewed to evaluate whether findings from these studies are compatible with A-bomb survivor-based models. In many cases, results of fitting models similar to those in this chapter have been published.

Risk estimates are subject to several sources of uncertainty due to inherent limitations in epidemiologic data and in our understanding of exactly how radiation exposure increases the risk of cancer. In addition to statistical uncertainty, the populations and exposures for which risk estimates are needed nearly always differ from those for whom epidemiologic data are available. This means that assumptions are required, many of which involve considerable uncertainty. Risk may depend on the type of cancer, the magnitude of the dose, the quality of the radiation, the dose-rate, the age and sex of the person exposed, exposure to other carcinogens such as tobacco, and other characteristics of the exposed individual. Despite the abundance of epidemiologic and experimental data on the health effects of exposure to radiation, data are not adequate to quantify these dependencies precisely. Uncertainties in the BEIR VII risk models are discussed, and a quantitative assessment of selected sources of uncertainty is made.

In recent years, several national and international organizations have developed models for estimating cancer risk from exposure to low levels of low-LET ionizing radiation. These include the work of the BEIR V committee (NRC 1990), the International Commission on Radiological Protection (ICRP 1991), the National Council on Radiation Protection and Measurements (NCRP 1993), the Environmental Protection Agency (EPA 1994, 1999), the United Nations Scientific Committee on the Effects of Atomic Radiation (UNSCEAR 2000b), and the National Institutes of Health (NIH 2003). The approaches used in these past assessments are described in Annex 12A.

DATA EVALUATED FOR BEIR VII MODELS

As in earlier BEIR reports addressing health effects from exposure to low-LET radiation, the committee's models for risk estimation are based primarily on the Life Span Study (LSS) cohort of survivors of the atomic bombings in Hiroshima and Nagasaki. As discussed in Chapter 6, the LSS cohort offers several advantages for developing quantitative estimates of risk from exposure to ionizing radiation. These include its large size, the inclusion of both sexes and all ages, a wide range of doses that have been estimated for individual subjects, and high-quality mortality and cancer incidence data. In addition, because the exposure was to the whole body, the LSS cohort offers the opportunity to assess risks for cancers of a large number of specific sites and to evaluate the comparability of site-specific risks.

Another consideration in the choice of data was that it was considered essential that the data used by the committee eventually be available to other investigators. The Radiation Effects Research Foundation (RERF) has developed a policy of making summarized data available to those who request it, thus enabling other investigators to analyze data used by the BEIR VII committee. This is not the case for data sets on most other radiation-exposed cohorts.

Although the committee's models have been developed from A-bomb survivor data, attention has been given to their compatibility with data from other cohorts. Fortunately, for most cohorts with suitable data for developing quantitative risk models, analyses based on models similar to those used by the committee have been conducted and published. This facilitated the committee's evaluation of data from other studies. Pooled analyses of thyroid cancer risks (Ron and others 1995a) and of breast cancer risks (Preston and others 2002a) were especially helpful in this regard, as were several meta-analyses by Little and colleagues. In addition, the many published analyses based on A-bomb survivor data have guided and facilitated the committee's efforts in its choice of models. The committee notes particularly the main publications on mortality (Preston and others 2003) and incidence data (Thompson and others 1994) and the models developed by UNSCEAR (2000b) and NIH (2003).

The use of data on persons exposed at low doses and low dose rates merits special mention. Of these studies, the most promising for quantitative risk assessment are the studies of nuclear workers who have been monitored for radiation exposure through the use of personal dosimeters. These studies, which are reviewed in Chapter 8, were not used as the primary source of data for risk modeling principally because of the imprecision of the risk estimates obtained. For example, in a large combined study of nuclear workers in three countries, the estimated relative risk per gray (ERR/Gy) for all cancers other than leukemia was negative, and the confidence interval included negative values and values larger than estimates based on A-bomb survivors (Cardis and others 1995).

Since the publication of BEIR V, data on cancer incidence in the LSS cohort from the Hiroshima Tumor Registry have become available, whereas previously only data from the Nagasaki Tumor Registry were available. Thus, the committee could use both incidence and mortality data to develop its models. The incidence data offer the advantages of including nonfatal cancers and of better diagnostic accuracy. However, the mortality data offer the advantages of covering a longer period (1950–2000) than the incidence data (1958–1998) and of including deaths of LSS members who migrated from Hiroshima and Nagasaki to other parts of Japan.

MEASURES OF RISK AND CHOICE OF CANCER END POINTS

To express the health impact of whole-body exposures to radiation, the lifetime risk of *total* cancer, without distinction as to site, is usually of primary concern. Estimates of risk for both mortality and incidence are of interest, the former because it is the most serious consequence of exposure to radiation and the latter because it reflects public health impact more fully. The time or age of cancer occurrence is also of interest, and for this reason, estimates of cancer mortality risks are sometimes accompanied by estimates of the years of life lost or years of life lost per death. Because leukemia exhibits markedly different patterns of

risk with time since exposure and other variables, and also because the excess relative risk for leukemia is clearly greater than that for solid cancers, all recent risk assessments have provided separate models and estimates for leukemia.

For exposure scenarios in which various tissues of the body receive substantially different doses, estimates of risks for cancers of specific sites are needed. Adjudication of compensation claims for possible radiation-related cancer, which is usually specific to organ site, also requires site-specific estimates. Furthermore, site-specific cancers vary in their causes and baseline risks, and it might thus be expected that models for estimating excess risks from radiation exposure could also vary by site. For this reason, even for estimating total cancer risk, it is desirable to estimate risks for each of several specific cancer sites and then sum the results.

The development of site-specific models is limited by data characteristics. For A-bomb survivor data on solid cancers, parameter estimates based on site-specific data are less precise than those based on all solid cancers analyzed as a group, particularly for less common cancers. It is especially difficult to detect and quantify the modifying effects of variables such as sex, age at exposure, and attained age for site-specific cancers. It was for these reasons that the BEIR V committee provided estimates for only five broad cancer categories.

In addition to statistical uncertainties, it has recently been recognized that estimates of the modifying effects of age at exposure based on A-bomb survivor data can be influenced strongly by secular trends in Japanese baseline rates (Pierce 2002; Preston and others 2003). This occurs because age at exposure in the LSS cohort is confounded with birth cohort, making it impossible to estimate their separate effects without additional information on the relation of baseline and radiation-related risks. (See Annex 12B for further discussion of this issue.) Japanese rates for several cancer sites changed over the period 1950–1998 as Japan became more Westernized, including rates for cancers of the stomach, colon, lung, and female breast. A related problem is that baseline risks for the United States and Japan differ substantially for many cancer sites, and it is unclear how to account for these differences in applying models developed from A-bomb survivor data to estimate risks for the U.S. population.

Pierce and colleagues (1996) and, more recently, Preston and colleagues (2003) found little evidence of heterogeneity among excess relative risk (ERR)[1] models developed for several specific cancer sites. Although these authors caution that this finding should be taken mainly as a warning against overinterpreting apparent differences in sites, some grouping of cancers seems justified. In developing its models, the committee has tried to strike a balance between allowing for differences among cancer sites and statistical precision. As discussed later in this chapter, most of the committee's mod-

---

[1]ERR is the rate of disease in an exposed population divided by the rate of disease in an unexposed population minus 1.0.

els for site-specific cancers make use of data on all solid cancers to estimate the modifying effects of age at exposure and attained age, but make use only of data for the site of interest to estimate the overall level of risk.

Considerations in deciding on the sites for which individual estimates should be provided are whether or not the cancer has been linked clearly with radiation exposure and the adequacy of the data for developing reliable risk estimates. On the first point, it can be argued that the range of uncertainty for risk of a particular cancer is of interest regardless of whether or not a statistically significant dose-response had been observed, a position taken by NIH (2003). Cancers of the salivary glands, stomach, colon, liver, lung, breast, bladder, ovary, and thyroid and nonmelanoma skin cancer have all been linked clearly with radiation exposure in A-bomb survivor data, with evidence somewhat more equivocal for a few additional sites such as esophagus, gall bladder, and kidney. Other studies support many of these associations, and bone cancer has been linked with exposure to α-irradiation from ^{224}Ra. Leukemia has been strongly linked with radiation exposure in several studies including those of atomic bomb survivors.

Another consideration in selecting sites for evaluation is the likelihood of exposure scenarios that will irradiate the site selectively. Here it is noted that inhalation exposures will selectively irradiate the lung, exposures from ingestion will selectively irradiate the digestive organs, exposure to strontium selectively irradiates the bone marrow, and exposure to uranium selectively irradiates the kidney.

Based on these considerations, the committee has provided models and mortality and incidence estimates for cancers of the stomach, colon, liver, lung, female breast, prostate, uterus, ovary, bladder, and all other solid cancers. Incidence estimates are also provided for thyroid cancer.

The inclusion of cancers of the prostate and uterus merits comment because these cancers are not usually thought to be radiation-induced and have not been evaluated separately in previous risk assessments. However, the committee did not want to include these cancers in the residual category of "all other solid cancers," particularly since prostate cancer is much more common in the United States than in Japan.

THE BEIR VII COMMITTEE'S PREFERRED MODELS

Approach to Analyses

This section describes the results of analyses of data on cancer incidence and mortality in the LSS cohort that were conducted by the committee with the help of RERF personnel acting as agents of the National Academies. Analyses of cancer incidence were based on cases diagnosed in the period 1958–1998. Analyses of cancer mortality from all solid cancers and from leukemia were based on deaths occurring in the period 1950–2000 (Preston and others 2004), whereas analyses of mortality from cancer of specific sites were based

on deaths occurring in the period 1950–1997 (Preston and others 2003). Both excess relative risk models and excess absolute risk (EAR)[2] models were evaluated. Methods were generally similar to those that have been used in recent reports by RERF investigators (Pierce and others 1996; Preston and others 2003) and were based on Poisson regression using the AMFIT module of the software package EPICURE (Preston and others 1991). Additional detail is given in Annex 12B.

All analyses were based on the newly implemented DS02 dose estimates. Doses were expressed in sieverts, with a constant weighting factor of 10 for the neutron dose; that is, the doses were calculated as γ-ray absorbed dose (Gy) + 10 × neutron absorbed dose (Gy). The DS02 system provides estimates of doses to several organs of the body. For site-specific estimates, the committee used dose to the organ being evaluated, with colon dose used for the residual category of "other" cancers. The weighted dose, d, to the colon was used for the combined category of all solid cancer or all solid cancers excluding thyroid and nonmelanoma skin cancer. Additional discussion of the doses used in the analyses is given in Annex 12B.

Models for All Solid Cancers

Risk estimates for all solid cancers were obtained by summing the estimates for cancers of specific sites. However, the general form of the model and the estimates of the parameters that quantify the modifying effects of age at exposure and attained age were (with some exceptions) based on analyses of data on all solid cancers. Such analyses offer the advantage of larger numbers of cancer cases and deaths, which increases statistical precision.

As discussed in Chapter 6, most recent analyses of data on the LSS cohort have been based on either ERR models, in which the excess risk is expressed relative to the background risk, or EAR models, in which the excess risk is expressed as the difference in the total risk and the background risk. With linear dose-response functions, the general models for the ERR and EAR are given below:

$$\lambda(c, s, a, b, d) = \lambda(c, s, a, b) [1 + \beta_s \, ERR(e, a)d]$$
or
$$\lambda(c, s, a, b, d) = \lambda(c, s, a, b) + \beta_s \, EAR(e, a)d,$$

where $\lambda(c, s, a, b)$ denotes the background rate at zero dose, and depends on city (c), sex (s), attained age (a), and birth cohort (b). The terms $\beta_s \, ERR(e, a)$ and $\beta_s \, EAR(e, a)$ are, respectively, the ERR and the EAR per unit of dose expressed in sieverts, which may depend on sex (s), age at exposure (e), and attained age (a).

---

[2]EAR is the rate of disease in an exposed population minus the rate of disease in an unexposed population.

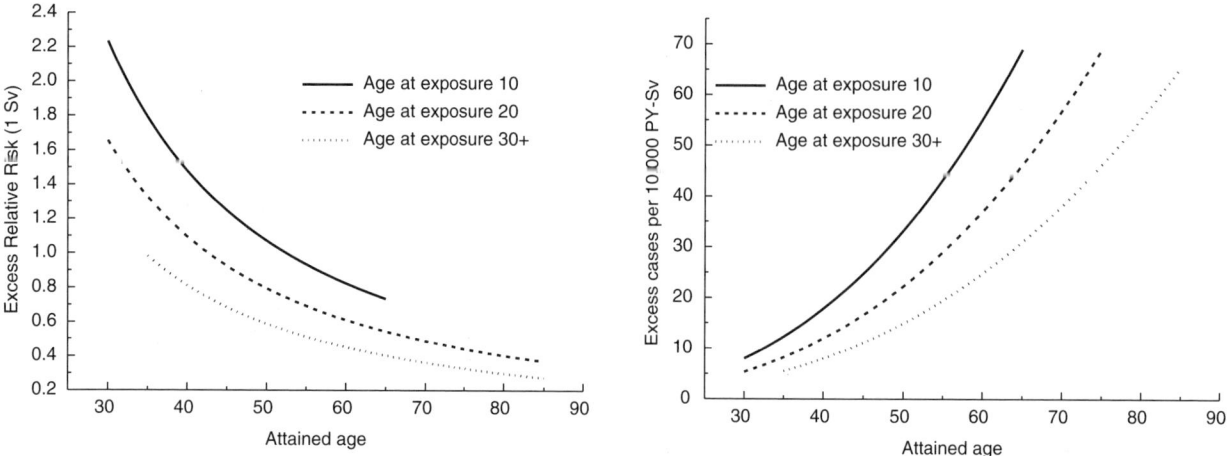

FIGURE 12-1 A Age-time patterns in radiation-associated risks for solid cancer incidence excluding thyroid and nonmelanoma skin cancer. Curves are sex-averaged estimates of the risk at 1 Sv for people exposed at age 10 (solid lines), age 20 (dashed lines), and age 30 or more (dotted lines). Estimates were computed using the parameter estimates shown in Table 12-1.

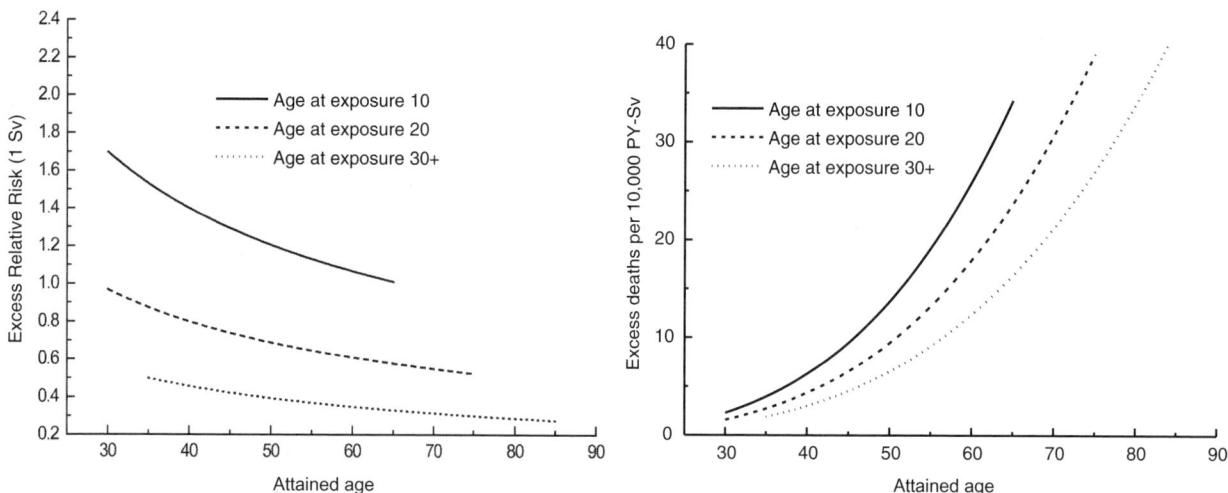

FIGURE 12-1B Age-time patterns in the radiation-associated risks for all solid cancer mortality. Curves are sex-averaged estimates of the risk at 1 Sv for people exposed at age 10 (solid lines), age 20 (dashed lines), and age 30 or more (dotted lines). Estimates were computed using the parameter estimates shown in Table 12-1.

The most recent analyses of A-bomb survivor cancer incidence and mortality data (*e.g.*, Preston and others 2003, 2004) are based on models in which ERR (*e, a*) and EAR (*e, a*) are of the form below:

RERF model:
$$\text{ERR}(e, a) \text{ or } \text{EAR}(e, a) = \exp(\gamma e) \, a^{\eta}. \quad (12\text{-}1)$$

The parameters γ and η quantify the dependence of the ERR or EAR on *e* and *a*. These models, with dependence on both age at exposure and attained age, were chosen because of

difficulties in distinguishing the fits of models with only one of those variables and because, with the incidence data, analyses of all solid cancers indicated dependence on both variables.

The committee's models were developed from analyses of both LSS incidence and LSS mortality data. Analyses of incidence data were based on the category consisting of all solid cancers excluding thyroid and nonmelanoma skin cancers. These exclusions were made because both thyroid cancer and nonmelanoma skin cancer exhibit exceptionally strong age-at-exposure dependencies that do not seem typical of cancer of other sites (Thompson and others 1994). Because the most recent mortality data (1950–2000) available to the committee did not include site-specific solid cancers and because thyroid cancer and nonmelanoma skin cancer are rarely fatal, analyses of mortality data were based on the category of all solid cancers. The committee's preferred models for estimating solid cancer risks are similar to the RERF model, except that the ERR and EAR depend on age at exposure only for exposure ages under 30 years and are constant for exposure ages over 30. That is,

BEIR VII model:
$$\text{ERR}(e, a) \text{ or } \text{EAR}(e, a) = \exp{(\gamma e^*)}\, a^\eta, \quad (12\text{-}2)$$

where e is age at exposure in years, e^* is equal to $e - 30$ when $e < 30$, and equal to zero when e 30, and a is attained age in years.

Figure 12-1A shows plots of the ERR and EAR for incidence of all solid cancers excluding thyroid cancer and nonmelanoma skin cancer as a function of exposure age and attained age using the BEIR VII model. Figure 12-1B shows similar plots for mortality from all solid cancers. Although the ERR and EAR models have the same form, the values and interpretation of the parameters are different. In particular, the ERR shows a decrease with attained age, whereas the EAR shows a strong increase with attained age. Both the ERR and the EAR decrease with increasing age at exposure for those exposed under age 30.

The committee chose the model shown in Equation (12-2) because it fitted both incidence and mortality data on all solid cancers excluding thyroid cancer and nonmelanoma skin cancer slightly better than the RERF model shown in Equation (12-1). There was no indication of a continued decrease with exposure age in the ERR or EAR after exposure age 30, and there was even a suggestion of an increase at older ages. Further discussion of the rationale for choosing the Equation (12-2) model, including a detailed description of analyses that were conducted by the committee, can be found in Annex 12B. In that annex, the committee evaluates several alternative model choices, including models that allow for dependence on age at exposure alone, on attained age alone, and on time since exposure instead of attained age. Also evaluated are models that use different functional forms to express the dependence on exposure age, attained age, or time since exposure. Although several alternative models provided reasonable descriptions of the data, the BEIR VII preferred model shown in Equation (12-2) provided the best fit.

Table 12-1 shows estimates of the parameters of the ERR and EAR models obtained from analyses of LSS incidence data (1958–1998) for all solid cancers excluding thyroid and nonmelanoma skin cancers and of LSS mortality data (1950–2000) for all solid cancers. Further description of these results and how they were obtained can be found in Annex 12B.

TABLE 12-1 ERR and EAR Models for Estimating Incidence of All Solid Cancers Excluding Thyroid and Nonmelanoma Skin Cancers and Mortality from All Solid Cancers[a,b]

ERR Models	No. of Cases or Deaths	ERR/Sv (95% CI) at Age 30 and Attained Age 60		Per-Decade Increase in Age at Exposure Over the Range 0–30 Years[c] (95% CI), γ	Exponent of Attained Age (95% CI), η
		Males (β_M)	Females (β_F)		
Incidence[d]	12,778	0.33 (0.24, 0.47)	0.57 (0.44, 0.74)	−0.30 (−0.51, −0.10)	−1.4 (−2.2, −0.7)
Mortality[e]	10,127	0.23 (0.15, 0.36)	0.47 (0.34, 0.65)	−0.56 (−0.80, −0.32)	−0.67 (−1.6, 0.26)
EAR Models		EAR per 10^4 PY-Sv (95% CI)			
		Males (β_M)	Females (β_F)		
Incidence[d]	12,778	22 (15, 30)	28 (22, 36)	−0.41 (−0.59, −0.22)	2.8 (2.15, 3.41)
Mortality[e]	10,127	11 (7.5, 17)	13 (9.8, 18)	−0.37 (−0.59, −0.15)	3.5 (2.71, 4.28)

NOTE: Estimated parameters with 95% CIs. PY = person-years.

[a]The ERR or EAR is of the form $\beta_s\, D \exp{(\gamma e^*)}\, (a/60)^\eta$, where D is the dose (Sv), e is age at exposure (years), e^* is $(e - 30)/10$ for $e < 30$ and zero for e 30, and a is attained age (years).

[b]The committee's preferred estimates of risks from all solid cancers are obtained as sums of estimates based on models for site-specific cancers (see Table 12-2 and text).

[c]Change in ERR/Sv or EAR per 10^4 PY-Sv (per-decade increase in age at exposure) is obtained as $1 - \exp{(\gamma)}$.

[d]Based on analyses of LSS incidence data 1958–1998 for all solid cancers excluding thyroid and nonmelanoma skin cancer.

[e]Based on analyses of LSS mortality data 1950–2000 for all solid cancers.

Models for Site-Specific Solid Cancers Other Than Breast and Thyroid

Although the committee provides risk estimates for both cancer incidence and mortality, models for site-specific cancers were based on cancer incidence data. This was done primarily because site-specific cancer incidence data are based on diagnostic information that is more detailed and accurate than death certificate data and because, for several sites, the number of incident cases is considerably larger than the number of deaths (see annex Table 12B-2). However, models developed from incidence data were checked for consistency with mortality data. Since there is little evidence that radiation-induced cancers are more rapidly fatal than cancer that occurs for other reasons, ERR models based on incidence data can be used directly to estimate risks of cancer mortality, whereas EAR models require adjustment. (See "Method of Calculating Lifetime Risks" for a description of how the models are used to estimate risks of cancer incidence and mortality.)

Models for estimating risks of solid cancers of specific sites other than breast and thyroid were also of the form shown in Equation (12-2). The committee's approach to quantifying the parameters γ and η was to use the estimates obtained from analyzing incidence data on all solid cancers excluding thyroid and nonmelanoma skin cancers (shown in Table 12-1) unless site-specific analyses indicated significant departure from these estimates. This approach is similar to that used by UNSCEAR (2000b) except that the committee estimated the parameters β_M and β_F separately for each site of interest.

The committee's preferred ERR and EAR models for site-specific cancer incidence and mortality are shown in Table 12-2. The estimates of β_M and β_F are for a person exposed at age 30 or older at an attained age of 60. Models for breast and thyroid cancer were based on published analyses that included data on medically exposed persons as discussed in the next two sections. For other sites, common values of the parameter γ indicating dependence on age at exposure could be used in all cases. With the ERR models, common values of the parameter indicating the dependence of risks on attained age (η) could be used in all cases except the category "all other solid cancers." With the EAR models, it was necessary to estimate the attained-age parameter, η, separately for cancers of the liver, lung, and bladder, which may reflect variation in the pattern of increase with age for site-specific baseline rates.

The committee emphasizes that there is considerable uncertainty in models for site-specific cancers. Statistical uncertainty in the estimates of the main effect parameter β_s is

TABLE 12-2 Committee's Preferred ERR and EAR Models for Estimating Site-Specific Solid Cancer Incidence and Mortality[a]

Cancer Site	No. of Cases	ERR Models				EAR Models			
		β_M[b] (95% CI)	β_F[b] (95% CI)	γ[c]	η[d]	β_M[e] (95% CI)	β_F[e] (95% CI)	γ[c]	η[d]
Stomach	3602	0.21 (0.11, 0.40)	0.48 (0.31, 0.73)	−0.30	−1.4	4.9 (2.7, 8.9)	4.9 (3.2, 7.3)	−0.41	2.8
Colon	1165	0.63 (0.37, 1.1)	0.43 (0.19, 0.96)	−0.30	−1.4	3.2 (1.8, 5.6)	1.6 (0.8, 3.2)	−0.41	2.8
Liver	1146	0.32 (0.16, 0.64)	0.32 (0.10, 1.0)	−0.30	−1.4	2.2 (1.9, 5.3)	1.0 (0.4, 2.5)	−0.41	4.1 (1.9, 6.4)
Lung	1344	0.32 (0.15, 0.70)	1.40 (0.94, 2.1)	−0.30	−1.4	2.3 (1.1, 5.0)	3.4 (2.3, 4.9)	−0.41	5.2 (3.8, 6.6)
Breast[f]	952	—	0.51 (0.28, 0.83)	0	−2.0	—	9.4 (6.7, 13.3)	−0.51	3.5, 1.1[g]
Prostate	281	0.12 (<0, 0.69)	—	−0.30	−1.4	0.11 (<0, 1.0)	—	−0.41	2.8
Uterus	875	—	0.055 (<0, 0.22)	−0.30	−1.4	—	1.2 (< 0, 2.6)	−0.41	2.8
Ovary	190	—	0.38 (0.10, 1.4)	−0.30	−1.4	—	0.70 (0.2, 2.1)	−0.41	2.8
Bladder	352	0.50 (0.18, 1.4)	1.65 (0.69, 4.0)	−0.30	−1.4	1.2 (0.4, 3.7)	0.75 (0.3, 1.7)	−0.41	6.0 (3.1, 9.0)
Other solid cancers	2969	0.27 (0.15, 0.50)	0.45 (0.27, 0.75)	−0.30	−2.8 (−4.1, −1.5)	6.2 (3.8, 10.0)	4.8 (3.2, 7.3)	−0.41	2.8
Thyroid[h]		0.53 (0.14, 2.0)	1.05 (0.28, 3.9)	−0.83	0				

NOTE: Estimated parameters with 95% CIs. PY = person-years.

[a]The ERR or EAR is of the form $\beta_s D \exp(\gamma e^*)(a/60)^\eta$, where D is the dose (Sv), e is age at exposure (years), e^* is $(e − 30)/10$ for $e < 30$ and zero for e 30, and a is attained age (years). Models for breast and thyroid cancer are based on e instead of e^*, although γ is still expressed per decade.

[b]ERR/Sv for exposure at age 30+ at attained age 60.

[c]Per-decade increase in age at exposure over the range 0–30 years (γ).

[d]Exponent of attained age (η).

[e]EAR per 10^4 PY-Sv for exposure at age 30+ and attained age 60; these values are for cancer incidence and must be adjusted as described in the text to estimate cancer mortality risks.

[f]Based on a pooled analysis by Preston and others (2002a). See text for details. Parameter estimates presented by Preston and colleagues were for exposure at age 25 at attained age 50, while estimates in this table are for exposure at age 30 at attained age 60.

[g]The first number is for attained ages less than 50; the second number is for attained ages 50 or greater.

[h]Based on a pooled analyses by Ron and others (1995a) and NIH (2003). Confidence intervals are based on standard errors of non-sex-specific estimates with allowance for heterogeneity among studies.

often large. Although the common values of the parameters γ and η that have been used to quantify the modifying effects of age at exposure and attained age are compatible with site-specific data, estimates of these parameters based on site-specific data are often quite different from the common values. Annex 12B shows the site-specific estimates of γ and η.

Models for Female Breast Cancer

The committee's preferred models for estimating breast cancer incidence and mortality are those developed by Preston and colleagues (2002a) from analyses of combined data on breast cancer incidence in several cohorts including the LSS. The LSS data used in these analyses were for the period 1958–1993, whereas the committee's analyses included data through 1998. Although these models were developed for estimating breast cancer incidence, they may also be used to estimate breast cancer mortality using the same approach as that for other site-specific solid cancers.

Preston and colleagues (2002a) found that common models could be used to describe data from the LSS cohort, the original Massachusetts tuberculosis fluoroscopy cohort and an extension of this cohort (Boice and others 1991b), and the Rochester infant thymus irradiation cohort (Hildreth and others 1989). Models for both the ERR and the EAR were developed for these cohorts. The ERR model was as follows:

$$\text{ERR/Sv} = \beta \, (a/60)^{-2},$$

where a is attained age. With this model, it was necessary to estimate β separately for the LSS and the remaining U.S. women. Parameter estimates were β = 1.46 for the LSS and 0.51 for the remaining U.S. cohorts. The committee's preferred ERR model for estimating risks for U.S. women uses β = 0.51. In the formulation above, the committee has parameterized the model so that β indicates the ERR at an attained age of 60 instead of 50 as given in Preston and colleagues. The pooled EAR model from Preston and colleagues (2002b) was as follows:

$$\text{EAR per } 10^4 \text{ woman-years per gray} = 9.4 \exp [-0.05 \, (e - 30)] \, (a/60)^{\eta},$$

where e is exposure age and a is attained age (years); η = 3.5 for $a < 50$ and η = 1 for $a \geq 50$. For the EAR, a common value of the overall level of risk (9.4) could be used for all four cohorts. Again, the model has been parameterized so that the value of 9.4 is for a woman exposed at age 30 at attained age 60 (instead of a woman exposed at age 25 at attained age 50 as in Preston and others).

Although the committee calculates lifetime risk estimates based on both the ERR and the EAR models described above, its preferred estimates are based on the EAR model. With this model the estimated main effect is more stable because it is based on both LSS and U.S. women. In addition, this model includes both age at exposure and attained age as

modifying factors and is thus more comparable to models used for other sites.

Model for Thyroid Cancer

The committee's preferred model for estimating thyroid cancer incidence is based on a pooled analysis of data from seven thyroid cancer incidence studies conducted by Ron and colleagues (1995a). The NIH (2003) adapted the results of data from five cohorts of persons exposed under age 15 to develop a thyroid cancer incidence model. The five studies were the A-bomb survivors (including only those exposed under age 15; Thompson and others 1994), the Rochester thymus study (Shore and others 1993b), the Israel tinea capitis study (Ron and others 1989), children treated for enlarged tonsils and other conditions (Pottern and others 1990; Schneider and others 1993), and an international childhood cancer study (Tucker and others 1991). Ron and colleagues found that the ERR/Gy for females was about twice that for males although the difference was not statistically significant. Although the NIH (2003) used a non-sex-specific model, for consistency with the treatment of cancers of other sites, the committee has used a sex-specific model. From data presented in NIH (2003, Table IV.D.8), it can be determined that the model takes the form ERR/Gy = 0.79 exp [−0.083 $(e − 30)$], where e is exposure age in years. The BEIR VII model is as follows:

$$\text{ERR/Gy} = 0.53 \exp [- 0.083 \, (e - 30)] \text{ for males,}$$
and
$$\text{ERR/Gy} = 1.05 \exp [- 0.083 \, (e - 30)] \text{ for females.}$$

The estimate of the ERR per Gy given by Ron and colleagues was 7.7 (95% CI 2.1, 29) in a model without modification by age at exposure. With the committee's model, this would be the ERR/Gy, averaged over the two sexes, for exposure at about 2.5 years of age, which was about the average exposure age in the data analyzed by Ron and colleagues.

Ron and colleagues (1995a) did not present results for ERR or EAR models that allowed for modification by both age at exposure and attained age.

Model for Leukemia

The committee's models for estimating leukemia risks were based on analyses of LSS leukemia mortality data for the period 1950–2000 (Preston and others 2004). The quality of diagnostic information for the non-type-specific leukemia mortality used in these analyses is thought to be high. Data on medically exposed cohorts have indicated that chronic lymphocytic leukemia (CLL) is not likely to be induced by radiation exposure (Boice and others 1987; Curtis and others 1994; Weiss and others 1995), but CLL is extremely rare in Japan. Details of the committee's leukemia analyses are given in Annex 12B.

Models used for estimating leukemia risks in the past have expressed the ERR (NRC 1990; NIH 2003) or EAR (ICRP 1991; UNSCEAR 2000b) as a linear-quadratic function of dose and have allowed for dependence on sex, age at exposure, and time since exposure. Both categorical and continuous treatments of age at exposure and time since exposure have been used. The BEIR VII committee models also express the ERR or EAR as a linear-quadratic function of dose with allowance for dependencies on sex, age at exposure, and time since exposure. The committee's preferred models are of the following form:

BEIR VII leukemia model:
EAR(D, s, e, t) or ERR(D, s, e, t) =
$\beta_s D (1 + \theta D) \exp [\gamma e^* + \delta \log (t / 25) + \phi e^* \log (t / 25)]$,
(12-3)

where D is dose (Sv), s is sex, and e^* is $(e - 30) / 10$ for $e < 30$ and 0 for $e \geq 30$ (e is age at exposure in years). Table 12-3 shows the parameter estimates, and Figure 12-2 depicts the dependence of the ERR or EAR on age at exposure and time since exposure. The parameter θ indicates the degree of curvature, which does not depend on sex, age at exposure, or time since exposure; β_M and β_F represent the ERR/Sv or the EAR (expressed as excess deaths per 10^4 PY-Sv, where PY = person-years), for exposure at age 30 or more at 25 years following exposure. This model was found to fit the data better than analogous models using e instead of e^*, or using t instead of $\log (t)$, and nearly as well as models with a

categorical treatment of age at exposure. It was also found to be necessary to allow the dependence on time since exposure to vary by age at exposure by including the term $e^* \log (t / 25)$. For the EAR model, there was no need to include a term for the main effect of time since exposure; note that with this parameterization, there is no decrease with time since exposure for those exposed at age 30 or more. For application of these models, the reader should consult the section "Use of the Committee's Preferred Models to Estimate Risks for the U.S. Population."

USE OF THE COMMITTEE'S PREFERRED MODELS TO ESTIMATE RISKS FOR THE U.S. POPULATION

To use models developed primarily from Japanese A-bomb survivor data for the estimation of lifetime risks for the U.S. population, several issues must be addressed. These include determining approaches for estimating risks at low doses and low dose rates, projecting risks over time, transporting risks from the Japanese to the U.S. population, and estimating risks from exposure to X-rays. This section describes the approach for addressing each of these issues, as well as the methodology used to estimate lifetime risk. More detailed discussion of some of the issues is given in Chapter 10, and the approach for quantifying the uncertainties associated with some of these issues is discussed later in this chapter.

Estimating Risks from Exposure to Low Doses and Low Dose Rates

The BEIR VII risk models have been developed primarily from analyses of data on the LSS cohort of Japanese A-bomb survivors. Although more than 60% of the exposed members of this cohort were exposed to relatively low doses (0.005–0.1 Sv), survivors who were exposed to doses exceeding 0.5 Gy are still influential in estimating the ERR/Sv. In addition, exposure of A-bomb survivors was at high dose rates, whereas exposure at low dose rates is of primary concern for risk assessment. Based on evidence from experimental data, ICRP (1991), NCRP (1993), EPA (1999), and UNSCEAR (2000b) recommended reducing linear estimates based on A-bomb survivor (or other high-dose-rate) exposure by a dose and dose-rate reduction factor (DDREF) of 2.0.

In Chapter 10, both data on solid cancer risks in the LSS cohort and experimental data pertinent to this issue are evaluated by the committee. Based on this evaluation, the committee found a believable range of DDREF values (for adjusting linear risk estimates based on the LSS cohort) to be 1.1 to 2.3. When a single value is needed, 1.5 (the median of the subjective probability distribution for the LSS DDREF) is used to estimate risk for solid tumors. To estimate the risk of leukemia, the BEIR VII model is linear-quadratic, since this model fitted the data substantially better than the linear model.

TABLE 12-3 Committee's Preferred ERR and EAR Models for Estimating Leukemia Incidence and Mortality[a,b,c]

Parameter	ERR Model	EAR Model
β_M	1.1 per Sv (0.1, 2.6)	1.62 deaths per 10^4 PY-Sv (0.1, 3.6)
β_F	1.2 per Sv (0.1, 2.9)	0.93 deaths per 10^4 PY-Sv (0.1, 2.0)
γ	–0.40 per decade (–0.78, 0.0)	0.29 per decade (0.0, 0.62)
δ	–0.48 (–1.1, 0.2)	0.0
ϕ	0.42 (0.0, 0.96)	0.56 (0.31, 0.85)
θ	0.87 per Sv (0.16, 15)	0.88 Sv^{-1} (0.16, 15)

NOTE: Estimated parameters with 95% CIs[d] based on likelihood ratio profile.

[a]The ERR or EAR is of the form $\beta_s(D + \theta D^2) \exp [\gamma e^* + \delta \log (t / 25) + \phi e^* \log (t / 25)]$, where D is the dose to the bone marrow (Sv), e is age at exposure (years), e^* is $(e - 30) / 10$ for $e < 30$ and zero for $e \geq 30$, and t is time since exposure (years).

[b]Based on analyses of LSS mortality data (1950–2000), with 296 deaths from leukemia.

[c]These models apply only to the period 5 or more years following exposure.

[d]Confidence intervals based on likelihood ratio profile.

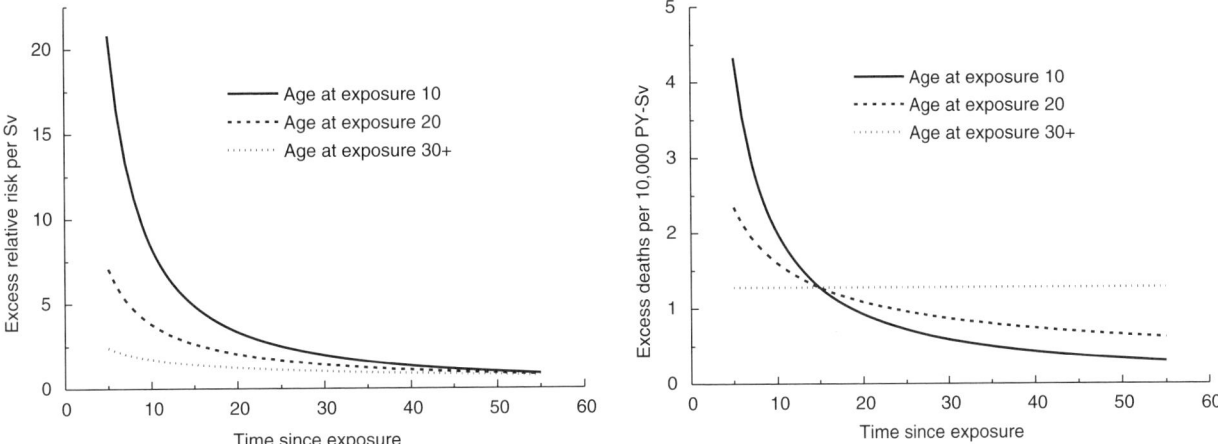

FIGURE 12-2 Age-time patterns in radiation-associated risks for leukemia mortality. Curves are sex-averaged estimates of the risk at 1 Sv for people exposed at age 10 (solid lines), age 20 (dashed lines), and age 30 or more (dotted lines). Estimates were computed using the parameter estimates shown in Table 12-3.

Projection of Risks over Time

The LSS cohort has now been followed for more than 50 years, so that lifetime follow-up is nearly complete for all but the youngest survivors (under age 20 at exposure). Although the extrapolation involved in estimating lifetime risks based on limited follow-up has been a major source of uncertainty in past risk assessments, it is now much less so. The BEIR VII models allow for dependencies of both the ERR and the EAR on attained age, and it is assumed that the identified patterns persist until the end of life for the youngest survivors. Additional discussion of this issue is found in Chapter 10.

For leukemia, the early years of follow-up also must be addressed. Ascertainment of leukemia cases for the LSS cohort did not begin until 1950, while data on medically exposed cohorts have demonstrated that excess leukemia cases can occur as early as a year or two after exposure (Boice and others 1987; Curtis and others 1992, 1994; Inskip and others 1993; Weiss and others 1994, 1995). In several of these studies, relative risks were highest in the period 1–5 years after exposure. In addition, a recent analysis of data on Mayak workers found that leukemia risks 3–5 years following external radiation exposure were more than an order of magnitude higher than risks for later periods (Shilnikova and others 2003). The UNSCEAR (2000b) committee addressed this problem by assuming that excess risks for the first 5 years after exposure were half those observed 5 years after exposure. The BEIR VII committee has instead assumed that excess absolute risk in the period 2–5 years following exposure is equal to that observed 5 years after exposure. Clearly there is uncertainty in the magnitude of the risk during the initial years following exposure.

Transport of Risks from a Japanese to a U.S. Population

Baseline risks for many site-specific cancers are different for the United States and Japan. For example, baseline risks for cancers of the colon, lung, and female breast are higher in the United States, whereas baseline risks for cancers of the stomach and liver are much higher in Japan. The BEIR V committee based its estimates on relative risk transport, where it is assumed that the excess risk due to radiation is proportional to baseline risks; that is, the ERR is the same for the United States and Japan. However, the BEIR III committee based its estimates on absolute risk transport, where it is assumed that the excess risk does not depend on baseline risks; that is, the EAR is the same for the United States and Japan. The EPA (1994) used the geometric mean of the two estimates, whereas UNSCEAR (2000b) presented estimates based on both approaches without indicating a preference. Estimates based on relative and absolute risk can differ substantially. For example, the UNSCEAR stomach cancer estimates for the U.S. population based on absolute risk transport are nearly an order of magnitude larger than those based on relative risk transport.

For breast and thyroid cancer, the committee's models are based on combined analyses that include Caucasian subjects. For other solid cancer sites including leukemia, the committee has calculated risks using both relative and absolute risk transport, which provides an indication of the uncertainty from this source. The recommended point estimates are weighted means of estimates obtained under the two models (adjusted by a DDREF of 1.5 as discussed above). For sites other than breast, thyroid, and lung, a weight of 0.7 is used for the estimate obtained using relative risk transport and a weight of 0.3 for the estimate obtained using absolute risk transport, with the weighting done on a logarithmic scale. This choice was made because, as discussed in Chapter 10, there is somewhat greater support for relative risk than for absolute risk transport. In addition, the ERR models used to obtain relative risk transport estimates may be less vulnerable to possible bias from underascertainment of cases. For lung cancer, the weighting scheme is reversed, and a weight of 0.7 is used for the absolute risk transport estimate and a weight of 0.3 for the relative risk transport estimate. This departure was made because of evidence that the interaction of radiation and smoking in A-bomb survivors is additive (Pierce and others 2003). Although it is likely that the correct transport model varies by cancer site, for sites other than breast, thyroid, and lung the committee judged that current knowledge was insufficient to allow the approach to vary by cancer site.

Transport has not generally been considered an important source of uncertainty for estimating leukemia risks. The committee has nevertheless developed both ERR and EAR models for leukemia and obtained estimates based on both relative and absolute risk transport. As shown later, the EAR model leads to substantially lower lifetime risks than the ERR model (Table 12-7). Since there is no reason to suspect underascertainment of leukemia deaths, apparently this comes about because baseline risks in the LSS cohort are different than those for a modern U.S. population. Because of the small number of deaths in the early period among those who were unexposed, it might be thought that the uncertainty in the estimated ERR/Sv would be large; however in fact, it is only slightly larger than that for the EAR model (Table 12-3).

Relative Effectiveness of X-Rays and γ-Rays

Risk estimates in this report have been developed primarily from data on A-bomb survivors and are thus directly relevant to exposure from high-energy photons. However, the report is concerned with low-LET radiation generally, which includes γ-rays, X-rays, and fast electrons. There is no principal difference between the action of these different types of radiation, because they all work through fast electrons that either are incident on the body or are released within the body by electrons or photons. The various types of low-LET radiation vary in their ability to penetrate to greater depths in the body. The more penetrating, high-energy radiation tends to produce electrons with linear energy transfer less than 1 keV / μm, while the softer X-rays release slower electrons with linear energy transfer up to several kiloelectronvolts per micrometer.

With regard to setting dose limits in radiation protection, γ-rays, fast electrons, and X-rays are all given the radiation weighting factor 1; that is, an absorbed dose of 1 Gy of these radiations is taken to be equal to the effective dose 1 Sv (ICRP 1991), which expresses the fact that the differences of effectiveness between different photon radiations are not considered of sufficient consequence to require explicit accounting in radiation protection regulations. However, the significant difference between the (dose average) unrestricted LET of ^{60}Co (about 0.4 keV / μm) or ^{137}Cs γ-rays (about 0.8 keV / μm) and that of 200 kVp X-rays (about 3.5 keV / μm) makes it clear that the relative biological effectiveness (RBE) at low doses can differ appreciably for γ-rays and X-rays. For actual risk estimates it is, therefore, necessary to consider these differences in terms of the radiobiological findings, the dosimetric and microdosimetric parameters of radiation quality, and the radioepidemiologic evidence.

As discussed in ICRP (2004) and in Chapters 1 and 3 of this report, there is evidence based on chromosomal aberration data and on biophysical considerations that, at low doses, the effectiveness per unit absorbed dose of standard X-rays may be about twice that of high-energy photons. The effectiveness of lower-energy X-rays may be even higher. How this translates into risks of late effects in man is an open question. Estimates based on studies of persons exposed to X-rays for medical reasons tend to be lower than those based on A-bomb survivors (Little 2001; ICRP 2004), but a number of other differences may confound these comparisons. In addition, doses in many medically exposed populations are higher than those at which the energy of the radiation (based on biophysical considerations) would be expected to be important.

Because of the lack of adequate epidemiologic data on this issue, the committee makes no specific recommendation for applying risk estimates in this report to estimate risk from exposure to X-rays. However, it may be desirable to increase risk estimates in this report by a factor of 2 or 3 for the purpose of estimating risks from low-dose X-ray exposure.

Relative Effectiveness of Internal Exposure

Internal exposure through inhalation or ingestion is also of interest. For example, internal exposure to ^{131}I, strontium, and cesium may occur from atmospheric fallout from nuclear weapons testing. Epidemiologic studies involving these exposures are reviewed in Chapter 9. Studies of thyroid cancer in relation to ^{131}I include those of persons exposed to atmospheric fallout in Utah, to releases from the Hanford plant, and as a result of the Chernobyl accident. There are also studies of persons exposed to cesium and strontium from releases from the Mayak nuclear facility in Russia into the Techa River. To date, these studies are not adequate to quantify carcinogenic risk reliably as a function of dose. Although there are no strong reasons to think that the dose-response from internal low-LET exposure would differ from that for external exposure, there is additional uncertainty in applying the BEIR VII risk models to estimate risks from internal exposure.

Method of Calculating Lifetime Risks

Several measures of lifetime risk have been used to express radiation risks and are discussed by Vaeth and Pierce (1990), Thomas and colleagues (1992), UNSCEAR (2000b), and Kellerer and colleagues (2001). The BEIR VII committee has chosen to use what Kellerer and coworkers refer to as the lifetime attributable risk (LAR), which was earlier called the risk of untimely death by Vaeth and Pierce (1990). The LAR is an approximation of the risk of exposure-induced death (REID), the measure used by UNSCEAR (2000b), which estimates the probability that an individual will die from (or develop) cancer associated with the exposure. Although the nomenclature is recent, the LAR was used by the BEIR III committee (1980b) and by the EPA (1994).

The LAR and the REID both differ from the excess lifetime risk (ELR) used by the BEIR V committee in that the former include deaths or incident cases of cancer that would have occurred without exposure but occurred at a younger age because of the exposure. As noted by Thomas and colleagues (1992) and earlier by Pierce and Vaeth (1989), the ratio of ELR to REID is approximately $1 - Q_c$ where Q_c is the lifetime risk of dying from the cause of interest. For example, the ELR for all cancer mortality would be about 20% lower than the REID. The LAR differs from the REID in that the survival function used in calculating the LAR does not take account of persons dying of radiation-induced disease, thus simplifying the computations. This difference may be important for estimating risks at higher doses (1+ Sv), but not at the low doses of interest for this report. Kellerer and colleagues show that the REID and the LAR are nearly identical at low doses and discuss other aspects of the LAR compared to the REID.

The LAR for a person exposed to dose D at age e is calculated as follows:

$$\text{LAR}(D, e) = {}_a M(D, e, a) S(a) / S(e), \quad (12-4)$$

where the summation is from $a = e + L$ to 100, where a denotes attained age (years) and L is a risk-free latent period ($L = 5$ for solid cancers; $L = 2$ for leukemia). The $M(D, e, a)$ is the EAR, $S(a)$ is the probability of surviving until age a, and $S(a) / S(e)$ is the probability of surviving to age a conditional on survival to age e. All calculations are sex-specific; thus, the dependence of all quantities on sex is suppressed.

The quantities $S(a)$ were obtained from a 1999 unabridged life table for the U.S. population (Anderson and DeTurk 2002). Lifetime risk estimates using relative risk transport were based on ERR models. For these calculations,

$$M(D, e, a) = \text{ERR}(D, e, a) \, \lambda_I^c(a)$$

for cancer incidence, and

$$M(D, e, a) = \text{ERR}(D, e, a) \, \lambda_M^c(a)$$

for cancer mortality. The ERR(D, e, a) was obtained from models shown in Tables 12-1, 12-2, and 12-3. The $\lambda_I^c(a)$ represents sex- and age-specific 1995–1999 U.S. cancer incidence rates from Surveillance Epidemiology, and End Results (SEER) registries, whereas the $\lambda_M^c(a)$ are sex- and age-specific 1995–1999 U.S. cancer mortality rates (http://seer.cancer.gov/csr/1975_2000), where c designates the cancer site or category. These rates were available for each 5-year age group with linear interpolation used to develop estimates for single years of age. With the exception of the category "all solid cancers," the same ERR models were used to estimate both cancer incidence and mortality.

Lifetime risk estimates using absolute risk transport were based on EAR models (see "Transport of Risks from a Japanese to a U.S. Population"). For estimating cancer incidence, $M(D, e, a)$ is taken to be the EAR(D, e, a) based on the models shown in Tables 12-1, 12-2, and 12-3. For estimating mortality from all solid cancers, the EAR mortality model shown in Table 12-1 was used directly. For estimating site-specific cancer mortality, it was necessary to adjust the EAR(D, e, a) from Tables 12-2 and 12-3 by multiplying by $\lambda_M^c(a) / \lambda_I^c(a)$, the ratio of the sex- and age-specific mortality and incidence rates for the U.S. population. That is, for site-specific mortality,

$$M(D, e, a) = \text{EAR}(D, e, a) \, \lambda_M^c(a) / \lambda_I^c(s, a).$$

Leukemia merits special comment. The approach for deriving incidence and mortality estimates based on relative and absolute risk transport is the same for leukemia as for other site-specific cancers, despite the fact that leukemia models were developed from LSS mortality data rather than incidence data as for other sites. This is because LSS leukemia data were obtained at a time when this disease was nearly always rapidly fatal, so that estimates of leukemia mortality should closely approximate those for leukemia incidence. In the last few decades, however, marked progress has been made in treating leukemia, and the disease is not always fatal. Thus, the committee has used the EAR model shown in Table 12-3 to estimate leukemia incidence, but has adjusted the EAR(D, s, e, a) from Table 12-3 in the manner described above to obtain estimates of leukemia mortality. In all cases, the U.S. leukemia baseline rates were for all leukemias excluding CLL.

Models for leukemia differ from those for solid cancers in that risk is expressed as a function of age at exposure (e) and time since exposure (t) instead of age at exposure and attained age (a). Since $t = a - e$, ERR(D, e, a) or EAR(D, e, a) is obtained by substituting $a - e$ for t in the models presented in Table 12-3. Note further that for the period 2–5 years after exposure, the EAR is assumed to be the same as that at 5 years after exposure. That is, for $a = e + 2$ to $e + 5$, $M(D, e, a) = M(D, e, e + 5)$.

The approach described above for obtaining estimates based on absolute transport differs from that used by UNSCEAR (2000b) and NIH (2003), where $M(D, e, a)$ for absolute risk transport was calculated by multiplying the ERR(D, e, a) estimated from LSS data by sex- and age-specific baseline risks for the 1985 population of Japan. Because Japanese rates for cancer of several sites changed in the period 1950–1985 (becoming more similar to U.S. rates), the committee's approach may reflect risks more truly in the LSS cohort than do 1985 baseline rates for Japan.

Another difference between the committee's approach and that of UNSCEAR is that for estimating cancer incidence, UNSCEAR lifetime risk calculations counted only first cancers. That is, once a person was diagnosed with cancer (baseline or radiation induced), that person was removed from the population at risk. By contrast, the committee's calculations count all primary cancers including those in persons previously diagnosed with another primary cancer.

To obtain estimates of risk for a population of mixed exposure ages, the age-at-exposure-specific estimates in Equation (12-4) were weighted by the fraction of the population in the age group based on the U.S. population in 1999 (http://wonder.cdc.gov/popu0.shtml). Estimates of chronic lifetime exposure are for a person at birth, with allowance for attrition of the population with age. These estimates are obtained by weighting the age-at-exposure-specific estimates by the probability of survival to each age, that is, $S(e)$. Similarly, estimates for chronic occupational exposure are for a person who enters the workforce at age 18 and continues to be exposed to age 65, again with allowance for attrition of the population with age. These estimates are obtained by weighting the age-at-exposure-specific estimates by the probability of survival to each age conditional on survival to age 18, that is, $S(e)$ / $S(18)$.

QUANTITATIVE EVALUATION OF UNCERTAINTY IN LIFETIME RISKS

Because of the various sources of uncertainty it is important to regard specific estimates of LAR with a healthy skepticism, placing more faith in a range of possible values. Although a confidence interval is the usual statistical device for doing so, the approach here also accounts for uncertainties external to the data, treating subjective probability distributions for these uncertainties *as if* they resulted from real data. The resulting range of plausible values for lifetime risk is consequently labeled a "subjective confidence interval" to emphasize its dependence on opinions in addition to direct numerical observation. Similar logic has been used in other uncertainty analyses (NCRP 1997; EPA 1999; UNSCEAR 2000b).

The quantitative analysis focuses on the three sources that are thought to matter most: (1) sampling variability in risk model parameter estimates from the LSS data, (2) the uncertainty about transport of risk from a Japanese (LSS) to a U.S. population (*i.e.*, whether ERR or EAR is transportable), and (3) the uncertainty in the appropriate value of a DDREF for adjusting low-dose risks based on linear-in-dose risk models

estimated from LSS data. The approach used is a conventional one that finds a variance for the estimated LAR (on the log scale) induced by the variances of these three sources. The computational approach for the subjective confidence intervals is detailed in Annex 12C. Additional sources of uncertainty that have not been quantified are discussed later in the chapter. For site-specific cancers other than leukemia, the assessment of sampling variability did not include uncertainty in the parameters quantifying the modifying effects of age at exposure and attained age. Although estimates of solid cancer risks are obtained as the sum of site-specific risks, the uncertainty in these estimates was evaluated using models for all solid cancers.

RESULTS OF RISK CALCULATIONS

Lifetime Risk Estimates for the U.S. Population

In this section, the committee's preferred estimates of the LAR are presented for several cancer categories. Estimates of the numbers of excess cancers or deaths due to cancer in a population of 100,000 exposed to 0.1 Gy are emphasized and are intended to apply to a population with an age composition similar to the 1999 U.S. population. In addition, estimates for all solid cancers and for leukemia are presented for three specific exposure ages (10, 30, and 50 years), for a population that is exposed throughout life to 1 mGy per year, and for a population that is exposed to 10 mGy per year from age 18 to 65. Additional examples are found in Annex 12D.

For perspective, Table 12-4 shows lifetime risks of cancer incidence and mortality in the absence of exposure. For

TABLE 12-4 Baseline Lifetime Risk Estimates of Cancer Incidence and Mortality

Cancer site	Incidence		Mortality	
	Males	Females	Males	Females
Solid cancer[a]	45,500	36,900	22,100 (11)	17,500 (11)
Stomach	1,200	720	670 (11)	430 (12)
Colon	4,200	4,200	2,200 (11)	2,100 (11)
Liver	640	280	490 (13)	260 (12)
Lung	7,700	5,400	7,700 (12)	4,600 (14)
Breast	—	12,000	—	3,000 (15)
Prostate	15,900	—	3,500 (8)	—
Uterus	—	3,000	—	750 (15)
Ovary	—	1,500	—	980 (14)
Bladder	3,400	1,100	770 (9)	330 (10)
Other solid cancer	12,500	8,800	6,800 (13)	5,100 (13)
Thyroid	230	550	40 (12)	60 (12)
Leukemia	830	590	710 (12)	530 (13)

NOTE: Number of estimated cancer cases or deaths in population of 100,000 (No. of years of life lost per death).

[a]Solid cancer incidence estimates exclude thyroid and nonmelanoma skin cancers.

nearly all sites other than breast, ovary, and thyroid, risks are higher for males than females, with especially large differences for cancers of the liver and bladder. In males, prostate cancer accounts for more than a third of the incident cases. In females, breast cancer accounts for about a third of the incident cases.

Tables 12-5A and 12-5B show estimates of the LAR for a population with an age composition similar to that of the U.S. population exposed to 0.1 Gy. Estimates of cancer incidence (Table 12-5A) and mortality (Table 12-5B) are shown for several site-specific solid cancers. The committee's preferred estimates are those in the third and sixth columns. These were obtained by calculating a weighted mean (on a logarithmic scale) of linear estimates based on relative and absolute risk transport (also shown) and then reducing them by DDREF of 1.5 as described earlier. The subjective confidence intervals reflect uncertainty due to sampling variability, transport, and DDREF. For most sites, these intervals cover at least an order of magnitude. For many sites, statistical uncertainty alone is large (see Table 12-2). For cancers of the stomach, liver, lung (females), prostate, and uterus, estimates based on relative and absolute risk differ by a factor of 2 or more, contributing substantially to the uncertainty

in estimates for these sites. It is perhaps surprising that the LAR for lung cancer is nearly twice as high for females as males even though the baseline risks show a reverse pattern. It is possible that this and other patterns for site-specific cancers reflect statistical anomalies or other biases in LARs estimated with high uncertainty.

The committee's preferred estimates for risk of all solid cancers can be obtained as the sums of the site-specific estimates and are shown in the next-to-the-last line of Tables 12-5A and 12-5B. These estimates are higher for females than males, even though the reverse is true for baseline risks (Table 12-4), a finding that comes about primarily because of the contribution of breast cancer and lung cancer (as noted above). For cancer mortality, the years of life lost per death are also of interest. For the sum of sites estimates, this was 14 per death for males and 15 per death for females.

The LAR for all cancer incidence is about twice that for cancer mortality. However, this ratio varies greatly by cancer site. The largest contribution to cancer incidence in males is from the residual category of "other solid cancers" followed by colon and lung cancer. These three categories are also the most important contributors to cancer mortality. Cancers of the lung, and breast and other solid cancers con-

TABLE 12-5A Lifetime Attributable Risk of Solid Cancer Incidence

Cancer Site	Males			Females		
	LAR Based on Relative Risk Transport[a]	LAR Based on Absolute Risk Transport[b]	Combined and Adjusted by DDREF[c] (Subjective 95% CI[d])	LAR Based on Relative Risk Transport[a]	LAR Based on Absolute Risk Transport[b]	Combined and Adjusted by DDREF[c] (Subjective 95% CI[d])
Incidence						
Stomach	25	280	34 (3, 350)	32	330	43 (5, 390)
Colon	260	180	160 (66, 360)	160	110	96 (34, 270)
Liver	23	150	27 (4, 180)	9	85	12 (1, 130)
Lung	250	190	140 (50, 380)	740	370	300 (120, 780)
Breast				510 Not used	460	310 (160, 610)
Prostate	190	6	44 (<0, 1860)			
Uterus				19	81	20 (<0, 131)
Ovary				66	47	40 (9, 170)
Bladder	160	120	98 (29, 330)	160	100	94 (30, 290)
Other	470	350	290 (120, 680)	490	320	290 (120, 680)
Thyroid	32	No model	21 (5, 90)	160	No model	100 (25, 440)
Sum of site-specific estimates	1400	1310[e]	800	2310[f]	2060[e]	1310
All solid cancer model[g]	1550	1250	970 (490, 1920)	2230	1880	1410 (740, 2690)

NOTE: Number of cases per 100,000 persons of mixed ages exposed to 0.1 Gy.

[a]Linear estimate based on ERR models shown in Table 12-2 with no DDREF adjustment.

[b]Linear estimate based on EAR models shown in Table 12-2 with no DDREF adjustment.

[c]Estimates obtained as a weighted average (on a logarithmic scale) of estimates based on relative and absolute risk transport. For sites other than lung, breast, and thyroid, relative risk transport was given a weight of 0.7 and absolute risk transport was given a weight of 0.3. These weights were reversed for lung cancer. Models for breast and thyroid cancer were based on data that included Caucasian subjects. The resulting estimates were reduced by a DDREF of 1.5.

[d]Including uncertainty from sampling variability, transport, and DDREF. Sampling uncertainty in the parameters that quantify the modifying effects of age at exposure and attained age is not included except for the all solid cancer model.

[e]Includes thyroid cancer estimate based on ERR model.

[f]Includes breast cancer estimate based on EAR model.

[g]Estimates based on model developed by analyzing LSS incidence data on all solid cancers excluding thyroid cancer and nonmelanoma skin cancer as a single category. See Table 12-1.

TABLE 12-5B Lifetime Attributable Risk of Solid Cancer Mortality

Cancer Site	Males			Females		
	LAR Based on Relative Risk Transport[a]	LAR Based on Absolute Risk Transport[b]	Combined and Adjusted by DDREF[c] (Subjective 95% CI[d])	LAR Based on Relative Risk Transport[a]	LAR Based on Absolute Risk Transport[b]	Combined and Adjusted by DDREF[c] (Subjective 95% CI[d])
Stomach	14	150	19 (2, 190)	19	190	25 (3, 220)
Colon	130	89	76 (32, 180)	78	50	46 (16, 130)
Liver	16	120	20 (3, 150)	8	84	11 (1, 130)
Lung	240	200	140 (52, 380)	620	340	270 (110, 660)
Breast				110 Not used	110	73 (37, 150)
Prostate	35	1	9 (<0, 300)			
Uterus				4	24	5 (<0, 38)
Ovary				37	34	24 (6, 98)
Bladder	34	31	22 (7, 73)	45	36	28 (10, 81)
Other	180	190	120 (54, 280)	200	180	132 (61, 280)
Sum of site-specific estimates	650	780	410	1120[e]	1050	610
All solid cancer model[f]	760	650	480 (240, 980)	1200	940	740 (370, 1500)

NOTE: Number of deaths per 100,000 exposed persons of mixed ages exposed to 0.1 Gy.

[a]Linear estimate based on ERR models shown in Table 12-2 with no DDREF adjustment.

[b]Linear estimate based on EAR models shown in Table 12-2 with no DDREF adjustment.

[c]Estimates obtained as a weighted average (on a logarithmic scale) of estimates based on relative and absolute risk transport. For sites other than lung, breast, and thyroid, relative risk transport was given a weight of 0.7 and absolute risk transport was given a weight of 0.3. These weights were reversed for lung cancer. Models for breast and thyroid cancer were based on data that included Caucasian subjects. The resulting estimates were reduced by a DDREF of 1.5.

[d]Including uncertainty from sampling variability, transport, and DDREF. Sampling uncertainty in the parameters that quantify the modifying effects of age at exposure and attained age is not included except for the all solid cancer model.

[e]Includes breast cancer estimate based on EAR model.

[f]Estimates based on model developed by analyzing LSS mortality data on all solid cancers as a single category. See Table 12-1.

tribute about equally to cancer incidence in females. Lung cancer is the most important contributor to cancer mortality in females.

Although the committee's preferred estimates for all solid cancers are the sums of the site-specific estimates, for comparison the last line of Tables 12-5 shows estimates based on models developed by analyzing LSS data incidence and mortality data on all solid cancers as a single category (see Table 12-1). These estimates are generally about 20% higher than those obtained using the sum-of-sites approach, a difference that comes about in part because of the weighting scheme used to combine estimates based on relative and absolute risk transport (particularly the greater weight given to absolute risk transport for lung cancer) and because of the use of the model developed by Preston and colleagues (2002a) for breast cancer, similar to assuming absolute risk transport for this site.

Table 12-6 shows estimates of the all solid cancer LARs for several exposure scenarios. In each case, these were obtained as the sum of the site-specific estimates. Additional detail is given in Annex 12D. Because models for most cancers allow for a decrease in both the ERR and the EAR with increasing age at exposure, estimates for persons exposed at age 10 are more than twice those for persons exposed at ages 30 or 50. However, because models allow for no further decrease after age 30, the difference in lifetime risk estimates

for persons exposed at ages 30 and 50 is not as great. Also shown are estimates of the LAR for chronic lifetime exposure to 1 mGy per year and of the LAR for an occupational scenario of exposure to 10 mGy per year from ages 18 to 65.

Table 12-7 shows estimates of the LARs for leukemia incidence and mortality for several exposure scenarios. The number of years of life lost per death was estimated to be 20 years for males and 21 years for females, values that are greater than those for solid cancers. Although the transport model has not been considered a major source of uncertainty in leukemia risk estimates (UNSCEAR 2000b; NIH 2003), Table 12-7 shows that LAR estimates based on relative risk transport are higher than those based on absolute risk transport, with the ratio ranging from about 1 to 3. This is not due to the contribution of CLL since that was excluded from the baseline rates used to calculate LARs based on relative risk transport. The committee's preferred estimates are based on a weighted mean of LAR estimates obtained from the two transport models as with most site-specific solid cancers, and the subjective confidence intervals include transport uncertainty. Unlike solid cancer models, the leukemia models (Table 12-3) are based on linear-quadratic functions of dose, so there is no need for further reduction by a DDREF. Uncertainty calculations include sampling uncertainty in both the linear coefficient and the curvature parameter. Previous risk assessments have considered leukemia incidence and

TABLE 12-6 Committee's Preferred Estimates of Lifetime Attributable Risk (LAR) of Solid Cancer Incidence and Mortality[a] (with 95% subjective CIs)[b]

Exposure Scenario	Incidence		Mortality	
	Males	Females	Males	Females
0.1 Gy to population of mixed ages	800 (400, 1590)	1310 (690, 2490)	410 (200, 830)	610 (300, 1230)
0.1 Gy at age 10	1330 (660, 2660)	2530 (1290, 4930)	640 (300, 1390)	1050 (470, 2330)
0.1 Gy at age 30	600 (290, 1260)	1000 (500, 2020)	320 (150, 650)	490 (250, 950)
0.1 Gy at age 50	510 (240, 1100)	680 (350, 1320)	290 (140, 600)	420 (210, 810)
1 mGy per year throughout life	550 (280, 1100)	970 (510, 1840)	290 (140, 580)	460 (230, 920)
10 mGy per year from ages 18 to 65	2600 (1250, 5410)	4030 (2070, 7840)	1410 (700, 2860)	2170 (1130, 4200)

NOTE: Number of cases or deaths per 100,000 exposed persons.

[a]These were obtained as the sum of site-specific LAR estimates. The site-specific estimates were obtained as a weighted average (on a logarithmic scale) of estimates based on relative and absolute risk transport. For sites other than lung, breast, and thyroid, relative risk transport was given a weight of 0.7 and absolute risk transport was given a weight of 0.3. These weights were reversed for lung cancer. Models for breast and thyroid cancer were based on data that included Caucasian subjects. The resulting linear estimates were reduced by a DDREF of 1.5.

[b]Including uncertainty from sampling variability, transport, and DDREF. The uncertainty evaluation was based on evaluation of estimates based on analyses of LSS cohort data on all solid cancers analyzed as a single category as described in Annex 12C.

TABLE 12-7 Lifetime Attributable Risk of Leukemia Incidence and Mortality[a]

Exposure Scenario	Males			Females		
	LAR Based on Relative Risk Transport[b]	LAR Based on Absolute Risk Transport[c]	Committee's Preferred Estimate[d] (Subjective 95% CI[e])	LAR Based on Relative Risk Transport[b]	LAR Based on Absolute Risk Transport[c]	Committee's Preferred Estimate[d] (Subjective 95% CI[e])
Incidence						
0.1 Gy to population of mixed ages	120	64	100 (33, 300)	94	38	72 (21, 250)
0.1 Gy at age 10	140	85	120 (40, 360)	110	50	86 (25, 300)
0.1 Gy at age 30	87	77	84 (31, 230)	69	49	62 (22, 170)
0.1 Gy at age 50	110	45	84 (24, 290)	84	30	62 (16, 230)
1 mGy per year throughout life	83	40	67 (19, 230)	68	26	51 (13, 200)
10 mGy per year from ages 18 to 65	430	240	360 (110, 1140)	340	160	270 (79, 920)
Mortality						
0.1 Gy to population of mixed ages	88	40	69 (22, 220)	71	25	52 (14, 190)
0.1 Gy at age 10	88	42	70 (21, 240)	71	26	53 (13, 210)
0.1 Gy at age 30	70	53	64 (23, 180)	59	36	51 (17, 150)
0.1 Gy at age 50	93	37	71 (20, 250)	74	26	54 (14, 210)
1 mGy per year throughout life	62	25	47 (13, 180)	53	17	38 (9, 160)
10 mGy per year from ages 18 to 65	350	170	290 (84, 970)	290	120	220 (61, 820)

NOTE: Number of cases or deaths per 100,000 exposed persons.

[a]All estimates are based on linear-quadratic model.

[b]Based on ERR model shown in Table 12-4.

[c]Based on EAR model shown in Table 12-4.

[d]Obtained as a weighted mean (on a logarithmic scale) with weights of 0.7 for the relative risk transport estimate and a weight of 0.3 for the absolute risk transport estimate.

[e]Including uncertainty from sampling variability and transport. Sampling uncertainty includes uncertainty in both the linear and the quadratic terms of the dose-response.

mortality to be very similar, and this was likely the case at the time many of the LSS leukemia data were obtained. However, currently leukemia is not always rapidly fatal, and the committee has thus reduced estimates based on the LSS cohort for estimating leukemia mortality (see "Methods of Calculating Lifetime Risks"). For a single exposure of a population of mixed ages to 0.1 Gy, leukemia mortality estimates are about 30% lower than those for leukemia incidence.

Detailed tables showing lifetime risk estimates are found in Annex 12D. Annex 12D also gives examples of the use of these tables to obtain risk estimates for specific exposure scenarios.

Comparison of BEIR VII Risk Estimates with Those from Other Sources

Tables 12-8 and 12-9 compare the BEIR VII committee's lifetime risk estimates with estimates recommended by other organizations in recent years. A description of the approaches used to obtain these earlier risk estimates is given in Annex 12A. The ICRP and EPA solid cancer estimates include reduction by a DDREF of 2 (except for the EPA estimates for breast and thyroid cancers, where linear estimates were used without reduction). Neither BEIR V nor UNSCEAR made specific recommendations regarding reduction of risks at low doses and low dose rates. Estimates from these organizations are shown with no reduction and, to facilitate comparison with BEIR VII estimates, are reduced by a DDREF of 1.5 with the latter shown in parentheses. UNSCEAR presents estimates for site-specific solid cancers based on both relative and absolute risk transport models without expressing a preference. Again to facilitate comparison, the UNSCEAR estimates in parentheses combine these estimates using the same approach adopted by the BEIR VII committee and reducing them by a DDREF of 1.5.

BEIR VII, BEIR V (NRC 1990), and UNSCEAR (2000b) present estimates that are sex-specific, whereas ICRP (1991) and EPA (1999) present a single estimate for both sexes.

TABLE 12-8 Comparison of BEIR VII Lifetime Cancer Mortality Estimates with Those from Other Reports

Cancer Category	BEIR V[a] (NRC 1990)	ICRP[b] (1991)	EPA[b] (1999)	UNSCEAR[c] (2000)		BEIR VII[d]
Leukemia[e]	95 50	56	50			61
All cancer except leukemia (sum)	700 (460)	450	520			
All solid cancers (sum)				1150, 780 1400[f], 1100[f]	(520)	510
Digestive cancers	230 (150)					
Esophagus		30	12	30, 60	(25)	
Stomach		110	41	15, 120	(18)	22
Colon		85	100	160, 50	(75)	61
Liver		15	15	20, 85	(20)	16
Respiratory cancer	170 (110)					
Lung		85	99	340, 210	(160)	210
Female breast[g]	35 (23)	20	51	280, 65	(43)	37
Bone		5	1	—		
Skin		2	1	—		
Prostate[g]						5
Uterus[g]						3
Ovary[g]		10	15			12
Bladder		30	24	40, 20	(22)	25
Kidney		—	5	—		
Thyroid		8	3	—		
Other cancers or other solid cancers[h]	260 (170)	50	150	280, 180	(160)	130

NOTE: Excess deaths for population of 100,000 of all ages and both sexes exposed to 0.1 Gy.

[a]Average of estimates for males and females. The measure used was the excess lifetime risk; unlike other estimates in this table, radiation-induced deaths in persons who would have died from the same cause at a later time in the absence of radiation exposure are excluded. The estimates are not reduced by a DDREF, but parentheses show the result that would be obtained if the DDREF of 1.5, used by the BEIR VII committee, had been employed.

[b]Except for the EPA breast and thyroid cancer estimates, the solid cancer estimates are linear estimates reduced by a DDREF of 2.

[c]Average of estimates for males and females. Except where noted otherwise, estimates are based on the attained-age model. The first estimate is based on relative risk transport; the second on absolute risk transport. The estimate in parentheses is a combined estimate (using the same weights as used by the BEIR VII committee applied on a logarithmic scale) reduced by a DDREF of 1.5, although these were not recommendations of the UNSCEAR committee.

[d]Average of the committee's preferred estimates for males and females from Table 12-5B.

[e]Estimates based on a linear-quadratic model.

[f]Estimates based on age-at-exposure model.

[g]These estimates are half those for females only.

[h]These estimates are for the remaining solid cancers.

TABLE 12-9 Comparison of BEIR VII Lifetime Sex-Specific Cancer Incidence and Mortality Estimates with Those from Other Reports

Cancer Category	Males			Females		
	BEIR V[a]	UNSCEAR[b]	BEIR VII[c]	BEIR V[a]	UNSCEAR[b]	BEIR VII[c]
Incidence						
Leukemia[d]	NA	50	100	NA	50	72
All solid cancer	NA	1330, 1160 (740) 2600,[e] 1700[e]	800	NA	3230, 1700 (910) 3800,[e] 2100[e]	1310
Mortality						
Leukemia[d]	110	50	69	80	60	52
All cancer except leukemia (sum)	660 (440)			730 (490)		
All solid cancers (sum of sites)		710, 620 (380) 900,[e] 900[e]	410		1580, 930 (660) 1900,[e] 1300[e]	610

NOTE: Excess deaths for population of 100,000 of all ages exposed to 0.1 Gy.

[a]The measure used was the ELR; unlike other estimates in this table, radiation-induced deaths in persons who would have died from the same cause at a later time in the absence of radiation exposure are excluded. The estimates are not reduced by a DDREF, but parentheses show the result that would be obtained if the DDREF of 1.5, used by the BEIR VII committee, had been employed.

[b]Except where noted otherwise, estimates are based on the attained-age model. The first estimate is based on relative risk transport; the second on absolute risk transport. The estimate in parentheses is a combined estimate (using the same weights as used by the BEIR VII committee applied on a logarithmic scale) reduced by a DDREF of 1.5, although these were not recommendations of the UNSCEAR committee.

[c]Estimates are from Tables 12-6 and 12-7, and are shown with 95% subjective confidence intervals.

[d]Estimates based on a linear-quadratic model.

[e]Estimates based on age-at-exposure model.

Table 12-8 addresses comparisons that include cancer mortality estimates developed by the ICRP and EPA. Thus, the estimates in this table from BEIR V, UNSCEAR (2000b), and BEIR VII are averages of estimates for males and females. The BEIR V leukemia estimates are higher than the other leukemia estimates presented, possibly because chronic lymphatic leukemia was included in applying its ERR model. Only BEIR VII adjusted the LSS data to account for the fact that leukemia is not always rapidly fatal. The estimates of mortality from all solid cancers are very similar if a DDREF of 1.5 is applied to the BEIR V and UNSCEAR estimates. The BEIR VII estimate is also similar to the ICRP and EPA estimates even though different DDREFs were used. BEIR V estimated the ELR, which can be expected to be smaller than estimates of REID or LAR; the all solid cancer ELR would be expected to be about 20% smaller than the REID or LAR.

There are several factors that account for variation in estimates for site-specific cancers, which include differences in the choice of transport model and differences in the data that were evaluated. Estimates by BEIR V, ICRP, and EPA were based mainly on LSS mortality data from 1950 to 1985. UNSCEAR evaluated LSS mortality data through 1990 and LSS cancer incidence data through 1987, whereas BEIR VII evaluated site-specific mortality data through 1997 and cancer incidence data through 1998. ICRP estimates were in-

tended to be relevant for a world population, whereas the other estimates were specifically for the U.S. population. To some extent, the variation in estimates of site-specific cancers simply reflects the general uncertainties in this process.

Table 12-9 shows the sex-specific estimates for cancer incidence and mortality recommended by BEIR V, UNSCEAR (2000b), and BEIR VII. BEIR V did not present lifetime cancer incidence estimates, although models were developed for estimating breast and thyroid cancer incidence. Sex-specific comparisons for cancer mortality follow the same patterns as the non-sex-specific estimates shown in Table 12-8, with similar estimates provided a DDREF is applied to BEIR V and UNSCEAR. The BEIR VII leukemia incidence estimates are larger than those of UNSCEAR. This is probably due primarily to the fact that the BEIR VII estimates are a weighted mean of estimates based on relative and absolute risk transport (using both ERR and EAR models), whereas UNSCEAR estimates are based entirely on absolute risk transport. The BEIR VII all solid cancer estimate for females is larger than the UNSCEAR estimate that would have been obtained if the same approach to transport and DDREF had been applied. Examining the site-specific incidence estimates (not shown) indicates that lung cancer and the residual category of other solid cancers are the strongest contributors to this difference. The committee notes once again that BEIR VII cancer incidence estimates were based on LSS data that included 11 more years of follow-up than the data analyzed by the UNSCEAR committee.

UNCERTAINTIES IN LIFETIME RISK ESTIMATES

As noted early in this chapter, quantitative estimates of cancer risk are subject to several sources of uncertainty, which come about because of limitations in epidemiologic data and because the populations and exposures for which risk estimates are needed nearly always differ from those for which epidemiologic data are available. Several organizations have conducted detailed uncertainty assessments, which are described in Annex 12A. The NCRP (1997) evaluated uncertainties in the lifetime risk of total cancer mortality, and EPA (1999) provides extensive discussion of sources of uncertainty and gives example quantitative evaluations for lung cancer and leukemia. The NIH (2003) conducted a comprehensive evaluation of uncertainty in the excess relative risk used to calculate the assigned share, and it would be possible to extend this to lifetime risk estimates.

Quantitative Evaluation of Uncertainty

The lifetime risk estimates shown in Tables 12-5, 12-6, and 12-7 are accompanied by subjective confidence intervals that quantify the most important uncertainty sources: (1) sampling variability in risk model parameter estimates from the LSS data, (2) the uncertainty about transport of risk from a Japanese (LSS) to a U.S. population, and (3) the uncertainty in the appropriate value of a DDREF for adjusting low-dose risks based on linear-in-dose risk models estimated from the LSS data. This section gives more details on the allocation of uncertainty by source and discusses sources of uncertainty that were not included in the committee's quantitative assessment.

As an example, Table 12-10 displays the estimated lifetime attributable risks of cancer incidence for various sites shown in Table 12-5A, corresponding to a population of persons of mixed ages exposed to 0.1 Gy. The confidence intervals in Table 12-5A were constructed from the standard error of the estimated logarithm of LAR. This standard error is conveyed in Table 12-10 as the coefficient of variation, which is the standard error of LAR as a percentage of the

TABLE 12-10 Estimated Lifetime Attributable Risks of Solid Cancer Incidence[a] for a Population of Mixed Ages Exposed to 0.1 Gy (Corresponding to Table 12-5A)

Site	LAR (per 10^5)	CV (%)	Variance of log (LAR)	Variance (Percentage) Due to		
				Estimation	Transport	DDREF
Males						
Stomach	34	176	1.41	0.10 (7)	1.22 (86)	0.09 (6)
Colon	154	46	0.19	0.15 (40)	0.08 (13)	0.09 (47)
Liver	27	127	0.96	0.14 (15)	0.73 (76)	0.09 (9)
Lung	138	55	0.26	0.16 (60)	0.01 (5)	0.09 (34)
Bladder	98	69	0.39	0.28 (72)	0.02 (5)	0.09 (23)
Other solid	285	46	0.19	0.09 (45)	0.02 (8)	0.09 (46)
All solid	669	36	0.12	0.02 (18)	0.01 (8)	0.09 (74)
Leukemia	101	61	0.32	0.23 (72)	0.09 (28)	—
Females						
Stomach	43	161	1.28	0.05 (4)	1.14 (89)	0.09 (7)
Colon	96	57	0.28	0.15 (54)	0.04 (14)	0.09 (32)
Liver	12	184	1.48	0.31 (21)	1.08 (73)	0.09 (6)
Lung	304	51	0.23	0.04 (16)	0.10 (44)	0.09 (39)
Breast	462	36	0.12	0.03 (25)	0.00 (0)	0.09 (75)
Ovary	60	85	0.54	0.42 (79)	0.02 (5)	0.09 (17)
Bladder	94	63	0.34	0.19 (58)	0.05 (15)	0.09 (27)
Other solid	288	45	0.19	0.06 (32)	0.04 (20)	0.09 (48)
All solid	1048	34	0.11	0.01 (11)	0.006 (6)	0.09 (83)
Leukemia	72	71	0.41	0.24 (58)	0.17 (42)	—

NOTE: Number of excess cases per 100,000 exposed.

[a]Also shown are the coefficients of variation (estimated standard deviation as a percentage of the estimated LAR value) and the variance of log (LAR) due to each of the three sources considered: sampling variability in the parameter estimates, uncertainty in the transport model (ERR or EAR), and presumed uncertainty in the DDREF. The approach for obtaining a single LAR and its uncertainty is detailed in Annex 12C.

quantity it estimates. Also shown are the contributions to the variance of the estimated LAR (on a log scale) due to each of the three sources considered: (1) the variability due to uncertain parameter estimates from LSS risk models, (2) the uncertainty due to choice of transport model, and (3) the uncertainty in the appropriate DDREF for low-dose adjustment. The percentages of overall uncertainty due to each of these three component sources are shown in parentheses. The effective value of the DDREF for values in Table 12-10 is 1.5, so *unadjusted* lifetime risks can be calculated by multiplying the second column by 1.5.

Uncertainty is largest for cancers of the stomach and liver, where the main contribution is from transport. Cancers of the bladder and ovary also have large uncertainties, but in this case the main contribution is from estimation (sampling variability). Female breast cancer and the combined category of all solid cancer (excluding thyroid and nonmelanoma skin cancer) have the least uncertainty. In both cases, the main contribution is from the DDREF. For leukemia, the variance due to estimation includes uncertainty in both the linear coefficient and the curvature parameter (Table 12-3) and, thus, can be considered as including uncertainty resulting from use of the LSS data for estimating risks at low doses and low dose rates.

Sources of Uncertainty Not Included in the Quantitative Assessment

Uncertainty sources that were not included in the quantitative assessment are discussed next. In general, sources of uncertainty can be broadly categorized as uncertainties in the estimated parameters that derive from limitations in the epidemiologic data, uncertainties in the models used to describe the LSS data, and uncertainties in applying these models to estimate risks from exposures at low doses and low dose rates to the U.S. population.

Uncertainties in Parameter Estimates Derived from Data on the LSS Cohort

The estimated parameters shown in Tables 12-1, 12-2, and 12-3 are subject to sampling variation that can be quantified objectively, and the above tables include confidence intervals. The lifetime risk estimates shown in Tables 12-5, 12-6, and 12-7 are also accompanied by subjective confidence intervals that include uncertainty from sampling variation.

Uncertainty in parameter estimates may also come about because of errors in the basic epidemiologic data used, including dose estimation errors and errors in disease detection and diagnosis. No epidemiologic study is free of such errors. For the LSS cohort, efforts have been made to quantify random errors in dose estimates, and analyses have been adjusted to account for these errors (Pierce and others 1990). However, there is uncertainty from this source because the

nature and magnitude of the random error are not known with certainty. Preston and colleagues (2004) acknowledge that the adjustments that have been used to account for random error in DS86 dose estimates may require modification for application to DS02 estimates.

Errors in disease detection and diagnosis can also bias parameter estimates, although this is probably not a serious source of uncertainty in risk estimates. Although a major strength of the LSS cohort is that mortality ascertainment is virtually complete, assignment of cause of death is not always accurate. Misclassification of cancer as noncancers will lead to underestimation of the EAR but should not bias estimates of the ERR since an RERF autopsy study by Sposto and coworkers (1992) found no evidence that ascertainment depends on dose. By contrast, misclassification of noncancers as cancers will lead to underestimation of the ERR but should not affect the EAR. Based on the study by Sposto and coworkers (1992), Pierce and colleagues (1996) estimate that the EAR for cancer mortality should be adjusted upward by about 16% to reflect errors in diagnostic misclassification, whereas the ERR for cancer mortality should be adjusted upward by about 12%. These results pertain to analyses of all cancer mortality. The magnitude of bias resulting from diagnostic misclassification undoubtedly varies by cancer site.

Cancer incidence data are probably much less subject to bias from underascertainment or from misclassification, and this was an important reason for the committee's decision to base models for site-specific cancers on incidence data. However incidence data are not available for survivors who migrated from Hiroshima and Nagasaki. Adjustments have been made to account for this (Sposto and others 1992), but there is likely some uncertainty in the adequacy of these adjustments.

A further source of uncertainty in parameter estimates arises because epidemiologic studies are not controlled experiments and thus are subject to potential bias from unmeasured factors that may differ by the level of exposure or dose. The LSS cohort is probably less subject to such bias than most other exposed cohorts since a primary determinant of dose is distance from the hypocenter, with a steep gradient of dose as a function of distance.

Uncertainty in the Selected Model for the Excess Relative Risk or Excess Absolute Risk

The committee has based its risk estimates for all solid cancers and for cancers of specific sites on models of the form shown in Equation (12-2). Although this model was chosen because it fitted the LSS cancer incidence and mortality data better than several alternative models that were evaluated, other models also fitted the data reasonably well. With mortality data, for example, a model in which the ERR or EAR depended only on age at exposure and not on attained age fitted the data nearly as well as the selected model (see annex Table 12B-2). Alternative models can lead to different lifetime risk estimates, particularly for persons exposed early in life. Furthermore, it was not feasible to evaluate all possible models that might be used to describe the LSS data.

The form of the model is particularly uncertain for cancers of specific sites. In most cases, the parameters that quantify the effects of age at exposure and attained age (see Equation 12-2) were taken to be those estimated in analyses of all solid cancers as a single outcome. However, for most sites, data were consistent with a wide range of values for these parameters. Even the form of the model might vary by cancer site. Although this was not investigated by the committee, it is doubtful that data for most specific sites would allow one to distinguish among various models. Finally, once again it should be noted that because all members of the LSS cohort were exposed at the same time, effects of age at exposure are confounded with secular trends (discussed further above and in Annex 12B).

Models based on either the ERR or the EAR as a function of dose, sex, age at exposure, and attained age can provide reasonable descriptions of the data, and the committee has presented risk estimates based on both choices. In its application, the differences in lifetime risks obtained for the two choices largely reflect differences in the method of transport to the U.S. population as discussed above. However, the two models could give somewhat different risk estimates even if applied to the LSS cohort. Preston and colleagues (1991) present lifetime risk estimates for solid cancer mortality in the LSS cohort. Estimates based on ERR and EAR models were similar for those exposed at ages of 30 or more, but for those exposed as children, estimates based on the EAR model were about 25% lower for men and 25% higher for women than estimates based an ERR model. (NOTE: The model used by Preston and others is the RERF model shown in Equation (12-2).)

Uncertainties in Use of the Model to Estimate Risks for the U.S. Population

The above section "Use of the Committee's Preferred Models to Estimate Risks for the U.S. Population" describes the committee's choices regarding several issues. Since data are inadequate to indicate clearly the correct choices, all are sources of uncertainty. The committee has quantified the uncertainty from its choice regarding transport of risks from a Japanese population to a U.S. population and from its choice regarding the DDREF for estimating risks from exposure to low doses and low dose rates. Additional sources of uncertainty which have not been quantified, are projection of risks over time, which is primarily important for persons exposed early in life, and estimating risks from low-energy X-rays, which is of importance in estimating risks from diagnostic medical procedures (for a discussion of this subject, see Chapter 1, "Different Effectiveness of γ-rays and X-rays").

COHERENCE OF MODELS WITH OTHER STUDIES

Comparison with Studies of Persons Exposed for Medical Reasons

Although the committee has analyzed only data from the LSS cohort, consideration has been given to published analyses of data from several medically exposed cohorts. For breast and thyroid cancers, the committee's recommended models are based on published analyses of pooled data from the LSS and from medically exposed persons. This section briefly describes results from relevant medical studies and their compatibility with BEIR VII models.

A number of studies involving radiation exposure for medical reasons are described and discussed in Chapter 7. Although these studies have increased our general knowledge of radiation risks, not all of them are suitable for quantitative risk assessment. Many studies lack the sample size and high-quality dosimetry that are necessary for precise estimation of risk as a function of dose, a point that is illustrated by the large confidence intervals for many of the risk estimates shown in Tables 7-2 to 7-6. Studies of therapeutic exposures often involve very large doses (5 Gy or more) where cell killing may lead to underestimation of the risk per unit dose. In addition, the presence of disease may modify radiation-related risk especially for organs directly affected by the disease, such as the lung in tuberculosis fluoroscopy patients and the breast in benign breast disease patients. Furthermore, studies frequently include only a limited range of exposure ages and thus provide little information on the modifying effect of this variable. For example, studies of persons treated with radiation for solid cancers are often limited to persons exposed at older ages; by contrast, most studies of thyroid cancer risk from external exposure involve exposure in childhood (Ron and others 1995a).

Often there is interest in comparing results from different studies to gain information on the modifying effects of factors that may differ among studies. For example, Chapter 10 ("Transport of Risks") discusses estimates from medical studies from the standpoint of comparing risks for cancer sites where baseline risks differ greatly for Japanese and Caucasian subjects. Most medically exposed cohorts differ in more than one way from the LSS cohort (e.g., baseline risks, size of doses, dose fractionation, age at exposure), making it difficult to interpret risk estimate comparisons. It must be acknowledged that data are inadequate to develop models that take account fully of the many factors that may influence risks. This is illustrated effectively in analyses by Preston and colleagues (2002a) of breast cancer incidence in eight cohorts, where it was not possible to find a common model that adequately described data from all eight cohorts.

In the material below, findings from selected studies are discussed that were *not* used in developing the BEIR VII models. The material is organized by cancer categories.

Breast Cancer

The BEIR VII committee's recommended model for breast cancer is the EAR model developed by Preston and colleagues (2002a), who found it possible to use a common model to describe data from female atomic bomb survivors, two cohorts of Massachusetts tuberculosis fluoroscopy patients, and the Rochester infant thymus irradiation cohort. Preston and colleagues (2002a) also analyzed data from additional cohorts: the New York acute postpartum mastitis cohort (Shore and others 1986), the Swedish benign breast disease cohort (Mattsson and others 1993), and two Swedish skin hemangioma cohorts exposed in infancy (Lundell and Holm 1996). These cohorts all exhibited patterns that were not compatible with the models noted in the previous paragraph and adopted by the committee. The reader should consult Preston and colleagues (2002a) for details on the differences, but they include lower risks for the skin hemangioma cohorts (possibly due to the lower dose rates at which they were exposed) and different age at exposure and attained age patterns for the New York postpartum mastitis and Swedish benign breast disease cohorts (possibly due to the existence of breast disease in these cohorts). The reasons for these differences are not understood, but remind us that our understanding of radiation risks is incomplete and that models used to describe radiation risks are likely to be oversimplifications.

Another study that was not used in the BEIR VII committee's breast cancer model is the Canadian tuberculosis fluoroscopy cohort, where breast cancer mortality has been evaluated. Howe and McLaughlin (1996) conducted combined analyses of this cohort and female atomic bomb survivors, and found it possible to describe both cohorts with common models, although it was necessary to exclude Nova Scotia women, who had risks (both ERR and EAR) that were much higher than non-Nova Scotia women. This study is also discussed in Chapter 10.

Thyroid Cancer

The committee's model for thyroid cancer risks was based on analyses of data from five studies of persons exposed under age 15 (Ron and others 1995a), as described earlier in this chapter. Although the pooled analyses did not include all studies addressing thyroid cancer risks from external radiation exposure, it included those considered most informative by the authors, who reviewed published studies of thyroid cancer and external radiation. Specifically, the analyses included cohort studies with at least 1000 irradiated subjects who had individual estimates of radiation dose to the thyroid and case-control studies with at least 20 thyroid cancer cases and adequate dose information.

Shore and Xue (1999) summarized data from several studies of thyroid cancer risks in persons exposed in childhood that were not included in the analyses by Ron and colleagues and found that the combined estimate of the ERR/Gy from these studies was similar to that obtained by Ron and colleagues. Shore and Xue also summarized data from studies involving adult exposure and confirmed the finding from A-bomb survivors that risks are much lower (and possibly nonexistent) among persons exposed as adults.

Site-Specific Solid Cancers Other Than Breast and Thyroid

Most medical exposure results in nonuniform doses to various organs of the body; thus, only site-specific estimates can be compared. As noted earlier, not all studies involving medical exposure have adequate dosimetry or sample sizes to obtain informative quantitative risk estimates. Furthermore, doses are often at a level where cell killing is likely to have reduced the risk per gray. Table 12-11 summarizes risk estimates for selected sites from six medically exposed cohorts where doses for individuals were estimated. The studies included are those of women treated for cervical cancer (Boice and others 1988), women treated for uterine bleeding with intrauterine radium capsules (Inskip and others 1990a) or X-irradiation (Darby and others 1994), ankylosing spondylitis patients (Weiss and others 1994), people treated for peptic ulcer (Carr and others 2002), and tuberculosis fluoroscopy patients (Howe 1995). The table is limited to sites for which (1) the BEIR VII committee provides lifetime risk estimates, (2) the investigators present estimates of the ERR/Gy (usually based on regression analyses), (3) the mean dose to the organ of interest is less than 4 Gy, and (4) the estimate is based on at least 30 cases. Further information on these studies is given in Chapter 7.

Also shown in Table 12-11 are BEIR VII sex-specific estimates of the ERR/Gy based on incidence data from the LSS cohort and taken from Table 12-2. Because the ankylosing spondylitis and peptic ulcer patients were predominantly male (sex-specific estimates were not presented), Table 12-11 compares estimates from these studies with those of male LSS cohort members. The estimates from medical studies can be considered an average over the exposure and attained ages of the study cohorts; in all cases, exposure occurred in adulthood. The LSS estimates are for exposure at age 30 or older at attained age 60, ages that seem likely to be reasonably appropriate for comparison with the medical studies.

In most cases, estimates from the medical studies are similar to those from the LSS cohort, especially if one considers statistical uncertainties reflected in the confidence intervals. The studies with mean organ doses exceeding 2 Gy (stomach cancer in ankylosing spondylitis patients and colon cancer in the U.K. uterine bleeding study) included many subjects with considerably higher doses, and this might have affected results. The colon cancer estimate based on women in the United Kingdom given X-ray therapy for uterine bleeding (Darby and others 1994) is lower than that for LSS females, but the two estimates are not significantly different. The estimate for women treated in the United States for this

TABLE 12-11 Comparison of the Estimated ERR/Gy from Selected Medical Studies to the ERR/Gy Used in the Committee's Preferred Models for Estimating Site-Specific Solid Cancer Incidence and Mortality

Cancer Site[a]	Medical Study	Sex	Mean Organ Dose (Gy)	Number of Exposed Cases	ERR/Gy Based on Medical Study (95% CI)	Comparable ERR/Gy from LSS Cohort[b] (95% CI)
Stomach	Cervical cancer[c]	Females	2	348	0.54 (0.05, 1.5)	0.48 (0.31, 0.73)
Stomach	Ankylosing spondylitis[d]	Males (83%)	2.5	127	−0.004 (< 0, 0.05)	0.21 (0.11, 0.40)
Colon	Uterine bleeding (US)[e]	Females	1.3	75	0.51 (<0, 5.6)	0.43 (0.19, 0.96)
Colon	Uterine bleeding (UK)[f]	Females	3.2	47	0.13 (0.01, 0.26)	0.43 (0.19, 0.96)
Lung	Peptic ulcer[g]	Males (80%)	1.8	125	0.24 (0.07, 0.44)	0.32 (0.15, 0.70)
Lung	Fluoroscopy[h]	Males	1.0	347	0.02 (< 0, 0.11)	0.32 (0.15, 0.70)
Lung	Fluoroscopy[h]	Females	1.0	108	−0.06 (< 0, 0.07)	1.40 (0.94, 2.1)
Prostate	Ankylosing spondylitis[d]	Males (83%)	1.5	88	0.14 (0.02, 0.28)	0.12 (<0, 0.69)
Bladder	Ankylosing spondylitis[d]	Males (83%)	1.5	71	0.24 (0.09, 0.41)	0.50 (0.18, 1.4)

NOTE: Estimated parameters with 95% CIs.

[a]Sites had to meet the following criteria: (1) the BEIR VII committee provides lifetime risk estimates, (2) the study investigators present estimates of the ERR/Gy, (3) the mean dose to the organ of interest is less than 4 Gy, and (4) the estimate is based on at least 30 exposed cases.
[b]For the LSS, estimates are sex-specific estimates from Table 12-3 (for the sex indicated in column 3) and are for exposure at age 30 at attained age 60.
[c]Boice and others (1988).
[d]Weiss and others (1994).
[e]Inskip and others (1990b).
[f]Darby and others (1994).
[g]Carr and others (2002).
[h]Howe (1995).

TABLE 12-12 Comparison of Estimated ERR/Gy from Selected Worker Studies with the ERR/Gy Used in the Committee's Preferred Models for Estimating Solid Cancer and Leukemia Mortality

	Age at Exposure		
	All Ages ERR per Gy (95% CI)	30+ ERR per Gy	20 ERR per Gy
All solid cancers (or all cancers but leukemia)			
Estimate from 3-country study[a]	−0.07 (−0.29, 0.30)		
Estimate from NRRW[b]	0.09 (−0.28, 0.52)		
BEIR VII estimate[c] reduced by a DDREF of 1.5			
Attained age 50		0.17	0.31
Attained age 60		0.15	0.27
Leukemia excluding CLL			
Estimate from 3-country study[a]	2.2 (0.1, 5.7)		
Estimate from NRRW[b]	2.6 (−0.03, 7.2)		
BEIR VII estimate[d] based on linear-quadratic function			
Time since exposure 5 years		2.4	6.4
Time since exposure 15 years		1.4	2.4
Time since exposure 25 years		1.1	1.6
Time since exposure 35 years		0.9	1.1

NOTE: Estimated parameters with 95% CIs.

[a]Cardis and others (1995).
[b]Muirhead and others (1999).
[c]Based on ERR model for cancer mortality in males shown in Table 12-1.
[d]Based on ERR model for leukemia mortality in males shown in Table 12-3.

disorder (Inskip and others 1990a) is higher and closer to that for LSS women. The most striking discrepancies are for stomach cancer in ankylosing spondylitis patients (Weiss and others 1994) and lung cancer in tuberculosis fluoroscopy patients (Howe 1995). In both cases, there is little evidence of risk in the medically exposed cohorts and estimates appear incompatible with those based on the LSS cohort. The stomach cancer discrepancy is especially striking if one considers that the baseline risk is much higher in the LSS cohort. Howe found no evidence of bias from several potential sources that were investigated in the fluoroscopy study and attributed this finding to the fractionated nature of the exposure. Nevertheless, modification of radiation-induced risk by the presence of lung disease (tuberculosis) in this cohort seems a reasonable and perhaps likely possibility.

Little (2001) has also made relevant comparisons. He compared estimates of the ERR/Gy from 65 studies of persons treated with radiation therapy for benign and malignant disease with estimates from LSS incidence (Thompson and others 1994) and mortality data (Pierce and others 1996). Little (2001) expanded on an earlier study by Little and colleagues (1999b). To address differences in ages at exposure and length of follow-up, Little derived estimates using only the portion of the LSS cohort corresponding to the age and follow-up period for each of the individual studies evaluated. A total of 116 cancer site-specific estimates were derived, including estimates for cancers of the salivary glands, esophagus, stomach, colon, rectum, liver, pancreas, larynx, lung, bone, nonmelanoma skin cancer, female breast, uterus, and ovary.

Little found that estimates of the ERR/Gy based on the medical studies were generally lower than those based on the LSS, although in most cases the differences were not statistically significant. He also found that the ratio of the medical and LSS results decreased with increasing dose and concluded that cell sterilization largely accounts for the discrepancy between estimates based on the LSS and the medical studies. Dose fractionation and differences in baseline risks were noted as additional contributing factors. The data used by Little included cancer cases through 1987 and cancer deaths through 1990, in contrast to 1995 for incidence data and 1998 for mortality data used by the BEIR VII committee. Also, using only subsets of the LSS data may result in less stable estimates than modeling age at exposure and time since exposure or attained age.

In addition to the overall level of risk, medical studies can potentially inform us regarding patterns of risk by sex, age at exposure, and time since exposure. However, many of the relevant studies (such as those included in Table 12-11) were primarily single-sex studies involving exposure in adulthood, thus providing little information on the modifying effects of these factors. Several studies have confirmed the persistence of excess risk 30 or more years after exposure. The study of ankylosing spondylitis patients (Weiss and others 1994) is noteworthy in that there is no evidence of excess lung cancer

risk 25 years or more after exposure. Other cancers in this cohort also exhibited a decline in risk with time since exposure, although there was still evidence of risk at a reduced level after 25 years. Little and colleagues (1998) used data on cancer incidence in the LSS cohort and in five studies of patients exposed for medical reasons in childhood to investigate the pattern of risk with time since exposure. They found no evidence of heterogeneity in the magnitude of the decrease in relative risk with time since exposure.

Leukemia

Little (2001) found particularly striking differences between LSS-based estimates of the ERR/Gy for leukemia and those based on medically exposed persons. In all 17 studies evaluated, the estimated ERR/Gy was lower than that based on a comparable subset of the LSS, and for many of the studies, the differences were statistically significant. He also found that the ratio of the LSS and medical study estimates showed a strong decrease with increasing dose. Little conducted additional analyses that took account of curvature in the dose-response, cell sterilization, and fractionation of dose. When these variables were accounted for, the differences in the LSS and medical study estimates largely disappeared. Little concluded that cell sterilization is the primary reason for differences in estimates of the ERR/Gy that do not account for this factor.

In an earlier paper, Little and colleagues (1999c) evaluated patterns in the ERR/Gy for leukemia with age at exposure, time since exposure, and attained age in the LSS cohort, women treated for cervical cancer, and patients treated for ankylosing spondylitis. They found that patterns varied by leukemia subtype. Preston (1995) also found evidence of heterogeneity among subtypes based on LSS leukemia incidence data alone, although these analyses were based on the EAR rather than the ERR. Within each type of leukemia, Little and colleagues found no indication that patterns varied among the three cohorts. However, analyses treating all non-CLL leukemia as a single category showed patterns that were cohort dependent. A limitation of these analyses was that interactions of age at exposure with time since exposure or attained age were not investigated, whereas analyses by Preston (1995; Preston and others 2003) and by the BEIR VII committee of the LSS data indicate a need to include such interactions. There also was no evaluation of the comparability of the EAR among studies and subtypes of leukemia.

The committee's leukemia models are based on combined analyses of all types of leukemia within the LSS cohort. This was done both to yield more stable risk estimates and because updated leukemia incidence data (which would allow distinctions by subtype) were not available. It is acknowledged that subtype-specific models might have advantages, particularly if the relative frequencies of leukemia subtypes differed for the LSS cohort and the general U.S. population.

Conclusions

For the most part, data from medically exposed cohorts support the BEIR VII models. Although there are a few estimates from medical studies that seem incompatible with BEIR VII estimates, the evidence is not sufficiently compelling and consistent to provide a basis for modifying models.

Comparison with Studies of Nuclear Workers Exposed at Low Doses and Low Dose Rates

As discussed in Chapter 8, the most promising studies for direct assessment of risk at low doses and low dose rates are those of nuclear workers who have been monitored for radiation exposure through the use of personal dosimeters. Currently, the most informative risk estimates based on workers are those from a combined analysis of workers in three countries (IARC 1995) and from an analysis of workers in the National Registry of Radiation Workers (NRRW) in the United Kingdom (Muirhead and others 1999). Estimates from these studies are summarized in Table 8-7.

Table 12-12 compares worker-based estimates of the ERR/Gy with estimates that form the basis of BEIR VII models. Specifically, the BEIR VII estimates for all solid cancers are based on the ERR mortality model shown in Table 12-1. The BEIR VII estimates for leukemia are based on the ERR model shown in Table 12-3. Table 12-12 shows estimates of the ERR/Gy for males because workers studies have involved predominantly male exposure. Because the BEIR VII models allow for dependencies on age at exposure, attained age (solid cancer model), and time since exposure (leukemia model), estimates for several values of these variables that might be typical of workers are shown. It would be expected that the average age of exposure for workers would be 30 or more, but BEIR VII values for persons exposed at age 20 are also shown. The average time since exposure for workers is likely to exceed 15 years. The only BEIR VII estimates that are outside the confidence intervals for the worker studies are those for exposure at age 20 and, for leukemia, 5 years after exposure. Although the comparison is not precise, the estimates from the three-country study and the NRRW seem reasonably compatible with BEIR VII models for solid cancer mortality among males, especially when the wide confidence intervals for the worker-based estimates are considered.

SUMMARY

As in past risk assessments, the LSS cohort of survivors of the atomic bombings in Hiroshima and Nagasaki plays a principal role in developing the committee's recommended cancer risk estimates. In contrast to previous BEIR reports, data on both cancer mortality and cancer incidence (from the Hiroshima and Nagasaki tumor registries) were available to the BEIR VII committee. This made it possible to give much

more detailed attention to cancer incidence (including non-fatal cancers) than in past evaluations. It also made it possible to develop more reliable estimates for site-specific cancers due to the higher-quality diagnostic information compared with that based on death certificates. The cancer incidence data analyzed by the committee included nearly 13,000 cases occurring in the period 1958–1998. In addition, the committee evaluated data on approximately 10,000 cancer deaths occurring in the period 1950–2000, in contrast to fewer than 6000 cancer deaths available to the BEIR V committee. The longer follow-up period and larger number of cancer deaths and cases allowed more precise evaluation of risk and also more reliable assessment of the long-term effects of radiation exposure.

Although the committee did not conduct its own analyses of data from studies other than the LSS, for most studies with suitable data the results of analyses based on models similar to those used by the committee were available and evaluated by the committee. For cancers of the breast and thyroid, several medically exposed groups offer quantitative data suitable for risk assessment, and the committee's recommended models for these sites are those developed in published combined analyses of data from the relevant studies. For other cancer sites, data suitable for quantitative risk assessment were limited; for example, medical exposures often involve large therapeutic doses.

To use models developed primarily from the LSS cohort to estimate lifetime risks for the U.S. population, it was necessary to make several assumptions. Because of inherent limitations in epidemiologic data and in our understanding of radiation carcinogenesis, these assumptions involve uncertainty. Two of the most important sources of uncertainty are (1) the possible reduction in risk for exposure at low doses and low dose rates (*i.e.*, the DDREF), and (2) the transport of risk estimates based on Japanese atomic bomb survivors to estimate risks for the U.S. population. With regard to the first issue, the committee evaluated both data on solid cancer risks in the LSS cohort and experimental animal data pertinent to this issue. Based on this evaluation, the committee concluded that linear risk estimates obtained from the LSS cohort should be reduced by a factor in the range 1.1 to 2.3 for estimating risks at low doses and low dose rates, and a value of 1.5 was used to estimate solid cancer risks. For estimating the risk of leukemia, the BEIR VII model is linear-quadratic, since this model fitted the data substantially better than the linear model. The use of data on Japanese A-bomb survivors for estimating risks for the U.S. population (transport of risks) is especially problematic for sites where baseline risks differ greatly between the two countries. For cancer sites other than breast and thyroid (where data on Caucasian subjects are available), the committee presents estimates based on the assumption that the excess risk due to radiation is proportional to baseline risks (relative risk transport) and also presents estimates based on the assumption that the excess risk is independent of baseline risks. As a central estimate, the committee recommends a weighted estimate of these two results with the ratio of the two used to reflect the uncertainty in transporting risks. For most sites, a weight of 0.7 is used for relative transport and a weight of 0.3 is used for absolute transport; the weighting is reversed for lung cancer.

The committee provides estimates of lifetime risks of both cancer incidence and mortality for leukemia, all solid cancers, and cancers of several specific sites (stomach, colon, liver, lung, female breast, prostate, uterus, ovary, bladder, and all other solid cancers). The committee's recommended models provide the basis for sex-specific estimates for exposure scenarios including single exposures at various ages, chronic exposure throughout life, or occupational exposure from age 18 to 65.

As an example, Table 12-13 shows the estimated number of incident cancer cases and deaths that would be expected to result if a population of 100,000 persons with an age distribution similar to that of the entire U.S. population were each exposed to 0.1 Gy, and also shows the numbers that would be expected in the absence of exposure. Results are shown for all solid cancers and for leukemia. The estimates are accompanied by 95% subjective confidence intervals that reflect the most important sources of uncertainty, namely, statistical variation, uncertainty in the factor used to adjust risk estimates for exposure at low doses and low dose rates, and uncertainty in the method of transport. Consideration of additional sources of uncertainty would increase the width of these intervals. Mortality estimates are reasonably compatible with those in previous risk assessments, particularly if uncertainties are considered. Previous risk assessments have paid much less attention to cancer incidence.

The committee also presents estimates for each of several specific cancer sites and for other exposure scenarios, although they are not shown here. For many cancer sites, uncertainty is very large, with subjective 95% confidence intervals covering greater than an order of magnitude.

ANNEX 12A: PREVIOUS MODELS FOR ESTIMATING CANCER RISKS FROM EXPOSURE TO LOW LEVELS OF LOW-LET IONIZING RADIATION

This annex briefly reviews models that have been used in recent years to estimate risks of cancer. All details of these models are not given, but the general approaches that have been used are described. The committee begins with mention of the BEIR IV model for estimating lung cancer risks from exposure to radon, which is important because it was the first major radiation risk assessment based on modeling ERR (NRC 1988). Specifically, the BEIR IV committee analyzed data on four cohorts of underground miners and developed expressions for the ERR of lung cancer as a function of working level months, time since exposure, and attained age.

BEIR V

The BEIR V committee (NRC 1990) used the same general approach initiated in BEIR IV and analyzed data to develop expressions for the ERR for estimating risks from low-LET radiation. At the time the BEIR V committee began its work, the analyses needed for ERR-based risk modeling were not available, so it was necessary for the committee to rely extensively on its own analyses. The BEIR V committee models express the ERR as a function of radiation dose, sex, age at exposure, and time since exposure. Separate models were developed for mortality from leukemia, breast cancer, respiratory cancer, digestive cancer, and all other cancers. With the exception of breast cancer, the BEIR V mortality models were derived from analyses of A-bomb survivor mortality data for the period 1950–1985 (Shimizu and others 1990). The model for breast cancer mortality was based on both A-bomb survivor data and Canadian fluoroscopy patients. Models were also developed for breast and thyroid cancer incidence, although no lifetime risk estimates based on these models were presented. The breast cancer incidence model was based on data from A-bomb survivors,

TABLE 12-13 Committee's Preferred Estimates of Lifetime Attributable Risk of Incidence and Mortality for All Solid Cancers and for Leukemia

	All Solid Cancer		Leukemia	
	Males	Females	Males	Females
Excess cases (including nonfatal cases) from exposure to 0.1 Gy	800 (400, 1600)	1300 (690, 2500)	100 (30, 300)	70 (20, 250)
Number of cases in the absence of exposure	45,500	36,900	830	590
Excess deaths from exposure to 0.1 Gy	410 (200, 830)	610 (300, 1200)	70 (20, 220)	50 (10, 190)
Number of deaths in the absence of exposure	22,100	17,500	710	530

NOTE: Number of cases or deaths per 100,000 exposed persons with 95% subjective CIs.

Massachusetts fluoroscopy patients (Hrubec and others 1989), and New York postpartum mastitis patients (Shore and others 1986). The thyroid cancer incidence model was based on children in the Israel Tinea Capitis Study (Ron and Modan 1984) and the Rochester Thymus Study (Hempelmann and others 1975).

For leukemia, the ERR was found to depend on a linear-quadratic function of dose with modification by age at exposure (20 and >20 years) and time since exposure (2–15 and 15–25 years for exposure under age 20; 2–25 and 25–30 years for exposure over age 20). For female breast cancer mortality, the ERR was expressed as a linear function of dose with modification by time since exposure (risks increase, then decrease) and age at exposure, with a decline starting at age 15. For digestive cancers, the ERR was expressed as a linear function of dose with modification by sex and age at exposure (25, 25–35, >35 years), with risks decreasing with increasing exposure age. For respiratory cancer, the ERR was expressed as a linear function of dose with modification by sex and time since exposure, with risks decreasing with increasing time since exposure. For the residual category of all other cancers, the ERR was expressed as a linear function of dose with modification by age at exposure, with a decline starting at age 10.

To estimate risks at low doses and low dose rates, BEIR V used a linear-quadratic model for leukemia, which reduced effects at low doses by a factor of 2 over estimates that would have been obtained from a linear model. For cancers other than leukemia, a linear model was used with a nonspecific recommendation to reduce the estimates obtained through linear extrapolation by a factor between 2 and 10 for doses received at low dose rates.

Demographic data for the 1980 U.S. population were used to calculate lifetime risk estimates. These estimates were based on a multiplicative transportation model in which relative risks were assumed to be the same for the U.S. population and for Japanese A-bomb survivors. The risk measure used was the excess lifetime risk, which excludes radiation-induced deaths in persons who would have died from the same cause at a later time in the absence of radiation exposure. The BEIR V report provides estimates for excess mortality from leukemia and all cancers except leukemia expected to result from a single exposure to 0.1 Sv, from continuous lifetime exposure to 1 mSv per year, and from continuous exposure to 0.01 Sv per year from age 18 until age 65 with separate estimates for males and females. Estimates of the number of excess deaths (with confidence intervals), the total years of life lost, and the average years of life lost per excess death were given. For the single exposure scenario, separate estimates were presented for leukemia, breast cancer, respiratory cancer, digestive cancer, and other cancers, with each presented for both sexes and nine age-at-exposure groups.

BEIR V used Monte Carlo simulations to evaluate statistical uncertainties in its lifetime risk estimates for leukemia

and all cancer excluding leukemia. Statistical uncertainties in ERR/Sv for specific disease categories were also shown for various ages at exposure, for time since exposure, and for the two sexes. In addition, BEIR V discusses several other sources of uncertainty and shows risk estimates based on alternative models.

ICRP

The ICRP (1991) reviewed estimates provided by UNSCEAR (1988) and by BEIR V (NRC 1990) and recommended the UNSCEAR estimates. The UNSCEAR (1988) report gave several estimates, but those recommended by the ICRP were obtained by applying a model developed from A-bomb survivor mortality data for the period 1950–1985 (Shimizu and others 1990) to demographic data for the 1982 population of Japan. The lifetime risk measure used was the risk of exposure-induced death.

The ICRP recommended estimate for leukemia was based on a model in which the EAR depended on age at exposure (separate estimates for three categories: 0–9, 10–19, and 20+ years) and in which risks were assumed to persist for 40 years after exposure. The recommended estimate for all cancers other than leukemia was based on a model in which the ERR depended on age at exposure (same three categories as for leukemia) and remained constant from 10 years after exposure to the end of life. The ICRP also recommended that for exposures below 0.2 Gy or below 0.1 Gy / h, the linear risk estimates obtained from high-dose data be reduced by a DDREF of 2. Based on this approach, about 500 cancer deaths would be predicted from exposure of 0.1 Gy to a population of 100,000 persons of all ages (5.0×10^{-2} Sv^{-1}). For a working population (excluding children), about 400 cancers would be predicted (4.0×10^{-2} Sv^{-1}).

The ICRP was especially concerned with developing weighting factors to indicate the relative sensitivity for different cancer sites. Although a major objective in developing these weighting factors was to estimate the detrimental effects of radiation exposures that deliver nonuniform doses to various organs of the body, they can also be used to obtain lifetime risks for site-specific cancers. This is done by multiplying these factors by the lifetime risk estimates for all cancers. To develop these weighting factors, ICRP made use of risk calculations by Land and Sinclair (1991), which were conducted specifically for the ICRP. Land and Sinclair estimated lifetime risks for several types of cancer using age-specific risk coefficients from Japanese A-bomb survivors (taken from Shimizu and others 1990). Because the ICRP wanted its factors to be useful for a world population, separate calculations were made for five reference countries (Japan, United States, United Kingdom, Puerto Rico, and China) and were based on three sets of assumptions for projecting risks over time and for transporting risks across countries. Final recommendations were based on results obtained by averaging results over countries and over two of the models: the relative model in which relative risks were assumed to be constant both over time and among populations, and the so-called NIH model in which relative risks were assumed to be constant over time, but absolute risks were assumed to be constant among populations. These two models represent relative and absolute transportation models. For cancers of the thyroid, bone surface, skin, and liver, the ICRP 60 considered sources of data other than the A-bomb survivors to determine estimates.

Although uncertainties were not addressed by the ICRP, a later report by the NCRP (1997) discusses sources of uncertainty in detail and quantifies uncertainties in the ICRP lifetime risk estimate for all fatal cancers. This is accomplished by specifying uncertainty distributions for each of several sources and then combining these distributions using Monte Carlo simulations.

NCRP

The NCRP (1993) undertook its own review of risk models provided in UNSCEAR (1998), and by the BEIR V committee. This review resulted in the NCRP's supporting the ICRP recommendations.

EPA

The EPA (1994, 1999) also reviewed the models noted above and, in addition, reviewed models provided by the National Radiological Protection Board (NRPB) in the United Kingdom (Strather 1988) and by the U.S. Nuclear Regulatory Commission (Gilbert 1991). For most cancers, the EPA used results from Land and Sinclair (1991); specifically, EPA used the geometric mean of lifetime risk estimates based on the relative and absolute transportation models for the U.S. population. An exception was breast cancer, where EPA used the NRC model, which was developed from data on Massachusetts fluoroscopy patients (Hrubec and others 1989) and New York postpartum mastitis patients (Shore and others 1986). The EPA developed its own estimate for kidney cancer based on A-bomb survivor data and made use of published results on studies other than A-bomb survivors for its estimates of mortality from cancers of the liver, bone, thyroid, and skin. The EPA accepted the ICRP recommendation of a DDREF of 2, except for breast cancer where it recommended a DDREF of 1.

UNSCEAR

The UNSCEAR (2000b) report presents lifetime risk estimates for mortality and incidence for leukemia; all solid cancer; cancers of the esophagus, stomach, colon, liver, lung, breast, bladder, and thyroid (incidence only); and all remaining solid cancers. The leukemia model was that developed by Preston and colleagues (1994) and based on A-bomb survivor leukemia incidence data for the period 1950–1987.

This model allows the EAR to vary as a linear-quadratic function of dose and allows both the overall level of risk and the dependence on time since exposure to vary by sex and age at exposure. That is,

$$EAR(d, s, e, t) = \beta_e (d + \theta d^2) \exp [\gamma_{female} + (\delta_e + \varepsilon_{female}) (t - 25)],$$

where d is dose in sieverts, s is sex; e is an index for three age-at-exposure categories: 0–19, 20–39, and 40+ years; and t is time since exposure in years. The parameter θ indicates the degree of curvature, which does not depend on sex or age at exposure; β_e is the EAR for males exposed at various ages 25 years following exposure; δ_e indicates the dependence on time since exposure for males exposed at various ages; and γ_{female} and ε_{female} express the dependence of these parameters on sex and do not vary by age at exposure. The parameter estimates were as follows: $\beta_e = 0.33, 0.48$, and 1.31 for the three respective age-at-exposure categories; $\theta = 0.79$; $\gamma_{female} = 0.69$; $\delta_e = -0.17, -0.13$, and -0.03 for the three respective age-at-exposure categories; and $\varepsilon_{female} = 0.10$. Preston and colleagues (1994) note that allowing overall modification by sex and age at exposure in an EAR model did not significantly improve the fit once time since exposure was included in the model, but that these factors significantly modified the effects of time since exposure. Specifically, risks for those exposed early in life decreased more rapidly than the risk for those exposed later, and the decrease was less rapid for women than for men. For the UNSCEAR (2000b) application, excess risks for the first 5 years after exposure were assumed to be half of those observed 5 years after exposure.

The UNSCEAR (2000b) models for solid cancer mortality were based on A-bomb survivor mortality data for the period 1950–1990 (Pierce and others 1996), and the models for solid cancer incidence were based on A-bomb survivor incidence data for 1958–1987 (Thompson and others 1994). Risk estimates based on the following two models were presented:

Age-at-exposure model:
$$ERR(d, s, e) = \beta d \exp [\phi s + \gamma(e - 30)],$$
and

Attained-age model:
$$ERR(d, s, a) = \beta d \exp (\phi s) a^\eta,$$

where d is dose, s is an indicator variable for sex, e is age at exposure, and a is attained age. The attained-age model generally gives lower lifetime risks because of the attenuation of risks as people age.

To obtain estimates of the parameters quantifying the modifying effects of sex (ϕ), age at exposure (γ), and attained age (η), an approach described by Pierce and colleagues (1996) was used. With this approach, the parameters ϕ, γ, and η were set equal to those for all solid cancers unless there was evidence of significant departure from these values. Even though there was little evidence of heterogeneity in the main effect parameters (β) by site, the individual estimates were retained.

UNSCEAR (2000b) presented lifetime risk estimates based on demographic data for the populations of China, Japan, Puerto Rico, the United Kingdom, and the United States. Estimates based on both relative and absolute transportation models were presented. With the absolute risk model, the absolute magnitude of the radiation risk is assumed not to depend on the baseline risk, whereas with the relative risk model, the magnitude of the radiation risk is assumed to be proportional to the baseline risk. No recommendations were made as to which approach is preferred. Because baseline risks for site-specific cancers vary considerably from country to country, estimates based on the two models can differ substantially. For leukemia, only absolute transportation was used, since differences in the two approaches were trivial. Table 12A-1 summarizes lifetime risk estimates from the reports discussed above.

NIH Radioepidemiologic Tables

The NIH was mandated in 1983 to "devise and publish radioepidemiologic tables that estimate the likelihood that persons who have or have had any of the radiation-related cancers and who have received specific doses prior to the onset of such disease developed cancer as a result of these doses." The mandate included a provision for periodic updating of the tables. The first NIH radioepidemiologic tables were published in 1985, and they have been updated recently (NIH 2003). Although these efforts were not directly addressed at estimating lifetime risks, developing estimates of the so-called probability of causation, more correctly referred to as the "assigned share," requires modeling the ERR as a function of dose and other factors. Specifically,

$$AS = \frac{\text{Risk due to radiation exposure}}{\text{Baseline risk} + \text{risk due to radiation exposure}}$$
$$= ERR / (1 + ERR).$$

TABLE 12A-1 Lifetime Risk Estimates from Several Reports

Cancer Category	BEIR V (1990)[a]	ICRP (1991)[b]	EPA (1999)[b]	UNSCEAR (2000)[c] Mortality	Incidence[d]
Leukemia[e]	95	50	56	60	50
All cancer except leukemia	695	900			
All solid cancers	1430		985	1400,[f] 1100,[f] 1050, 780	3200,[f] 1900,[f] 2280,
Digestive cancers	230				
Esophagus		60	23	30, 60	15, 30
Stomach		220	81	15, 115	20, 170
Colon		170	208	160, 50	185, 160
Liver		30	30	20, 85	15, 320
Respiratory cancer	170				
Lung		170	198	335, 205	635, 150
Female breast[g]	35	40	51	280, 65	785, 260
Bone		10	2	—	—
Skin		4	2	—	—
Ovary[g]		20	30	—	—
Bladder		60	48	40, 20	75, 80
Kidney		—	10	—	—
Thyroid	16	3	—	50, 55	
Other cancers or other solid cancers[h]	260	100	299	275, 175	500, 205

NOTE: Excess deaths for population of 100,000 of all ages and both sexes exposed to 0.1 Gy. Estimates are based on linear models with no modification for low doses and low dose rates, although in some cases reduction by a factor of 2 or so was recommended.

[a]Estimates are the average of estimates for males and females. The measure used was ELR; unlike other estimates in this table, radiation-induced deaths in persons who would have died from the same cause at a later time in the absence of radiation exposure are excluded.

[b]Except for leukemia (see footnote e) and the EPA breast and thyroid cancer estimates, these estimates would be reduced by about a factor of 2 for exposures at low doses and low dose rates. No specific recommendations for such modification were made for BEIR V and UNSCEAR estimates.

[c]Average of estimates for males and females. Except where noted otherwise, estimates are based on the attained-age model. The first estimate is based on relative risk transportation; the second on absolute risk transportation.

[d]Excess cases instead of excess deaths.

[e]Estimates based on a linear-quadratic model and would not be further reduced for exposures at low doses and low dose rates.

[f]Estimates based on age-at-exposure model.

[g]Estimates are half those for females only.

[h]Estimates are for remaining cancers; the group differs for the various reports.

Thus, the models developed for the NIH report could be used to estimate lifetime risks.

Most NIH models were developed by analyzing A-bomb survivor cancer incidence data for the period 1958–1987 (Thompson and others 1994). Exceptions were thyroid cancer, where models were based on a pooled analysis of data from six different study populations by Ron and colleagues (1995a). Nonmelanoma skin cancer risks were estimated from a special A-bomb survivor data set used by Ron and colleagues (1998a). Models for leukemia were based on data from Preston and coworkers (1994).

Because adjudication of compensation claims for possibly radiation-related cancer is almost always specific to organ site, the list of sites for which models were provided was more extensive than most previous risk assessments. The NIH developed models for all sites with 50 or more incident cases among A-bomb survivors exposed to at least 5 mSv and, unlike most previous assessments, included site-specific cancers that have not been clearly linked with radiation exposure. The rationale for this was that the range of uncertainty is of interest regardless of whether or not a statistically significant dose-response association has been observed.

Although most previous leukemia models have been based on the EAR, NIH models were based on the ERR. Separate models were developed for leukemias of all types and for specific types of leukemia (acute lymphocytic leukemia, acute myelogenous leukemia, and chronic myelogenous leukemia) with different dependencies of the ERR on sex, age at exposure, and time since exposure. All leukemia models were based on a linear-quadratic function of dose, with equal contributions of the linear and quadratic terms.

For solid cancers other than thyroid and nonmelanoma skin cancer, the following linear dose-response function was used to model the ERR:

$$\text{ERR}(D, s, e, a) = \beta D \exp [\phi I_s(sex) + \gamma f(e) + \eta g(a)],$$

where D is dose in sieverts; $I_s(sex) = 1$ for females and $= 0$ for males; e is age at exposure in years; a is attained age in years; f and g are specified functions of e and a, respectively; and β, N, and O are unknown parameters. The choice of a model that included both age at exposure and attained age was based in part on the knowledge that models being applied at RERF to updated mortality and incidence data include both variables. Several specifications for the functions $f(e)$ and $g(a)$ were evaluated, with final models based on

$$f(e) = -15 \text{ for } e < 15, = e - 30 \text{ for } e \text{ between 15 and 30,}$$
$$\text{and } = 0 \text{ for } e > 30; g(a) = \log (a / 50) \text{ for}$$
$$0 < a < 50, \text{ and } = 0 \text{ for } a \ge 50.$$

This model has the property that for fixed attained age, the ERR/Sv is constant (at different levels) for exposure ages less than 15 years and greater than 30, but decreases between ages 15 and 30; the estimated ERR/Sv for ages less than 15 years is about 2.2 times that for ages greater than 30. For fixed exposure ages, the ERR/Sv declines up to an attained age of 50 and then remains constant; the ERR/Sv for attained age 30 is about 2.3 times that at attained age 50 or more. This model was chosen because it provided a slightly better fit to the data than a model that allowed risks to vary over the full range of exposure and attained ages [*i.e.*, $f(e) = e - 30$; $g(a) = \log (a / 50)$] and because it allowed more statistically stable estimates at the extremes of the exposure ages and attained ages.

The parameters ϕ, γ, and η were estimated from an analysis of all solid cancers excluding thyroid and nonmelanoma skin cancer, although cancers that occur in only one sex were excluded in estimating N; the estimated values of ϕ, γ, and η were, respectively, 0.84, –0.053, and –1.63. For cancer of a specific site, only data for that site were used to estimate β, and an approach similar to that used by Pierce and colleagues (1996) and by UNSCEAR (2000b) was used to estimate the parameters ϕ, γ, and η. With this approach, the common values noted above were used unless there was evidence that the site-specific values differed significantly from these common values. In the NIH application, the common values of γ and 0 were used for all specific sites other than lung and the category female genital cancers other than ovary, where these parameters were set equal to zero. The common value for ϕ was used for all sites except liver cancer, where the ERR/Sv for the two sexes was assumed to be equal; this choice was based on an analysis of liver cancer by Cologne and colleagues (1999). The assessment of uncertainty in the estimated parameters, some of which were site-specific and some of which were common to several sites, was complex and made use of an approach known as joint analysis (Pierce and Preston 1993); joint analysis allows some parameters to depend on cancer site whereas others are assumed to be common to several sites.

Although the models for solid cancer ERRs were based on linear dose-response functions, these estimates were reduced by a DDREF for estimating risks at low doses and low dose rates. Uncertainty in the DDREF was reflected in distributions that included values ranging from 0.5 to 5; different distributions were used for breast and thyroid cancer (more weight was given to linearity for these cancers). The DDREF is applied to all chronic exposures, whereas for acute exposures, the DDREF is phased in as the dose is decreased.

The NIH undertook a comprehensive uncertainty assessment. In fact, uncertainty was a fundamental part of the process in that the emphasis was not on determining single point estimates, but on estimating the uncertainty distribution. Uncertainty distributions for each of several sources were developed and Monte Carlo simulations were carried out to obtain overall uncertainty distributions for both the estimated ERR and the assigned share. A software tool Interactive RadioEpidemiological Program (IREP) was developed to carry out the simulations. Output from IREP gives several percentiles of the distribution for both the ERR and the assigned share. The following uncertainty sources were included: sampling variability in the estimated ERRs (including uncertainty in parameters expressing modification by age at exposure, sex, etc.), correction for random and systematic errors in A-bomb survivor dosimetry, extrapolation of risk from high to low doses and low dose rates (expressed as uncertainty in the DDREF), transfer of risk estimates from A-bomb survivors to a U.S. population, and modification by smoking history (lung cancer only). With the exception of sampling variability, the uncertainty distributions for the individual sources were based on informed but nevertheless subjective judgments.

ANNEX 12B: COMMITTEE ANALYSES OF DATA ON THE LSS COHORT TO DEVELOP BEIR VII MODELS FOR ESTIMATING CANCER RISKS

Chapter 12 describes models that were used by the committee to estimate cancer risks. These models were based primarily on analyses of A-bomb survivor cancer incidence data (1958–1995) and, to a lesser extent, mortality data (1950–1997), with consideration of published analyses of data from selected studies involving medical exposures. This annex presents details of analyses of data from the LSS cohort of atomic bomb survivors that were conducted to develop these models. Analyses of cancer mortality data were conducted by the committee. Because the most recent cancer incidence data were not yet available outside of RERF, analyses of these data were conducted under the direction of the committee by RERF investigators who served as agents of the Academy.

The committee's selected models for estimating solid cancer risks allow the ERR or EAR to depend on both age at exposure and attained age. Both the ERR and the EAR decline with increasing age at exposure. The ERR also declines with increasing attained age, while the EAR increases with increasing attained age. The BEIR VII models are similar to the model used in recent analyses of atomic bomb survivor data by RERF investigators, except that with the BEIR VII model, the ERR and EAR decrease with age at exposure only over the range 0–30 years with no further decrease after age 30. The material that follows describes analyses that were conducted to evaluate several possible models for solid cancer risks, including models that allow for dependence on age at exposure alone, on attained age alone, on time since exposure instead of attained age, and on the use of different functional forms to express these dependencies. Also evaluated are several models for leukemia risks.

Aproach to Analyses

Analyses of cancer incidence were based on cases diagnosed in the period 1958–1998. Analyses of mortality from all solid cancers and from leukemia were based on deaths occurring in the period 1950–2000 (Preston and others 2004), whereas analyses of mortality from cancer of specific sites were based on deaths occurring in the period 1950–1997 (Preston and others 2003). Both ERR models and EAR models were evaluated. Methods were generally similar to those used in recent reports by RERF investigators (Pierce and others 1996; Preston and others 2003) and were based on Poisson regression using the AMFIT module of the software package EPICURE (Preston and others 1991). Confidence intervals (95%) were usually calculated as the estimate plus and minus 1.96 times the standard error. For estimates of linear coefficients of dose, these were calculated on a logarithmic scale. Occasionally (as noted) confidence intervals were calculated using the likelihood profile. All p-values were based on chi-square approximations of likelihood ratio tests. These are based on differences in the maximized log likelihood statistics, often referred to as deviances.

To fit ERR models, baseline risks were handled by stratifying on gender, city of exposure (Hiroshima or Nagasaki), age at exposure, and attained age as described by Pierce and colleagues (1996). To fit EAR models, baseline risks were modeled using the parametric model described by Preston and coworkers (2003). For leukemia, the parametric model is that described by Preston and coworkers (2004). The risk of radiation-induced cancer was modeled as described in the sections that follow.

All analyses were based on newly implemented DS02 dose estimates. Doses were expressed in sieverts with a constant weighting factor of 10 for the neutron dose; that is, the doses were calculated as γ-ray absorbed dose (grays) + 10 × neutron absorbed dose (grays). Analyses were adjusted for random errors in doses using an approach described by Pierce and colleagues (1990) and developed from DS86 dose estimates. Preston and colleagues (2004) note that it has not yet been determined if modification of these methods is needed for DS02 dose estimates. Unless stated otherwise, doses are truncated to correspond to the 4 Gy kerma level.

The DS02 system provides estimates of doses to several organs of the body. For site-specific cancers the committee used dose to the organ being evaluated, with colon dose used for the residual category of "other" cancers. The weighted dose, d, to the colon was used for the combined category of all solid cancers or all solid cancers excluding thyroid and nonmelanoma skin cancer. This choice was made to achieve comparability with analyses by RERF investigators. Reference to an average organ dose—approximated, say, by the dose to the liver—might be more realistic for the analysis of solid cancers combined and would likely lead to about a 10% increase in the values of the weighted dose, d, and thus a reduction of about 10% in the risk coefficients (Kellerer and others 2001). However, the committee's estimates of the risk for all solid cancers are obtained by summing estimates for individual organ sites (based on doses to these organs), and thus should not be subject to this bias.

It has also been suggested that a weighting factor of roughly 30 for the neutron absorbed dose might be a better choice than 10. The higher value corresponds more closely to the radiation weighting factor recommended by the ICRP (1991) for fission neutrons. However, with the higher value it becomes critical that the weighting factor stands for the neutron low-dose RBE (*i.e.*, the ratio of the linear coefficients in the dose relations for neutrons and for γ-rays). The weighted dose, *d*, is then relevant only to the linear term in the dose-effect relation, while the dose-square term, which corresponds to a DDREF, has to contain the γ-ray absorbed dose alone. This change in the analysis might, in the case DDREF = 1, reduce the estimated ERR/Sv by roughly another 15% (Kellerer and others 2001).

General Considerations in Describing Dependencies of Solid Cancer Risks on Exposure Age and Attained Age

A decline in the solid cancer ERR with increasing exposure age has been demonstrated in several epidemiologic studies (UNSCEAR 2000b), and most models for estimating risks of solid cancers in the last decade have included a term that allowed for such a decline. Analyses of A-bomb survivor incidence and mortality data from the recent past (Thompson and others 1994; Pierce and others 1996) emphasized models of the form

$$\text{Exposure age model:}$$
$$\text{ERR} = \beta_s D \exp(\gamma e), \qquad (12B\text{-}1)$$

where *D* is dose in sieverts, β_M and β_F are sex-specific estimates of the ERR/Sv for exposure at age 30, and *e* is age at exposure in years. This model is often parameterized so that the β_s are the risks at an exposure age of 30, that is, by replacing *e* by (*e* − 30).

Although allowing for dependence of the ERR on exposure age seems appropriate, quantifying this dependence is subject to considerable uncertainty, especially for cancers of specific sites. Most medically exposed cohorts involve limited ranges of exposure age, and there is no medically exposed cohort that covers the full range of exposure ages from early childhood to old age. Thus, statistical power for evaluating the effects of exposure age within any single cohort is usually low.

The LSS cohort of Japanese A-bomb survivors is unique in providing data on persons exposed at all ages and, for this reason, has been used in many past risk assessments to quantify the effects of age at exposure (BEIR V, UNSCEAR). Reasonably precise estimates can be obtained when all solid cancers are analyzed as a single outcome. However, sample sizes for individual cancer sites are usually too small to quantify the effects of exposure age precisely. Estimates of the parameter γ vary widely among sites, but it is not possible to determine the extent to which this variation reflects real differences and the extent to which it reflects statistical variation.

An additional problem in quantifying the effect of age at exposure is that since all A-bomb survivors were exposed at the same time, the effects of exposure age are confounded with birth cohort effects. Japanese baseline rates for cancers of many specific sites show strong secular trends, which probably result at least in part from changes in life-style that have come about with the Westernization of Japan. For example, baseline rates for cancers of the colon, lung, and female breast have increased over the past 50 years so that early birth cohorts have lower baseline risks than later birth cohorts. This means that the appropriate way to estimate the effects of exposure age depends on how the factors responsible for secular trends affect radiation risks. If these factors increase or decrease radiation risks to the same extent that they increase or decrease baseline risks (a multiplicative relationship), then estimation of the effect of age at exposure should be based on modeling the ERR. However, if the factors responsible for secular trends in baseline risks have no effect on radiation risks (an additive relationship), then estimation of the effects of age at exposure should be based on modeling the EAR. If the chosen model is not correct, then estimated exposure age effects may be influenced by secular trends in Japanese baseline rates and may not be applicable to populations other than the LSS cohort. Further discussion of secular trends and their influence on estimating the effects of age at exposure can be found in Preston and colleagues (2003).

With the Equation (12B-1) model, the ERR is assumed to be constant over the follow-up period for fixed exposure age. (This is likely to be an oversimplification since Little and others 1991, Thompson and others 1994, and Pierce and others 1996 have all demonstrated that the ERR declines with increasing attained age, or time since exposure, at least for those exposed early in life [under age 20]). In addition, it is now recognized that some and perhaps all of the decline in the ERR with exposure age can also be described as a decline in the ERR with attained age (Kellerer and Barclay 1992; Preston and others 2002a). Pierce (2002) describes the age-time patterns in A-bomb survivor cancer incidence data and discusses difficulties in interpreting them. He also discusses a possible biological rationale for a model in which the ERR decreases with attained age.

As noted in Annex 12A, UNSCEAR (2000b) provided two solid cancer models—one based on age at exposure (as shown by Equation (12B-1)) and one based on attained age. The UNSCEAR attained-age model is of the form

$$\text{Attained age model: ERR} = \beta_s D a^\eta, \qquad (12B\text{-}2)$$

where *D* is dose in sieverts, β_M and β_F are sex-specific estimates of the ERR/Sv, and *a* is attained age in years. This model is often parameterized so that β_s represents the ERR/Sv at attained age 60, that is, by replacing *a* by (*a* / 60).

Even when the exposure-age and attained-age models provide comparable fits to the data, estimated lifetime risks based on the two models are not the same, especially for specific age-at-exposure groups such as persons exposed early in life. For example, UNSCEAR estimates for persons exposed at age 10 based on the exposure-age model are about twice those based on the attained-age model.

The most recent analyses of A-bomb survivor incidence (R13 incidence report) and mortality data (Preston and others 2003) have emphasized models that allow the ERR to depend on both age at exposure and attained age. That is,

$$\text{RERF model: ERR} = \beta_s D \exp(\gamma e) a^{\eta}. \quad (12B-3)$$

This choice was made in part because of difficulties in distinguishing the fits of the two models above and because, with the incidence data, analyses of all solid cancers indicated a need for modification by both exposure age and attained age. It is expected that analyses of updated cancer incidence data will allow for dependencies on both exposure age and attained age.

Recent A-bomb survivor reports also show results based on models for the EAR. These models are of the same form as given above, although the parameters have different interpretations. In particular, the parameter that quantifies the dependence on attained age describes the strong increase in excess risk with this variable.

The models developed in the following two sections allow for dependencies on both exposure age and attained age. Although the RERF model is evaluated, consideration has also been given to other forms for the dependencies on exposure age and attained age. Both ERR and EAR models are evaluated. Because sample sizes for individual cancer sites are usually too small to quantify precisely the effects of either age at exposure or attained age, the parameters that quantify these effects are in most cases obtained from analyses of all solid cancers. As shown later, with ERR models there are few instances in which the site-specific estimates of these parameters differ significantly from the common values. However, with EAR models there is evidence that the dependence on attained age varies by site.

In the material that follows, the committee first describes analyses conducted to determine the basic form of the preferred model. It then describes analyses of site-specific cancers that were used to confirm the committee's model choice and to evaluate the appropriateness of using common parameters.

Analyses of Incidence Data on All Solid Cancers Excluding Thyroid and Nonmelanoma Skin Cancer and of Mortality Data on All Solid Cancers

The analyses of cancer incidence data described in this section were based on the category of all solid cancers excluding thyroid cancer and nonmelanoma skin cancer. These exclusions were made primarily because both thyroid cancer and nonmelanoma skin cancer exhibit exceptionally strong age dependencies that do not seem to be typical of cancers of other sites (Thompson and others 1994). With the incidence

data, there were 12,778 solid cancer cases occurring in the periods 1958–1998 after the exclusion of 401 thyroid cancers and 275 nonmelanoma skin cancers. Because the most recent mortality data (1950–2000) available to the committee did not include site-specific solid cancer, analyses of mortality data were based on all solid cancers (including thyroid and nonmelanoma skin cancer). There were 10,127 solid cancer deaths occurring in the period 1950–2000. The number of thyroid and nonmelanoma skin cancers included in this group is likely to have been small. Of the 9399 solid cancer deaths occurring in the period 1950–1997 (Preston and others 2003), only 64 (0.7%) were due to thyroid cancer and 32 (0.3%) to nonmelanoma skin cancer. Table 12B-1A shows the distribution of cases (1958–1998) and deaths (1950–2000) by sex and dose category. Table 12B-1B shows the distribution of site-specific cancers by sex, with the num-

TABLE 12B-1A Number of Incidence Cases of Solid Cancer Excluding Thyroid and Nonmelanoma Skin Cancer and Number of Deaths from Solid Cancer by Sex and Colon Dose

Colon Dose (Sv)	No. of Cases (1958–1998)			No. of Deaths (1950–2000)		
	Males	Females	Total	Males	Females	Total
< 0.005	2,504	2,855	5,359	2,089	2,181	4,270
0.005 – 0.1	1,900	2,295	4,195	1,603	1,784	3,387
0.1 – 0.2	379	547	926	307	425	732
0.2 – 0.5	473	602	1,075	379	436	815
0.5 – 1.0	294	348	642	241	242	483
1.0 – 2.0	199	219	418	160	166	326
2.0 +	74	89	163	51	63	114
Total	5,823	6,955	12,778	4,830	5,297	10,127

TABLE 12B-1B Number of Incidence Cases and Number of Deaths by Cancer Site and Sex

Cancer Site	No. of cases (1958–1998)			No. of Deaths (1950–1997)		
	Males	Females	Total	Males	Females	Total
Stomach	1,899	1,703	3,602	1,555	1,312	2,867
Colon	547	618	1,165	206	272	478
Liver	676	470	1,146	722	514	1,236
Lung	770	574	1,344	716	548	1,264
Breast	7	847	854	3	272	275
Prostate	281	0	281	104	0	104
Ovary	0	190	190	0	136	136
Uterus	0	875	875	0	518	518
Bladder	227	125	352	83	67	150
Other solid	1,416	1,553	2,969	1,036	1,175	2,211
Total	5,823	6,955	12,778	4,425	4,814	9,239

ber of deaths based on the period 1950–1997 rather than 1950–2000, the period used in Table 12B-1A.

The following general linear dose-response function was used to model the ERR or EAR:

$$\text{ERR}(D, s, e, a) \text{ or } \text{EAR}(D, s, e, a) = \beta_s D \exp[h(e, a)], \quad (12\text{B-4})$$

where D is dose in sieverts, β_M and β_F are sex-specific estimates of the ERR/Sv, e is age at exposure in years, and a is attained age in years. The function h includes parameters to be estimated. Most commonly, h is of the form

$$h(e, a) = \gamma f(e) + \eta g(a). \quad (12\text{B-5})$$

As noted above, recent analyses by RERF investigators of A-bomb survivor solid cancer mortality (Preston and others 2003) and incidence data have taken $f(e) = e$ and $g(a) = \log a$; note that $\exp(\eta \log a) = a^\eta$. Others (Kellerer and Barclay 1992) have developed models with $g(a) = a$.[3] Some past risk assessments (BEIR V) have taken h to be a function of sex, age at exposure, and time since exposure (t). Note that any two of the variables e, a, and t determine the third ($t = a - e$) so models based on e and t are included in the equation (12B-4) specification.

In recent analyses conducted for the purpose of updating radioepidemiologic tables (NIH 2003), the NIH evaluated models of the form indicated above, but the ERR was allowed to vary over only a limited range of exposure ages or attained ages. The models evaluated by the NIH included those in which the ERR varied with age at exposure only over the ranges 0–30 and 15–30 years, and in which the ERR varied with attained age only over the range 0–50 years. Stated mathematically, these models for age at exposure are as follows:

$$f(e) = \min(e - 30, 0) = e\_u30$$
$$f(e) = \min[\max(-15, e - 30), 0] = e\_15to30,$$

where min = minimum and max = maximum. Note that the variable e_u30 is equal to $e - 30$ for $e < 30$, and equal to zero for older ages. That is, it allows for modification of the ERR for exposure ages between 0 and 30, but allows no further modification after age 30. The variable e_15to30 allows the ERR to change over the interval 15–30 years, whereas the ERR is constant (at different levels) for exposure ages under 15 or over 30. As seen later, e_u30 is selected for use in the committee's preferred models. Thus, the simpler notation $e^* = e\_u30$ is introduced.

The alternative for attained age was

$$g(a) = \min[\log(a/50), 0] = \log(a)\_u50.$$

The variable $\log(a)\_u50$ allows for modification of the ERR for attained ages under 50, but allows no further modification after age 50. The final model used in the NIH report (2003) was based on e_15to30 and $\log(a)\_u50$. With the LSS cancer incidence data available at the time (1958–1987), these choices gave a slightly better fit to the data than the alternatives that were evaluated.

To decide on the preferred BEIR VII model, the committee evaluated several alternative choices for $f(e)$ and $g(a)$ using data both on incidence of all solid cancers excluding thyroid cancer and nonmelanoma skin cancer and on mortality from all solid cancers. The committee conducted a series of analyses of all solid cancers excluding thyroid cancer and nonmelanoma skin cancer with several alternative choices for $f(e)$ and $g(a)$; it also evaluated models based on time since exposure instead of attained age. As noted earlier, in fitting ERR models, baseline risks were handled by stratifying on sex, city of exposure (Hiroshima or Nagasaki), age at exposure, and attained age as described by Pierce and colleagues (1996). More limited analyses were conducted using an EAR model with the same form as the ERR model. To fit the EAR models, it was necessary to use a parametric baseline. The committee used a similar parametric model to that described by Preston and colleagues (2003). To evaluate comparability with stratified results, a limited number of parametric ERR models were also fitted.

Table 12B-2 shows the drop in deviance for each of the models compared to a model with no modification by age at exposure, attained age, or time since exposure; such a model is not realistic but facilitates comparison among models. The deviance differences, which follow (approximately) chi-square distributions with the number of degrees of freedom indicated, can be regarded as a measure of the improvement in fit brought about by use of the indicated function of e, a, and t. In general, the greater the deviance difference, the better is the fit of the model. Comparison of these deviance differences is most appropriate among models based on the same data and of the same type (e.g., comparisons among stratified ERR models for incidence data).

All models were of the form indicated in Equation (12B-5), and most (models 1–13) were ERR models with a stratified background. Model 1 is the RERF model given by Equation (12B-3), while models 2 and 3 included only one of the variables e or $\log(a)$. With the incidence data, the RERF model fitted the data significantly better than model 2 with only e ($p < .001$) or model 3 with only $\log(a)$ ($p = .013$). With the updated incidence data, models that include only exposure age (or a function of exposure age) or only attained age (or a function of attained age) do not provide an adequate fit to the data. With the mortality data, the RERF model fitted the data slightly better than model 2 with only e ($p = .08$), and much better than model 3 with only $\log(a)$ ($p < .001$).

Models 4–6 were addressed at evaluating alternative choices for $f(e)$ and $g(a)$, specifically the alternatives that

---

[3]In general, information published since the BEIR V report (1990).

TABLE 12B-2 Comparison of Fits of Several Models (As Measured by the Deviance): Estimated Parameters and Deviances for Several Models Expressing Dependence of Risk of Solid Cancer Incidence and Mortality on Age at Exposure (*e*), Attained Age (*a*), and Time Since Exposure (*t*)[a]

Model Number	Model Type[b]	*f(e)*	*g(a)* or *g(t)*	Difference in Deviance for This Model and Model with No Modifiers (degrees of freedom)	
				Incidence Data[c]	Mortality Data[d]
1-RERF	ERR-S	*e*	log (*a*)	50.2 (2)	37.2 (2)
2	ERR-S	E	None	32.9 (1)	34.2 (1)
3	ERR-S	None	log (*a*)	44.0 (1)	17.5 (1)
4–BEIR VII	ERR-S	*e*e*	log (*a*)	53.1 (2)	40.2 (2)
5	ERR-S	*e**	None	39.4 (1)	38.1 (1)
6	ERR-S	*e_15to30*[f]	log (*a*)	48.1 (2)	33.8 (2)
7	ERR-S	*e**	log (*a*)_u50[g]	49.2 (2)	39.1 (2)
8	ERR-S	*e**	*a*	51.4 (2)	40.2 (2)
9	ERR-S	*e**	*t*	51.1 (2)	39.4 (2)
10	ERR-S	*e**	log (*t*)	49.1 (2)	38.5 (2)
11	ERR-S	*e*	*a*	48.2 (2)	37.1 (2)
12	ERR-S	*e*	*t*	48.2 (2)	37.1 (2)
13	ERR-S	*e*	log (*t*)	46.3 (2)	35.6 (2)
1B-RERF	ERR-P	*e*	log (*a*)	50.55[h] (2)	37.0[i] (2)
4B-BEIR VII	ERR-P	*e**	log (*a*)	53.85 (2)	41.1 (2)
1C-RERF	EAR-P	*e*	log (*a*)	80.34[h] (2)	113.0[i] (2)
4C-BEIR VII	EAR-P	*e**	log (*a*)	84.21 (2)	115.5 (2)

[a]Based on models in which the ERR or EAR is given by $\beta_s D \exp[\gamma e^* + \eta \log(a/60)]$.
[b]ERR-S: stratified excess relative risk model; ERR-P: parametric excess relative risk model; EAR-P: parametric excess absolute risk model.
[c]Based on analyses of solid cancers excluding thyroid and nonmelanoma skin cancers, 1958–1998 (12,778 incident cases).
[d]Based on analyses of all solid cancers, 1950–2000 (10,127 deaths).
[e]e^* is $(e - 30)/10$ for $e < 30$ and zero for $e \geq 30$, where e is age at exposure in years.
[f]e; e_15to30 is -1.5 for $e < 15$, $(e - 30)/10$ for e between 15 and 30, and 0 for $e \geq 30$, where e is exposure age in years.
[g]$\log(a)\_u50 = \log(a/50)$ for $a < 50$ and 0 for $a \geq 50$, where a is attained age in years.
[h]The deviance based on incidence data for model 4B was 20,377.0; that for model 4C was 20,388.3.
[i]The deviance based on incidence data for model 4B was 13,766.8; that for model 4C was 13,769.0.

were evaluated by NIH (2003). Of these choices, model 4 resulted in the best fit (greatest deviance difference) for both incidence and mortality data, although differences between models 1 and 4 were not great. This model allows for variation in the ERR with age at exposure only over the range 0 to 30 years, but allows for variation in attained age over the full range. The model, which is as follows,

$$\text{BEIR VII Model: ERR}(D, s, e, a) = \beta_s\, D \exp(e^*)\, a^\eta, \quad (12\text{B-6})$$

is referred to as the BEIR VII ERR model. With the incidence data, the BEIR VII ERR model fitted the data significantly better than model 3 with only log (*a*) ($p = .0025$) or model 5 with only e^* ($p < .001$). With the mortality data, this model fitted the data much better than model 3 with only log (*a*) ($p < .001$), and slightly better than model 5 with only e^* ($p = .15$). With both incidence and mortality data, the BEIR VII model fitted the data somewhat better than the RERF model. Comparison of the BEIR VII model with model 6 indicates that e^* is a better choice than e_15to30 (particu-

larly for the mortality data), whereas comparison with model 7 indicates that log (*a*) is a slightly better choice than log (*a*)_u50. However, the fits of models 1, 4, and 7 do not differ greatly.

The committee also evaluated the use of *a*, *t*, and log (*t*) as substitutes for log (*a*) in models that also included e^* (respectively, models 8, 9, and 10) or that also included *e30* (models 11, 12, and 13). None of these models fitted the data as well as the BEIR VII model, although the differences were not great. Models 11 and 12 are different parameterizations of the same model.

For reasons that are discussed later in this annex, the committee fitted models that added exposure age variables $e\_over30 = \max(0, e - 30)$ and $e\_over55 = \max(0, e - 55)$. When e_over30 was added to the BEIR VII model, the estimated coefficients were positive. There was little evidence of improvement in fit ($p > .50$ for both incidence and mortality data), but this shows that there is no evidence of a decline in risk after age 30. However, with the incidence data, when e_over55 was added to the BEIR VII model, the *p*-value for improvement in fit was .044; when added to the RERF model, the *p*-value was .015. The *e_over55* parameter was positive, indicating an increase in risks for those exposed at older ages. This finding is discussed later in this section. For now, note only that there is no evidence that risks continue to decline for exposure ages greater than 30; thus, the BEIR VII model seems a better choice than the RERF model. With mortality data, there was little indication that adding *e_over55* improved the fit (*p* = .46).

The committee also evaluated whether using sex-specific estimates of the modifying effects of *e*_u30* or log (*a*) would substantially improve the fit of the BEIR VII model. There was no evidence of improvement in fit with the sex-specific model (*p* > .5 for both incidence and mortality) resulting in a deviance of 11,825.63, only slightly lower than model 1 (difference in deviances = 1.10; *p* > .5).

Models 1B and 4B were comparable to models 1 and 4 except that they were based on parametric modeling of the baseline risks. The estimated coefficients were similar to those obtained for the comparable models 1 and 4 based on stratified ERR models. These analyses also support the use of *e** instead of *e*.

Models 1C and 4C are EAR models. Model 1C has the same functional form as the RERF model shown in Equation (12B-3), whereas model 4C has the same functional form as the BEIR VII model shown in Equation (12B-6). The same variables were used to model the baseline risk as in respective models 1B and 4B. Again, model 4C (with *e**) provides a somewhat better fit than does model 1C (with *e*). The deviances for the ERR models are slightly lower than the corresponding EAR models. Model 1C is subsequently referred to as the "RERF ERR model," and model 4C as "the BEIR VII EAR model."

As another approach to evaluating alternative models, separate ERRs per sievert and EARs per 10^4 PY-Sv were estimated for each of five groups defined by age at exposure

(<15, 15–30, 30–45, 45–60 and 60+ years). That is, the following model was fitted:

$$ERR(D, s, e, a) \text{ or } EAR(D, s, e, a) = \beta_j [1 + \theta s] D (a / 60)^\eta, \qquad (12B\text{-}7)$$

where *j* indexes the five age-at-exposure categories. The estimated β_js, averaged for the two sexes, are shown in Table 12B-3. For both incidence and mortality data, we see the expected decline in risk for exposure ages under 60, with a stronger decline for mortality data than for incidence data. Somewhat surprisingly, the estimates for the 60+ exposure age group are three to four times those in the next-oldest age group for the incidence data, and about twice those in next-oldest age group for the mortality data. The difference between the coefficients β_j for the two oldest age-at-exposure groups was statistically significant for the incidence data (*p* = .04 for the ERR model; *p* = .03 for the EAR model) but not for the mortality data (*p* > .3). Because there was not an *a priori* hypothesis that the ERR/Sv would increase for the oldest exposure age category, a more appropriate test may be to compare the deviance for the BEIR VII model shown in Equation (12B-6) to that of the model shown below:

$$ERR(D, s, e, a) = \beta_j [1 + \theta s] D \exp [e\_u30] (a / 60)^\eta, \qquad (12B\text{-}8)$$

in which the single parameter β is replaced by five separate parameters for the five age-at-exposure categories. This four-degree-of-freedom test resulted in *p*-values that exceeded .15 for the incidence data and exceeded .45 for the mortality data.

Further exploration of the cancer incidence data revealed that the elevation of both ERR and EAR for the oldest exposure age category was strongest for stomach and liver cancers; for these cancers, β_js for the 60+ exposure age group

TABLE 12B-3 Sex-Averaged Estimates of ERR/Sv and EAR per 10^4 PY-Sv by Age-at-Exposure Categories for All Solid Cancers Excluding Thyroid and Nonmelanoma Skin[a]

Data Used	Age at Exposure				
	<15 years	15–30 years	30–45 years	45–60 years	60+ years
Incidence data[b]					
Number of cases	2044	3465	4417	2526	326
ERR/Sv (95% CI)	0.78 (0.58, 1.06)	0.63 (0.49, 0.81)	0.42 (0.28, 0.62)	0.43 (0.23, 0.79)	1.7 (0.76, 3.8)
EAR per 10^4 PY-Sv (95% CI)	57 (43, 76)	40 (30, 48)	23 (16, 33)	20 (11, 36)	67 (35, 131)
Mortality data[c]					
Number of deaths	1220	2188	3679	2572	468
ERR/Sv (95% CI)	1.12 (0.80, 1.58)	0.63 (0.46, 0.84)	0.35 (0.22, 0.55)	0.25 (0.10, 0.55)	0.55 (0.19, 1.7)
EAR per PY-Sv (95% CI)	29 (21, 39)	18 (14, 25)	12 (7.9, 19)	8.4 (3.7, 19)	17 (6.1, 45)

[a]Based on models in which the ERR or EAR is of the form $\beta_j [1 + \theta s] D \exp [\eta \log (a)]$, where *j* indexes age-at-exposure categories, *s* is sex, and *a* is attained age in years.

[b]Based on solid cancers excluding thyroid and nonmelanoma skin cancers, 1958–1998.

[c]Based on all solid cancers, 1950–2000.

were more than eight times those for the 45–60 exposure age group ($p = .02$), while for the remaining solid cancers this ratio was less than 2 and did not differ significantly from unity ($p > .5$).

The increased ERR/Sv and EAR per 10^4 PY-Sv for the oldest age-at-exposure group was one of the reasons the committee selected the BEIR VII model with no decline with exposure age after age 30 in preference to the RERF model with a decline throughout the entire range of exposure age. The committee notes particularly that stomach and liver cancers, for which this effect was strongest, are far more prevalent in Japan than in the United States. With the incidence data, about 37% of the cancers in the solid cancer category that the committee analyzed were cancers of the stomach and liver; by contrast, SEER data for the United States (see Table 12-3) indicate that only about 3% of incident cancers are of these types. Furthermore, risks for stomach and liver cancers may be affected by infectious agents such as *Helicobacter pylori* for stomach cancer and the hepatitis virus for liver cancer (Parsonnet and others 1994; Aromaa and others 1996; Goldstone and others 1996). Infection rates might differ by birth cohort (and thus exposure age), which could affect risks due to radiation in ways that are not currently understood. Although the reason for the relatively high ERR/Sv among those exposed at older ages is not fully understood the committee does not think that this effect is likely to generalize to a modern U.S. population.

Based on the analyses of A-bomb survivor data described above, the committee has selected the model shown in Equation (12B-6) as its preferred model for estimating solid cancer risks. However, several alternative choices, including the RERF model shown in Equation (12B-3), fitted the data nearly as well and would also have been reasonable choices. Both ERR and EAR models are evaluated. Table 12B-4 shows the estimated parameters (with 95% confidence intervals) for ERR and EAR models obtained from both incidence and mortality data. With the ERR models, the effect of exposure age is stronger for mortality than for incidence data, while the effect of attained age is weaker. The two EAR models show similar exposure age effects, but the rate of increase with attained age is greater for the mortality data than for the incidence data.

The committee also evaluated mortality data on all solid cancers to compare the use of 5- and 10-year minimal latent periods. This was done by fitting the BEIR VII ERR model, and estimating the ERR/Sv separately for the period 5–9 years following exposure and for the period 10 or more years following exposure. Although the estimate for the 5–9-year period was not quite statistically significant with a two-sided test ($p = .10$), there was no evidence that it differed from the estimate for the later follow-up period ($p = .44$). The committee accordingly has used a minimal latent period of 5 years in its calculations of lifetime risks.

TABLE 12B-4 ERR and EAR Models for Estimating Incidence of All Solid Cancers Excluding Thyroid and Nonmelanoma Skin Cancers and Mortality from All Solid Cancers[a,b]

ERR Models	No. of Cases or Deaths	ERR/Sv (95% CI) at Age 30 and Attained Age 60		Per-Decade Increase in Age at Exposure Over the Range 0–30 Years[c] (95% CI), γ	Exponent of Attained Age (95% CI), η
		Males (β_M)	Females (β_F)		
Incidence[d]	12,778	0.33 (0.24, 0.47)	0.57 (0.44, 0.74)	−0.30 (−0.51, −0.10)	−1.4 (−2.2, −0.7)
Mortality[e]	10,127	0.23 (0.15, 0.36)	0.47 (0.34, 0.65)	−0.56 (−0.80, −0.32)	−0.67 (−1.6, 0.26)
EAR models		EAR per 10^4 PY-Sv (95% CI)			
		Males (β_M)	Females (β_F)		
Incidence[d]	12,778	22 (15, 30)	28 (22, 36)	−0.41 (−0.59, −0.22)	2.8 (2.15, 3.41)
Mortality[e]	10,127	11 (7.5, 17)	13 (9.8, 18)	−0.37 (−0.59, −0.15)	3.5 (2.71, 4.28)

NOTE: Estimated parameters with 95% CIs. PY = person-years.

[a]The ERR or EAR is of the form $\beta_s D \exp(\gamma e^*)(a/60)^\eta$, where D is the dose (Sv), e is age at exposure (years), e^* is $(e − 30)/10$ for $e < 30$ and zero for $e \geq 30$, and a is attained age (years).

[b]The committee's preferred estimates of risks from all solid cancers are obtained as sums of estimates based on models for site-specific cancers (see Table 12-2 and text).

[c]Change in ERR/Sv or EAR per 10^4 PY-Sv (per-decade increase in age at exposure) is obtained as $1 − \exp(\gamma)$.

[d]Based on analyses of LSS incidence data 1958–1998 for all solid cancers excluding thyroid and nonmelanoma skin cancer.

[e]Based on analyses of LSS mortality data 1950–2000 for all solid cancers.

Analyses of Incidence and Mortality Data on Site-Specific Solid Cancers

Although the committee provides risk estimates for both cancer incidence and mortality, models for site-specific cancers were based mainly on cancer incidence data. This was done primarily because site-specific cancer incidence data are based on diagnostic information that is more detailed and accurate than death certificate data and because, for several sites, the number of incident cases is considerably larger than the number of deaths. For cancers of the colon, breast, prostate, and bladder, the number of cases in the LSS cohort is more than double the number of deaths (Table 12B-1B). In addition, mortality data may be more subject than incidence data to changes over time brought about because of improved survival. Models developed from incidence data were however evaluated for consistency with mortality data. Since there is little evidence that radiation-induced cancers are more rapidly fatal than cancer that occurs for other reasons, ERR models based on incidence data can be used directly to estimate risks of cancer mortality. EAR models require adjustment as discussed in the chapter.

Models for site-specific cancers were based on the BEIR VII model indicated by Equation (12B-6). The committee's approach to quantifying the parameters γ and η was to use the estimates obtained from analyzing incidence data on all solid cancers excluding thyroid and nonmelanoma skin cancers unless site-specific analyses indicated significant departure from these estimates. Table 12B-5A shows the results of fitting ERR site-specific models to the incidence data. Results are shown for a model in which all four of the parameters β_M, β_F, γ, and η were estimated and are also shown for a model in which the parameters quantifying the modifying effects of age of exposure and attained age γ and η were set equal to the values obtained from analysis of the category all solid cancers excluding thyroid and nonmelanoma skin cancers; these values are referred to subsequently as the "common values." The final column gives the deviance difference between the two models and the resulting p-value based on a two-degree-of-freedom test comparing the fits of the two models. This test does not take account of uncertainty in the estimates of the common values of γ and η. In addition, the committee fitted models in which just one of the parameters γ and η was fixed, with the other estimated allowing a one-degree-of-freedom test for each of the parameters.

The only sites with even modest evidence ($p < .10$) of departure from the fixed values of γ and η were cancer of the uterus and the category "all other solid cancers." For cancer of the uterus, the estimated ERR/Sv was very small and nonsignificant so that it was not possible to obtain stable estimates of the modifying parameters; thus the common values were used. For other solid cancers, a test for the parameter η alone resulted in a p-value of .025; thus, results are also

TABLE 12B-5A Results of Fitting Stratified ERR Models to Site-Specific Cancer Incidence Data Using the Model ERR(D, s, e, a) = $\beta_s D \exp [\gamma e^* + \eta \log (a / 60)]^a$

Cancer Site	No. of Cases	All Parameters Estimated				Fixed Parameters: $\gamma = -0.30$; $\eta = -1.4$		Deviance Difference[b] (p-value)
		β_M	β_F	γ	η	β_M	β_F	
Solid cancer[c]	12,778	0.33	0.57	−0.30	−1.4	0.33	0.57	
Stomach	3602	0.25	0.54	−0.13	−1.9	0.21	0.48	0.5 (> 0.5)
Colon	1165	0.72	0.54	−0.16	−3.1	0.63	0.43	1.0 (> 0.5)
Liver	1146	0.40	0.36	−0.15	−1.5	0.32	0.32	0.2 (> 0.5)
Lung	1344	0.39	1.68	0.05	−1.1	0.32	1.40	2.9 (0.23)
Breast	847	—	1.19	−0.04	−2.0	—	0.91	2.4 (0.34)
Prostate[d]	281	—	—	—	—	0.12	—	—
Uterus	875	—	0.027	−2	5.6	—	0.055	5.8 (0.055)
Ovary	190	—	0.47	−0.13	−1.6	—	0.38	0.05 (> 0.5)
Bladder	352	0.51	1.62	−0.04	0.28	0.50	1.65	2.7 (0.26)
Other solid cancers	2969	0.27	0.45	−0.29	−2.8	0.33	0.51	5.0 (0.081)
Other solid cancers (alternative)	2969	0.27	0.45	Fixed at −0.30	−2.8			0.003[e] (>0 .5)

[a]D is dose (Sv); $e^* = (e - 30) / 10$ for $e < 30$, where e is age at exposure (years); $e^* = 0$ for e 30; and a is attained age (years). β_M and β_F are the ERR/Sv for males and females exposed at age 30 at attained age 60, γ is expressed per decade increase in age at exposure over the range 0–30 years, and a is the exponent of attained age.

[b]Difference in deviance for model shown in columns 7 and 8 and model shown in columns 3–6.

[c]Solid cancer excluding thyroid and nonmelanoma skin cancers.

[d]Model with all parameters estimated would not converge.

[e]Difference in deviance for this model and that shown in columns 3–6 in the row immediately above.

shown for an alternative model with η estimated separately for this category.

Table 12B-5B shows results based on mortality data on site-specific cancers. As in Table 12B-5A, columns 3–6 show results with all four of the parameters β_M, β_F, γ, and η estimated using data on that site alone. Columns 7 and 8 show the results of testing the compatibility of these models with models developed from the incidence data with γ and η fixed as indicated in columns 7 and 8 of Table 12B-5A. Column 7 is based on analyses in which γ was set equal to –0.30 per decade and η was set equal to –1.4, and the parameters β_M and β_F were estimated, and thus tests whether the fixed values of data γ and η are compatible with the mortality data. Column 8 is based on analyses in which all four of the parameters β_M, β_F, γ and η were set equal to the values estimated from the incidence data (Table 12B-5A). The alternative model for "all other solid cancers," based on the incidence data, was also evaluated. Because of difficulties in fitting four-parameter models for cancers of the prostate and uterus, these sites are not shown in Table 12B-4B. Only for colon cancer and for all other solid cancers was there a suggestion ($p < .10$) that the models based on incidence data did not fit the mortality data. Because there was no evidence against using the common values of η and γ for colon cancer based on the incidence data, the committee chose to use the common values for this site. For all other solid cancers, the alternative model developed from the incidence data was also more compatible with the mortality data, and this was chosen as the preferred model.

Table 12B-5C shows results of fitting EAR models to the cancer incidence data and is analogous to Table 12B-5A for the ERR models. There is clear evidence that common values of the parameters γ and η are not appropriate for cancers of lung, breast, and bladder. For all three of these sites, and also for liver cancer (see below), alternative models in which η was estimated and γ was set at the common value (–0.41) provided acceptable fits.

For breast cancer, the committee fitted additional EAR models with separate parameters for attained ages under 50 and over 50, similar to the model used by Preston and colleagues (2002a) in a pooled analysis of breast cancer incidence data from several cohorts including the LSS data. This model (labeled alternative 2) provided a significantly better fit ($p < .001$) than did the model with a single parameter for attained age. As discussed in this chapter, the committee's preferred models for breast cancer were based on pooled analyses by Preston and colleagues (2002a). However, it was of interest to compare these results with those obtained from models based on the same approach as most other cancer sites.

Table 12B-5D shows results of fitting EAR models to the mortality data. All but the last column are analogous to those in Table 12B-4C for the ERR models. The last column of Table 12B-5D shows the deviance differences for models based on the mortality data and the alternative models shown in Table 12B-5C. Only for cancers of the liver, lung, breast, and bladder was there evidence ($p < .10$) of departure from the main incidence models. However, for these sites, there

TABLE 12B-5B Results of Fitting Stratified ERR Models to Site-Specific Cancer Mortality Data Using the Model ERR(D, s, e, a) $= \beta_s D \exp [\gamma e^* + \eta \log (a / 60)]^a$

Cancer Site	No. of Deaths	All Parameters Estimated				Fixed Parameters $\gamma = -0.30$; $\eta = -1.4$			
		β_M	β_F	γ	η	β_M	β_F	Deviance Difference for Testing γ and η^b (p-value)	Deviance Difference for Testing β_M, β_F, γ, and η (p-value)^c
Stomach	2,867	0.11	0.41	–0.65	0.29	0.14	0.46	2.6 (0.28)	3.3 (>0.5)
Colon	478	0.65	0.79	–0.19	–5.3	0.68	0.68	4.8 (0.09)	5.8 (0.22)
Liver	1,236	0.23	0.25	–0.51	0.82	0.28	0.29	1.8 (0.40)	2.0 (>0.50)
Lung	1,264	0.36	0.80	–0.36	0.34	0.45	0.93	3.0 (0.23)	6.6 (0.16)
Breast	272	—	0.56	–0.72	–1.5	—	0.94	1.9 (0.38)	2.0 (>0.5)
Ovary	136	—	0.34	–0.10	–5.1	—	0.65	1.3 (> 0.5)	2.1 (>0.5)
Bladder	150	1.27	1.65	0.10	–0.65	0.90	1.18	3.3 (0.20)	3.8 (0.44)
All other solid cancers	2,211	0.24	0.30	–0.68	–1.7	0.35	0.53	5.1 (0.079)	5.1 (0.28)
All other solid cancer (alternative)		Fixed at –0.30	Fixed at –2.8			0.32	0.44	3.3 (0.20)	3.5 (0.48)

$^a D$ is dose (Sv); $e^* = (e - 30) / 10$ for $e < 30$, where e is age at exposure (years); $e^* = 0$ for e 30; and a is attained age (years). β_M and β_F are the ERR/Sv for males and females exposed at age 30 at attained age 60, γ is expressed per decade increase in age at exposure over the range 0–30 years, and η is the exponent of attained age.

^{b} Difference in deviance for model shown in columns 7 and 8 (with $\gamma = -0.30$ and $\eta = -1.4$) and model shown in columns 3–6 (2 degrees of freedom).

^{c} Difference in deviance for model shown in columns 7 and 8 of Table 12B-5A and model shown in columns 3–6 of this table (4 degrees of freedom for cancers occurring in both sexes; 3 degrees of freedom for cancers of the breast, prostate, uterus, and ovary).

TABLE 12B-5C Results of Fitting Parametric EAR Models to Site-Specific Cancer Incidence Data Using the Model
$EAR(D, s, e, a) = \beta_s D \exp [\gamma e^* + \eta \log (a / 60)]^a$

| Cancer Site | No. of Cases | All Parameters Estimated | | | | Fixed Parameters: $\gamma = -4.1; \eta = 2.8$ | | Deviance Difference[b] (p-value) |
		β_M	β_F	γ	η	β_M	β_F	
Solid cancer[c]	12,778	22	28	−0.41	2.8	22	28	
Stomach	3,602	7.0	7.1	0.002	1.8	4.9	4.9	3.4 (0.18)
Colon	1,165	2.2	0.84	−1.0	5.7	3.2	1.6	4.0 (0.14)
Liver	1,146	1.8	0.81	−0.64	4.8	1.9	0.83	1.9 (0.39)
Liver[d] (alternative)	1,146	2.2	1.0	Fixed at −0.41	4.1			0.3[e] (> 0.5)
Lung	1,344	3.1	4.6	−0.3	4.4	1.5	3.3	15.4 (<0.001)
Lung (alternative)	1,344	2.3	3.4	Fixed at −0.41	5.2			2.0[e] (0.16)
Breast	847	—	5.6	−0.51	1.5	—	6.3	16.5 (<0.001)
Breast (alternative 1)	847	—	6.1	Fixed at −0.41	1.3			0.42[e] (> 0.5)
Breast (alternative 2)	847	—	5.9	Fixed at −0.41	3.4, −2.4[f]			−13.9[g] (<0.001)
Prostate[h]	281	—	—	—	—	0.11	—	—
Uterus	875	—	0.28	−1.6	6.3	—	1.2	2.7 (0.27)
Ovary	190	—	0.50	−0.66	2.7	—	0.7	1.2 (> 0.5)
Bladder	352	1.3	0.88	−0.23	5.6	1.1	0.62	6.4 (0.04)
Bladder (alternative)	352	1.2	0.75	Fixed at −0.41	6.0			0.1[e] (>0.5)
Other solid cancers	2,969	5.1	4.2	−0.39	1.9	6.2	4.8	3.1 (0.22)

[a]D is dose (Sv); $e^* = (e - 30) / 10$ for $e < 30$, where e is age at exposure in years; $e^* = 0$ for e 30; and a is attained age in years. β_M and β_F are the number of excess cases per 10^4 PY-Sv for males and females exposed at age 30 at attained age 60, γ is expressed per decade increase in age at exposure over the range 0–30 years, and a is the exponent of attained age.
[b]Difference in deviance for model shown in columns 7 and 8 and model shown in columns 3–6.
[c]Solid cancer excluding thyroid and nonmelanoma skin cancers.
[d]This alternative was developed to obtain a model that was consistent with mortality data.
[e]Difference in deviance for this model and that shown in columns 3–6 in the row immediately above.
[f]The first coefficient is for attained age under 50; the second coefficient is for attained age over 50.
[g]Difference in deviance for alternative 1 breast model and this model.
[h]Model with all parameters estimated would not converge.

was no evidence of departure from the alternate incidence models. In fact, the alternative liver cancer model was developed because of the large attained age effect identified in the mortality data. In general, the numbers of excess deaths per 10^4 PY-Sv would be expected to be less than the numbers of excess cases; thus, it was not sensible to evaluate the compatibility of the estimated β_M and β_F as was done for the ERR models. However, for sites common to both sexes, the committee tested whether or not the ratio β_F / β_M estimated from the mortality data was compatible with that estimated from the incidence data (with the latter treated as a fixed value). The p-values for the sites tested, based on a single-degree-of-freedom test, were as follows: stomach ($p = .19$), colon ($p = .35$), liver ($p > .5$), lung ($p = .28$), and all other solid cancers ($p > .5$).

The analyses of site-specific cancer presented in the last few paragraphs address the use of common parameters to quantify the modifying effects of age at exposure and at-

tained age, but do not address the possibility of common parameters for the overall level of the ERR or EAR (β_M and β_F). Because at least some of the variation among cancer sites in these estimated parameters is due to sampling variation, one might consider using common parameters for sites where there is no evidence of statistical differences. The committee chose not to use such an approach because it seems likely that there are true differences among the sites and because it was considered desirable to use site-specific data to reflect the uncertainty in site-specific estimates. A promising approach for the future is to use methods that draw both on data for individual sites and on data for the combined category of all solid cancers. With this approach, the variance of the site-specific estimate and the degree of deviation from the all-solid-cancer estimate are considered in developing site-specific estimates that draw both on data for the specific individual site and on data for all solid cancers. The National Research Council (2000) gives a simple il-

TABLE 12B-5D Results of Fitting Parametric EAR Models to Site-Specific Cancer Mortality Data Using the Model $EAR(D, s, e, a) = \beta_s\, D \exp[\gamma\, e^* + \eta \log (a / 60)]^a$

Cancer Site	No. of Deaths	All Parameters Estimated				Fixed Parameters: $\gamma = -4.1$; $\eta = 2.8$		
		β_M	β_F	γ	η	β_M	β_Γ	Deviance Difference[b] (p-value)
Stomach	2867	2.6	4.3	0.008	2.7	1.4	2.8	2.8 (0.25)
Colon	478	0.82	0.66	−0.66	3.6	0.96	0.83	0.6 (> 0.5)
Liver	1236	0.61	0.30	−1.2	7.9	1.1	0.56	6.9 (0.033)
Liver (alternative)				Fixed at −0.41	4.1	1.7	0.72	3.0 (0.23)
Lung	1264	2.1	1.8	−0.36	6.1	1.2	1.4	19.3 (<0 .001)
Lung (alternative)				Fixed at −0.41	Fixed at 5.2	2.1	1.9	1.8 (0.41)
Breast	272	—	0.90	−0.90	2.8	—	1.5	5.1 (0.077)
Breast (alternative 2)		—	2.0	−0.60	6.5, −2.9[c]	—	2.0	3.2[d] (0.36)
Ovary	136	—	0.78	−0.19	2.0	0.66	0.2 (> 0.5)	
Bladder	150	0.76	0.21	0.76	6.7	0.20	< 0	6.6 (0.037)
Bladder (alternative)		0.53	0.13	Fixed at −0.41	Fixed at 6.0	2.7 (0.26)		
All other solid cancers	2211	2.2	2.0	−0.61	2.9	2.9	2.6	0.8 (>0.5)

[a]D is dose (Sv); $e^* = (e - 30) / 10$ for $e < 30$, where e is age at exposure (years); $e^* = 0$ for e 30; and a is attained (years). β_M and β_F are the number of excess cases per 10^4 PY-Sv for males and females exposed at age 30 at attained age 60, γ is expressed per decade increase in age at exposure over the range 0–30 years, and η is the exponent of attained age.

[b]Difference in deviance for models shown in columns 7 and 8 (with $\gamma = -0.41$ and $\eta = 2.8$) and model shown in columns 3–6 (2 degrees of freedom).

[c]The first parameter is for attained age under 50; the second coefficient is for attained age over 50.

[d]Difference in deviance for alternative 2 breast model with $\gamma = -0.41$ and the two attained age parameters set at the values shown in Table 12B-5C and the model shown in columns 3–6 of this table (3 degrees of freedom).

lustration of this approach, using methods described in DerSimonian and Laird (1986) for estimating site-specific excess relative risks for the purpose of developing radio-epidemiologic tables.

The committee's preferred models for estimating site-specific cancer incidence and mortality are summarized in Table 12-2. With the exception of the category of all other solid cancers, the ERR models are based on common values of the parameters γ and η that quantify the modifying effects of age at exposure and attained age. For the EAR models, the preferred models are based on site-specific estimates of η for cancers of the liver, lung, and bladder; for the remaining sites (other than breast), common values of γ and η were used. For breast and thyroid cancers, models developed by Preston and colleagues (2002a) and by Ron and coworkers (1995a) are used as discussed in this chapter. The EAR coefficients β_M and β_F shown in Table 12-2 can be used directly only for cancer incidence and must be adjusted as described in this chapter for cancer mortality.

As stated earlier, the committee's models for mortality from all solid cancers were based on mortality data. An alternative might have been to use incidence data for this purpose as was done for site-specific cancers. However, the two main reasons for using incidence data for estimating

mortality from site-specific data were the better diagnostic quality and the larger number of cases for several cancer sites. These considerations do not apply when evaluating risks for the broad category of all solid cancers. In addition, the mix of cancers is different for incidence and mortality data so that one might expect greater differences than for site-specific data as evidenced from the parameter estimates shown in Table 12B-4. Nevertheless, the committee conducted analyses of the solid cancer mortality data with parameters set equal to the estimates obtained from the incidence data (as in columns 7 and 8 of Tables 12B-5B and 12B-5D). With the solid cancer ERR model, a joint test of $\gamma = -0.30$ per decade and $\eta = -1.4$ (the values from the incidence data) resulted in a p-value of .06. However, there was no evidence of further differences when main effects parameters β_M and β_F were set equal to those for the incidence data ($\beta_M = 0.33$; $\beta_F = 0.57$).

With EAR models, the estimated main effects (β_M and β_F) based on the incidence data were about twice those based on mortality data, reflecting the fact that not all cancers are fatal. The estimates of γ, the parameter quantifying the effects of age at exposure, were similar, whereas the increase with attained age (quantified by η) was stronger for the mortality data than for the incidence data. When mortality data were analyzed with the parameters γ and η set equal to the values estimated from incidence data, the joint test resulted in a p-value of .041; the evidence for differences came about mainly from differences in the attained age parameter η ($p = .047$) with little evidence of differences in the exposure age parameter γ ($p > .5$).

Analyses of Data on Leukemia

The committee's model for estimating leukemia risks is based on analyses of LSS leukemia mortality data for the period 1950–2000. Recent LSS leukemia incidence data based on DS02 doses are not yet available. The quality of diagnostic information for non-type-specific leukemia mortality is thought to be much better than for most site-specific solid cancers. Although Preston and colleagues (1994) used incidence data to develop separate models for all types of leukemia—acute lymphocytic leukemia, acute myelogenous leukemia, chronic myelogenous leukemia, and other leukemias—in Hiroshima, the models in most past risk assessments (NRC 1990; ICRP 1991; UNSCEAR 2000b) have been based on leukemias of all types, and the BEIR VII committee has followed the same practice. Data on medically exposed cohorts indicate that CLL is not likely to be induced by radiation exposure (Boice and others 1987; Curtis and others 1994; Weiss and others 1995) but CLL is rare in Japan.

The committee began by considering the model used in a recent report on cancer mortality (Preston and others 2004). This model allows the EAR to vary as a linear-quadratic function of dose and allows both the overall level of risk and the dependence on time since exposure to vary by age at exposure:

$$\text{RERF leukemia model: EAR}(d, s, e, t) = \\ \beta_s (D + \theta D^2) \exp [\gamma_e + \delta_e \log (t / 25)], \quad (12B\text{-}9)$$

where D is dose in sieverts; s is sex; e is an index for three age-at-exposure categories: 0–19, 20–39, and 40+ years with γ_{20-39} fixed at 0; and t is time since exposure in years. The parameter θ indicates the degree of curvature, which does not depend on sex or age at exposure; β_M and β_F are the EAR at exposure ages 20–39 and 25 years following exposure (expressed as excess deaths per 10^4 PY-Sv for males and females, respectively); and δ_e indicates the dependence on time since exposure for each of the three age groups. Parameter estimates for this model are given by Preston and colleagues (2002b).

The committee also considered the UNSCEAR (2000) model, which was developed by Preston and colleagues (2004) and based on A-bomb survivor leukemia incidence data for the period 1950–1987. This model, which is described in Annex 12A, and is similar to the RERF model above except that $t - 25$ replaces $\log (t / 25)$ and the parameters δ_e are allowed to depend on sex.

Although the committee could have used the RERF or the UNSCEAR model, it was judged desirable to develop alternative models with the EAR and ERR expressed as continuous functions of age at exposure and without dependence of the modifying effect of time since exposure on sex (as in the UNSCEAR model). The committee thus analyzed the same leukemia mortality data (1950–2000) used by Preston and colleagues (2004), using the same model for baseline leukemia rates, and evaluated models of the following form:

$$\text{ERR}(D, s, e, t) \text{ or EAR}(D, s, e, t) = \\ \beta_s(D + \theta D^2) \exp [\gamma f(e) + \\ \delta g(t) + \phi f(e) g(t)], \quad (12B\text{-}10)$$

where e is age at exposure in years and t is time since exposure in years. The functions of age at exposure evaluated were $f(e) = e$; $f(e) = e^* = (e - 30) / 10$ for $e < 30$, and 0 for e 0; and the RERF model in which $f(e)$ was an indicator for one of the three categories: $e < 20$, 20 $e < 40$, and e 40. The functions of time since exposure evaluated were $g(t) = \log (t)$ and $g(t) = t$. The committee also fitted ERR models for leukemia of the form shown in Equation (12B-10).

Table 12B-6 shows the drop in deviance (compared to a model with no modification by e or t) for both the EAR and the ERR models. For comparisons among different models of the same type (EAR or ERR), the greater the drop in deviance, the better is the fit. Because it is not meaningful to compare the drop in deviance for an EAR model to that for an ERR model, the total deviances are also shown. In general, models in which age at exposure was treated as a continuous variable fitted the data nearly as well even though they have fewer parameters. Comparing the use of e and e^* in models that are otherwise the same resulted in very similar fits, with slightly better fits with e^*. The use of $\log (t)$ resulted in better fits that the use of t.

For the EAR models using e^* and $\log (t)$ (models 5–7), the interaction term $[e^* \times \log (t)]$ was clearly needed ($p < .001$), but the main effect for $\log (t)$ was not ($p > .5$). With the main effect for $\log (t)$ in the model (model 5), the EAR decreases with time since exposure for those exposed under about age 25, but increases slightly with time since exposure at older exposure ages. Without the main effect (model 7), the EAR remains constant with time since exposure for those exposed over age 30 and decreases with time since exposure for those exposed under age 30, with a stronger decrease at the youngest ages. The latter model is the committee's preferred EAR model for estimating leukemia risks. With this model, there was no need for an interaction of sex and time since exposure ($p = .23$), which was included in the UNSCEAR (2000b) leukemia model.

The committee's preferred ERR leukemia model is model 5. With this model, the ERR decreases with time since exposure regardless of age at exposure, although the decrease is not as strong at older ages. Again, there was no strong evidence of a need for an interaction of sex and time since expo-

TABLE 12B-6 Comparison of Fits of Several Models (as Measured by the Deviance) Expressing the Dependence of Risk of Leukemia Mortality on Age at Exposure (*e*) and Time Since Exposure (*t*)[a]

Model Number	Age at Exposure (*e*), *f(e)*	Time Since Exposure (*t*) or Attained Age (*a*), *g(t)* or *g(a)*	Difference in Deviance for This Model and Model with No Modification by *e* or *t* (degrees of freedom)		Deviance	
			EAR Model	ERR Model	EAR Model	ERR Model
1	Categorical[b]	log (*t*): full model	21.3 (5)	21.4 (5)	2254.9	2258.7
2	*e* − 30	log (*t*): full model	20.1 (3)	22.5 (3)	2256.1	2257.6
3	*e* − 30	log (*t*): main effect only (φ = 0)	9.4 (2)	20.2 (2)	2266.8	2259.8
4	*e* − 30	log (*t*): interaction only (δ = 0)	19.5 (2)	13.3 (2)	2256.7	2266.8
5	*e**[c]	log (*t*): full model	21.1 (3)	24.9 (3)	2255.1	2255.1
6	*e**	log (*t*): main effect only (φ = 0)	9.4 (2)	21.9 (2)	2266.9	2258.2
7	*e**	log (*t*): interaction only (δ = 0)	20.4 (2)	22.9 (2)	2255.8	2257.2
8	Categorical	*t*	17.7 (5)	19.9 (5)	2258.5	2260.2
9	*e* − 30	*t*: full model	15.9 (3)	21.0 (3)	2260.3	2259.1
10	*e**	*t*: full model	18.2 (3)	23.9 (3)	2258.1	2256.1

[a]Based on analyses of leukemia mortality (1950–2000) using models in which the EAR or ERR is given by $\beta_s(d + \theta d^2) \exp[\gamma f(e) + \delta g(t) + \phi f(e) g(t)]$.
[b]Separate estimates for *e* < 20, 20 ≤ *e* < 40, *e* ≥ 40.
[c]*e** is min[(*e* − 30) / 10, 0], where *e* is age at exposure in years.

sure (*p* = .15). The total deviances for the preferred EAR and ERR leukemia models were nearly identical.

Thus, the committee's preferred models for the EAR and the ERR are as follows, with δ = 0 for the EAR model:

BEIR VII leukemia model: EAR(*d, s, e, t*) or
ERR(*d, s, e, t*) = $\beta_s(D + \theta D^2) \exp [\gamma e^* +$
$\delta \log (t / 25) + \phi e^* \log (t / 25)]$.   (12B-11)

The parameter estimates for the committee's preferred leukemia models are listed in Table 12-3 in the main chapter. Figure 12-2 shows both the ERR and the EAR as a function of time since exposure for exposure ages of 10, 20, and 30+ years. The ERR model is similar to that used for all leukemia by NIH (2003), although its leukemia model was based on *e* instead of *e**, and on *t* instead of log (*t*), and did not allow for the dependence of the ERR on sex. Although there was no indication that the ERR depended on sex, this was included for compatibility with models for site-specific solid cancers.

ANNEX 12C: DETAILS OF LAR UNCERTAINTY ANALYSIS

Uncertainty Due to Sampling Variability

The approximate variance of the estimated LAR due to the uncertainty in LSS estimated linear models can be derived with the "delta method" (Feinberg 1988). As an example, the estimated LAR based on relative risk transport for solid cancer (for males or females) is calculated as

$$LAR = \hat{\beta} D \times \exp[\hat{\gamma} e^*] \sum_{a=e+5}^{100} \exp[\hat{\eta} \log(a / 60)]B(a)S(a) / S(e),$$

(12C-1)

where *e** = *e* − 30 if *e* (exposure age) is less than 30 years and 0 otherwise; *B(a)* is the age-specific baseline rate at age *a* for the cancer of interest; *S(a)* is the probability of survival (in the 1999 U.S. population) to age *a*; and the Greek letters with hats represent the estimated coefficients in the excess relative risk model. The logarithm of Equation (12C-1) gives

$$\log(LAR) = \log(D) + \hat{\beta}^* + \hat{\gamma} e^* + \log\left\{ \sum_{a=e+5}^{100} \exp[\hat{\eta} \log(a / 60)]B(a)S(a) / S(e) \right\},$$

where $\hat{\beta}^* = \log(\hat{\beta})$.

The result of a first-order Taylor's approximation about $\hat{\eta} = \eta$ is

$$\log(LAR) \approx \log(D) + \hat{\beta}^* + \hat{\gamma} e^* + \log\left\{ \sum_{a=e+5}^{100} \exp[\eta \log(a / 60)]B(a)S(a) / S(e) \right\} + \frac{\sum_{a=e+5}^{100} \exp[\eta \log(a / 60)][B(a)S(a) / S(e)]\log(a / 60)}{\sum_{a=e+5}^{100} \exp[\eta \log(a / 60)][B(a)S(a) / S(e)]}$$

so that the estimate of log (LAR) is a constant plus $A^T \hat{\theta}$, where

$$A^T =$$

$$\left(1, \ e^*, \ \frac{\sum_{a=e+5}^{100} \exp[\hat{\eta}\log(a/60)][B(a)S(a)/S(e)]\log(a/60)}{\sum_{a=e+5}^{100} \exp[\hat{\eta}\log(a/60)][B(a)S(a)/S(e)]} \right),$$

and $\hat{\theta}^T = (\hat{\beta}^*, \hat{\gamma}, \hat{\eta})$. Then var[log(*LAR*)] may be estimated by

$$\text{var}[\log(LAR)] = A^T V A, \qquad (12\text{-}C2)$$

where V is the estimated variance-covariance matrix of $(\hat{\beta}^*, \hat{\gamma}, \hat{\eta})$, which is available as a component of the output from the computer program used to estimate the risk models. The standard error of the log of estimated LAR is the square root of the estimate of this variance. A 95% confidence interval for log (LAR) is obtained as the estimate of log (LAR) plus and minus 1.96 times the standard error, and the confidence interval for LAR is obtained by taking the antilogarithm of these end points.

The LAR based on absolute risk transport is

$$LAR = \hat{\beta}D \times \exp[\hat{\gamma}e^*]\sum_{a=e+5}^{100} \exp[\hat{\eta}\log(a/60)]S(a)/S(e).$$

The issues and computations involve only slight modifications of what has been described above. For scenarios that involve a weighted average of different ages at exposure and for relative and absolute risk models for leukemia, which involve quadratic-in-dose terms and different modifiers including interactions, the computations differ but the ideas behind the delta method calculations are the same as above.

The confidence intervals in Tables 12-5A and 12-5B for risks of cancer incidence and mortality at specific sites were based on the same procedure as above, but without accounting for the uncertainty in γ and η, since, with a few exceptions, these quantities were fixed at their values estimated from all solid cancers combined (although the values of γ and η used in site-specific models were compatible with data for each site, the fixed values cannot be considered unbiased estimates of the correct values). For most sites, uncertainty in the estimated coefficient of dose (β) is quite large and is expected to dominate the uncertainty in the estimated LAR.

Combining Several Sources of Uncertainty

A single estimate of LAR is obtained from estimates based on ERR and EAR transport models as a combination on the log scale: log (LAR) = [p (log (LAR$_{ERR}$) + (1 − p) log (LAR$_{EAR}$)], where LAR$_{ERR}$ and LAR$_{EAR}$ are the estimates based on ERR and EAR transport, respectively, and p is a number between 0 and 1, reflecting the relative strength of belief in the two transport models. For most cancers, a value of .7 was taken for p. Exceptions were lung cancer, where $p = .3$, and thyroid cancer, where only an ERR model developed from data on Caucasian women was available. A further

adjustment to the single estimate of LAR, due to the presumed curvature in the dose-response, is obtained by dividing this combined estimate by the presumed DDREF. A value of 1.5 was used for DDREF, which is an estimate of the median of the Bayesian posterior probability distribution for DDREF, as discussed in the chapter.

The uncertainty analysis here arrives at an approximate variance for log (LAR), emanating from the individual variances in LAR$_{ERR}$ and LAR$_{EAR}$ (sampling variability from the LSS risk model estimation, as discussed above), p (uncertainty in the knowledge of whether absolute risk or excess risk is transportable from Japanese A-bomb survivors to the U.S. population), and DDREF (uncertainty in estimating dose-response curvature from animal studies and uncertainty with which the animal curvature applies to humans).

To accomplish this, the model above is written more formally as depending on four sets of unknown quantities: θ_R, the parameters in the relevant ERR model; θ_A, the parameters in the EAR model; I_R, an indicator variable that takes on the value 1 if the ERR model is the correct one for transport and 0 if the EAR model is the correct one; and θ_{DDREF}, the unknown DDREF. The LAR associated with an acute radiation dose D at age e may be written as

$$LAR(e, D; \theta_R, \theta_A, I_R, \theta_{DDREF}) =$$
$$LAR_R(e, D; \theta_R)^{I_R} LAR_A(e, D; \theta_A)^{1-I_R} / \theta_{DDREF},$$

where LAR$_R$(e, D; θ_R) and LAR$_A$(e, D; θ_A) are the LARs based on EAR and ERR transport, prior to DDREF adjustment, and θ_{DDREF} is the correct DDREF value. Notice that if the ERR model is the correct one for transport, then I_R is 0 and the LAR expression above reduces to LAR$_A$(e, D; θ_A / θ_{DDREF}. Similarly, if the relative risk model is the correct one for transport, then the LAR expression reduces to the excess relative risk LAR with DDREF adjustment.

The *estimated* LAR can be expressed by the same formula, but with the known parameters replaced by their estimators: LAR($e, D; \hat{\theta}_R, \hat{\theta}_A, \hat{I}_E, \hat{\theta}_{DDREF}$), where $\hat{\theta}_R$ and are parameter estimates for the ERR and EAR models; $\hat{I}_R$ is the (subjective) probability that the relative risk model is the correct one for transport; and $\hat{\theta}_{DDREF}$ is the (subjective) estimate of DDREF. Every quantity with a "hat" on it is an uncertain estimator and has a variance associated with it. The variance in the estimated LAR, consequently, is that which is propagated by the variances of these estimators.

Statistically, it is best to consider this propagation on the log scale:

$$\log LAR(e, D; \hat{\theta}_R, \hat{\theta}_A, \hat{I}_R, \hat{\theta}_{DDREF}) = \log LAR_A(e, D; \hat{\theta}_A) +$$
$$\hat{I}_A \log[LAR_R(e, D; \hat{\theta}_R)/LAR_A(e, D; \hat{\theta}_A)] - \log\hat{\theta}_{DDREF}.$$

With the simplifying approximation that the "hats" can be dropped from $\hat{\theta}_A$ and $\hat{\theta}_R$ in the middle term and the assumption that the uncertainties due to risk model estimation, subjective assessment of DDREF, and subjective assessment of transport model are independent of one another, the variance of the log of the estimated LAR is the sum of three pieces:

$$\mathrm{var}[\log \mathrm{LAR}(e, D; \hat{\theta}_A, \hat{\theta}_R, \hat{I}_A, \hat{\theta}_{\mathrm{DDREF}})] =$$

$$\mathrm{var}[\log \mathrm{LAR}_A(e, D; \hat{\theta}_A)] + \{\log[\mathrm{LAR}_R(e, D; \hat{\theta}_R)/$$

$$\mathrm{LAR}_A(e, D; \hat{\theta}_A)]\}^2 \, \mathrm{var}(\hat{I}_R) + \mathrm{var}(\log \hat{\theta}_{\mathrm{DDREF}}),$$

which are due, respectively, to the variability in the parameter estimators in the EAR model, the uncertainty in the transport model, and the uncertainty in the DDREF. It is a fairly simple matter to estimate the variance of the log (LAR) from these quantities. The variance of log (LAR), with a normal approximation to the sampling distribution of log (LAR), leads directly to the coefficient of variation in Table 12-10 and the subjective confidence intervals in Tables 12-6 and 12-7.

The simplifying approximation mentioned above amounts to assuming that $\log[\mathrm{LAR}_A(e, D; \hat{\theta}_A)]$ and $\log[\mathrm{LAR}_E(e, D; \hat{\theta}_R)]$ have equal variances and a correlation of 1 or, in other words, that the variance of an average of these two quantities is the same as the variance of either one individually. The effect of inaccuracies in this assumption is expected to be small relative to the overall variability. Furthermore, because the first term in the variance expression represents the variance of the estimated LAR for either transport model, a weighted average of $\mathrm{var}[\log \mathrm{LAR}_R(e, D; \hat{\theta}_R)]$ and $\mathrm{var}[\log \mathrm{LAR}_R(e, D; \hat{\theta}_A)]$ is used to estimate it (with the weight corresponding to the strength of belief in the relative risk transport model).

The approach for estimating the variances of the sampling distributions of the estimated LARs is discussed in the first section of this annex. The variance of $\hat{I}_R$ is taken to be Bernoulli variance. If, for example, the probability that the relative risk transport is correct is taken to be .7, then the variance of $\hat{I}_R$ is $.7 \times 0.3$. The Bernoulli variance tends to be larger than a variance from a uniform distribution (for a model in which the correct transport is some completely unknown combination of relative and absolute risk) or from a beta distribution (for a model in which the correct transport is some unknown combination, but with more specific information about the possible combination). In the absence of any real knowledge about which of these is correct, the committee has elected to use the more conservative approach, which leads to somewhat wider confidence intervals.

As discussed in Annex 11B, the DDREF analysis is necessarily rough and the variance of the uncertainty distribution described there is, if anything, misleadingly small. For the uncertainty analysis considered here, therefore, the variance representing the uncertainty in log (DDREF) was inflated by 50%, using 0.09 as the variance of $\mathrm{var}(\log \hat{\theta}_{\mathrm{DDREF}})$, rather than the derived posterior variance 0.06.

ANNEX 12D: ADDITIONAL EXAMPLES OF LIFETIME RISK ESTIMATES BASED ON BEIR VII PREFERRED MODELS

Tables 12D-1 and 12D-2 show lifetime risk estimates for cancer incidence and mortality resulting from a single dose of 0.1 Gy at several specific ages. Estimates are shown for all cancer, leukemia, all solid cancer, and cancer of several specific sites. Table 12D-3 shows analogous lifetime risk estimates for exposure to 1 mGy per year throughout life and to 10 mGy per year from ages 18 to 65. The examples below illustrate how these tables may be used to obtain estimates for other exposure scenarios. For clarity of presentation, the committee has generally shown more decimal places than are justified.

Example 1: A 10-year-old male receives a dose of 0.01 Gy (10 mGy) to the colon from a computed tomography (CT) scan. Table 12D-1 shows the estimated lifetime risk of being diagnosed with colon cancer for a male exposed to 0.1 Gy at age 10 as 241 per 100,000. The estimate for a male exposed at 0.01 Gy is obtained as $(0.01 / 0.1) \times 241 = 24.1$ per 100,000 (about 1 in 4000). An estimate of the lifetime risk of dying of colon cancer can also be obtained using Table 12D-2, and is $(0.01 / 0.1) \times 117 = 11.7$ per 100,000 (about 1 in 8500).

Example 2: A 45-year-old woman receives a dose of 0.001 Gy (1 mGy) to the breast from a mammogram. Table 12D-1 shows an estimated lifetime risk of being diagnosed with breast cancer for a female exposed to 0.1 Gy at age 40 as 141 per 100,000; the comparable estimate for exposure at age 50 is 70 per 100,000. Using linear interpolation, the risk from exposure to 0.1 Gy at age 45 is $(141 + 70) / 2 = 105.5$ per 100,000. The risk from exposure to 0.001 Gy is estimated as $(0.001 / 0.1) \times 105.5 = 1.055$ per 100,000. A rough estimate of the risk from repeated annual mammograms could be obtained by adding estimates obtained from receiving a mammogram at ages 45, 46, 47, 48, and so forth. For most purposes, such an estimate will be reasonable, although this approach does not account for the possibility of dying before subsequent doses are received.

Example 3: A female is exposed to high natural background of 0.004 Gy (4 mGy) per year throughout life. Lifetime risk estimates for exposure to 0.001 Gy (1 mGy) per year throughout life are shown in columns 2 (incidence) and 4 (mortality) of Table 12D-3. To obtain estimates for exposure to 4 mGy throughout life, these estimates must be multiplied by 4. For example, the estimated risk of a female being diagnosed with a solid cancer would be 3872 (4×968), per 100,000 whereas the risk of being diagnosed with leukemia would be 204 (4×51) per 100,000, yielding a total risk of being diagnosed with cancer of 4076 per 100,000 (about 1 in 25). The risk of dying of cancer can be obtained in a similar manner and would be 1988 per 100,000 (about 1 in 50).

TABLE 12D-1 Lifetime Attributable Risk of Cancer Incidence[a]

Cancer Site	Age at Exposure (years)										
	0	5	10	15	20	30	40	50	60	70	80
Males											
Stomach	76	65	55	46	40	28	27	25	20	14	7
Colon	336	285	241	204	173	125	122	113	94	65	30
Liver	61	50	43	36	30	22	21	19	14	8	3
Lung	314	261	216	180	149	105	104	101	89	65	34
Prostate	93	80	67	57	48	35	35	33	26	14	5
Bladder	209	177	150	127	108	79	79	76	66	47	23
Other	1123	672	503	394	312	198	172	140	98	57	23
Thyroid	115	76	50	33	21	9	3	1	0.3	0.1	0.0
All solid	2326	1667	1325	1076	881	602	564	507	407	270	126
Leukemia	237	149	120	105	96	84	84	84	82	73	48
All cancers	2563	1816	1445	1182	977	686	648	591	489	343	174
Females											
Stomach	101	85	72	61	52	36	35	32	27	19	11
Colon	220	187	158	134	114	82	79	73	62	45	23
Liver	28	23	20	16	14	10	10	9	7	5	2
Lung	733	608	504	417	346	242	240	230	201	147	77
Breast	1171	914	712	553	429	253	141	70	31	12	4
Uterus	50	42	36	30	26	18	16	13	9	5	2
Ovary	104	87	73	60	50	34	31	25	18	11	5
Bladder	212	180	152	129	109	79	78	74	64	47	24
Other	1339	719	523	409	323	207	181	148	109	68	30
Thyroid	634	419	275	178	113	41	14	4	1	0.3	0.0
All solid	4592	3265	2525	1988	1575	1002	824	678	529	358	177
Leukemia	185	112	86	76	71	63	62	62	57	51	37
All cancers	4777	3377	2611	2064	1646	1065	886	740	586	409	214

NOTE: Number of cases per 100,000 persons exposed to a single dose of 0.1 Gy.

[a]These estimates are obtained as combined estimates based on relative and absolute risk transport and have been adjusted by a DDREF of 1.5, except for leukemia, which is based on a linear-quadratic model.

TABLE 12D-2 Lifetime Attributable Risk of Cancer Mortality[a]

Cancer Site	Age at Exposure (years)										
	0	5	10	15	20	30	40	50	60	70	80
Males											
Stomach	41	34	30	25	21	16	15	13	11	8	4
Colon	163	139	117	99	84	61	60	57	49	36	21
Liver	44	37	31	27	23	16	16	14	12	8	4
Lung	318	264	219	182	151	107	107	104	93	71	42
Prostate	17	15	12	10	9	7	6	7	7	7	5
Bladder	45	38	32	27	23	17	17	17	17	15	10
Other	400	255	200	162	134	94	88	77	58	36	17
All solid	1028	781	641	533	444	317	310	289	246	181	102
Leukemia	71	71	71	70	67	64	67	71	73	69	51
All cancers	1099	852	712	603	511	381	377	360	319	250	153
Females											
Stomach	57	48	41	34	29	21	20	19	16	13	8
Colon	102	86	73	62	53	38	37	35	31	25	15
Liver	24	20	17	14	12	9	8	8	7	5	3
Lung	643	534	442	367	305	213	212	204	183	140	81
Breast	274	214	167	130	101	61	35	19	9	5	2
Uterus	11	10	8	7	6	4	4	3	3	2	1
Ovary	55	47	39	34	28	20	20	18	15	10	5
Bladder	59	51	43	36	31	23	23	22	22	19	13
Other	491	287	220	179	147	103	97	86	69	47	24
All solid	1717	1295	1051	862	711	491	455	415	354	265	152
Leukemia	53	52	53	52	51	51	52	54	55	52	38
All cancers	1770	1347	1104	914	762	542	507	469	409	317	190

NOTE: Number of deaths per 100,000 persons exposed to a single dose of 0.1 Gy.

[a]These estimates are obtained as combined estimates based on relative and absolute risk transport and have been adjusted by a DDREF of 1.5, except for leukemia, which is based on a linear-quadratic model.

TABLE 12D-3 Lifetime Attributable Risk of Solid Cancer Incidence and Mortality[a]

Cancer site	Incidence: Exposure Scenario		Mortality: Exposure Scenario	
	1 mGy per Year throughout Life	10 mGy per Year from Ages 18 to 65	1 mGy per Year throughout Life	10 mGy per Year from Ages 18 to 65
Males				
Stomach	24	123	13	66
Colon	107	551	53	273
Liver	18	93	14	72
Lung	96	581	99	492
Prostate	32	164	6.3	32
Bladder	69	358	16	80
Other	194	801	85	395
Thyroid	14	28		
All solid	554	2699	285	1410
Leukemia	67	360	47	290
All cancers	621	3059	332	1700
Females				
Stomach	32	163	19	94
Colon	72	368	34	174
Liver	8.7	44	8	40
Lung	229	1131	204	1002
Breast	223	795	53	193
Uterus	14	19	3.5	18
Ovary	29	140	18	91
Bladder	71	364	21	108
Other	213	861	98	449
Thyroid	75	139		
All solid	968	4025	459	2169
Leukemia	51	270	38	220
All cancers	1019	4295	497	2389

NOTE: Number of cases or deaths per 100,000 persons exposed to 1 mGy per year throughout life or to 10 mGy per year from ages 18 to 64.

[a]These estimates are obtained as combined estimates based on relative and absolute risk transport and have been adjusted by a DDREF of 1.5, except for leukemia, which is based on a linear-quadratic model.

13

Summary and Research Needs

The research needs stated here relate to the committee's primary task: "To develop the best possible risk estimate for exposure to low-dose, low-LET [linear energy transfer] radiation in human subjects."

EVIDENCE FROM BIOLOGY

Molecular and Cellular Responses to Ionizing Radiation

This chapter discusses the biological effects of the ranges of radiation dose that are most relevant for the committee's deliberations on the shapes of dose-response relationships. Considering the levels of background radiation, the maximal permissible levels of exposure of radiation workers now in effect, and the fact that much of the epidemiology of low-dose exposures includes people who in the past have received up to 500 mGy, the committee has focused on evaluating radiation effects in the low-dose range of <100 mGy, with emphasis on the lowest doses when relevant data are available. Effects that may occur as the radiation is delivered chronically over several months to a lifetime are thought to be most relevant.

At low doses, damage is caused by the passage of single particles that can produce multiple, locally damaged sites leading to DNA double-strand breaks (DSBs). DNA DSBs in the low-dose range can be quantified by a number of novel techniques, including immunofluorescence, comet assay, chromosome aberrations, translocation, premature chromosome condensation, and others. Some of these indicators of DSBs show linearity down to doses of 5 to 10 mGy.

In vitro data on the introduction of gene mutations by low-LET ionizing radiation are consistent with knowledge of DNA damage response mechanisms and imply a non-threshold low-dose response for mutations involved in cancer development. Experiments that quantified DNA breakage, chromosomal aberrations, or gene mutations induced by low total doses or low doses per fraction suggest that the dose-response over the range of 20 to 100 mGy is linear. Limited data indicate that the dose-response for DNA breakage is linear down to 1 mGy, and biophysical arguments suggest that the response should be linear between zero and 5 mGy.

In vitro studies of gene mutation induction provide evidence for a dose and dose rate effectiveness factor (DDREF) in the range of 2–4. The DDREF has been used in past estimates of risk to adjust data obtained from acute exposures at Hiroshima and Nagasaki to the expected lower risk posed by chronic low-dose exposures that the general population might experience.

Research Need 1. Determination of the level of various molecular markers of DNA damage as a function of low-dose ionizing radiation

Currently identified molecular markers of DNA damage and other biomarkers that can be identified in the future should be used to quantify low levels of DNA damage and to identify the chemical nature and repair characteristics of the damage to the DNA molecule. These biomarkers have to be evaluated fully to understand their biological significance for radiation damage and repair and for radiation carcinogenesis.

Most studies suggest that the repair of ionizing radiation damage occurs through nonhomologous end joining and related pathways that are constitutive in nature, occur in excess, and are not induced to higher levels by low radiation doses.

Data from animal models of radiation tumorigenesis were evaluated with respect to the cellular mechanisms involved. For animal models of radiation carcinogenesis that are dependent on cell killing, there tend to be threshold-like dose-responses and high values of DDREF; therefore, less weight was placed on these data. Once cell-killing dependence is excluded, animal data are not inconsistent with a linear nonthreshold (LNT) dose response, and DDREF values are in the range 2–3 for solid cancers and somewhat higher for acute myeloid leukemia.

Research Need 2. Determination of DNA repair fidelity, especially as regards double- and multiple-strand breaks at low doses, and determination of whether repair capacity is independent of dose

Repair capacity at low levels of damage must be investigated, especially in light of conflicting evidence for stimulation of repair at low doses. In such studies the accuracy of DNA sequences rejoined by these pathways has to be determined, and the mechanisms of error-prone repair of radiation lesions must be elucidated. Identification of critical genetic alterations that can be characteristic of radiation exposure would be important.

Consideration of Phenomena That Might Affect Risk Estimates for Carcinogenesis at Very Low Doses

A number of biological phenomena that could conceivably affect risk estimates at very low radiation doses have been reported. These phenomena include the existence of radiation-sensitive human subpopulations, hormetic or adaptive effects, bystander effects, low-dose hyperradiosensitivity, and genomic instability.

Radiation-Sensitive Subpopulations

Epidemiologic, clinical, and experimental data provide clear evidence that genetic factors can influence radiation cancer risk. Strongly expressing human mutations of this type are rare and are not expected to influence significantly the development of estimates of population-based, low-dose risks. They are, however, potentially important in the context of high-dose medical exposures. Evidence for the complex interaction of weakly expressing genetic factors in cancer risk is growing, but current understanding is insufficient for a detailed consideration of the potential impact on population risk.

Adaptive Response

Adaptive responses have been well documented in bacteria, where exposures to radiation or chemicals induce subsequent resistance to these agents by inducing expression of DNA damage repair genes. This induced expression of repair genes does not occur to a significant extent in human cells, although changes in signal transduction do take place. A type of apparent adaptive response, however, has been documented for the induction of chromosomal aberrations in human lymphocytes stimulated to divide.

In most studies, a priming or adaptive dose of about 10 mGy significantly reduces the frequency of chromosomal aberrations and mutations induced a few hours later by 1000–3000 mGy. Similar effects are sometimes seen with other end points. However, priming doses less than 5 mGy or greater than ~200 mGy generally give very little, if any, adaptation, and adaptation has not been reported for challenge doses of less than about 1000 mGy. To have relevance for risk assessment, the adaptive response has to be demonstrated for both priming and challenging doses of 1–50 mGy.

Furthermore, the induction and magnitude of the adaptive response in human lymphocytes are highly variable, with much heterogeneity demonstrated among different individuals. The adaptive response could not be induced when noncycling lymphocytes were given the priming dose. Although inhibitor and electrophoretic studies suggest that alterations in messenger RNA transcription and protein synthesis are involved in the adaptive response in lymphocytes, no specific signal transduction or repair pathways have been identified. At this time, the assumption that any stimulating effects from low doses of ionizing radiation will have a significant effect in reducing long-term deleterious effects of radiation on humans is unwarranted.

Bystander Effects

The bystander effect that results from irradiated cells' reacting with nearby nonirradiated cells could influence dose-response relationships. Such an effect might come into play at low-LET doses below 1–5 mGy, where some cells of the body would not be irradiated. Current limitations of low-LET bystander studies include the lack of demonstrated bystander effects below 50 mGy and uncertainties about whether the effect occurs *in vivo*. Another complication is that both beneficial and detrimental effects have been postulated for bystander effects by different investigators. Until molecular mechanisms are elucidated, especially as they relate to an intact organism, and until reproducible bystander effects are observed for low-dose low-LET radiation in the dose range of 1–5 mGy, where an average of less than 1 electron tracks traverse the nucleus, the assumption should be made that bystander effects will not influence the shape of the low-dose, low-LET dose-response relationship.

Hyperradiosensitivity for Low Doses

In some cell lines, hyperradiosensitivity (HRS) has been reported for cell lethality induced by low-LET radiation at doses less than 100–200 mGy. In this dose range, survival decreases to 85–90%, which is significantly lower that projected from data obtained above 1–2 Gy. It is not known whether HRS for cell lethality would cause an increase in deleterious effects in surviving cells or would actually decrease deleterious effects by increased killing of damaged cells. Until molecular mechanisms responsible for HRS that may or may not play a role in carcinogenesis are understood, the extrapolation of data for HRS for cell lethality to the dose-response for carcinogenesis in the 0–100 mGy range is not warranted.

Genomic Instability

During the last decade, evidence has accumulated that under certain experimental conditions, the progeny of cells surviving radiation appear to express new chromosomal aberrations and gene mutations over many postirradiation cell generations. This feature is termed radiation-induced persistent genomic instability. Some inconsistencies were identified in the data that describe the diverse manifestation of induced genomic instability, and clear evidence of its general involvement in radiation-induced cancer is lacking. Although developing data on the various phenomena classified as genomic instability may eventually provide useful insights into the mechanisms of carcinogenesis, it is not possible to predict whether induced genomic instability will influence low-dose, low-LET response relationships.

Research Need 3. Evaluation of the relevance of adaptation, low-dose hypersensitivity, bystander effects, and genomic instability for radiation carcinogenesis

Mechanistic data are needed to establish the relevance of these processes to low-dose radiation exposure (*i.e.*, <100 mGy). Relevant end points should include not only chromosomal aberrations and mutations but also genomic instability and induction of cancer. *In vitro* and *in vivo* data are needed for delivery of low doses over several weeks or months at very low dose rates or with fractionated exposures. The cumulative effect of multiple low doses of less than 10 mGy delivered over extended periods has to be explored further. The development of *in vitro* transformation assays utilizing nontransformed human diploid cells is judged to be of special importance.

Hormesis

The possibility that low doses of radiation may have beneficial effects (a phenomenon often referred to as "hormesis") has been the subject of considerable debate. Evidence for hormetic effects was reviewed, with emphasis on material published since the 1990 BEIR V study on the health effects of exposure to low levels of ionizing radiation. Although examples of apparent stimulatory or protective effects can be found in cellular and animal biology, the preponderance of available experimental information does not support the contention that low levels of ionizing radiation have a beneficial effect. The mechanism of any such possible effect remains obscure. At this time, the assumption that any stimulatory hormetic effects from low doses of ionizing radiation will have a significant health benefit to humans that exceeds potential detrimental effects from radiation exposure at the same dose is unwarranted.

Research Need 4. Identification of molecular mechanisms for postulated hormetic effects at low doses

Definitive experiments that identify molecular mechanisms are necessary to establish whether hormetic effects exist for radiation-induced carcinogenesis.

Radiation-Induced Cancer: Mechanism, Quantitative Experimental Studies, and the Role of Molecular Genetics

A critical conclusion on mechanisms of radiation tumorigenesis is that the data reviewed greatly strengthen the view that there are intimate links between the dose-dependent induction of DNA damage in cells, the appearance of gene or chromosomal mutations through DNA damage misrepair, and the development of cancer. Although less well established, the data available point toward a single-cell (monoclonal) origin for induced tumors and suggest that low-dose radiation acts predominantly as a tumor-initiating agent. These data also provide some evidence on candidate, radiation-associated mutations in tumors. These mutations are predominantly loss-of-function DNA deletions, some of which are represented as segmental loss of chromosomal material (*i.e.*, multigene deletions).

This form of tumorigenic mechanism is broadly consistent with the more firmly established *in vitro* processes of DNA damage response and mutagenesis considered in Chapters 1 and 2. Thus, if as judged in Chapters 1 and 2, error-prone repair of chemically complex DNA double-strand damage is the predominant mechanism for radiation-induced gene or chromosomal mutation, there can be no expectation of a low-dose threshold for the mutagenic component of radiation cancer risk.

One mechanistic caveat explored was that novel forms of cellular damage response, collectively termed induced genomic instability, might contribute significantly to radiation cancer risk. The cellular data reviewed in Chapter 2 identified uncertainties and some inconsistencies in the expression of this multifaceted phenomenon. However, telomere-associated mechanisms did provide a coherent explanation for some *in vitro* manifestations of induced genomic instability. The data considered did not reveal consistent evidence for the involvement of induced genomic instability in radiation tumorigenesis, although telomere-associated processes may account for some tumorigenic phenotypes. A further conclusion was that there is little evidence of specific tumorigenic signatures of radiation causation, but rather that radiation-induced tumors develop in a tumor-specific multistage manner that parallels that of tumors arising spontaneously.

Quantitative animal data on dose-response relationships provide a complex picture for low-LET radiation, with some tumor types showing linear or linear-quadratic relationships while other studies are suggestive of a low-dose threshold, particularly for thymic lymphoma and ovarian cancer. Since, however, the induction or development of these two cancer types is believed to proceed via atypical mechanisms involving cell killing, it was judged that the threshold-like responses observed should not be generalized.

Radiation-induced life shortening in mice is largely a reflection of cancer mortality, and the data reviewed generally support the concept of a linear dose-response at low doses and low dose rates. Other dose-response data for animal tumorigenesis, together with cellular data, contributed to the judgments developed and the choice of a DDREF for use in the interpretation of epidemiologic information on cancer risk.

Adaptive responses for radiation tumorigenesis have been investigated in quantitative animal studies, and recent information is suggestive of adaptive processes that increase tumor latency but not lifetime risk. However, these data are difficult to interpret, and the implications for radiological protection remain most uncertain.

Research Need 5. Tumorigenic mechanisms

Further cytogenetic and molecular genetic studies are needed to reduce current uncertainties about the specific role of radiation in multistage radiation tumorigenesis; such investigations would include studies with radiation-associated tumors of humans and experimental animals.

The review of cellular, animal, and epidemiologic or clinical studies on the role of genetic factors in radiation tumorigenesis suggests that many of the known strongly expressing cancer-prone human genetic disorders are likely to show an elevated risk of radiation-induced cancer, probably with a high degree of organ specificity. Cellular and animal studies suggest that the molecular mechanisms underlying these genetically determined radiation effects largely mirror those that apply to spontaneous tumorigenesis and are consistent with knowledge of somatic mechanisms of tumorigenesis. In particular, evidence was obtained that major deficiencies in DNA damage response and tumor-suppressor-type genes can serve to elevate radiation cancer risk.

Limited epidemiologic data from follow-up of second cancers in gene carriers receiving radiotherapy were supportive of the above conclusions, but quantitative judgments about the degree of increased cancer risk remain uncertain. However, since major germline deficiencies in the genes of interest are known to be rare, it has been possible to conclude from published analyses that they are most unlikely to create a significant distortion of population-based estimates of cancer risk. The major practical issue associated with these strongly expressing cancer genes is judged to be the risk of radiotherapy-related cancer.

A major theme developing in cancer genetics is the interaction and potential impact of more weakly expressing variant cancer genes that may be relatively common in human populations. The animal genetic data provide proof-of-principle evidence of how such variant genes with functional polymorphisms can influence cancer risk, including limited data on radiation tumorigenesis. Attention was also given to human molecular epidemiology data on associations between functional polymorphisms and cancer risk, particularly with respect to DNA damage response genes.

Given that functional gene polymorphisms associated with cancer risk may be relatively common, the potential for significant distortion of population-based risk was explored with emphasis on the organ specificity of the genes of interest. An interim conclusion was that common polymorphisms of DNA damage response genes associated with organ-wide radiation cancer risk would be the most likely source of major interindividual differences in radiation response.

Research Need 6. Genetic factors in radiation cancer risk

Further work is needed in humans and mice on gene mutations and functional polymorphisms that influence the risk of radiation-induced cancers. Where possible, human molecular genetic studies should be coupled with epidemiologic investigations.

GENETIC EFFECTS OF RADIATION ON HUMAN POPULATIONS

As noted in BEIR V, heritable effects of radiation are estimated using what is referred to as the "doubling dose method" and expressed in terms of increases in the frequencies of genetic diseases in the population over and above those that occur as a result of spontaneous mutations. The doubling dose (DD) is the amount of radiation required to produce as many mutations as those that occur spontaneously in a generation and is calculated as a ratio of the average rates of spontaneous and induced mutations in defined genes. If the DD is small, the relative mutation risk per unit dose (*i.e.*, 1/DD) is high, and if DD is large, the relative mutation risk is low. The DD, therefore, provides a convenient yardstick to express risks and a perspective of whether the predicted increases are trivial, small, or substantial relative to the baseline.

Revision of the Conceptual Basis for Calculating the DD

In the BEIR V report, mouse data on both spontaneous and induced mutation rates were used for DD calculations. A reassessment of the assumptions underlying this procedure revealed that the use of mouse data for spontaneous mutation rates can no longer be considered appropriate and that reverting to the use of human data on spontaneous mutation rates for DD calculations, as was first done in the 1972 BEIR report, is correct. The DD calculated is 1 Gy and is the same as the one based entirely on mouse data.

Revision of the Baseline Frequencies of Mendelian Diseases in Humans

The baseline frequencies of genetic diseases constitute an important quantity in risk estimation. While there is no reason to consider revision of the baseline frequencies of congenital abnormalities (6%) and chronic diseases (65%), these two classes together constitute what are referred to as "multifactorial diseases" because of the multiple factors involved in their etiology. Advances in human genetics now suggest that the frequencies of Mendelian diseases (*i.e.*, those that are due to mutations in single genes and show simple and predictable patterns of inheritance) have to be revised upwards from the 1.25% used previously (based on estimates made in the mid-1970s) to 2.40% at this time.

Delineation of a New Concept—The Concept of Potential Recoverability Correction Factor

Mouse data on rates of radiation-induced mutations constitute the primary basis for estimating the risk of radiation-inducible genetic diseases in humans. Advances in the molecular biology of human genetic diseases and in studies of radiation-induced mutations in experimental systems show that mouse mutation rates cannot readily be converted into rates of genetic disease in human live births and that a correction factor, the potential recoverability correction factor (PRCF), is required to make the transition from induced mutations in mice to inducible genetic disease in humans. A framework and methods have been developed to estimate PRCFs for Mendelian and chronic multifactorial diseases.

Introduction of the Concept That Adverse Hereditary Effects of Radiation Are Likely to Be Manifest as Multisystem Developmental Abnormalities

The adverse hereditary effects of radiation are more likely to be manifest as multisystem developmental abnormalities than as Mendelian diseases. This concept incorporates elements of current knowledge of the mechanisms of radiation-induced genetic damage, the molecular nature of radiation-induced mutations, the phenotypic manifestations of naturally occurring multigene deletions in humans, empirical observations in mice on the phenotypic effects of radiation-induced multigene deletions, and the enormous number and distribution of genes involved in development in nearly all the human chromosomes. Appropriate mouse data that can serve as a basis for a preliminary estimate of radiation-induced adverse developmental effects have been identified and used.

Risk estimates have been made only for the first two post-irradiation generations. The population genetic theory of equilibrium between mutation and selection (*i.e.*, the equilibrium theory) underlies the DD method that is used to estimate genetic risks of radiation. This theory postulates that the stability of mutant gene frequencies (and therefore of disease frequencies) in a population is a reflection of the existence of a balance between the rates at which spontaneous mutations arise in every generation and enter the gene pool and the rates at which they are eliminated by natural selection. When such an "equilibrium population" sustains radiation exposure generation after generation, additional mutations are introduced into the gene pool, and these are

also subject to the action of natural selection. The prediction is that a new equilibrium between mutation and selection will be reached. The time it takes in terms of generations to attain the new equilibrium, the rate of approach to it, and the magnitude of increase in mutant (and disease) frequencies are dependent on the induced mutation rate, the intensity of selection, and the type of disease.

Research Need 7. Heritable genetic effects of radiation

Further work is necessary to establish (1) the potential roles of DNA DSB repair processes in the origin of deletions in irradiated stem cell spermatogonia and oocytes (the germ cell stages of importance in risk estimation) in mice and humans and (2) the extent to which large, radiation-induced deletions in mice are associated with multisystem development defects. In humans, the problem can be explored using genomic databases and knowledge of mechanisms of the origin of radiation-induced deletions to predict regions that may be particularly prone to such deletions. These predictions can subsequently be tested in the mouse, these tests can also provide insights into the potential phenotypes associated with such deletions in humans.

With respect to epidemiology, studies on the genetic effects of radiotherapy for childhood cancer, of the type that have been under way in the United States and Denmark since the mid-1990s, should be encouraged, especially when they can be coupled with modern molecular techniques (such as array-based comparative genomic hybridization). These techniques enable one to screen the whole genome for copy number abnormalities (*i.e.*, deletions and duplications of genomic segments) with a resolution beyond the level of a light microscope.

EPIDEMIOLOGIC STUDIES OF POPULATIONS EXPOSED TO IONIZING RADIATION

Atomic Bomb Survivor Studies

The Life Span Study (LSS) cohort of survivors of the atomic bombings in Hiroshima and Nagasaki continues to serve as a major source of information for evaluating health risks from exposure to ionizing radiation, and particularly for developing quantitative estimates of risk. Its advantages include its large size, the inclusion of both sexes and all ages, a wide range of doses that have been estimated for individual subjects, and high-quality mortality and cancer incidence data. In addition, the whole-body exposure received by this cohort offers the opportunity to assess risks for cancers of a large number of specific sites and to evaluate the comparability of site-specific risks.

As an illustration, Figure 13-1 shows estimated ERRs of solid cancer versus dose (averaged over sex and standardized to represent individuals exposed at age 30 at attained

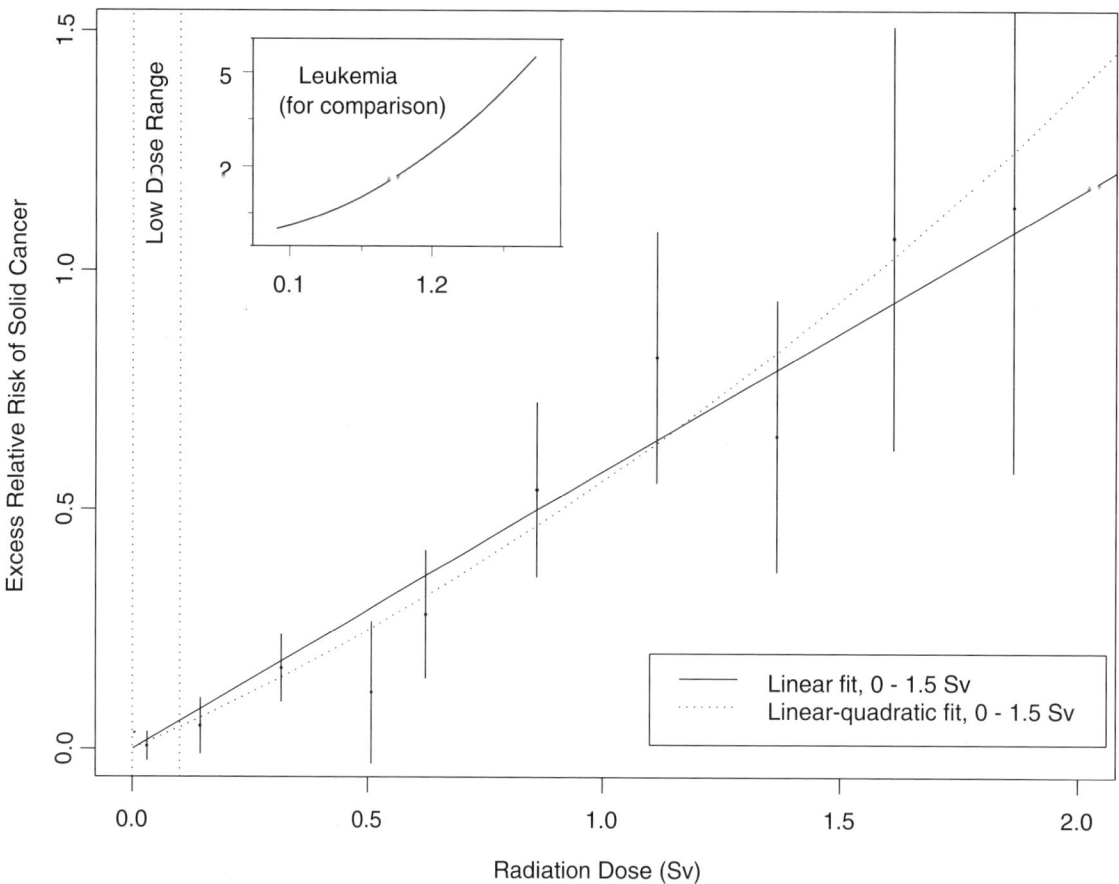

FIGURE 13-1 Excess relative risks of solid cancer for Japanese atomic bomb survivors. The insert shows the fit of a linear-quadratic model for leukemia, to illustrate the greater degree of curvature observed for that cancer.

age 60) for atomic bomb survivors with doses in each of 10 dose intervals less than 2.0 Sv. This plot helps convey the overall dose-response relationship from the LSS cohort and its role in low-dose risk estimation. Specific models are detailed in Chapter 6. It is important to note that the difference between the linear and linear-quadratic models in the low-dose ranges is small relative to the error bars; therefore, the difference between these models is small relative to the uncertainty in the risk estimates produced from them. For solid cancer incidence the linear-quadratic model did not offer statistically significant improvement in the fit, so the linear model was used. For leukemia, a linear-quadratic model (insert in Figure 13-1) was used because it fitted the data significantly better than the linear model.

Plotted points are the estimated ERRs of solid cancer incidence (averaged over sex and standardized to represent individuals exposed at age 30 attained age 60) for atomic bomb survivors with doses in each of 10 dose intervals, plotted above the midpoints of the dose intervals. If $R(d)$ represents the age-specific instantaneous risk at some dose d, then the *excess relative risk* at dose d is $[R(d) - R(0)] / R(0)$ (which is necessarily zero when dose is zero). Vertical

lines are approximate 95% confidence intervals. Solid and dotted lines are estimated linear and linear-quadratic models for ERR, estimated from all subjects with doses in the range 0 to 1.5 Sv. (These are not estimated from the points, but from the lifetimes and doses of individual survivors, using statistical methods discussed in Chapter 6.) A linear-quadratic model will always fit the data better than a linear model, since the linear model is a restricted special case with quadratic coefficient equal to zero. For solid cancer incidence, however, there is no *statistically significant* improvement in fit due to the quadratic term. It should also be noted that in the low-dose range of interest the difference between the estimated linear and linear-quadratic models is small relative to the 95% CIs.

The full LSS cohort consists of approximately 120,000 persons who were identified at the time of the 1950 census. However, most recent analyses have been restricted to approximately 87,000 survivors who were in the city at the time of the bombings and for whom it is possible to estimate doses. Special studies of subgroups of the LSS have provided clinical data, biological measurements, and information on potential confounders or modifiers.

The availability of high-quality cancer incidence data has resulted in several analyses and publications addressing specific cancer sites. These analyses often include special pathological review of the cases and sometimes include data on additional variables (such as smoking for the evaluation of lung cancer risks). Papers focusing on the following cancer sites have been published in the last decade: female breast cancer, thyroid cancer, salivary gland cancer, liver cancer, lung cancer, skin cancer, and central nervous system tumors. Special analyses have also been conducted of cancer mortality in survivors who were exposed either *in utero* or during the first 5 years of life.

Health end points other than cancer have been linked with radiation exposure in the LSS cohort. Of particular note, a dose-response relationship with mortality from nonneoplastic disease was demonstrated in 1992, and subsequent analyses in 1999 and 2003 have strengthened the evidence for this association. Statistically significant associations were seen for the categories of heart disease, stroke, and diseases of the digestive, respiratory, and hematopoietic systems. The data were inadequate to distinguish between a linear dose-response, a pure quadratic response, or a dose-response with a threshold as high as 0.5 Sv.

Medical Radiation Studies

Published studies on the health effects of medical exposures were reviewed to identify those that provide information for quantitative risk estimation. Particular attention was focused on estimating risks of leukemia and of lung, breast, thyroid, and stomach cancer in relation to radiation dose for comparison with estimates derived from other exposed populations, particularly the atomic bomb survivors. The possible association between radiation exposure and cardiovascular mortality and morbidity was also reviewed.

For lung cancer, the ERRs per Gy (ERRs/Gy) from the studies of acute high-dose-rate exposures are statistically compatible and in the range 0.1–0.4. It is difficult to evaluate the effects of age at exposure or of exposure protraction based on these studies because only one study (the hemangioma cohort) is available in which exposure occurred at very young ages and protracted low-dose-rate exposures were received. The study of tuberculosis patients, however, appears to indicate that substantial fractionation of exposure leads to a reduction of risk.

For breast cancer, excess absolute risks (EARs) appear to be similar—of the order of 9.9 per 10^4 person-years (PY) per gray at age 50—following acute and fractionated moderate-to-high-dose-rate exposure. Effects of attained age and age at exposure are important modifiers of risk. The excess risks appear to be higher in populations of women treated for benign breast conditions, suggesting that these women may be at an elevated risk of radiation-induced breast cancer. The hemangioma cohorts showed lower risks, suggesting a possible reduction of risks following protracted low-dose-rate exposures.

For thyroid cancer, all of the studies providing quantitative information about risks are studies of children who received radiotherapy for benign conditions. A combined analysis of data from some of these cohorts and data from atomic bomb survivors and from two case-control studies of thyroid cancer nested within the International Cervical Cancer Survivor Study and the International Childhood Cancer Survivor Study provides the most comprehensive information about thyroid cancer risks. For subjects exposed below the age of 15, a linear dose-response was seen, with a leveling or decrease in risk at the higher doses used for cancer therapy. The pooled ERR was 7.7 Gy^{-1}, and the EAR was 4.4 per 10^4 PY-Gy. Both estimates were significantly affected by age at exposure, with a strong decrease in risk with increasing age at exposure and little apparent risk for exposures after age 20. The ERR appeared to decline over time about 30 years after exposure but was still elevated at 40 years.

Little information on thyroid cancer risk in relation to exposure in childhood to iodine-131 was available. Studies of the effects of ^{131}I exposure later in life provide little evidence of an increased risk of thyroid cancer following ^{131}I exposure after childhood.

For leukemia, ERR estimates from studies with average doses ranging from 0.1 to 2 Gy are relatively close, in the range 1.9 to 5 Gy^{-1}, and are statistically compatible. Estimates of EAR are also similar across studies, ranging from 1 to 2.6 per 10^4 PY-Gy. Little information is available on the effects of age at exposure or of exposure protraction.

For stomach cancer, the estimates of ERR range from negative to 1.3 Gy^{-1}. The confidence intervals are wide, however, and they all overlap, indicating that these estimates are statistically compatible.

Finally, studies of patients having undergone radiotherapy for Hodgkin's disease or breast cancer suggest that there may be some risk of cardiovascular morbidity and mortality for very high doses and high-dose-rate exposures. The magnitude of the radiation risk and the shape of the dose-response curve for these outcomes are uncertain.

Research Need 8. Future medical radiation studies

Most studies of medical radiation should rely on exposure information collected prospectively, including cohort studies as well as nested case-control studies. Future studies should continue to include individual dose estimation to the site of interest, as well as an evaluation of the uncertainty in dose estimation. Ideally, where population-based cancer registries do not exist to establish cohorts of cancer survivors, hospital-based registries can be established to identify cohorts of exposed patients whose mortality and morbidity can be followed. If these registries can be linked to appropriate radiation therapy or diagnostic records, they can be used as a basis for nested case-control studies of specific outcomes, and detailed exposure estimation for the site of interest can be undertaken.

Studies of populations with high- and moderate-dose medical exposures are particularly important for the study of modifiers of radiation risks. Because of the high level of radiation exposure in these populations, they are also ideally suited to study the effects of gene-radiation interactions that may render particular subsets of the population more sensitive to radiation-induced cancer. Genes of particular interest include BRCA1, BRCA2, ATM, CHEK2, NBS1, XRCC1, and XRCC3. These are among the most important genes known to be involved in detection and repair of radiation-induced DNA damage.

Of concern for radiological protection is the increasing use of computed tomography (CT) scans and diagnostic X-rays. Epidemiologic studies of these exposures would be particularly useful if they are feasible, particularly the following: (1) follow-up studies of cohorts of persons receiving CT scans, especially children; and (2) studies of infants who experience diagnostic exposures related to cardiac catheterization, those who have recurrent exposures to follow their clinical status, and premature babies monitored for pulmonary development with repeated X-rays.

The widespread use of interventional radiological procedures in the heart, lungs, abdomen, and many vascular beds, with extended fluoroscopic exposure times of patients and operators, emphasizes the need for recording of dose and later follow-up studies of potential radiation effects among these populations. There is a need to organize worldwide consortia that would use similar methods in data collection and follow-up. These consortia should record delivered doses and technical data from all X-ray or isotope-based imaging approaches including CT, positron emission tomography, and single photon emission computed tomography.

Occupational Radiation Studies

The risk of cancer among physicians and other persons exposed to ionizing radiation in the workplace has been a subject of study since the 1940s, when increased mortality from leukemia was reported among radiologists in comparison to mortality among other medical specialists. Since then, numerous studies have considered the mortality and cancer incidence of various occupationally exposed groups in medicine, industry, defense, research, and aviation industries.

Studies of occupationally exposed groups are, in principle, well suited for direct estimation of the effects of low doses and low dose rates of ionizing radiation. The most informa-

tive studies at present are those of nuclear industry workers (including the workers at Mayak in the former USSR), for whom individual real-time estimates of doses have been collected over time with the use of personal dosimeters. More than 1 million workers have been employed in this industry since its beginning in the early 1950s. However, studies of individual worker cohorts are limited in their ability to estimate precisely the potentially small risks associated with low levels of exposure. Risk estimates from these studies are variable, ranging from no risk to risks an order of magnitude or more than those seen in atomic bomb survivors.

Combined analyses of data from multiple cohorts offer an opportunity to increase the sensitivity of such studies and provide direct estimates of the effects of long-term, low-dose, low-LET radiation. The most comprehensive and precise estimates to date are those derived from the U.K. National Registry of Radiation Workers and the three-country study (Canada-U.K.-U.S.), which have provided estimates of leukemia and all cancer risks. Although the estimates are lower than the linear estimates obtained from studies of atomic bomb survivors, they are compatible with a range of possibilities, from a reduction of risk at low doses to risks twice those on which current radiation protection recommendations are based. Overall, there is no suggestion that the current radiation risk estimates for cancer at low levels of exposure are appreciably in error. Uncertainty regarding the size of this risk remains as indicated by the width of the confidence intervals.

Because of the absence of individual dose estimates in most of the cohorts, studies of occupational exposures in medicine and aviation provide minimal information useful for the quantification of these risks.

Because of the uncertainty in occupational risk estimates and the fact that errors in doses have not formally been taken into account in these studies, the committee concluded that the occupational studies were not suitable for the projection of population-based risks. These studies, however, provide a comparison to the risk estimates derived from atomic bomb survivors.

Research Need 9. Future occupational radiation studies
Studies of occupational radiation exposures, particularly among nuclear industry workers, including nuclear power plant workers, are well suited for direct assessment of the carcinogenic effects of long-term, low-level radiation exposure in humans. Ideally, studies of occupational radiation should be prospective in nature and rely on individual real-time estimates of radiation doses. Where possible, national registries of radiation exposure of workers should be established and updated as additional radiation exposure is accumulated and as workers change employers. These registries should include at least annual estimates of whole-body radiation dose from external photon exposure. These exposure registries should be linked

with mortality registries and, where they exist, with national tumor (and other disease) registries. Where national dose registries cannot be set up, cohort studies based on records of nuclear installations are a useful alternative. It is noted that the power of individual cohort studies at the local and even national levels is limited. To maximize the information about the effects of low-dose, protracted exposures from these studies, it is therefore necessary to combine data across cohorts and countries. Most studies published to date have been based on relatively short follow-up periods, and the majority of workers were still young at the end of follow-up. Extended mortality follow-up over the next decades—and, where possible, cancer morbidity follow-up—of these workers, as they enter an age range when cancer incidence and mortality rates increase, will provide useful improvements of the direct cancer risk estimates drawn from these studies of exposure to low-dose, low-LET radiation. It is also important to continue follow-up of workers exposed to relatively high doses, that is, workers at the Mayak nuclear facility and workers involved in the Chernobyl cleanup.

Environmental Radiation Studies

Ecologic studies of populations living around nuclear facilities and of other environmentally exposed populations do not contain individual estimates of radiation dose or provide a direct quantitative estimate of risk in relation to radiation dose. This limits the interpretation of these data.

Several cohort studies have reported health outcomes among persons exposed to environmental radiation. No consistent or generalizable information is contained in these studies. Four ecologic studies of populations exposed to natural background did not find any association between disease rates and indicators of high background levels of radiation exposure. Ecologic studies of children of adults exposed to radiation while working at the Sellafield nuclear facility in Great Britain have suggested some increased risk of leukemia and lymphoma associated with individual dose, but the findings are based on small numbers of cases and the results across studies are not consistent.

Evidence from ecologic studies does not indicate an increased risk of leukemia among persons exposed *in utero* to radiation from Chernobyl or an increase in rates of childhood leukemia. In contrast to a considerable body of evidence regarding the risk of thyroid cancer in persons exposed to external radiation, there is relatively little information regarding the risk of thyroid cancer in humans exposed internally to ^{131}I. There is some evidence of a small increase in thyroid cancer associated with exposure to ^{131}I from therapeutic and diagnostic uses, but the findings are inconsistent and the small increases in thyroid cancer observed in some studies

may be due to the underlying thyroid condition and not radiation exposure.

Results from external environmental exposures to ^{131}I have been inconsistent. The most informative findings are from studies of individuals exposed to radiation after the Chernobyl accident. Recent evidence indicates that exposure to radiation from Chernobyl is associated with an increased risk of thyroid cancer and that the relationship is dose dependent. The quantitative estimate of excess thyroid cancer risk is generally consistent with estimates from other radiation-exposed populations and is observed in both males and females. Iodine deficiency appears to be an important modifier of risk, enhancing the risk of thyroid cancer following radiation exposure.

Ecologic studies of persons exposed to environmental sources of ionizing radiation have not been useful in developing risk estimates. Exposure levels are low, the studies relate to exposure of populations rather than individuals, and there is minimal possibility of follow-up of exposed individuals. The few exceptions to these circumstances are populations where there is unusual exposure because of accidents involving radiation exposure or long-term releases of relatively high levels of ionizing radiation (*e.g.*, Chernobyl, Hanford).

Research Need 10. Future environmental radiation studies

In general, additional ecologic studies of persons exposed to low levels of radiation from environmental sources are not recommended. However, if disasters occur in which a local population is exposed to unusually high levels of radiation, it is important that there be a rapid response not only for the prevention of further exposure but also for the establishment of scientific evaluation of the possible effects of exposure. The data collected should include basic demographic information on individuals, estimates of acute and possible continuing exposure, the nature of the ionizing radiation, and the means of following these individuals for many years. The possibility of enrolling a comparable nonexposure population should be considered. Studies of persons exposed environmentally as a result of the Chernobyl disaster or as a result of releases from the Mayak nuclear facility should continue.

INTEGRATION OF BIOLOGY AND EPIDEMIOLOGY

This chapter highlights the ways in which cellular, molecular, and animal data can be integrated with epidemiologic findings in order to develop coherent judgments on the health effects of low-LET radiation. Emphasis is placed on data integration for the purposes of modeling these health risks. The principal conclusions from this work are the following:

• Current knowledge on the cellular and molecular mechanisms of radiation tumorigenesis tends to support the application of models that incorporate the excess relative risk projection over time.

• The choice of models for the transport of cancer risk from Japanese A-bomb survivors to the U.S. population is influenced by mechanistic knowledge and information on the etiology of different cancer types.

• A combined Bayesian analysis of A-bomb epidemiologic information and experimental data has been employed to provide an estimate of the DDREF for cancer risk.

• Knowledge of adaptive responses, genomic instability, and bystander signaling between cells that may act to alter radiation cancer risk was judged to be insufficient to be incorporated in a meaningful way into the modeling of epidemiologic data. The same judgment is made with respect to the possible contribution to cancer risk of postirradiation genomic instability and bystander signaling between cells.

• Genetic variation in the population is a potentially important factor in the estimation of radiation cancer risk. Strongly expressing cancer-predisposing mutations are judged from modeling studies to be too rare to distort population-based estimates of risk appreciably, but they are a significant issue in some medical radiation settings. The position regarding potentially more common variant genes that express only weakly remains uncertain.

• Estimation of the heritable effects of radiation takes advantage of new information on human genetic disease and on mechanisms of radiation-induced germline mutation. The application of a new approach to genetic risk estimation leads the committee to conclude that low-dose induced genetic risks are very small compared to baseline risks in the population.

• The committee judges that the balance of evidence from epidemiologic, animal, and mechanistic studies tends to favor a simple proportionate relationship at low doses between radiation dose and cancer risk. Uncertainties in this judgment are recognized and noted.

MODELS FOR ESTIMATING THE LIFETIME RISK OF CANCER

As in past risk assessments, the LSS cohort of survivors of the atomic bombings of Hiroshima and Nagasaki plays a principal role in developing the committee's recommended cancer risk estimates. In contrast to previous BEIR reports, data on both cancer mortality and cancer incidence (from the Hiroshima and Nagasaki tumor registries) were available to the committee. The cancer incidence data analyzed by the committee included nearly 13,000 cases occurring in the period 1958–1998. In addition, the committee evaluated data on approximately 10,000 cancer deaths occurring in the period 1950–2000, in contrast to fewer than 6000 cancer deaths available to the BEIR V committee.

Although the committee did not conduct its own analyses of data from studies other than the LSS, for most studies with suitable data, results of analyses based on models similar to those used by the committee were available and were evaluated. For cancers of the breast and thyroid, several medically exposed groups offer quantitative data suitable for risk assessment, and the recommended models for these sites are those developed in published combined analyses of data from the relevant studies.

To use models developed primarily from the LSS cohort for the estimation of lifetime risks for the U.S. population, it was necessary to make several assumptions. Because of inherent limitations in epidemiologic data and in our understanding of radiation carcinogenesis, these assumptions involve uncertainty. Two important sources of uncertainty are (1) the possible reduction in risk for exposure at low doses and low-dose rates (*i.e.*, the DDREF), and (2) the "transport" of risk estimates based on Japanese atomic bomb survivors to use in estimating risks for the U.S. population. With regard to the DDREF, the committee concluded that linear risk estimates obtained from the LSS cohort should be reduced by a factor of 1.1 to 2.3 for estimating risks at low doses and low dose rates, and the BEIR VII committee used a value of 1.5 to estimate solid cancer risks. To estimate the risk of leukemia, the BEIR VII model is linear-quadratic, since this model fitted the data substantially better than the linear model. The use of data on Japanese A-bomb survivors to estimate risks for the U.S. population (transport) is problematic for sites where baseline risks differ greatly between the two countries. For cancer sites other than breast and thyroid (where data on Caucasian subjects are available), the committee presents estimates based on the assumption that the excess risk due to radiation is proportional to baseline risks (relative risk transport) and also presents estimates based on the assumption the excess risk is independent of baseline risks. As a central estimate, the committee recommends a weighted estimate of these two results, with the ratio of the two used to reflect the uncertainty in transporting risks. For most sites, a weight of 0.7 is used for relative transport and a weight of 0.3 is used for absolute transport; the weighting is reversed for lung cancer.

The committee provides estimates of lifetime risks of both cancer incidence and mortality for leukemia, all solid cancers, and cancers of several specific sites: stomach, colon, liver, lung, female breast, prostate, uterus, ovary, bladder, and all other solid cancers. The committee's models provide the basis for sex-specific estimates for exposure scenarios including single exposures at various ages, chronic exposure throughout life, or occupational exposure from age 18 to 65. These models are based primarily on the LSS study, with additional use of medical data for breast and thyroid.

As an example, Table 13-1 shows the estimated number of incident cancer cases and deaths expected to result if a population of 100,000 persons with an age distribution simi-

TABLE 13-1 Committee's Preferred Estimates of the Lifetime Attributable Risk of Incidence and Mortality for All Solid Cancers and for Leukemia

	All Solid Cancer		Leukemia	
	Males	Females	Males	Females
Excess cases (including nonfatal cases) from exposure to 0.1 Gy	800 (400, 1600)	1300 (690, 2500)	100 (30, 300)	70 (20, 250)
Number of cases in the absence of exposure	45,500	36,900	830	590
Excess deaths from exposure to 0.1 Gy	410 (200, 830)	610 (300, 1200)	70 (20, 220)	50 (10, 190)
Number of deaths in the absence of exposure	22,100	17,500	710	530

NOTE: Number of cases or deaths per 100,000 exposed persons.

lar to that of the entire U.S. population were each exposed to 0.1 Gy; also shown are the numbers that would be expected in the absence of exposure. Results are shown for all solid cancers and for leukemia. The estimates are accompanied by 95% subjective confidence intervals that reflect the most important uncertainty sources—namely, statistical variation, uncertainty in the factor used to adjust risk estimates for exposure at low doses and low dose rates, and uncertainty in the method of transport. Additional sources of uncertainty would increase the width of these intervals. Mortality estimates are reasonably compatible with those in previous risk assessments, particularly if uncertainties are considered.

The committee also presents estimates for each of several specific cancer sites and for other exposure scenarios, although they are not shown here. For many cancer sites, uncertainty is very large, with subjective 95% confidence intervals covering more than an order of magnitude.

In general the magnitude of estimated risks for total cancer mortality or leukemia has not changed greatly from estimates provided in past reports such as BEIR V, those of the United Nations Scientific Committee on the Effects of Atomic Radiation, and those of the International Commission on Radiological Protection. New data and analyses have reduced sampling uncertainty, but uncertainties related to estimating risk for exposure at low doses and low dose rates and to transporting risks from Japanese A-bomb survivors to the U.S. population remain large. Uncertainties in estimating risks of site-specific cancers are especially large.

Research Need 11. Japanese atomic-bomb survivor studies

The LSS cohort of Japanese A-bomb survivors has played a central role in BEIR VII and past risk assessments. It is thus important that follow-up for mortality and cancer incidence continue for the 45% of the cohort who remained alive at the end of 2000.

In the near future, an uncertainty evaluation of the DS02 dosimetry system is expected to become available. Dose-response analyses that make use of this evaluation should thus be conducted to account for dosimetry uncertainties.

Development and application of analytic methods that allow more reliable site-specific estimates are also needed. Specifically, methods that draw on both data for the specific site and data on broader cancer categories could be useful.

Research Need 12. Epidemiologic studies in general

Data from the LSS should be supplemented with data on populations exposed to low doses and/or low dose rates, especially those with large enough doses to allow risks to be estimated with reasonable precision. Studies of nuclear industry workers and careful studies of persons exposed in countries of the former Soviet Union are particularly important in this regard.

Studies in non-Japanese populations are also important, especially for estimating risks of cancers in organs where baseline risks vary widely. Studies that elucidate the relationship of radiation and other risk factors (for example, smoking) are needed, possibly by conducting nested case-control studies within cohorts currently under study.

Combined analyses of data from several cohorts have been used successfully in the past and are encouraged to provide a unified treatment of data from the LSS and other studies.

Development and application of analytic methods that take account of dosimetry uncertainties are encouraged for all studies. For the LSS, analyses that make use of the uncertainty evaluation of the DS02 dosimetry system, which is expected to become available in the near future, are needed.

CONCLUSION

The committee concludes that the current scientific evidence is consistent with the hypothesis that there is a linear, no-threshold dose-response relationship between exposure to ionizing radiation and the development of cancer in humans.

Appendixes

A

Basic Biological and Genetic Concepts

DNA, Genes, and Chromosomes

The genetic material of living organisms, DNA, is contained in chromosomes, which are present in the nuclei of cells. Chromosomes contain genes, which are the basic units of inheritance. Humans have 23 pairs of chromosomes: one member of each pair derived from the father and the other from the mother. Males have 22 pairs of *autosomes* and an X and a Y chromosome (the latter two are called *sex chromosomes*). Females have 22 pairs of autosomes and two X chromosomes. Ordinary body cells (*somatic cells*) contain the full complement of 23 pairs of chromosomes (referred to as the *diploid* number), whereas the mature *germ cells*—sperm and ova—contain only half the diploid number of chromosomes (referred to as the *haploid* number) that consists of 3×10^9 base pairs (bp) of DNA. Each of the genes occupies a specific position in a specific chromosome called the *locus* (plural *loci*). The two genes at each locus, one paternal and one maternal, are called *alleles*. The totality of all the genes is the *genotype* of the individual, and their manifestation is the phenotype.

Most eukaryotic (including human) genes are made up of sequences (exons) that code for amino acid sequences in proteins and noncoding intervening sequences (introns). Genes differ not only in the DNA sequences that specify the amino acids of the proteins they encode but also in their structures. A few human genes, such as histone genes, interferon genes, and mitochondrial genes, do not contain introns; some contain a considerable number of introns whose lengths vary from a few bases to several kilobases (kb; *e.g.*, the dystrophin gene, DMD, mutations in which result in Duchenne's and Becker's muscular dystrophies, is 2400 kb long and contains 79 introns).

The 5′ end of the gene is marked by the translational start site (the ATG codon). Upstream from this are a number of noncoding sequences referred to as promoters; further upstream are a number of *cis*-acting regulatory elements of defined sequence (TATAAA and CCAAAT motifs), which play a role in constitutive gene expression, and enhancers,

which respond to particular proteins in a tissue-specific manner by increasing transcription. At the 3′ end is the termination codon (*e.g.*, TAA, TAG, TGA) and a poly-A tail.

The process by which genetic information in DNA is used to produce amino acids and proteins is called transcription. During this process, the entire unit of both introns and exons is transcribed into precursor messenger RNA (mRNA). The region of the precursor mRNA transcribed from the introns is then excised and removed and does not form the definitive mRNA. Precursor mRNA from the exons is spliced together to form the definitive mRNA, which specifies the primary structure of the gene product. The definitive mRNA is then transported to the cytoplasm, where protein synthesis occurs.

Mutations and Their Effects on the Phenotype

Mutations are permanent heritable changes that occur in the genetic material. They arise spontaneously and can be induced by exposure to radiation or chemical mutagens. When mutations arise or are induced in somatic cells, there is a very small probability that they will cause cancer, but somatic mutations are not transmitted to progeny. If mutations occur or are induced in germ cells, they can be transmitted to progeny and they may result in genetic (hereditary) diseases. Mutations are classified as *dominant* or *recessive*, depending on their effects on the phenotype (physical appearance of the organism). In the case of a dominant mutation, a single mutant allele inherited from either parent is sufficient to cause an altered phenotype; the organism has one mutant and one normal allele of the gene in question and is called a *heterozygote* with respect to that gene. In the case of a recessive mutation, two mutant alleles of the same gene—one from each parent—are required to produce a mutant phenotype; the organism is called a *homozygote* for the gene. In general, mutations in genes that code for structural proteins are dominant, and those in genes that code for enzymatic proteins are recessive.

Genetic Diseases

Genetic diseases are traditionally classified as Mendelian or multifactorial diseases. Mendelian diseases are due to mutations in single genes; multifactorial diseases arise as a result of the joint action of multiple genetic and environmental factors.

Molecular analyses have revealed that a wide variety of mutational changes underlie Mendelian diseases: "microlesions," such as single base-pair substitutions, deletions, insertions, or duplications involving one to a few base pairs; and "gross lesions," such as whole-gene or multigene deletions, complex rearrangements, and large insertions and duplications. Microlesions dominate the spectrum of Mendelian diseases (Krawczak and Cooper 1997).

At the functional level, mutations can be classified as causing either a loss of function or the gain of a new function. Normal gene function can be abolished by some types of point mutations, partial or total gene deletions, disruption of the gene structure by translocations or inversions of the genetic material, and so on. In most cases, loss-of-function mutations in enzyme-coding genes are recessive, because 50% of the gene product is usually sufficient for normal functioning. Loss-of-function mutations in genes that code for structural or regulatory proteins, however, result in dominant phenotypes through haploinsufficiency (a 50% reduction in the gene product in the heterozygote is insufficient for normal functioning but is compatible with viability) or through dominant negative effects (the product of the mutant gene not only loses its own function but also prevents the product of the normal allele from functioning in a heterozygous organism). Dominant negative effects are seen particularly in the case of genes whose products function as aggregates (dimers and multimers).

In contrast, gain of function is likely when only specific changes cause a given disease phenotype. Gains of truly novel functions are not common except in cancer, but in inherited diseases, gain of function usually means that the mutant gene is expressed at the wrong time in development, in the wrong tissue, in response to wrong signals, or at an inappropriately high level. The spectrum of gain-of-function mutations would therefore be more restricted, and deletion or disruption of the gene would not produce the disease.

Genetic Effects of Radiation

Exposure of cells and organisms to ionizing radiation causes DNA damage. The cellular processing of radiation-induced damage to DNA by enzymes may result in a return to normal sequence and structure (Lobrich and others 1995), or processing may fail or may cause alterations in DNA that lead to lethality or heritable changes (mutations and chromosomal aberrations) in surviving cells. Heritable changes induced in reproductive (germ) cells can be transmitted to the following generations and cause genetic disease of one kind or another (a concept that lies at the core of estimation of the genetic risks posed by radiation). Changes induced in nonreproductive (somatic) cells have a small but finite probability of contributing to the complex process of carcinogenesis.

The types of mutational changes induced by radiation are broadly similar to the types that occur naturally, but the proportions of the different types are not the same. The results of molecular studies of radiation-induced germ cell mutations in experimental organisms and in mammalian somatic cells support the view that most radiation-induced mutations involve changes in large segments of the DNA, such as deletions that often encompass more than one gene. Hence, radiation readily induces the kinds of molecular changes that can derange a genome and lead to cancer. Conversely, many of those changes, if they occur in germ cells, are incompatible with embryo development and result in developmental abnormalities or lethal mutations in the germline, which would result in nonviable progeny.

B

Commentary on "Radiation from Medical Procedures in the Pathogenesis of Cancer and Ischemic Heart Disease: Dose-Response Studies with Physicians per 100,000 Population"

A monograph authored by Dr. John W. Gofman and dated 1999 was submitted to the committee for its consideration. Dr. Gofman is professor emeritus of molecular and cell biology at the University of California, Berkeley.

In his monograph, Dr. Gofman uses two databases: (1) the database for age-adjusted mortality rates derived from U.S. age-adjusted mortality rates in the decade years from 1940 to approximately 1990—these data are grouped into nine census divisions—and (2) the database for physicians per 100,000 population according to census division obtained from records maintained by the American Medical Association.

Dr. Gofman argues that the number of physicians per 100,000 population may be used as a surrogate for the average dose of medical radiation to the population of each census division. However, no data are presented to support this argument.

In his analyses, Dr. Gofman regresses cause-specific mortality rates on physician population values. Three major causes of death are used: all cancers combined, ischemic heart disease, and all other causes. He demonstrates a positive association of physician population values with all cancer and ischemic heart disease and an inverse association with all other causes. He argues that this evidence ". . . strongly indicates that over 50% of the death-rate from cancer today, and over 60% of the death-rate from Ischemic Heart Disease today, are xray-induced as defined and explained in Part 5 of the Introduction."

Comment

The data used by Dr. Gofman share certain limitations with the data used in the committee's evaluation of environmental radiation and in the consideration of the existence of hormesis in relation to radiation. The primary issue is that so-called ecologic data are used, that is, data on populations rather than data on individuals.

A second limitation of the data used by Dr. Gofman is the assumption that the number of physicians per 100,000 population is a surrogate for the dose of medical radiation received by the population. It is not possible to verify the quantitative nature of this assumption.

Summary

The interpretation that medical radiation has been a major contributor to death from cancer and ischemic heart disease in the United States during the period 1940–1990 is not shared by the committee. There are insufficient data on dose and disease in individuals to lead to this conclusion.

C

Issues Raised by the
Institute for Energy and Environment Research (IEER)

A letter dated September 3, 1999, and authored by Ms. Lisa Ledwidge and Dr. Arjun Makhijani on behalf of IEER and a number of other signatories requested that the BEIR VII committee address six issues. The committee's response to these issues follows.

1. Effects of Radionuclides That Cross the Placenta

In Chapter 8 the committee considers post-Chernobyl data on the excess papillary thyroid cancers arising in radioiodine-exposed children, some of whom received their exposure *in utero*. With respect to carbon-14 and tritium, brief comments are made in response to issue 3. The committee recommends that this issue be addressed as part of a larger review of maternal exposures in humans that may affect the fetus.

2. Effects of Radiation on Female Fetuses

In Chapters 6 and 7, the committee considers the effects of *in utero* radiation, including medical radiation and radiation from the atomic bombs. In the recent paper by Delong-champ and colleagues (1997), nine cancer deaths among females exposed *in utero* to the atomic bombs were noted in comparison to only one among males. Minimal information exists in the medical literature with respect to sex-specific effects, and none reports a gender-specific association between radiation and cancer. Because of the current practice of minimizing radiation exposure to pregnant women, the committee considers it unlikely that this issue will be able to be addressed by future epidemiologic studies.

3. Effects of Organically Bound Radionuclides

Cellular and animal data are available for the development of judgments on the tumorigenic, genetic, and developmental effects of tritiated water and organically bound tri-

tium (Straume 1991; Straume and Carsten 1993). The tritium effects observed do not differ qualitatively from those resulting from external irradiation with X-rays or γ-rays. The evidence available indicates that the relative biological effectiveness of β-irradiation from tritium is generally greater (by two- to threefold) than that of γ-irradiation and similar to or slightly greater (one- to twofold) than X-irradiation. Higher effectiveness is seen *in vitro* in cellular studies when tritium is incorporated into DNA (*e.g.*, as tritiated thymidine). Although the observed effects of tritium are largely attributable to ionization damage from beta particles, transmutation of incorporated tritium to helium also has the potential to damage DNA (NCRP 1979; Hill and Johnson 1993). However, following ingestion of organically bound tritium (OBT, including tritiated thymidine) the *in vivo* activity of digestive metabolic processes means that only a very small fraction of tritium is incorporated into cellular DNA. Thus, the predominant *in vivo* source of DNA damage from OBT is β-particle ionization, not transmutation. The observed *in vivo* effects of tritium will, in any event, include any contribution from transmutational damage to DNA. The same general principles also apply to *in vivo* effects from organically bound carbon-14.

It is important to point out that the committee was not constituted to review the biokinetic aspects of doses from internal radionuclides such as tritium, carbon-14, strontium-90, radiocesium, and radioiodine. Nevertheless the BEIR VII committee considered potentially informative epidemiologic data that relate to risks from internal radiation as part of its brief to review risks at low doses of low-LET (linear energy transfer) radiation.

4. Synergistic Effects

This issue has been comprehensively addressed in Annex H of UNSCEAR (2000b). The BEIR committee endorses the recommendations made on page 217 of that report.

5. Data Integrity and Quality

This issue is addressed in Chapter 8 on occupational radiation studies. The committee acknowledges that there is imprecision in exposure estimates of all epidemiologic studies, especially in retrospective studies of occupational groups. In general, however, studies of workers exposed to radiation tend to have better exposure data than studies of workers exposed to chemicals because of the concurrent estimation of exposure through the use of radiation badges.

The committee notes that imprecision in the estimation of radiation exposure will tend toward an underestimation of any true association between radiation and health effects. To the extent that models based on these data are utilized to set standards of population exposure, the standards will tend to be lower than those that would be based on completely accurate data.

6. Effects on Various Populations

The atomic bomb data are based on two populations in Japan at one point of time. The relation of radiation exposure to age at exposure and gender has been extensively studied and is summarized in Chapter 6. Data on occupational and medical exposure to radiation are available for a number of populations throughout the world for many decades. However, few details are presented in these studies on age at exposure and sex, except of course for sex-specific studies.

The committee recommends that future studies of populations exposed to ionizing radiation include not only information on factors that may interact with radiation exposure, but also information on possible risks present in persons with varying demographic characteristics.

D

Hormesis

Hormesis has been defined as "the stimulating effect of small doses of substances which in larger doses are inhibitory." As stated by Wolff (1989) the meaning has been modified in recent times to refer not only to a stimulatory effect but also to a beneficial effect. In other words, hormesis now connotes a value judgment whereby a low dose of a noxious substance is considered beneficial to an organism.

The committee has reviewed evidence for "hormetic effects" after radiation exposure, with emphasis on material published since the previous BEIR study on the health effects of exposure to low levels of ionizing radiation. Historical material relating to this subject has been reviewed by the United Nations Scientific Committee on the Effects of Atomic Radiation (UNSCEAR 1994), and a special edition of Health Physics on hormesis is available (Sagan 1987). A recent publication reviews data for and against the concept of hormesis (Upton 2000), while noting that further research needs to be done at low-dose and low-dose-rate exposures to resolve the issue. Another recent review argues against the validity of a linear no-threshold model in the low-dose region (Cohen 2002). The committee also reviewed a compilation of materials submitted by Radiation, Science, and Health Inc., entitled Low Level Radiation Health Effects: Compiling the Data and materials provided by Dr. Edward J. Calabrese including the Belle Newsletter Vol. 8, no. 2, December 1999 and the article; Hormesis: a highly generalizable and reproducible phenomenon with important implications for risk assessment (Calabrese and coworkers 1999).

Much of the historical material on radiation hormesis relates to plants, fungi, algae, protozoans, insects, and nonmammalian vertebrates (Calabrese and Baldwin 2000). For the purposes of this report on human health effects, the committee focused on recent information from mammalian cell and animal biology and from human epidemiology. In this context, some investigators have suggested that radiation exposure may enhance immune response (Luckey 1996; Liu 1997) or DNA repair processes (see "Adaptive Response" below). It has been postulated that such stimulation might result in a net health benefit after exposure, and these observations are sometimes offered as mechanisms for hormesis.

Theoretical Considerations

Pollycove and Feinendegen have made a theoretical argument that the hazards of radiation exposure are negligible in comparison to DNA damage that results from oxidative processes during normal metabolism. They argue that endogenous processes, autoxidation, depurination, and/or deamination can lead to cellular DNA damage resembling that produced by ionizing radiation. Oxidative damage is much more complex than they appreciate and involves predominantly proteins and mitochondrial targets associated with transcription, protein trafficking, and vacuolar functions (Thorpe and others 2004). The identity of the particular radical species generated endogenously in undamaged cells is unknown, and therefore yields of endogenous single-strand breaks (SSBs) and double-strand breaks (DSBs) cannot be estimated reliably a priori. Direct measurements of SSBs in unirradiated cells indicate levels several orders of magnitude less than that estimated by Pollycove and Feinendegen. The authors' hypothesis that endogenous processes within cells give rise to significant levels of DSBs from SSBs in close proximity is speculative and not supported by current experimental information. Exposure of cells to high levels of hydrogen peroxide, for example, produces high frequencies of SSBs but no DSBs, suggesting that overlap of SSBs does not occur to a significant extent experimentally (Ward and others 1985). They also hypothesize that low-dose radiation induces a specific repair mechanism that then acts to reduce both spontaneous and radiation-induced damage to below spontaneous levels, thus causing a hormetic effect. The evidence for such a repair mechanism is weak and indirect and is contradicted by direct measures of DSB repair foci at low doses (Rothkamm and Lobrich 2003).

Evidence from Cell Biology

Possible stimulatory effects have been reported for radiation exposure, such as mobilization of intracellular calcium (Liu 1994), gene activation (Boothman and others 1993), activation of signal transduction pathways (Liu 1994; Ishii and others 1997), increase in antioxidants such as reduced glutathione (GSH; Kojima and others 1997), increase in lipoperoxide levels (Petcu and others 1997), and increase in circulating lymphocytes (Luckey 1991). The general thesis presented is that stress responses activated by low doses of radiation, particularly those that would increase immunological responses, are more beneficial than any deleterious effects that might result from the low doses of ionizing radiation. Although evidence for stimulatory effects from low doses has been presented, little if any evidence is offered concerning the ultimate deleterious effects that may occur. In the section of this report on observed dose-response relationships at low doses, bystander effects and hyper radiation sensitivity for low-dose deleterious effects in mammalian cells have been observed for doses in the 10–100 mGy range. End points for these deleterious effects include mutations, chromosomal aberrations, oncogenic transformation, genomic instability, and cell lethality. These deleterious effects have been observed for cells irradiated *in vivo* as well as *in vitro*.

Adaptive Response

The radiation-adaptive response in mammalian cells was demonstrated initially in human lymphocyte experiments (Olivieri and others 1984) and has been associated in recent years with the older concept of radiation hormesis. A more extensive treatment of adaptive effects is discussed in another section of this report. Radiation adaptation, as it was initially observed in human lymphocytes, is a transient phenomenon that occurs in some (but not all) individuals when a conditioning radiation dose lowers the biological effect of a subsequent (usually higher) radiation exposure. In lymphocyte experiments, this reduction occurs under defined temporal conditions and at specific radiation dose levels and dose rates (Shadley and others 1987; Shadley and Wiencke 1989). However, priming doses less than 5 mGy or greater than ~200 mGy generally result in very little if any adaptation, and adaptation has not been reported for challenge doses of less than about 1000 mGy. Furthermore, the induction and magnitude of the adaptive response in human lymphocytes is highly variable (Bose and Olivieri 1989; Hain and others 1992; Vijayalaxmi and others 1995), with a great deal of heterogeneity demonstrated between different individuals (Upton 2000). Also, the adaptive response could not be induced when the lymphocytes were given the priming dose during G_0. Although inhibitor and electrophoretic studies suggest that alterations in transcribing messenger RNA and

synthesis of proteins are involved in the adaptive response in lymphocytes, no specific signal transduction or repair pathways have been identified. A recent study (Barquinero and others 1995), which reported that chronic average occupational exposure of about 2.5 mSv per year over 7 to 21 years induced an adaptive response for radiation-induced chromosomal aberrations in human lymphocytes, also reported that the spontaneous level of aberrations was elevated significantly, presumably by the occupational exposure. (See Barquinero and others [1995] for references to six other reports that basal levels of chromosome abnormalities are in general higher in exposed human populations.) These results suggest that occupational exposure may have induced chromosomal damage in the worker population while protecting lymphocytes from a subsequent experimental radiation exposure administered years after initiation of the chronic exposure. It is unclear whether such competing events would result in a net gain, net loss, or no change in health status.

In general, to observe hormetic effects the spontaneous levels of these effects have to be rather high. The committee notes in the Biology section that a very low radiation dose was reported to cause a reduction in transformation *in vitro* below a relatively high spontaneous transformation frequency. However, problems and possible artifacts of the assay system employed are also discussed. When radioresistance is observed after doses that cause some cell lethality—for example, after chronic doses that continually eliminate cells from the population—the radioresistance that emerges may be caused either (1) by some inductive phenomenon or (2) by selecting for cells that are intrinsically radioresistant. Either process 1 or process 2 could occur as the radiosensitive cells are selectively killed and thus eliminated from the population as the chronic irradiation is delivered. In the end, an adaptive or hormetic response in the population may appear to have occurred, but this would be at the expense of eliminating the sensitive or weak components in the population.

In chronic low-dose experiments with dogs (75 mGy/d for the duration of life), vital hematopoietic progenitors showed increased radioresistance along with renewed proliferative capacity (Seed and Kaspar 1992). Under the same conditions, a subset of animals showed an increased repair capacity as judged by the unscheduled DNA synthesis assay (Seed and Meyers 1993). Although one might interpret these observations as an adaptive effect at the cellular level, the exposed animal population experienced a high incidence of myeloid leukemia and related myeloproliferative disorders. The authors concluded that "the acquisition of radioresistance and associated repair functions under the strong selective and mutagenic pressure of chronic radiation is tied temporally and causally to leukemogenic transformation by the radiation exposure" (Seed and Kaspar 1992).

Evidence from Animal Experiments

Life Span Data

In contrast to experiments showing that radiation shortens the life span, some early publications reported apparent radiation-induced life lengthening following exposure to low levels of single or protracted doses of radiation (Lorenz 1950; Lorenz and others 1954). Statistical analyses of the distribution of deaths in these studies indicate control animals usually show a greater variance around the mean survival time than groups exposed to low doses of radiation. In addition, the longer-living irradiated animals generally have a reduced rate of intercurrent mortality from nonspecific and infectious diseases during their early adult life, followed by a greater mortality rate later in life. Since these investigations were conducted under conditions in which infectious diseases made a significant contribution to overall mortality, the interpretation of these studies with respect to radiation-induced cancer or other chronic diseases in human populations must be viewed with caution.

Problems with variability in controls was a major difficulty in the early studies before animal maintenance and heath care issues were dealt with by transitioning to the use of specific pathogen-free (SPF) facilities; this change to SPF facilities substantially reduced interexperimental variability. For example, the cited data of Lorenz (1950) show a small difference in life span in mice exposed to 0.11 r/d compared to controls; the irradiated group lived somewhat longer than the unirradiated group, but the difference was not significant. A French study (Caratero and others 1998) shows life lengthening in irradiated mice compared to controls; unfortunately, the control life spans were significantly shorter by 100–150 d than any in other published data for this mouse strain (Sacher 1955; Congdon 1987).

Tumor Incidence Data

Two studies have reported a significant reduction in tumor incidence of lymphoma in animals that have a high spontaneous tumor incidence (>40%; Covelli and others 1989; Ishii and others 1996). A paper by Ishii and colleagues (1996) describes a reduction in lymphoma incidence after chronic, fractionated, low-dose total-body irradiation of AKR mice with a spontaneous lymphoma incidence of 80.5%. The spontaneous lymphoma incidence was decreased significantly (to 48.6%) by 150 mGy X-irradiation delivered twice a week for 40 weeks. A protocol of 50 mGy three times a week gave a smaller (not statistically significant) decrease to 67.5% lymphoma incidence. The mean survival time was significantly prolonged from 283 d for the control animals to 309 d with the three-exposure-per-week protocol and to 316 d with the twice-a-week protocol.

In a study by Covelli and colleagues (1989), a decrease in incidence of malignant lymphoma at low doses of radiation (46 and 52% age-adjusted incidence at X-ray exposures of 500 and 1000 mGy versus 57% incidence in control animals) shows a reduction in tumor incidence relative to the control frequency. After peaking at 60% lymphoma incidence (3000 mGy), the frequencies decline, "possibly due to cell inactivation becoming predominant at higher doses over the initial transforming events."

The reduction in spontaneous tumors noted in the previous two studies may in some way be related to the high spontaneous lymphoma incidence in this mouse strain. In the Ishii study, the authors speculate that possible mechanisms may include augmentation of the immune system or initiation of an "adaptive response." One might also consider that the substantial doses delivered to the animals in this study (6000 and 12,000 mGy) are effectively acting as radiotherapy in the reduction of spontaneous tumor incidence. Human populations, which have a wider spectrum of "spontaneous" tumors occurring at a lower incidence, may not be expected to respond to radiation in the same way as mouse strains with high lymphoma incidence.

HORMESIS AND EPIDEMIOLOGY

The term *hormesis* is not commonly used in the epidemiologic literature. Rather, epidemiologists discuss associations between exposure and disease. A *positive association* is one in which the rate of disease is higher among a group exposed to some substance or condition than among those not exposed, and a *negative (or inverse) association* is one in which the rate of disease is lower among the exposed group. If an association is judged to be causal, a positive association may be termed a *causal effect* and a negative association could be termed a *protective effect*.

One type of epidemiologic study that has been used to evaluate the association between exposure to radiation and disease is the *"ecologic"* study in which data on populations, rather than data on individuals, are compared. These data have been used to argue for the existence of radiation hormesis.

Another example of an ecologic study is the evaluation of geographic areas with high background levels of radiation compared to areas with "normal" background levels. The fact that cancer rates in these high-background-radiation geographic regions are not elevated is sometimes cited as evidence against a linear no-threshold model (Jaworowowski 1995).

It is also true that certain populations residing in high-background areas, such as occur at high altitudes, have lower levels of health problems than those residing at lower altitudes. This observation has been interpreted by some as evidence for a hormetic effect of radiation. BEIR V discussed the effect of confounders and the ecological fallacy[2] in the evaluation of high-background-radiation areas and concluded that "these two problems alone are enough to make such studies essentially meaningless" (NRC 1990).

Another important consideration is the expected magnitude of the increase in health effect induced by excess background radiation. If one assumes a linear no-threshold response, a calculation can be made for expected cancers induced by excess radiation in a high-background-radiation area. As an example, consider the elevated levels of gamma radiation in Guodong Province, Peoples' Republic of China (PRC). In this study, a population receiving 3–4 mGy per year was compared to an adjacent control population receiving 1 mGy per year. No difference in cancers was noted between the high-background area and the control area (NRC 1990). One can estimate the expected excess percentage of cancers resulting from the 2–3 mGy difference in exposure per year using a linear nonthreshold model and the lifetime risk estimates developed in this report. A calculation by this committee indicated that the expected percentage of cancers induced by the excess background radiation would be 1–2% above the cancers occurring from all other causes in a lifetime. Even if all confounding factors were accounted for, it is questionable whether one could detect an excess cancer rate of 1–2%. Excess cancers may indeed be induced by elevated radiation exposure in high-background areas, but the excess may not be detectable given the high lifetime occurrence of cancer from all causes.

Ordinarily, epidemiologists do not consider ecologic data such as this as being sufficient for causal interpretations. Since the data are based on populations, no information is available on the exposure and disease status of individuals. Such data cannot be controlled adequately for confounding factors or for selection bias. Although ecologic data may be consistent with an inverse association between radiation and cancer, they may not be used to make causal inferences.

A second type of epidemiologic study that has been used to evaluate the association between exposure to radiation and disease is the *retrospective cohort* study. Persons who have had past exposure to radiation are followed forward in time, and the rate of disease is compared between exposed and nonexposed subjects or between exposed subjects and the general population. Especially valuable are occupational studies that include both unexposed and exposed subjects, so that a dose-response evaluation can be made of the relation between radiation exposure and health outcome. Typically, study populations in retrospective cohort studies include persons who have worked with radiation in medical facilities or in the nuclear industry or patients with cancer or other disease who have been treated with radiation.

It is common in cohort studies of occupational populations to observe that the overall mortality rate is lower than that of the general population, commonly about 15%. This is not interpreted to mean that work *per se* reduces the risk of mortality, but rather that healthy persons start to work more often than unhealthy persons (Monson 1990). The term "healthy worker effect" (HWE) is commonly used to describe this observation. Diseases such as cancer that develop in later life ordinarily have less of an HWE than noncancerous diseases. The HWE is observed in most occupational studies, including those of radiation workers, and should not be interpreted to mean that low doses of radiation prevent death from cancer or other causes.

A third type of epidemiologic study that has been used to evaluate the association between exposure to radiation and disease is the *case-control* study. Persons with a specific disease are compared to a control group of persons without the disease with respect to their past exposure to radiation. This type of study is unusual in radiation epidemiology, in that most general populations have relatively low exposure to radiation.

While no phenomenon similar to the HWE is observed in case-control studies, the play of chance is always operative, as it is in cohort studies. Thus, if some exposure does not cause cancer and if a number of case-control studies are conducted, there will be a normal distribution observed in the odds ratios that describe the association between exposure and disease. Some studies will have an odds ratio that is less than 1.0; others will have an odds ratio greater than 1.0. In interpreting these studies, it is inappropriate to select only those that are consistent with an excess or deficit of disease. Rather, the entire distribution must be examined to assess the likely relationship between exposure and disease.

The studies discussed here illustrate the variability that is inherent in all epidemiologic studies and the need to evaluate the entire body of relevant literature in order to assess possible associations between radiation and disease, be they positive or negative. In its evaluation of the literature and in its discussions, the committee has found no consistent evidence in the epidemiologic literature that low doses of ionizing radiation lower the risk of disease or death. Some studies show isolated positive associations between radiation exposure and disease, and some show isolated negative associations. However, the weight of the evidence does not lead to the interpretation that low doses of radiation exert what in biological terms is called hormesis.

Summary

The committee concludes that the assumption that any stimulatory hormetic effects from low doses of ionizing radiation will have a significant health benefit to humans that exceeds potential detrimental effects from the radiation exposure is unwarranted at this time.

---

[2]*Ecological fallacy:* two populations differ in many factors other than those being evaluated, and one or more of these may be the underlying reason for any difference noted in their morbidity or mortality experience (Lilienfeld and Stolley 1994).

E

Fifteen-Country Workers Study

After the BEIR VII committee's draft report had been reviewed by the Review Panel, an additional paper was submitted for its consideration (Cardis 2005). This publication provides results from a multinational collaborative study coordinated by the International Agency for Research on Cancer (IARC) and updates and supersedes the three-country study referred to in this report (IARC 1995). This new IARC 15-country study included almost 600,000 individually monitored workers from 15 countries (Australia, Belgium, Canada, Finland, France, Hungary, Japan, Korea, Lithuania, Slovakia, Spain, Sweden, Switzerland, the United Kingdom, and the United States). The main analyses included 407,391 nuclear industry workers who were employed for at least one year in a participating facility and who were monitored individually for external radiation. The total duration of follow-up was 5,192,710 person-years, and the total collective recorded dose was 7892 Sv, almost exclusively from external photon exposure. Most workers in the study were men (90%), who received 98% of the collective dose. The overall average cumulative recorded dose was 19.4 mSv. The distribution of recorded doses was skewed: 90% of workers received cumulative doses of less than 50 mSv, and less than 0.1% received cumulative doses of greater than 500 mSv. The study included a comprehensive study of errors in doses. The major sources of errors were quantified and taken into account in the risk estimates.

The excess relative risk estimate for all cancers excluding leukemia was reported as 0.97 Gy^{-1} (95% CI 0.14, 1.97) and for all solid cancers 0.87 Gy^{-1} (95% CI 0.03, 1.88). These estimates are somewhat higher than, but statistically compatible with, the estimates on which current radiation protection recommendations are based. Analyses of smoking- and non-smoking-related causes of death indicate that although confounding by smoking may be present, it is unlikely to explain all of this increased risk.

The excess relative risk estimate for leukemia excluding chronic lymphocytic leukemia was reported as 1.93 Gy^{-1} (95% CI <0, 8.47). This estimate, although not statistically significantly elevated, is close to that observed in previous nuclear workers studies.

References

Abramson, D.H., M.R. Melson, L.J. Dunkel, and C.M. Frank. 2001. Third (fourth and fifth) tumours in survivors of retinoblastoma. Ophthalmology 108:1868-1876.

Acquavella, J.F., L.D. Wiggs, R.J. Waxweiler, D.G. Macdonell, G.L. Tietjen, and G.S. Wilkinson. 1985. Mortality among workers at the Pantex weapons facility. Health Phys 48:735-746.

Adams, N., and W.A. Langmead. 1962. An Investigation into the Accuracy Attained in Routine Film-Badge Dosimetry at UKAEA Establishments. Authority Health and Safety Branch Report. Harwell, UK: United Kingdom Atomic Energy Authority.

Aghamohammadi, S.Z., and J.R. Savage. 1991. A BrdU pulse double-labelling method for studying adaptive response. Mutat Res 251:133-141.

Aghamohammadi, S.Z., T. Morris, D.L. Stevens, and J. Thacker. 1992. Rapid screening for deletion mutations in the hprt gene using the polymerase chain reaction: x-ray and alpha-particle mutant spectra. Mutat Res 269:1-7.

Alam, N.A., S. Bevan, M. Churchman, E. Barclay, K. Barker, E.E. Jaeger, H.M. Nelson, E. Healy, A.C. Pembroke, P.S. Friedmann, K. Dalziel, E. Calonje, J. Anderson, P.J. August, M.G. Davies, R. Felix, C.S. Munro, M. Murdoch, J. Rendall, S. Kennedy, I.M. Leigh, D.P. Kelsell, I.P. Tomlinson, and R.S. Houlston. 2001. Localization of a gene (MCUL1) for multiple cutaneous leiomyomata and uterine fibroids to chromosome 1q42.3-q43. Am J Hum Genet 68:1264-1269.

Allan, J.M., C.P. Wild, S. Rollinson, E.V. Willett, A.V. Moorman, G.J. Dovey, P.L. Roddam, E. Roman, R.A. Cartwright, and G.J. Morgan. 2001. Polymorphism in glutathione S-transferase P1 is associated with susceptibility to chemotherapy-induced leukemia. Proc Natl Acad Sci USA 98:11592-11597.

Alper, T., C. Mothersill, and C.B. Seymour. 1988. Lethal mutations attributable to misrepair of Q-lesions. Int J Radiat Biol 54:525-530.

Amundson, S.A., and D.J. Chen. 1996. Inverse dose-rate effect for mutation induction by gamma-rays in human lymphoblasts. Int J Radiat Biol 69:555-563.

Amundson, S.A., and H.L. Liber. 1991. A comparison of induced mutations at homologous alleles of the tk locus in human cells. Mutat Res 247:19-27.

Amundson, S.A., M. Bittner, Y. Chen, J. Trent, P. Meltzer, and A.J. Fornace Jr. 1999a. Fluorescent cDNA microarray hybridization reveals complexity and heterogeneity of cellular genotoxic stress responses. Oncogene 18:3666-3672.

Amundson, S.A., K.T. Do, and A.J. Fornace Jr. 1999b. Induction of stress genes by low doses of gamma rays. Radiat Res 152:225-231.

Anderson, R.M., S.J. Marsden, E.G. Wright, M.A. Kadhim, D.T. Goodhead, and C.S. Griffin. 2000. Complex chromosome aberrations in peripheral blood lymphocytes as a potential biomarker of exposure to high-LET alpha-particles. Int J Radiat Biol 76:31-42.

Anderson, R.N., and P.B. DeTurk. 2002. United States Life Tables, National Vital Statistics Reports, Vol. 50, Number 6, 1-12. Available online at www.cdc.gov/nchs/data/nvsr/nvsr50/nvsr50_06.pdf.

Andersson, M., G. Engholm, K. Ennow, K.A. Jessen, and H.H. Storm. 1991. Cancer risk among staff at two radiotherapy departments in Denmark. Brit J Radiol 64:455-460.

Angel, J.M., N. Popova, N. Lanko, V.S. Turusov, and J. DiGiovanni. 2000. A locus that influences susceptibility to 1,2-dimethylhydrazine-induced colon tumors maps to the distal end of mouse chromosome 3. Mol Carcinogen 27:47-54.

Antoniou, A.C., P.D. Pharoah, G. McMullan, N.E. Day, M.R. Stratton, J. Peto, B.J. Ponder, and D.F. Easton. 2002. A comprehensive model for familial breast cancer incorporating BRCA1, BRCA2 and other genes. Brit J Cancer 86:76-83.

Appleby, J.M., J.B. Barber, E. Levine, J.M. Varley, A.M. Taylor, T. Stankovic, J. Heighway, C. Warren, and D. Scott. 1997. Absence of mutations in the ATM gene in breast cancer patients with severe responses to radiotherapy. Brit J Cancer 76:1546-1549.

Arai, T., T. Nakano, K. Fukuhisa, T. Kasamatsu, R. Tsunematsu, K. Masubuchi, K. Yamauchi, T. Hamada, T. Fukuda, H. Noguchi, and M. Murata. 1991. Second cancer after radiation therapy for cancer of the uterine cervix. Cancer 67:394-405.

Armitage, P. 1985. Biometry and medical statistics. Biometrics 41:823-833.

Armitage, P., and R. Doll. 1954. The age distribution of cancer and a multistage theory of carcinogenesis. Brit J Cancer 8:1-12.

Armitage, P., and R. Doll. 1957. A two-stage theory of carcinogenesis in relation to the age distribution of human cancer. Brit J Cancer 11:161-169.

Armitage, P., and R. Doll. 1961. Stochastic models for carcinogenesis. Pp. 18-38 in Proceedings of the Fourth Berkeley Symposium on Mathematical Statistics and Probability, Vol. 4. Berkeley, CA: University of California Press.

Armour, J.A., M.H. Brinkworth, and A. Kamischke. 1999. Direct analysis by small-pool PCR of MS205 minisatellite mutation rates in sperm after mutagenic therapies. Mutat Res 445:73-80.

Aromaa, A., T.U. Kosunen, P. Knekt, J. Maatela, L. Teppo, O.P. Heinonen, M. Harkonen, and M.K. Hakama. 1996. Circulating anti-Helicobacter pylori immunoglobulin A anitbodies and low serum pepsinogen I level are associated with increased risk of gastric cancer. Am J Epidemiol 144:142-149.

Arseneau, J. C., R. W. Sponzo, D. L. Levin, L. E. Schnipper, H. Bonner, R. C. Young, G. P. Canellos, R. E. Johnson, and V. T. DeVita. 1972. Nonlymphomatous malignant tumors complicating Hodgkin's disease. Possible association with intensive therapy. N Engl J Med 287: 1119-22.

Artalejo, F.R., S.C. Lara, B. de Andres Manzano, M.G. Ferruelo, L.I. Martin, and J.R. Calero. 1997. Occupational exposure to ionising radiation and mortality among workers of the former Spanish Nuclear Energy Board. Occup Environ Med 54:202-208.

Artandi, S.E., S. Chang, S.L. Lee, S. Alson, G.J. Gottlieb, L. Chin, and R.A. DePinho. 2000. Telomere dysfunction promotes non-reciprocal translocations and epithelial cancers in mice. Nature 406:641-645.

Ashizawa, K., Y. Shibata, S. Yamashita, H. Namba, M. Hoshi, N. Yokoyama, M. Izumi, and S. Nagataki. 1997. Prevalence of goiter and urinary iodine excretion levels in children around Chernobyl. J Clin Endocr Metab 82:3430-3433.

Ashmore, J.P., D. Krewski, J.M. Zielinski, H. Jiang, R. Semenciw, and P.R. Band. 1998. First analysis of mortality and occupational radiation exposure based on the National Dose Registry of Canada. Am J Epidemiol 148:564-574.

Astakhova, L.N., L.R. Anspaugh, G.W. Beebe, A. Bouville, V.V. Drozdovitch, V. Garber, Y.I. Gavrilin, V.T. Khrouch, A.V. Kuvshinnikov, Y.N. Kuzmenkov, V.P. Minenko, K.V. Moschik, A.S. Nalivko, J. Robbins, E.V. Shemiakina, S. Shinkarev, S.I. Tochitskaya, and M.A. Waclawiw. 1998. Chernobyl-related thyroid cancer in children of Belarus: a case-control study. Radiat Res 150:349-356.

Aszterbaum, M., J. Epstein, A. Oro, V. Douglas, P.E. LeBoit, M.P. Scott, and E.H. Epstein Jr. 1999. Ultraviolet and ionizing radiation enhance the growth of BCCs and trichoblastomas in patched heterozygous knockout mice. Nat Med 5:1285-1291.

Atkinson, W.D., D.V. Law, K.J. Bromley, and H.M. Inskip. 2004. Mortality of employees of the United Kingdom Atomic Energy Authority, 1946-97. Occup Environ Med 61:577-585.

Austin, D.F., and P. Reynolds. 1997. Investigation of an excess of melanoma among employees of the Lawrence Livermore National Laboratory. Am J Epidemiol 145:524-531.

Auvinen, A., M. Hakama, H. Arvela, T. Hakulinen, T. Rahola, M. Suomela, B. Soderman, and T. Rytomaa. 1994. Fallout from Chernobyl and incidence of childhood leukaemia in Finland, 1976-92. Brit Med J 309: 151-154.

Auvinen, A., E. Pukkala, H. Hyvonen, M. Hakama, and T. Rytomaa. 2002. Cancer incidence among Finnish nuclear reactor workers. J Occ Env Med 44:634-638.

Azzam, E.I., G.P. Raaphorst, and R.E.J. Mitchel. 1994. Radiation-induced adaptive response for protection against micronucleus formation and neoplastic transformation in C3H 10T1/2 mouse embryo cells. Radiat Res 138:S28-S31.

Azzam, E.I., S.M. de Toledo, G.P. Raaphorst, and R.E.J. Mitchel. 1996. Low-dose ionizing radiation decreases the frequency of neoplastic transformation to a level below the spontaneous rate in C3H 10T1/2 cells. Radiat Res 146:369-373.

Azzam, E.I., S.M. de Toledo, T. Gooding, and J.B. Little. 1998. Intercellular communication is involved in the bystander regulation of gene expression in human cells exposed to very low fluences of alpha particles. Radiat Res 150:497-504.

Azzam, E.I., S.M. de Toledo, D.R. Spitz, and J.B. Little. 2002. Oxidative metabolism modulates signal transduction and micronucleus formation in bystander cells from alpha-particle-irradiated normal human fibroblast cultures. Cancer Res 62:5436-5442.

Bacchetti, S. 1996. Telomere maintenance in tumour cells. Cancer Surv 28:197-216.

Backlund, M.G., S.L. Trasti, D.C. Backlund, V.L. Cressman, V. Godfrey, and B.H. Koller. 2001. Impact of ionizing radiation and genetic background on mammary tumorigenesis in p53-deficient mice. Cancer Res 61:6577-6582.

Bacq, Z.M., and P. Alexander. 1961. Fundamentals of Radiobiology. New York: Pergamon Press.

Bailey, S.M., J. Meyne, D.J. Chen, A. Kurimasa, G.C. Li, B.E. Lehnert, and E.H. Goodwin. 1999. DNA double-strand break repair proteins are required to cap the ends of mammalian chromosomes. Proc Natl Acad Sci USA 96:14899.

Bailey, S.M., M.A. Brenneman, J. Halbrook, J.A. Nickoloff, R.L. Ullrich, and E.H. Goodwin. 2004a. The kinase activity of DNA-PK is required to protect mammalian telomeres. DNA Repair (Amst) 3:225-233.

Bailey, S.M., M.N. Cornforth, R.L. Ullrich, and E.H. Goodwin. 2004b. Dysfunctional mammalian telomeres join with DNA double-strand breaks. DNA Repair (Amst) 3:349-357.

Baird, P.A., T.W. Anderson, H.B. Newcombe, and R.B. Lowry. 1988. Genetic disorders in children and young adults: a population study. Am J Hum Genet 42:677-693.

Baker, T.G. 1971. Comparative aspects of the effects of radiation during oogenesis. Mutat Res 11:9-22.

Bakkenist, C.J., and M.B. Kastan. 2003. DNA damage activates ATM through intermolecular autophosphorylation and dimer dissociation. Nature 421:499-506.

Balmain, A., and H. Nagase. 1998. Cancer resistance genes in mice: models for the study of tumour modifiers. Trends Genet 14:139-144.

Balmain, A., J. Gray, and B. Ponder. 2003. The genetics and genomics of cancer. Nat Genet 33(Suppl):238-244.

Barber, R., M. Plumb, A.G. Smith, C.E. Cesar, E. Boulton, A.J. Jeffreys, and Y.E. Dubrova. 2000. No correlation between germline mutation at repeat DNA and meiotic crossover in male mice exposed to x-rays or cisplatin. Mutat Res 457:79-91.

Barber, R., M.A. Plumb, E. Boulton, I. Roux, and Y.E. Dubrova. 2002. Elevated mutation rates in the germ line of first- and second-generation offspring of irradiated male mice. Proc Natl Acad Sci USA 99:6877-6882.

Baria, K., C. Warren, S.A. Roberts, C.M. West, and D. Scott. 2001. Chromosomal radiosensitivity as a marker of predisposition to common cancers? Brit J Cancer 84:892-896.

Barlow, C., S. Hirotsune, R. Paylor, M. Liyanage, M. Eckhaus, F. Collins, Y. Shiloh, J.N. Crawley, T. Reid, D. Tagle, and A. Wynshaw-Boris. 1996. Atm-deficient mice: a paradigm of ataxia-telangiectasia. Cell 86:159-171.

Barlow, C., P.A. Dennery, M.K. Shigenaga, M.A. Smith, J.D. Morrow, L.J. Roberts 2nd, A. Wynshaw-Boris, and R.L. Levine. 1999. Loss of the ataxia-telangiectasia gene product causes oxidative damage in target organs. Proc Natl Acad Sci USA 96:9915-9919.

Barquinero, J.F., L. Barrios, M.R. Caballin, R. Miro, M. Ribas, A. Subias, and J. Egozcue. 1995. Occupational exposure to radiation induces an adaptive response in human lymphocytes. Int J Radiat Biol 67:187-191.

Bartsch, H., M. Hollstein, R. Mustonen, J. Schmidt, A. Spiethoff, H. Wesch, T. Wiethege, and K.M. Muller. 1995. Screening for putative radon-specific p53 mutation hotspot in German uranium miners. Lancet 346:121.

Basco, V.E., A.J. Coldman, J.M. Elwood, and M.E. Young. 1985. Radiation dose and second breast cancer. Brit J Cancer 52:319-325.

Bauchinger, M., E. Schmid, S. Streng, and J. Dresp. 1983. Quantitative analysis of the chromosome damage at first division of human lymphocytes after ^{60}CO gamma-irradiation. Radiat Environ Bioph 22:225-229.

Baumann, P., and S.C. West. 1998. DNA end-joining catalyzed by human cell-free extracts. Proc Natl Acad Sci USA 95:14066-14070.

Baverstock, K., B. Egloff, A. Pinchera, C. Ruchti, and D. Williams. 1992. Thyroid cancer after Chernobyl. Nature 359:21-22.

Beauchesne, P.D., S. Bertrand, R. Branche, S.P. Linke, R. Revel, J.F. Dore, and R.M. Pedeux. 2003. Human malignant glioma cell lines are sensitive to low radiation doses. Int J Cancer 105:33-40.

Bebeshko, V.G., E.M. Bruslova, V.I. Klimenko, and others. 1997. Leukemias and lymphomas in Ukraine population exposed to chronic low dose irradiation. Pp. 337-338 in Low Doses of Ionizing Radiation: Biological Effects and Regulatory Control. Contributed papers. International Conference held in Seville, Spain. IAEA-TECDOC-976.

Beckman, K.B., and B.N. Ames. 1997. Oxidative decay of DNA. J Biol Chem 272:19633-19636.

Bedford, J.S., and W.C. Dewey. 2002. Radiation Research Society. 1952-2002. Historical and current highlights in radiation biology: has anything important been learned by irradiating cells? Radiat Res 158:251-291.

Bell, G.I., S. Horita, and J.H. Karam. 1984. A polymoprhic locus near the human insulin gene is associated with insulin-dependent diabetes mellitus. Diabetes 31:176-183.

Belyaev, I.Y., and M. Harms-Ringdahl. 1996. Effects of gamma rays in the 0.5-50-cGy range on the conformation of chromatin in mammalian cells. Radiat Res 145:687-693.

Belyakov, O.V., A.M. Malcolmson, M. Folkard, K.M. Prise, and B.D. Michael. 2001. Direct evidence for a bystander effect of ionizing radiation in primary human fibroblasts. Brit J Cancer 84:674-679.

Bender, M.A., H.G. Griggs, and J.S. Bedford. 1974. Mechanisms of chromosomal aberration production. 3. Chemicals and ionizing radiation. Mutat Res 23:197-212.

Bennett, R.A., D.M. Wilson 3rd, D. Wong, and B. Demple. 1997. Interaction of human apurinic endonuclease and DNA polymerase beta in the base excision repair pathway. Proc Natl Acad Sci USA 94:7166-7169.

Bennett, S.T., A.M. Lucassen, S.C. Gough, E.E. Powell, D.E. Undlien, L.E. Pritchard, M.E. Merriman, Y. Kawaguchi, M.J. Dronsfield, F. Pociot, and 1 other. 1995. Susceptibility to human type 1 diabetes at IDDM2 is determined by tandem repeat variation at the insulin gene minisatellite locus. Nat Genet 9:284-292.

Bennett, W.P., M.C. Alavanja, B. Blomeke, K.H. Vahakangas, K. Castren, J.A. Welsh, E.D. Bowman, M.A. Khan, D.B. Flieder, and C.C. Harris. 1999. Environmental tobacco smoke, genetic susceptibility, and risk of lung cancer in never-smoking women. J Natl Cancer Inst 91:2009-2014.

Beral, V., H. Inskip, P. Fraser, M. Booth, D. Coleman, and G. Rose. 1985. Mortality of employees of the United Kingdom Atomic Energy Authority, 1946-1979. Brit Med J (Clin Res Ed) 291:440-447.

Beral, V., P. Fraser, L. Carpenter, M. Booth, A. Brown, and G. Rose. 1988. Mortality of employees of the Atomic Weapons Establishment, 1951-1982. Brit Med J 297:757-770.

Berrington, A., S.C. Darby, H.A. Weiss, and R. Doll. 2001. 100 years of observation on British radiologists: mortality from cancer and other causes 1897-1997. Brit J Radiol 74:507-519.

Bertoni, L., C. Attolini, L. Tessera, E. Mucciolo, and E. Giulotto. 1994. Telomeric and nontelomeric (TTAGGG) n sequences in gene amplification and chromosome stability. Genomics 24:53-62.

Bettega, D., P. Calzolari, A. Ottolenghi, E. Rimoldi, and L. Tallone Lombardi. 1989. Cell density dependence of transformation frequencies in C3H10T1/2 cells exposed to x-rays. Int J Radiat Biol 56:989-998.

Billen, D. 1990. Spontaneous DNA damage and its significance for the "negligible dose" controversy in radiation protection. Radiat Res 124:242-245.

Binks, K., D.I. Thomas, and D. McElvenny. 1989. Mortality of workers at Chapelcross plant of British Nuclear Fuels Limited. Cancer 67:615-624.

Birdwell, S.H., S.L. Hancock, A. Verghese, R.S. Cox, and R.T. Hoppe. 1997. Gastrointestinal cancer after treatment of Hodgkin's disease. Int J Radiat Oncol 37:67-73.

Birrell, G.W., G. Giaever, A.M. Chu, R.W. Davis, and J.M. Brown. 2001. A genome-wide screen in *Saccharomyces cerevisiae* for genes affecting UV radiation sensitivity. Proc Natl Acad Sci USA 98:12608-12613.

Birrell, G.W., J.A. Brown, H.I. Wu, G. Giaever, A.M. Chu, R.W. Davis, and J.M. Brown. 2002. Transcriptional response of *Saccharomyces cerevisiae* to DNA-damaging agents does not identify the genes that protect against these agents. Proc Natl Acad Sci USA 99:8778-8783.

Bishayee, A., D.V. Rao, and R.W. Howell. 1999. Evidence for pronounced bystander effects caused by nonuniform distributions of radioactivity using a novel three-dimesional tissue culture model. Radiat Res 152: 88-97.

Bishop, D.T. 1990. Multifactorial inheritance. Pp. 165-174 in Principles and Practice of Medical Genetics. New York: Churchill Livingstone.

Bishop, D.T., and J. Hopper. 1997. AT-tributable risks? Nat Genet 15:226.

Bishop, J.M. 1991. Molecular themes in oncogenesis. Cell 64:235-248.

Bithell, J.F. 1989. Epidemiological studies of children irradiated in utero. Pp. 77-87 in Low Dose Radiation: Biological Bases of Risk Assessment, K.F. Baverstock and J.W. Stather, eds. London: Taylor & Francis.

Bithell, J.F. 1990. An application of density estimation to geographical epidemiology. Stat Med 9:691-701.

Bithell, J.F. 1993. Statistical issues in assessing the evidence associating obstetric irradiation and childhood malignancy. Pp. 53-60 in Neue Bewertung des Strahlenriskos: Niedrigdosis-Strahlung und Gesundheit, E. Lengfelder and H. Wendhausen, eds. Munich: MMV Medizin Verlag.

Bithell, J.F., and A.M. Stewart. 1975. Pre-natal irradiation and childhood malignancy: a review of British data from the Oxford Survey. Brit J Cancer 31:271-287.

Bithell, J.F., S.J. Dutton, G.J. Draper, and N.M. Neary. 1994. Distribution of childhood leukaemias and non-Hodgkin's lymphomas near nuclear installations in England and Wales. Brit Med J 309:501-505.

Black, R.J., J.D. Urquhart, S.W. Kendrick, K.J. Bunch, J. Warner, and D.A. Jones. 1992. Incidence of leukaemia and other cancers in birth and schools cohorts in the Dounreay area. Brit Med J 304:1401-1405.

Black, R.J., L. Sharp, E.F. Harkness, and P.A. McKinney. 1994a. Leukaemia and non-Hodgkin's lymphoma: incidence in children and young adults resident in the Dounreay area of Caithness, Scotland in 1968-91. J Epidemiol Commun H 48:232-236.

Black R.J., L. Sharp, A.R. Finlayson, E.F. Harkness. 1994b. Cancer incidence in a population potentially exposed to radium-226 at Dalgety Bay, Scotland. Brit J Cancer 69:140-143.

Blackburn, E.H. 2000. Telomere states and cell fates. Nature 408:53-56.

Blaisdell, J.O., and S.S. Wallace. 2001. Abortive base-excision repair of radiation-induced clustered DNA lesions in *Escherichia coli*. Proc Natl Acad Sci USA 98:7426-7430.

Blake, J.A., J.T. Eppig, J.E. Richardson, and M.T. Davisson. 2000. The Mouse Genome Database (MGD): expanding genetic and genomic resources for the laboratory mouse. The Mouse Genome Database Group. Nucleic Acids Res 28:108-111.

Blattner, C., A. Knebel, A. Radler-Pohl, C. Sachsenmaier, P. Herrlich, and H.J. Rahmsdorf. 1994. DNA damaging agents and growth factors induce changes in the program of expressed gene products through common routes. Environ Mol Mutagen 24:3-10.

Blattner, C., E. Tobiasch, M. Litfen, H.J. Rahmsdorf, and P. Herrlich. 1999. DNA damage induced p53 stabilization: no indication for an involvement of p53 phosphorylation. Oncogene 18:1723-1732.

Blettner, M., and H. Zeeb. 1999. Epidemiological studies among pilots and cabin crew. Radiat Prot Dosimet 86:269-273.

Blettner, M., B. Grosche, and H. Zeeb. 1998. Occupational cancer risk in pilots and flight attendants: current epidemiological knowledge. Radiat Environ Bioph 37:75-80.

Bleuer, J.P., Y.I. Averkin, and T. Abelin. 1997. Chernobyl-related thyroid cancer: what evidence for role of short-lived iodines. Environ Health Persp 105:1483-1486.

BNL (Brookhaven National Laboratory). Conard, R.A., D.E. Peglia, P.R. Larson, and others. 1980. Review of Medical Findings in a Marshallese Population Twenty-Six Years after Accidental Exposure to Radioactive Fallout. BNL 51261. Upton, NY: Brookhaven National Laboratory.

Boice, J.D., Jr., and R.W. Miller. 1992. Risk of breast cancer in ataxia-telangiectasia. New Engl J Med 326:1357-1358; discussion 1360-1361.

Boice, J.D., Jr., and R.W. Miller. 1999. Childhood and adult cancer after intrauterine exposure to ionizing radiation. Teratology 59:227-233.

Boice, J.D., Jr., M. Rosenstein, and E.D. Trout. 1978. Estimation of breast doses and breast cancer risk associated with repeated fluoroscopic chest examinations of women with tuberculosis. Radiat Res 73:373-390.

Boice, J.D., Jr., N.E. Day, A. Andersen, L.A. Brinton, R. Brown, N.W. Choi, E.A. Clarke, M.P. Coleman, R.E. Curtis, J.T. Flannery, M. Hakama, T. Hakulinen, G.R. Howe, O.M. Jensen, R.A. Kleinerman, D. Magnin, K. Magnus, K. Mäkelä, B. Malker, A.B. Miller, N. Nelson, C.C. Patterson, F. Pettersson, V. Pompe-Kirn, M. Primic-Zakelj, P. Prior, B. Ravnihar, R.G. Skeet, J.E. Skjerven, P.G. Smith, M. Sok, R.F. Spengler, H.H. Storm, M. Stovall, G.W.O. Tomkins, and C. Wall. 1985. Second cancers following radiation treatment for cervical cancer. An international collaboration among cancer registries. J Natl Cancer Inst 74:955-975.

Boice, J.D., Jr., M. Blettner, R.A. Kleinerman, M. Stovall, W.C. Moloney, G. Engholm, D.F. Austin, A. Bosch, D.L. Cookfair, E.T. Krementz, H.B. Latourette, L.J. Peters, M.D. Schulz, M. Lundell, F. Pettersson, H.H. Storm, J. Bell, M.P. Coleman, P. Fraser, M. Palmer, P. Prior, N.W. Choi, T.G. Hilsop, M. Kock, D. Robb, D. Robson, R.F. Spengler, M. von Fournier, R. Frischkorn, H. Lochmuller, V. Pompe-Kirn, A. Rimpela, M. Kjorstad, H. Pejovic, K. Sigurdsson, P. Pisani, H. Kucera, and G.B. Hutchison. 1987. Radiation dose and leukemia risk in patients treated for cancer of the cervix. J Natl Cancer Inst 79:1295-1311.

Boice, J.D., Jr., G. Engholm, R.A. Kleinerman, M. Blettner, M. Stovall, H. Lisco, W.C. Moloney, D.F. Austin, A. Bosch, D.L. Cookfair, E.T. Krementz, H.B. Latourette, J.A. Merrill, L.J. Peters, M.D. Schulz, H.H. Storm, E. Björkholm, F. Pettersson, C.M.J. Bell, M.P. Coleman, P. Fraser, F.E. Neal, P. Prior, N.W. Choi, T.G. Hislop, M. Koch, N. Kreiger, D. Robb, D. Robson, D.H. Thomson, H. Lochmüller, D. von Fournier, R. Frischkorn, K.E. Kjørstad, A. Rimpela, M.H. Pejovic, V. Pompe Kirn, H. Stankusova, F. Berrino, K. Sigurdsson, G.B. Hutchison, and B. MacMahon. 1988. Radiation dose and second cancer risk in patients treated for cancer of the cervix. Radiat Res 116:3-55.

Boice, J.D., Jr., M. Blettner, R.A. Kleinerman, G. Engholm, M. Stovall, H. Lisco, D.F. Austin, A. Bosch, L. Harlan, E.T. Krementz, and J. Wactawski-Wende. 1989. Radiation dose and breast cancer risk in patients treated for cancer of the cervix. Int J Cancer 44:7-16.

Boice, J.D., Jr., M.M. Morin, A.G. Glass, G.D. Friedman, M. Stovall, R.N. Hoover, and J.F. Fraumeni. 1991a. Diagnostic x-ray procedures and risk of leukemia, lymphoma, and multiple myeloma. J Am Med Assoc 265:1290-1294.

Boice, J.D., Jr., D. Preston, F.G. Davis, and R.R. Monson. 1991b. Frequent chest x-ray fluoroscopy and breast cancer incidence among tuberculosis patients in Massachusetts. Radiat Res 125:214-222.

Boice, J.D., Jr., E.B. Harvey, M. Blettner, M. Stovall, and J.T. Flannery. 1992. Cancer in the contralateral breast after radiotherapy for breast cancer. N Engl J Med 326:781-785.

Boice, J.D., Jr., M. Blettner, and A. Auvinen. 2000. Epidemiologic studies of pilots and aircrew. Health Phys 79:576-584.

Bois, P., J.D. Stead, S. Bakshi, J. Williamson, R. Neumann, B. Moghadaszadeh, and A.J. Jeffreys. 1998a. Isolation and characterization of mouse minisatellites. Genomics 50:317-330.

Bois, P., J. Williamson, J. Brown, Y.E. Dubrova, and A.J. Jeffreys. 1998b. A novel unstable mouse VNTR family expanded from SINE B1 elements. Genomics 49:122-128.

Boivin, J., and G.B. Hutchison. 1982. Coronary heart disease mortality after irradiation for Hodgkin's disease. Cancer 49:2470-2475.

Boivin, J.F., G.B. Hutchison, M. Lyden, J. Godbold, J. Chorosh, and D. Schottenfeld. 1984. Second primary cancers following treatment of Hodgkin's disease. J Natl Cancer Inst 72:233-241.

Boland, C.R., J. Sato, H.D. Appelman, R.S. Bresalier, and A.P. Feinberg. 1995. Microallelotyping defines the sequence and tempo of allelic losses at tumour suppressor gene loci during colorectal cancer progression. Nat Med 1:902-909.

Bonassi, S., A. Abbondandolo, L. Camurri, L. Dalpra, M. DeFerrari, F. Degrassi, L. Lamberti, C. Lando, P. Padovani, I. Sbrana, D. Vecchio, and R. Puntoni. 1995. Are chromosome aberrations in circulating lymphocytes predictive of future cancer onset in humans. Cancer Genet Cytogen 79:133-135.

Bond, V.P., C.B. Meinhold, and H.H. Rossi. 1978. Low-dose RBE and Q for x-ray compared to gamma-ray radiations. Health Phys 34:433-438.

Bondy, M.L., A.P. Kyritsis, J. Gu, M. de Andrade, J. Cunningham, V.A. Levin, J.M. Bruner, and Q. Wei. 1996. Mutagen sensitivity and risk of gliomas: a case-control analysis. Cancer Res 56:1484-1486.

Bongarzone, I., M.G. Butti, L. Fugazzola, F. Pacini, A. Pinchera, T.V. Vorontsova, E.P. Demidchik, and M.A. Pierotti. 1997. Comparison of the breakpoint regions of ELE1 and RET genes involved in the generation of RET/PTC3 oncogene in sporadic and in radiation-associated papillary thyroid carcinomas. Genomics 42:252-259.

Bonser, G., and J. Jull. 1977. Tumors of the ovary. Pp. 129-184 in The Ovary, L. Zuckerman and B. Weir, eds. New York: Academic Press.

Boothman, D.A., M. Meyers, N. Fukunaga, and S.W. Lee. 1993. Isolation of x-ray inducible transcripts from radioresistant human melanoma cells. Proc Natl Acad Sci USA 90:7200-7204.

Borek, C., E.J. Hall, and M. Zaider. 1983. X rays may be twice as potent as gamma rays for malignant transformation at low doses. Nature 301:156-158.

Bose, A., and G. Olivieri. 1989. Variability of the adaptive response to ionizing radiations in humans. Mutat Res 211:13-17.

Boudaiffa, B., P. Cloutier, D. Hunting, M.A. Huels, and L. Sanche. 2000. Resonant formation of DNA strand breaks by low-energy (3 to 20 eV) electrons. Science 287:1658-1660.

Bouffler, S.D. 1998. Involvement of telomeric sequences in chromosomal aberrations. Mutat Res 404:199-204.

Bouffler, S.D., A. Silver, D. Papworth, J. Coates, and R. Cox. 1991. Murine radiation myeloid leukemogenesis: relationship between interstitial telomere like sequences and chromosome 2 fragile sites. Gene Chromosome Canc 6:90-106.

Bouffler, S.D., C.J. Kemp, A. Balmain, and R. Cox. 1995. Spontaneous and ionizing radiation-induced chromosomal abnormalities in p53-deficient mice. Cancer Res 55:3883-3889.

Bouffler, S.D., G. Breckon, and R. Cox. 1996. Chromosomal mechanisms in murine radiation acute myeloid leukaemogenesis. Carcinogenesis 17:655-659.

Bouffler, S.D., E.I. Meijne, D.J. Morris, and D. Papworth. 1997. Chromosome 2 hypersensitivity and clonal development in murine radiation acute myeloid leukaemia. Int J Radiat Biol 72:181-189.

Bouffler, S.D., J.W. Haines, A.A. Edwards, J.D. Harrison, and R. Cox. 2001. Lack of detectable transmissible chromosomal instability after in vivo or in vitro exposure of mouse bone marrow cells to ^{224}Ra alpha particles. Radiat Res 155:345-352.

Boulton, E., H. Cleary, D. Papworth, and M. Plumb. 2001. Susceptibility to radiation-induced leukaemia/lymphoma is genetically separable from sensitivity to radiation-induced genomic instability. Int J Radiat Biol 77:21-29.

Bounacer, A., R. Wicker, B. Caillou, A.F. Cailleux, A. Sarasin, M. Schlumberger, and H.G. Suarez. 1997. High prevalence of activating ret proto-oncogene rearrangements, in thyroid tumors from patients who had received external radiation. Oncogene 15:1263-1273.

Boutou, O., A.V. Guizard, R. Slama, D. Pottier, and A. Spira. 2002. Population mixing and leukaemia in young people around the La Hague nuclear waste reprocessing plant. Brit J Cancer 87:740-745.

Box, H.C., H.G. Freund, E.E. Budzinski, J.C. Wallace, and A.E. Maccubbin. 1995. Free radical-induced double base lesions. Radiat Res 141:91-94.

Bradley, W.E., A. Belouchi, and K. Messing. 1988. The aprt heterozygote/hemizygote system for screening mutagenic agents allows detection of large deletions. Mutat Res 199:131-138.

Branch, P., D.C. Bicknell, A. Rowan, W.F. Bodmer, and P. Karran. 1995. Immune surveillance in colorectal carcinoma. Nat Genet 9:231-232.

Breckon, G., and R. Cox. 1990. Alpha particle leukaemogenesis. Lancet 335:656-657.

Breckon, G., D. Papworth, and R. Cox. 1991. Murine radiation myeloid leukemogenesis: a possible role for radiation sensitive sites on chromosome 2. Gene Chromosome Canc 3:367-375.

Brenner, D.J. 1999. Does fractionation decrease the risk of breast cancer induced by low-LET radiation? Radiat Res 151:225-229.

Brenner, D.J., and C.D. Elliston. 2004. Estimated radiation risks potentially associated with full-body CT screening. Radiology 232:735-738.

Brenner, D.J., and J.F. Ward. 1995. Constraints on energy deposition and target size of multiple-damaged sites associated with DNA DSB. Pp. 443-446 in Radiat Res 1895-1995, Proceedings of the Tenth International Congress of Radiation Research, Vol. 2, U. Hagen and others, eds.

Brenner, D.J., P. Hahnfeldt, S.A. Amundson, and R.K. Sachs. 1996. Interpretation of inverse dose-rate effects for mutagenesis by sparsely ionizing radiation. Int J Radiat Biol 70:447-458.

Brenner, D.J., C. Elliston, E. Hall, and W. Berdon. 2001. Estimated risks of radiation-induced fatal cancer from pediatric CT. Am J Roentgenology 176:289-296.

Brewen, J.G., H.S. Payne, and R.J. Preston. 1976. X-ray-induced chromosome aberrations in mouse dictyate oocytes. I. Time and dose relationships. Mutat Res 35:111-120.

Brewer, C., S. Holloway, P. Zawalnyski, A. Schinzel, and D. FitzPatrick. 1998. A chromosomal deletion map of human malformations. Am J Hum Genet 63:1153-1159.

Bridges, B.A. 2001. Radiation and germline mutation at repeat sequences: are we in the middle of a paradigm shift? Radiat Res 156:631-641.

Brill, A.G., M. Tomonaga, and R.M. Heyssel. 1962. Leukemia in man following exposure to ionizing radiation. A summary of the findings in Hiroshima and Nagasaki, and a comparison with other human experience. Ann Intern Med 56:590-609.

Broeks, A., N.S. Russell, A.N. Floore, J.H. Urbanus, E.C. Dahler, M.B. van't Veer, A. Hagenbeek, E.M. Noordijk, M.A. Crommelin, F.E. van Leeuwen, and L.J. van't Veer. 2000. Increased risk of breast cancer following irradiation for Hodgkin's disease is not a result of ATM germline mutations. Int J Radiat Biol 76:693-698.

Broerse, J.J., and G.B. Gerber. 1982. Neutron carcinogenesis. Commission of the European Communities Directorate-General for Science, Research and Development Luxembourg.

Broome, E.J., D.L. Brown, and R.E. Mitchel. 1999. Adaptation of human fibroblasts to radiation alters biases in DNA repair at the chromosomal level. Int J Radiat Biol 75:681-690.

Broome, E.J., D.L. Brown, and R.E. Mitchel. 2002. Dose responses for adaption to low doses of (60)Co gamma rays and (3)H beta particles in normal human fibroblasts. Radiat Res 158:181-186.

Brown, D.P., and T. Bloom. 1987. Cincinnati, OH: National Institute for Occupational Safety and Health. National Technical Information Services, Springfield, VA.

Bryant, P.E. 1984. Enzymatic restriction of mammalian cell DNA using Pvu II and Bam H1: evidence for the double-strand break origin of chromosomal aberrations. Int J Radiat Biol Re 46:57-65.

Buard, J., A. Bourdet, J. Yardley, Y. Dubrova, and A.J. Jeffreys. 1998. Influences of array size and homogeneity on minisatellite mutation. EMBO J 17:3495-3502.

Buard, J., A. Collick, J. Brown, and A.J. Jeffreys. 2000. Somatic versus germline mutation processes at minisatellite CEB1 (D2S90) in humans and transgenic mice. Genomics 65:95-103.

Buchholz, T.A., X. Wu, A. Hussain, S.L. Tucker, G.B. Mills, B. Haffty, S. Bergh, M. Story, F.B. Geara, and W.A. Brock. 2002. Evidence of haplotype insufficiency in human cells containing a germline mutation in BRCA1 or BRCA2. Int J Cancer 97:557-561.

Buchhop, S., M.K. Gibson, X.W. Wang, P. Wagner, H.W. Sturzbecher, and C.C. Harris. 1997. Interaction of p53 with the human Rad51 protein. Nucleic Acids Res 25:3868-3874.

Budarf, M.L., and B.S. Emanuel. 1997. Progress in the autosomal segmental aneusomy syndromes (SASs): single or multi-locus disorders? Hum Mol Genet 6:1657-1665.

Burak, L.E., Y. Kodama, M. Nakano, K. Ohtaki, M. Itoh, N.D. Okladnikova, E.K. Vasilenko, J.B. Cologne, and N. Nakamura. 2001. FISH examination of lymphocytes from Mayak workers for assessment of translocation induction rate under chronic radiation exposures. Int J Radiat Biol 77:901-908.

Burki, H.J. 1980. Ionizing radiation-induced 6-thioguanine-resistant clones in synchronous CHO cells. Radiat Res 81:76-84.

Burma, S., B.P. Chen, M. Murphy, A. Kurimasa, and D.J. Chen. 2001. ATM phosphorylates histone H2AX in response to DNA double-strand breaks. J Biol Chem 276:42462-42467.

Burnet, N.G., J. Johansen, I. Turesson, J. Nyman, and J.H. Peacock. 1998. Describing patients' normal tissue reactions: concerning the possibility of individualising radiotherapy dose prescriptions based on potential predictive assays of normal tissue radiosensitivity. Steering Committee of the BioMed2 European Union Concerted Action Programme on the Development of Predictive Tests of Normal Tissue Response to Radiation Therapy. Int J Cancer 79:606-613.

Burns, F.J., R.E. Albert, I.P. Sinclair, and P. Bennett. 1973. The effect of fractionation on tumor induction and hair follicle damage in rat skin. Radiat Res 53:235-240.

Burns, F.J., R.E. Albert, I.P. Sinclair, and M. Vanderlaan. 1975. The effect of 24-hour fractionation interval on the induction of rat skin tumors by electron radiation. Radiat Res 62:478-487.

Burns, F.J., R.E. Albert, and S.J. Garte. 1989a. Radiation-induced cancer in rat skin. Carcinog Compr Surv 11:293-319.

Burns, F.J., R.E. Albert, and S.J. Garte. 1989b. Multiple stages in radiation carcinogenesis of rat skin. Environ Health Persp 81:67-72.

Burns, F.J., R.N. Shore, C. Loomis, and P. Zhao. 2002. PTCH (patched) and XPA genes in radiation-induced basal cell carcinomas. Pp. 175-178 in Radiation and Homeostasis, T. Sugahara, O. Nikaido, and O. Niwa, eds. Amsterdam: Elsevier.

Byrne, J., S.A. Rasmussen, S.C. Steinhorn, R.R. Connelly, M.H. Myers, C.F. Lynch, J. Flannery, D.F. Austin, F.F. Holmes, G.E. Holmes, L.C. Strong, and J.J. Mulvihill. 1998. Genetic disease in offspring of long-term survivors of childhood and adolescent cancer. Am J Hum Genet 62:45-52.

Calabrese, E.J., and L.A. Baldwin. 2000. Radiation hormesis: its historical foundations as a biological hypothesis. Hum Exp Toxicol 19:41-75.

Calabrese, E.J., and L.A. Baldwin. 2003. Toxicology rethinks its central belief: hormesis demands a reappraisal of the way risks are assessed. Nature 421:691-692.

Calabrese, E.J., L.A. Baldwin, and C.D. Holland. 1999. Hormesis: A highly generalizable and reproducible phenomenon with important implications for risk assessment. Risk Anal 19:261-281.

Cao, J., R.L. Wells, and M.M. Elkind. 1992. Enhanced sensitivity to neoplastic transformation by ^{137}Cs gamma-rays of cells in the G2-/M-phase age interval. Int J Radiat Biol 62:191-199.

Cao, J., R.L. Wells, and M.M. Elkind. 1993. Neoplastic transformation of C3H mouse embryo cells, 10T1/2: cell-cycle dependence for 50 kV X-rays and UV-B light. Int J Radiat Biol 64:83-92.

Caratero, A., M. Courtade, L. Bonnet, H. Planel, and C. Caratero. 1998. Effect of a continuous gamma irradiation at a very low dose on the life span of mice. Gerontology 44:272-276.

Cardis, E., and A.E. Okeanov. 1996. What's feasible and desirable in the epidemiologic follow-up of Chernobyl. First International Conference of the European Commission, Belarus, the Russian Federation, and the Ukraine on the Radiological Consequences of the Chernobyl Accident. Minsk, Belarus.

Cardis, E., and D. Richardson. 2000. Invited editorial: health effects of radiation exposure at uranium processing facilities. J Radiol Prot 20: 95-97.

Cardis, E., E.S. Gilbert, L. Carpenter, G. Howe, I. Kato, B.K. Armstrong, V. Beral, G. Cowper, A. Douglas, J. Fix, F.A. Fry, J. Kaldor, C. Lave, L. Salmon, P.G. Smith, G.L. Voelz, and L.D. Wiggs. 1995. Effects of low doses and low dose rates of external ionizing radiation: cancer mortality among nuclear industry workers in three countries. Radiat Res 142:117-132.

Cardis, E., A.E. Okeanov, L. Anspaugh, V.K. Ivanov, I. Likthariev, A.E. Okeanov, and A. Prisyazhniuk. 1996. Estimated long term health effects of the Chernobyl accident. Pp. 241-279 in Proceedings of the Joint EU, IAEA and WHO International Conference One Decade after Chernobyl. Vienna, Austria.

Cardis, E., D. Richardson, and A. Kesminiene. 2000. Radiation Risk Estimates in the Beginning of the 21st Century. Bethesda, MD: National Council on Radiation Protection and Measurements.

Cardis, E., M. Vrijheid, M. Blettner, E. Gilbert, M. Hakama, C. Hill, G. Howe, J. Kaldor, C.R. Muirhead, M. Schubauer-Berigan, Yoshimura, F. Bermann, G. Cowper, J. Fix, C. Hacker, B. Heinmiller, M. Marshall, I. Thierry-Chef, D. Utterback, Y.O. Ahn, E. Amoros, P. Ashmore, A. Auvinen, J.M. Bae, J. Bernar Solano, A. Biau, E. Combalot, P. Deboodt, A. Diez Sacristan, M. Eklof, H. Engels, G. Engholm, G. Gulis, R. Habib, K. Holan, H. Hyvonen, A. Kerekes, I. Kurtinaitis, H. Malker, M. Martuzzi, A. Mastauskas, A. Monnet, M. Moser, M. Murata, M.S. Pearce, D.B. Richardson, F. Rodriguez-Artalejo, A. Rogel, H. Tardy, M. Telle-Lamberton, I. Turai, M. Usel, and K. Veress. 2005a. Cancer risk following low doses of ionising radiation—a 15-country study. Brit Med J.

Cardis, E., A. Kesminiene, V. Ivanov, I. Malakhova, Y. Shibata, V. Khrouch, V. Drozdovitch, E. Maceika, I. Zvonova, O. Vlassov, A. Bouville, G. Goulko, M. Hoshi, A. Abrosimov, J. Anoshko, L. Astakhova, S. Chekin, E. Demidchik, R. Galanti, M. Ito, E. Korobova, E. Lushnikov, M. Maksioutov, V. Masyakin, A. Nerovnia, V. Parshin, E. Parshkov, N. Piliptsevich, A. Pinchera, S. Polyakov, N. Shabeka, E. Suonio, V. Tenet, A. Tsyb, S. Yamashita, and D. Williams. 2005b. Risk of thyroid cancer after exposure to ^{131}I in childhood. J Natl Cancer Inst 97:724-732.

Carnes, B.A., and T.E. Fritz. 1991. Responses of the beagle to protracted irradiation. I. Effect of total dose and dose rate. Radiat Res 128:125-132.

Carnes, B.A., D. Grahn, and J.F. Thomson. 1989. Dose-response modeling of life shortening in a retrospective analysis of the combined data from the JANUS program at Argonne National Laboratory. Radiat Res 119:39-56.

Carnes, B.A., S.J. Olshansky, and D. Grahn. 1998. An interspecies prediction of the risk of radiation-induced mortality. Radiat Res 149:487-492.

Carnes, B.A., N. Gavrilova, and D. Grahn. 2002. Pathology effects at radiation doses below those causing increased mortality. Radiat Res 158:187-194.

Carnes, B.A., D. Grahn, and D. Hoel. 2003. Mortality of atomic bomb survivors predicted from laboratory animals. Radiat Res 160:159-167.

Carpenter, L., P. Fraser, M. Booth, C. Higgins, and V. Beral. 1989. Smoking habits and radiation exposure. J Radiol Prot 9:286-287.

Carpenter, L., V. Beral, P. Fraser, and M. Booth. 1990. Health related selection and death rates in the United Kingdom Atomic Energy Authority workforce. Brit J Ind Med 47:248-258.

Carpenter, L., C. Higgins, A. Douglas, P. Fraser, V. Beral, and P. Smith. 1994. Combined analysis of mortality in three United Kingdom nuclear industry workforces, 1946-1988. Radiat Res 138:224-238.

Carpenter, L.M., C.D. Higgins, A.J. Douglas, N.E. Maconochie, R.Z. Omar, P. Fraser, V. Beral, and P.G. Smith. 1998. Cancer mortality in relation to monitoring for radionuclide exposure in three UK nuclear industry workforces. Brit J Cancer 78:1224-1232.

Carr, Z.A., R.A. Kleinerman, M. Stovall, R.M. Weinstock, M.L. Griem, and C.E. Land. 2002. Malignant neoplasms after radiation therapy for peptic ulcer. Radiat Res 157:668-677.

Carrano, A.V., and J.A. Heddle. 1973. The fate of chromosome aberrations. J Theor Biol 38:289-304.

Carter, C.O. 1961. The inheritance of pyloric stenosis. Brit Med Bull 17:251-254.

Carter, C.O. 1976a. Genetics of common single malformations. Brit Med Bull 32:21-26.

Carter, C.O. 1976b. Monogenic disorders. J Med Genet 14:316-320.

Cattanach, B.M., and H. Moseley. 1974. Sterile period, translocation and specific locus mutation in the mouse following fractionated x-ray treatments with different fractionation intervals. Mutat Res 25:63-72.

Cattanach, B.M., and C. Rasberry. 1994. Enhanced specific-locus mutation response of 101/H male mice to single, acute X-irradiation. Mutat Res 311:77-84.

Cattanach, B.M., C. Jones, and D.G. Papworth. 1985. Specific-locus mutation response to unequal, 1 + 9 Gy X-ray fractionations at 24-h and 4-day fraction intervals. Mutat Res 149:105-118.

Cattanach, B.M., M.D. Burtenshaw, C. Rasberry, and E.P. Evans. 1993. Large deletions and other gross forms of chromosome imbalance compatible with viability and fertility in the mouse. Nat Genet 3:56-61.

Cattanach, B.M., E.P. Evans, C. Rasberry, and M.D. Burtenshaw. 1996. Incidence and distribution of radiation-induced large deletions in the mouse. Pp. 531-534 in Radiation Research 1895-1995, Vol. 2, Proceedings of the 10th International Congress of Radiation Research, Wurzburg, August 27-September 1, 1995.

Chakraborty, R., M.P. Little, and K. Sankaranarayanan. 1997. Cancer predisposition, radiosensitivity and the risk of radiation-induced cancers. III. Effects of incomplete penetrance and dose-dependent radiosensitivity on cancer risks in populations. Radiat Res 147:309-320.

Chakraborty, R., M.P. Little, and K. Sankaranarayanan. 1998a. Cancer predisposition, radiosensitivity and the risk of radiation-induced cancers. IV. Prediction of risks in relatives of cancer-predisposed individuals. Radiat Res 149:493-507.

Chakraborty, R., N. Yasuda, C. Denniston, and K. Sankaranarayanan. 1998b. Ionizing radiation and genetic risks. VII. The concept of mutation component and its use in risk estimation for Mendelian diseases. Mutat Res 400:541-552.

Chan, T.A., H. Hermeking, C. Lengauer, K.W. Kinzler, and B. Vogelstein. 1999. 14-3-3 Sigma is required to prevent mitotic catastrophe after DNA damage. Nature 401:616-620.

Chang, B.D., K. Watanabe, E.V. Broude, J. Fang, J.C. Poole, T.V. Kalinichenko, and I.B. Roninson. 2000. Effects of p21Waf1/Cip1/Sdi1 on cellular gene expression: implications for carcinogenesis, senescence, and age-related diseases. Proc Natl Acad Sci USA 97:4291-4296.

Chang, B.H., L.C. Shimmin, S.K. Shyue, D. Hewett-Emmett, and W.H. Li. 1994. Weak male-driven molecular evolution in rodents. Proc Natl Acad Sci USA 91:827-831.

Chang, S., C. Khoo, and R.A. DePinho. 2001. Modeling chromosomal instability and epithelial carcinogenesis in the telomerase-deficient mouse. Semin Cancer Biol 11:227-239.

Chang, W.P., and J.B. Little. 1992. Evidence that DNA double-strand breaks initiate the phenotype of delayed reproductive death in Chinese hamster ovary cells. Radiat Res 131:53-59.

Charles, D.J., and W. Pretsch. 1986. Enzyme-activity mutations detected in mice after paternal fractionated irradiation. Mutat Res 160:243-248.

Chauveinc, L., M. Ricoul, L. Sabatier, G. Gaboriaud, A. Srour, X. Bertagna, and B. Dutrillaux. 1997. Dosimetric and cytogenetic studies of multiple radiation-induced meningiomas for a single patient. Radiother Oncol 43:285-288.

Checkoway, H., R.M. Mathew, C.M. Shy, J.E. Watson Jr., W.G. Tankersley, S.H. Wolf, J.C. Smith, and S.A. Fry. 1985. Radiation, work experience, and cause specific mortality among workers at an energy research laboratory. Brit J Ind Med 42:525-533.

Checkoway, H., N. Pearce, D.J. Crawford-Brown, and D.L. Cragle. 1988. Radiation doses and cause-specific mortality among workers at a nuclear materials fabrication plant. Am J Epidemiol 127:255-266.

Chehab, N.H., A. Malikzay, M. Appel, and T.D. Halazonetis. 2000. Chk2/hCds1 functions as a DNA damage checkpoint in G(1) by stabilizing p53. Gene Dev 14:278-288.

Chenevix-Trench, G., A.B. Spurdle, M. Gatei, H. Kelly, A. Marsh, X. Chen, K. Donn, M. Cummings, D. Nyholt, M.A. Jenkins, C. Scott, G.M. Pupo, T. Dork, R. Bendix, J. Kirk, K. Tucker, M.R. McCredie, J.L. Hopper, J. Sambrook, G.J. Mann, and K.K. Khanna. 2002. Dominant negative ATM mutations in breast cancer families. J Natl Cancer Inst 94:205-215.

Cherbonnel-Lasserre, C., S. Gauny, and A. Kronenberg. 1996. Suppression of apoptosis by Bcl-2 or Bcl-xL promotes susceptibility to mutagenesis. Oncogene 13:1489-1497.

Childs, J.D. 1981. The effect of a change in mutation rate on the incidence of dominant and X-linked recessive disorders in man. Mutat Res 83:145-158.

Choi, H., S. Kim, P. Mukhopadhyay, S. Cho, J. Woo, G. Storz, and S. Ryu. 2001. Structural basis of the redox switch in the OxyR transcription factor. Cell 105:103-113.

Chuang, Y.Y., and H.L. Liber. 1996. Effects of cell cycle position on ionizing radiation mutagenesis. I. Quantitative assays of two genetic loci in a human lymphoblastoid cell line. Radiat Res 146:494-500.

Clark, D.J., E.I.M. Meijne, S.D. Bouffler, R. Huiskamp, C.J. Skidmore, R. Cox, and A.R.J. Silver. 1996. Microsatellite analysis of recurrent chromosome 2 deletions in acute myeloid leukaemia induced by radiation in F_1 hybrid mice. Gene Chromosome Canc 16:238-246.

Clayton, D., and M. Hills. 1993. Statistical Methods in Epidemiology. New York: Oxford Press.

Cleaver, J.E. 1968. Defective repair replication of DNA in xeroderma pigmentosum. Nature 218:652-656.

Clifton, K.H. 1996. Comments on the evidence in support of the epigenetic nature of radiogenic initiation. Mutat Res 350:77-80.

Clutton, S.M., K.M.S. Townsend, D.T. Goodhead, J.D. Ansell, and E.G. Wright. 1996a. Differentiation and delayed cell death in embryonal stem cells exposed to low doses of ionising radiation. Cell Death Differ 3: 141-148.

Clutton, S.M., K.M.S. Townsend, C. Walker, J.D. Ansell, and E.G. Wright. 1996b. Radiation-induced genomic instability and persisting oxidative stress in primary bone marrow cultures. Carcinogenesis 17:1633-1639.

Cohen, B.L. 2002. Cancer risk from low-level radiation. Am J Roentgenology 179:1137-1143.

Collis, S.J., J.M. Schwaninger, A.J. Ntambi, T.W. Keller, W.G. Nelson, L.E. Dillehay, and T.L. Deweese. 2004. Evasion of early cellular response mechanisms following low level radiation-induced DNA damage. J Biol Chem 279:49624-49632.

Cologne, J.B., and D.L. Preston. 2000. Longevity of atomic-bomb survivors. Lancet 356:303-307.

Cologne, J.B., S. Tokuoka, G.W. Beebe, T. Fukuhara, and K. Mabuchi. 1999. Effects of radiation on incidence of primary liver cancer among atomic bomb survivors. Radiat Res 152:364-373.

Conard, R.A. 1980. Review of Medical Findings in a Marshallese Population Twenty-Six Years after Accidental Exposure to Radioactive Fallout. Upton, NY: Brookhaven National Laboratory.

Conard, R.A. 1984. Late radiation effects in Marshall Islanders exposed to fallout 28 years ago. Pp. 57-71 in Radiation Carcinogenesis: Epidemiology and Biologic Significance, J.D. Boice Jr. and J.F. Fraumeni Jr., eds. New York: Raven Press.

Congdon, C.C. 1987. A review of certain low-level ionizing radiation studies in mice and guinea pigs. Health Phys 52:593-597.

Connor, F., D. Bertwistle, P.J. Mee, G.M. Ross, S. Swift, E. Grigorieva, V.L. Tybulewicz, and A. Ashworth. 1997. Tumorigenesis and a DNA repair defect in mice with a truncating Brca2 mutation. Nat Genet 17:423-430.

Cormier, R.T., A. Bilger, A.J. Lillich, R.B. Halberg, K.H. Hong, K.A. Gould, N. Borenstein, E.S. Lander, and W.F. Dove. 2000. The Mom1AKR intestinal tumor resistance region consists of Pla2g2a and a locus distal to D4Mit64. Oncogene 19:3182-3192.

Cornforth, M.N., and J.S. Bedford. 1983. X-ray-induced breakage and rejoining of human interphase chromosomes. Science 222:1141-1143.

Cornforth, M.N., and J.S. Bedford. 1987. A quantitative comparison of potentially lethal damage repair and the rejoining of interphase chromosome breaks in low passage normal human fibroblasts. Radiat Res 111:385-405.

Cornforth, M.N., and J.S. Bedford. 1993. Ionizing radiation damage and its early development in chromosomes. Adv Radiat Biol 423:423-496.

Cornforth, M.N., S.M. Bailey, and E.H. Goodwin. 2002. Dose responses for chromosome aberrations produced in noncycling primary human fibroblasts by alpha particles and by gamma rays delivered at sublimiting low dose rates. Radiat Res 158:43-53.

Correa, C.R., and V.G. Cheung. 2004. Genetic variation in radiation-induced expression phenotypes. Am J Hum Genet 75:885-890.

Court Brown, W.M., and R. Doll. 1958. Expectation of life and mortality from cancer among British radiologists. Brit Med J:181-187.

Coussens, L.M., and Z. Werb. 2002. Inflammation and cancer. Nature 420:860-867.

Covelli, V., V. Di Majo, M. Coppola, and S. Rebessi. 1989. The dose-response relationships for myeloid leukemia and malignant lymphoma in BC3F1 mice. Radiat Res 119:553-561.

Cox, R., and A.A. Edwards. 2002. Comments on the paper: Microsatellite instability in acute myelocytic leukaemia developed from A-bomb survivors—and related cytogenic data. Int J Radiat Biol 78:443-445.

Cox, R., and W.K. Masson. 1978. Do radiation-induced thioguanine-resistant mutants of cultured mammalian cells arise by HGPRT gene mutation or X-chromosome rearrangement? Nature 276:629-630.

Cox, R., and W.K. Masson. 1979. Mutation and inactivation of cultured mammalian cells exposed to beams of accelerated heavy ions. III. Human diploid fibroblasts. Int J Radiat Biol Re 36:149-160.

Cox, R., J. Thacker, and D.T. Goodhead. 1977. Inactivation and mutation of cultured mammalian cells by aluminium characteristic ultrasoft x-rays. II. Dose-responses of Chinese hamster and human diploid cells to aluminium x-rays and radiations of different LET. Int J Radiat Biol Re 31:561-576.

Cragle, D.L., R.W. McLain, J.R. Qualters, J.L. Hickey, G.S. Wilkinson, W.G. Tankersley, and C.C. Lushbaugh. 1988. Mortality among workers at a nuclear fuels production facility. Am J Ind Med 14:379-401.

Cragle, D.L., K. Robertson-Demers, and J.P. Watkins. 1994. Mortality among Workers at a Nuclear Fuels Production Facility: The Savannah River Site, 1952-1986. Center for Epidemiologic Research. Oak Ridge, TN: Oak Ridge Institute for Science and Education.

Critchlow, S.E., and S.P. Jackson. 1998. DNA end-joining: from yeast to man. Trends Biochem Sci 23:394-398.

Crompton, N.E.A., B. Barth, and J. Kiefer. 1990. Inverse dose-rate effect for the induction of 6-thioguanine-resistant mutants in Chinese hamster V79-S cells by ^{60}Co gamma rays. Radiat Res 124:300-308.

Crow, J.F. 1993. How much do we know about spontaneous human mutation rates? Environ Mol Mutagen 21:122-129.

Crow, J.F. 1997. The high spontaneous mutation rate: is it a health risk? Proc Natl Acad Sci USA 94:8380-8386.

Crow, J.F. 1999. Spontaneous mutation in man. Mutat Res 437:5-9.

Crow, J.F. 2001. The beanbag lives on. Nature 409:771.

Crow, J.F., and C. Denniston. 1981. The mutation component of genetic damage. Science 212:888-893.

Crow, J.F., and C. Denniston. 1985. Mutations in human populations. Adv Hum Genet 14:59-123.

Curtis, R.E., J.D. Boice Jr., M. Stovall, L. Bernstein, R.S. Greenberg, J.T. Flannery, A.G. Schwartz, P. Weyer, W.C. Moloney, and R.N. Hoover. 1992. Risk of leukemia after chemotherapy and radiation treatment for breast cancer. N Engl J Med 326:1745-1751.

Curtis, R.E., J.D. Boice Jr., M. Stovall, L. Bernstein, E. Holowaty, S. Karjalainen, F. Langmark, P.C. Nasca, A.G. Schwartz, and M.J. Schymura. 1994. Relationship of leukemia risk to radiation dose following cancer of the uterine corpus. J Natl Cancer Inst 86:1315-1324.

Czeizel, A., and K. Sankaranarayanan. 1984. The load of genetic and partially genetic disorders in man. I. Congenital anomalies: estimates of detriment in terms of years of life lost and years of impaired life. Mutat Res 128:73-103.

Czeizel, A., and G. Tusnady. 1984. Aetiological Studies of Isolated Common Congenital Abnormalities in Hungary. Budapest: Akademiai Kiado.

Czeizel, A., K. Sankaranayanan, A. Losonci, T. Rudas, and M. Keresztes. 1988. The load of genetic and partially genetic diseases in man. II. Some selected common multifactorial diseases: estimates of detriment in terms of years lost and impaired life. Mutat Res 196:259-292.

Czene, K., P. Lichtenstein, and K. Hemminki. 2002. Environmental and heritable causes of cancer among 9.6 million individuals in the Swedish Family-Cancer Database. Int J Cancer 99:260-266.

Dalager, N.A., H.K. Kang, and C.M. Mahan. 2000. Cancer mortality among the highest exposed US atmospheric nuclear test participants. J Occup Environ Med 42:798-805.

Damber, L., L.G. Larsson, L. Johansson, and T. Norin. 1995. A cohort study with regard to the risk of haematological malignancies in patients treated with x-rays for benign lesions in the locomotor system. I. Epidemiological analyses. Acta Oncol 34:713-719.

Darby, S.C., G.M. Kendall, T.P. Fell, R. Doll, A.A. Goodill, A.J. Conquest, D.A. Jackson, and R.G. Haylock. 1993. Further follow up of mortality and incidence of cancer in men from the United Kingdom who participated in the United Kingdom's atmospheric nuclear weapon tests and experimental programmes. Brit Med J 307:1530-1535.

Darby, S.C., G. Reeves, T. Key, R. Doll, and M. Stovall. 1994. Mortality in a cohort of women given x-ray therapy for metropathia haemorrhagica. Int J Cancer 56:793-801.

Darby, S., P. McGale, R. Peto, F. Granath, P. Hall, and A. Ekbom. 2003. Mortality from cardiovascular disease more than 10 years after radiotherapy for breast cancer: nationwide cohort study of 90,000 Swedish women. Brit Med J 326:256-257.

Darroudi, F., J. Fomina, M. Meijers, and A.T. Natarajan. 1998. Kinetics of the formation of chromosome aberrations in x-irradiated human lymphocytes, using PCC and FISH. Mutat Res 404:55-65.

Davis, F., J. Boice, Z. Hrubec, and R. Monson. 1989. Cancer mortality in a radiation-exposed cohort of Massachusetts. Cancer Res 49:6130-6136.

Davis, S., K.J. Kopecky, and T.E. Hamilton. 2001. Final Report. Hanford Thyroid Disease Study.

Davis S., K.J. Kopecky, T.E. Hamilton, and L.E. Onstad. 2004a. Thyroid neoplasia, autoimmune thyroiditis, and hypothyroidism in persons exposed to I-131 from the Hanford Nuclear Site. J Am Med Assoc 292:2600-2613.

Davis, S., V. Stepanenko, N. Rivkind, K.J. Kopecky, P. Voilleque, V. Shakhtarin, E. Parshkov, S. Kulikov, E. Lushnikov, A. Abrosimov, V. Troshin, G. Romanova, V. Doroschenko, A. Proshin, and A. Tsyb. 2004b. Risk of thyroid cancer in the Bryansk Oblast of the Russian Federation after the Chernobyl Power Station accident. Radiat Res 162:241-248.

Day, J.P., C.L. Limoli, and W.F. Morgan. 1998. Recombination involving interstitial telomere repeat-like sequences promotes chromosomal instability in Chinese hamster cells. Carcinogenesis 19:259-265.

Daza, P., S. Reichenberger, B. Gottlich, M. Hagmann, E. Feldmann, and P. Pfeiffer. 1996. Mechanisms of nonhomologous DNA end-joining in frogs, mice and men. Biol Chem 377:775-786.

de Vathaire, F., P. Francois, C. Hill, O. Schweisguth, C. Rodary, D. Sarrazin, O. Oberlin, C. Beurtheret, A. Dutreix, and R. Flamant. 1989. Role of radiotherapy and chemotherapy in the risk of second malignant neoplasms after cancer in childhood. Brit J Cancer 59:792-796.

de Vathaire, F., M. Schlumberger, M.J. Delisle, C. Francese, C. Challeton, E. de la Genardiere, F. Meunier, C. Parmentier, C. Hill, and H. Sancho-Garnier. 1997. Leukaemias and cancers following iodine-131 administration for thyroid cancer. Brit J Cancer 75:734-739.

de Vathaire, F., M. Hawkins, S. Campbell, O. Oberlin, M.A. Raquin, J.Y. Schlienger, A. Shamsaldin, I. Diallo, J. Bell, E. Grimaud, C. Hardiman, J.L. Lagrange, N. Daly-Schveitzer, X. Panis, J.M. Zucker, H. Sancho-Garnier, F. Eschwege, J. Chavaudra, and J. Lemerle. 1999. Second malignant neoplasms after a first cancer in childhood: temporal pattern of risk according to type of treatment. Brit J Cancer 79:1884-1893.

Degg, N.L., M.M. Weil, A. Edwards, J. Haines, M. Coster, J. Moody, M. Ellender, R. Cox, and A. Silver. 2003. Adenoma multiplicity in irradiated Apc(Min) mice is modified by chromosome 16 segments from BALB/c. Cancer Res 63:2361-2363.

Delongchamp, R.R., K. Mabuchi, Y. Yoshimoto, and D.L. Preston. 1997. Cancer mortality among atomic bomb survivors exposed in utero or as young children, October 1950-May 1992. Radiat Res 147:385-395.

Demple, B. 1991. Regulation of bacterial oxidative stress genes. Ann Rev Genet 25:315-337.

Deng, C.X., and S.G. Brodie. 2001. Knockout mouse models and mammary tumorigenesis. Semin Cancer Biol 11:387-394.

Denniston, C. 1983. Are human studies possible? Some thoughts on the mutation component and population monitoring. Environ Health Persp 52:41-44.

Denniston, C., R. Chakraborty, and K. Sankaranarayanan. 1998. Ionizing radiation and genetic risks. VIII. The concept of mutation component and its use in risk estimation for multifactorial diseases. Mutat Res 405:57-79.

Dent, P., D.B. Reardon, J.S. Park, G. Bowers, C. Logsdon, K. Valerie, and R. Schmidt-Ullrich. 1999. Radiation-induced release of transforming growth factor alpha activates the epidermal growth factor receptor and mitogen-activated protein kinase pathway in carcinoma cells, leading to increased proliferation and protection from radiation-induced cell death. Mol Biol Cell 10:2493-2506.

DerSimonian, R., and N. Laird. 1986. Meta-analysis in clinical trials. Control Clin Trials 7:177-188.

Dewey, W.C., R.M. Humphrey, and B.A. Jones. 1965. Comparisons of tritiated thymidine, tritiated water, and cobalt-60 gamma rays in inducing chromosomal aberrations. Radiat Res 24:214-238.

Dewey, W.C., S.C. Furman, and H.H. Miller. 1970. Comparison of lethality and chromosomal damage induced by x-rays in synchronized Chinese hamster cells in vitro. Radiat Res 43:561-581.

Dianov, G.L., R. Prasad, S.H. Wilson, and V.A. Bohr. 1999. Role of DNA polymerase beta in the excision step of long patch mammalian base excision repair. J Biol Chem 274:13741-13743.

Dickinson, H.O., L. Parker, K. Binks, R. Wakeford, and J. Smith. 1996. The sex ratio of children in relation to paternal preconceptional radiation dose: a study in Cumbria, Northern England. J Epidemiol Commun H 50:645-652.

Difilippantonio, M.J., S. Petersen, H.T. Chen, R. Johnson, M. Jasin, R. Kanaar, T. Ried, and A. Nussenzweig. 2002. Evidence for replicative repair of DNA double-strand breaks leading to oncogenic translocation and gene amplification. J Exp Med 196:469-480.

Dizdaroglu, M. 1992. Oxidative damage to DNA in mammalian chromatin. Mutat Res 275:331-342.

Dizdaroglu, M., M.L. Dirksen, H.X. Jiang, and J.H. Robbins. 1987. Ionizing-radiation-induced damage in the DNA of cultured human cells. Identification of 8,5-cyclo-2-deoxyguanosine. Biochem J 241:929-932.

Dizdaroglu, M., Z. Nackerdien, B.C. Chao, E. Gajewski, and G. Rao. 1991. Chemical nature of in vivo DNA base damage in hydrogen peroxide-treated mammalian cells. Arch Biochem Biophys 285:388-390.

Dobson, R.L., T. Straume, A.V. Carrano, J.L. Minkler, L.L. Deaven, L.G. Littlefield, and A.A. Awa. 1991. Biological effectiveness of neutrons from Hiroshima bomb replica: results of a collaborative cytogenetic study. Radiat Res 128:143-149.

Dobyns, B.M., G.E. Sheline, J.B. Workman, E.A. Tompkins, W.M. McConahey, and D.V. Becker. 1974. Malignant and benign neoplasms of the thyroid in patients treated for hyperthyroidism: a report of the cooperative thyrotoxicosis therapy follow-up study. J Clin Endocr Metab 38:976-998.

Doll, R., and R. Wakeford. 1997. Risk of childhood cancer from fetal irradiation. Brit J Radiol 70:130-139.

Dondon, M.G., F. de Vathaire, A. Shamsaldin, F. Doyon, I. Diallo, L. Ligot, C. Paoletti, M. Labbe, M. Abbas, J. Chavaudra, M.F. Avril, P. Fragu, and F. Eschwege. 2004. Cancer mortality after radiotherapy for a skin hemangioma during childhood. Radiother Oncol 72:87-93.

Dong, C., and K. Hemminki. 2001. Modification of cancer risks in offspring by sibling and parental cancers from 2,112,616 nuclear families. Int J Cancer 92:144-150.

Doody, M.M., J.S. Mandel, J.H. Lubin, and J.D. Boice Jr. 1998. Mortality among United States radiologic technologists, 1926-90. Cancer Cause Control 9:67-75.

Doody, M.M., J.E. Lonstein, M. Stoval, D.G. Hacker, N. Luckyanov, and C.E. Land. 2000. Breast cancer mortality after diagnostic radiography: findings from the US Scoliosis Cohort Study. Spine 25:2052-2063.

Dores, G.M., C. Metayer, R.E. Curtis, C.F. Lynch, E.A. Clarke, B. Glimelius, H. Storm, E. Pukkala, F.E. van Leeuwen, E.J. Holowaty, M. Andersson, T. Wiklund, T. Joensuu, M.B. van't Veer, M. Stovall, M. Gospodarowicz, and L.B. Travis. 2002. Second malignant neoplasms among long-term survivors of Hodgkin's disease: a population-based evaluation over 25 years. J Clin Oncol 20:3484-3494.

Douglas, A.J., R.Z. Omar, and P.G. Smith. 1994. Cancer mortality and morbidity among workers at the Sellafield plant of British Nuclear Fuels. Brit J Cancer 70:1232-1243.

Doyle, P., N. Maconochie, E. Roman, G. Davies, P.G. Smith, and V. Beral. 2000. Fetal death and congenital malformation in babies born to nuclear industry employees: report from the nuclear industry family study. Lancet 356:1293-1299.

Draper, G.J., M.P. Little, T. Sorahan, L.J. Kinlen, K.J. Bunch, A.J. Conquest, G.M. Kendall, G.W. Kneale, R.J. Lancashire, C.R. Muirhead, C.M. O'Connor, and T.J. Vincent. 1997. Cancer in the offspring of radiation workers: a record linkage study. Brit Med J 315:1181-1188.

Drost, J.B., and W.R. Lee. 1995. Biological basis of germline mutation: comparisons of spontaneous germline mutation rates among drosophila, mouse, and human. Environ Mol Mutagen 25(Suppl 26):48-64.

Dublin, L.I., and M. Spigelman. 1948. Mortality among medical specialists. J Am Med Assoc 137:1519-1524.

Dubrova, Y.E., A.J. Jeffreys, and A.M. Malashenko. 1993. Mouse minisatellite mutations induced by ionizing radiation. Nat Genet 5: 92-94.

Dubrova, Y.E., V.N. Nesterov, N.G. Krouchinsky, V.A. Ostapenko, R. Neumann, D.L. Neil, and A.J. Jeffreys. 1996. Human minisatellite mutation rate after the Chernobyl accident. Nature 380:683-686.

Dubrova, Y.E., V.N. Nesterov, N.G. Krouchinsky, V.A. Ostapenko, G. Vergnaud, F. Giraudeau, J. Buard, and A.J. Jeffreys. 1997. Further evidence for elevated human minisatellite mutation rate in Belarus eight years after the Chernobyl accident. Mutat Res 381:267-278.

Dubrova, Y.E., M. Plumb, J. Brown, J. Fennelly, P. Bois, D. Goodhead, and A.J. Jeffreys. 1998a. Stage specificity, dose response, and doubling dose for mouse minisatellite germ-line mutation induced by acute radiation. Proc Natl Acad Sci USA 95:6251-6255.

Dubrova, Y.E., M. Plumb, J. Brown, and A.J. Jeffreys. 1998b. Radiation-induced germline instability at minisatellite loci. Int J Radiat Biol 74:689-696.

Dubrova, Y.E., M. Plumb, J. Brown, E. Boulton, D. Goodhead, and A.J. Jeffreys. 2000a. Induction of minisatellite mutations in the mouse germline by low-dose chronic exposure to gamma-radiation and fission neutrons. Mutat Res 453:17-24.

Dubrova, Y.E., M. Plumb, B. Gutierrez, E. Boulton, and A.J. Jeffreys. 2000b. Transgenerational mutation by radiation. Nature 405:37.

Dubrova, Y.E., R.I. Bersimbaev, L.B. Djansugurova, M.K. Tankimanova, Z.Zh. Mamyrbaeva, R. Mustonen, C. Lindholm, M. Hulten, and S. Salomaa. 2002a. Nuclear weapons tests and human germline mutation rate. Science 295:1037.

Dubrova, Y.E., G. Grant, A.A. Chumak, V.A. Stezhka, and A.N. Karakasian. 2002b. Elevated minisatellite mutation rate in the post-Chernobyl families from Ukraine. Am J Hum Genet 71:801-809.

Ducray, C., J.P. Pommier, L. Martins, F.D. Boussin, and L. Sabatier. 1999. Telomere dynamics, end-to-end fusions and telomerase activation during the human fibroblast immortalization process. Oncogene 18:4211-4223.

Duell, E.J., J.K. Wiencke, T.J. Cheng, A. Varkonyi, Z.F. Zuo, T.D. Ashok, E.J. Mark, J.C. Wain, D.C. Christiani, and K.T. Kelsey. 2000. Polymorphisms in the DNA repair genes XRCC1 and ERCC2 and biomarkers of DNA damage in human blood mononuclear cells. Carcinogenesis 21:965-971.

Duell, E.J., R.C. Millikan, G.S. Pittman, S. Winkel, R.M. Lunn, C.K. Tse, A. Eaton, H.W. Mohrenweiser, B. Newman, and D.A. Bell. 2001. Polymorphisms in the DNA repair gene XRCC1 and breast cancer. Cancer Epidemiol Biomarkers Prev 10:217-222.

Duensing, S., and K. Munger. 2001. Centrosome abnormalities, genomic instability and carcinogenic progression. Biochim Biophys Acta 1471:M81-M88.

Duensing, S., A. Duensing, C.P. Crum, and K. Munger. 2001. Human papillomavirus type 16 E7 oncoprotein-induced abnormal centrosome synthesis is an early event in the evolving malignant phenotype. Cancer Res 61:2356-2360.

Dugan, L.C., and J.S. Bedford. 2003. Are chromosomal instabilities induced by exposure of cultured normal human cells to low- or high-LET radiation? Radiat Res 159:301-311.

Dummer, T.J., H.O. Dickinson, M.S. Pearce, M.E. Charlton, J. Smith, J. Salotti, and L. Parker. 1998. Stillbirth rates around the nuclear installation at Sellafield, North West England: 1950-1989. Int J Epidemiol 27:74-82.

Duncan, K.P., and R.W. Howell. 1970. Health workers in the United Kingdom Atomic Energy Authority. Health Phys 19:285-291.

Dunning, A.M., C.S. Healey, P.D. Pharoah, M.D. Teare, B.A. Ponder, and D.F. Easton. 1999. A systematic review of genetic polymorphisms and breast cancer risk. Cancer Epidemiol Biomarkers Prev 8:843-854.

Dupree, E.A., D.L. Cragle, R.W. McLain, D.J. Crawford-Brown, and M.J. Teta. 1987. Mortality among workers at a uranium processing facility, the Linde Air Products Company Ceramics Plant, 1943-1949. Scand J Work Env Health 13:100-107.

Dupree, E.A., J.P. Watkins, J.N. Ingle, P.W. Wallace, C.M. West, and W.G. Tankersley. 1995. Uranium dust exposure and lung cancer risk in four uranium processing operations. Epidemiology 6:370-375.

Easton, D.F. 1994. Cancer risks in A-T heterozygotes. Int J Radiat Biol 66:S177-S182.

Eckardt-Schupp, F., and C. Klaus. 1999. Radiation inducible DNA repair processes in eukaryotes. Biochimie 81:161-171.

Edwards, A.A. 1992. Low Dose and Low Dose Rate Effects in Laboratory Animals, Technical Memorandum 1(92). Chilton, UK: National Radiological Protection Board.

Edwards, A.A., and R. Cox. 2000. Commentary on the Second Event Theory of Busby. Int J Radiat Biol 76:119-122.

Edwards, A.A., and J.R. Savage. 1999. Is there a simple answer to the origin of complex chromosome exchanges? Int J Radiat Biol 75:19-22.

Edwards, A.A., D.C. Lloyd, R.J. Purrott, and J.C. Prosser. 1982. The dependence of chromosome aberration yields on dose rate and radiation quality. In: Research and Development Report, 1979-1981, R&D 4. Chilton, UK: National Radiological Protection Board.

Edwards, A.A., D.C. Lloyd, and J.S. Prosser. 1989. Chromosome aberrations in human lymphocytes—a radiobiological review. Pp. 423-432 in Low Dose Radiation: Biological Bases of Risk Assessment, K.F. Baverstock and J.W. Stather, eds. London: Taylor and Francis.

Eeles, R., B. Bonder, D. Easton, and A. Horwich. 1996. Genetic Predisposition to Cancer. London: Chapman and Hall.

Ehling, U.H. 1965. The frequency of x-ray-induced dominant mutations affecting the skeleton in mice. Genetics 51:723-732.

Ehling, U.H. 1966. Dominant mutations affecting the skeleton in offspring of x-irradiated male mice. Genetics 54:1381-1389.

Ehling, U.H. 1985. Induction and manifestation of hereditary cataracts. Basic Life Sci 33:345-367.

Elkind, M.M. 1996. Guest editorial: enhanced risks of cancer from protracted exposures to x- or gamma-rays: a radiobiological model of radiation-induced breast cancer. Brit J Cancer 73:133-138.

Ellegren, H. 2000. Microsatellite mutations in the germline: implications for evolutionary inference. Trends Genet 16:551-558.

Ellender, M., S.M. Larder, and J.D. Harrison. 1997. Radiation-induced intestinal neoplasia in a genetically-predisposed mouse (*Min*). Radioprotection 32:C1-C287.

Elson, A., Y. Wang, C.J. Daugherty, C.C. Morton, F. Zhou, J. Campos-Torres, and P. Leder. 1996. Pleiotropic defects in ataxia-telangiectasia protein-deficient mice. Proc Natl Acad Sci USA 93:13084-13089.

Elston, R.C., and J. Stewart. 1971. A general model for the genetic analysis of pedigree data. Hum Hered 21:523-542.

Endlich, B., I.R. Radford, H.B. Forrester, and W.C. Dewey. 2000. Computerized video time-lapse microscopy studies of ionizing radiation-induced rapid-interphase and mitosis-related apoptosis in lymphoid cells. Radiat Res 153:36-48.

Eng, C., F.P. Li, D.H. Abramson, R.M. Ellsworth, F.L. Wong, M.B. Goldman, J. Seddon, N. Tarbell, and J.D. Boice Jr. 1993. Mortality from second tumors among long-term survivors of retinoblastoma. J Natl Cancer Inst 85:1121-1128.

Engels, H., and A. Wambersie. 1998. Relative biological effectiveness of neutrons for cancer induction and other late effects: a review of radiobiological data. Recent Res Cancer 150:54-87.

EPA (Environmental Protection Agency). 1994. Estimating Radiogenic Cancer Risks, EPA Report 402-R-93-076. Washington DC: U.S. Environmental Protection Agency.

EPA (Environmental Protection Agency). 1999. Estimating Radiogenic Cancer Risks. Addendum: Uncertainty Analysis. Washington DC: U.S. Environmental Protection Agency.

Epstein, C.J. 1995. The new dysmorphology: application of insights from basic developmental biology to the understanding of human birth defects. Proc Natl Acad Sci USA 92:8566-8573.

ESGNWJ (Epidemiological Study Group of Nuclear Workers [Japan]). 1997. First analysis of mortality of nuclear industry workers in Japan, 1986-1992. J Health Phys 32:173-184.

Falconer, D.S. 1965. The inheritance liability to certain diseases, estimated from the incidence among relatives. Ann Hum Genet 29:51-76.

Falconer, D.S. 1967. The inheritance of liability to diseases with variable age of onset, with particular reference to diabetes mellitus. Ann Hum Genet 31:1-20.

Fan, Y.J., Z. Wang, S. Sadamoto, Y. Ninomiya, N. Kotomura, K. Kamiya, K. Dohi, R. Kominami, and O. Niwa. 1995. Dose-response of a radiation induction of a germline mutation at a hypervariable mouse minisatellite locus. Int J Radiat Biol 68:177-183.

Favor, J. 1989. Risk estimation based on germ-cell mutations in animals. Genome 31:844-852.

FCRGERG (FANTOM Consortium and Riken Genome Exploration Research Group Phase I and II Teams). 2002. Analysis of the mouse transcriptome based on functional annotation of 60,770 full-length cDNAs. Nature 420:563-573.

Federal Register. 1987. Radiation protection guidance to federal agencies for occupational exposure. Washington, DC: Federal Register 52:2822-2834.

Feinberg, A.P. 1993. Genomic imprinting and gene activation in cancer. Nat Genet 4:110-113.

Feinberg, A.P. 2004. The epigenetics of cancer etiology. Semin Cancer Biol 14:427-432.

Feinberg, S.E. 1988. Method of Statistical Differentials. New York: Wiley-Interscience.

Finnon, P., D.C. Lloyd, and A.A. Edwards. 1995. Fluorescence in situ hybridization detection of chromosomal aberrations in human lymphocytes: applicability to biological dosimetry. Int J Radiat Biol 68:429-435.

Finnon, P., J.E. Moquet, A.A. Edwards, and D.C. Lloyd. 1999. The 60Co gamma ray dose-response for chromosomal aberrations in human lymphocytes analysed by FISH; applicability to biological dosimetry. Int J Radiat Biol 75:1215-1222.

Finnon, R., J. Moody, E. Meijne, J. Haines, D. Clark, A. Edwards, R. Cox, and A. Silver. 2002. A major breakpoint cluster domain in murine radiation-induced acute myeloid leukemia. Mol Carcinogen 34:64-71.

Finucane, D.M., E. Bossy-Wetzel, N.J. Waterhouse, T.G. Cotter, and D.R. Green. 1999. Bax-induced caspase activation and apoptosis via cytochrome c release from mitochondria is inhibitable by Bcl-xL. J Biol Chem 274:2225-2233.

Fishel, R., and R.D. Kolodner. 1995. Identification of mismatch repair genes and their role in the development of cancer. Curr Opin Genet Dev 5:382-395.

Fisher, B., H. Rockette, E.R. Fisher, D.L. Wickerham, C. Redmond, and A. Brown. 1985. Leukemia in breast cancer patients following adjuvant chemotherapy or postoperative radiation: the NSABP experience. J Clin Oncol 3:1640-1658.

Fisher, R.A. 1918. The correlation between relatives on the supposition of Mendelian inheritance. Trans Roy Soc Edin 52:399-433.

Fix, J.J., E.S. Gilbert, and W.V. Baumgartner. 1994. An Assessment of Bias and Uncertainty in Recorded Dose from External Sources of Radiation for Workers at the Hanford Site. PNL-10066. Richland, WA: Pacific Northwest Laboratory.

Fix, J.J., L. Salmon, G. Cowper, and E. Cardis. 1997. A retrospective evaluation of the dosimetry employed in an international combined epidemiological study. Radiat Prot Dosim 74:39-53.

Fjalling, M., L.E. Tisell, S. Carlsson, G. Hansson, L.M. Lundberg, and A. Oden. 1986. Benign and malignant thyroid nodules after neck irradiation. Cancer 58:1219-1224.

Floyd, R.A. 1995. Measurement of oxidative stress in vivo. Pp. 89-103 in The Oxygen Paradox, K.J.A. Davies and F. Ursini, eds. Padova, Italy: CLEUP University Press.

Folkman, J. 1995. Angiogenesis in cancer, vascular, rheumatoid and other disease. Nat Med 1:27-31.

Folley, J.H., W. Borges, and T. Yamasaki. 1952. Incidence of leukemia in survivors of the atom bomb in Hiroshima and Nagasaki, Japan. Am J Med 13:311-321.

Ford, H.L., and A.B. Pardee. 1999. Cancer and the cell cycle. J Cell Biochem 32-33 (Suppl):166-172.

Forrester, H.B., C.A. Vidair, N. Albright, C.C. Ling, and W.C. Dewey. 1999. Using computerized video time lapse for quantifying cell death of x-irradiated rat embryo cells transfected with c-myc or c-Ha-ras. Cancer Res 59:931-939.

Fortini, P., B. Pascucci, E. Parlanti, R.W. Sobol, S.H. Wilson, and E. Dogliotti. 1998. Different DNA polymerases are involved in the short- and long-patch base excision repair in mammalian cells. Biochemistry-US 37:3575-3580.

Fouladi, B., L. Sabatier, D. Miller, G. Pottier, and J.P. Murnane. 2000. The relationship between spontaneous telomere loss and chromosome instability in a human tumor cell line. Neoplasia 2:540-554.

Foulds, L. 1975. Neoplastic Development. New York: Acedemic Press.

Franklyn, J.A., P. Maisonneuve, M. Sheppard, J. Betteridge, and P. Boyle. 1999. Cancer incidence and mortality after radioiodine treatment for hyperthyroidism: a population-based cohort study. Lancet 353:2111-2115.

Fraser, P., L. Carpenter, N. Maconochie, C. Higgins, M. Booth, and V. Beral. 1993. Cancer mortality and morbidity in employees of the United Kingdom Atomic Energy Authority, 1946-86. Brit J Cancer 67:615-624.

Freudenreich, C.H., S.M. Kantrow, and V.A. Zakian. 1998. Expansion and length-dependent fragility of CTG repeats in yeast. Science 279:853-856.

Friedberg, W., D.N. Faulkner, L. Snyder, E.B. Darden Jr., and K. O'Brien. 1989. Galactic cosmic radiation exposure and associated health risks for air carrier crewmembers. Aviat Space Envir Md 60:1104-1108.

Frome, E.L., D. Cragle, and R. McLain. 1990. Poisson regression analysis of the mortality among a cohort of World War II nuclear industry workers. Radiat Res 123:138-152.

Frome, E.L., D.L. Cragle, J.P. Watkins, S. Wing, C.M. Shy, W.G. Tankersley, and C.M. West. 1997. A mortality study of employees of the nuclear industry in Oak Ridge, Tennessee. Radiat Res 148:64-80.

Fry, R.J., J.B. Storer, and F.J. Burns. 1986. Radiation induction of cancer of the skin. Brit J Radiol 19(Suppl):58-60.

Fu, Y.P., J.C. Yu, T.C. Cheng, M.A. Lou, G.C. Hsu, C.Y. Wu, S.T. Chen, H.S. Wu, P.E. Wu, and C.Y. Shen. 2003. Breast cancer risk associated with genotypic polymorphism of the nonhomologous end-joining genes: a multigenic study on cancer susceptibility. Cancer Res 63:2440-2446.

Fuchs, C.S., and R.J. Mayer. 1995. Gastric carcinoma. N Engl J Med 333:32-41.

Fujiwara, S., R. Sposto, H. Ezaki, S. Akiba, K. Neriishi, K. Kodama, Y. Hosoda, and K. Shimaoka. 1992. Hyperparathyroidism among atomic bomb survivors in Hiroshima. Radiat Res 130:372-378.

Fujiwara, S., S. Kusumi, J. Cologne, M. Akahoshi, K. Kodama, and H. Yoshizawa. 2000. Prevalence of anti-hepatitis C virus antibody and chronic liver disease among atomic bomb survivors. Radiat Res 154: 12-19.

Fukasawa, K., F. Wiener, G.F. Vande Woude, and S. Mai. 1997. Genomic instability and apoptosis are frequent in p53 deficient young mice. Oncogene 15:1295-1302.

Furre, T., M. Koritzinsky, D.R. Olsen, and E.O. Pettersen. 1999. Inverse dose-rate effect due to pre-mitotic accumulation during continuous low dose-rate irradiation of cervix carcinoma cells. Int J Radiat Biol 75: 699-707.

Gajewski, E., G. Rao, Z. Nackerdien, and M. Dizdaroglu. 1990. Modification of DNA bases in mammalian chromatin by radiation-generated free radicals. Biochemistry 29:7876-7882.

Game, J.C., G.W. Birrell, J.A. Brown, T. Shibata, C. Baccari, A.M. Chu, M.S. Williamson, and J.M. Brown. 2003. Use of a genome-wide approach to identify new genes that control resistance of *Saccharomyces cerevisiae* to ionizing radiation. Radiat Res 160:14-24.

Gardner M.J., A.J. Hall, S. Downes, and J.D. Terrell. 1987. Follow up study of children born to mothers resident in Seascale, West Cumbria (birth cohort). Brit Med J 295:822-827.

Gardner, M.J., A.J. Hall, M.P. Snee, S. Downes, C.A. Powell, and J.D. Terrell. 1990a. Methods and basic data of case-control study of leukaemia and lymphoma among young people near Sellafield nuclear plant in West Cumbria. Brit Med J 300:429-434.

Gardner, M.J., M.P. Snee, A.J. Hall, C.A. Powell, S. Downes, and J.D. Terrell. 1990b. Results of case-control study of leukaemia and lymphoma among young people near Sellafield nuclear plant in West Cumbria. Brit Med J 300:423-429.

Gellon, L., R. Barbey, P. Auffret van der Kemp, D. Thomas, and S. Boiteux. 2001. Synergism between base excision repair, mediated by the DNA glycosylases Ntg1 and Ntg2, and nucleotide excision repair in the removal of oxidatively damaged DNA bases in *Saccharomyces cerevisiae*. Mol Genet Genomics 265:1087-1096.

Geoffroy-Perez, B., N. Janin, K. Ossian, A. Lauge, M.F. Croquette, C. Griscelli, M. Debre, B. Bressac-de-Paillerets, A. Aurias, D. Stoppa-Lyonnet, and N. Andrieu. 2001. Cancer risk in heterozygotes for ataxia-telangiectasia. Int J Cancer 93:288-293.

Gibbs, M., A. Collick, R.G. Kelly, and A.J. Jeffreys. 1993. A tetranucleotide repeat mouse minisatellite displaying substantial somatic instability during early preimplantation development. Genomics 17:121-128.

Gibbs, R.A., J. Camakaris, G.S. Hodgson, and R.F. Martin. 1987. Molecular characterization of ^{125}I decay and X-ray-induced HPRT mutants in CHO cells. Int J Radiat Biol Re 51:193-199.

Gilbert, E.S. 1989. Issues in analysing the effects of occupational exposure to low levels of radiation. Stat Med 8:173-187.

Gilbert, E.S. 1991. Chapter 3: Late somatic effects. In Health Effects Models for Nuclear Power Plant Accident Consequence Analysis. Modifications of Models Resulting from Recent Reports on Health Effects of Ionizing Radiation, Low LET Radiation, Part II: Scientific Bases for Health Effects Models, S. Abrahamson and others, eds. NUREG/CR-4214, Rev. 1, Part II, Addendum 1, LMF-132.

Gilbert, E.S. 1998. Accounting for errors in dose estimates used in studies of workers exposed to external radiation. Health Phys 74:22-29.

Gilbert, E.S., and J.J. Fix. 1995. Accounting for bias in dose estimates in analyses of data from nuclear worker mortality studies. Health Phys 68:650-660.

Gilbert, E.S., S.A. Fry, L.D. Wiggs, G.L. Voelz, D.L. Cragle, and G.R. Petersen. 1989. Analyses of combined mortality data on workers at the Hanford Site, Oak Ridge National Laboratory, and Rocky Flats Nuclear Weapons Plant. Radiat Res 120:19-35.

Gilbert, E.S., D.L. Cragle, and L.D. Wiggs. 1993a. Updated analyses of combined mortality data for workers at the Hanford Site, Oak Ridge National Laboratory, and Rocky Flats Weapons Plant. Radiat Res 136:408-421.

Gilbert, E.S., E. Omohundro, J. Buchanna, and N. Holter. 1993b. Mortality of workers at the Hanford Site: 1945-1986. Health Phys 64:577-590.

Gilbert, E.S., N.A. Koshurnikova, M. Sokolnikov, V.F. Khokhryakov, S. Miller, D.L. Preston, S.A. Romanov, N.S. Shilnikova, K.G. Suslova, and V.V. Vostrotin. 2000. Liver cancers in Mayak workers. Radiat Res 154:246-252.

Gilbert, E.S., M. Stovall, M. Gospodarowicz, F.E. Van Leeuwen, M. Andersson, B. Glimelius, T. Joensuu, C.F. Lynch, R.E. Curtis, E. Holowaty, H. Storm, E. Pukkala, M.B. van't Veer, J.F. Fraumeni, J.D. Boice Jr., E.A. Clarke, and L.B. Travis. 2003. Lung cancer after treatment for Hodgkin's disease: focus on radiation effects. Radiat Res 159:161-173.

Gilman, E.G., A.M. Stewart, E.G. Knox, and G.W. Kneale. 1989. Trends in obstetric radiography, 1939-81. J Radiol Prot 9:93-101.

Gimm, O., H. Dziema, J. Brown, C. Hoang-Vu, R. Hinze, H. Dralle, L.M. Mulligan, and C. Eng. 2001. Over-representation of a germline variant in the gene encoding RET co-receptor GFRalpha-1 but not GFRalpha-2 or GFRalpha-3 in cases with sporadic medullary thyroid carcinoma. Oncogene 20:2161-2170.

Gisselsson, D., L. Pettersson, M. Hoglund, M. Heidenblad, L. Gorunova, J. Wiegant, F. Mertens, P. Dal Cin, F. Mitelman, and N. Mandahl. 2000. Chromosomal breakage-fusion-bridge events cause genetic intratumor heterogeneity. Proc Natl Acad Sci USA 97:5357-5362.

Goldman, M.B., F. Maloof, R.R. Monson, A. Aschengrau, D. Cooper, and E. Ridgway. 1988. Radioactive iodine therapy and breast cancer: a follow-up study of hyperthyroid women. Am J Epidemiol 127:969-980.

Goldstone, A.R., P. Quirke, and M.F. Dixon. 1996. *Helicobacter pylori* infection and gastric cancer. J Pathol 179:129-137.

Goode, E.L., C.M. Ulrich, and J.D. Potter. 2002. Polymorphisms in DNA repair genes and associations with cancer risk. Cancer Epidemiol Biomarkers Prev 11:1513-1530.

Goodhead, D.T. 1994. Initial events in the cellular effects of ionizing radiations: clustered damage in DNA. Int J Radiat Biol 65:7-17.

Gorman, M.A., S. Morera, D.G. Rothwell, E. de La Fortelle, C.D. Mol, J.A. Tainer, I.D. Hickson, and P.S. Freemont. 1997. The crystal structure of the human DNA repair endonuclease HAP1 suggests the recognition of extra-helical deoxyribose at DNA abasic sites. EMBO J 16:6548-6558.

Gowen, L.C., A.V. Avrutskaya, A.M. Latour, B.H. Koller, and S.A. Leadon. 1998. BRCA1 required for transcription-coupled repair of oxidative DNA damage. Science 281:1009-1012.

Greaves, M.F., and J. Wiemels. 2003. Origins of chromosome translocations in childhood leukaemia. Nat Rev Cancer 3:639-649.

Green, P.M., S. Saad, C.M. Lewis, and F. Giannelli. 1999. Mutation rates in humans. I. Overall and sex-specific rates obtained from a population study of hemophilia B. Am J Hum Genet 65:1572-1579.

Greenberg, E.R., B. Rosner, C. Hennekens, R. Rinsky, and T. Colton. 1985. An investigation of bias in a study of nuclear shipyard workers. Am J Epidemiol 121:301-308.

Greenblatt, M.S., W.P. Bennett, M. Hollstein, and C.C. Harris. 1994. Mutations in the p53 tumor suppressor gene: clues to cancer etiology and molecular pathogenesis. Cancer Res 54:4855-4878.

Greider, C.W. 1996. Telomere length regulation. Annu Rev Biochem 65:337-365.

Greinert, R., E. Detzler, and D. Harder. 2000. The kinetics of postirradiation chromatin restitution as revealed by chromosome aberrations detected by premature chromosome condensation and fluorescence in situ hybridization. Radiat Res 154:87-93.

Gribbin, M.A., J.L. Weeks, and G.R. Howe. 1993. Cancer mortality (1956-1985) among male employees of Atomic Energy of Canada Limited with respect to occupational exposure to external low-linear-energy-transfer ionizing radiation. Radiat Res 133:375-380.

Griem, M.L., R.A. Kleinerman, J.D. Boice, M. Stovall, D. Shefner, and J.H. Lubin. 1994. Cancer following radiotherapy for peptic ulcer. J Natl Cancer Inst 86:842-849.

Griffin, C.S., S.J. Marsden, D.L. Stevens, P. Simpson, and J.R. Savage. 1995. Frequencies of complex chromosome exchange aberrations induced by ^{238}Pu alpha-particles and detected by fluorescence in situ hybridization using single chromosome-specific probes. Int J Radiat Biol 67:431-439.

Griffin, C.S., A. Neshasateh-Riz, R.J. Mairs, E.G. Wright, and T.E. Wheldon. 2000. Absence of delayed chromosomal instability in a normal human fibroblast cell line after ^{125}I iododeoxyuridine. Int J Radiat Biol 76:963-969.

Grosovsky, A.J. 1999. Radiation-induced mutations in unirradiated DNA. Proc Natl Acad Sci USA 96:5346-5347.

Grosovsky, A.J., and J.B. Little. 1985. Evidence for linear response for the induction of mutations in human cells by x-ray exposures below 10 rads. Proc Natl Acad Sci USA 82:2092-2095.

Grosovsky, A.J., K.K. Parks, C.R. Giver, and S.L. Nelson. 1996. Clonal analysis of delayed karyotypic abnormalities and gene mutations in radiation-induced genetic instability. Mol Cell Biol 16:6252-6262.

Guerin, S., A. Dupuy, H. Anderson, A. Shamsaldin, G. Svahn-Tapper, T. Moller, E. Quiniou, S. Garwicz, M. Hawkins, M.F. Avril, O. Oberlin, J. Chavaudra, and F. de Vathaire. 2003. Radiation dose as a risk factor for malignant melanoma following childhood cancer. Eur J Cancer 39:2379-2386.

Guizard, A.V., O. Boutou, D. Pottier, X. Troussard, D. Pheby, G. Launoy, R. Slama, and A. Spira. 2001. The incidence of childhood leukaemia around the La Hague nuclear waste reprocessing plant (France): a survey for the years 1978-1998. J Epidemiol Commun H 55:469-474.

Gulis, G. 2003. Cancer occurrence among radiation workers at Jaslovske Bohunice nuclear power plant. Cent Eur J Public Health 11:91-97.

Gulis, G., and O. Fitz. 1998. Cancer incidence around the nuclear power plant Jaslovske Bohunice. Cent Eur J Public Health 6:183-187.

Gunay, U., A. Meral, and B. Sevinir. 1996. Pediatric malignancies in Bursa, Turkey. J Environ Pathol Toxicol Oncol 15:263-265.

Guo, M., C. Chen, C. Vidair, S. Marino, W.C. Dewey, and C.C. Ling. 1997. Characterization of radiation-induced apoptosis in rodent cell lines. Radiat Res 147:295-303.

Gusev, B.I., Z.N. Abylkassimova, and K.N. Apsalikov. 1997. The Semipalatinsk nuclear test site: a first assessment of the radiological situation and the test-related radiation doses in the surrounding territories. Radiat Environ Bioph 36:201-204.

Haber, J.E. 1998. The many interfaces of Mre11. Cell 95:583-586.

Hadjimichael, O.C., A.M. Ostfeld, D.A. D'Atri, and R.E. Brubaker. 1983. Mortality and cancer incidence experience of employees in a nuclear fuels fabrication plant. J Occup Med 25:48-61.

Hahn, H., L. Wojnowski, A.M. Zimmer, J. Hall, G. Miller, and A. Zimmer. 1998. Rhabdomyosarcomas and radiation hypersensitivity in a mouse model of Gorlin syndrome. Nat Med 4:619-622.

Haimovitz-Friedman, A. 1998. Radiation-induced signal transduction and stress response. Radiat Res 150:S102-S108.

Hain, J., R. Jaussi, and W. Burkart. 1992. Lack of adaptive response to low doses of ionizing radiation in human lymphocytes from five different donors. Mutat Res 283:137-144.

Haines, J., R. Dunford, J. Moody, M. Ellender, R. Cox, and A. Silver. 2000. Loss of heterozygosity in spontaneous and x-ray-induced intestinal tumors arising in F_1 hybrid min mice: evidence for sequential loss of apc(+) and dpc4 in tumor development. Gene Chromosome Canc 28:387-394.

Hall, E.J. 2000. Radiation, the two-edged sword: cancer risks at high and low doses. Cancer J 6:343-350.

Hall, P., L.E. Holm, G. Lundell, G. Bjelkengren, L.G. Larsson, S. Lindberg, J. Tennvall, H. Wicklund, and J.D. Boice Jr. 1991. Cancer risks in thyroid cancer patients. Brit J Cancer 64:159-163.

Hall, P., J.D. Boice Jr., G. Berg, G. Bjelkengren, U.B. Ericsson, A. Hallquist, M. Lidberg, G. Lundell, A. Mattsson, J. Tennvall, K. Wiklund, and L.E. Holm. 1992. Leukaemia incidence after iodine-131 exposure. Lancet 340:1-4.

Hall, P., A. Mattsson, and J.D. Boice. 1996. Thyroid cancer after diagnostic administration of iodine-131. Radiat Res 145:86-92.

Hamilton, T.E., G. van Belle, and J.P. LoGerfo. 1987. Thyroid neoplasia in Marshall Islanders exposed to nuclear fallout. J Am Med Assoc 258:629-635.

Hammer, G.P., H. Zeeb, U. Tveten, and M. Blettner. 2000. Comparing different methods of estimating cosmic radiation exposure of airline personnel. Radiat Environ Bioph 39:227-231.

Han, A., C.K. Hill, and M.M. Elkind. 1980. Repair of cell killing and neoplastic transformation at reduced dose rates of ^{60}Co gamma-rays. Cancer Res 40:3328-3332.

Hancock, S.L., R.S. Cox, and I.R. McDougall. 1991. Thyroid diseases after treatment of Hodgkin's disease. N Engl J Med 325:599-605.

Hancock, S.L., M.A. Tucker, and R.T. Hoppe. 1993a. Breast cancer after treatment of Hodgkin's disease. J Natl Cancer Inst 85:25-31.

Hancock, S.L., M.A. Tucker, and R.T. Hoppe. 1993b. Factors affecting late mortality from heart disease after treatment of Hodgkin's disease. J Am Med Assoc 270:1949-1955.

Hancock, S.L., S.S. Donaldson, and R.T. Hoppe. 1993c. Cardiac disease following treatment of Hodgkin's disease in children and adolescents. J Clin Oncol 11:1209-1215.

Hanel, G., B. Gstir, S. Denifl, P. Scheier, M. Probst, B. Farizon, M. Farizon, E. Illenberger, and T.D. Mark. 2003. Electron attachment to uracil: effective destruction at subexcitation energies. Phys Rev Lett 90:188104.

Haran-ghera, N. 1976. Pathways in murine radiation leukemogenesis-coleukemogeneisis. Pp. 245-260 in Biology of Radiation Carcinogenesis, J.M. Yuhas, J.D. Regan, and R.W. Tennant, eds. New York: Raven Press.

Harley, C.B., and B. Villeponteau. 1995. Telomeres and telomerase in aging and cancer. Curr Opin Genet Dev 5:249-255.

Hartwell, L. 1992. Defects in a cell cycle checkpoint may be responsible for the genomic instability of cancer cells. Cell 71:543-546.

Hartwell, L., T. Weinert, L. Kadyk, and B. Garvik. 1994. Cell cycle checkpoints, genomic integrity, and cancer. Cold Spring Harb Symp Quant Biol 59:259-263.

Harvey, D.M., and A.J. Levine. 1991. p53 alteration is a common event in the spontaneous immortalization of primary BALB/c murine embryo fibroblasts. Genes Dev 5:2375-2385.

Hatch, M. 1992. Childhood leukemia around nuclear facilities. Sci Total Environ 127:37-42.

Hatch, M., and M. Susser. 1990. Background gamma radiation and childhood cancers within ten miles of a US nuclear plant. Int J Epidemiol 19:546-552.

Hatch, M.C., J. Beyea, J.W. Nieves, and M. Susser. 1990. Cancer near the Three Mile Island Nuclear Plant: radiation emissions. Am J Epidemiol 132:397-412.

Hawkins, M.M. 1990. Second primary tumors following radiotherapy for childhood cancer. Int J Radiat Oncol 19:1297-1301.

Hawkins, M.M., G.J. Draper, and J.E. Kingston. 1987. Incidence of second primary tumours among childhood cancer survivors. Brit J Cancer 56:339-347.

Hawkins, M.M., L.M. Wilson, M.A. Stovall, H.B. Marsden, M.H. Potok, J.E. Kingston, and J.M. Chessells. 1992. Epipodophyllotoxins, alkylating agents, and radiation and risk of secondary leukaemia after childhood cancer. Brit Med J 304:951-958.

Hawkins, M.M., L.M. Wilson, H.S. Burton, M.H. Potok, D.L. Winter, H.B. Marsden, and M.A. Stovall. 1996. Radiotherapy, alkylating agents, and risk of bone cancer after childhood cancer. J Natl Cancer Inst 88:270-278.

Hayata, I., M. Seki, K. Yoshida, K. Hirashima, T. Sado, J. Yamagiwa, and T. Ishihara. 1983. Chromosomal aberrations observed in 52 mouse myeloid leukemias. Cancer Res 43:367-373.

Hazelton, W.D., E.G. Luebeck, W.F. Heidenreich, and S.H. Moolgavkar. 2001. Analysis of a historical cohort of Chinese tin miners with arsenic, radon, cigarette smoke, and pipe smoke exposures using the biologically based two-stage clonal expansion model. Radiat Res 156:78-94.

Heidenreich, P.A., S.L. Hancock, B.K. Lee, C.S. Mariscal, and I. Schnittger. 2003. Asymptomatic cardiac disease following mediastinal irradiation. J Am Coll Cardiol 42:743-749.

Heidenreich, W.F., and H.G. Paretzke. 2001. The two-stage clonal expansion model as an example of a biologically based model of radiation-induced cancer. Radiat Res 156:678-681.

Heidenreich, W., P. Jacob, and H.G. Paretzke. 1997. Exact solutions of the clonal expansion model and their application to the incidence of solid tumors of the atomic bomb survivors. Radiat Environ Bioph 36:45-58.

Heidenreich, W.F., P. Jacob, H.G. Paretzke, F.T. Cross, and G.E. Dagle. 1999. Two-step model for the risk of fatal and incidental lung tumors in rats exposed to radon. Radiat Res 151:209-217.

Heidenreich, W.F., T.I. Bogdanova, P. Jacob, A.G. Biryukov, and N.D. Tronko. 2000. Age and time patterns in thyroid cancer after the Chernobyl accidents in the Ukraine. Radiat Res 154:731-732; discussion 734-735.

Heidenreich, W.F., E.G. Luebeck, W.D. Hazelton, H.G. Paretzke, and S.H. Moolgavkar. 2002a. Multistage models and the incidence of cancer in the cohort of atomic bomb survivors. Radiat Res 158:607-614.

Heidenreich, W.F., J. Wellmann, P. Jacob, and H.E. Wichmann. 2002b. Mechanistic modelling in large case-control studies of lung cancer risk from smoking. Stat Med 21:3055-3070.

Hempelmann, L.H., W.J. Hall, M. Phillips, R.A. Cooper, and W.R. Ames. 1975. Neoplasms in persons treated with x-rays in infancy: fourth survey in 20 years. J Natl Cancer Inst 55:519-530.

Henderson, S.D., B.F. Kimler, and M.F. Scanlan. 1982. Interaction of hyperthermia and radiation on the survival of synchronous 9L cells. Radiat Res 92:146-159.

Herrlich, P., H. Ponta, and H.J. Rahmsdorf. 1992. DNA damage-induced gene expression: signal transduction and relation to growth factor signaling. Rev Physiol Biochem Pharmacol 119:187-223.

Heyer, B.S., A. MacAuley, O. Behrendtsen, and Z. Werb. 2000. Hypersensitivity to DNA damage leads to increased apoptosis during early mouse development. Genes Dev 4:2072-2084.

Heyn, R., V. Haeberlen, W.A. Newton, A.H. Ragab, R.B. Raney, M. Tefft, M. Wharam, L.G. Ensign, and H.M. Maurer. 1993. Second malignant neoplasms in children treated for rhabdomyosarcoma. Intergroup Rhabdomyosarcoma Study Committee. J Clin Oncol 11:262-270.

Hickman, A.W., R.J. Jaramillo, J.F. Lechner, and N.F. Johnson. 1994. Alpha-particle-induced p53 protein expression in a rat lung epithelial cell strain. Cancer Res 54:5797-5800.

Hildreth, N.G., R.E. Shore, L.H. Hempelmann, and M. Rosenstein. 1985. Risk of extrathyroid tumors following radiation treatment in infancy for thymic enlargement. Radiat Res 102:378-391.

Hildreth, N.G., R.E. Shore, and P.M. Dvoretsky. 1989. The risk of breast cancer after irradiation of the thymus in infancy. N Engl J Med 321:1281-1284.

Hill, A.B. 1966. Principles of Medical Statistics. New York: Oxford University Press.

Hill, C.K., F.M. Buonaguro, C.P. Myers, A. Han, and M.M. Elkind. 1982. Fission-spectrum neutrons at reduced dose rates enhance neoplastic transformation. Nature 298:67-69.

Hill, C.K., A. Han, and M.M. Elkind. 1984. Fission-spectrum neutrons at a low dose rate enhance neoplastic transformation in the linear, low dose region (0-10 cGy). Int J Radiat Biol Re 46:11-15.

Hill, R.L., and J.R. Johnson. 1993. Metabolism and dosimetry of tritium. Health Phys 65:628-647.

Hinds, P.W., and R.A. Weinberg. 1994. Tumor suppressor genes. Curr Opin Genet Dev 4:135-141.

Hino, O., A.J. Klein-Szanto, J.J. Freed, J.R. Testa, D.Q. Brown, M. Vilensky, R.S. Yeung, K.D. Tartof, and A.G. Knudson. 1993. Spontaneous and radiation-induced renal tumors in the Eker rat model of dominantly inherited cancer. Proc Natl Acad Sci USA 90:327-331.

Hino, O., H. Mitani, and J. Sakaurai. 2002. "Second hit" of Tsc2 gene in radiation induced renal tumours of Eker rat model. Pp. 175-178 in Radiation and Hoomeostasis, T. Sugahara and others, eds. Amsterdam: Elsevier.

Hjalmars, U., M. Kulldorff, and G. Gustafsson. 1994. Risk of acute childhood leukaemia in Sweden after the Chernobyl reactor accident. Swedish Child Leukaemia Group. Brit Med J 309:154-157.

Hlatky, L., R. Sachs, and P. Hahnfeldt. 1991. Reaction kinetics for the development of radiation-induced chromosome aberrations. Int J Radiat Biol 59:1147-1172.

Hoeijmakers, J.H. 2001. Genome maintenance mechanisms for preventing cancer. Nature 411:366-374.

Hoel, D.G., and P. Li. 1998. Threshold models in radiation carcinogenesis. Health Phys 75:241-250.

Hoffman, D.A. 1984. Late effects of I-131 therapy in the United States. Pp. 273-280 in Radiation Carcinogenesis: Epidemiology and Biological Significance, J.D. Boice Jr. and J.F. Fraumeni Jr., eds. New York: Raven Press.

Holley, W.R., and A. Chatterjee. 1996. Clusters of DNA damage induced by ionizing radiation: formation of short DNA fragments. 1. Theoretical modelling. Radiat Res 145:188-199.

Hollowell, J.G., Jr., and L.G. Littlefield. 1968. Chromosome damage induced by plasma of x-rayed patients: an indirect effect of x ray. Proc Soc Exp Biol Med 129:240-244.

Holm, L.E. 1984. Malignant disease following iodine-131 therapy in Sweden. Pp. 263-271 in J.D. Boice Jr. and J.F. Fraumeni Jr., eds. New York: Raven Press.

Holm, L.E., I. Dahlqvist, A. Israelsson, and G. Lundell. 1980a. Malignant thyroid tumors after iodine-131 therapy: a retrospective cohort study. N Engl J Med 303:188-191.

Holm L.E., G. Lundell, and G. Wallinder. 1980b. Incidence of malignant thyroid tumors in humans after exposure to diagnostic doses of iodine-131. I. Retrospective cohort study. J Natl Cancer Inst 64:1055-1059.

Holm, L.E., K.E. Wiklund, G.E. Lundell, N.A. Bergman, G. Bjelkengren, E.S. Cederquist, U.B. Ericsson, L.G. Larsson, M.E. Lindberg, R.S. Lindberg, H.V. Wicklund, and J.D. Boice Jr. 1988. Thyroid cancer after diagnostic doses of iodine-131: a retrospective cohort study. J Natl Cancer Inst 80:1132-1138.

Holm, L.E., K.E. Wiklund, G.E. Lundell, N.A. Bergman, G. Bjelkengren, U.B. Ericsson, E.S. Cederquist, M.E. Lidberg, R.S. Lindberg, and H.V. Wicklund. 1989. Cancer risk in population examined with diagnostic doses of ^{131}I. J Natl Cancer Inst 81:302-306.

Holm, L.E., P. Hall, K. Wiklund, G. Lundell, G. Berg, G. Bjelkengren, E. Cederquist, U.B. Ericsson, A. Hallquist, and L.G. Larsson. 1991. Cancer risk after iodine-131 therapy for hyperthyroidism. J Natl Cancer Inst 83:1072-1077.

Holmberg, K., A.E. Meijer, M. Harms-Ringdahl, and B. Lambert. 1998. Chromosomal instability in human lymphocytes after low dose rate gamma-irradiation and delayed mitogen stimulation. Int J Radiat Biol 73:21-34.

Hoover, R.N. 2000. Cancer—nature, nurture, or both. N Engl J Med 343:135-136.

Hoshi, Y., H. Tanooka, K. Miyazaki, and H. Wakasugi. 1997. Induction of thioredoxin in human lymphocytes with low-dose ionizing radiation. Biochim Biophys Acta 1359:65-70.

Hoshino, H., and H. Tanooka. 1975. Interval effect of beta-irradiation and subsequent 4-nitroquinoline 1-oxide painting on skin tumor induction in mice. Cancer Res 35:3663-3666.

Host, H., and M. Loeb. 1986. Post-operative radiotherapy in breast cancer: long-term results from the Oslo study. Int J Radiat Oncol 12:727-732.

Houlston, R.S., and I.P. Tomlinson. 2000. Detecting low penetrance genes in cancer: the way ahead. J Med Genet 37:161-167.

Houlston, R.S., and I.P. Tomlinson. 2001. Polymorphisms and colorectal tumor risk. Gastroenterology 121:282-301.

Howe, G.R. 1995. Lung cancer mortality between 1950 and 1987 after exposure to fractionated moderate-dose-rate ionizing radiation in the Canadian fluoroscopy cohort study and a comparison with lung cancer mortality in the atomic bomb survivors study. Radiat Res 142:295-304.

Howe, G.R., and J. McLaughlin. 1996. Breast cancer mortality between 1950 and 1987 after exposure to fractionated moderate-dose-rate ionizing radiation in the Canadian fluoroscopy cohort study and a comparison with breast cancer mortality in the atomic bomb survivors study. Radiat Res 145:694-707.

Howe, G.R., J.L. Weeks, A.B. Miller, A.M. Chiarelli, and J. Etezadi-Amoli. 1987. AECL-9442, Open Literature Report.

Howe, G.R., A.M. Chiarelli, and J.P. Lindsay. 1988. Components and modifiers of the healthy worker effect: evidence from three occupational cohorts and implications for industrial compensation. Am J Epidemiol 128:1364-1375.

Howe, G.R., L.B. Zablotska, J.J. Fix, J. Egel, and J. Buchanan. 2004. Analysis of the mortality experience amongst U.S. nuclear power industry workers following chronic low-dose exposure to ionizing radiation. Radiat Res 162:517-526.

Hrubec, Z., J.D. Boice Jr., R.R. Monson, and M. Rosenstein. 1989. Breast cancer after multiple chest fluoroscopies: second follow-up of Massachusetts women with tuberculosis. Cancer Res 49:229-234.

Hu, J.J., T.R. Smith, M.S. Miller, H.W. Mohrenweiser, A. Golden, and L.D. Case. 2001. Amino acid substitution variants of APE1 and XRCC1 genes associated with ionizing radiation sensitivity. Carcinogenesis 22:917-922.

Hut, H.M.J., W. Lemstra, E.H. Blaauw, G.W.A. van Cappellen, H.H. Kampinga, and O.C.M. Sibon. 2003. Centrosomes split in the presence of impaired DNA integrity during mitosis. Mol Biol of the Cell 14:1993-2004.

Hutchinson, F. 1985. Chemical changes induced in DNA by ionizing radiation. Prog Nucleic Acid Res 32:115-154.

IARC (International Agency for Research on Cancer). 1994. Direct estimates of cancer mortality due to low doses of ionising radiation: an international study. IARC Study Group on Cancer Risk among Nuclear Industry Workers. Lancet 344:1039-1043.

IARC (International Agency for Research on Cancer). 1995. Combined Analyses of Cancer Mortality Among Nuclear Industry Workers in Canada, the United Kingdom and the United States of America. Lyon, France: IARC Press.

IARC (International Agency for Research on Cancer). 2000. Ionizing Radiation, Part I: X- and Gamma (γ) Radiation and Neutrons. IARC Monographs ,Vol. 75. Lyon, France: IARC Press.

IARC (International Agency for Research on Cancer). 2002. Cancer Incidence in Five Continents, Vol. VIII, D.M. Parkin and others, eds. Lyon, France: IARC Press.

IARC (International Agency for Research on Cancer). 2003. World Cancer Report, B.W. Stewart and P.E. Kleihues, eds. Lyon, France: IARC Press.

IARC (International Agency for Research on Cancer). 2004. Monographs on the Evaluation of Carcinogenic Risks to Humans. IARC Monographs Vol. 83. Tobacco Smoke and Involuntary Smoking. Lyon, France: IARC Press.

Ichimaru, M., T. Ishimaru, and J.L. Belsky. 1978. Incidence of leukemia in atomic bomb survivors belonging to a fixed cohort in Hiroshima and Nagasaki, 1950-71. Radiation dose, years after exposure, age at exposure, and type of leukemia. J Radiat Res 19:262-282.

ICRP (International Commission on Radiological Protection). 1990. ICRP Publication 60: 1990 Recommendations of the ICRP. Oxford, UK: Pergamon, Elsevier Science.

ICRP (International Commission on Radiological Protection). 1998. ICRP Publication 79: Genetic susceptibility to cancer. Ann ICRP 28:1-157.

ICRP (International Commission on Radiological Protection). 1999. Risk estimation for multifactorial diseases. A report of the International Commission on Radiological Protection. Ann ICRP 29:1-144.

ICRP (International Commission on Radiological Protection). 2003. ICRP Publication 92: Relative biological effectiveness (RBE), quality factor (QF), and radiation weighting factor (WR). A report of the International Commission on Radiological Protection. Ann ICRP 33:1-117.

ICRU (International Commission on Radiation Units and Measurements). 1970. Linear Energy Transfer. Washington, DC: ICRU Publications.

ICRU (International Commission on Radiation Units and Measurements). 1986. The Quality Factor in Radiation Protection. ICRU Report 40. Bethesda, MD.

Ikushima, T., H. Aritomi, and J. Morisita. 1996. Radioadaptive response: efficient repair of radiation-induced DNA damage in adapted cells. Mutat Res 358:193-198.

Inoue, M., G.P. Shen, M.A. Chaudhry, H. Galick, J.O. Blaisdell, and S.S. Wallace. 2004. Expression of the oxidative base excision repair enzymes is not induced in TK6 human lymphoblastoid cells after low doses of ionizing radiation. Radiat Res 161:409-417.

Inskip, H., V. Beral, P. Fraser, M. Booth, D. Coleman, and A. Brown. 1987. Further assessment of the effects of occupational radiation exposure in the United Kingdom Atomic Energy Authority mortality study. Brit J Ind Med 44:149-160.

Inskip, P.D., R.R. Monson, J.K. Wagoner, M. Stovall, F.G. Davis, R.A. Kleinerman, and J.D. Boice. 1990a. Cancer mortality following radium treatment for uterine bleeding. Radiat Res 123:331-344.

Inskip, P.D., R.R. Monson, J.K. Wagoner, M. Stovall, F.G. Davis, R.A. Kleinerman, and J.D. Boice. 1990b. Leukemia following radiotherapy for uterine bleeding. Radiat Res 122:107-119.

Inskip, P.D., R.A. Kleinerman, M. Stovall, D.L. Cookfair, O. Hadjimichael, W.C. Moloney, R.R. Monson, W.D. Thompson, J. Wactawski-Wende, J.K. Wagoner, and J.D. Boice. 1993. Leukemia, lymphoma, and multiple myeloma after pelvic radiotherapy for benign disease. Radiat Res 135:108-124.

Inskip, P.D., M. Stovall, and J. Flannery. 1994. Lung cancer risk and radiation dose among women treated for breast cancer. J Natl Cancer Inst 86:983-988.

Inskip, P.D., A. Ekbom, M.R. Galanti, L. Grimelius, and J.D. Boice. 1995. Medical diagnostic x-rays and thyroid cancer. J Natl Cancer Inst 87:1613-1621.

Ishii, K., Y. Hosoi, S. Yamada, T. Ono, and K. Sakamoto. 1996. Decreased incidence of thymic lymphoma in AKR mice as a result of chronic, fractionated low-dose total-body x irradiation. Radiat Res 146:582-585.

Ishii, K., Y. Hoshi, T. Iwasaki, and M. Watanabe. 1997. Participation of intracellular communication and intracellular signal transduction in radio-adaptive response of human fibroblastic cells. Pp. 410-413 in Low Doses of Ionizing Radiation: Biological Effects and Regulatory Control. IAEA-CN-67/128.

Issa, J.P., and S.B. Baylin. 1996. Epigenetics and human disease. Nat Med 2:281-282.

Ivanov, E., G. Tolochko, L. Shuvaeva, S. Becker, E. Nekolla, and A. Kellerer. 1996. Childhood leukemia in Belarus before and after the Chernobyl accident. Radiat Environ Bioph 35:75-80.

Ivanov, E.P., G. Tolochko, V.S. Lazarev, and L. Shuvaeva. 1993. Child leukaemia after Chernobyl. Nature 365:702.

Ivanov, V.K., and A.F. Tsyb. 1996. National Chernobyl Registry: radiation risk analysis. Proceedings of the EC/WHO/IAEA International Conference: One Decade after Chernobyl. Vienna.

Ivanov, V.K., E.V. Nilova, V.A. Efendiev, A.I. Gorskii, V.A. Pishkevich, S.I. Leshakov, and V.I. Shiriaev. 1997a. Oncoepidemiologic situation in the Kaluga region of the Russian Federation ten years after the Chernobyl accident. Voprosy Onkologii 43:143-150.

Ivanov, V.K., A.F. Tsyb, E.V. Nilova, V.F. Efendiev, A.I. Gorsky, V.A. Pitkevich, S. Leshakov, and V.I. Shiryaev. 1997b. Cancer risks in the Kaluga Oblast of the Russian Federation 10 years after the Chernobyl accident. Radiat Environ Bioph 36:161-167.

Ivanov, V.K., E.M. Rastopchin, A.I. Gorsky, and V.B. Ryvkin. 1998. Cancer incidence among liquidators of the Chernobyl accident: solid tumors, 1986-1995. Health Phys 74:309-315.

Ivanov, V. K., A. I. Gorsky, A. F. Tsyb, M. A. Maksyutov, and E. M. Rastopchin. 1999. Dynamics of thyroid cancer incidence in Russia following the Chernobyl accident. J Radiol Prot 19:305-18.

Ivanov, V.K., A.F. Tsyb, A.V. Petrov, M.A. Maksioutov, T.P. Shilyaeva, and E.V. Kochergina. 2002. Thyroid cancer incidence among liquidators of the Chernobyl accident. Absence of dependence of radiation risks on external radiation dose. Radiat Environ Bioph 41:195-198.

Ivanov, V.K., A.I. Gorski, M.A. Maksioutov, O.K. Vlasov, A.M. Godko, A.F. Tsyb, M. Tirmarche, M. Valenty, and P. Verger. 2003. Thyroid cancer incidence among adolescents and adults in the Bryansk region of Russia following the Chernobyl accident. Health Phys 84:46-60.

Ivanov, V.K., A.I. Gorski, A.F. Tsyb, S.I. Ivanov, R.N. Naumenko, and L.V. Ivanova. 2004a. Solid cancer incidence among the Chernobyl emergency workers residing in Russia: estimation of radiation risks. Radiat Environ Bioph 43:35-42.

Ivanov, V.K., L. Ilyin, A. Gorski, A. Tukov, and R. Naumenko. 2004b. Radiation and epidemiological analysis for solid cancer incidence among nuclear workers who participated in recovery operations following the accident at the Chernobyl NPP. J Radiat Res 45:41-44.

Iwamoto, K.S., S. Fujii, A. Kurata, M. Suzuki, T. Hayashi, Y. Ohtsuki, Y. Okada, M. Narita, M. Takahashi, S. Hosobe, K. Doishita, T. Manabe, S. Hata, I. Murakami, S. Hata, S. Itoyama, S. Akatsuka, N. Ohara, K. Iwasaki, H. Akabane, M. Fujihara, T. Seyama, and T. Mori. 1999. p53 mutations in tumor and non-tumor tissues of thorotrast recipients: a model for cellular selection during radiation carcinogenesis in the liver. Carcinogenesis 20:1283-1291.

Iyer, R., and B.E. Lehnert. 2000. Factors underlying the cell growth-related bystander responses to alpha particles. Cancer Res 60:1290-1298.

Jablon, S., Z. Hrubec, and J.D. Boice. 1991. Cancer in populations living near nuclear facilities. A survey of mortality nationwide and incidence in two states. J Am Med Assoc 265:1403-1408.

Jacks, T., and R.A. Weinberg. 1998. The expanding role of cell cycle regulators. Science 280:1035-1036.

Jacob, P., G. Goulko, W.F. Heidenreich, I. Likhtarev, I. Kairo, N.D. Tronko, T.I. Bogdanova, J. Kenigsberg, E. Buglova, V. Drozdovitch, A. Golovneva, E.P. Demidchik, M. Balonov, I. Zvonova, and V. Beral. 1998. Thyroid cancer risk to children calculated. Nature 392:31-32.

Jacob, P., Y. Kenigsberg, I. Zvonova, G. Goulko, E. Buglova, W.F. Heidenreich, A. Golovneva, A.A. Bratilova, V. Drozdovitch, J. Kruk, G.T. Pochtennaja, M. Balonov, E.P. Demidchik, and H.G. Paretzke. 1999. Childhood exposure due to the Chernobyl accident and thyroid cancer risk in contaminated areas of Belarus and Russia. Brit J Cancer 80:1461-1469.

Jacob, P., Y. Kenigsberg, G. Goulko, E. Buglova, F. Gering, A. Golovneva, J. Kruk, and E.P. Demidchik. 2000. Thyroid cancer risk in Belarus after the Chernobyl accident: comparison with external exposures. Radiat Environ Bioph 39:25-31.

Jaffe, D.R., and G.T. Bowden. 1986. Ionizing radiation as an initiator in the mouse two-stage model of skin tumor formation. Radiat Res 106: 156-165.

Jaworowowski, Z. 1995. Beneficial radiation. Nukleonika 40:3-12.

Jeffreys, A.J., and Y.E. Dubrova. 2001. Monitoring spontaneous and induced human mutation by RAPD-PCR: a response to Weinberg et al. Proc R Soc Lond B Biol Sci 268:2493-2494.

Jeffreys, A.J., and R. Neumann. 1997. Somatic mutation processes at a human minisatellite. Hum Mol Genet 6:129-132; 134-136.

Jeffreys, A.J., V. Wilson, and S.L. Thein. 1985. Hypervariable "minisatellite" regions in human DNA. Nature 314:67-73.

Jeffreys, A.J., N.J. Royle, V. Wilson, and Z. Wong. 1988. Spontaneous mutation rates to new length alleles at tandem-repetitive hypervariable loci in human DNA. Nature 332:278-281.

Jeffreys, A.J., A. MacLeod, K. Tamaki, D.L. Neil, and D.G. Monckton. 1991. Minisatellite repeat coding as a digital approach to DNA typing. Nature 354:204-209.

Jeffreys, A.J., K. Tamaki, A. MacLeod, D.G. Monckton, D.L. Neil, and J.A. Armour. 1994. Complex gene conversion events in germline mutation at human minisatellites. Nat Genet 6:136-145.

Jeffreys, A.J., M.J. Allen, J.A. Armour, A. Collick, Y. Dubrova, N. Fretwell, T. Guram, M. Jobling, C.A. May, D.L. Neil, and R. Neumann. 1995. Mutation processes at human minisatellites. Electrophoresis 16:1577-1585.

Joenje, H. 1989. Genetic toxicology of oxygen. Mutat Res 219:193-208.

Joiner, M.C., P. Lambin, E.P. Malaise, T. Robson, J.E. Arrand, K.A. Skov, and B. Marples. 1996. Hypersensitivity to very-low single radiation doses: its relationship to the adaptive response and induced radioresistance. Mutat Res 358:171-183.

Jones, I.M., H. Galick, P. Kato, R.G. Langlois, M.L. Mendelsohn, G.A. Murphy, P. Pleshanov, M.J. Ramsey, C.B. Thomas, J.D. Tucker, L. Tureva, I. Vorobtsova, and D.O. Nelson. 2002. Three somatic genetic biomarkers and covariates in radiation-exposed Russian cleanup workers of the Chernobyl nuclear reactor 6-13 years after exposure. Radiat Res 158:424-442.

Jones, P.A., W.M. Rideout, J.C. Shen, C.H. Spruck, and Y.C. Tsai. 1992. Methylation, mutation and cancer. Bioessays 14:33-36.

Jostes, R.F., K.M. Bushnell, and W.C. Dewey. 1980. X-ray induction of 8-azaguanine-resistant mutants in synchronous Chinese hamster ovary cells. Radiat Res 83:146-161.

Jostes, R.F., E.W. Fleck, T.L. Morgan, G.L. Stiegler, and F.T. Cross. 1994. Southern blot and polymerase chain reaction exon analyses of HPRT-mutations induced by radon and radon progeny. Radiat Res 137: 371-379.

Kaatsch, P., U. Kaletsch, R. Meinert, and J. Michaelis. 1998. An extended study on childhood malignancies in the vicinity of German nuclear power plants. Cancer Cause Control 9:529-533.

Kadhim, M.A., D.A. Macdonald, D.T. Goodhead, S.A. Lorimore, S.J. Marsden, and E.G. Wright. 1992. Transmission of chromosomal instability after plutonium alpha-particle irradiation. Nature 355:738-740.

Kadhim, M.A., C.A. Walker, M.A. Plumb, and E.G. Wright. 1996. No association between p53 status and alpha-particle-induced chromosomal instability in human lymphoblastoid cells. Int J Radiat Biol 69:167-174.

Kai, M., E.G. Luebeck, and S.H. Moolgavkar. 1997. Analysis of the incidence of solid cancer among atomic bomb survivors using a two-stage model of carcinogenesis. Radiat Res 148:348-358.

Kaldor, J.M., N.E. Day, E.A. Clarke, F.E. Van Leeuwen, M. Henry-Amar, M.V. Fiorentino, J. Bell, D. Pedersen, P. Band, D. Assouline, M. Koch, W. Choi, P. Prior, V. Blair, F. Langmark, V. Pompe Kirn, F. Neal, D. Peters, R. Pfeiffer, S. Karjalainen, J. Cuzick, S. Sutcliffe, R. Somers, B. Pellae-Cosset, G. Pappagallo, P. Fraser, H. Storm, and M. Stovall. 1990a. Leukemia following Hodgkin's disease. N Engl J Med 322:7-13.

Kaldor, J.M., N.E. Day, F. Pettersson, E.A. Clarke, D. Pedersen, W. Mehnert, J. Bell, H. Host, P. Prior, S. Karjalainen, F. Neal, M. Koch, P. Band, W. Choi, V. Pompe Kirn, A. Arslan, B. Zaren, A.R. Belch, H. Storm, B. Kittelmann, P. Fraser, and M. Stovall. 1990b. Leukemia following chemotherapy for ovarian cancer. N Engl J Med 322:1-6.

Kaldor, J.M., N.E. Day, J. Bell, E.A. Clarke, F. Langmark, S. Karjalainen, P. Band, D. Pedersen, W. Choi, V. Blair, M. Henry-Amar, P. Prior, D. Assouline, V. Pompe-Kirn, R.A. Cartwright, M. Koch, A. Arslan, P. Fraser, S.B. Sutcliffe, H. Host, M. Hakama, and M. Stovall. 1992. Lung cancer following Hodgkin's disease: a case-control study. Int J Cancer 52:677-681.

Kamijo, T., F. Zindy, M.F. Roussel, D.E. Quelle, J.R. Downing, R.A. Ashmun, G. Grosveld, and C.J. Sherr. 1997. Tumor suppression at the mouse INK4a locus mediated by the alternative reading frame product p19ARF. Cell 91:649-659.

Kang, J., R.T. Bronson, and Y. Xu. 2002. Targeted disruption of NBS1 reveals its roles in mouse development and DNA repair. EMBO J 21:1447-1455.

Kaplan, H.S. 1950. Influence of ovarian function on incidence of radiation-induced ovarian tumors. J Natl Cancer Inst 11:125-132.

Kaplan, H.S. 1964. The role of radiation on experimental leukaemogenesis. Natl Cancer Inst Monogr 14:207-220.

Kaplan, H.S. 1967. On the natural history of the murine leukemias: presidential address. Cancer Res 27:1325-1340.

Kaplan, M.I., and W.F. Morgan. 1998. The nucleus is the target for radiation-induced chromosomal instability. Radiat Res 150:382-390.

Karlsson, P., E. Holmberg, L.M. Lundberg, C. Nordborg, and A. Wallgren. 1997. Intracranial tumors after radium treatment for skin hemangioma during infancy—a cohort and case-control study. Radiat Res 148: 161-167.

Karlsson, P., E. Holmberg, M. Lundell, A. Mattsson, L.E. Holm, and A. Wallgren. 1998. Intracranial tumors after exposure to ionizing radiation during infancy: a pooled analysis of two Swedish cohorts of 28,008 infants with skin hemangioma. Radiat Res 150:357-364.

Karp, J.E., and S. Broder. 1995. Molecular foundations of cancer: new targets for intervention. Nat Med 1:309-320.

Kasatkina, E.P., D.E. Shilin, A.L. Rosenbloom, M.I. Pykov, G.V. Ibragimova, V.N. Sokolovskaya, A.N. Matkovskaya, T.N. Volkova, E.A. Odoud, M.I. Bronshtein, A.M. Poverenny, and N.M. Mursankova. 1997. Effects of low level radiation from the Chernobyl accident in a population with iodine deficiency. Eur J Pediatr 156:916-920.

Kastan, M.B., and D.S. Lim. 2000. The many substrates and functions of ATM. Nat Rev Mol Cell Biol 1:179-186.

Kawamura, S., F. Kasagi, K. Kodama, S. Fujiwara, M. Yamada, K. Ohama, and K. Oto. 1997. Prevalence of uterine myoma detected by ultrasound examination in the atomic bomb survivors. Radiat Res 147:753-758.

Kazakov, V. S., E. P. Demidchik, and L. N. Astakhova. 1992. Thyroid cancer after Chernobyl. Nature 359:21.

Kellerer, A.M. 2002. Electron spectra and the RBE of x-rays. Radiat Res 158:13-22.

Kellerer, A.M., and M. Barclay. 1992. Age dependencies in the modeling of radiation carcinogenesis. Radiat Prot Dosim 41:273-281.

Kellerer, A.M., and E. Nekolla. 1997. Neutron versus gamma-ray risk estimates. Inferences from the cancer incidence and mortality data in Hiroshima. Radiat Environ Bioph 36:73-83.

Kellerer, A.M., and H.H. Rossi. 1972a. The theory of dual radiation action. Curr Top Radiat Res Q 8:85-158.

Kellerer, A.M., and H.H. Rossi. 1972b. Dependence of RBE on neutron dose. Brit J Radiol 45:626.

Kellerer, A.M., E.A. Nekolla, and L. Walsh. 2001. On the conversion of solid cancer ERR into lifetime attributable risk. Radiat Environ Bioph 40:249-257.

Kelley, M.R., Y.W. Kow, and D.M. Wilson 3rd. 2003. Disparity between DNA base excision repair in yeast and mammals: translational implications. Cancer Res 63:549-554.

Kelly, R., G. Bulfield, A. Collick, M. Gibbs, and A.J. Jeffreys. 1989. Characterization of a highly unstable mouse minisatellite locus: evidence for somatic mutation during early development. Genomics 5:844-856.

Kelsey, K.T., A. Memisoglu, D. Frenkel, and H.L. Liber. 1991. Human lymphocytes exposed to low doses of X-rays are less susceptible to radiation-induced mutagenesis. Mutat Res 263:197-201.

Kemp, C.J., T. Wheldon, and A. Balmain. 1994. p53-deficient mice are extremely susceptible to radiation-induced tumorigenesis. Nat Genet 8:66-69.

Kendall, G.M., C.R. Muirhead, B.H. MacGibbon, J.A. O'Hagan, A.J. Conquest, A.A. Goodill, B.K. Butland, T.P. Fell, D.A. Jackson, M.A. Webb, R.G.E. Haylock, J.M. Thomas, and T.J. Silk. 1992a. Mortality and occupational exposure to radiation: first analysis of the National Registry for Radiation Workers. Brit Med J 304:220-225.

Kendall, G.M., C.R. Muirhead, B.H. MacGibbon, J.A. O'Hagan, A.J. Conquest, A.A. Goodill, B.K. Butland, T.P. Fell, D.A. Jackson, M.A. Webb, R.G.E. Haylock, J.M. Thomas, and T.J. Silk. 1992b. First analysis of the National Registry for Radiation Workers: occupational exposure to ionising radiation and mortality. Chilton, UK.

Kendler, K.S., and K.K. Kidd. 1986. Recurrence risks in an oligogenic threshold model: the effect of alterations in allele frequency. Ann Hum Genet 50(Pt 1):83-91.

Kennedy, A.R., M. Fox, G. Murphy, and J.B. Little. 1980. Relationship between x-ray exposure and malignant transformation in C3H 10T1/2 cells. Proc Natl Acad Sci USA 77:7262-7266.

Kennedy, G.C., M.S. German, and W.J. Rutter. 1995. The minisatellite in the diabetes susceptibility locus IDMM2 regulates insulin transcription. Nat Genet 9:293-298.

Kerber, R.A., J.E. Till, S.L. Simon, J.L. Lyon, D.C. Thomas, S. Preston-Martin, M.L. Rallison, R.D. Lloyd, and W.S. Stevens. 1993. A cohort study of thyroid disease in relation to fallout from nuclear weapons testing. J Am Med Assoc 270:2076-2082.

Kesminiene, A., J. Kurtinaitis, and G. Rimdeika. 1997. The study of Chernobyl clean-up workers from Lithuania. Acta Med Lituanica 2: 55-61.

Khanna, K.K. 2000. Cancer risk and the ATM gene: a continuing debate. J Natl Cancer Inst 92:795-802.

Khanna, K.K., and S.P. Jackson. 2001. DNA double-strand breaks: signaling, repair and the cancer connection. Nat Genet 27:247-254.

Kinashi, Y., R. Okayasu, G.E. Iliakis, H. Nagasawa, and J.B. Little. 1995. Induction of DNA double-strand breaks by restriction enzymes in x-ray-sensitive mutant Chinese hamster ovary cells measured by pulsed-field gel electrophoresis. Radiat Res 141:153-159.

Kinlen, L.J. 1993a. Can paternal preconceptional radiation account for the increase of leukaemia and non-Hodgkin's lymphoma in Seascale. Brit Med J 306:1718-1721.

Kinlen, L.J. 1993b. Childhood leukaemia and non-Hodgkin's lymphoma in young people living close to nuclear reprocessing sites. Biomed Pharmacol 47:429-434.

Kinzler, K.W., and B. Vogelstein. 1996. Life (and death) in a malignant tumour. Nature 379:19-20.

Kinzler, K.W., and B. Vogelstein. 1997. Cancer-susceptibility genes. Gatekeepers and caretakers. Nature 386:761-763.

Kinzler, K.W., and B. Vogelstein. 1998. Landscaping the cancer terrain. Science 280:1036-1037.

Kirk, K.M., and M.F. Lyon. 1984. Induction of congenital malformations in the offspring of male mice treated with x-rays at pre-meiotic and post-meiotic stages. Mutat Res 125:75-85.

Kirk, M., and M.F. Lyon. 1982. Induction of congenital anomalies in offspring of female mice exposed to varying doses of x-rays. Mutat Res 106:73-83.

Kiuru, A., A. Auvinen, M. Luokkamaki, K. Makkonen, T. Veidebaum, M. Tekkel, M. Rahu, T. Hakulinen, K. Servomaa, T. Rytomaa, and R. Mustonen. 2003. Hereditary minisatellite mutations among the offspring of Estonian Chernobyl cleanup workers. Radiat Res 159:651-655.

Kleinerman, R.A., J.D. Boice, H.H. Storm, P. Sparen, A. Andersen, E. Pukkala, C.F. Lynch, B.F. Hankey, and J.T. Flannery. 1995. Second primary cancer after treatment for cervical cancer. An international cancer registries study. Cancer 76:442-452.

Kliauga, P., and R. Dvorak. 1978. Microdosimetric measurements of ionization by monoenergetic photons. Radiat Res 73:1-20.

Klugbauer, S., E. Lengfelder, E.P. Demidchik, and H.M. Rabes. 1995. High prevalence of RET rearrangement in thyroid tumors of children from Belarus after the Chernobyl reactor accident. Oncogene 11:2459-2467.

Klugbauer, S., P. Pfeiffer, H. Gassenhuber, C. Beimfohr, and H.M. Rabes. 2001. RET rearrangements in radiation-induced papillary thyroid carcinomas: high prevalence of topoisomerase I sites at breakpoints and microhomology-mediated end joining in ELE1 and RET chimeric genes. Genomics 73:149-160.

Klungland, A., and T. Lindahl. 1997. Second pathway for completion of human DNA base excision-repair: reconstitution with purified proteins and requirement for DNase IV (FEN1). EMBO J 16:3341-3348.

Kneale, G.W., and A.M. Stewart. 1993. Reanalysis of Hanford data: 1944-1986 deaths. Am J Ind Med 23:371-389.

Kneale, G.W., T.F. Mancuso, and A.M. Stewart. 1981. Hanford radiation study III: a cohort study of the cancer risks from radiation to workers at Hanford (1944-77 deaths) by the method of regression models in life-tables. Brit J Ind Med 38:156-166.

Knox, E.G., A.M. Stewart, G.W. Kneale, and E.A. Gilman. 1987. Ten-day rule. Lancet 2:1404.

Knudson, A.G., Jr. 1986. Genetics of human cancer. Annu Rev Genet 20:231-251.

Kodaira, M., C. Satoh, K. Hiyama, and K. Toyama. 1995. Lack of effects of atomic bomb radiation on genetic instability of tandem-repetitive elements in human germ cells. Am J Hum Genet 57:1275-1283.

Kodama, K., S. Fujiwara, M. Yamada, F. Kasagi, Y. Shimizu, and I. Shigematsu. 1996. Profiles of non-cancer diseases in atomic bomb survivors. World Health Stat Q 49:7-16.

Kodell, R.L., D. Krewski, and J.M. Zielinski. 1991. Additive and multiplicative relative risk in the two-stage clonal expansion model of carcinogenesis. Risk Anal 11:483-490.

Kofler A, T.H. Abelin, I. Prudyves, and others. 1999. Factors related to latency period in post Chernobyl carcinogenesis. Pp. 123-129 in Radiation and Thyroid Cancer. Proceedings of an Internal Seminar held in St John's College, Cambridge, UK, July 20-23, 1998, G. Thomas, A. Karaoglou, and E.D. Williams, eds. Singapore: World Scientific.

Kojima, S., O. Matsuki, T. Nomura, M. Takahasi, and K. Yamaoka. 1997. Effect of small doses of gamma-rays on the glutathione synthesis in mouse. Low Doses of Ionizing Radiation: Biological Effects and Regulatory Control, IAEA-TECDOC976: IAEA-CN-67/130.

Kominami, R., Y. Saito, T. Shinbo, A. Matsuki, Y. Ochiai, Y. Kodama, Y. Wakabayashi, Y. Takahashi, Y. Mishima, and O. Niwa. 2002. Genetic analysis of radiation-induced thymic lymphoma. Pp. 143-150 in Radiation and Homeostasis, T. Sugahara and others, eds. Amsterdam.

Kony, S.J., F. de Vathaire, A. Chompret, A. Shamsaldim, E. Grimaud, M.A. Raquin, O. Oberlin, L. Brugieres, J. Feunteun, F. Eschwege, J. Chavaudra, J. Lemerle, and C. Bonaiti-Pellie. 1997. Radiation and genetic factors in the risk of second malignant neoplasms after a first cancer in childhood. Lancet 350:91-95.

Kopecky, K.J., E. Nakashima, T. Yamamoto, and H. Kato. 1986. Lung Cancer, Radiation, and Smoking Among A-bomb Survivors, Hiroshima and Nagasaki. Hiroshima, Japan: Radiation Effects Research Foundation.

Korshurnikova, N., and N. Shilnikova. 1996. The risk of cancer among nuclear workers at the "Mayak" production association: preliminary results of an epidemiological study. NCRP Proceedings 113:113-122.

Koshurnikova, N.A., E.S. Gilbert, M. Sokolnikov, V.F. Khokhryakov, S. Miller, D.L. Preston, S.A. Romanov, N.S. Shilnikova, K.G. Suslova, and V.V. Vostrotin. 2000. Bone cancers in Mayak workers. Radiat Res 154:237-245.

Koshurnikova, N.A., G.S. Mushkacheva, N.S. Shilnikova, E.I. Rabinovich, N.P. Petrushkina, P. Hall, M.G. Bolotnikova, D.L. Preston, and E. Ron. 2002. Studies on the Ozyorsk population: health effects. Radiat Environ Bioph 41:37-39.

Kossenko, M.M. 1996. Cancer mortality among Techa River residents and their offspring. Health Phys 71:77-82.

Kossenko, M.M., P.V. Izhevsky, M.O. Degteva, A.V. Akleev, and O.V. Vyushkova. 1994. Pregnancy outcome and early health status of children born to the Techa River population. Sci Total Environ 142:91-100.

Kossenko, M.M., Y. Ostroumova, A. Akleyev, N. Startsev, M. Degteva, F. Granath, and P. Hall. 2000. Mortality in the offspring of individuals living along the radioactively contaminated Techa River: a descriptive analysis. Radiat Environ Bioph 39:219-225.

Kountouras, J., P. Boura, and N.J. Lygidakis. 2001. Liver regeneration after hepatectomy. Hepatogastroenterology 48:556-562.

Kratochvilova, J., and U.H. Ehling. 1979. Dominant cataract mutations induced by gamma-irradiation of male mice. Mutat Res 63:221-223.

Krawczak, M., and D.N. Cooper. 1997. The human gene mutation database. Trends Genet 13:121-122.

Kreisheimer, M., N.A. Koshurnikova, E. Nekolla, V.F. Khokhryakov, S.A. Romanow, M.E. Sokolnikov, N.S. Shilnikova, P.V. Okatenko, and A.M. Kellerer. 2000. Lung cancer mortality among male nuclear workers of the Mayak facilities in the former Soviet Union. Radiat Res 154:3-11.

Krontiris, T.G., B. Devlin, D.D. Carp, N.J. Robert, and N. Risch. 1993. An association between the risk of cancer and mutations in the HRAS1 minisatellite locus. N Engl J Med 329:517-523.

Kubota, Y., R.A. Nash, A. Klungland, P. Schar, D.E. Barnes, and T. Lindahl. 1996. Reconstitution of DNA base excision-repair with purified human proteins: interaction between DNA polymerase beta and the XRCC1 protein. EMBO J 5:6662-6670.

Kumpusalo, L., E. Kumpusalo, S. Soimakallio, S. Salomaa, W. Paile, S. Kolmakow, G. Zhukowsky, I. Ilchenko, and A. Nissinen. 1996. Thyroid ultrasound findings 7 years after the Chernobyl accident. A comparative epidemiological study in the Bryansk region of Russia. Acta Radiol 37:904-909.

Kuraoka, I., C. Bender, A. Romieu, J. Cadet, R.D. Wood, and T. Lindahl. 2000. Removal of oxygen free-radical-induced 5′,8-purine cyclodeoxynucleosides from DNA by the nucleotide excision-repair pathway in human cells. Proc Natl Acad Sci USA 97:3832-3837.

Kuschel, B., A. Auranen, S. McBride, K.L. Novik, A. Antoniou, J.M. Lipscombe, N.E. Day, D.F. Easton, B.A. Ponder, P.D. Pharoah, and A. Dunning. 2002. Variants in DNA double-strand break repair genes and breast cancer susceptibility. Hum Mol Genet 11:1399-1407.

Laake, K., P. Vu, T.I. Andersen, B. Erikstein, R. Karesen, P.E. Lonning, E. Skovlund, and A.L. Borresen-Dale. 2000. Screening breast cancer patients for Norwegian ATM mutations. Brit J Cancer 83:1650-1653.

Labhart, P. 1999. Nonhomologous DNA end joining in cell-free systems. Eur J Biochem 265:849-861.

Lafuma, J., D. Chmelevsky, J. Chameaud, M. Morin, R. Masse, and A.M. Kellerer. 1989. Lung carcinomas in Sprague-Dawley rats after exposure to low doses of radon daughters, fission neutrons, or gamma rays. Radiat Res 118:230-245.

Lambin, P., B. Marples, B. Fertil, E.P. Malaise, and M.C. Joiner. 1993. Hypersensitivity of a human tumour cell line to very low radiation doses. Int J Radiat Biol 63:639-650.

Land, C.E., and W.K Sinclair. 1991. The relative contributions of different organ sites to the total cancer mortality associated with low-dose radiation exposure. Ann ICRP 22:31-57.

Land, C.E., J.D. Boice Jr., R.E. Shore, J.E. Norman, and M. Tokunaga. 1980. Breast cancer risk from low-dose exposures to ionizing radiation: results of parallel analysis of three exposed populations of women. J Natl Cancer Inst 65:353-376.

Land, C.E., N. Hayakawa, S.G. Machado, Y. Yamada, M.C. Pike, S. Akiba, and M. Tokunaga. 1994a. A case-control interview study of breast cancer among Japanese A-bomb survivors. I. Main effects. Cancer Cause Control 5:157-165.

Land, C.E., N. Hayakawa, S.G. Machado, Y. Yamada, M.C. Pike, S. Akiba, and M. Tokunaga. 1994b. A case-control interview study of breast cancer among Japanese A-bomb survivors. II. Interactions with radiation dose. Cancer Cause Control 5:167-176.

Land, C.E., T. Saku, Y. Hayashi, O. Takahara, H. Matsuura, S. Tokuoka, M. Tokunaga, and K. Mabuchi. 1996. Incidence of salivary gland tumors among atomic bomb survivors, 1950-1987. Evaluation of radiation-related risk. Radiat Res 146:28-36.

Land, C.E., M. Tokunaga, K. Koyama, M. Soda, D.L. Preston, I. Nishimori, and S. Tokuoka. 2003. Incidence of female breast cancer among atomic bomb survivors, Hiroshima and Nagasaki, 1950-1990. Radiat Res 160:707-717.

Lane, D.P. 1992. Cancer: p53, guardian of the genome. Nature 358:15-16.

Lazjuk, G.I., D.L. Nikolaev, and I.V. Novikova. 1997. Changes in registered congenital anomalies in the Republic of Belarus after the Chernobyl accident. Stem Cells 15:255-260.

Le, X.C., J.Z. Xing, J. Lee, S.A. Leadon, and M. Weinfeld. 1998. Inducible repair of thymine glycol detected by an ultrasensitive assay for DNA damage. Science 280:1066-1069.

Lea, D.E.A. 1946. Actions of Radiations on Living Cells. Cambridge, UK: University Press.

Lee, G.H., H. Matsushita, and T. Kitagawa. 2001. Fine chromosomal localization of the mouse Par2 gene that confers resistance against urethane-induction of pulmonary adenomas. Oncogene 20:3979-3985.

Lengauer, C., K.W. Kinzler, and B. Vogelstein. 1998. Genetic instabilities in human cancers. Nature 396:643-649.

Leonhardt, E.A., M. Trinh, H.B. Forrester, R.T. Johnson, and W.C. Dewey. 1997. Comparisons of the frequencies and molecular spectra of HPRT mutants when human cancer cells were x-irradiated during G_1 or S phase. Radiat Res 148:548-560.

Le Rhun, Y., J.B. Kirkland, and G.M. Shah. 1998. Cellular responses to DNA damage in the absence of poly(ADP-ribose) polymerase. Biochem Bioph Res Commun 245:1-10.

Levine, A.J. 1993. The tumor suppressor genes. Ann Rev Biochem 62: 623-651.

Levine, A.J. 1997. p53, the cellular gatekeeper for growth and division. Cell 88:323-331.

Levine, A.J., and J.R. Broach. 1995. Oncogenes and cell proliferation. Curr Opin Genet Dev 5:1-4.

Levy, D.B., K.J. Smith, Y. Beazer-Barclay, S.R. Hamilton, B. Vogelstein, and K.W. Kinzler. 1994. Inactivation of both APC alleles in human and mouse tumors. Cancer Res 54:5953-5958.

Lewis, S.E., and F.M. Johnson. 1986. The nature of spontaneous and induced electrophoretically detected mutations in the mouse. Pp. 359-365 in Genetic Toxicology of Environmental Chemicals, C. Ramel, B. Lambert, and J. Magnusson, eds. New York: Alan Liss.

Li, W.H., D.L. Ellsworth, J. Krushkal, B.H. Chang, and D. Hewett-Emmett. 1996. Rates of nucleotide substitution in primates and rodents and the generation-time effect hypothesis. Mol Phylogenet Evol 5:182-187.

Lichtenstein, P., N.V. Holm, P.K. Verkasalo, A. Iliadou, J. Kaprio, M. Koskenvuo, E. Pukkala, A. Skytthe, and K. Hemminki. 2000. Environmental and heritable factors in the causation of cancer—analyses of cohorts of twins from Sweden, Denmark, and Finland. N Engl J Med 343:78-85.

Likhtarev, I.A., B.G. Sobolev, I.A. Kairo, N.D. Tronko, T.I. Bogdanova, V.A. Oleinic, E.V. Epshtein, and V. Beral. 1995. Thyroid cancer in the Ukraine. Nature 375:365.

Lilienfeld, D.E., and P.D. Stolley. 1994. Foundations of Epidemiology. Third Edition. New York: Oxford University Press.

Limoli, C.L., M.I. Kaplan, J. Corcoran, M. Meyers, D.C. Boothman, and W.F. Morgan. 1997a. Chromosomal instability and its relationship to other end points of genomic instability. Cancer Res 57:5557-5563.

Limoli, C.L., M.I. Kaplan, J.W. Phillips, G.M. Adair, and W.F. Morgan. 1997b. Differential induction of chromosomal instability by DNA strand-breaking agents. Cancer Res 57:4048-4056.

Limoli, C.L., J.J. Corcoran, J.R. Milligan, J.F. Ward, and W.F. Morgan. 1999. Critical target and dose and dose-rate responses for the induction of chromosomal instability by ionizing radiation. Radiat Res 151: 677-685.

Limoli, C.L., M.I. Kaplan, E. Giedzinski, and W.F. Morgan. 2001. Attenuation of radiation-induced genomic instability by free radical scavengers and cellular proliferation. Free Radical Bio Med 31:10-19.

Lindahl, T. 1993. Instability and decay of the primary structure of DNA. Nature 362:709-715.

Lindahl, T., and R.D. Wood. 1999. Quality control by DNA repair. Science 286:1897-1905.

Lindberg, S., P. Karlsson, B. Arvidsson, E. Holmberg, L.M. Lundberg, and A. Wallgren. 1995. Cancer incidence after radiotherapy for skin haemangioma during infancy. Acta Oncol 34:735-740.

Little, J.B. 1994. Failla Memorial Lecture. Changing views of cellular radiosensitivity. Radiat Res 140:299-311.

Little, J.B. 1998. Radiation-induced genomic instability. Int J Radiat Biol 74:663-671.

Little, J.B. 2000. Challenging the standard paradigms for radiation carcinogenesis. Radiation Research 2, M. Moriarty and others, eds. Lawrence: Allen Press.

Little, J.B. 2003. Genomic instability and bystander effects: a historical perspective. Oncogene 22:6978-6987.

Little, J.B., H. Nagasawa, T. Pfenning, and H. Vetrovs. 1997. Radiation-induced genomic instability: delayed mutagenic and cytogenetic effects of x rays and alpha particles. Radiat Res 148:299-307.

Little, M.P. 2001. Comparison of the risks of cancer incidence and mortality following radiation therapy for benign and malignant disease with the cancer risks observed in the Japanese A-bomb survivors. Int J Radiat Biol 77:431-464.

Little, M.P. 2002. Comments on the paper: Microsatellite instability in acute myelocytic leukaemia developed from A-bomb survivors. Int J Radiat Biol 78:441-443.

Little, M.P., and J.D. Boice Jr. 1999. Comparison of breast cancer incidence in the Massachusetts tuberculosis fluoroscopy cohort and in the Japanese atomic bomb survivors. Radiat Res 151:218-224.

Little, M.P., and C.R. Muirhead. 1997. Curvilinearity in the dose-response curve for cancer in Japanese atomic bomb survivors. Environ Health Persp 105:1505-1509.

Little, M.P., M.M. Hawkins, R.E. Shore, M.W. Charles, and N.G. Hildreth. 1991. Time variations in the risk of cancer following irradiation in childhood. Radiat Res 126:304-316.

Little, M.P., G.M. Kendall, C.R. Muirhead, B.H. MacGibbon, R.G.E. Haylock, J.M. Thomas, and A.A. Goodill. 1993. Further analysis, incorporating assessment of the robustness of risks of cancer mortality in the National Registry for Radiation Workers. J Radiol Prot 13:95-108.

Little, M.P., F. de Vathaire, M.W. Charles, M.M. Hawkins, and C.R. Muirhead. 1998. Variations with time and age in the risks of solid cancer incidence after radiation exposure in childhood. Stat Med 17:1341-1355.

Little, M.P., C.R. Muirhead, and M.W. Charles. 1999a. Describing time and age variations in the risk of radiation-induced solid tumour incidence in the Japanese atomic bomb survivors using generalized relative and absolute risk models. Stats Med 18:17-33.

Little, M.P., C.R. Muirhead, R.G. Haylock, and J.M. Thomas. 1999b. Relative risks of radiation-associated cancer: comparison of second cancer in therapeutically irradiated populations with the Japanese atomic bomb survivors. Radiat Environ Bioph 38:267-283.

Little, M.P., H.A. Weiss, J.D. Boice Jr., S.C. Darby, N.E. Day, and C.R. Muirhead. 1999c. Risks of leukemia in Japanese atomic bomb survivors, in women treated for cervical cancer, and in patients treated for ankylosing spondylitis. Radiat Res 152:280-292.

Littlefield, L.G., J.G. Hollowell Jr., and W.H. Pool Jr. 1969. Chromosomal aberrations induced by plasma from irradiated patients: an indirect effect of x radiation. Further observations and studies of a control population. Radiology 93:879-886.

Liu, S.Z. 1994. Multilevel mechanisms of stimulatory effect of low dose radiation on immunity. Low Dose Irradiation and Biological Defense Mechanisms, T.L. Sugahara, ed. Amsterdam: Exerpta Medica.

Liu, S.Z. 1997. Cellular and molecular basis of the stimulator effect of low-dose radiation on immunity. Pp. 341-353 in High Levels of Natural Radiation, L. Wei, T. Sugahara, and Z. Tao, eds. Beijing: Elsevier.

Liu, R.S., W.L. Chen, and F.D. Chen. 2002. Health examination and chromosome aberration analysis of residents living in ^{60}Co-contaminated rebar buildings. Int J Radiat Biol 78:635-639.

Livshits, L.A., S.G. Malyarchuk, S.A. Kravchenko, G.H. Matsuka, E.M. Lukyanova, Y.G. Antipkin, L.P. Arabskaya, E. Petit, F. Giraudeau, P. Gourmelon, G. Vergnaud, and B. Le Guen. 2001. Children of Chernobyl cleanup workers do not show elevated rates of mutations in minisatellite alleles. Radiat Res 155:74-80.

Lloyd, D.C., A.A. Edwards, A. Leonard, G.L. Deknudt, L. Verschaeve, A.T. Natarajan, F. Darroudi, G. Obe, F. Palitti, C. Tanzarella, and E.J. Tawn. 1992. Chromosomal aberrations in human lymphocytes induced in vitro by very low doses of x-rays. Int J Radiat Biol 61:335-343.

Lo, A.W., C.N. Sprung, B. Fouladi, M. Pedram, L. Sabatier, M. Ricoul, G.E. Reynolds, and J.P. Murnane. 2002a. Chromosome instability as a result of double-strand breaks near telomeres in mouse embryonic stem cells. Mol Cell Biol 22:4836-4850.

Lo, A.W., L. Sabatier, B. Fouladi, G. Pottier, M. Ricoul, and J.P. Murnane. 2002b. DNA amplification by breakage/fusion/bridge cycles initiated by spontaneous telomere loss in a human cancer cell line. Neoplasia 4:531-538.

Lo, Y.M., S. Darby, L. Noakes, E. Whitley, P.B. Silcocks, K.A. Fleming, and J.I. Bell. 1995. Screening for codon 249 p53 mutation in lung cancer associated with domestic radon exposure. Lancet 345:60.

Lobrich, M., B. Rydberg, and P.K. Cooper. 1995. Repair of x-ray-induced DNA double-strand breaks in specific NotI restriction fragments in human fibroblasts: joining of correct and incorrect ends. Proc Natl Acad Sci USA 92:12050-12054.

Lobrich, M., P.K. Cooper, and B. Rydberg. 1996. Non-random distribution of DNA double-strand breaks induced by particle irradiation. Int J Radiat Biol 70:493-503.

Loeb, L.A. 1991. Mutator phenotype may be required for multistage carcinogenesis. Cancer Res 51:3075-3079.

Loeb, L.A. 1994. Microsatellite instability: marker of a mutator phenotype in cancer. Cancer Res 54:5059-5063.

Logue, J.N., M.K. Barrick, and G.L. Jessup Jr. 1986. Mortality of radiologists and pathologists in the Radiation Registry of Physicians. J Occup Med 28:91-99.

Lohrer, H.D., H. Braselmann, H.E. Richter, G. Jackl, J. Herbeck, L. Hieber, A.M. Kellerer, and M. Bauchinger. 2001. Instability of microsatellites in radiation-associated thyroid tumours with short latency periods. Int J Radiat Biol 77:891-899.

Lomat, L., G. Galburt, M.R. Quastel, S. Polyakov, A. Okeanov, and S. Rozin. 1997. Incidence of childhood disease in Belarus associated with the Chernobyl accident. Environ Health Persp 105:1529-1532.

Loomis, D.P., and S.H. Wolf. 1996. Mortality of workers at a nuclear materials production plant at Oak Ridge, Tennessee, 1947-1990. Am J Ind Med 29:131-141.

Loomis, D.P., and S.H. Wolf. 1997. Influence of social class on SMRs among Y-12 workers. Am J Ind Med 31:122.

Lorenz, E. 1950. Some biologic effects of long continued irradiation. Am J Roentgenol Ra 63:176-185.

Lorenz, E., W. Heston, A. Eschenbrennen, and M. Deringer. 1947. Biological studies in the tolerance range. Radiology 49:274-287.

Lorenz, E., L.O. Jacobson, W.E. Heston, M. Shimkin, A.B. Eschenbrenner, M.K. Deringer, J. Doniger, and R. Schweisthal. 1954. Effects of long-continued total-body gamma irradiation on mice, guinea pigs, and rabbits. III. Effects on life span, weight, blood picture, and carcinogenesis and the role of the intensity of radiation. P. 24 in Biological Effects of External X and Gamma Radiation, R.E. Zirkle, ed. National Nuclear Energy Series, Div. IV, Vol. 22B. New York: McGraw-Hill.

Lorenz, R., W. Deubel, K. Leuner, T. Gollner, E. Hochhauser, and K. Hempel. 1994. Dose and dose-rate dependence of the frequency of hprt deficient T lymphocytes in the spleen of the ^{137}Cs gamma-irradiated mouse. Int J Radiat Biol 66:319-326.

Lorimore, S.A., M.A. Kadhim, D.A. Pocock, D. Papworth, D.L. Stevens, D.T. Goodhead, and E.G. Wright. 1998. Chromosomal instability in the descendants of unirradiated surviving cells after alpha-particle irradiation. Proc Natl Acad Sci USA 95:5730-5733.

Lorimore, S.A., P.J. Coates, and E.G. Wright. 2003. Radiation-induced genomic instability and bystander effects: inter-related nontargeted effects of exposure to ionizing radiation. Oncogene 22:7058-7069.

Loucas, B.D., and M.N. Cornforth. 2001. Complex chromosome exchanges induced by gamma rays in human lymphocytes: an mFISH study. Radiat Res 155:660-671.

Loucas, B.D., R. Eberle, S.M. Bailey, and M.N. Cornforth. 2004. Influence of dose rate on the induction of simple and complex chromosome exchanges by gamma rays. Radiat Res 162:339-349.

Louis, E.L., M.J. McLoughlin, and G. Wortzman. 1974. Chronic damage to medium and large arteries following irradiation. J Can Assoc Radiol 25:94-104.

Lubin, J.H., J.D. Boice Jr., C. Edling, R.W. Hornung, G.R. Howe, E. Kunz, R.A. Kusiak, H.I. Morrison, E.P. Radford, and J.M. Samet. 1995. Lung cancer in radon-exposed miners and estimation of risk from indoor exposure. J Natl Cancer Inst 87:817-827.

Luckey, T.D. 1991. Radiation Hormesis. Boca Raton: CRC Press.

Luckey, T.D. 1996. The evidence for radiation hormesis. 21st Century Sci Technol 9:12-20.

Ludwig, D.L., M.A. MacInnes, Y. Takiguchi, P.E. Purtymun, M. Henrie, M. Flannery, J. Meneses, R.A. Pedersen, and D.J. Chen. 1998. A murine AP-endonuclease gene-targeted deficiency with post-implantation embryonic progression and ionizing radiation sensitivity. Mutat Res 409:17-29.

Luebeck, E.G., S.B. Curtis, F.T. Cross, and S.H. Moolgavkar. 1996. Two-stage model of radon-induced malignant lung tumors in rats: effects of cell killing. Radiat Res 145:163-173.

Luebeck, E.G., W.F. Heidenreich, W.D. Hazelton, H.G. Paretzke, and S.H. Moolgavkar. 1999. Biologically based analysis of the data for the Colorado uranium miners cohort: age, dose and dose-rate effects. Radiat Res 152:339-351.

Lundell, M., and L.E. Holm. 1995. Risk of solid tumors after irradiation in infancy. Acta Oncologie 34:735-740.

Lundell, M., and L.E. Holm. 1996. Mortality from leukemia after irradiation in infancy for skin hemangioma. Radiat Res 145:595-601.

Lundell, M., T. Hakulinen, and L.E. Holm. 1994. Thyroid cancer after radiotherapy for skin hemangioma in infancy. Radiat Res 140:334-339.

Lundell, M., A. Mattsson, T. Hakulinen, and L.E. Holm. 1996. Breast cancer after radiotherapy for skin hemangioma in infancy. Radiat Res 145:225-230.

Lundell, M., A. Mattsson, P. Karlsson, E. Holmberg, A. Gustafsson, and L.E. Holm. 1999. Breast cancer risk after radiotherapy in infancy: a pooled analysis of two Swedish cohorts of 17,202 infants. Radiat Res 151:626-632.

Lüning, K.G., and A.G. Searle. 1971. Estimates of the genetic risks from ionizing irradiation. Mutat Res 12:291-304.

Luongo, C., and W.F. Dove. 1996. Somatic genetic events linked to the Apc locus in intestinal adenomas of the Min mouse. Gene Chromosome Canc 17:194-198.

Lu-xin W, Jian-zhi W. Estimate of cancer risk for a large population continuously exposed to higher background radiation in Yangjiang, China. Chinese Med J 107(7):541-544, 1994.

Lyon, M.F., and T. Morris. 1969. Gene and chromosome mutation after large fractionated, or unfractionated radiation dose to mouse spermatogonia. Mutat Res 8:191-198.

Lyon, M.F., and R. Renshaw. 1984. Induction of congenital malformations in the offspring of mutagen-treated mice. Prog Clin Biol Res 209B: 449-458.

Lyon, M.F., R.J.S. Phillips, and A.G. Searle. 1964. The overall rate of dominant and recessive lethal and visible mutations induced by spermatological x-irradiation of mice. Gen Res Cambridge 5:448-467.

Lyon, M.F., R.J. Phillips, and H.J. Bailey. 1972. Mutagenic effects of repeated small radiation doses to mouse spermatogonia. I. Specific-locus mutation rates. Mutat Res 15:185-190.

Mabuchi, K., M. Soda, E. Ron, M. Tokunaga, S. Ochikubo, S. Sugimoto, T. Ikeda, M. Terasaki, D.L. Preston, and D.E. Thompson. 1994. Cancer incidence in atomic bomb survivors. Part I: Use of the tumor registries in Hiroshima and Nagasaki for incidence studies. Radiat Res 137: S1-S16.

Mackey, Z.B., C. Niedergang, J.M. Murcia, J. Leppard, K. Au, J. Chen, G. de Murcia, and A.E. Tomkinson. 1999. DNA ligase III is recruited to DNA strand breaks by a zinc finger motif homologous to that of poly(ADP-ribose) polymerase. Identification of two functionally distinct DNA binding regions within DNA ligase III. J Biol Chem 274:21679-21687.

MacMahon, B. 1962. Prenatal x-ray exposure and childhood cancer. J Natl Cancer Inst 28:1173-1191.

MacMahon, B., T. Pugh, and J. Ipsen. 1960. Epidemiologic Methods. Boston: Little, Brown.

Maconochie, N., P. Doyle, E. Roman, G. Davies, P.G. Smith, and V. Beral. 1999. The nuclear industry family study: linkage of occupational exposures to reproduction and child health. Brit Med J 318:1453-1454.

Maffei, F., S. Angelini, G.C. Forti, F.S. Violante, V. Lodi, S. Mattioli, and P. Hrelia. 2004. Spectrum of chromosomal aberrations in peripheral lymphocytes of hospital workers occupationally exposed to low doses of ionizing radiation. Mutat Res 547:91-99.

Malaise, E.P., P. Lambin, and M.C. Joiner. 1994. Radiosensitivity of human cell lines to small doses. Are there some clinical implications? Radiat Res 138:S25-S27.

Malkin, D. 1993. p53 and the Li-Fraumeni syndrome. Cancer Genet Cytogen 66:83-92.

Mao, J.H., J. Perez-Losada, D. Wu, R. Delrosario, R. Tsunematsu, K.I. Nakayama, K. Brown, S. Bryson, and A. Balmain. 2004. Fbxw7/Cdc4 is a p53-dependent, haploinsufficient tumour suppressor gene. Nature 432:775-779.

March, H.C. 1944. Leukemia in radiologists. Radiology 43:275-278.

Marder, B.A., and W.F. Morgan. 1993. Delayed chromosomal instability induced by DNA damage. Mol Cell Biol 13:6667-6677.

Marintchev, A., M.A. Mullen, M.W. Maciejewski, B. Pan, M.R. Gryk, and G.P. Mullen. 1999. Solution structure of the single-strand break repair protein XRCC1 N-terminal domain. Nat Struct Biol 6:884-893.

Mark, R.J., J.W. Bailet, J. Poen, L.M. Tran, T.C. Calcaterra, E. Abemayor, Y.S. Fu, and R.G. Parker. 1993. Postirradiation sarcoma of the head and neck. Cancer 72:887-893.

Marnett, L.J., and P.C. Burcham. 1993. Endogenous DNA adducts: potential and paradox. Chem Res Toxicol 6:771-785.

Marples, B., and M.C. Joiner. 1995. The elimination of low-dose hypersensitivity in Chinese hamster V79-379A cells by pretreatment with x rays or hydrogen peroxide. Radiat Res 141:160-169.

Marples, B., and K.A. Skov. 1996. Small doses of high-linear energy transfer radiation increase the radioresistance of Chinese hamster V79 cells to subsequent x irradiation. Radiat Res 146:382-387.

Marples, B., N.E. Cann, C.R. Mitchell, P.J. Johnston, and M.C. Joiner. 2002. Evidence for the involvement of DNA-dependent protein kinase in the phenomena of low dose hyper-radiosensitivity and increased radioresistance. Int J Radiat Biol 78:1139-1147.

Marples, B., B.G. Wouters, and M.C. Joiner. 2003. An association between the radiation-induced arrest of G_2-phase cells and low-dose hyper-radiosensitivity: a plausible underlying mechanism? Radiat Res 160:38-45.

Marples, B., B.G. Wouters, S.J. Collis, A.J. Chalmers, and M.C. Joiner. 2004. Low-dose hyper-sensitivity: a consequence of ineffective cell cycle arrest of radiation-damaged G_2-phase cells. Radiat Res 161:247-255.

Marshall, J.H., and P.G. Groer. 1977. A theory of the induction of bone cancer by alpha radiation. Radiat Res 71:149-192.

Matanoski, G.M., R. Seltser, P.E. Sartwell, E.L. Diamond, and E.A. Elliott. 1975a. The current mortality rates of radiologists and other physician specialists: specific causes of death. Am J Epidemiol 101:199-210.

Matanoski, G.M., R. Seltser, P.E. Sartwell, E.L. Diamond, and E.A. Elliott. 1975b. The current mortality rates of radiologists and other physician specialists: deaths from all causes and from cancer. Am J Epidemiol 101:188-198.

Matanoski, G.M., A. Sternberg, and E.A. Elliott. 1987. Does radiation exposure produce a protective effect among radiologists. Health Phys 52:637-643.

Mathieu, N., L. Pirzio, M.A. Freulet-Marriere, C. Desmaze, and L. Sabatier. 2004. Telomeres and chromosomal instability. Cell Mol Life Sci 61: 641-656.

Matsumoto, Y., and K. Kim. 1995. Excision of deoxyribose phosphate residues by DNA polymerase beta during DNA repair. Science 269: 699-702.

Matsuo, T., M. Tomonaga, J.M. Bennett, K. Kuriyama, F. Imanaka, A. Kamada, N. Kuramoto, M. Ichimaru, S.S. Finch, A.V. Pisciotta, and T. Ishimaru. 1988. Reclassification of leukemia among A-bomb survivors in Nagasaki using French-American-British (FAB) classification for acute leukemia. Jpn J Clini Oncol 18:91-96.

Mattsson, A., B.I. Ruden, N. Wilking, and L.E. Rutqvist. 1993. Radiation-induced breast cancer: long-term follow-up of radiation therapy for benign breast disease. J Natl Cancer Inst 85:1679-1685.

Mattsson, A., B.I. Ruden, J. Palmgren, and L.E. Rutqvist. 1995. Dose- and time-response for breast cancer risk after radiation therapy for benign breast disease. Brit J Cancer 72:1054-1061.

Mattsson, A., P. Hall, B.I. Ruden, and L.E. Rutqvist. 1997. Incidence of primary malignancies other than breast cancer among women treated with radiation therapy for benign breast disease. Radiat Res 148: 152-160.

May, C.A., A.J. Jeffreys, and J.A. Armour. 1996. Mutation rate heterogeneity and the generation of allele diversity at the human minisatellite MS205 (D16S309). Hum Mol Genet 5:1823-1833.

May, C.A., K. Tamaki, R. Neumann, G. Wilson, G. Zagars, A. Pollack, Y.E. Dubrova, A.J. Jeffreys, and M.L. Meistrich. 2000. Minisatellite mutation frequency in human sperm following radiotherapy. Mutat Res 453:67-75.

McGeoghegan, D., and K. Binks. 1999. The mortality and cancer morbidity experience of employees at the Chapelcross plant of British Nuclear Fuels Ltd, 1955-1995. Pp. 261-264 in Proceedings of the 6th International Symposium on Achievements and Challenges: Advancing Radiation Protection into the 21st Century, Southport, M.C. Thorne, ed. London: Society for Radiological Protection.

McGeoghegan, D., and K. Binks. 2000a. The mortality and cancer morbidity experience of workers at the Springfields uranium production facility, 1946-95. J Radiol Prot 20:111-137.

McGeoghegan, D., and K. Binks. 2000b. The mortality and cancer morbidity experience of workers at the Capenhurst uranium enrichment facility 1946-95. J Radiol Prot 20:381-401.

McKusick, V.A. 1998. Mendelian Inheritance in Man. Tenth Edition. Baltimore: Johns Hopkins University Press.

McKusick, V.A. 2000. Mendelian Inheritance in Man. Available online at http://www.ncbi.nlm.nih.gov/omim.

McLaughlin, J.R., E.A. Clarke, E.D. Nishri, and T.W. Anderson. 1993a. Childhood leukemia in the vicinity of Canadian nuclear facilities. Cancer Cause Control 4:51-58.

McLaughlin J.R., W.D. King, T.W. Anderson, E.A. Clarke, and J.P. Ashmore. 1993b. Paternal radiation exposure and leukemia in offspring: the Ontario case-control study. Brit Med J 307:959-966.

Meadows, A.T. 2001. Second tumours. Eur J Cancer 37:2074-2079; discussion 2079-2081.

Meijers-Heijboer, H., A. van den Ouweland, J. Klijn, M. Wasielewski, A. de Snoo, R. Oldenburg, A. Hollestelle, M. Houben, E. Crepin, M. van Veghel-Plandsoen, F. Elstrodt, C. van Duijn, C. Bartels, C. Meijers, M. Schutte, L. McGuffog, D. Thompson, D. Easton, N. Sodha, S. Seal, R. Barfoot, J. Mangion, J. Chang-Claude, D. Eccles, R. Eeles, D.G. Evans, R. Houlston, V. Murday, S. Narod, T. Peretz, J. Peto, C. Phelan, H.X. Zhang, C. Szabo, P. Devilee, D. Goldgar, P.A. Futreal, K.L. Nathanson, B. Weber, N. Rahman, and M.R. Stratton. 2002. Low-penetrance susceptibility to breast cancer due to CHEK2*1100delC in noncarriers of BRCA1 or BRCA2 mutations. Nat Genet 31:55-59.

Meistrich, M.L., and J. Byrne. 2002. Genetic disease in offspring of long-term survivors of childhood and adolescent cancer treated with potentially mutagenic therapies. Am J Hum Genet 70:1069-1071.

Melendez, B., J. Santos, and J. Fernandez-Piqueras. 1999. Loss of heterozygosity at the proximal-mid part of mouse chromosome 4 defines two novel tumor suppressor gene loci in T-cell lymphomas. Oncogene 18:4166-4169.

Menu-Branthomme, A., C. Rubino, A. Shamsaldin, M.M. Hawkins, E. Grimaud, M.G. Dondon, C. Hardiman, G. Vassal, S. Campbell, X. Panis, N. Daly-Schveitzer, J.L. Lagrange, J.M. Zucker, J. Chavaudra, O. Hartman, and F. de Vathaire. 2004. Radiation dose, chemotherapy and risk of soft tissue sarcoma after solid tumours during childhood. Int J Cancer 110:87-93.

Merlo, A., J.G. Herman, L. Mao, D.J. Lee, E. Gabrielson, P.C. Burger, S.B. Baylin, and D. Sidransky. 1995. 5'-CpG island methylation is associated with transcriptional silencing of the tumour suppressor p16/CDKN2/MTS1 in human cancers. Nat Med 1:686-692.

Mettler, F.H., Jr., D.V. Becker, B.W. Wachholz, and A.C. Bouville. 1996. Chernobyl: 10 years later. J Nucl Med 37:24N, 26N-27N.

Meyer, K.R., and R.L. Ullrich. 1981. Effects of x rays and fission neutrons on an induced proliferative response in lung type 2 epithelial cells. Radiat Res 85:380-389.

Meyer, K.R., H. Witschi, and R.L. Ullrich. 1980. Proliferative responses of type 2 lung epithelial cells after x rays and fission neutrons. Radiat Res 82:559-569.

MGSC (Mouse Genome Sequencing Consortium). 2002. Initial sequencing and comparative analysis of the mouse genome. Nature 420:520-562.

Michaelis, J., B. Keller, G. Haaf, and P. Kaatsch. 1992. Incidence of childhood malignancies in the vicinity of West German nuclear power plants. Cancer Cause Control 3:255-263.

Michaelis, U., U. Kaletsch, W. Burkahrt, and B. Grosche. 1997. Infant leukemia after the Chernobyl accident. Nature 387:246.

Miles, C., G. Sargent, G. Phear, and M. Meuth. 1990. DNA sequence analysis of gamma radiation-induced deletions and insertions at the APRT locus of hamster cells. Mol Carcinog 3:233-242.

Mill, A.J., D. Frankenberg, D. Betega, L. Hieber, A. Saran, L.A. Allen, P. Calzolari, M. Frankenberg-Schwager, M.M. Lehane, G.R. Morgan, L. Pariset, S. Pazzaglia, C.J. Roberts, and L. Tallone. 1998. Transformation of C3H 10T1/2 cells by low doses of ionizing radiation: a collaborative study by six European laboratories strongly supporting a linear dose-response relationship. J Radiol Prot 18:79-100.

Miller, A.B., G. Howe, G. Sherman, J. Lindsay, M. Yaffe, P. Dinner, H. Risch, and D. Preston. 1989. Mortality from breast cancer after irradiation during fluoroscopic examinations in patients being treated for tuberculosis. N Engl J Med 321:1285-1289.

Miller, M.C. 3rd, H.W. Mohrenweiser, and D.A. Bell. 2001. Genetic variability in susceptibility and response to toxicants. Toxicol Lett 120:269-280.

Miller, R.C., R.C. Geard, M.J. Geard, and E.J. Hall. 1992. Rapid communication. Cell-cycle-dependent radiation-induced oncogenic transformation of C3H 10T1/2 cells. Radiat Res 130:129-133.

Milligan, J.R., J.Y. Ng, J.A. Aguilera, J.F. Ward, Y.W. Kow, S.S. Wallace, and R.P. Cunningham. 1996. Methylperoxyl radicals as intermediates in DNA damage by ionizing radiation. Radiat Res 146:436-443.

Mills, K.D., D.O. Ferguson, J. Essers, M. Eckersdorff, R. Kanaar, and F.W. Alt. 2004. Rad54 and DNA Ligase IV cooperate to maintain mammalian chromatid stability. Gene Dev 18:1283-1292.

Mitchel, R.E.J., J.S. Jackson, R.A. McCann, and D.R. Boreham. 1999. The adaptive response modifies latency for radiation-induced myeloid leukemia in CBA/H mice. Radiat Res 152:273-279.

Mitchel, R.E., J.S. Jackson, D.P. Morrison, and S.M. Carlisle. 2003. Low doses of radiation increase the latency of spontaneous lymphomas and spinal osteosarcomas in cancer-prone, radiation-sensitive Trp53 heterozygous mice. Radiat Res 159:320-327.

Mitchell, C.R., M. Folkard, and M.C. Joiner. 2002. Effects of exposure to low-dose-rate (60)Co gamma rays on human tumor cells in vitro. Radiat Res 158:311-318.

Mitchell, J.B., J.S. Bedford, and S.M. Bailey. 1979. Dose-rate effects in mammalian cells in culture. III. Comparison of cell killing and cell proliferation during continuous irradiation for six different cell lines. Radiat Res 79:537-551.

Mitchell, T.J., G. Ostrouchov, E.L. Frome, and G.D. Kerr. 1997. A method for estimating occupational radiation dose to individuals, using weekly dosimetry data. Radiat Res 147:195-207.

Mitelman, F., F. Mertens, and B. Johansson. 1997. A breakpoint map of recurrent chromosomal rearrangements in human neoplasia. Nat Genet 15:417-474.

Mizuta, R., J.M. LaSalle, H.L. Cheng, A. Shinohara, H. Ogawa, N. Copeland, N.A. Jenkins, M. Lalande, and F.W. Alt. 1997. RAB22 and RAB163/mouse BRCA2: proteins that specifically interact with the RAD51 protein. Proc Natl Acad Sci USA 94:6927-6932.

Modan, B., A. Chetrit, E. Alfandary, and L. Katz. 1989. Increased risk of breast cancer after low-dose irradiation. Lancet 8639:629-631.

Modan, B., A. Chetrit, E. Alfandary, A. Tamir, A. Lusky, M. Wolf, and O. Shpilberg. 1998. Increased risk of salivary gland tumors after low-dose irradiation. Laryngoscope 108:1095-1097.

Mohan, N., and M.L. Meltz. 1994. Induction of nuclear factor kappa B after low-dose ionizing radiation involves a reactive oxygen intermediate signaling pathway. Radiat Res 140:97-104.

Mohrenweiser, H.W., and I.M. Jones. 1998. Variation in DNA repair is a factor in cancer susceptibility: a paradigm for the promises and perils of individual and population risk estimation? Mutat Res 400:15-24.

Moiseenko, V.V., A.A. Edwards, H. Nikjoo, and W.V. Prestwich. 1997. The influence of track structure on the understanding of relative biological effectiveness for induction of chromosomal exchanges in human lymphocytes. Radiat Res 147:208-214.

Mole, R.H., and I.R. Major. 1983. Myeloid leukaemia frequency after protracted exposure to ionizing radiation: experimental confirmation of the flat dose-response found in ankylosing spondylitis after a single treatment course with x-rays. Leuk Res 7:295-300.

Mole, R.H., D.G. Papworth, and M.J. Corp. 1983. The dose-response for x-ray induction of myeloid leukaemia in male CBA/H mice. Brit J Cancer 47:285-291.

Monsieurs, M.A., H.M. Thierens, A.M. Vral, C. Van De Wiele, L.I. De Ridder, and R.A. Dierckx. 2000. Adaptive response in patients treated with ^{131}I. J Nucl Med 41:17-22.

Monson, R.R. 1990. Occupational Epidemiology. Boca Raton, FL: CRC Press.

Monson, R.R., and B. MacMahon. 1984. Prenatal x-ray exposure and cancer in children. Pp. 97-105 in Radiation Carcinogenesis: Epidemiology and Biological Significance. J.D. Boice Jr. and J.F. Fraumeni Jr., eds. New York: Raven Press.

Moolgavkar, S.H. 1991. Carcinogenesis models: an overview. Basic Life Sci 58:387-396; discussion 396-399.

Moolgavkar, S.H., and A.G. Knudson Jr. 1981. Mutation and cancer: a model for human carcinogenesis. J Natl Cancer Inst 66:1037-1052.

Moolgavkar, S.H., and E.G. Luebeck. 1990. Two-event model for carcinogenesis: biological, mathematical and statistical considerations. Risk Anal 10:323-341.

Moolgavkar, S.H., and E.G. Luebeck. 2003. Multistage carcinogenesis and the incidence of human cancer. Gene Chromosome Canc 38:302-306.

Moolgavkar, S.H., and Venzon, D.J. 1987. General relative risk regression models for epidemiologic studies. Am J Epidemiol 126:949-961.

Moolgavkar, S.H., N.E. Day, and R.G. Stevens. 1980. Two-stage model for carcinogenesis: epidemiology of breast cancer in females. J Natl Cancer Inst 65:559-569.

Moolgavkar, S.H., E.G. Luebeck, D. Krewski, and J.M. Zielinski. 1993. Radon, cigarette smoke, and lung cancer: a re-analysis of the Colorado Plateau uranium miners' data. Epidemiology 4:204-217.

Moolgavkar, S.H., D. Krewski, and M. Schwarz. 1999. Mechanisms of carcinogenesis and biologically based models for estimation and prediction of risk. IARC Sci Publ 131:179-237.

Moore, D.H. 2nd, H.W. Patterson, F. Hatch, D. Discher, J.S. Schneider, D. Bennett, and M.L. Mendelsohn. 1997. Case-control study of malignant melanoma among employees of the Lawrence Livermore National Laboratory. Am J Ind Med 32:377-391.

Morey, N.J., P.W. Doetsch, and S. Jinks-Robertson. 2003. Delineating the requirements for spontaneous DNA damage resistance pathways in genome maintenance and viability in *Saccharomyces cerevisiae*. Genetics 164:443-545.

Morgan, W.F. 2003a. Non-targeted and delayed effects of exposure to ionizing radiation: I. Radiation-induced genomic instability and bystander effects in vivo. Radiat Res 159:567 580.

Morgan, W.F. 2003b. Non-targeted and delayed effects of exposure to ionizing radiation: II. Radiation-induced genomic instability and bystander effects in vivo, clastogenic factors and transgenerational effects. Radiat Res 159:581-596.

Morgan, W.F., and R.A. Winegar. 1990. The use of restriction endonucleases to study the mechanisms of chromosome damage. Pp. 70-78 in Chromosomal Aberrations: Basic and Applied Aspects, G. Obe and A.T. Natarajan, eds. Berlin: Springer-Verlag.

Morgan, W.F., J.P. Day, M.I. Kaplan, E.M. McGhee, and C.L. Limoli. 1996. Genomic instability induced by ionizing radiation. Radiat Res 146:247-258.

Mori, N., Y. Matsumoto, M. Okumoto, N. Suzuki, and J. Yamate. 2001. Variations in Prkdc encoding the catalytic subunit of DNA-dependent protein kinase (DNA-PKcs) and susceptibility to radiation-induced apoptosis and lymphomagenesis. Oncogene 20:3609-3619.

Morimura, K., A. Romanenko, W. Min, E.I. Salim, A. Kinoshita, H. Wanibuchi, A. Vozianov, and S. Fukushima. 2004. Possible distinct molecular carcinogenic pathways for bladder cancer in Ukraine, before and after the Chernobyl disaster. Oncol Rep 11:881-886.

Morris, T., and J. Thacker. 1993. Formation of large deletions by illegitimate recombination in the HPRT gene of primary human fibroblasts. Proc Natl Acad Sci USA 90:1392-1396.

Morrison, C., G.C. Smith, L. Stingl, S.P. Jackson, E.F. Wagner, and Z.Q. Wang. 1997. Genetic interaction between PARP and DNA-PK in V(D)J recombination and tumorigenesis. Nat Genet 17:479-482.

Morton, N.E., and C.J. MacLean. 1974. Analysis of family resemblance. 3. Complex segregation of quantitative traits. Am J Hum Genet 26:489-503.

Moser, A.R., L.F. Hegge, and R.D. Cardiff. 2001. Genetic background affects susceptibility to mammary hyperplasias and carcinomas in Apc(min)/+ mice. Cancer Res 61:3480-3485.

Mothersill, C., and C.B. Seymour. 1997. Survival of human epithelial cells irradiated with cobalt 60 as microcolonies or single cells. Int J Radiat Biol 72:597-606.

Mothersill, C., and C.B. Seymour. 1998a. Cell-cell contact during gamma irradiation is not required to induce a bystander effect in normal human keratinocytes: evidence for release during irradiation of a signal controlling survival into the medium. Radiat Res 149:256-262.

Mothersill, C., and C.B. Seymour. 1998b. Mechanisms and implications of genomic instability and other delayed effects of ionizing radiation exposure. Mutagenesis 13:421-426.

Mothersill, C., T.D. Stamato, M.L. Perez, R. Cummins, R. Mooney, and C.B. Seymour. 2000a. Involvement of energy metabolism in the production of "bystander effects" by radiation. Brit J Cancer 82:1740-1746.

Mothersill, C., T.D. Stamato, M.L. Perez, R. Mooney, R. Cummins, and C.B. Seymour. 2000b. A role for mitochondria in the induction of radiation-induced bystander effects. Brit J Cancer 82:1740-1746.

Mothersill, C., C.B. Seymour, and M.C. Joiner. 2002. Relationship between radiation-induced low-dose hypersensitivity and the bystander effect. Radiat Res 157:526-532.

Motulsky, A.G., and J.D. Brunzell. 1992. The genetics of coronary atherosclerosis. Pp. 150-169 in The Genetic Basis of Common Diseases, A.G. Motulsky, ed. Oxford, UK: Oxford University Press.

MRC (Medical Research Council). 1958. The Hazards to Man of Nuclear and Allied Radiations. London: HMS Office.

Muirhead, C.R., and G.W. Kneale. 1989. Prenatal irradiation and childhood cancer. J Radiol Prot 9:209-212.

Muirhead, C.R., A.A. Goodill, R.G. Haylock, J. Vokes, M.P. Little, D.A. Jackson, J.A. O'Hagan, J.M. Thomas, G.M. Kendall, T.J. Silk, D. Bingham, and G.L. Berridge. 1999. Occupational radiation exposure and mortality: second analysis of the National Registry for Radiation Workers. J Radiol Prot 19:3-26.

Muirhead, C.R., D. Bingham, R.G. Haylock, J.A. O'Hagan, A.A. Goodill, G.L. Berridge, M.A. English, N. Hunter, and G.M. Kendall. 2003. Follow up of mortality and incidence of cancer 1952-98 in men from the UK who participated in the UK's atmospheric nuclear weapon tests and experimental programmes. Occup Environ Med 60:165-172.

Muller, H.J. 1927. Artificial transmutation of the gene. Science 66:84-87.

Muller, H.J. 1951. Radiation damage to the genetic material. Pp. 93-177 in Science in Progress, G.A. Baitsell, ed. New Haven: Yale University Press.

Muller, H.J. 1954. The manner of dependence of the permissible dose of radiation on the amount of genetic damage. Acta Radiol 41:5-20.

Muller, H.J. 1959. Advances in radiation mutagenesis through studies on *Drosophila*. Pp. 146-160 in Progress in Nuclear Energy Series, J.C. Burger, ed. New York: Pergamon Press.

Muller, W.U., C. Streffer, and F. Niedereichholz. 1992. Adaptive response in mouse embryos? Int J Radiat Biol 62:169-175.

Mulvihill, J.J. 1999. Catalog of human cancer genes. McKusick's Mendelian Inheritance in Man for Clinical and Research Oncologists (OncoMIN). Baltimore, MD: Johns Hopkins University Press.

Murnane, J.P. 1990. The role of recombinational hotspots in genome instability in mammalian cells. Bioessays 12:577-581.

Murnane, J.P. 1996. Role of induced genetic instability in the mutagenic effects of chemicals and radiation. Mutat Res 367:11-23.

Murnane, J.P., and L. Sabatier. 2004. Chromosome rearrangements resulting from telomere dysfunction and their role in cancer. Bioessays 26:1164-1174.

Myrianthopoulos, N.C., and C.S. Chung. 1974. Congenital malformations in singletons: epidemiologic survey. Report from the Collaborative Perinatal project. Birth Defects Orig Artic Ser 10:1-58.

Nackerdien, Z., G. Rao, M.A. Cacciuttolo, E. Gajewski, and M. Dizdaroglu. 1991. Chemical nature of DNA-protein cross-links produced in mammalian chromatin by hydrogen peroxide in the presence of iron or copper ions. Biochemistry 30:4873-4879.

Nackerdien, Z., R. Olinski, and M. Dizdaroglu. 1992. DNA base damage in chromatin of gamma-irradiated cultured human cells. Free Radic Res Commun 16:259-273.

Nagasawa, H., and J.B. Little. 1992. Induction of sister chromatid exchanges by extremely low doses of alpha-particles. Cancer Res 52:6394-6396.

Nagasawa, H., and J.B. Little. 1999. Unexpected sensitivity to the induction of mutations by very low doses of alpha-particle radiation: evidence for a bystander effect. Radiat Res 152:552-557.

Nagasawa, H., A. Cremesti, R. Kolesnick, Z. Fuks, and J.B. Little. 2002. Involvement of membrane signaling in the bystander effect in irradiated cells. Cancer Res 62:2531-2534.

Nagase, H., J.H. Mao, J.P. de Koning, T. Minami, and A. Balmain. 2001. Epistatic interactions between skin tumor modifier loci in interspecific (*spretus/musculus*) backcross mice. Cancer Res 61:1305-1308.

Nagataki, S., Y. Shibata, S. Inoue, N. Yokoyama, M. Izumi, and K. Shimaoka. 1994. Thyroid diseases among atomic bomb survivors in Nagasaki. J Am Med Assoc 272:364-370.

Nair, M.K., K.S. Nambi, N.S. Amma, P. Gangadharan, P. Jayalekshmi, S. Jayadevan, V. Cherian, and K.N. Reghuram. 1999. Population study in the high natural background radiation area in Kerala, India. Radiat Res 152:S145-S148.

Najarian, T., and T. Colton. 1978. Mortality from leukaemia and cancer in shipyard nuclear workers. Lancet 1:1018-1020.

Nakanishi, M., K. Tanaka, T. Shintani, T. Takahashi, and N. Kamada. 1999. Chromosomal instability in acute myelocytic leukemia and myelodysplastic syndrome patients among atomic bomb survivors. J Radiat Res 40:159-167.

Nakanishi, M., K. Tanaka, T. Takahashi, T. Kyo, H. Dohy, M. Fujiwara, and N. Kamada. 2001. Microsatellite instability in acute myelocytic leukaemia developed from A-bomb survivors. Int J Radiat Biol 77: 687-694.

Nambi, K.S.V., and S.D. Soman. 1990. Further observations on environmental radiation and cancer in India. Health Phys 59:339-344.

Nambi, K.S., Y.S. Mayya, and S.D. Soman. 1991. A cancer mortality study in Bombay-based atomic energy community: 1975-1987. Indian J Cancer 28:61-69.

Nambi, K.S., Y.S. Mayya, D.D. Rao, and S.D. Soman. 1992. A study on cancer mortality in Tarapur-based atomic energy community. Arch Environ Health 47:155-157.

Narayanan, P.K., E.H. Goodwin, and B.E. Lehnert. 1997. Alpha particles initiate biological production of superoxide anions and hydrogen peroxide in human cells. Cancer Res 57:3963-3971.

Narayanan, P.K., K.E. LaRue, E.H. Goodwin, and B.E. Lehnert. 1999. Alpha particles induce the production of interleukin-8 by human cells. Radiat Res 152:57-63.

Natarajan, A.T., and G. Obe. 1996. Radiation induced chromosomal aberrations. Introduction. Mutat Res 366:65-67.

Nathanson, K.L., and B.L. Weber. 2001. "Other" breast cancer susceptibility genes: searching for more holy grail. Hum Mol Genet 10:715-720.

Nathrath, M.H., V. Kuosaite, M. Rosemann, M. Kremer, C. Poremba, S. Wakana, M. Yanagi, W.B. Nathrath, H. Hofler, K. Imai, and M.J. Atkinson. 2002. Two novel tumor suppressor gene loci on chromosome 6q and 15q in human osteosarcoma identified through comparative study of allelic imbalances in mouse and man. Oncogene 21:5975-5980.

NCRP (National Council on Radiation Protection and Measurements). 1979. Tritium and Other Radionuclide Labelled Organic Compounds Incorporated in Genetic Material. Bethesda, MD.

NCRP (National Council on Radiation Protection and Measurements). 1980. Influence of Dose and Its Distribution in Time on Dose-Response Relationships for Low-LET Radiations, NCRP Report No. 64. Bethesda, MD.

NCRP (National Council on Radiation Protection and Measurements). 1990. The Relative Biological Effectiveness of Radiations of Different Quality, NCRP Report No. 104. Bethesda, MD.

NCRP (National Council on Radiation Protection and Measurements). 1993. Limitating of Exposure to Ionizing Radiation, NCRP Report No. 116. Bethesda, MD.

NCRP (National Council on Radiation Protection and Measurements). 1995. Radiation Exposure and High Altitude Flight, NCRP Commentary No. 12. Bethesda, MD.

NCRP (National Council on Radiation Protection and Measurements). 1997. Uncertainties in Fatal Cancer Risk Estimates Used in Radiation Protection, NCRP Report No. 126. Bethesda, MD.

Neel, J.V. 1998. Genetic studies at the Atomic Bomb Casualty Commission—Radiation Effects Research Foundation: 1946-1997. Proc Natl Acad Sci USA 95:5432-5436.

Neel, J.V., and S.E. Lewis. 1990. The comparative radiation genetics of humans and mice. Annu Rev Genet 24:327-362.

Neel, J.V., and W.J. Schull. 1991. The Children of the Atomic Bomb Survivors: A Genetic Study. Washington, DC: National Academy Press.

Neel, J.V., H. Kato, and W.J. Schull. 1974. Mortality in the children of atomic bomb survivors and controls. Genetics 76:311-336.

Neel, J.V., W.J. Schull, and M. Otake. 1982. Current status of the genetic followup studies Hiroshima and Nagasaki. Pp. 39-51 in Progress in Mutation Research, K.C. Bora and others, eds. Amsterdam: Elsevier Biomedical.

Neel, J.V., W.J. Schull, A.A. Awa, C. Satoh, H. Kato, M. Otake, and Y. Yoshimoto. 1990. The children of parents exposed to atomic bombs: estimates of the genetic doubling dose of radiation for humans. Am J Hum Genet 46:1053-1072.

Neriishi, K., S. Akiba, T. Amano, T. Ogino, and K. Kodama. 1995. Prevalence of hepatitis B surface antigen, hepatitis Be antigen and antibody, and antigen subtypes in atomic bomb survivors. Radiat Res 144: 215-221.

Nichols, K.E., S. Levitz, K.E. Shannon, D.C. Wahrer, D.W. Bell, G. Chang, S. Hegde, D. Neuberg, T. Shafman, N.J. Tarbell, P. Mauch, C. Ishioka, D.A. Haber, and L. Diller. 1999. Heterozygous germline ATM mutations do not contribute to radiation-associated malignancies after Hodgkin's disease. J Clin Oncol 17:1259.

NIH (National Institutes of Health). 1985. Report of the National Institutes of Health Ad Hoc Working Group to Develop Radioepidemiological Tables. Bethesda, MD.

NIH (National Institutes of Health). 2003. Report of the NCI-CDC Working Group to Revise the 1985 NIH Radioepidemiological Tables. NIH Publication No. 03-5387. Bethesda, MD.

Nikiforov, Y.E., M.N. Nikiforova, D.R. Gnepp, and J.A. Fagin. 1996. Prevalence of mutations of ras and p53 in benign and malignant thyroid tumors from children exposed to radiation after the Chernobyl nuclear accident. Oncogene 13:687-693.

Nikjoo, H., P. O'Neill, D.T. Goodhead, and M. Terrissol. 1997. Computational modelling of low-energy electron-induced DNA damage by early physical and chemical events. Int J Radiat Biol 71:467-483.

Nikjoo, H., P. O'Neill, W.E. Wilson, D.T. Goodhead. 2001. Computational approach for determining the spectrum of DNA damage by ionizing radiation. Radiat Res 156:577-583.

Nikjoo, H., C.E. Bolton, R. Watanabe, M. Terrissol, P. O'Neill, and D.T. Goodhead. 2002. Modelling of DNA damage induced by energetic electrons (100 eV to 100 keV). Radiat Prot Dosim 99:77-80.

Niwa, O., and R. Kominami. 2001. Untargeted mutation of the maternally derived mouse hypervariable minisatellite allele in F_1 mice born to irradiated spermatozoa. Proc Natl Acad Sci USA 98:1705-1710.

Niwa, O., Y.J. Fan, M. Numoto, K. Kamiya, and R. Kominami. 1996. Induction of a germline mutation at a hypervariable mouse minisatellite locus by 252Cf radiation. J Radiat Res 37:217-224.

Nomura, T. 1982. Parental exposure to x rays and chemicals induces heritable tumours and anomalies in mice. Nature 296:575-577.

Nomura, T. 1988. X-ray- and chemically-induced germ-line mutation causing phenotypical anomalies in mice. Mutat Res 198:309-320.

Nomura, T. 1989. Congenital malformations as a consequence of parental exposure to radiation and chemicals in mice. J UOEH 11(Suppl): 406-415.

Nomura, T. 1994. Male-mediated teratogenesis: Ionizing radiation/ ethylnitrosourea studies. Pp. 117-128 in Male Mediated Developmental Toxicity, A.F. Olshan and D.R. Mattison, eds. New York: Plenum Press.

Noshchenko, A.G., K.B. Moysich, A. Bondar, P.V. Zamostyan, V.D. Drosdova, and A.M. Michalek. 2001. Patterns of acute leukaemia occurrence among children in the Chernobyl region. Int J Epidemiol 30:125-129.

Noshchenko, A.G., P.V. Zamostyan, O.Y. Bondar, and V.D. Drozdova. 2002. Radiation-induced leukemia risk among those aged 0-20 at the time of the Chernobyl accident: a case-control study in the Ukraine. Int J Cancer 99:609-618.

NRC (National Research Council). 1947. Genetics Conference, Committee on Atomic Casualties. Genetic effects of the atomic bombs in Hiroshima and Nagasaki. Science 106:331-333.

NRC (National Research Council). 1956. Biological Effects of Atomic Radiation (BEAR I). Washington, DC: National Academy Press.

NRC (National Research Council). 1972. The Effects on Populations of Exposure to Low Levels of Ionizing Radiation (BEIR I). Washington, DC: National Academy Press.

NRC (National Research Council). 1980. The Effects on Populations of Exposure to Low Levels of Ionzing Radiation (BEIR III). Washington, DC: National Academy Press.

NRC (National Research Council). 1988. Health Effects of Radon and Other Internally Deposited Alpha-Emitters (BEIR IV). Washington, DC: National Academy Press.

NRC (National Research Council). 1990. Health Effects of Exposure to Low Levels of Ionzing Radiation (BEIR V). Washington, DC: National Academy Press.

NRC (National Research Council). 1999. Health Effects of Exposure to Radon (BEIR VI). Washington, DC: National Academy Press.

NRC (National Research Council). 2000. A Review of the Draft Report of the NCI-CDC Working Group to Revise the 1985 Radioepidemiological Tables. Washington, DC: National Academy Press.

NRPB (National Radiological Protection Board). 1999. Report of an Advisory Group on Ionising Radiation.

NRPB (National Radiological Protection Board). 2000. Report of an Advisory Group on Ionising Radiation.

Oakberg, E.F., and E. Clark. 1964. Species comparisons of radiation response of the gonads. Pp. 11-24 in Effects of Radiation on the Reproductive System, W.D. Carlson and F.X. Gassner, eds. Oxford: Pergamon Press.

Obe, G., F. Palitti, C. Tanzarella, F. Degrassi, and R. De Salvia. 1985. Chromosomal aberrations induced by restriction endonucleases. Mutat Res 150:359-368.

O'Connor, M.M., and M.R. Mayberg. 2000. Effects of radiation on cerebral vasculature: a review. Neurosurgery 46:138-151.

Okayasu, R., K. Suetomi, Y. Yu, A. Silver, J.S. Bedford, R. Cox, and R.L. Ullrich. 2000. A deficiency in DNA repair and DNA-PKcs expression in the radiosensitive BALB/c mouse. Cancer Res 60:4342-4345.

Okeanov A.E., S.M. Polyakov, H.H. Storm, A. Sobolev, and R. Winkelman. 1996. Development of cancer registration system in Belarus. The Radiological Consequences of the Chernobyl Accident. Proceedings of the First International Conference, Minsk, Belarus, A. Karaoglou, G. Desmet, G.N. Kelly, and others, eds. ECSC-EC-EAEC, Brussels, Luxembourg.

Olivieri, G., J. Bodycote, and S. Wolff. 1984. Adaptive response of human lymphocytes to low concentrations of radioactive thymidine. Science 223:594-597.

Olsen, J.H., J.M. Hahnemann, A.L. Borresen-Dale, K. Brondum-Nielsen, L. Hammarstrom, R. Kleinerman, H. Kaariainen, T. Lonnqvist, R. Sankila, N. Seersholm, S. Tretli, J. Yuen, J.D. Boice Jr., and M. Tucker. 2001. Cancer in patients with ataxia-telangiectasia and in their relatives in the Nordic countries. J Natl Cancer Inst 93:121-127.

Omar, R.Z., J.A. Barber, and P.G. Smith. 1999. Cancer mortality and morbidity among plutonium workers at the Sellafield plant of British Nuclear Fuels. Brit J Cancer 79:1288-1301.

Orkin, S.H. 1996. Development of the hematopoietic system. Curr Opin Genet Dev 6:597-602.

Osechinsky I.V., and A.R. Martirosov. 1995. Haematological diseases in the Belarus Republic after the Chernobyl accident. Presented at the International Conference on Health Consequences of the Chernobyl and Other Radiological Accidents, World Health Organization, Geneva.

Otake, M., W.J. Schull, and J.V. Neel. 1990. The effects of parental exposure to the atomic bombings of Hiroshima and Nagasaki on congenital malformations, still births, and early mortality among their children: a reanalysis. Radiat Res 122:1-11.

Pacini, F., T. Vorontsova, E.P. Demidchik, E. Molinaro, L. Agate, C. Romei, E. Shavrova, E.D. Cherstvoy, Y. Ivashkevitch, E. Kuchinskaya, M. Schlumberger, G. Ronga, M. Filesi, and A. Pinchera. 1997. Post-Chernobyl thyroid carcinoma in Belarus children and adolescents: comparison with naturally occurring thyroid carcinoma in Italy and France. J Clin Endocr Metab 82:3563-3569.

Pacini, F., T. Vorontsova, E. Molinaro, E. Kuchinskaya, L. Agate, E. Shavrova, L. Astachova, L. Chiovato, and A. Pinchera. 1998. Prevalence of thyroid autoantibodies in children and adolescents from Belarus exposed to the Chernobyl radioactive fallout. Lancet 352:763-766.

Padovani, L., M. Appolloni, P. Anzidei, B. Tedeschi, D. Caporossi, P. Vernole, and F. Mauro. 1995. Do human lymphocytes exposed to the fallout of the Chernobyl accident exhibit an adaptive response? 1. Challenge with ionizing radiation. Mutat Res 332:33-38.

Pala, F.S., J.E. Moquet, A.A. Edwards, and D.C. Lloyd. 2001. In vitro transmission of chromosomal aberrations through mitosis in human lymphocytes. Mutat Res 474:139-146.

Parikh, S.S., C.D. Mol, G. Slupphaug, S. Bharati, H.E. Krokan, and J.A. Tainer. 1998. Base excision repair initiation revealed by crystal structures and binding kinetics of human uracil-DNA glycosylase with DNA. EMBO J 17:5214-5226.

Parker, L., A.W. Craft, J. Smith, H. Dickinson, R. Wakeford, K. Binks, D. McElvenny, L. Scott, and A. Slovak. 1993. Geographical distribution of preconceptional radiation doses to fathers employed at the Sellafield nuclear installation, West Cumbria. Brit Med J 307:966-971.

Parker, L., M.S. Pearce, H.O. Dickinson, M. Aitkin, and A.W. Craft. 1999. Stillbirths among offspring of male radiation workers at Sellafield nuclear reprocessing plant. Lancet 354:1407-1414.

Parkin, D.M., E. Cardis, E. Masuyer, H.P. Friedl, H. Hansluwka, D. Bobev, E. Ivanov, J. Sinnaeve, J. Augustin, I. Plesko, and others. 1993. Childhood leukaemia following the Chernobyl accident. The European Leukaemia-Lymphoma Incidence Study (ECLIS). Eur J Cancer 29A:87-95.

Parkin, D.M., D. Clayton, R.J. Black, E. Masuyer, H.P. Friedl, E. Ivanov, J. Sinnaeve, C.G. Tzvetansky, E. Geryk, H.H. Storm, M. Rahu, E. Pukkala, J.L. Bernard, P.M. Carli, M.C. L'Huilluier, F. Menegoz, P. Schaffer, S. Schraub, P. Kaatsch, J. Michaelis, E. Apjok, D. Schuler, P. Crosignani, C. Magnani, B.G. Bennett, and others. 1996. Childhood leukaemia in Europe after Chernobyl: 5 year follow-up. Brit J Cancer 73:1006-1012.

Parshad, R., F.M. Price, V.A. Bohr, K.H. Cowans, J.A. Zujewski, and K.K. Sanford. 1996. Deficient DNA repair capacity, a predisposing factor in breast cancer. Brit J Cancer 74:1-5.

Parsonnet, J., S. Hansen, L. Rodriguez, A.B. Gelb, R.A. Warnke, E. Jellum, N. Orentreich, J.H. Vogelman, and G.D. Friedman. 1994. *Helicobacter pylori* infection and gastric lymphoma. N Engl J Med 330:1267-1271.

Parvin, J.D. 2001. BRCA1 at a branch point. Proc Natl Acad Sci USA 98:5952-5954.

Pazzaglia, S., M. Mancuso, M.J. Atkinson, M. Tanori, S. Rebessi, V.D. Majo, V. Covelli, H. Hahn, and A. Saran. 2002. High incidence of medulloblastoma following x-ray-irradiation of newborn Ptc1 heterozygous mice. Oncogene 21:7580-7584.

Peissel, B., D. Zaffaroni, S. Pazzaglia, G. Manenti, N. Zanesi, I. Zedda, S. Rebessi, V. Covelli, T.A. Dragani, and A. Saran. 2001. Use of intercross outbred mice and single nucleotide polymorphisms to map skin cancer modifier loci. Mamm Genome 12:291-294.

Petcu, I., D. Moisoi, and G.J. Koteles. 1997. In vitro and in vivo effects of low-dose HTO contamination modulated by dose rate. Pp. 312-315 in Low Doses of Ionizing Radiation: Biological Effects and Regulatory Control, IAEA-CN-67/95: IAEA-TECDOC-976.

Petersen, G.R., E.S. Gilbert, J.A. Buchanan, and R.G. Stevens. 1990. A case-cohort study of lung cancer, ionizing radiation, and tobacco smoking among males at the Hanford Site. Health Phys 58:3-11.

Peto, J., and T.M. Mack. 2000. High constant incidence in twins and other relatives of women with breast cancer. Nat Genet 26:411-414.

Petridou, E., C. Proukakis, D. Tong, D. Kassimos, F. Athanassiadou-Piperopoulou, S. Haidas, M. Kalmanti, D. Koliouskas, H. Kosmidis, A. Louizi, and 1 other. 1994. Trends and geographical distribution of childhood leukemia in Greece in relation to the Chernobyl accident. Scand J Soc Med 22:127-131.

Petridou, E., D. Trichopoulos, N. Dessypris, V. Flytzani, S. Haidas, M. Kalmanti, D. Koliouskas, H. Kosmidis, F. Piperopoulou, and F. Tzortzatou. 1996. Infant leukaemia after in utero exposure to radiation from Chernobyl. Nature 382:352-353.

Petrini, J.H. 1999. The mammalian Mre11-Rad50-nbs1 protein complex: integration of functions in the cellular DNA-damage response. Am J Hum Genet 64:1264-1269.

Pettersson, F., S. Fotiou, N. Einhorn, and C. Silfversward. 1985. Cohort study of the long-term effect of irradiation for carcinoma of the uterine cervix. Second primary malignancies in the pelvic organs in women irradiated for cervical carcinoma at Radiumhemmet 1914-1965. Acta Radiol Oncol 24:145-151.

Phelan, C.M., T.R. Rebbeck, B.L. Weber, P. Devilee, M.H. Rutledge, H.T. Lynch, G.M. ReLenoir, M.R. Stratton, D.F. Eason, and S.A. Narod. 1996. Ovarian cancer risk in BRCA1 carriers is modified by the HRAS1 variable number of tandem repeat (VNTR) locus. Nat Genet 12: 309-311.

Phillips, J.W., and W.F. Morgan. 1994. Illegitimate recombination induced by DNA double-strand breaks in a mammalian chromosome. Mol Cell Biol 14:5794-5803.

Phillips, R.J.S. 1961. A comparison of the mutations induced by acute and chronic gamma irradiation. Brit J Radiol 34:261-264.

Pierce, D.A. 2002. Age-time patterns of radiogenic cancer risk: their nature and likely explanations. J Radiol Prot 22:A147-A154.

Pierce, D.A., and M.L. Mendelsohn. 1999. A model for radiation-related cancer suggested by atomic bomb survivor data. Radiat Res 152:642-654.

Pierce, D.A., and D.L. Preston. 1993. Joint analysis of site-specific cancer risks for the atomic bomb survivors. Radiat Res 134:134-142.

Pierce, D.A., and D.L. Preston. 2000. Radiation-related cancer risks at low doses among atomic bomb survivors. Radiat Res 154:178-186.

Pierce, D.A., and M. Vaeth. 1989. Cancer Risk Estimation from the Atomic Bomb Survivors: Extrapolation to Low Doses, Use of Relative Risk Models, and Other Uncertainties. London: Taylor and Francis.

Pierce, D.A., and M. Vaeth. 1991. The shape of the cancer mortality dose-response curve for the A-bomb survivors. Radiat Res 126:36-42.

Pierce, D.A., D.O. Stram, and M. Vaeth. 1990. Allowing for random errors in radiation dose estimates for the atomic bomb survivor data. Radiat Res 123:275-284.

Pierce, D.A., Y. Shimizu, D.L. Preston, M. Vaeth, and K. Mabuchi. 1996. Studies of the mortality of atomic bomb survivors. Report 12, Part I. Cancer: 1950-1990. Radiat Res 146:1-27.

Pierce, D.A., G.B. Sharp, and K. Mabuchi. 2003. Joint effects of radiation and smoking on lung cancer risk among atomic bomb survivors. Radiat Res 159:511-520.

Pierce, L.J., M. Strawderman, S.A. Narod, I. Oliviotto, A. Eisen, L. Dawson, D. Gaffney, L.J. Solin, A. Nixon, J. Garber, C. Berg, C. Isaacs, R. Heimann, O.I. Olopade, B. Haffty, and B.L. Weber. 2000. Effect of radiotherapy after breast-conserving treatment in women with breast cancer and germline BRCA1/2 mutations. J Clin Oncol 18:3360-3369.

Pihan, G.A., A. Purohit, J. Wallace, H. Knecht, B. Woda, P. Quesenberry, and S.J. Doxsey. 1998. Centrosome defects and genetic instability in malignant tumors. Cancer Res 58:3974-3985.

Pitkevitch, V.A., V.K. Ivanov, A.F. Tsyb, M.A. Maksyoutov, V.A. Matiash, and N.V. Shchukina. 1997. Exposure levels for persons involved in recovery operations after the Chernobyl accident. Statistical analysis based on the data of the Russian National Medical and Dosimetric Registry (RNMDR). Radiat Environ Biophys 36:149-160.

Pobel, D., and J.F. Viel. 1997. Case-control study of leukaemia among young people near La Hague nuclear reprocessing plant: the environmental hypothesis revisited. Brit Med J 314:101-106.

Pohl-Ruling, J., P. Fischer, O. Haas, G. Obe, A.T. Natarajan, P.P. van Buul, K.E. Buckton, N.O. Bianchi, M. Larramendy, M. Kucerova, Z. Polikova, A. Leonard, L. Fabry, F. Palitti, T. Sharma, W. Binder, R.N. Mukherjee, and U. Mukherjee. 1983. Effect of low-dose acute x-irradiation on the frequencies of chromosomal aberrations in human peripheral lymphocytes in vitro. Mutat Res 110:71-82.

Polednak, A.P., and E.L. Frome. 1981. Mortality among men employed between 1943 and 1947 at a uranium-processing plant. J Occup Med 23:169-178.

Pollycove, M., and L.E. Feinendegen. 2003. Radiation-induced versus endogenous DNA damage: possible effect of inducible protective responses in mitigating endogenous damage. Hum Exp Toxicol 22:290-306; discussion 307, 315-317, 319-323.

Ponder, B.A. 2001. Cancer genetics. Nature 411:336-341.

Ponnaiya, B., M.N. Cornforth, and R.L. Ullrich. 1997. Radiation-induced chromosomal instability in BALB/c and C57BL/6 mice: the difference is as clear as black and white. Radiat Res 147:121-125.

Pope, C.A., 3rd, R.T. Burnett, M.J. Thun, E.E. Calle, D. Krewski, K. Ito, and G.D. Thurston. 2002. Lung cancer, cardiopulmonary mortality, and long-term exposure to fine particulate air pollution. J Am Med Assoc 287:1132-1141.

Potten, C.S., and J. Hendry, eds. 1997. Radiation and Gut. Amsterdam: Elsevier Science.

Pottern, L.M., M.M. Kaplan, P.R. Larsen, J.E. Silva, R.J. Koenig, J.H. Lubin, M. Stovall, and J.D. Boice Jr. 1990. Thyroid nodularity after childhood irradiation for lymphoid hyperplasia: a comparison of questionnaire and clinical findings. J Clin Epidemiol 43:449-460.

Poupon, M.F., K.A. Smith, O.B. Chernova, C. Gilbert, and G.R. Stark. 1996. Inefficient growth arrest in response to dNTP starvation stimulates gene amplification through bridge-breakage-fusion cycles. Mol Biol Cell 7:345-354.

Powell, S.N., and L.A. Kachnic. 2003. Roles of BRCA1 and BRCA2 in homologous recombination, DNA replication fidelity and the cellular response to ionizing radiation. Oncogene 22:5784-5791.

Prentice, R.L., and M.W. Mason. 1986. On the application of linear relative risk regression models. Biometrics 42:109-120.

Prentice, R.L., Y. Yoshimoto, and M.W. Mason. 1983. Relationship of cigarette smoking and radiation exposure to cancer mortality in Hiroshima and Nagasaki. J Natl Cancer Inst 70:611-622.

Preston, D. 1995. Cancer risks and biomarker studies in the atomic bomb survivors. Stem Cells 13:40-48.

Preston, D.L., J.H. Lubin, and D.A. Pierce. 1991. EPICURE User's Guide. Seattle: Hirosoft International.

Preston, D.L., S. Kusumi, M. Tomonaga, S. Izumi, E. Ron, A. Kuramoto, N. Kamada, H. Dohy, T. Matsuo, H. Nonaka, D.E. Thompson, M. Soda, and K. Mabuchi. 1994. Cancer incidence in atomic bomb survivors. Part III. Leukemia, lymphoma and multiple myeloma, 1950-1987. Radiat Res 137:S68-S97.

Preston, D.L., A. Mattsson, E. Holmberg, R. Shore, N.G. Hildreth, and J.D. Boice Jr. 2002a. Radiation effects on breast cancer risk: a pooled analysis of eight cohorts. Radiat Res 158:220-235.

Preston, D.L., E. Ron, S. Yonehara, T. Kobuke, H. Fujii, M. Kishikawa, M. Tokunaga, S. Tokuoka, and K. Mabuchi. 2002b. Tumors of the nervous system and pituitary gland associated with atomic bomb radiation exposure. J Natl Cancer Inst 94:1555-1563.

Preston, D.L., Y. Shimizu, D.A. Pierce, A. Suyama, and K. Mabuchi. 2003. Studies of mortality of atomic bomb survivors. Report 13: Solid cancer and noncancer disease mortality: 1950-1997. Radiat Res 160:381-407.

Preston, D.L., D.A. Pierce, Y. Shimizu, H.M. Cullings, S. Fujita, S. Funamoto, and K. Kodama. 2004. Effect of recent changes in atomic bomb survivor dosimetry on cancer mortality risk estimates. Radiat Res 162:377-389.

Preston-Martin, S., D.C. Thomas, S.C. White, and D. Cohen. 1988. Prior exposure to medical and dental x-rays related to tumors of the parotid gland. J Natl Cancer Inst 80:943-949.

Pretsch, W., J. Favor, W. Lehmacher, and A. Neuhauser-Klaus. 1994. Estimates of the radiation-induced mutation frequencies to recessive visible, dominant cataract and enzyme-activity alleles in germ cells of AKR, BALB/c, DBA/2 and (102xC3H)F1 mice. Mutagenesis 9: 289-294.

Price, E.A., S.L. Bourne, R. Radbourne, P.A. Lawton, J. Lamerdin, L.H. Thompson, and J.E. Arrand. 1997. Rare microsatellite polymorphisms in the DNA repair genes XRCC1, XRCC3 and XRCC5 associated with cancer in patients of varying radiosensitivity. Somat Cell Mol Genet 23:237-247.

Prise, K.M., M. Folkard, H.C. Newman, and B.D. Michael. 1994. Effect of radiation quality on lesion complexity in cellular DNA. Int J Radiat Biol 66:537-542.

Prise, K.M., O.V. Belyakov, M. Folkard, and B.D. Michael. 1998. Studies of bystander effects in human fibroblasts using a charged particle microbeam. Int J Radiat Biol 74:793-798.

Prise, K.M., M. Folkard, and B.D. Michael. 2003. Bystander responses induced by low LET radiation. Oncogene 22:7043-7049.

Prisyazhiuk, A., O.A. Pjatak, V.A. Buzanov, G.K. Reeves, and V. Beral. 1991. Cancer in the Ukraine, post-Chernobyl. Lancet 338:1334-1335.

Prisyazhniuk, A., V. Gristchenko, V. Zakordonets, N. Fouzik, Y. Slipeniuk, and I. Ryzhak. 1995. The time trends of cancer incidence in the most contaminated regions of the Ukraine before and after the Chernobyl accident. Radiat Environ Biophys 34:3-6.

Prisyazhniuk, A., Z. Fedorenko, A. Okaenov, V. Ivanov, V. Starinsky, V. Gristchenko, and L. Remennik. 1996. Epidemiology of cancer in populations living in contaminated territories of Ukraine, Belarus, Russia after the Chernobyl accident. Pp. 909-921 in The Radiological Consequences of the Chernobyl Accident. Brussels: European Commission, EUR 16544.

Prysyazhnyuk A.Y., L.O. Gulak, V.G. Gristchyenko, and Z.P. Fedorenko. 2002. Cancer incidence in Ukraine after the Chernobyl accident. In Chernobyl: Message for the 21st Century. Proceedings of the Sixth Chernobyl Sasakawa Medical Cooperation Symposium. New York: Elsevier.

Puck, T.T., and C.A. Waldren. 1987. Mutation in mammalian cells: theory and implications. Somat Cell Mol Genet 13:405-409.

Puck, T.T., R. Johnson, and S. Rasumussen. 1997. A system for mutation measurement in mammalian cells: application to gamma-irradiation. Proc Natl Acad Sci USA 94:1218-1223.

Rabbitts, T.H. 1994. Chromosomal translocations in human cancer. Nature 372:143-149.

Rabes, H.M., E.P. Demidchik, J.D. Sidorow, E. Lengfelder, C. Beimfohr, D. Hoelzel, and S. Klugbauer. 2000. Pattern of radiation-induced RET and NTRK1 rearrangements in 191 post-Chernobyl papillary thyroid carcinomas: biological, phenotypic, and clinical implications. Clin Cancer Res 6:1093-1103.

Rafii, S., P. O'Regan, G. Xinarianos, I. Azmy, T. Stephenson, M. Reed, M. Meuth, J. Thacker, and A. Cox. 2002. A potential role for the XRCC2 R188H polymorphic site in DNA-damage repair and breast cancer. Hum Mol Genet 11:1433-1438.

Rahu, M., M. Tekkel, T. Veidebaum, E. Pukkala, T. Hakulinen, A. Auvinen, T. Rytomaa, P.D. Inskip, and J.D. Boice. 1997. The Estonian study of Chernobyl cleanup workers: II. Incidence of cancer and mortality. Radiat Res 147:653-657.

Rallison, M.L., B.M. Dobyns, F.R. Keating Jr., J.E. Rall, and F.H. Tyler. 1975. Thyroid nodularity in children. J Am Med Assoc 233:1069-1072.

Ramsay, C.N., P.M. Ellis, and H. Zealley. 1991. Down's syndrome in the Lothian region of Scotland—1978 to 1989. Biomed Pharmacother 45:267-272.

Ranier, S., L.A. Johnson, C.J. Dobry, A.J. Ping, P.E. Grundy, and A.P. Feinberg. 1993. Relaxation of imprinted genes in human cancer. Nature 362:749-751.

Redpath, J.L., and R.J. Antoniono. 1998. Induction of an adaptive response against spontaneous neoplastic transformation in vitro by low-dose gamma radiation. Radiat Res 149:517-520.

Redpath, J.L., D. Liang, T.H. Taylor, C. Christie, and E. Elmore. 2001. The shape of the dose-response curve for radiation-induced neoplastic transformation in vitro: evidence for an adaptive response against neoplastic transformation at low doses of low-LET radiation. Radiat Res 156:700-707.

Remennik, L.V., V.V. Starinsky, V.I. Chissov, V.D. Mokina, and M.M. Rubtsova. 1996. Malignant neoplasm of thyroid in children and in adults on the territories suffered from the Chernobyl accident. Pp. 519-520 in International Conference One Decade After Chernobyl: Summing up the Consequences of the Accident, Book of Extended Synopses. Vienna: International Atomic Energy Agency.

Ribeiro, R.C., F. Sandrini, B. Figueiredo, G.P. Zambetti, E. Michalkiewicz, A.R. Lafferty, L. DeLacerda, M. Rabin, C. Cadwell, G. Sampaio, I. Cat, C.A. Stratakis, and R. Sandrini. 2001. An inherited p53 mutation that contributes in a tissue-specific manner to pediatric adrenal cortical carcinoma. Proc Natl Acad Sci USA 98:9330-9335.

Richardson, D.B., and S. Wing. 1998. Methods for investigating age differences in the effects of prolonged exposures. Am J Ind Med 33:123-130.

Richardson, D.B., and S. Wing. 1999a. Greater sensitivity to ionizing radiation at older age: follow-up of workers at Oak Ridge National Laboratory through 1990. Int J Epidemiol 28:428-436.

Richardson, D.B., and S. Wing. 1999b. Radiation and mortality of workers at Oak Ridge National Laboratory: positive associations for doses received at older ages. Environ Health Persp 107:649-656.

Richardson, S., C. Monfort, M. Green, G. Draper, and C. Muirhead. 1995. Spatial variation of natural radiation and childhood leukaemia incidence in Great Britain. Stat Med 14:2487-2501.

Richter, H.E., H.D. Lohrer, L. Hieber, A.M. Kellerer, E. Lengfelder, and M. Bauchinger. 1999. Microsatellite instability and loss of heterozygosity in radiation-associated thyroid carcinomas of Belarussian children and adults. Carcinogenesis 20:2247-2252.

Rigaud, O., A. Laquerbe, and E. Moustacchi. 1995. DNA sequence analysis of HPRT-mutants induced in human lymphoblastoid cells adapted to ionizing radiation. Radiat Res 144:181-189.

Rinsky, R.A., R.D. Zumwalde, R.J. Waxweiler, W.E. Murray Jr., P.J. Bierbaum, P.J. Landrigan, M. Terpilak, and C. Cox. 1981. Cancer mortality at a naval nuclear shipyard. Lancet 1:231-235.

Rinsky, R.A., J.M. Melius, R.W. Hornung, R.D. Zumwalde, R.J. Waxweiler, P.J. Landrigan, P.J. Bierbaum, and W.E. Murray Jr. 1988. Case-control study of lung cancer in civilian employees at the Portsmouth Naval Shipyard, Kittery, Maine. Am J Epidemiol 127:55-64.

Ritz, B. 1999. Radiation exposure and cancer mortality in uranium processing workers. Epidemiology 10:531-538.

Ritz, B., H. Morgenstern, J. Froines, and B.B. Young. 1999a. Effects of exposure to external ionizing radiation on cancer mortality in nuclear workers monitored for radiation at Rocketdyne/Atomics International. Am J Ind Med 35:21-31.

Ritz, B., H. Morgenstern, and J. Moncau. 1999b. Age at exposure modifies the effects of low-level ionizing radiation on cancer mortality in an occupational cohort. Epidemiology 10:135-140.

Ritz, B., H. Morgenstern, D. Crawford-Brown, and B. Young. 2000. The effects of internal radiation exposure on cancer mortality in nuclear workers at Rocketdyne/Atomics International. Environ Health Persp 108:743-751.

Roberts, S.A., A.R. Spreadborough, B. Bulman, J.B. Barber, D.G. Evans, and D. Scott. 1999. Heritability of cellular radiosensitivity: a marker of low-penetrance predisposition genes in breast cancer? Am J Hum Genet 65:784-794.

Robison, L.L., and A. Mertens. 1993. Second tumors after treatment of childhood malignancies. Hematol Oncol Clin North Am 7:401-415.

Robson, T.A., H. Lohrer, J.R. Bailie, D.G. Hirst, M.C. Joiner, and J.E. Arrand. 1997. Gene regulation by low-dose ionizing radiation in a normal human lung epithelial cell line. Biochem Soc Trans 25:335-342.

Robson, T., M.C. Joiner, G.D. Wilson, W. McCullough, M.E. Price, I. Logan, H. Jones, S.R. McKeown, and D.G. Hirst. 1999. A novel human stress response-related gene with a potential role in induced radioresistance. Radiat Res 152:451-461.

Robson, T., M.E. Price, M.L. Moore, M.C. Joiner, V.J. McKelvey-Martin, S.R. McKeown, and D.G. Hirst. 2000. Increased repair and cell survival in cells treated with DIR1 antisense oligonucleotides: implications for induced radioresistance. Int J Radiat Biol 76:617-623.

Roderick, T.H. 1963. The response of twenty-seven inbred strains of mice to daily doses of whole body x-irradiation. Radiat Res 20:613-614.

Roesch, W.C., Radiation Effects Research Foundation, and National Academy of Sciences (U.S.). 1987. US-Japan joint reassessment of atomic bomb radiation dosimetry in Hiroshima and Nagasaki. Minami-ku, Hiroshima: Radiation Effects Research Foundation.

Rogel A, N. Carré, E. Amoros, M. Bonnet-Belfais, M. Goldberg, E. Imbernon, T. Calvez, and C. Hill. 2005. Mortality of workers exposed to ionizing radiation at the French National Electricity Company. Am J Ind Med 47:72–82.

Roman, E., A. Watson, V. Beral, S. Buckle, D. Bull, K. Baker, H. Ryder, and C. Barton. 1993. Case-control study of leukaemia and non-Hodgkin's lymphoma among children aged 0-4 years living in west Berkshire and north Hampshire health districts. Brit Med J 306: 615-621.

Roman, E., P. Doyle, N. Maconochie, G. Davies, P.G. Smith, and V. Beral. 1999. Cancer in children of nuclear industry employees: report on children aged under 25 years from nuclear industry family study. Brit Med J 318:1443-1450.

Romanenko, A., L. Morell-Quadreny, V. Nepomnyaschy, A. Vozianov, and A. Llombart-Bosch. 2000. Pathology and proliferative activity of renal-cell carcinomas (RCCS) and renal oncocytomas in patients with different radiation exposure after the Chernobyl accident in Ukraine. Int J Cancer 87:880-883.

Romanenko, A., K. Morimura, M. Wei, W. Zaparin, A. Vozianov, and S. Fukushima. 2002. DNA damage repair in bladder urothelium after the Chernobyl accident in Ukraine. J Urol 168:973-977.

Romanenko, A., K. Morimura, H. Wanibuchi, M. Wei, W. Zaparin, W. Vinnichenko, A. Kinoshita, A. Vozianov, and S. Fukushima. 2003. Urinary bladder lesions induced by persistent chronic low-dose ionizing radiation. Cancer Sci 94:328-333.

Romney, C.A., J.D. Paulauskis, H. Nagasawa, and J.B. Little. 2001. Multiple manifestations of x-ray-induced genomic instability in Chinese hamster ovary (CHO) cells. Mol Carcinog 32:118-127.

Ron, E. 1998. Ionizing radiation and cancer risk: evidence from epidemiology. Radiat Res 150:S30-S41.

Ron, E., and B. Modan. 1984. Thyroid and other neoplasms following childhood scalp irradiation. Radiation Carcinogenesis: Epidemiology and Biological Significance, J.D. Boice and J.F Fraumeni, eds. New York: Raven Press.

Ron, E., B. Modan, J.D. Boice, E. Alfandary, M. Stovall, A. Chetrit, and L. Katz. 1988a. Tumors of the brain and nervous system after radiotherapy in childhood. N Engl J Med 319:1033-1039.

Ron, E., B. Modan, and J.D. Boice Jr. 1988b. Mortality after radiotherapy for ringworm of the scalp. Am J Epidemiol 127:713-725.

Ron, E., B. Modan, D. Preston, E. Alfandary, M. Stovall, and J.D. Boice Jr. 1989. Thyroid neoplasia following low-dose radiation in childhood. Radiat Res 120:516-531.

Ron, E., B. Modan, D. Preston, S. Hamburger, and M. Stovall. 1994. Mortality following radiation treatment for infertility of hormonal origin or amenorrhoea. Int J Epidemiol 23:1165-1173.

Ron, E., J.H. Lubin, R.E. Shore, K. Mabuchi, B. Modan, L. Pottern, A.B. Schneider, M.A. Tucker, and J.D. Boice. 1995a. Thyroid cancer after exposure to external radiation: a pooled analysis of seven studies. Radiat Res 141:259-277.

Ron, E., F.L. Wong, and K. Mabuchi. 1995b. Incidence of benign gastrointestinal tumors among atomic bomb survivors. Am J Epidemiol 142:68-75.

Ron, E., M.M. Doody, D.V. Becker, A.B. Brill, R.E. Curtis, M.B. Goldman, B.S. Harris 3rd, D.A. Hoffman, W.M. McConahey, H.R. Maxon, S. Preston-Martin, M.E. Warshauer, F.L. Wong, and J.D. Boice Jr. 1998a. Cancer mortality following treatment for adult hyperthyroidism. Cooperative Thyrotoxicosis Therapy Follow-Up Study Group. J Am Med Assoc 280:347-355.

Ron, E., D.L. Preston, M. Kishikawa, T. Kobuke, M. Iseki, S. Tokuoka, M. Tokunaga, and K. Mabuchi. 1998b. Skin tumor risk among atomic-bomb survivors in Japan. Cancer Cause Control 9:393-401.

Rooney, C., V. Beral, N. Maconochie, P. Fraser, and G. Davies. 1993. Case-control study of prostatic cancer in employees of the United Kingdom Atomic Energy Authority. Brit Med J 307:1391-1397.

Roots, R., and S. Okada. 1972. Protection of DNA molecules of cultured mammalian cells from radiation-induced single-strand scissions by various alcohols and SH compounds. Int J Radiat Biol Res 21:329-342.

Rosemann, M., M. Lintrop, J. Favor, and M.J. Atkinson. 2002. Bone tumorigenesis induced by alpha-particle radiation: mapping of genetic loci influencing predisposition in mice. Radiat Res 157:426-434.

Rosenquist, T.A., E. Zaika, A.S. Fernandes, D.O. Zharkov, H. Miller, and A.P. Grollman. 2003. The novel DNA glycosylase, NEIL1, protects mammalian cells from radiation-mediated cell death. DNA Repair 2:581-591.

Rossi, H.H., and M. Zaider. 1996. Microdosimetry and Its Applications. Berlin: Springer-Verlag.

Roth, S.Y. 1996. Something about silencing. Nat Genet 14:3-4.

Rothkamm, K., and M. Lobrich. 2003. From the cover: evidence for a lack of DNA double-strand break repair in human cells exposed to very low x-ray doses. Proc Natl Acad Sci USA 100:5057-5062.

Rothkamm, K., I. Kruger, L.H. Thompson, and M.Lobrich. 2003. Pathways of DNA double-strand break repair during the mammalian cell cycle. Mol Cell Biol 23:5706-5715.

Rubino, C., E. Adjadj, S. Guerin, C. Guibout, A. Shamsaldin, M.G. Dondon, D. Valteau-Couanet, O. Hartmann, M. Hawkins, and F. de Vathaire. 2003. Long-term risk of second malignant neoplasms after neuroblastoma in childhood: role of treatment. Int J Cancer 107:791-796.

Rudolph, K.L., M. Millard, M.W. Bosenberg, and R.A. DePinho. 2001. Telomere dysfunction and evolution of intestinal carcinoma in mice and humans. Nat Genet 28:155-159.

Russell, L.B. 1999. Significance of the perigametic interval as a major source of spontaneous mutations that result in mosaics. Environ Mol Mutagen 34:16-23.

Russell, L.B., and W.L. Russell. 1996. Spontaneous mutations recovered as mosaics in the mouse specific-locus test. Proc Natl Acad Sci USA 93:13072-13077.

Russell, W.L. 1965. Studies in mammalian radiation genetics. Nucleonics 23:53-56, 62.

Russell, W.L. 1968. Recent studies on the genetic effects of radiation in mice. Pediatrics 41:223-230.

Russell, W.L., L.B. Russell, and E.M. Kelly. 1958. Radiation dose rate and mutation frequency. Science 128:1546-1550.

Russell, W.L., L.B. Russell, and M.B. Cupp. 1959. Dependence of mutation frequency on radiation dose rate in female mice. Proc Natl Acad Sci USA 45:18-23.

Russell, W.L., J.W. Bangham, and L.B. Russell. 1998. Differential response of mouse male germ-cell stages to radiation-induced specific-locus and dominant mutations. Genetics 148:1567-1578.

Rutledge, J.C., K.T. Cain, L.A. Hughes, P.W. Braden, and W.M. Generoso. 1986. Difference between two hybrid stocks of mice in the incidence of congenital abnormalities following x-ray exposure of stem-cell spermatogonia. Mutat Res 163:299-302.

Ryberg, M., M. Lundell, B. Nilsson, and F. Pettersson. 1990. Malignant disease after radiation treatment of benign gynaecological disorders. A study of a cohort of metropathia patients. Acta Oncol 29:563-567.

Rydberg, B. 1996. Clusters of DNA damage induced by ionizing radiation: formation of short DNA fragments. 2. Experimental detection. Radiat Res 145:200-209.

Sabatier, L., B. Dutrillaux, and M.B. Martin. 1992. Chromosomal instability. Nature 357:548.

Sabatier, L., M. Ricoul, G. Pottier, N. Mathieu, and J.P. Murnane. 2005. The loss of a single telomere can result in instability of multiple chromosomes in a human tumor cell line. Mol Can Res 3:139-150.

Sacher, G.A. 1955. A comparative analysis of radiation lethality in mammals exposed at constant average intensity for the duration of life. J Natl Cancer Inst 15:1125-1144.

Sachs, R.K., A.M. Chen, and D.J. Brenner. 1997. Review: proximity effects in the production of chromosome aberrations by ionizing radiation. Int J Radiat Biol 71:1-19.

Sadamoto, S., S. Suzuki, K. Kamiya, R. Kominami, K. Dohi, and O. Niwa. 1994. Radiation induction of germline mutation at a hypervariable mouse minisatellite locus. Int J Radiat Biol 65:549-557.

Sadekova, S., S. Lehnert, and T.Y. Chow. 1997. Induction of PBP74/mortalin/Grp75, a member of the hsp70 family, by low doses of ionizing radiation: a possible role in induced radioresistance. Int J Radiat Biol 72:653-660.

Sadetzki, S., P. Flint-Richter, T. Ben-Tal, and D. Nass. 2002. Radiation-induced meningioma: a descriptive study of 253 cases. J Neurosurg 97:1078-1082.

Sadetzki, S., A. Chetrit, L. Freedman, M. Stovall, B. Modan, and I. Novikov. 2005. Long-term follow-up for brain tumor development after childhood exposure to ionizing radiation for tinea capitis. Radiat Res 163:424-432.

Saenger, E.L., G.E. Thoma, and E.A. Tompkins. 1968. Incidence of leukemia following treatment of hyperthyroidism. Preliminary report of the Cooperative Thyrotoxicosis Therapy Follow-up Study. J Am Med Assoc 205:855-862.

Safa, A.M., O.P. Schumacher, and A. Rodriguez-Antunez. 1975. Long-term follow-up results in children and adolescents treated with radioactive iodine (^{131}I) for hyperthyroidism. N Engl J Med 292:167-171.

Sagan, L.A. 1987. Health Physics, Vol. 52: Special Issue on Radiation Hormesis. New York: Pergamon Press.

Saito, Y., Y. Ochiai, Y. Kodama, Y. Tamura, T. Togashi, H. Kosugi-Okano, T. Miyazawa, Y. Wakabayashi, K. Hatakeyama, S. Wakana, O. Niwa, and R. Kominami. 2001. Genetic loci controlling susceptibility to gamma-ray-induced thymic lymphoma. Oncogene 20:5243-5247.

Sankaranarayanan, K. 1976. Evaluation and re-evaluation of genetic radiation hazards in man. III. Other relevant data and risk assessment. Mutat Res 35:387-414.

Sankaranarayanan, K. 1991. Ionizing radiation and genetic risks. III. Nature of spontaneous and radiation-induced mutations in mammalian in vitro systems and mechanisms of induction of mutations by radiation. Mutat Res 258:75-97.

Sankaranarayanan, K. 1998. Ionizing radiation and genetic risks. IX. Estimates of the frequencies of Mendelian diseases and spontaneous mutation rates in human populations: a 1998 perspective. Mutat Res 411:129-178.

Sankaranarayanan, K. 1999. Ionizing radiation and genetic risks. X. The potential "disease phenotypes" of radiation-induced genetic damage in humans: perspectives from human molecular biology and radiation genetics. Mutat Res 429:45-83.

Sankaranarayanan, K., and R. Chakraborty. 2000a. Ionizing radiation and genetic risks. XIII. Summary and synthesis of papers VI to XII and estimates of genetic risks in the year 2000. Mutat Res 453:183-197.

Sankaranarayanan, K., and R. Chakraborty. 2000b. Ionizing radiation and genetic risks. XII. The concept of "potential recoverability correction factor" (PRCF) and its use for predicting the risk of radiation-inducible genetic disease in human live births. Mutat Res 453:129-181.

Sankaranarayanan, K., A.V. Duyn, M.J. Loos, and A.T. Natarajan. 1989. Adaptive response of human lymphocytes to low-level radiation from radioisotopes or x-rays. Mutat Res 211:7-12.

Sankaranarayanan, K., N. Yasuda, R. Chakraborty, G. Tusnady, and A.E. Czeizel. 1994. Ionizing radiation and genetic risks. V. Multifactorial diseases: a review of epidemiological and genetic aspects of congenital abnormalities in man and of models on maintenance of quantitative traits in populations. Mutat Res 317:1-23.

Sankaranarayanan, K., R. Chakraborty, and E.A. Boerwinkle. 1999. Ionizing radiation and genetic risks. VI. Chronic multifactorial diseases: a review of epidemiological and genetical aspects of coronary heart disease, essential hypertension and diabetes mellitus. Mutat Res 436:21-57.

Santos, J., M. Herranz, M. Fernandez, C. Vaquero, P. Lopez, and J. Fernandez-Piqueras. 2001. Evidence of a possible epigenetic inactivation mechanism operating on a region of mouse chromosome 19 in gamma-radiation-induced thymic lymphomas. Oncogene 20:2186-2189.

Sasaki, M.S. 1991. Primary damage and fixation of chromosomal DNA as probed by monochromatic soft x-rays and low-energy neutrons. Pp. 369-384 in The Early Effects of Radiation on DNA, E.M. Fielden and P. O'Neil, eds. NATO ASI Series, Vol. H54. Berlin: Springer-Verlag.

Sasaki, M.S. 1995. On the reaction kinetics of the radioadaptive response in cultured mouse cells. Int J Radiat Biol 68:281-291.

Sasaki, M.S., K. Kobayashi, K. Hieda, Y. Yamada, Y. Ejima, H. Maezawa, Y. Furusawa, T. Ito, and S. Okada. 1989. Induction of chromosome aberrations in human lymphocytes by monochromatic x-rays of quantum energy between 4.8 and 14.6 keV. Int J Radiat Biol 56:975-988.

Satoh, C., and M. Kodaira. 1996. Effects of radiation on children. Nature 383:226.

Satoh, C., N. Takahashi, J. Asakawa, M. Kodaira, R. Kuick, S.M. Hanash, and J.V. Neel. 1996. Genetic analysis of children of atomic bomb survivors. Environ Health Persp 104(Suppl 3):511-519.

Saul, R.L., and B.N. Ames. 1986. Background levels of DNA damage in the population. Basic Life Sci 38:529-535.

Savage, J.R. 1996. Insight into sites. Mutat Res 366:81-95.

Savitsky, K., S. Sfez, D.A. Tagle, Y. Ziv, A. Sartiel, F.S. Collins, Y. Shiloh, and G. Rotman. 1995. The complete sequence of the coding region of the ATM gene reveals similarity to cell cycle regulators in different species. Hum Mol Genet 4:2025-2032.

Sawant, S.G., G. Randers-Pehrson, C.R. Geard, D.J. Brenner, and E.J. Hall. 2001a. The bystander effect in radiation oncogenesis: I. Transformation in C3H 10T1/2 cells in vitro can be initiated in the unirradiated neighbors of irradiated cells. Radiat Res 155:397-401.

Sawant, S.G., G. Randers-Pehrson, N.F. Metting, and E.J. Hall. 2001b. Adaptive response and the bystander effect induced by radiation in C3H 10T(1/2) cells in culture. Radiat Res 156:177-180.

Sawant, S.G., W. Zheng, K.M. Hopkins, G. Randers-Pehrson, H.B. Lieberman, and E.J. Hall. 2002. The radiation-induced bystander effect for clonogenic survival. Radiat Res 157:361-364.

Schappi-Buchi, C. 1994. On the genetic background of the adaptive response to x-rays in Drosophila melanogaster. Int J Radiat Biol 65:427-435.

Schiestl, R.H., F. Khogali, and N. Carls. 1994. Reversion of the mouse pink-eyed unstable mutation induced by low doses of x-rays. Science 266:1573-1576.

Schinzel, A. 1988. Microdeletion syndromes, balanced translocations, and gene mapping. J Med Genet 25:454-462.

Schmickel, R.D. 1986. Contiguous gene syndromes: a component of recognizable syndromes. J Pediatr 109:231-241.

Schmid, E., D. Regulla, S. Guldbakke, D. Schlegel, and M. Bauchinger. 2000. The effectiveness of monoenergetic neutrons at 565 keV in producing dicentric chromosomes in human lymphocytes at low doses. Radiat Res 154:307-312.

Schmid, E., D. Regulla, H.M. Kramer, and D. Harder. 2002. The effect of 29 kV x rays on the dose response of chromosome aberrations in human lymphocytes. Radiat Res 158: 771-777.

Schmidt-Kastner, P.K., K. Jardine, M. Cormier, and M.W. McBurney. 1998. Absence of p53-dependent cell cycle regulation in pluripotent mouse cell lines. Oncogene 16:3003-3011.

Schmutte, C., and R. Fishel. 1999. Genomic instability: first step to carcinogenesis. Anticancer Res 19:4665-4696.

Schneider, A.B., E. Shore-Freedman, U.Y. Ryo, C. Bekerman, M. Favus, and S. Pinsky. 1985. Radiation-induced tumors of the head and neck following childhood irradiation. Prospective studies. Medicine 64:1-15.

Schneider, A.B., E. Ron, J. Lubin, M. Stovall, and T.C. Gierlowski. 1993. Dose-response relationships for radiation-induced thyroid cancer and thyroid nodules: evidence for the prolonged effects of radiation on the thyroid. J Clin Endocrinol Metab 77:362-369.

Schollnberger, H., R.E. Mitchel, E.I. Azzam, D.J. Crawford-Brown, and W. Hofmann. 2002. Explanation of protective effects of low doses of gamma-radiation with a mechanistic radiobiological model. Int J Radiat Biol 78:1159-1173.

Schull, W.J., M. Otake, and J.V. Neel. 1981. Genetic effects of the atomic bomb: a reappraisal. Science 213:1220-1227.

Schull, W.J., J.V. Neel, M. Otake, A. Awa, C. Satoh, and H.B. Hamilton. 1982. Hiroshima and Nagasaki: three and a half decades of genetic screening. Pp. 687-700 in Environmental Mutagens and Carcinogens, T. Sugimura, S. Kondo, and H. Takebe, eds. New York: Alan R. Liss.

Scott, D. 1980. Molecular mechanisms of chromosome structural changes. Pp. 101-113 in Progress in Environmental Mutagenesis, M. Alcevic, ed. Amsterdam: Elsevier.

Scott, D., A. Spreadborough, E. Levine, and S.A. Roberts. 1994. Genetic predisposition in breast cancer. Lancet 344:1444.

Scott, D., J.B. Barber, E.L. Levine, W. Burrill, and S.A. Roberts. 1998. Radiation-induced micronucleus induction in lymphocytes identifies a high frequency of radiosensitive cases among breast cancer patients: a test for predisposition? Brit J Cancer 77:614-620.

Scully, R., J. Chen, R.L. Ochs, K. Keegan, M. Hoekstra, J. Feunteun, and D.M. Livingston. 1997. Dynamic changes of BRCA1 subnuclear location and phosphorylation state are initiated by DNA damage. Cell 90:425-435.

Searle, A.G. 1967. Progress in mammalian radiation genetics. Pp. 469-481 in Proceedings III International Congress of Radiation Research Cortina d'Ampezzo, 1966, Radiation Research, G. Silini, ed. Amsterdam: North Holland.

Searle, A.G. 1976. Use of the Doubling Doses for the Estimation of Genetic Risks. Orsay, France: Euratom.

Searle, A.G., and C. Beechey. 1986. The role of dominant visibles in mutagenicity testing. Prog Clin Biol Res 209B:511-518.

Seed, T.M., and L.V. Kaspar. 1992. Acquired radioresistance of hematopoietic progenitors (granulocyte/monocyte colony-forming units) during chronic radiation leukemogenesis. Cancer Res 52:1469-1476.

Seed, T.M., and S.M. Meyers. 1993. Chronic radiation-induced alteration in hematopoietic repair during preclinical phases of aplastic anemia and myeloproliferative disease: assessing unscheduled DNA synthesis responses. Cancer Res 53:4518-4527.

Seelentag, W.W., W. Panzer, G. Drexler, L. Platz, and F. Pfantner. 1979. A Catalogue of Spectra for the Calibration of Dose Meters, GSF-Report 560. Neuherberg/Munich.

Selby, P.B. 1998a. Discovery of numerous clusters of spontaneous mutations in the specific-locus test in mice necessitates major increases in estimates of doubling doses. Genetica 102-103:463-487.

Selby, P.B. 1998b. Major impacts of gonadal mosaicism on hereditary risk estimation, origin of hereditary diseases, and evolution. Genetica 102-103:445-462.

Selby, P.B., and P.R. Selby. 1977. Gamma-ray-induced dominant mutations that cause skeletal abnormalities in mice. I. Plan, summary of results and discussion. Mutat Res 43:357-375.

Selby, P.B., and P.R. Selby. 1978. Gamma-ray-induced dominant mutations that cause skeletal abnormalities in mice. III. Description of presumed mutations. Mutat Res 50:341-351.

Sever, L.E., E.S. Gilbert, N.A. Hessol, and J.M. McIntyre. 1988. A case-control study of congenital malformations and occupational exposure to low-level ionizing radiation. Am J Epidemiol 127:226-242.

Seymour, C.B., and C. Mothersill. 1997. Delayed expression of lethal mutations and genomic instability in the progeny of human epithelial cells which survived in a bystander killing environment. Radiat Oncol Investi 5:106-110.

Seymour, C.B., and C. Mothersill. 2000. Relative contribution of bystander and targeted cell killing to the low-dose region of the radiation dose-response curve. Radiat Res 153:508-511.

Shadley, J.D., and J.K. Wiencke. 1989. Induction of the adaptive response by x-rays is dependent on radiation intensity. Int J Radiat Biol 56:107-118.

Shadley, J.D., V. Afzal, and S. Wolff. 1987. Characterization of the adaptive response to ionizing radiation induced by low doses of x rays to human lymphocytes. Radiat Res 111:511-517.

Shafman, T.D., S. Levitz, A.J. Nixon, L.A. Gibans, K.E. Nichols, D.W. Bell, C. Ishioka, K.J. Isselbacher, R. Gelman, J. Garber, J.R. Harris, and D.A. Haber. 2000. Prevalence of germline truncating mutations in ATM in women with a second breast cancer after radiation therapy for a contralateral tumor. Gene Chromosome Canc 27:124-129.

Shakhtarin, V.V., A.F. Tsyb, V.F. Stepanenko, M.Y. Orlov, K.J. Kopecky, and S. Davis. 2003. Iodine deficiency, radiation dose, and the risk of thyroid cancer among children and adolescents in the Bryansk region of Russia following the Chernobyl power station accident. Int J Epidemiol 32:584-591.

Sharan, S.K., M. Morimatsu, U. Albrecht, D.S. Lim, E. Regel, C. Dinh, A. Sands, G. Eichele, P. Hasty, and A. Bradley. 1997. Embryonic lethality and radiation hypersensitivity mediated by Rad51 in mice lacking Brca2. Nature 386:804-810.

Shellabarger, C.J., D. Chmelevsky, and A.M. Kellerer. 1980. Induction of mammary neoplasms in the Sprague-Dawley rat by 430keV neutrons and x-rays. J Natl Cancer Inst 64:821-833.

Shen, M.R., I.M. Jones, and H. Mohrenweiser. 1998. Nonconservative amino acid substitution variants exist at polymorphic frequency in DNA repair genes in healthy humans. Cancer Res 58:604-608.

Shibata, Y., S. Yamashita, V.B. Masyakin, G.D. Panasyuk, and S. Nagataki. 2001. 15 years after Chernobyl: new evidence of thyroid cancer. Lancet 358:1965-1966.

Shields, L.M., W.H. Wiese, B.J. Skipper, B. Charley, and L. Benally. 1992. Navajo birth outcomes in the Shiprock uranium mining area. Health Phys 63:542-551.

Shields, P.G., and C.C. Harris. 2000. Cancer risk and low-penetrance susceptibility genes in gene-environment interactions. J Clin Oncol 18:2309-2315.

Shilnikova, N.S., D.L. Preston, E. Ron, E.S. Gilbert, E.K. Vassilenko, S.A. Romanov, I.S. Kuznetsova, M.E. Sokolnikov, P.V. Okatenko, V.V. Kreslov, and N.A. Koshurnikova. 2003. Cancer mortality risk among workers at the Mayak Nuclear Complex. Radiat Res 159:787-798.

Shiloh, Y. 2001. ATM and ATR: networking cellular responses to DNA damage. Curr Opin Genet Dev 11:71-77.

Shiloh, Y. 2004. Bridge over broken ends. The cellular response to DNA breaks in health and disease. DNA Repair 3:779-1254.

Shimizu, Y., H. Kato, and W.J. Schull. 1990. Studies of the mortality of A-bomb survivors. 9. Mortality, 1950-1985: Part 2. Cancer mortality based on the recently revised doses (DS86). Radiat Res 121:120-141.

Shimizu, Y., H. Kato, W.J. Schull, and K. Mabuchi. 1992. Dose-response analysis among atomic-bomb survivors exposed to low-level radiation. Pp. 71-74 in Low Dose Irradiation and Biological Defense Mechanisms, T. Sugahara, L. Sagan, and T. Aoyama, eds. Amsterdam: Exerpta Medica.

Shimizu, Y., D.A. Pierce, D.L. Preston, and K. Mabuchi. 1999. Studies of the mortality of atomic bomb survivors. Report 12, Part II. Noncancer mortality: 1950-1990. Radiat Res 152:374-389.

Shippey, C.A., M. Layton, and L.M. Secker-Walker. 1990. Leukemia characterized by multiple sub-clones with unbalanced translocations involving different telomeric segments: case report and review of the literature. Gene Chromosome Canc 2:14-17.

Shizuma, K., K. Iwatani, H. Hasai, M. Hoshi, T. Oka, and H. Morishima. 1993. Residual ^{152}Eu and ^{60}Co activities induced by neutrons from the Hiroshima atomic bomb. Health Phys 65:272-282.

Shore, R.E. 1992. Issues and epidemiological evidence regarding radiation-induced thyroid cancer. Radiat Res 131:98-111.

Shore, R.E. 2001. Radiation-induced skin cancer in humans. Med Pediatr Oncol 36:549-554.

Shore, R.E., and X. Xue. 1999. Comparative thyroid cancer risk of childhood and adult radiation exposure and estimation of lifetime risk. Radiation and Thyroid Cancer. Singapore: World Scientific Publishing.

Shore, R.E., R.E. Albert, M. Reed, N. Harley, and B.S. Pasternack. 1984. Skin cancer incidence among children irradiated for ringworm of the scalp. Radiat Res 100:192-204.

Shore, R.E., E. Woodard, N. Hildreth, P. Dvoretsky, L. Hempelmann, and B. Pasternack. 1985. Thyroid tumors following thymus irradiation. J Natl Cancer Inst 74:1177-1184.

Shore, R.E., N. Hildreth, E. Woodard, P. Dvoretsky, L. Hempelmann, and B. Pasternack. 1986. Breast cancer among women given x-ray therapy for acute postpartum mastitis. J Natl Cancer Inst 77:689-696.

Shore, R.E., N. Hildreth, P. Dvoretsky, E. Andresen, M. Moseson, and B. Pasternack. 1993a. Thyroid cancer among persons given x-ray treatment in infancy for an enlarged thymus gland. Am J Epidemiol 137:1068-1080.

Shore, R.E., N. Hildreth, P. Dvoretsky, B. Pasternack, and E. Andresen. 1993b. Benign thyroid adenomas among persons x-irradiated in infancy for enlarged thymus glands. Radiat Res 134:217-223.

Short, S.C., J. Kelly, C.R. Mayes, M. Woodcock, and M.C. Joiner. 2001. Low-dose hypersensitivity after fractionated low-dose irradiation in vitro. Int J Radiat Biol 77:655-664.

Short, S.C., M. Woodcock, B. Marples, and M.C. Joiner. 2003. Effects of cell cycle phase on low-dose hyper-radiosensitivity. Int J Radiat Biol 79:99-105.

Sidransky, D. 1996. Is human patched the gatekeeper of common skin cancers? Nat Genet 14:7-8.

Sielken, R.L., Jr., R.S. Bretzlaff, and D.E. Stevenson. 1994. Incorporating additional biological phenomena into two-stage cancer models. Prog Clin Biol Res 387:237-260.

Silver, A., J. Moody, R. Dunford, D. Clark, S. Ganz, R. Bulman, S. Bouffler, P. Finnon, E. Meijne, R. Huiskamp, and R. Cox. 1999. Molecular mapping of chromosome 2 deletions in murine radiation-induced AML localizes a putative tumor suppressor gene to a 1.0 cM region homologous to human chromosome segment 11p11-12. Gene Chromosome Canc 24:95-104.

Sinclair, W.K. 1985. Experimental RBE values of high LET radiations at low doses and the implications for quality factor assignment. Radiat Prot Dosim 13:319-326.

Sinclair, W.K., and R.A. Morton. 1963. Variations in x-ray response during the division cycle of partially synchronized Chinese hamster cells in culture. Nature 199:1158-1160.

Singer, M.F. 1982. Highly repeated sequences in mammalian genomes. Int Rev Cytol 76:67-112.

Skov, K.A. 1999. Radioresponsiveness at low doses: hyper-radiosensitivity and increased radioresistance in mammalian cells. Mutat Res 430:241-253.

Skuse, G.R., and J.W. Ludlow. 1995. Tumour suppressor genes in disease and therapy. Lancet 345:902-906.

Sloan, S.R., E.W. Newcomb, and A. Pellicer. 1990. Neutron radiation can activate K-ras via a point mutation in codon 146 and induces a different spectrum of ras mutations than does gamma radiation. Mol Cell Biol 10:405-408.

Smida, J., H. Zitzelsberger, A.M. Kellerer, H. Lehmann, G. Minkus, T. Negele, F. Spelsberg, L. Hieber, E.P. Demidchik, E. Lengfelder, and M. Bauchinger. 1997. p53 mutations in childhood thyroid tumours from Belarus and in thyroid tumours without radiation history. Int J Cancer 73:802-807.

Smida, J., H. Zitzelsberger, L. Lengfelder, and others. 1999a. Molecular genetic analysis of RET rearrangements in papillary thyroid carcinomas from BelaRussian children and adults. Pp. 263-267 in Radiation and Thyroid Cancer, G. Thomas, E.D. Williams, and A. Karaoglou, eds. Luxembourg: World Scientific Publishing.

Smida, J., K. Salassidis, L. Hieber, H. Zitzelsberger, A.M. Kellerer, E.P. Demidchik, T. Negele, F. Spelsberg, E. Lengfelder, M. Werner, and M. Bauchinger. 1999b. Distinct frequency of RET rearrangements in papillary thyroid carcinomas of children and adults from Belarus. Int J Cancer 80:32-38.

Smith, C. 1975. Quantitative inheritance. Pp. 382-441 in Text Book of Human Genetics, G. Fraser and O. Mayo, eds. London: Blackwell.

Smith, J.C., C.R. Newton, A. Alves, R. Anwar, D. Jenner, and A.F. Markham. 1990. Highly polymorphic minisatellite DNA probes. Further evaluation for individual identification and paternity testing. J Forensic Sci Soc 30:3-18.

Smith, J.W., and H. Inskip. 1985. Estimation of below measurement threshold doses following the remeasurement of a sample of old films. J Soc Radiol Prot 5:159-164.

Smith, L.G., R.C. Miller, M. Richards, D.J. Brenner, and E.J. Hall. 1999. Investigation of hypersensitivity to fractionated low-dose radiation exposure. Int J Radiat Oncol 45:187-191.

Smith, P.G., and R. Doll. 1981. Mortality from cancer and all causes among British radiologists. Brit J Radiol 54:187-194.

Smith, P.G., and A.J. Douglas. 1986. Mortality of workers at the Sellafield plant of British Nuclear Fuels. Brit Med J (Clin Res Ed) 293:845-854.

Smith, W.M., X.P. Zhou, K. Kurose, X. Gao, F. Latif, T. Kroll, K. Sugano, S.A. Cannistra, S.K. Clinton, E.R. Maher, T.W. Prior, and C. Eng. 2001. Opposite association of two PPARG variants with cancer: overrepresentation of H449H in endometrial carcinoma cases and underrepresentation of P12A in renal cell carcinoma cases. Hum Genet 109:146-151.

Sobol, R.W., J.K. Horton, R. Kuhn, H. Gu, R.K. Singhal, R. Prasad, K. Rajewsky, and S.H. Wilson. 1996. Requirement of mammalian DNA polymerase-beta in base-excision repair. Nature 379:183-186.

Sobolev, B., W.F. Heidenreich, I. Kairo, P. Jacob, G. Goulko, and I. Likhtarev. 1997. Thyroid cancer incidence in the Ukraine after the Chernobyl accident: comparison with spontaneous incidences. Radiat Environ Bioph 36:195-199.

Sofer, T., J.R. Goldsmith, I. Nusselder, and L. Katz. 1991. Geographical and temporal trends of childhood leukemia in relation to the nuclear plant in the Negev, Israel, 1960-1985. Public Health Rev 19:191-198.

Sonoda, E., M.S. Sasaki, J.M. Buerstedde, O. Bezzubova, A. Shinohara, H. Ogawa, M. Takata, Y. Yamaguchi-Iwai, and S. Takeda. 1998. Rad51-deficient vertebrate cells accumulate chromosomal breaks prior to cell death. EMBO J 17:598-608.

Sont, W.N., J.M. Zielinski, J.P. Ashmore, H. Jiang, D. Krewski, M.E. Fair, P.R. Band, and E.G. Letourneau. 2001. First analysis of cancer incidence and occupational radiation exposure based on the National Dose Registry of Canada. Am J Epidemiol 153:309-318.

Sorahan, T., and P.J. Roberts. 1993. Childhood cancer and paternal exposure to ionizing radiation: preliminary findings from the Oxford Survey of Childhood Cancers. Am J Ind Med 23:343-354.

Sorensen, K.J., C.M. Attix, A.T. Christian, A.J. Wyrobek, and J.D. Tucker. 2002. Adaptive response induction and variation in human lymphoblastoid cell lines. Mutat Res 519:15-24.

Southwick, G.J., and R.A. Schwartz. 1979. The basal cell nevus syndrome: disasters occurring among a series of 36 patients. Cancer 44:2294-2305.

Sposto, R., and D.L. Preston. 1992. Correction for catchment area nonresidency in studies based on tumor-registry data. Hiroshima, Japan: Radiation Effects Research Foundation.

Sposto, R., D.L. Preston, Y. Shimizu, and K. Mabuchi. 1992. The effect of diagnostic misclassification on non-cancer and cancer mortality dose response in A-bomb survivors. Biometrics 48:605-617.

Spring, K., F. Ahangari, S.P. Scott, P. Waring, D.M. Purdie, P.C. Chen, K. Hourigan, J. Ramsay, P.J. McKinnon, M. Swift, and M.F. Lavin. 2002. Mice heterozygous for mutation in ATM, the gene involved in ataxia-telangiectasia, have heightened susceptibility to cancer. Nat Genet 32:185-190.

Stadler, L.J. 1928a. Genetic effects of x-rays in maize. Proc Natl Acad Sci USA 14:69-75.

Stadler, L.J. 1928b. Mutations in barley induced by x-rays and radium. Science 68:186.

Stead, J.D., and A.J. Jeffreys. 2000. Allele diversity and germline mutation at the insulin minisatellite. Hum Mol Genet 9:713-723.

Stecca, C., and G.B. Gerber. 1998. Adaptive response to DNA-damaging agents: a review of potential mechanisms. Biochem Pharmacol 55:941-951.

Steiner, M., W. Burkart, B. Grosche, U. Kaletsch, and J. Michaelis. 1998. Trends in infant leukaemia in West Germany in relation to in utero exposure due to the Chernobyl accident. Radiat Environ Bioph 37:87-93.

Stern, F.B., R.A. Waxweiler, J.J. Beaumont, S.T. Lee, R.A. Rinsky, R.D. Zumwalde, W.E. Halperin, P.J. Bierbaum, P.J. Landrigan, and W.E. Murray Jr. 1986. A case-control study of leukemia at a naval nuclear shipyard. Am J Epidemiol 123:980-992.

Stewart, A.M., and G.W. Kneale. 1996. Relations between age at occupational exposure to ionising radiation and cancer risk. Occup Environ Med 53:225-230.

Stewart, A., J. Webb, and D. Hewitt. 1958. A survey of childhood malignancies. Brit Med J 30:1495-1508.

Stewart, J.R., and L.F. Fajardo. 1984. Radiation-induced heart disease: an update. Prog Cardiovasc Dis 27:173-194.

Stewart, J.R., L.F. Fajardo, S.M. Gillette, and L.S. Constine. 1995. Radiation injury to the heart. Int J Radiat Oncol 31:1205-1211.

Storer, J.B., and R.L. Ullrich. 1983. Life shortening in BALB/c mice following brief, protracted, or fractionated exposures to neutrons. Radiat Res 96:335-347.

Storer, J.B., L.J. Serrano, E.B. Darden Jr., M.C. Jernigan, and R.L. Ullrich. 1979. Life shortening in RFM and BALB/c mice as a function of radiation quality, dose, and dose rate. Radiat Res 78:122-161.

Storer, J.B., T.J. Mitchell, and R.L. Ullrich. 1982. Causes of death and their contribution to radiation-induced life shortening in intact and ovariectomized mice. Radiat Res 89:618-643.

Storer, J.B., T.J. Mitchell, and R.J. Fry. 1988. Extrapolation of the relative risk of radiogenic neoplasms across mouse strains and to man. Radiat Res 114:331-353.

Storm, H.H., M. Andersson, J.D. Boice, M. Blettner, M. Stovall, H.T. Mouridsen, P. Dombernowsky, C. Rose, A. Jacobsen, and M. Pedersen. 1992. Adjuvant radiotherapy and risk of contralateral breast cancer. J Natl Cancer Inst 84:1245-1250.

Storm, H.H., A.E. Prisyazhniuk, A.E. Okeanov, V.K. Ivanov, and L. Gulak. 1996. Development of infrastructure for epidemiological studies in the three CIS republics. The Radiological Consequences of the Chernobyl Accident. Proceedings of the First International Conference, Minsk, Belarus, A. Karaoglou, G. Desmet, G.N. Kelly, and others, eds. ECSC-EC-EAEC, Brussels, Luxembourg.

Strasser, A., A.W. Harris, T. Jacks, and S. Cory. 1994. DNA damage can induce apoptosis in proliferating lymphoid cells via p53-independent mechanisms inhibitable by Bcl-2. Cell 79:329-339.

Strather, J.W., C.R. Muirheard, A.A. Edwards, J.D. Harrison, D.C. Lloyd, and N.R. Wood. 1988. Health Effects Models Developed from the 1988 UNSCEAR Report. NRPBR226. Chilton, UK: National Radiological Protection Board.

Straume, T. 1991. Health Risks from Exposure to Tritium. Lawrence Livermore National Laboratory Report.

Straume, T. 1996. Risk implications of the neutron discrepancy in the Hiroshima DS86 dosimetry system. Radiat Prot Dosim 67:9-12.

Straume, T., and A.L. Carsten. 1993. Tritium radiobiology and relative biological effectiveness. Health Phys 65:657-672.

Straume, T., S.D. Egbert, W.A. Woolson, R.C. Finkel, P.W. Kubik, H.E. Gove, P. Sharma, and M. Hoshi. 1992. Neutron discrepancies in the DS86 Hiroshima dosimetry system. Health Phys 63:421-426.

Strong, L.C. 1977. Genetic and environmental interactions. Cancer 40:1861-1866.

Strong, L.C., and W.R. Williams. 1987. The genetic implications of long-term survival of childhood cancer. A conceptual framework. Am J Pediat Hematol 9:99-103.

Stsjazhko, V.A., A.F. Tsyb, N.D. Tronko, G. Souchkevitch, and K.F. Baverstock. 1995. Childhood thyroid cancer since accident at Chernobyl. Brit Med J 310:801.

Sturzbecher, H.W., B. Donzelmann, W. Henning, U. Knippschild, and S. Buchhop. 1996. p53 is linked directly to homologous recombination processes via RAD51/RecA protein interaction. EMBO J 15:1992-2002.

Sugenoya, A., K. Asanuma, Y. Hama, H. Masuda, G.S. Skidanenko, A.T. Anatoliebna, K. Koike, A. Komiyama, and F. Iida. 1995. Thyroid abnormalities among children in the contaminated area related to the Chernobyl accident. Thyroid 5:29-33.

Suzuki, K., R. Takahara, S. Kodama, and M. Watanabe. 1998. In situ detection of chromosome bridge formation and delayed reproductive death in normal human embryonic cells surviving X irradiation. Radiat Res 150:375-381.

Swanson, R.L., N.J. Morey, P.W. Doetsch, and S. Jinks-Robertson. 1999. Overlapping specificities of base excision repair, nucleotide excision repair, recombination, and translesion synthesis pathways for DNA base damage in Saccharomyces cerevisiae. Mol Cell Biol 19:2929-2935.

Swerdlow, A.J., M.J. Schoemaker, R. Allerton, A. Horwich, J.A. Barber, D. Cunningham, T.A. Lister, A.Z. Rohatiner, G. Vaughan Hudson, M.V. Williams, and D.C. Linch. 2001. Lung cancer after Hodgkin's disease: a nested case-control study of the relation to treatment. J Clin Oncol 19:1610-1618.

Szymanska, H., M. Sitarz, E. Krysiak, J. Piskorowska, A. Czarnomska, H. Skurzak, A.A. Hart, D. de Jong, and P. Demant. 1999. Genetics of susceptibility to radiation-induced lymphomas, leukemias and lung tumors studied in recombinant congenic strains. Int J Cancer 83:674-678.

Taalman, R.D., N.G. Jaspers, J.M. Scheres, J. de Wit, and T.W. Hustinx. 1983. Hypersensitivity to ionizing radiation, in vitro, in a new chromosomal breakage disorder, the Nijmegen Breakage Syndrome. Mutat Res 112:23-32.

Takahashi, T., M.J. Schoemaker, K.R. Trott, S.L. Simon, K. Fujimori, N. Nakashima, A. Fukao, and H. Saito. 2003. The relationship of thyroid cancer with radiation exposure from nuclear weapon testing in the Marshall Islands. J Epidemiol 13:99-107.

Tamaki, K., C.A. May, Y.E. Dubrova, and A.J. Jeffreys. 1999. Extremely complex repeat shuffling during germline mutation at human minisatellite B6.7. Hum Mol Genet 8:879-888.

Tanaka, K., N.J. Tchaijunusova, T. Takatsuji, B.I. Gusev, A.K. Sakerbaev, M. Hoshi, and N. Kamada. 2000. High incidence of micronuclei in lymphocytes from residents of the area near the Semipalatinsk nuclear explosion test site. J Radiat Res 41:45-54.

Tankersley, W.G., C.M. West, J.E. Watson, and J.L. Reagan. 1996. Retrospective assessment of radiation exposures at or below the minimum detectable level at a federal nuclear reactor facility. App Occ Env Hyg 11:330-333.

Tauchi, H., S. Endo, K. Eguchi-Kasai, Y. Furusawa, M. Suzuki, S. Matsuura, K. Ando, N. Nakamura, S. Sawada, and K. Komatsu. 1999. Cell cycle and LET dependence for radiation-induced mutation: a possible mechanism for reversed dose-rate effect. J Radiat Res 40(Suppl):45-52.

Tawn, E.J., C.A. Whitehouse, D. Holdsworth, S. Morris, and R.E. Tarone. 2000a. Chromosome analysis of workers occupationally exposed to radiation at the Sellafield nuclear facility. Int J Radiat Biol 76:355-365.

Tawn, E.J., C.A. Whitehouse, and F.A. Martin. 2000b. Sequential chromosome aberration analysis following radiotherapy—no evidence for enhanced genomic instability. Mutat Res 465:45-51.

Tawn, E.J., C.A. Whitehouse, and R.E. Tarone. 2004. FISH chromosome aberration analysis on retired radiation workers from the Sellafield nuclear facility. Radiat Res 162:249-256.

Taylor, A.M.R. 1983. The effect of radiation on the chromosomes of patients with an unusual cancer susceptibility. Pp. 166-169 in Radiation Induced Chromosome Damage in Man, T. Ishikara and M. Sasaki, eds. New York: A.R. Liss.

Taylor, A.M.R., D. Scott, C.F. Arlett, and J. Cole. 1994a. Ataxia-telangiectasia: the effect of a pleiotropic gene. Int J Radiat Biol 66:S5-S201.

Taylor, J.A., M.A. Watson, T.R. Devereux, R.Y. Michels, G. Saccomanno, and M. Anderson. 1994b. p53 mutation hotspot in radon-associated lung cancer. Lancet 343:86-87.

Taylor, N.A. 1991. Estimation of dose received when dosemeter results are recorded below a threshold level. J Radiol Prot 3:191-198.

Tedeschi, B., D. Caporossi, P. Vernole, L. Padovani, M. Appolloni, P. Anzidei, and F. Mauro. 1995. Do human lymphocytes exposed to the fallout of the Chernobyl accident exhibit an adaptive response? 2. Challenge with bleomycin. Mutat Res 332:39-44.

Tedeschi, B., D. Caporossi, P. Vernole, L. Padovani, and F. Mauro. 1996. Do human lymphocytes exposed to the fallout of the Chernobyl accident exhibit an adaptive response? III. Challenge with bleomycin in lymphocytes from children hit by the initial acute dose of ionizing radiation. Mutat Res 354:77-80.

Telle, M.A. 1995. Historique des pratiques de dosimetrie au CEA: metrologie, enreistrement des doses et criteres de surveillance. Institut de Protection et de Surete Nucleaire. Fontenay-aux-Roses, France.

Telle-Lamberton, M.A., E. Auperin, J. Chemali, and D. Bergot. 1998. Reconstitution a posteriori des expositions professionnelles aux rayonnements ionisants: Incertitudes liees aux seuils de detection des dosimetres. Revue de Medecine du Travail 25:37-41.

Telle-Lamberton, M., D. Bergot, M. Gagneau, E. Samson, J.M. Giraud, M.O. Neron, and P. Hubert. 2004. Cancer mortality among French Atomic Energy Commission workers. Am J Ind Med 45:34-44.

Teraoka, S.N., K.E. Malone, D.R. Doody, N.M. Suter, E.A. Ostrander, J.R. Daling, and P. Concannon. 2001. Increased frequency of ATM mutations in breast carcinoma patients with early onset disease and positive family history. Cancer 92:479-487.

Terasima, T., and L.J. Tolmach. 1963. X-ray sensitivity and DNA synthesis in synchronous populations of HeLa cells. Science 140:490-492.

Thacker, J. 1986. The nature of mutants induced by ionising radiation in cultured hamster cells. III. Molecular characterization of HPRT-deficient mutants induced by gamma-rays or alpha-particles showing that the majority have deletions of all or part of the hprt gene. Mutat Res 160:267-275.

Thacker, J. 1992. Radiation-induced mutation in mammalian cells at low doses and dose rates. Adv Radiat Biol 16:77-124.

Thacker, J., and R. Cox. 1983. The relationship between specific chromosome aberrations and radiation-induced mutations in cultured mammalian cells. Pp. 235-275 in Radiation-Induced Chromosome Damage in Man, T. Ishihara and M.S. Sasaki, eds. New York: Liss.

Thacker, J., A. Stretch, and M.A. Stephens. 1979. Mutation and inactivation of cultured mammalian cells exposed to beams of accelerated heavy ions. II. Chinese hamster V79 cells. Int J Radiat Biol Re 36:137-148.

Thacker, J., A.N. Ganesh, A. Stretch, D.M. Benjamin, A.J. Zahalsky, and E.A. Hendrickson. 1994. Gene mutation and V(D)J recombination in the radiosensitive irs lines. Mutagenesis 9:163-168.

Thomas, G.A., and E.D. Williams. 1991. Evidence for and possible mechanisms of non-genotoxic carcinogenesis in the rodent thyroid. Mutat Res 248:357-370.

Thomas, D., S. Darby, F. Fagnani, P. Hubert, M. Vaeth, and K. Weiss. 1992. Definition and estimation of lifetime detriment from radiation exposures: principles and methods. Health Phys 63:259-272.

Thompson, D.E., K. Mabuchi, E. Ron, M. Soda, M. Tokunaga, S. Ochikubo, S. Sugimoto, T. Ikeda, M. Terasaki, S. Izumi, and D.L. Preston. 1994. Cancer incidence in atomic bomb survivors. Part II: Solid tumors, 1958-1987. Radiat Res 137:S17-S67.

Thompson, L.H. 1996. Evidence that mammalian cells possess homologous recombinational repair pathways. Mutat Res 363:77-88.

Thompson, L.H., and R.M. Humphrey. 1968. Response of mouse L-P59 cells to x-irradiation in the G_2 phase. Int J Radiat Biol 15:181-184.

Thompson, L.H., and D. Schild. 1999. The contribution of homologous recombination in preserving genome integrity in mammalian cells. Biochimie 81:87-105.

Thorpe, G.W., C.S. Fong, N. Alic, V.J. Higgins, and I.W. Dawes. 2004. Cells have distinct mechanisms to maintain protection against different reactive oxygen species: oxidative-stress-response genes. Proc Natl Acad Sci USA 101:6564-6569.

Tishkoff, D.X., N. Filosi, G.M. Gaida, and R.D. Kolodner. 1997. A novel mutation avoidance mechanism dependent on S. cerevisiae RAD27 is distinct from DNA mismatch repair. Cell 88:253-263.

Tokunaga, M., C.E. Land, Y. Aoki, T. Yamamoto, M. Asano, E. Sato, S. Tokuoka, G. Sakamoto, and D.L. Page. 1993. Proliferative and nonproliferative breast disease in atomic bomb survivors. Results of a histopathologic review of autopsy breast tissue. Cancer 72:1657-1665.

Tokunaga, M., C.E. Landd, S. Tokuoka, I. Nishimori, M. Soda, and S. Akiba. 1994. Incidence of female breast cancer among atomic bomb survivors, 1950-1985. Radiat Res 138:209-223.

Tomlinson, I., and W. Bodmer. 1999. Selection, the mutation rate and cancer: ensuring that the tail does not wag the dog. Nat Med 5:11-12.

Tomonaga, M.T., R.L. Matsuo, J.M. Carter, K. Bennett, F. Kuriyama, F. Imanaka, S. Kusumi, K. Mabuci, A. Kuramoto, N. Kamada, M. Ichimaru, A.V. Pisciotta, and C.C. Finch. 1991. Differential effects of atomic bomb irradiation in inducing major leukemia types: analysis of open-city cases including the Life Span Study cohort based upon updated diagnostic systems and the Dosimetry System 1986 (DS86). Hiroshima, Japan: Radiation Effects Research Foundation.

Tondel, M., G. Carlsson, L. Hardell, M. Eriksson, S. Jakobsson, U. Flodin, A. Skoldestig, and O. Axelson. 1996. Incidence of neoplasms in ages 0-19 y in parts of Sweden with high ^{137}Cs fallout after the Chernobyl accident. Health Phys 71:947-950.

Travis, L.B., R.E. Curtis, J.D. Boice, B.F. Hankey, and J.F. Fraumeni. 1991. Second cancers following non-Hodgkin's lymphoma. Cancer 67:2002-2009.

Travis, L.B., R.E. Curtis, H. Storm, P. Hall, E. Holowaty, F.E. van Leeuwen, B.A. Kohler, E. Pukkala, C.F. Lynch, M. Andersson, K. Bergfeldt, E.A. Clarke, T. Wiklund, G. Stoter, M. Gospodarowicz, J. Sturgeon, J.F. Fraumeni Jr., and J.D. Boice Jr. 1997. Risk of second malignant neoplasms among long-term survivors of testicular cancer. J Natl Cancer Inst 89:1429-1439.

Travis, L.B., E.J. Holowaty, K. Bergfeldt, C.F. Lynch, B.A. Kohler, T. Wiklund, R.E. Curtis, P. Hall, M. Andersson, E. Pukkala, J. Sturgeon, and M. Stovall. 1999. Risk of leukemia after platinum-based chemotherapy for ovarian cancer. N Engl J Med 340:351-357.

Travis, L.B., M. Andersson, M. Gospodarowicz, F.E. van Leeuwen, K. Bergfeldt, C.F. Lynch, R.E. Curtis, B.A. Kohler, T. Wiklund, H. Storm, E. Holowaty, P. Hall, E. Pukkala, D.T. Sleijfer, E.A. Clarke, J.D. Boice Jr., M. Stovall, and E. Gilbert. 2000. Treatment-associated leukemia following testicular cancer. J Natl Cancer Inst 92:1165-1171.

Travis, L.B., M. Gospodarowicz, R.E. Curtis, E.A. Clarke, M. Andersson, B. Glimelius, T. Joensuu, C.F. Lynch, F.E. van Leeuwen, E. Holowaty, H. Storm, I. Glimelius, E. Pukkala, M. Stovall, J.F. Fraumeni Jr., J.D. Boice Jr., and E. Gilbert. 2002. Lung cancer following chemotherapy and radiotherapy for Hodgkin's disease. J Natl Cancer Inst 94:182-192.

Travis, L.B, D.A. Hill, G.M. Dores, M. Gospodarowicz, F.E. van Leeuwen, E. Holowaty, B. Glimelius, M. Andersson, T. Wiklund, C.F. Lynch, M.B. van't Veer, I. Glimelius, H. Storm, E. Pukkala, M. Stovall, R. Curtis, J.D. Boice Jr., and E. Gilbert. 2003. Breast cancer following radiotherapy and chemotherapy among young women with Hodgkin disease. J Am Med Assoc 290:465-475; erratum in J Am Med Assoc 290:1318.

Trenz, K., A. Rothfuss, P. Schutz, and G. Speit. 2002. Mutagen sensitivity of peripheral blood from women carrying a BRCA1 or BRCA2 mutation. Mutat Res 500:89-96.

Trepicchio, W.L., and T.G. Krontiris. 1992. Members of the rel/NF_B family of transcriptional regulatory proteins bind the HRAS1 minisatellite DNA sequence. Nucleic Acids Res 20:2427-2434.

Trimble, B.K., and J.H. Doughty. 1974. The amount of hereditary disease in human populations. Ann Hum Genet 38:199-223.

Tripodis, N., A.A. Hart, R.J. Fijneman, and P. Demant. 2001. Complexity of lung cancer modifiers: mapping of thirty genes and twenty-five interactions in half of the mouse genome. J Natl Cancer Inst 93:1484-1491.

Trivedi, A., and M.A. Hannan. 2004. Radiation and cardiovascular diseases. J Environ Pathol Toxicol Oncol 23:99-106.

Tronko, M.D., T.I. Bogdanova, I.V. Komissarenko, O.V. Epstein, V. Oliynyk, A. Kovalenko, I.A. Likhtarev, I. Kairo, S.B. Peters, and V.A. LiVolsi. 1999. Thyroid carcinoma in children and adolescents in Ukraine after the Chernobyl nuclear accident: statistical data and clinicomorphologic characteristics. Cancer 86:149-156.

Tronko, N.D., T.I. Bogdanova, O.V. Epstein, V.A. Oleynyk, I.V. Komissarenko, S.I. Rybakov, A.E. Kovalenko, V.P. Tereshchenko, I.A. Likhtarev, I.A. Kairo, V.M. Shpak, and N.I. Chepurnoy. 2002. Thyroid cancer in children and adolescents of Ukraine having been exposed as a result of the Chernobyl accident (15-year expertise of investigations). Int J Radiat Med 4:222-232.

Tsyb, A.F., A.N. Dedenkov, V.K. Ivanov, V.F. Stepanenko, and V.V. Pozhidaev. 1989. The development of an all-Union registry of persons exposed to radiation resulting from the accident at the Chernobyl atomic power station. Med Radiol 34:3-6.

Tucker, J.D., K.J. Sorensen, C.S. Chu, D.O. Nelson, M.J. Ramsey, C. Urlando, and J.A. Heddle. 1998. The accumulation of chromosome aberrations and Dlb-1 mutations in mice with highly fractionated exposure to gamma radiation. Mutat Res 400:321-335.

Tucker, M.A., G.J. D'Angio, J.D. Boice Jr., L.C. Strong, F.P. Li, M. Stovall, B.J. Stone, D.M. Green, F. Lombardi, W. Newton, R.N. Hoover, and J.F. Fraumeni Jr. 1987a. Bone sarcomas linked to radiotherapy and chemotherapy in children. N Engl J Med 317:588-593.

Tucker, M.A., A.T. Meadows, J.D. Boice Jr., M. Stovall, O. Oberlin, B.J. Stone, J. Birch, P.A. Voute, R.N. Hoover, and J.F. Fraumeni Jr. 1987b. Leukemia after therapy with alkylating agents for childhood cancer. J Natl Cancer Inst 78:459-464.

Tucker, M.A., P.H. Morris Jones, J.D. Boice Jr., L.L. Robison, B.J. Stone, M. Stovall, R.D. Jenkin, J.H. Lubin, E.S. Baum, S.E. Siegel, A.T. Meadows, R.N. Hoover, and J.F. Fraumeni Jr. 1991. Therapeutic radiation at a young age is linked to secondary thyroid cancer. Cancer Res 51:2885-2888.

Tukiendorf, A., G. Krasowski, and Z. Rybak. 2003. Thyroid cancer morbidity in Opole Province, Poland, after Chernobyl disaster. Cent Eur J Public Health 11:98-101.

Tusher, V.G., R. Tibshirani, and G. Chu. 2001. Significance analysis of microarrays applied to the ionizing radiation response. Proc Natl Acad Sci USA 98:5116-5121.

Ueno, A.M., D.B. Vannais, D.L. Gustafson, J.C. Wong, and C.A. Waldren. 1996. A low, adaptive dose of gamma-rays reduced the number and altered the spectrum of S1 mutants in human-hamster hybrid AL cells. Mutat Res 358:161-169.

Ullrich, R.L. 1983. Tumor induction in BALB/c female mice after fission neutron or gamma irradiation. Radiat Res 93:506-515.

Ullrich, R.L. 1984. Tumor induction in BALB/c mice after fractionated or protracted exposures to fission-spectrum neutrons. Radiat Res 97:587-597.

Ullrich, R.L. 1986. The rate of progression of radiation-transformed mammary epithelial cells is enhanced after low-dose-rate neutron irradiation. Radiat Res 105:68-75.

Ullrich, R.L. 1999. Risks for radiation-induced breast cancer: the debate continues. Radiat Res 151:123-124.

Ullrich, R.L., and C.M. Davis. 1999. Radiation-induced cytogenetic instability in vivo. Radiat Res 152:170-173.

Ullrich, R.L., and B. Ponnaiya. 1998. Radiation-induced instability and its relation to radiation carcinogenesis. Int J Radiat Biol 74:747-754.

Ullrich, R.L., and R.J. Preston. 1987. Myeloid leukemia in male RFM mice following irradiation with fission spectrum neutrons or gamma rays. Radiat Res 109:165-170.

Ullrich, R.L., and J.B. Storer. 1979a. Influence of gamma irradiation on the development of neoplastic disease in mice. I. Reticular tissue tumors. Radiat Res 80:303-316.

Ullrich, R.L., and J.B. Storer. 1979b. Influence of gamma irradiation on the development of neoplastic disease in mice. II. Solid tumors. Radiat Res 80:317-324.

Ullrich, R.L., and J.B. Storer. 1979c. Influence of gamma irradiation on the development of neoplastic disease in mice. III. Dose-rate effects. Radiat Res 80:325-342.

Ullrich, R.L., M.C. Jernigan, G.E. Cosgrove, L.C. Satterfield, N.D. Bowles, and J.B. Storer. 1976. The influence of dose and dose rate on the incidence of neoplastic disease in RFM mice after neutron irradiation. Radiat Res 68:115-131.

Ullrich, R.L., M.C. Jernigan, L.C. Satterfield, and N.D. Bowles. 1987. Radiation carcinogenesis: time-dose relationships. Radiat Res 111:179-184.

UNSCEAR (United Nations Scientific Committee on the Effects of Atomic Radiation). 1958. The 1958 Report to the General Assembly with Annexes. New York: United Nations.

UNSCEAR (United Nations Scientific Committee on the Effects of Atomic Radiation). 1962. The 1962 Report to the General Assembly with Annexes. New York: United Nations.

UNSCEAR (United Nations Scientific Committee on the Effects of Atomic Radiation). 1972. The 1972 Report to the General Assembly with Annexes. New York: United Nations.

UNSCEAR (United Nations Scientific Committee on the Effects of Atomic Radiation). 1977. The 1977 Report to the General Assembly with Annexes. New York: United Nations.

UNSCEAR (United Nations Scientific Committee on the Effects of Atomic Radiation). 1982. Ionizing Radiation: Sources and Biological Effects. The 1982 Report to the General Assembly with Annexes. New York: United Nations.

UNSCEAR (United Nations Scientific Committee on the Effects of Atomic Radiation). 1986. Genetic and Somatic Effects of Ionizing Radiation. The 1986 Report to the General Assembly with Annexes. New York: United Nations.

UNSCEAR (United Nations Scientific Committee on the Effects of Atomic Radiation). 1988. Sources, Effects, and Risks of Ionizing Radiation. The 1988 Report to the General Assembly with Annexes. New York: United Nations.

UNSCEAR (United Nations Scientific Committee on the Effects of Atomic Radiation). 1993. Sources and Effects of Ionizing Radiation. 1993 Report to the General Assembly, with Scientific Annexes. New York: United Nations.

UNSCEAR (United Nations Scientific Committee on the Effects of Atomic Radiation). 1994. Sources and Effects of Ionizing Radiation. 1994 Report to the General Assembly, with Scientific Annexes. Annex B Adaptive Responses to Radiation in Cells and Organisms. New York: United Nations.

UNSCEAR (United Nations Scientific Committee on the Effects of Atomic Radiation). 1996. Sources and Effects of Ionizing Radiation. 1996 Report to the General Assembly with Scientific Annex. New York: United Nations.

UNSCEAR (United Nations Scientific Committee on the Effects of Atomic Radiation). 2000a. Sources and Effects of Ionizing Radiation. UNSCEAR Report to the General Assembly, Volume I: Sources. New York: United Nations.

UNSCEAR (United Nations Scientific Committee on the Effects of Atomic Radiation). 2000b. Sources and Effects of Ionizing Radiation. UNSCEAR Report to the General Assembly, Volume II: Effects. New York: United Nations.

UNSCEAR (United Nations Scientific Committee on the Effects of Atomic Radiation). 2001. United Nations Scientific Committee on the Effects of Atomic Radiation. Hereditary Effects of Radiation. The 2002 Report to the General Assembly with Scientific Annex. New York: United Nations.

Upton, A.C. 2000. Why the concept of hormesis has not been incorporated into mainstream radiation health theory: radiation perspective. Hum Ecol Risk Assess 6:249-271.

Upton, A.C., M.L. Randolph, J.W. Conklin, M.A. Kastenbaum, M. Slater, G.S. Melville Jr., F.P. Conte, and J.A. Sproul Jr. 1970. Late effects of fast neutrons and gamma-rays in mice as influenced by the dose rate of irradiation: induction of neoplasia. Radiat Res 41:467-491.

Urquhart, J.D., R.J. Black, M.J. Muirhead, L. Sharp, M. Maxwell, O.B. Eden, and D.A. Jones. 1991. Case-control study of leukaemia and non-Hodgkin's lymphoma in children in Caithness near the Dounreay nuclear installation. Brit Med J 302:687-692; erratum in Brit Med J 302:818.

Vaeth, M., and D.A. Pierce. 1990. Calculating excess lifetime risk in relative risk models. Environ Health Persp 87:83-94.

Vahakangas, K.H., J.M. Samet, R.A. Metcalf, J.A. Welsh, W.P. Bennett, D.P. Lane, and C.C. Harris. 1992. Mutations of p53 and ras genes in radon-associated lung cancer from uranium miners. Lancet 339:576-580.

van der Houven van Oordt, C.W., R. Smits, T.G. Schouten, J.J. Houwing-Duistermaat, S.L. Williamson, A. Luz, P. Meera Khan, A.J. van der Eb, M.L. Breuer, and R. Fodde. 1999. The genetic background modifies the spontaneous and x-ray-induced tumor spectrum in the Apc1638N mouse model. Gene Chromosomes Cancer 24:191-198.

van Leeuwen, F.E., W.J. Klokman, M. Stovall, A. Hagenbeek, A.W. van den Belt-Dusebout, R. Noyon, J.D. Boice Jr., J.M. Burgers, and R. Somers. 1995. Roles of radiotherapy and smoking in lung cancer following Hodgkin's disease. J Natl Cancer Inst 87:1530-1537.

van Leeuwen, F.E., W.J. Klokman, M. Stovall, E.C. Dahler, M.B. van't Veer, E.M. Noordijk, M.A. Crommelin, B.M. Aleman, A. Broeks, M. Gospodarowicz, L.B. Travis, and N.S. Russell. 2003. Roles of radiation dose, chemotherapy, and hormonal factors in breast cancer following Hodgkin's disease. J Natl Cancer Inst 95:971-980.

van Wezel, T., A.P. Stassen, C.J. Moen, A.A. Hart, M.A. van der Valk, and P. Demant. 1996. Gene interaction and single gene effects in colon tumour susceptibility in mice. Nat Genet 14:468-470.

Venitt, S., and P.J. Biggs. 1994. Radon, mycotoxins, p53, and uranium mining. Lancet 343:795.

Verger, P., O. Catelinois, M. Tirmarche, L. Cherie-Challine, P. Pirard, M. Colonna, and P. Hubert. 2003. Thyroid cancers in France and the Chernobyl accident: risk assessment and recommendations for improving epidemiological knowledge. Health Phys 85:323-329.

Vergnaud, G., and F. Denoeud. 2000. Minisatellites: mutability and genome architecture. Genome Res 10:899-907.

Vermiglio, F., M.G. Castagna, E. Volnova, V.P. Lo Presti, M. Moleti, M.A. Violi, A. Artemisia, and F. Trimarchi. 1999. Post-Chernobyl increased prevalence of humoral thyroid autoimmunity in children and adolescents from a moderately iodine-deficient area in Russia. Thyroid 9: 781-786.

Viel, J.F., D. Pobel, and A. Carre. 1995. Incidence of leukaemia in young people around the La Hague nuclear waste reprocessing plant: a sensitivity analysis. Stat Med 14:2459-2472.

Vijayalaxmi, B.Z. Leal, T.S. Deahl, and M.L. Meltz. 1995. Variability in adaptive response to low dose radiation in human blood lymphocytes: consistent results from chromosome aberrations and micronuclei. Mutat Res 348:45-50.

Vilenchik, M.M., and A.G. Knudson Jr. 2000. Inverse radiation dose-rate effects on somatic and germ-line mutations and DNA damage rates. Proc Natl Acad Sci USA 97:5381-5386.

Villeneuve, P.J., and H.I. Morrison. 1997. Coronary heart disease mortality among Newfoundland fluorspar miners. Scan J Work Env Health 23:221-226.

Voehringer, D.W., D.L. Hirschberg, J. Xiao, Q. Lu, M. Roederer, C.B. Lock, L.A. Herzenberg, L. Steinman, and L.A. Herzenberg. 2000. Gene microarray identification of redox and mitochondrial elements that control resistance or sensitivity to apoptosis. Proc Natl Acad Sci USA 97:2680-2685.

Voelz, G.L., J.N. Lawrence, and E.R. Johnson. 1997. Fifty years of plutonium exposure to the Manhattan Project plutonium workers: an update. Health Phys 73:611-619.

Vogel, F. 1979. Genetics of retinoblastoma. Hum Genet 52:1-54.

Vogel, F., and Motulsky, A.G. 1997. Human Genetics: Problems and Approaches. Berlin: Springer.

Vogelstein, B., and K.W. Kinzler. 1993. The multistep nature of cancer. Trends Genet 9:138-141.

Vykhovanets, E.V., V.P. Chernyshov, I.I. Slukvin, Y.G. Antipkin, A.N. Vasyuk, H.F. Klimenko, and K.W. Strauss. 1997. I-131 dose-dependent thyroid autoimmune disorders in children living around Chernobyl. Clin Immunol Immnop 84:251-259

Wakeford, R., and M.P. Little. 2003. Risk coefficients for childhood cancer after intrauterine irradiation: a review. Int J Radiat Biol 79:293-309.

Wakeford, R., and L. Parker. 1996. Leukaemia and non-Hodgkin's lymphoma in young persons resident in small areas of West Cumbria in relation to paternal preconceptional irradiation. Brit J Cancer 73: 672-679.

Waldren, C.A. 2004. Classical radiation biology dogma, bystander effects and paradigm shifts. Hum Exp Toxicol 23:95-100.

Waller, L.A., B.W. Turnbull, G. Gustafsson, U. Hjalmars, and B. Andersson. 1995. Detection and assessment of clusters of disease: an application to nuclear power plant facilities and childhood leukaemia in Sweden. Stat Med 14:3-16.

Wang, B., H. Ohyama, T. Nose, H. Itsukaichi, T. Nakajima, O. Yukawa, T. Odaka, K. Tanaka, E. Kojima, T. Yamada, and I. Hayata. 1998. Adaptive response in embryogenesis: I. Dose and timing of radiation for reduction in prenatal death and congenital malformation during the late period of organogenesis. Radiat Res 150:120-122.

Wang, J.X., J.D. Boice Jr., B.X. Li, J.Y. Zhang, and J.F. Fraumeni Jr. 1988. Cancer among medical diagnostic x-ray workers in China. J Natl Cancer Inst 80:344-350.

Wang, Y., D. Cortez, P. Yazdi, N. Neff, S.J. Elledge, and J. Qin. 2000. BASC, a super complex of BRCA1-associated proteins involved in the recognition and repair of aberrant DNA structures. Genes Dev 14: 927-939.

Wang, Z.Y., J.D. Boice, L.X. Wei, G.W. Beebe, Y.R. Zha, M.M. Kaplan, Z.F. Tao, H.R. Maxon III, S.Z. Zhang, A.B. Schneider, B. Tan, T.A. Wesseler, D. Chen, A.G. Ershow, R.A. Kleinerman, L.G. Littlefield, and D. Preston. 1990. Thyroid nodularity and chromosome aberrations among women in areas of high background radiation in China. J Natl Cancer Inst 82:478-485.

Ward, J.F. 1981. Some biochemical consequences of the spatial distribution of ionizing radiation-produced free radicals. Radiat Res 86:185-195.

Ward, J.F. 1988. DNA damage produced by ionizing radiation in mammalian cells: identities, mechanism of formation, and repairability. Prog Nucleic Acid Res 35:95-125.

Ward, J.F. 1994. The complexity of DNA damage—relevance to biological consequences. Int J Radiat Biol 66:427-432.

Ward, J.F. 2002. The radiation-induced lesions which trigger the bystander effect. Mutat Res 499:151-154.

Ward, J.F., W.F. Blakely, and E.I. Joner. 1985. Mammalian cells are not killed by DNA single-strand breaks caused by hydroxyl radicals from hydrogen peroxide. Radiat Res 103:383-392.

Ward, J.F., C.L. Limoli, P.M. Calabro-Jones, and J. Aguilera. 1991. An examination of the repair saturation hypothesis for describing shouldered survival curves. Radiat Res 127:90-96.

Watanabe, M., and M. Horikawa. 1980. Analyses of differential sensitivities of synchronized HeLa S3 cells to radiations and chemical carcinogens during the cell cycle. Par V. Radiation- and chemical carcinogen-induced mutagenesis. Mutat Res 71:219-231.

Waters, T.R., P. Gallinari, J. Jiricny, and P.F. Swann. 1999. Human thymine DNA glycosylase binds to apurinic sites in DNA but is displaced by human apurinic endonuclease 1. J Biol Chem 274:67-74.

Watkins, J.P., D.L. Cragle, E.L. Frome, J.L. Reagan, C.M. West, D. Crawford-Brown, and W.G. Tankersley. 1997. Collection, validation, and treatment of data for mortality study of nuclear industry workers. Appl Occ Env Hyg 12:195-205.

Watson, G.E., S.A. Lorimore, and E.G. Wright. 1996. Long-term in vivo transmission of alpha-particle-induced chromosomal instability in murine haemopoietic cells. Int J Radiat Biol 69:175-182.

Watson, G.E., D.A. Pocock, D. Papworth, S.A. Lorimore, and E.G. Wright. 2001. In vivo chromosomal instability and transmissible aberrations in the progeny of haemopoietic stem cells induced by high- and low-LET radiations. Int J Radiat Biol 77:409-417.

Weil, M.M., C. Xia, X. Xia, X. Gu, C.I. Amos, and K.A. Mason. 2001. A chromosome 15 quantitative trait locus controls levels of radiation-induced jejunal crypt cell apoptosis in mice. Genomics 72:73-77.

Weinberg, II.S., A.B. Korol, V.M. Kirzhner, T. Fahima, E. Nevo, S. Shapiro, G. Rennert, O. Piatak, E.I. Stepanova, and E. Skvarskaja. 2001. Very high mutation rate in offspring of Chernobyl accident liquidators. Proc R Soc Lond 268:1001-1005.

Weinberg, R.A. 1994. Oncogenes and tumor suppressor genes. CA-Cancer J Clin 44:160-170.

Weiss, H.A., S.C. Darby, and R. Doll. 1994. Cancer mortality following x-ray treatment for ankylosing spondylitis. Int J Cancer 59:327-338.

Weiss, H.A., S.C. Darby, T. Fearn, and R. Doll. 1995. Leukemia mortality after x-ray treatment for ankylosing spondylitis. Radiat Res 142:1-11.

White, E., and C. Prives. 1999. DNA damage enables p73. Nature 399:734-735, 737.

White, E., S.K. Chiou, L. Rao, P. Sabbatini, and H.J. Lin. 1994. Control of p53-dependent apoptosis by E1B, Bcl-2, and Ha-ras proteins. Cold Spring Harb Symp 59:395-402.

Whitehouse, C.A., and E.J. Tawn. 2001. No evidence for chromosomal instability in radiation workers with in vivo exposure to plutonium. Radiat Res 156:467-475.

Whitehouse, C.J., R.M. Taylor, A. Thistlethwaite, H. Zhang, F. Karimi-Busheri, D.D. Lasko, M. Weinfeld, and K.W. Caldecott. 2001. XRCC1 stimulates human polynucleotide kinase activity at damaged DNA termini and accelerates DNA single-strand break repair. Cell 104:107-117.

Whittemore, A. S.1978. Quantitative theories of oncogenesis. Adv Cancer Res, v. 27: 55-88.

WHO (World Health Organization). 1998. The public health impact of nuclear weapons testing in Kazakhstan. World Health Organization Report.

Wiggs, L.D., C.A. Cox-DeVore, and G.L. Voelz. 1991a. Mortality among a cohort of workers monitored for ^{210}Po exposure: 1944-1972. Health Phys 61:71-76.

Wiggs, L.D., C.A. Cox-DeVore, G.S. Wilkinson, and M. Reyes. 1991b. Mortality among workers exposed to external ionizing radiation at a nuclear facility in Ohio. J Occup Med 33:632-637.

Wiggs, L.D., E. Johnson, C. Cox-Devore, and G. Voelz. 1994. Mortality through 1990 among white male workers at the Los Alamos National Laboratory: considering exposures to plutonium and external ionizing radiation. Health Phys 67:577-588.

Wilkinson, G.S., G.L. Tietjen, L.D. Wiggs, W.A. Galke, J.F. Acquavella, M. Reyes, G.L. Voelz, and R.J. Waxweiler. 1987. Mortality among plutonium and other radiation workers at a plutonium weapons facility. Am J Epidemiol 125:231-250.

Williams, E.D. 1997. Thyroid Cancer and the Chernobyl Accident. Health Effects of Low Dose Radiation. London: British Nuclear Energy Society.

Williams, E.D. 2003. Chernobyl, 15 years later, correlation of clinical, epidemiological and molecular outcomes. Ann Endocrinol 64:72.

Wilson, D.M., 3rd, and L.H. Thompson. 1997. Life without DNA repair. Proc Natl Acad Sci USA 94:12754-12757.

Wing, S., C. Shy, J. Wood, S. Wolf, D. Cragle, and E. Frome. 1991. Mortality among workers at Oak Ridge National Laboratory. Evidence of radiation effects in follow-up through 1984. J Am Med Assoc 265:1397-1402.

Wing, S., C. Shy, J. Wood, S. Wolf, D. Cragle, W. Tankersley, and E. Frome. 1993. Job factors, radiation and cancer mortality at Oak Ridge National Laboratory: follow-up through 1984. Am J Ind Med 23:265-279.

Witt, K.L., and J.B. Bishop. 1996. Mutagenicity of anticancer drugs in mammalian germ cells. Mutat Res 355:209-234.

Wojcik, A., K. Bonk, W.U. Muller, C. Streffer, U. Weissenborn, and G. Obe. 1992. Absence of adaptive response to low doses of x-rays in pre-implantation embryos and spleen lymphocytes of an inbred mouse strain as compared to human peripheral lymphocytes: a cytogenetic study. Int J Radiat Biol 62:177-186.

Wolf, C., J. Lafuma, R. Masse, M. Morin, and A.M. Kellerer. 2000. Neutron RBE for induction of tumors with high lethality in Sprague-Dawley rats. Radiat Res 154:412-420.

Wolff, S. 1989. Are radiation-induced effects hormetic? Science 245: 575, 621.

Wolff, S. 1992a. Low dose exposures and the induction of adaptation. Pp. 21-28 in Low Dose Irradiation and Biological Defense Mechanisms, T. Sugahara and others, eds. Amsterdam, London, New York, Tokyo: Excerpta Medica.

Wolff, S. 1992b. Failla Memorial Lecture. Is radiation all bad? The search for adaptation. Radiat Res 131:117-123.

Wolff, S. 1996. Aspects of the adaptive response to very low doses of radiation and other agents. Mutat Res 358:135-142.

Wong, F.L., M. Yamada, H. Sasaki, K. Kodama, S. Akiba, K. Shimaoka, and Y. Hosoda. 1993. Noncancer disease incidence in the atomic bomb survivors: 1958-1986. Radiat Res 135:418-430.

Wong, F.L., J.D. Boice Jr., D.H. Abramson, R.E. Tarone, R.A. Kleinerman, M. Stovall, M.B. Goldman, J.M. Seddon, N. Tarbell, J.F. Fraumeni Jr., and F.P. Li. 1997a. Cancer incidence after retinoblastoma. Radiation dose and sarcoma risk. J Am Med Assoc 278:1262-1267.

Wong, F.L., J.D. Boice Jr., D.H. Abramson, R.E. Tarone, R.A. Kleinerman, M. Stovall, M.B. Goldman, J.M. Seddon, N. Tarbell, J.F. Fraumeni Jr., and F.P. Li. 1997b. Cancer incidence after retinoblastoma. Radiation dose and sarcoma risk. J Am Med Assoc 278:1262-1267.

Wong, F.L., M. Yamada, H. Sasaki, K. Kodama, and Y. Hosoda. 1999. Effects of radiation on the longitudinal trends of total serum cholesterol levels in atomic bomb survivors. Radiat Res 151:736-746.

Wood, R.D., M. Mitchell, J. Sgouros, and T. Lindahl. 2001. Human DNA repair genes. Science 291:1284-1289.

Worgul, B.V., L. Smilenov, D.J. Brenner, A. Junk, W. Zhou, and E.J. Hall. 2002. ATM heterozygous mice are more sensitive to radiation-induced cataracts than are their wild-type counterparts. Proc Natl Acad Sci USA 99:9836-9839.

Wouters, B.G., A.M. Sy, and L.D. Skarsgard. 1996. Low-dose hypersensitivity and increased radioresistance in a panel of human tumor cell lines with different radiosensitivity. Radiat Res 146:399-413.

Wright, E.G. 2000. Genomic instability manifestations and mechanisms. Pp. 483-487 in Radiation Research 2, M. Moriarty and others, eds. Lawrence: Allen Press.

Wu, H.I., J.A. Brown, M.J. Dorie, L. Lazzeroni, J.M. Brown. 2004. Genome-wide identification of genes conferring resistance to the anticancer agents cisplatin, oxaliplatin, and mitomycin C. Cancer Res 64:3940-3948.

Wu, L.J., G. Randers-Pehrson, A. Xu, C.A. Waldren, C.R. Geard, Z. Yu, and T.K. Hei. 1999. Targeted cytoplasmic irradiation with alpha particles induces mutations in mammalian cells. Proc Natl Acad Sci USA 96:4959-4964.

Wu, X., J. Li, X. Li, C.L. Hsieh, P.M. Burgers, and M.R. Lieber. 1996. Processing of branched DNA intermediates by a complex of human FEN-1 and PCNA. Nucleic Acids Res 24:2036-2043.

Wyllie, A. 1998. Apoptosis. An endonuclease at last. Nature 391:20-21.

Xanthoudakis, S., R.J. Smeyne, J.D. Wallace, and T. Curran. 1996. The redox/DNA repair protein, Ref-1, is essential for early embryonic development in mice. Proc Natl Acad Sci USA 93:8919-8923.

Xia, F., X. Wang, Y.H. Wang, N.M. Tsang, D.W. Yandell, K.T. Kelsey, and H.L. Liber. 1995. Altered p53 status correlates with differences in sensitivity to radiation-induced mutation and apoptosis in two closely related human lymphoblast lines. Cancer Res 55:12-15.

Xu, Y., T. Ashley, E.E. Brainerd, R.T. Bronson, M.S. Meyn, and D. Baltimore. 1996. Targeted disruption of ATM leads to growth retardation, chromosomal fragmentation during meiosis, immune defects, and thymic lymphoma. Genes Dev 10:2411-2422.

Yang, C.R., K. Leskov, K. Hosley-Eberlein, T. Criswell, J.J. Pink, T.J. Kinsella, and D.A. Boothman. 2000. Nuclear clusterin/XIP8, an x-ray-induced Ku70-binding protein that signals cell death. Proc Natl Acad Sci USA 97:5907-5912.

Yang, N., H. Galick, and S.S. Wallace. 2004. Attempted base excision repair of ionizing radiation damage in human lymphoblastoid cells produces lethal and mutagenic double strand breaks. DNA Repair 3:1323-1334.

Yauk, C.L., Y.E. Dubrova, G.R. Grant, and A.J. Jeffreys. 2002. A novel single molecule analysis of spontaneous and radiation-induced mutation at a mouse tandem repeat locus. Mutat Res 500:147-156.

Yin, E., D.O. Nelson, M.A. Coleman, L.E. Peterson, and A.J. Wyrobek. 2003. Gene expression changes in mouse brain after exposure to low-dose ionizing radiation. Int J Radiat Biol 79:759-775.

Yokoro, K., M. Nakano, A. Ito, K. Nagao, and Y. Kodama. 1977. Role of prolactin in rat mammary carcinogenesis: detection of carcinogenicity of low-dose carcinogens and of persisting dormant cancer cells. J Natl Cancer Inst 58:1777-1783.

Yoshimoto, Y., H. Kato, and W.J. Schull. 1988. Risk of cancer among children exposed in utero to A-bomb radiations, 1950-84. Lancet 2:665-669.

Yoshinaga, S., T. Aoyama, Y. Yoshimoto, and T. Sugahara. 1999. Cancer mortality among radiological technologists in Japan: updated analysis of follow-up data from 1969 to 1993. J Epidemiol 9:61-72.

Young, H.E. 2004. Existence of reserve quiescent stem cells in adults, from amphibians to humans. Curr Top Microbiol 280:71-109.

Youngblom, J.H., J.K. Wiencke, and S. Wolff. 1989. Inhibition of the adaptive response of human lymphocytes to very low doses of ionizing radiation by the protein synthesis inhibitor cycloheximide. Mutat Res 227:257-261.

Yu, Y., C.Y. Li, and J.B. Little. 1997. Abrogation of p53 function by HPV16 E6 gene delays apoptosis and enhances mutagenesis but does not alter radiosensitivity in TK6 human lymphoblast cells. Oncogene 14:1661-1667.

Yu, Y., R. Okayasu, M.M. Weil, A. Silver, M. McCarthy, R. Zabriskie, S. Long, R. Cox, and R.L. Ullrich. 2001. Elevated breast cancer risk in irradiated BALB/c mice associates with unique functional polymorphism of the Prkdc (DNA-dependent protein kinase catalytic subunit) gene. Cancer Res 61:1820-1824.

Yuhas, J.M. 1974. Recovery from radiation-carcinogenic injury to the mouse ovary. Radiat Res 60:321-332.

Zablotska, L.B., J.P. Ashmore, and G.R. Howe. 2004. Analysis of mortality experience amongst Canadian nuclear power industry workers following chronic low-dose exposure to ionizing radiation. Radiat Res 161:633-641.

Zaridze, D.G., N. Li, T. Men, and S.W. Duffy. 1994. Childhood cancer incidence in relation to distance from the former nuclear testing site in Semipalatinsk, Kazakhstan. Int J Cancer 59:471-475.

Zhang, S., E.S. Ramsay, and B.A. Mock. 1998. Cdkn2a, the cyclin-dependent kinase inhibitor encoding p16INK4a and p19ARF, is a candidate for the plasmacytoma susceptibility locus, Pctr1. Proc Natl Acad Sci USA 95:2429-2434.

Zhao, S., Y.C. Weng, S.S. Yuan, Y.T. Lin, H.C. Hsu, S.C. Lin, E. Gerbino, M.H. Song, M.Z. Zdzienicka, R.A. Gatti, J.W. Shay, Y. Ziv, Y. Shiloh, and E.Y. Lee. 2000. Functional link between ataxia-telangiectasia and Nijmegen breakage syndrome gene products. Nature 405:473-477.

Zheng, N., D.G. Monckton, G. Wilson, F. Hagemeister, R. Chakraborty, T.H. Connor, M.J. Siciliano, and M.L. Meistrich. 2000. Frequency of minisatellite repeat number changes at the MS205 locus in human sperm before and after cancer chemotherapy. Environ Mol Mutagen 36:134-145.

Zhou, P.K., and O. Rigaud. 2001. Down-regulation of the human CDC16 gene after exposure to ionizing radiation: a possible role in the radioadaptive response. Radiat Res 155:43-49.

Zhou, P.K., X.Y. Liu, W.Z. Sun, Y.P. Zhang, and K. Wei. 1993. Cultured mouse SR-1 cells exposed to low doses of gamma-rays become less susceptible to the induction of mutagenesis by radiation as well as bleomycin. Mutagenesis 8:109-111.

Zhou, H., G. Randers-Pehrson, C.A. Waldren, D. Vannais, E.J. Hall, and T.K. Hei. 2000. Induction of a bystander mutagenic effect of alpha particles in mammalian cells. Proc Natl Acad Sci USA 97:2099-2104.

Zielinski, J., R. Kodell, and D. Krewski. 2001. Interaction between two carcinogens in the two-stage clonal expansion model of carcinogenesis. J Epidemiol Biostat 6:219-228.

Glossary

Absolute risk (AR). The rate of disease among a population.

Absorbed dose (D). The mean energy imparted by ionizing radiation to a medium per unit mass. Units: gray (Gy), rad. 1 Gy = 1 J/kg = 100 rads. 1 rad = 0.01 J/kg = 100 ergs/g.

Activity. The amount of radioactivity defined in terms of the mean number of decays per unit time. Units: becquerel (Bq), curie (Ci). $1 \text{ Bq} = 1 \text{ s}^{-1} = 2.7 \times 10^{-11} \text{ Ci}$. $1 \text{ Ci} = 3.7 \times 10^9 \text{ Bq}$.

Additive effect. When two agents do not interact, the combined effect is equal to the sum of the effects of the two agents acting alone.

Apoptosis. Programmed cell death. The cell death is characterized by a distinctive fragmentation of DNA that is regulated by cellular functions.

Attributable risk (AR). The estimated rate of a disease (such as lung cancer) that could, in theory, be prevented if all exposures to a particular causative agent (such as radon) were eliminated.

Background radiation. The radiation to which a member of the population is exposed from natural sources, such as terrestrial radiation due to naturally occurring radionuclides in the soil, cosmic radiation originating in outer space, and naturally occurring radionuclides in the human body.

Baseline rate of cancer. The annual cancer incidence observed in a population in the absence of the specific agent being studied; the baseline rate includes cancers from a number of other causes, such as smoking, occupational exposures, and so forth.

Becquerel (Bq). SI unit of activity (see *Units*). $1 \text{ Bq} = 1 \text{ s}^{-1} = 2.7 \times 10^{-11} \text{ Ci}$.

BEIR V. Report of the fifth National Research Council Committee on the Biological Effects of Ionizing Radiation; the report was published in 1990.

Beta particle. A charged particle emitted from a nucleus during radioactive decay that, if negatively charged, is identical to an electron.

Bias. Factors that influence the outcome of data collection, such as causing certain measurements to have a greater chance of being included than others.

Cancer. A malignant tumor of potentially unlimited growth, capable of invading surrounding tissue or spreading to other parts of the body by metastasis.

Carcinogen. An agent that can cause cancer. Ionizing radiation is a physical carcinogen; there are also chemical and biological carcinogens; biological carcinogens may be extrinsic (*e.g.*, viruses) or intrinsic (genetic defects).

Carcinoma. A malignant tumor (cancer) of epithelial origin.

Case-control study. An epidemiologic study in which people with disease and a similarly composed control group are compared in terms of exposures to a putative causative agent.

Cell culture. The growing of cells *in vitro* (in a glass or plastic container, or in suspension) in such a manner that the cells are no longer organized into tissues.

Cohort study. An epidemiologic study in which groups of people (the cohort) are identified with respect to the presence or absence of exposure to a disease-causing agent, and in which the outcomes of disease rates are compared; also called a follow-up study.

Collective effective dose. Number of persons times average effective dose (see *Effective dose*). Unit: person-Sv (special name used with collective dose).

Competing risks. Causes other than the agent under study that contribute to the mortality rate. The mortality rate from these other causes is not included in the risk of dying from the factor under study.

Confidence interval (CI). An interval estimate of an unknown parameter, such as a risk. A 95% confidence interval, as an example, is constructed from a procedure that is theoretically successful in capturing the parameter of interest in 95% of its applications. Confidence limits are the end points of a confidence interval.

Constant relative risk (CRR). A risk model that assumes the ratio of the risk at a specific dose and the risk in the absence of the dose remains constant after a certain time.

Curie (Ci). Former special unit of activity (see *Units*). 1 Ci = 3.7×10^{10} Bq.

DEF. The reduction in risk for low doses.

DNA. Deoxyribonucleic acid; the genetic material of cells.

Deletions. Type of mutation in which sections of DNA are removed; term can refer to the removal of a single base or many bases.

Dose. Short name for absorbed dose (1 Gy = 1 J/kg) and also for equivalent dose, effective dose, and weighted dose (1 Sv = 1 J/kg). Definitions of low, medium, and high doses vary widely in the literature. For the purposes of this report, dose levels have been defined as follows:

Low dose: 0-100 mGy (mSv)

Medium dose: In excess of 100 mGy up to a maximum of 1 Gy

High dose: In excess of 1 Gy up to the very high total doses used in radiation therapy (on the order of 20–60 Gy)

Dose-effect (dose-response) model. A mathematical formulation and description of the way the effect (or biological response) depends on dose.

Dose rate. The absorbed dose delivered per unit time.

Dose-rate effectiveness factor (DREF). The factor by which the effect caused by a specific type of radiation changes at low doses or low dose rates (protracted or fractionated delivery of dose) as compared to high doses delivered at high (or acute) dose rates.

Dose and dose-rate effectiveness factor (DDREF). A judged factor by which the radiation effect, per unit of dose, caused by a given high or moderate dose of radiation received at high dose rates is reduced when doses are low or are received at low dose rates.

Dosimetric model. A method for estimating risk based on the use of physical models for doses to target cells and the use of results from epidemiologic studies of exposures to humans from other types of radiations.

Ecological fallacy. The fact that two populations differ in many factors other than the one being evaluated and that one or more of these other factors may be the underlying reason for any difference noted in their morbidity or mortality experience.

Ecologic study. A method of epidemiologic study in which rates of health effects outcome based on population rather than individual data are related to the measure of population radiation exposure.

Effective dose. Sum over the absorbed doses to different organs from different radiation types multiplied by organ weighting factors and radiation weighting factors, as defined by the International Commission for Radiation Protection (ICRP). Unit: 1 Sv = 1 J/kg = 100 rem. Equal effective doses are meant to correspond—apart from age- and sex-dependent differences—to roughly the same overall risk. For a uniform whole-body exposure by a specified radiation type the effective dose equals the absorbed dose times the radiation weighting factor.

Electron volt (eV). A special unit of energy: 1 eV = 1.6×10^{-19} J = 1.6×10^{-12} erg; 1 eV is equivalent to the energy gained by an electron in passing through a potential difference of 1 V; 1 keV = 1000 eV; 1 MeV = 1,000,000 eV.

Empirical model. A model that is derived from measurements in populations, as opposed to a theoretical model.

Epidemiology. The study of the determinants of the frequency of disease in humans. The two main types of epidemiologic studies of chronic disease are cohort (or follow-up) studies and case-control studies.

Equivalent dose. Absorbed dose multiplied by the *quality factor, Q,* which represents, for the purposes of radiation protection and control, the effectiveness of the radiation relative to sparsely ionizing radiation (see *Quality factor*). Units: 1 Sv = 1 J/kg = 100 rem. 1 rem = 0.01 Sv.

Etiology. The science or description of cause(s) of disease.

Excess absolute risk (EAR). The rate of disease in an exposed population minus the rate of disease in an unexposed population. Also termed "attributable risk" or "risk difference."

Excess relative risk (ERR). The rate of disease in an exposed population divided by the rate of disease in an unexposed population minus 1.0.

Exposure. The condition of having contact with a physical or chemical agent.

Fibrosis. Damage to normal tissue that results in a modification of tissue structure but is not cancer.

Fractionation. The delivery of a given dose of radiation as several smaller doses separated by intervals of time.

Gamma radiation. Also gamma rays; short-wavelength electromagnetic radiation of nuclear origin, similar to X-rays but usually of higher energy (100 keV to several million electronvolts).

Geometric mean. The geometric mean of a set of positive numbers is the exponential of the arithmetic mean of their logarithms. The geometric mean of a lognormal distribution is the exponential of the mean of the associated normal distribution.

Geometric standard deviation (GSD). The geometric standard deviation of a lognormal distribution is the exponential of the standard deviation of the associated normal distribution.

Germ cells. Reproductive cells such as the sperm and egg and their progenitor cells.

Gray (Gy). Special name of the SI unit of absorbed dose (see *Units*). 1 Gy = 1 J/kg = 100 rads.

Half-life, biological. Time required for the body to eliminate half of an administered dose of any substance by metabolic processes of elimination; it is approximately the same for both stable and radioactive isotopes of a particular element.

Half-life, radioactive. Time required for a radioactive substance to lose 50% of its activity by decay.

ICRP (International Commission on Radiological Protection). An independent international organization that provides recommendations and guidance on protection against ionizing radiation.

ICRU (International Commission on Radiation Units and Measurements). An independent international organization that provides recommendations and guidance on radiation quantities, units, and measurements.

Incidence. Also, incidence rate; the rate of occurrence of a disease within a specified period of time, often expressed as a number of cases per 100,000 individuals per year.

In utero. In the womb (*i.e.*, before birth).

Inverse dose-rate effect. An effect in which, for a given exposure, the probability of effect increases as the dose rate is lowered.

In vitro. Cell culture conditions in glass or plastic containers.

In vivo. In the living organism.

Ionizing radiation. Radiation sufficiently energetic to dislodge electrons from an atom, thereby producing an ion pair. Ionizing radiation includes X- and gamma radiation, electrons (beta radiation), alpha particles (helium nuclei), and heavier charged atomic nuclei. Neutrons ionize indirectly by first colliding with components of atomic nuclei.

Kerma (kinetic energy released in material). The kinetic energy transferred to charged particles per unit mass of irradiated medium by indirectly ionizing (uncharged) particles, such as photons or neutrons. Unit: gray (Gy). 1 Gy = 1 J/kg. If all of the kinetic energy is absorbed "locally," the kerma is equal to the absorbed dose.

Latent period. The time between exposure and expression of the disease. After exposure to a dose of radiation, there typically is a delay of several years (the latent period) before any cancer is observed.

Life table. A table showing the number of persons who, of a given number born or living at a specified age, live to attain successivly higher ages, together with the numbers who die in each interval.

Linear energy transfer (LET). Mean energy lost by charged particles in electronic collisions per unit track length. Unit: keV/mm.

High-LET radiation. Neutrons or heavy, charged particles, such as protons or alpha particles, that produce ionizing events densely spaced on a molecular scale (*e.g.*, $L > 10$ keV/μm; see *Unrestricted LET*).

Low-LET radiation. X-rays and gamma rays or light, charged particles, such as electrons, that produce sparse ionizing events far apart on a molecular scale (*e.g.*, $L < 10$ keV/μm).

Restricted LET (L_Δ). The mean energy lost per unit track length in electronic collisions with energy transfer not larger than Δ.

Unrestricted LET (*L*) or *Total collision stopping power*, the mean energy lost per unit track length in all electronic collisions.

Linear (L) model or relationship (also linear dose-effect relationship). The linear model is a special case of the linear-quadratic model, with the quadratic coefficient equal to zero; the linear model expresses the effect (*e.g.*, cancer or mutation) as proportional to the dose (linear function of the dose).

Linear-quadratic (LQ) model. Also, *linear-quadratic dose-effect relationship*; expresses the effect (*e.g.*, cancer) as the sum of two components, one proportional to the dose (linear term) and one proportional to the square of the dose (quadratic term). The linear term predominates at low doses; the quadratic term, at high doses.

LNT model. Linear no-threshold dose-response for which any dose greater than zero has a positive probability of producing an effect (*e.g.*, mutation or cancer). The probability is calculated either from the slope of a linear (L) model or from the limiting slope, as the dose approaches zero, of a linear-quadratic (LQ) model.

Lognormal distribution. When the logarithms of a randomly distributed quantity have a normal (Gaussian) distribution.

LSS (Life Span Study). Long-term study of health effects in the Hiroshima and Nagasaki atomic bomb survivors.

Mechanistic basis. An explanation derived from a knowledge of the individual stages leading to an effect.

Meta-analysis. An analysis of epidemiologic data from several studies based on data included in publications.

Model. A schematic description of a system, theory, or phenomenon that accounts for its known or inferred properties and may be used for further study of its characteristics.

Monte Carlo calculation. The method for evaluation of a probability distribution by means of random sampling.

Mortality (rate). The frequency at which people die from a disease (*e.g.*, a specific cancer), often expressed as the number of deaths per 100,000 population per year.

Multiplicative effects. The combined effect of two agents is equal to the product of the effects of the two agents acting alone.

NCRP (National Council on Radiation Protection and Measurements). U.S. Council commissioned to formulate and disseminate information, guidance, and recommendations on radiation protection and measurements.

Neoplasm. Any new and abnormal growth, such as a tumor; neoplastic disease refers to any disease that forms tumors, whether malignant or benign.

Nonstochastic. A description of effects whose severity is a function of dose; for these, a threshold may occur; some examples of somatic effects believed to be nonstochastic are cataract induction, nonmalignant damage to the skin, hematological deficiencies, and impairment of fertility.

Normal distribution. The so-called bell-shaped curve of randomly distributed quantities; also referred to as a "Gaussian distribution."

Odds ratio (OR). The odds of being exposed among diseased persons divided by the odds of being exposed among nondiseased persons.

Oncogenes. Genes that encode the potential for cancer.

Phenotype. The genetically and environmentally determined physical appearance of an organism.

Photon. An electromagnetic quantum whose energy (E_{ph}) equals the product of the Planck constant (h) and its frequency (n). With the convenient units eV and s, and with the wave length λ in μm: $E_{ph} = 4.136 \times 10^{-15}$ v = $1.24/\lambda$.

Pooled analysis. An analysis of epidemiologic data from several studies based on original data from the studies.

Prevalence. The number of cases of a disease in existence at a given time per unit of population, usually 100,000 persons.

Probability of causation. A number that expresses the probability that a given cancer, in a specific tissue, has been caused by a previous exposure to a carcinogenic agent, such as radiation.

Projection model. A mathematical model that simultaneously describes the excess cancer risk at different levels of some factor such as dose, time after exposure, or baseline level of risk, in terms of a parametric function of that factor. It becomes a projection model when data in a particular range of observations are used to assign values to the parameters in order to estimate (or project) excess risk for factor values outside that range.

Promoter. An agent that is not by itself carcinogenic but can amplify the effect of a true carcinogen by increasing the probability of late-stage cellular changes necessary to complete the carcinogenic process.

Proof of principle. Proof-of-principle studies are those that extend or strengthen the validity of some elements of a hypothesis or a model using a system that is different from the one that provided the basis for the hypothesis or model.

Proportional mortality ratio. The ratio of the percentage of a specific cause of death among all deaths in the population being studied divided by the comparable percentage in a standard population.

Protraction. The spreading out of a radiation dose over time by continuous delivery at a lower dose rate.

Quadratic-dose model. A model that assumes that the excess risk is proportional to the square of the dose.

Quality factor (Q). An LET-dependent factor by which the absorbed dose is multiplied to obtain (for radiation protection purposes) the dose equivalent that corresponds roughly to the absorbed dose from X- or gamma rays that causes the same degree of biologic effect. Absorbed dose (Gy) $\times Q$ = dose equivalent (Sv).

Rad. A special unit of absorbed dose, now replaced by the SI unit gray (see *Units*). 1 rad = 0.01 Gy = 100 erg/g.

Radiation. Energy emitted in the form of waves or particles by radioactive atoms as a result of radioactive decay or produced by artificial means, such as X-ray generators.

Radioactivity. The property of nuclide decay in which particles or gamma radiations are usually emitted.

   Artificial radioactivity. Man-made radioactivity produced by fission, fusion, particle bombardment, or electromagnetic irradiation.

   Natural radioactivity. The property of radioactivity exhibited by more than 50 naturally occurring radionuclides.

Radiogenic. Caused by radiation.

Radioisotope. A radioactive atomic species of an element with the same atomic number and usually identical chemical properties.

Radionuclide. A radioactive species of an atom characterized by the constitution of its nucleus.

Relative biologic effectiveness (RBE). The ratio D_{ref}/D, where D is the absorbed dose of a specified radiation and D_{ref} is the absorbed dose of a sparsely ionizing reference radiation (γ-rays or X-rays) that produces the same level of effect. When the magnitude of the dose D is not specified, the RBE is meant to be the low-dose limit of the ratio D_{ref}/D (this low-dose *RBE* equals the low-dose effectiveness (initial slope) of the specified radiation to that of the reference radiation).

Relative risk (RR). The rate of disease in an exposed population divided by the rate of disease in an unexposed population. Also termed "rate ratio."

Rem (rad equivalent man). A special unit of dose equivalent, now replaced by the SI unit sievert (see *Units*). 1 rem = 0.01 Sv.

Retinoblastoma. An eye tumor that is an example of an inherited malignant tumor with a dominant autosomal gene inheritance pattern.

Risk. A chance of injury, loss, or detriment; a measure of the deleterious effects that may be expected as the result of an action or inaction.

Risk assessment. The process by which the risks associated with an action or inaction are identified and quantified.

Risk coefficient. The increase in the annual incidence or mortality rate per unit dose: (1) absolute risk coefficient is the increase of the incidence or mortality rate per unit dose; (2) relative risk coefficient is the fractional increase above the baseline incidence or mortality rate per unit dose.

Risk estimate. The increment of the incidence or mortality rate projected to occur in a specified exposed population per unit dose for a specified exposure regime and expression period.

Sievert (Sv). Special name of the SI unit of dose equivalent (see *Units*). 1 Sv = 1 J/kg = 100 rem.

SI units. Units of the International System of Units as defined by the General Conference of Weights and Measures in 1960. They are the base units, such as meter (m), kilogram (kg), second (s), and their combinations, which have special names (e.g., the unit of energy, 1 J = 1 kg m^2/s^2, or absorbed dose, 1 Gy = 1 J/kg = 1 m^2/s^2 (see *Units*).

Solid cancers. Solid cancers include all malignant neoplasms other than those of the lymphatic and hematopoietic tissue.

Somatic cells. Nonreproductive cells.

Specific activity. Activity of a given nuclide per unit mass of a compound, element, or radioactive nuclide.

Specific energy (z). The energy per unit mass actually deposited in a microscopic volume in a single energy deposition event or at a given absorbed dose. This is a stochastic quantity as opposed to its average, the absorbed dose, *D*. The mean energy imparted by ionizing radiation to a medium per unit mass. Unit: 1 Gy = 1 J/kg.

Standardized morbidity ratio or *Standardized mortality rate (SMR).* The ratio (multiplied by 100) of the mortality rate from a disease in the population being studied divided by the comparable rate in a standard population. The ratio is similar to a relative risk times 100.

Stochastic. Effects whose probability of occurrence in an exposed population (rather than severity in an affected individual) depends on dose; stochastic effects are commonly regarded as having no threshold; hereditary effects are stochastic; some somatic effects, especially cancers, are regarded as being stochastic.

Suppressor gene. A gene that can suppress another gene such as an oncogene. Changes in suppressor genes can lead to expression by genes such as oncogenes.

Synergistic effect. Increased effectiveness results from an interaction between two agents, so that the total effect is greater than the sum of the effects of the two agents acting alone.

Target cells. Cells in a tissue that have been determined to be the key cells in which changes occur in order to produce an end point such as cancer.

Threshold hypothesis. The assumption that no radiation injury occurs below a specified dose.

Transformed cells. Tissue culture cells changed from growing in an orderly pattern exhibiting contact inhibition to growing in a pattern more like that of cancer cells.

Uncertainty. The range of values within which the true value is estimated to lie. It is a best estimate of possible inaccuracy due to both random and systemic errors.

> *Random Errors.* Errors that vary in a nonreproducible way around a limiting mean. These errors can be treated statistically by use of the laws of probability.

> *Systemic Errors.* Errors that are reproducible and tend to bias a result in one direction. Their causes can be assigned, at least in principle, and they can have constant and variable components. Generally, these errors cannot be treated statistically.

Units of dose. Also known as dosimetric units.

Unit[a]	Symbol	Conversion Factors
Becquerel (SI)	Bq	1 disintegration/s = 2.7×10^{-11} Ci
Curie	Ci	3.7×10^{10} disintegrations/s = 3.7×10^{10} Bq
Gray (SI)	Gy	1 J/kg = 100 rad
Rad	rad	0.01 Gy = 100 erg/g
Sievert (SI)	Sv	1 J/kg = 100 rem
Rem	rem	0.01 Sv

[a]International Units are designated SI.

NOTE: Equivalent dose equals absorbed dose times *Q* (quality factor). Gray is the special name of the unit (J/kg) to be used with absorbed dose; sievert is the special name of the unit (J/kg) to be used with equivalent dose.

UNSCEAR (United Nations Scientific Committee on the Effects of Atomic Radiation). A UN committee that publishes periodic reports on sources and effects of ionizing radiation.

Variability. The variation of a property or quantity among members of a population. Such variation is inherent in nature and is often assumed to be random; it can then be represented by a frequency distribution.

Weighted dose (d). The dose to A-bomb survivors, roughly adjusted to account for the increased effectiveness of the small neutron absorbed dose contribution. The weighted dose equals the gamma-ray absorbed dose to a specified organ plus the neutron absorbed dose multiplied by a weighting factor that has usually been set equal to 10 in analyses by the Radiation Effects Research Foundation (RERF). Unit: 1 Sv = 1 J/kg.

X-radiation. Also X-rays; penetrating electromagnetic radiation, usually produced by bombarding a metallic target with fast electrons in a high vacuum.

Committee Biographies

Richard R. Monson, M.D., Sc.D. (chair), is associate dean for professional education and professor of epidemiology at the Harvard University School of Public Health. He earned an M.D. from Harvard Medical School and an Sc.D. in epidemiology and biostatistics from the Harvard School of Public Health. He was a Fulbright scholar. His research interests in epidemiology include evaluation of the risks of exposure to radiation, health effects of exposure to machining fluids, and potential health effects of environmental exposure to chemicals. From 1992 through 2004, he was principal investigator of a Superfund program project award from the National Institute of Environmental Health Sciences (NIEHS). Dr. Monson has served on numerous national and international advisory groups, including two committees of the National Research Council (NRC), and on several public and private advisory committees that addressed issues concerning exposure to radiation and the potential health effects of environmental exposure to chemicals. He served as chair for the multinational peer review of the epidemiology program of the U.S.-Japan Radiation Effects Research Foundation (RERF). Dr. Monson was a member of the scientific committee on Effects of Radiation on the Embryo-Fetus, of the National Council on Radiation Protection and Measurements (NCRP). He served as editor of *Cancer Causes and Control* from 1993 to 1998. Dr. Monson has more than 150 publications in radiation and occupational epidemiology, including the text *Occupational Epidemiology*.

James E. Cleaver, Ph.D. (vice chair), is a professor of dermatology at the University of California, San Francisco Cancer Center, with a joint appointment in the Department of Pharmaceutical Chemistry. His research interests have focused on the effects of ultraviolet light, internal emitters (tritium and carbon-14), radon, and chemical carcinogens on human cells in culture from patients with mutations in genes involved in DNA repair and genetic instability. He is best known for the discovery that the human disease xeroderma pigmentosum is defective in DNA nucleotide excision repair. Dr. Cleaver is a member of several professional societies including Radiation Research, the American Society for Photobiology, the American Society for Human Genetics, the Environmental Mutagen Society, and the American Association for Cancer Research. He received the Radiation Research Society Research Award, the Senior Investigator Award from the American Society of Photobiology, and the Lila Gruber Award for Cancer Research from the American Academy of Dermatology, the Luigi Provasoli Award from the Phycological Society of America, and the John B. Little Award for radiation research from Harvard University; he was the 47th Faculty Research Lecturer for the University of California San Francisco in 2004. He was elected to the Roll of Honor of the International Union for Research on Cancer, Geneva, in 1997. He was president of the Genetic and Toxicology Association of the Bay area for 1999-2000. He was a member of the NRC Committee on Risk Assessment of Exposure to Radon in Drinking Water. Dr. Cleaver is a member of the National Academy of Sciences (NAS).

Herbert L. Abrams, M.D., a faculty member at Stanford Medical School from 1951 to 1967, served as professor of radiology from 1960 to 1967, when he was appointed Philip H. Cook Professor and Chairman of Radiology at Harvard Medical School, as well as chief of radiology at Brigham and Women's Hospital and Dana Farber Cancer Institute (1967–1985). He returned to Stanford as professor in 1985. Long interested in the applications and effects of ionizing radiation, he was the first to demonstrate the striking influence of age on the lethality of total-body radiation in mammals (*Proc Soc* 1951) and, together with H.S. Kaplan, explored the mitigating effects of regional shielding on radiation mortality. His studies of the incidence and distribution of metastases in carcinoma; of the role of lymphography in detection of metastases and of the effects of lymphatic obstruction (*NEJM* 1965, 1967, 1968); of the altered response to pharmacologic agents of the tumor vascular bed in animals and humans (*Nature* 1964; *Cancer Research* 1979); of renal, adrenal, hepatic, pancreatic, colon, and splenic tumors (1951, 1964, 1965, 1980, 1982, 1983); and of the reasons for and effects of overutilization (*NEJM* 1979) and underutilization of X-rays (*NEJM* 1984) have appeared in major journals over the years. Dr. Abrams served as a member of the Radiation Study Section of the National Institutes of Health (NIH); a member of the Institute of Medicine (IOM) Planning Committee for the Symposium on the Medical Implications of Nuclear War (1985); as one of the four physician members of the congressionally mandated National Council on Health Care Technology and as co-chair of its Methods Section (1986-1990); as chairman of the NIH Consensus Conference on Magnetic Resonance Imaging (1987); as a consultant to the Presidential Advisory Committee on Human Radiation Experiments; and as one of two American members of the International Blue Ribbon Panel on the future of the Radiation Effects Research Foundation in Hiroshima (1996). His examinations of the "Medical Problem of Survivors of Nuclear War" (*NEJM* 1981); "Medical Resources After Nuclear War: Availability vs Need" (*JAMA* 1984); "Nuclear Radiation in Warfare" (XV Berzelius Symposium 1988); "The Fallout from Chernobyl" (*The Bulletin*); "Human Reliability and Safety in the Handling of Nuclear Weapons" (*Science and Global Security* 1991); and "Security Issues in the Handling and Disposition of Fissionable Materials" (1993) and other related subjects have been published in numerous journals and four multiauthored books. A member of the IOM since 1980, he is also an Honorary Fellow of the Royal Society of Radiology of Great Britain and the Royal College of Surgeons of Ireland. The author of more than 200 articles, he has also written and/or edited eight books on health policy, technology assessment, and the clinical and pathophysiologic aspects of cardiovascular disease. The Crookshank Lecturer of the Royal College of Radiology in 1981 and the Caldwell Lecturer of the American Roentgen Ray Society in 1982, he was also the recipient of the Gold Medal of the Association of University Radiologists in 1984 and the Gold Medal of the Radiological Society of North America in 1996. In 1998, he presented the Charles Dotter Memorial Lecture of the American Heart Association. He was the founding vice-president of International Physicians for the Prevention of Nuclear War, recipient of the 1985 Nobel Peace Prize. Currently, his time is divided between his work in radiology at the medical school and his activities as a member-in-residence of the Stanford Center for International Security and Cooperation.

Eula Bingham, Ph.D., is a professor of environmental health at the University of Cincinnati. Dr. Bingham's interests include risk assessment, regulatory toxicology, environmental carcinogenesis, and occupational health surveillance. She was a volunteer investigator at NIEHS and an Assistant Secretary of Labor, Occupational Safety and Health Administration. She was the first recipient of the William Lloyd Award for occupational safety. Throughout her career, Dr. Bingham has served on numerous national and international advisory groups, including advisory committees of the NRC, the NAS, the Food and Drug Administration, Department of Labor, National Institute for Occupational Safety and Health, NIH, Natural Resources Defense Council, and the International Agency for Research on Cancer (IARC). The committees addressed issues concerning research needs in health risk assessment and the potential health effects of environmental exposure to chemicals. In 1989, Dr. Bingham was elected to the IOM.

Patricia A. Buffler, Ph.D., is professor of epidemiology and the Kenneth and Marjorie Kaiser Chair of Cancer Epidemiology at the University of California, Berkeley, School of Public Health. Her current research interests in epidemiology include studies of leukemia in children, health effects of exposure to tobacco smoke, and health effects of nonionizing radiation. She has served on numerous national and international advisory groups including advisory committees to the Department of Energy, the Department of Defense, the Department of Health and Human Services, the Environmental Protection Agency, the University of California Office of the President, the National Research Council and the World Health Organization (WHO). From 1996 to 2002 she served as a visiting director for the U.S.-Japan RERF. She has served as president of the Society of Epidemiologic Research, the American College of Epidemiology, and the International Society for Environmental Epidemiology and as an officer of the Medical Sciences Council of the American Association for the Advancement of Science (AAAS). She was awarded the American College of Epidemiology Lilienfeld Award in 1996 and the James Bruce Award in Preventive Medicine from the American College of Physicians in 1999. She is a fellow of both the American College of Epidemiology and AAAS and is a member of the IOM.

Elisabeth Cardis, Ph.D., currently runs the Radiation Group at the IARC, where she was previously chief of the Unit of Radiation and Cancer and head of the Radiation Programme. She is responsible for the planning, conduct, and analyses of numerous epidemiologic studies of cancer in relation to exposure to radiation—both ionizing and nonionizing. She has been consultant to the United Nations Scientific Committee on the Effects of Atomic Radiation (UNSCEAR). She is currently a member of the Steering Committee for the EMF-Net project; of the International Advisory Committee for the WHO International EMF Project; and of the Scientific Council of the French Agence Française pour la Sécurité Sanitaire et Environnementale. She is a fellow of the Institute of Physics. She was a member of the International Commission for Non-ionizing Radiation Protection Standing Committee on Epidemiology from 1998 to 2002 and has been a corresponding member since then. She is also a member of the Scientific Council for the Joint Congress of the International Society for Environmental Epidemiology and the International Society for Exposure Assessment (ISEE-ISEA), Paris 2006, and was a member of the Scientific Council for the 6th International Conference on High Levels of Natural Radiation and Radon Areas (Osaka, September 2004) and of the International Programme Committee for the 11th International Radiation Protection Association (IRPA) Congress, Madrid 2004. She has been responsible for numerous projects supported currently or previously under the European Commission's Quality of Life, Radiation Protection, and INCO-Copernicus Programmes.

Roger Cox, Ph.D., is director of the National Radiological Protection Board (NRPB) at Chilton, Oxfordshire. He was formerly division head and head of the Radiation Effects Department. After his Ph.D. research in microbial genetics, he joined the MRC Radiobiology Unit at Harwell in 1971 where he was involved in cellular and molecular research relating to postirradiation repair, mutagenesis, and tumor development. At NRPB (1990–) he has continued to pursue personal research interests in the mechanisms and genetics of radiation tumorigenesis. Roger Cox is involved in the work of a variety of national and international committees considering radiation effects and radiological protection. These include the NRPB Advisory Group on Ionizing Radiation (1995–), the International Commission on Radiological Protection (ICRP; 1989–), and UNSCEAR (1989–1993 and 1996–).

Scott Davis, Ph.D., is professor and chairman of the Department of Epidemiology in the School of Public Health and Community Medicine at the University of Washington and a full member in the Program in Epidemiology of the Division of Public Health Sciences at the Fred Hutchinson Cancer Research Center, Seattle, Washington. He obtained his undergraduate degree in biology and chemistry from the University of New Mexico, a master of science in community health from the University of Rochester, and a Ph.D. in epidemiology from the University of Washington. He served as a research associate in epidemiology at RERF in Hiroshima, Japan, from 1983 to 1985. His primary research focus is radiation epidemiology. For more than a decade he has directed two major research activities investigating the effects of ionizing radiation on human health. One is a series of studies in the Russian Federation of the effects of exposure to radiation from the Chernobyl Power Station. These studies have focused on the risk of thyroid cancer and leukemia among children in the Bryansk Oblast. The second is a long-term follow-up study of thyroid disease in persons exposed to atmospheric releases of radiation from the Hanford Site in eastern Washington State (the Hanford Thyroid Disease Study). He has conducted several epidemiologic studies of the possible health effects associated with exposure to power frequency magnetic fields, focusing on the risk of leukemia

and breast cancer. Recently this work has expanded to include investigations of the effects of exposure to light-at-night and circadian disruption on melatonin and reproductive hormones important in the etiology of breast and other hormone-related cancers. He has also maintained a long-standing interest in the etiology of leukemias and lymphomas and has directed epidemiologic studies of Hodgkin's disease, non-Hodgkin's lymphoma, and multiple myeloma. Dr. Davis was a Special Fellow of the Leukemia Society of America from 1986 to 1987 and the recipient of a Research Career Development Award from the National Cancer Institute (NCI) from 1988 to 1993. He is an elected member of the American Epidemiological Society and a fellow of the American College of Epidemiology. He was recently elected a member (academician) of the Russian Academy of Medical Sciences.

William C. Dewey, Ph.D., is emeritus professor of radiation oncology, University of California, San Francisco (UCSF). From 1981 to 2004 when he retired, he was director of the Radiation Oncology Research Laboratory at UCSF. Dr. Dewey earned his doctorate in radiation biology from the University of Rochester in 1958. He was a faculty member of the University of Texas Graduate School of Biomedical Sciences at M.D. Anderson Hospital in Houston, Texas (1958–1965), and of the Department of Radiology and Radiation Biology, Colorado State University (1965–1981). He served as president of the Radiation Research Society in 1979 and was Failla lecturer of that society in 1989. Dr. Dewey was program chairman for the Ninth International Congress of Radiation Research in 1991, and chairman and organizer of Third International Symposium: Cancer Therapy by Hyperthermia, Drugs and Radiation in 1980. He is the author and coauthor of 264 publications mainly on the effects of radiation and hyperthermia on mammalian cells in culture. He received the Andrew G. Clark research award in 1977 and was an American Society for Therapeutic Radiology and Oncology (ASTRO) gold medal recipient in 1998. Dr. Dewey is known, in part, for his studies of the effects of radiation and hyperthermia on synchronous cell populations and for the number of investigators in the radiation biology community who trained in his laboratory.

Ethel S. Gilbert, Ph.D., is a biostatistician in the Radiation Epidemiology Branch of NCI. She holds a B.A. in mathematics from Oberlin College and an M.P.H. and Ph.D. in biostatistics from the University of Michigan. Her current research includes studies of workers at the Mayak nuclear plant in Russia, studies of second cancers after radio- and chemotherapy, and radiation risk assessment. Formerly, Dr. Gilbert spent several years as a senior staff scientist at Pacific Northwest National Laboratories, where her research focused on epidemiologic studies of nuclear workers, including combined analyses of national and international data. Dr. Gilbert is a fellow of the American Statistical Association, and a member of the NCRP, and she served on the BEIR VI committee.

Albrecht Kellerer, Ph.D., is professor emeritus at the University of Munich. He was formerly professor of radiation biophysics at Columbia University in New York, and subsequently professor and chief of the institute for medical radiation research at the University of Wurzburg. Dr. Kellerer's research specialties include microdosimetry, radiation risk assessment, and radiobiology. Dr. Kellerer was a member of the German National Commission for Radiation Protection and a member of committees of ICRU and ICRP.

Daniel Krewski, Ph.D., M.H.A., is professor of medicine and of epidemiology and community medicine at the University of Ottawa, where he serves as director of the McLaughlin Centre for Population Health Risk Assessment. He received his M.Sc. and Ph.D. in mathematics and statistics from Carleton University and his M.H.A. in health administration from the University of Ottawa. Dr. Krewski served on the NRC's Committee on the Biological Effects of Ionizing Radiation (BEIR VI) and is a member of the Board on Radiation Effects Research. He chaired the NRC's Committee on Acute Exposure Guideline Levels for Highly Hazardous Substances from 1998 to 2004, and currently chairs the NRC's Committee on Toxicity Testing and Assessment of Environmental Agents. Dr. Krewski has contributed to the publication of more than 500 scientific and technical articles in the areas of population health risk assessment, epidemiology, and biostatistics.

Tomas Lindahl, M.D., obtained his early training in medicine and research at the Karolinska Institute, Stockholm. After postdoctoral training at Princeton and Rockefeller Universities, he returned to Stockholm in 1969 as an assistant professor at the Karolinska Institute. Dr. Lindahl is currently director of the Cancer Research UK Clare Hall Laboratories in London. The laboratories have 12 research groups working on DNA repair, recombination, and replication, as well as cell cycle control and transcription. Dr. Lindahl's main contributions have been in the field of DNA repair, where he discovered the family of DNA glycosylases and the base excision-repair pathway as well as the unique suicide enzyme O^6-methylguanine-DNA methyltransferase and DNA dioxygenases for reversal of DNA damage. He also discovered the mammalian DNA ligases and defined their distinct roles. Dr. Lindahl is a fellow of the Royal Society, a member of the Royal Swedish Academy of Sciences, a foreign member of the Norwegian Academy of Science and Letters, a member of the European Molecular Biology Organization, and a member of the Academia Europea.

Katherine E. Rowan, Ph.D. (Purdue University, 1985), is professor and associate chair of the Department of Communication at George Mason University, Fairfax, Virginia. Her research concerns science and risk communication, particularly effective methods for earning trust and explaining complex science. Her scholarship on these topics appears in more than 40 journals and edited books. She has presented lectures, briefings, and workshops on science communication and risk communication for organizations such as the Institute of Medicine, the National Library of Medicine, the staff of the U.S. Congress, the U.S. Department of Agriculture, the National Academy of Sciences, Health Canada, Dow Elanco, Portland Cement, and the U.S. Environmental Protection Agency. Her risk communication consulting clients have included the NIH, the National Library of Medicine, the National Safety Council, and the University of Wisconsin in connection with a grant from the U.S. Nuclear Regulatory Commission. In 1999, she was selected as a journalism fellow by the American Society of Newspaper Editors.

K. Sankaranarayanan, Ph.D., is a professor emeritus in the Department of Toxicogenetics, Leiden University Medical Centre, Sylvius Laboratories, Leiden, in the Netherlands. Dr. Sankaranarayanan's research activities include understanding the adverse effects of radiation and chemical mutagens and the use of this knowledge for the assessment of biological, especially hereditary, risks associated with the exposure of humans to these agents. He was a consultant in genetics to UNSCEAR (1970–2001) and a member of Committee 1 of the ICRP (1975–2004). In the late 1990s, he chaired an ICRP Task Group on Risk Estimation for Multifactorial Diseases. Since the late 1960s, he has served the journal *Mutation Research* in several editorial capacities (including executive managing editor) until he stepped down from the editorship in 2000.

Daniel W. Schafer, Ph.D., received his master's and Ph.D. in statistics from the University of Chicago. Dr. Schafer is a professor in the Department of Statistics at Oregon State University. His expertise includes measurement errors and generalized linear models. Dr. Schafer is a fellow of the American Statistical Association and a past president of the Oregon Chapter of the American Statistical Association. He was scientific adviser to the NCI and the Centers for Disease Control and Prevention for updating radiologic properties of causation tables (1998–2000). Dr. Schafer is coauthor of the textbook *The Statistical Sleuth, A Course in Methods of Data Analysis*.

Robert Ullrich, Ph.D., is the Barbara Cox Anthony University Chair in Oncology at Colorado State University, Departments of Environmental and Radiological Health Sciences and of Clinical Sciences. Dr. Ullrich earned his Ph.D. from the University of Rochester School of Medicine and Dentistry. He was formerly head of the Radiation Carcinogenesis Unit at the Oak Ridge National Laboratory. He is the current president of the Radiation Research Society. He received the R.J. Michael Fry Research Award of the Radiation Research Society in 1987. Dr. Ullrich has served on the NRC Panel on Space Radiation Effects, a number of NIH advisory committees, and ICRP Committee 1 and the Task Group on Estimates of Radiation-Induced Cancer at Low Doses.

Index

A

Abscopal effect. *See* Bystander effects
Absolute risk. *See also* Excess absolute risk
    defined, 132, 373
Absorbed dose
    atomic bomb survivors, 20, 27, 142
    and biological effect, 20
    defined, xi, 373
    neutron weighting factor, 296-297
    photon energy fluence and, 22
Achondroplasia, 92, 98, 125
Activity
    defined, 373
    units of, 373
Acute myocardial infarction, 95
ada gene, 36, 37
Adaptive response
    animal studies, 51, 52, 53, 55, 78-79, 251
    bacteria, 50
    bystander effect and, 55
    and cancer risk assessment, 250-251
    cell cycle alterations, 51, 53
    for cell lethality, 51, 52, 55, 78
    in Chernobyl children, 51
    to chromosomal aberrations, 51, 53, 55
    cumulative effect of multiple low doses, 53
    DDREF and, 78-79
    to DNA damage and repair, 39, 50, 51, 52,
        53, 251
    dose fractionation and, 78
    dose-response relationships, 45, 50-53, 62,
        89
    genotype and, 53
    and hormesis, 333
    in humans, 50
    to iodine-131, 51
    in lymphocytes, 51, 53
    malignant transformation as, 51-52, 62, 78-
        79
    in mammalian cells in vitro, 51, 52, 62
    mechanisms, 11, 29, 39, 51-53
    to medical uses of radiation, 51
    in mice, 51, 52, 53, 55, 78
    microarray expression studies, 39, 53

    modeling, 250-251
    mutagenesis as, 51
    to occupational exposures, 51, 53
    to oxidative stress, 50
    priming dose and, 51-53, 55, 78, 251
    radiation resistance as, 37, 51
    research needs, 53
    signal transduction pathways, 51, 53
    in tumorigenesis, 12, 51-52, 78-79, 250-251
    variation in, 51
Additive effects, 148, 150, 297, 373
Adenine phosphoribosyltransferase, 46
Adenocarcinomas, 50, 74, 76
Adenomas, 69, 151, 168
Adrenal cortical carcinoma, 86, 161
Adriamycin, 130
Adult Health Study, 142, 150, 151, 153
Aflatoxins, 242
Air and space travel, 3, 19
    occupational exposures, 204
Alcohol consumption, 242
All-Union Distributed Registry, 202
Alpha particles
    and bone cancer, 87, 269
    bystander effect, 53-54, 55
    cell killing/lethality, 54
    chromosome aberrations, 53
    genetic susceptibility, 53, 87
    and genomic instability, 70, 71
    internal contamination, 199-200
    LET value, 21
    and leukemia, 71
    and liver cancer, 68
    and lung cancer, 242
    mutagenesis, 53, 54, 68
    RBE, 71
    and tumorigenesis, 70
Amelogenesis imperfecta, 98
American Registry of Radiologic Technologists,
    205
Animal studies. *See also specific animals*
    adaptive response, 78-79, 251
    bone cancer, 74
    breast/mammary cancer, 74, 82, 83, 88
    DDREF from, 77-79, 254-258

    DNA damage and repair mechanisms, 34, 36,
        42, 52, 69, 71
    dose-response relationships, 73-75, 89, 245
    extrapolation to humans, 73, 96-98, 109-111,
        114, 115
    genetic effects of radiation, 12, 68-70, 73,
        82-83, 92, 96-97
    genetic susceptibility to cancer, 68-70, 71,
        73, 82-83, 87-88, 89
    of growth and developmental effects, 115
    of hormesis, 334
    leukemia, 68-70, 71, 72, 73-74, 87
    life-shortening effects of radiation, 246
    limitations of, 245
    lung cancer, 28, 50, 74, 76, 87
    lymphoma, 68, 73, 74, 76, 78
    mutation rates, 68-70, 96-98
    neutron RBE, 28-29, 68, 126
    predictive value, 73, 75, 241
    radiosensitivity, 34, 69, 71, 82-83
    skin cancer, 74, 75, 76, 87
    tumorigenesis, radiation-induced, 11, 12, 67,
        68-70, 73-79, 82-83, 89, 240, 241, 246
    X-ray exposure, 68, 69, 100-101
Aniridia, 98
Ankylosing spondylitis cohorts
    bone cancer, 164
    breast cancer, 164-165, 176, 177
    cancer mortality, 164, 165
    cohort description, 164
    colon cancer, 164
    dosimetry, 164, 165
    esophageal cancer, 164
    kidney cancer, 164
    leukemia, 164, 165, 183, 289
    lung cancer, 164, 174
    malignant lymphoma, 164
    modeling cancer risk in, 287
    multiple myeloma, 164
    pancreatic cancer, 164
    prostate cancer, 164, 288
    radiotherapy-related cancer risks, 164-165,
        174, 176, 177, 185, 287, 288, 289
    stomach cancer, 164, 185, 287, 288, 289
    urinary tract cancer, 164, 288

Fenton oxidants, 30
Fernald facility, 191, 200
Fetal exposure to radiation
    atomic bomb survivors, 151
    and cancer in childhood, 6
    IEER issues, 330
    and mental retardation, 1
Fifteen-Country Workers Study, 336
Finite-locus threshold model, 105-108, 124
Fluorescence *in situ* hybridization (FISH)
    method of chromosome painting, 45-46,
    57
Fluoroscopy
    and breast cancer, 26, 170, 176, 177, 180, 243
    Canadian epidemiological study, 176, 287
    dosimetry, 170, 176
    and lung cancer, 174, 175, 176, 288, 289
    Massachusetts tuberculosis cohort, 170-171,
    174, 176, 243, 273, 287, 292
    and skin cancer, 2
14-3-3 proteins, 39
Free radicals, DNA damage from, 19, 29-30, 239
French Commissariat a l'Energie Atomique, 28,
    190

G

GADD45 gene, 39
Gall bladder cancer, 147, 148, 149, 269
Gamma rays
    bystander effects, 54-55
    cancer risk estimates, 276
    carcinogenesis, 24-26, 28
    cobalt-60, 19 n.1, 21, 22, 24
    damage mechanisms, 26-27
    DDREF, 61
    defined, 374
    dose-rate effect, 28
    dosimetry for atomic bomb survivors, 6, 142
    environmental exposures, 212
    extrapolation to X-ray exposures, 24
    high-energy, 24
    mutations in mice from, 126
    occupational exposures, 204
    photon energies, 19, 20, 21, 22, 24, 276
    RBE, 21, 22, 24, 25, 28, 126, 276
    signal transduction study, 56
    terrestrial, 30
Gap junctions, 39, 54, 55
Gastric ulcer, 153, 241
Gastrointestinal cancers, 218
Gene amplification, 47, 49, 72
Gene transcription, 55, 113
Gene-environment interactions, 12, 86, 87, 88
Genes. *See also specific genes*
    antibody, 34
    autosomal recessive disorders, 80
    basic concepts, 327
    breast cancer, 67, 79, 80, 82, 83, 84, 85-86,
    243
    caretaker, 67
    checkpoint kinase, 86
    deletions, 11, 58, 61, 66, 67, 86, 109, 110,
    111, 112, 119, 124, 125
    DNA repair, 36, 37, 39, 40, 42, 71, 86, 87-88

epigenetic silencing events, 66, 67
functional polymorphisms, 12, 86-88, 113, 114
gatekeeper, 66-67, 69, 81
haploinsufficiency, 125
human counterparts in mice, 98-99
interactions (epistasis), 12, 70, 108
low-penetrance, 85-88
oxidative damage resistance, 40-42
proto-oncogenes, 81-82
radiation resistance, 30, 40-42
radiation-responsive, 39
recoverability of induced mutations in live
    births, 109-111, 124, 125
tumor-supressor, 39, 65, 66, 69, 80, 80, 81
X-linked HPRT, 47, 54
Genetic diseases. *See also* Genetic susceptibility
    to cancer; *specific diseases*
    absolute risk, 110
    autosomal dominant, 79, 80, 81, 92, 94, 95,
    96, 97-98, 101, 102, 103-105, 110, 112,
    115, 116, 117, 125
    autosomal recessive, 79, 80, 81, 92-93, 94-95,
    96, 101, 104, 105, 110-111, 115, 117-118
    baseline frequencies, 12, 92, 94-95, 96, 115,
    117-118
    basic concepts, 328
    chromosomal, 93, 96, 115, 119
    chronic, 111, 115, 116, 119
    congenital, 93, 95, 105-106, 112, 115, 117, 119
    defined, 92
    experimental data, 112-113
    frequency, 105
    inborn errors of metabolisms, 113
    marker, 125
    Mendelian, 92-93, 94-95, 96, 98, 105, 111,
    112, 113, 115, 119
    MIM maps, 125
    minisatellite polymorphisms and, 113, 114
    multifactorial, 8, 93, 95, 96, 101, 112, 115,
    117, 119
    multisystem developmental abnormalities,
    12, 112
    mutation component, 98, 101-105, 117
    potential phenotypes, 111-113
    PRCF estimates, 110-111
    prevalence, 80, 96, 98, 110
    "radiation-inducible," 92
    risk estimation by class of, 94, 115, 116
    spontaneous mutation rates, 97-98, 109
    sporadic cases, 108
    UNSCEAR estimates, 94-95, 96
    X-linked, 92, 93, 94, 95, 96, 97, 98, 101,
    104-105, 110, 115, 116, 125
Genetic effects of radiation. *See also*
    Chromosome aberrations; DNA damage;
    Mutations
    animal studies, 12, 68-70, 73, 82-83, 92
    background data from humans, 8
    basic concepts, 328
    BEIR I estimates, 94
    cardiovascular disease, 95, 96
    in children, 8-9, 68, 114, 118, 161
    detection, 8
    disease liability concept, 107, 120-121
    germ-cell mutations at ESTR loci and, 113-
    114, 125-130

heritability concept, 121-122
    at human minisatellite loci, 126, 128-130
    indicators of risk, 8, 118
    models of multifactorial diseases, 120-122,
    252
    multifactorial threshold model, 120-121, 252
Genetic risk assessment. *See also* Mutation
    component of genetic diseases
    advances since BEIR V, 115-116
    in atomic bomb survivors, 8-9, 91, 92, 114,
    115, 118, 131, 252
    autosomal dominant disorders, 94, 111, 115,
    116, 117
    autosomal recessive diseases, 94, 112, 115
    back calculation, 94, 117
    baseline disease frequencies, 12, 94-96, 115,
    117, 252
    bases for, 8, 96, 115-116
    BEIR III estimates, 94-95
    BEIR V estimates, 12, 94-95, 96, 115-116,
    117-118, 252
    calculations, 109, 115, 116
    chromosomal diseases, 115, 117
    chronic multifactorial diseases, 115, 116, 117
    by class of disease, 94, 117
    congenital abnormalities, 115, 116, 117, 120
    current estimates, 116, 118-120
    DDREF and, 92, 246
    direct method, 8, 112
    doubling dose method, 8, 12, 93-101, 102,
    111, 115, 117, 118, 119, 252
    endpoints, 118, 130, 131
    equilibrium effects, 94, 117
    extrapolation of animal data to humans, 96-
    97, 109-111, 114
    first postradiation generation, 94, 116, 117,
    118
    framework, 92
    germ cell stages and, 92
    goal, 92
    indirect method, 8, 93-94
    methods, 91, 92
    mouse data, 92, 96-97, 98-101, 109-111,
    112-113, 114, 115-116, 252
    mutation component, 12, 81, 94, 101-113,
    115, 116, 117, 119, 252
    overlap in estimates, 119-120
    potential recoverability correction factor, 12,
    109-110, 115, 119, 252
    reconciliation of present and past estimates,
    117, 118
    reproductive outcomes, 252
    second-generation progeny, 116
    sex differences, 119, 120
    spontaneous mutations, 8, 96-101
    UNSCEAR estimates, 92, 94, 95, 96, 109,
    111, 112, 115, 118
    X-linked disorders, 94, 111, 115, 116, 117
Genetic susceptibility to cancer. *See also*
    Radiosensitivity; Tumorigenesis,
    radiation induced
    age of onset in carriers, 81
    alpha particles and, 53, 87
    animal studies, 68-70, 71, 73, 82-83, 87-88,
    89
    autosomal dominant disorders, 66, 79-81, 85

Tumor registries, 141, 142, 147-148, 150, 151, 152
Tumor suppressor
    disorders, 80, 84
    genes, 39, 65, 66, 67-68, 69, 80, 80, 81, 82-83
Tumorigenesis, radiation induced
    adaptive responses, 12, 51, 52, 62, 78-79, 250-251
    aggressive, 72
    alpha particles and, 70
    animals studies, 11, 12, 67, 68-70, 73-79, 82-83, 89, 240
    cell killing and, 12, 74, 75, 76, 82
    chromatid instability and, 70-73
    chromosome aberrations and, 24, 46, 66, 68-69, 70, 72-73, 74, 82, 86
    confounding stress factors, 71
    DDREF, 77-79, 246-250
    diet and, 242
    DNA damage response and, 6, 11, 65, 68, 69, 70, 73, 82, 83, 239, 227-228, 239, 246
    dose-response relationships, 12, 45, 50, 55, 59, 60, 61-62, 68, 70, 73-77, 84, 89, 140, 245-246
    epidemiologic studies, 68
    etiology at different histologic sites, 241-245
    fractionation kinetics, 73, 75-76, 78-79, 182
    genetic susceptibility, 12, 65-66, 79-90
    genomic instability and, 11-12, 39, 49, 65, 66, 67, 69, 70-73
    in hematopoietic cells, 70-71
    human data, 68, 71, 83-85
    latency, 65, 68, 78
    life-span shortening, 76-77
    lymphoma and leukemia, 68-70, 71, 73-74
    malignant transformation, 51-52, 62, 78-79
    mammary epithelial cells, 71-73
    mechanisms, 11-12, 27, 65, 66-70, 74, 75, 76, 81-82, 89, 245
    monoclonal origins, 11, 245, 253-254
    mouse models, 68-69, 70, 71
    mutations and, 11, 65, 66-70
    persistence of initiated cells postirradiation, 76
    phases and process, 11, 66, 240
    in progeny, 70
    quantitative studies, 73-79, 89
    RBE for, 24, 28, 29
    solid tumors, 74-75
    spontaneous mechanisms compared, 11, 12, 66-67, 70, 239
    target for, 54
    telomere sequence instability and, 12, 71-73
    temporal projections of risk, 239-240
    UV radiation and, 67-68
Twin studies
    of genetic susceptibility to cancer, 88
    of multifactorial diseases, 93

U

Ulcer patients
    dose-response relationship, 153
    radiotherapy-related cancer risk, 163

Ultraviolet light
    DNA damage, 40
    skin cancer, 75, 151
    tumorigenesis, 67-68, 75
Uncertainties. See also Bias; Confounding factors
    from age-related effects, 297
    in atomic bomb survivor data, 130, 131, 141, 147, 172, 285-286
    cancer risk estimates, 25-26, 147, 174, 251, 268, 272-273, 275, 276, 278, 279, 280, 284-286, 297, 308-310
    in Chernobyl data, 128, 129, 202-203
    chromosomal radiosensitivity, 82
    combining sources of, 309-310
    in cross-population transport, 279, 284, 285, 286, 296
    in DDREF, 279, 284, 285, 286, 295, 296, 310
    defined, 377
    in diagnostic radiation risk estimates, 286
    in dose-response relationships, 246
    in dosimetry, 3, 14, 128, 129, 139, 142, 169, 198-200, 233, 241, 266, 285
    in doubling dose estimates, 98-99, 130, 131
    EPA assessment of, 284
    in epidemiological studies, 133, 284
    in error correction, 296
    in exposure estimates, 114, 174
    in genetic risk estimates, 98-99, 114
    joint analysis, 295
    LAR analysis, 278, 279, 284, 308-310
    in lifetime cancer risk, 278, 279, 284-286
    Monte Carlo analysis, 293, 295
    NCRP assessment of, 284
    NIH assessment of, 284, 295
    occupational exposures, 14, 194, 198
    in population effects, 286
    procedures for addressing, 308-310
    quantitative evaluation, 278, 284-285
    in radiation exposure data, 3, 169
    in RBE, 29, 286
    in risk estimates, 29, 284, 286
    from sampling variability, 278, 279, 280, 284, 285, 296, 308-309
    sources, 133, 285, 295-296
    statistical, 29
    time since exposure, 275
Underground miners. See Uranium miners
United Nations Scientific Committee on the Effects of Atomic Radiation (UNSCEAR), 91, 93
    cancer risk model, 138, 240, 268, 272, 274, 275, 277-278, 282-283, 293-294, 297-298, 307
    congenital abnormality frequencies, 112
    DDREF, 131, 246, 249
    defined, 377
    genetic risk calculations, 92, 94, 95, 96, 109, 111, 112, 118, 131
    lifetime cancer risk estimates, 294
    mutation rate estimation, 109, 111
    occupational exposure studies, 190
United Nuclear Corporation, 190
Units of dose, xi, 2
Uracil, 30

Uranium, 2, 3, 20, 269
Uranium miners, 190
    lung cancer, 68, 138, 242
    risk modeling, 262
Urinary tract cancer. See also Bladder cancer
    ankylosing spondylitis cohort, 164
    atomic bomb survivors, 148, 149
    cervical cancer survivors, 157, 158
    Chernobyl accident and, 223, 227-228
U.S. Department of Energy, nuclear facilities, 190, 198
U.S. Environmental Protection Agency (EPA), 274, 275, 282, 283, 293, 294
U.S. Food and Drug Administration, 5
U.S. Scoliosis Cohort Study, 172
Uterine bleeding disorders (benign), 163-164, 183, 287, 288
Uterine cancer
    in atomic bomb survivors, 147, 148, 149, 269, 298, 303-306
    radiotherapy-related, 157, 158, 159, 162, 163, 287
    risk models and estimates, 272, 278, 279, 280, 282, 303-306
Uterine fibroma, 86
Uterine myoma, 151-152, 153

V

Vaginal cancer, 157
Varicose veins, 95
VHL gene, 66, 80, 81
Vinblastine, 130
Vincristine, 130
Von Hippel-Lindau syndrome, 80, 125

W

Warthin's tumor, 150
Water, electron ionization, 21, 29-32
World Health Organization, 128
WT-1 gene, 66, 80

X

Xenon, 209
Xeroderma pigmentosum, 79, 80, 81
X-linked disorders, 79
XP-A to XP-G gene, 67, 68, 80
XPV gene, 80
X-rays. *See also* Radiotherapy studies
    animal studies, 68, 69, 100-101
    atomic bomb explosion, 22
    bystander effects, 54
    cancer risk estimates, 171, 276
    childhood exposures, 211
    chromosome aberrations, 48, 86
    correction to gamma rays, 24
    defined, 378
    diagnostic, 3, 4, 5, 6, 21, 22, 156, 171; *see also* Mammography
    discovery and early studies, 2
    DNA damage, 30, 32, 33, 36
    dose-response relationships, 49, 56, 61, 62